# MEASUREMENTS USED IN MOLECULAR AND CELL BIOLOGY

| Length | meter (m) | 1 m = 39.37 inches |
|---|---|---|
| | centimeter (cm) | $1\text{ cm} = 10^{-2}\text{ m}$ |
| | millimeter (mm) | $1\text{ mm} = 10^{-3}\text{ m}$ |
| | micrometer ($\mu$m) | $1\ \mu\text{m} = 10^{-6}\text{ m}$ |
| | nanometer (nm) | $1\text{ nm} = 10^{-9}\text{ m}$ |
| | picometer (pm) | $1\text{ pm} = 10^{-12}\text{ m}$ |
| | Ångström (Å) | 1 Å = 0.1 nm |
| Volume | liter (l) | 1 l = 1.057 quarts |
| | milliliter (ml) | $1\text{ ml} = 10^{-3}\text{ l}$ |
| | microliter ($\mu$l) | $1\ \mu\text{l} = 10^{-6}\text{ l}$ |
| | nanoliter (nl) | $1\text{ nl} = 10^{-9}\text{ l}$ |
| | picoliter (pl) | $1\text{ pl} = 10^{-12}\text{ l}$ |
| Mass | kilogram (kg) | 1 kg = 2.2 pounds |
| | gram (g) | $1\text{ g} = 10^{-3}\text{ kg}$ |
| | milligram (mg) | $1\text{ mg} = 10^{-3}\text{ g}$ |
| | microgram ($\mu$g) | $1\ \mu\text{g} = 10^{-6}\text{ g}$ |
| | nanogram (ng) | $1\text{ ng} = 10^{-9}\text{ g}$ |
| | picogram (pg) | $1\text{ pg} = 10^{-12}\text{ g}$ |
| Heat | calorie (c) | 1 c = heat needed to raise the temperature of 1 g of water by 1°C |
| | kilocalorie (kcal) | $1\text{ kcal} = 10^{3}\text{ c}$ |
| | joule (J) | 1 J = 0.2389 c |
| Molecular mass | dalton (d) | 1 d = approximately the mass of a hydrogen atom $(1.66 \times 10^{-24}\text{ g})$ |
| | kilodalton (kd) | $1\text{ kd} = 10^{3}\text{ d}$ |
| Molecular length | kilobase (kb) | 1 kb = 1000 base pairs of double-stranded nucleic acid or 1000 bases of single-stranded nucleic acid |
| | megabase (Mb) | $1\text{ Mb} = 10^{3}\text{ kb}$ |

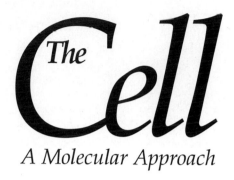

# The Cell

*A Molecular Approach*

# The Cell

*A Molecular Approach*

## Geoffrey M. Cooper

Dana-Farber Cancer Institute
Harvard Medical School

ASM Press   Sinauer Associates, Inc.
Washington, D.C. Sunderland, Massachusetts

*The Cover*
*Nautilus Shell*, 1927. Photograph by Edward Weston.
Copyright © 1981 Center for Creative Photography, Arizona Board of Regents.

*Part One opener image*
Courtesy of the President and Fellows of Harvard College, Gift of Mark Rothko.
© 1997 Kate Rothko - Prizel & Christopher Rothko/Artists Rights Society, N.Y.

*Part Two opener image*
Courtesy of the National Museum of the American Indian Smithsonian Institution
# 22/7882.

*Part Three opener image*
Courtesy of Bruce Heinemann.

*Part Four opener image*
Courtesy of The Metropolitan Museum of Art, The Alfred Steiglitz Collection, 1950.
Photograph by Malcolm Varon. © 1997. The Georgia O'Keeffe Foundation/Artists
Rights Society, N.Y.

**The Cell: A Molecular Approach**

Copyright © 1997 by Geoffrey M. Cooper.  All rights reserved.
This book may not be reproduced in whole or in part without permission.

Address editorial correspondence to ASM Press, c/o The American Society for
Microbiology, 1325 Massachusetts Avenue NW, Washington, DC 20005 U.S.A.

Address orders and requests for examination copies
to Sinauer Associates, Inc., PO Box 407,
23 Plumtree Road, Sunderland, MA 01375 U.S.A.

FAX: 413-549-1118
email: publish@sinauer.com

**Library of Congress Cataloging-in-Publication Data**

Cooper, Geoffrey M.
   The cell : a molecular approach / Geoffrey M. Cooper
        p.      cm.
Includes bibliographical references and index
ISBN 0–87893–119–8  (cloth : alk. paper)
1. Cytology   2. Molecular biology.   I. Title.
QH581.2.C66   1996
574.87--dc20                                          96–31922
                                   CIP

Printed in U.S.A.

5 4 3 2 1

*This book is dedicated to*
*Howard M. Temin (1934–1994),*
*whom I was privileged to know as*
*a teacher, mentor, and friend.*

# The Author

Geoffrey M. Cooper is Professor of Pathology at the Dana-Farber Cancer Institute and Harvard Medical School. He received his B.S. from the Massachusetts Institute of Technology in 1969, and his Ph.D. in Biochemistry from the University of Miami in 1973. He then pursued postdoctoral work with Howard Temin at the University of Wisconsin, where he developed gene transfer assays to characterize the proviral DNAs of Rous sarcoma virus and related retroviruses. After joining the faculty of Dana-Farber Cancer Institute and Harvard Medical School in 1975, he extended these studies to the identification of oncogenes in human tumors.

Dr. Cooper's current research is focused on understanding the roles of oncogene proteins in the signaling pathways that regulate cell proliferation, differentiation, and programmed cell death. He has published over 80 research papers in journals such as *Science*, *Nature*, and *Cell*, and received the U.S. Steel Award from the National Academy of Sciences in 1984 for "the identification and characterization of cellular oncogenes." In addition to training graduate students, medical students, and postdoctoral fellows, he has taught undergraduate courses in Harvard University's Biochemistry Department and served on the Board of Tutors in Biochemical Sciences. Dr. Cooper is the author of two textbooks on cancer, *Elements of Human Cancer* and *Oncogenes*, as well as *The Cancer Book*, written for the general public. He also edited, together with Rayla Greenberg Temin and Bill Sugden, the commemorative volume *The DNA Provirus: Howard Temin's Scientific Legacy*.

Dr. Cooper lives with his wife Ann and their family in Bedford, Massachusetts, where they enjoy a rural setting that provided a tranquil and diverse biological environment for writing *The Cell*.

# Contents in Brief

# Contents

## PART I  Introduction

### 1

#### An Overview of Cells and Cell Research  3

### 2

#### The Chemistry of Cells  39

### 3

#### Fundamentals of Molecular Biology  87

# PART II *The Flow of Genetic Information*

# PART III Cell Structure and Function

# Preface

*The Cell* is a basic text for undergraduate or medical students who are taking a first course on cell and molecular biology. When I initially considered this project in 1991, I had just completed my advanced text *Oncogenes* and a general undergraduate text on cancer biology. While working on these books, I became aware of the need for a text that would provide undergraduate students with an accessible introduction to contemporary cell biology. To be sure, there were then and are now some outstanding cell biology texts—books that are accurate, detailed, and comprehensive. But there was also a need for another kind of book—a book that could be more easily approached and mastered by undergraduate students, while still conveying the excitement and challenges of research in this dynamic area of the biological sciences.

My goals from the beginning of this project were twofold. First, *The Cell* was designed to be an approachable and readable text that undergraduates could understand and master. At the same time, the book was planned to be intellectually gratifying and to convey not only the facts, but also a sense of the excitement of modern molecular and cellular biology.

To accomplish these goals, I felt the cohesiveness of a single-authored text was important. The book was then focused on the molecular biology of cells as a unifying theme, with specialized topics discussed throughout the book as examples of more general principles. Aspects of developmental biology, the immune system, the nervous system, and plant biology are thus discussed in their broader biological context in chapters covering areas such as gene expression, DNA rearrangements, the plasma membrane, cell signaling, and the cell cycle. This organization has helped keep *The Cell* to a manageable length, while still allowing coverage of some of the exciting areas at the frontiers of contemporary research.

Some of the most dramatic advances in recent years have come from understanding the molecular and cellular basis of human diseases, in some cases allowing the development of new strategies for prevention and treatment. Examples of such relationships between cell biology and medical practice are therefore discussed throughout the text, as well as being highlighted in the Molecular Medicine essays that are included as special features in each chapter. By illustrating the growing impact of molecular and cellular biology on human health, I hope these discussions will stimulate as well as inform the reader.

*The Cell* is necessarily an unfinished work, because the science upon which it is based is not a fixed, static set of facts. To the contrary, this book deals with one of the most rapidly progressing areas of biology. It is thus critical for students to recognize the experimental nature of cell and molecular biology, not only in order to understand the current status of our knowledge but also to appreciate and hopefully contribute to the advances that will continue to be made in coming years. Although it was impossible to fully discuss experimental details in a book of this length, I have taken two approaches to introduce the reader to the experimental foundations of contemporary cell biology. First, critical experiments are briefly discussed throughout the text to illustrate the kinds of contributions that have moved the field. Second, each chapter contains a Key Experiment essay that describes a seminal paper and its background in detail, with the intent of giving the reader a sense of "doing science. Together, I hope these approaches impart a flavor for the ways in which progress in this field has been and continues to be made.

Finally, I have considered it nearly as important to tell the reader what is not known as what is known. In so doing, I have tried to point out areas of controversy, gaps in our knowledge, and scientific frontiers awaiting exploration. I hope that this approach will convey not only the science, but also the challenges and excitement, of understanding the workings of the cell.

ORGANIZATION AND FEATURES OF
# *The Cell*

*The Cell* has been designed to be an approachable and teachable text that can be covered in a single semester while allowing students to master the material in the entire book. It is assumed that most students will have had introductory biology and general chemistry courses, but will not have had previous courses in organic chemistry, biochemistry, or molecular biology. Several aspects of the organization and features of the book will help students to approach and understand its subject matter.

## ORGANIZATION

*The Cell* is divided into four parts, each of which is self-contained so that the order and emphasis of topics can be easily varied according to the needs of individual courses. In covering this vast subject matter, however, I developed an organizational overview of the book, as described below.

Part I of the book provides background chapters on the evolution of cells, methods for studying cells, the chemistry of cells, and the fundamentals of modern molecular biology. For those students who have a strong background from either a comprehensive introductory biology course or a previous course in molecular biology, various parts of these chapters can be skipped or used for review.

Part II focuses on the molecular biology of cells and contains chapters dealing with genome organization; DNA replication, repair, and recombination; transcription and RNA processing; and the synthesis, processing, and regulation of proteins. The order of chapters follows the flow of genetic information (DNA → RNA → protein) and provides a concise but up to date overview of these topics.

Part III contains the core block of chapters on cell structure and function, including chapters on the nucleus, cytoplasmic organelles, the cytoskeleton, and the cell surface. This part of the book starts with coverage of the nucleus, which puts the molecular biology of Part II within the context of the eukaryotic cell, and then works outward through cytoplasmic organelles and the cytoskeleton to the plasma membrane. These chapters are relatively self-contained, however, and could be used in a different order should that be more appropriate for a particular course.

Finally, Part Four focuses on the exciting and fast-moving area of.cell regulation, including coverage of topics such as cell signaling, the cell cycle, and programmed cell death. This part of the book concludes with a chapter on cancer, my own field of research, which synthesizes the consequences of defects in basic cell regulatory mechanisms.

## FEATURES

Several pedagogical features have been incorporated into *The Cell* in order to help students master and integrate its contents. These features are reviewed below as a guide to students studying from this book.

*Chapter organization.* Each chapter is divided into four or five major sections, which are further divided into a similar number of subsections. An outline listing the major sections at the beginning of each chapter provides a brief overview of its contents.

*Key terms and Glossary.* Key terms are identified as boldfaced words when they are introduced in each chapter. These key terms are reiterated in the chapter summary and defined in the glossary at the end of the book.

*Illustrations and micrographs.* An illustration program of full-color art and micrographs has been carefully developed to complement and visually reinforce the text.

*Key Experiment and Molecular Medicine essays.* Each chapter contains one Key Experiment and one Molecular Medicine feature. These essays are designed to provide the student with a sense of both the experimental basis of cell and molecular biology and its applications to modern medicine.

*Chapter summaries.* Chapter summaries are organized in outline form corresponding to the major sections and subsections of each chapter. This section-by-section format is coupled with a list of the key terms introduced in each section, providing a succinct but comprehensive review of the material.

*Questions and Answers.* Questions at the end of each chapter (with answers in the back of the book) are designed to further facilitate review by calling for students to understand and integrate the material presented in the chapter and to use this material to predict or interpret experimental results.

*References.* Comprehensive lists of references at the end of each chapter provide access to both reviews and selected papers from the primary literature. In order to help the student identify articles of interest, the references are organized according to chapter sections. Review articles and primary papers are distinguished by [R] and [P] designations, respectively.

## SUPPLEMENTS

In addition to these features of *The Cell* itself, two supplements are provided to facilitate study from the text and to help the reader keep up with the rapid progress in this area of science.

**CD-ROM.** Each copy of *The Cell* includes a CD-ROM designed to accompany the text. The disc contains minicourses that consist of an integrated series of text and figure entries reviewing selected topics covered in most chapters of *The Cell*. Many of the entries in these minicourses are illustrated by animations, which are particularly valuable in allowing the user to appreciate the dynamic nature of cell behavior. In addition, the disc includes an extensive encyclopedia of cell and molecular biology, again with many animated entries. We think the CD-ROM will not only be useful for review and lecture organization, but will also provide students with a novel visual presentation of many key processes in cell biology.

**Newsletter.** This book deals with an active, rapidly moving area of science. To help keep *The Cell* current, we will provide updates in the form of newsletters published twice a year. Each newsletter will contain summaries of recent papers representing major advances, as well as new Molecular Medicine features when appropriate. These newsletter updates will be keyed to appropriate sections of *The Cell* and written so as to help the reader integrate and appreciate current advances in the context of the text. In addition to providing updated information, we believe these newsletters will further help convey the excitement of research in this field of science, where there are still so many key questions to be answered and so many beautiful experiments to be done.

# Acknowledgments

Any book, but especially one of this magnitude, represents the efforts of many people, and it is both a privilege and a pleasure for me to thank my colleagues, friends, and family for their contributions to *The Cell*.

At ASM Press, Patrick Fitzgerald deserves special thanks for helping to guide and develop this project from beginning to end. Working with Pat over the last five years has been both fun and enlightening, and Pat's ideas have contributed in many ways to the organization and features of this book.

A book dealing with a field as vast as cell biology is beyond the immediate expertise of any one individual, so I have relied heavily on the input of colleagues who are experts in their various fields of research. Seventy-one scientists (listed on the following pages) read and critiqued draft chapters of *The Cell*, giving generously of their time and expertise. I am deeply grateful for their efforts and advice. Among this group, Thomas Roberts (Florida State University) and Roger Sloboda (Dartmouth College) were particularly helpful in reading and providing general advice on almost the entire book from their viewpoints not only as scientists, but also as active and interested teachers. I am also grateful to Brian Storrie (Virginia Tech) for his help, perspectives, and advice during the early stages of development of this project. Finally, thanks are due to Yale Altman and Karen Jones of ASM Press for their efforts in obtaining and organizing the many reviews that were so critical to the development of the book.

The production of *The Cell* has been primarily undertaken by Andy Sinauer and his colleagues at Sinauer Associates. Andy and his team have been efficient, professional, and a pleasure to work with. Andy personally oversaw the entire project with attention to the smallest details, Dean Scudder provided valuable advice on a variety of issues, Carol Wigg deftly coordinated all the complex parts of editing and collating the text and artwork, Jane Potter did a wonderful job of obtaining the many micrographs included in the book, and Christopher Small put the many parts together to produce the completed volume. The artwork itself was done by John Woolsey and Patrick Lane of the J/B Woolsey studio. Illustrations are a critical part of the book, and John and Patrick did a marvelous job of generating an outstanding art program based on my rough sketches. I am also grateful to Susan Schmidler for the attractive interior and cover design of *The Cell*.

Many authors acknowledge their families for their patience during the long hours of work required to write a book, and I am gratified to do so as well. However, my family also played a more active role in producing *The Cell*, with everyone contributing to its completion. My wife Ann, who is also a cell biologist, read and corrected multiple versions of the manuscript. The task of library research and compiling and copying the many papers I used as references was undertaken largely by Gwen and Ryan. At a later stage of production, Allison helped collate many chapters of the book during last minute dashes to meet production deadlines. And finally, the extensive glossary at the end of the book was drafted by Rachel, who brought a recent biology graduate's perspective to that undertaking.

# Reviewers

Gerald M. Adair
M. D. Anderson Cancer Center
University of Texas

Ibrahim Ades
University of Maryland, College Park

Ronald Berezney
State University of New York at Buffalo

Niels Bols
University of Waterloo
Ontario, Canada

Joan S. Brugge
ARIAD Pharmaceuticals, Inc.
Cambridge, MA

Brian Burke
The University of Calgary
Alberta, Canada

David A. Clayton
Stanford University School of Medicine

Roger I. Davis
Howard Hughes Medical Institute
University of Massachusetts Medical
Center

Michael Edidin
Johns Hopkins University

Gary Felsenfeld
National Institutes of Health
Bethesda, MD

Michael Forman
Purdue University

Errol C. Friedberg
University of Texas Southwestern
Medical Center

David Fromson
California State University, Fullerton

Larry Gerace
Scripps Research Institute
La Jolla, CA

Reid Gilmore
University of Massachusetts
Medical Center

Dennis Goode
University of Maryland

Michael M. Gottesman
National Cancer Institute
National Institutes of Health

Michael E. Greenberg
Children's Hospital
Harvard Medical School

Barry Gumbiner
Memorial Sloan-Kettering Cancer Center

Alan Hall
MRC Laboratory for Molecular Cell Biology
University College London

Joyce L. Hamlin
University of Virginia, Charlottesville

Ulla Hansen
Dana-Farber Cancer Institute
Harvard Medical School

F. Ulrich Hartl
Howard Hughes Medical Institute
Memorial Sloan-Kettering Cancer Center

Lawrence Hightower
University of Connecticut. Storrs

Tony Hunter
Salk Institute, La Jolla, CA

James. N. Ihle
St. Jude's Children's Research Hospital
Memphis, TN

Judith A. Jaehning
Indiana University, Bloomington

Haig H. Kazazian, Jr.
University of Pennsylvania

Kenneth Keegstra
MSU-DOE Plant Research Laboratory
Michigan State University

Robert A. Koch
California State University, Fullerton

Elaine Lai
Brandeis University

Jerry B. Lingrel
University of Cincinnati Medical Center

Elizabeth J. Luna
Worcester Foundation for Experimental
Biology
Shrewsbury, MA

Carolyn Machamer
Johns Hopkins University School of
Medicine

Umadas Maitra
Albert Einstein College of Medicine

Russell Malmberg
University of Georgia, Athens

James Manley
Columbia University

J. Richard McIntosh
University of Colorado, Boulder

Donald Miles
University of Missouri, Columbia

David A. Mullin
Tulane University

Eva Neer
Harvard Medical School

Carol S. Newlon
UMDNJ/New Jersey Medical School

John Newport
University of California, San Diego

Danton O'Day
University of Toronto, Erindale Campus

Donata Oertel
University of Wisconsin, Madison

Barbara Pearse
MRC Laboratory of Molecular Biology
Cambridge, UK

Hugh Pelham
Medical Research Council
Cambridge, UK

Jonathon Pines
The Wellcome Cancer Research Centre
Cambridge, UK

T. A. Rapoport
Max-Delbrück-Centrum
Berlin, Germany

Marilyn D. Resh
Memorial Sloan-Kettering Cancer Center

Joel D. Richter
Worcester Foundation for Biomedical
Research
Shrewsbury, MA

David A. Rintoul
Kansas State University, Manhattan

Thomas M. Roberts
Florida State University, Tallahassee

Alan Sachs
University of California, Berkeley

Shelley Sazer
Baylor College of Medicine

Gottfried Schatz
University of Basel
Switzerland

Randy Schekman
University of California, Berkeley

Tom Shrader
Albert Einstein College of Medicine

David Schlessinger
Washington University, St. Louis

Kai Simons
European Molecular Biology Laboratory
Heidelberg, Germany

Roger Sloboda
Dartmouth College

Timothy A. Springer
Harvard Medical School

Joseph M. Steffen
University of Louisville

Thomas P. Stossel
Brigham and Women's Hospital
Harvard Medical School

Suresh Subramani
University of California, San Diego

John Taylor
Wayne State University

Patricia Wadsworth
University of Massachusetts, Amherst

Jonathan R. Warner
Albert Einstein College of Medicine

Larry G. Williams
Kansas State University, Manhattan

Fred Wilt
University of California, Berkeley

Debra J. Wolgemuth
Columbia University

a critical part of the book, and John and Patrick did a marvelous job of generating an outstanding art program based on my rough sketches. I am also grateful to Susan Schmidler for the attractive interior and cover design of *The Cell*.

Many authors acknowledge their families for their patience during the long hours of work required to write a book, and I am gratified to do so as well. However, my family also played a more active role in producing *The Cell*, with everyone contributing to its completion. My wife Ann, who is also a cell biologist, read and corrected multiple versions of the manuscript. The task of library research and compiling and copying the many papers I used as references was undertaken largely by Gwen and Ryan. At a later stage of production, Allison helped collate many chapters of the book during last minute dashes to meet production deadlines. And finally, the extensive glossary at the end of the book was drafted by Rachel, who brought a recent biology graduate's perspective to that undertaking.

# Acknowledgments

Any book, but especially one of this magnitude, represents the efforts of many people, and it is both a privilege and a pleasure for me to thank my colleagues, friends, and family for their contributions to *The Cell*.

At ASM Press, Patrick Fitzgerald deserves special thanks for helping to guide and develop this project from beginning to end. Working with Pat over the last five years has been both fun and enlightening, and Pat's ideas have contributed in many ways to the organization and features of this book.

A book dealing with a field as vast as cell biology is beyond the immediate expertise of any one individual, so I have relied heavily on the input of colleagues who are experts in their various fields of research. Seventy-one scientists (listed on the following pages) read and critiqued draft chapters of *The Cell*, giving generously of their time and expertise. I am deeply grateful for their efforts and advice. Among this group, Thomas Roberts (Florida State University) and Roger Sloboda (Dartmouth College) were particularly helpful in reading and providing general advice on almost the entire book from their viewpoints not only as scientists, but also as active and interested teachers. I am also grateful to Brian Storrie (Virginia Tech) for his help, perspectives, and advice during the early stages of development of this project. Finally, thanks are due to Yale Altman and Karen Jones of ASM Press for their efforts in obtaining and organizing the many reviews that were so critical to the development of the book.

The production of *The Cell* has been primarily undertaken by Andy Sinauer and his colleagues at Sinauer Associates. Andy and his team have been efficient, professional, and a pleasure to work with. Andy personally oversaw the entire project with attention to the smallest details, Dean Scudder provided valuable advice on a variety of issues, Carol Wigg deftly coordinated all the complex parts of editing and collating the text and artwork, Jane Potter did a wonderful job of obtaining the many micrographs included in the book, and Christopher Small put the many parts together to produce the completed volume. The artwork itself was done by John Woolsey and Patrick Lane of the J/B Woolsey studio. Illustrations are

Mark Rothko, *Mural for Holyoke Center,* 1961

# PART I  *Introduction*

# 1

# An Overview of Cells and Cell Research

UNDERSTANDING THE MOLECULAR BIOLOGY OF CELLS is an active area of research that is fundamental to all of the biological sciences. This is true not only from the standpoint of basic science, but also with respect to a growing number of practical applications in agriculture, biotechnology, and medicine. Medical applications provide some particularly exciting examples, with new approaches to prevention and treatment being made possible by an increased understanding of the cellular and molecular basis of many human diseases.

Because cell and molecular biology is a rapidly growing field of research, this chapter will focus on how cells are studied, as well as reviewing some of their basic properties. Appreciating the similarities and differences between cells is particularly important to understanding cell biology. The first section of this chapter therefore discusses both the unity and the diversity of present-day cells in terms of their evolution from a common ancestor. On the one hand, all cells share common fundamental properties that have been conserved throughout evolution. For example, all cells employ DNA as their genetic material, are surrounded by plasma membranes, and use the same basic mechanisms for energy metabolism. On the other hand, present-day cells have evolved a variety of different lifestyles. Many organisms, such as bacteria, amoebas, and yeasts, consist of single cells that are capable of independent self-replication. More complex organisms are composed of collections of cells that function in a coordinated manner, with different cells specialized to perform particular tasks. The human body, for example, is composed of more than 200 different kinds of cells, each specialized for such distinctive functions as memory, sight, movement, and digestion. The diversity exhibited by the many different kinds of cells is striking; for example, consider the differences between bacteria and the cells of the human brain.

The fundamental similarities between different types of cells provide a unifying theme to cell biology, allowing the basic principles learned from experiments with one kind of cell to be extrapolated and generalized to other cell types. Several kinds of cells and organisms are widely used to study different aspects of cell and molecular biology; the second section of

3

this chapter discusses some of the properties of these cells that make them particularly valuable as experimental models. Finally, it is important to recognize that progress in cell biology depends heavily on the availability of experimental tools that allow scientists to make new observations or conduct novel kinds of experiments. This introductory chapter therefore concludes with a discussion of some of the experimental approaches used to study cells, as well as a review of some of the major historical developments that have led to our current understanding of cell structure and function.

## THE ORIGIN AND EVOLUTION OF CELLS

Cells are divided into two main classes, initially defined by whether they contain a nucleus. **Prokaryotic cells** (bacteria) lack a nuclear envelope; **eukaryotic cells** have a nucleus in which the genetic material is separated from the cytoplasm. Prokaryotic cells are generally smaller and simpler than eukaryotic cells; in addition to the absence of a nucleus, their genomes are less complex and they do not contain cytoplasmic organelles or a cytoskeleton (Table 1.1). In spite of these differences, the same basic molecular mechanisms govern the lives of both prokaryotes and eukaryotes, indicating that all present-day cells are descended from a single primordial ancestor. How did this first cell develop? And how did the complexity and diversity exhibited by present-day cells evolve?

### The First Cell

Life is thought to have originated between 3.8 billion and 3.5 billion years ago, approximately a billion years after Earth was formed (Figure 1.1). The placement of the origin of life within this relatively narrow period of time (in the evolutionary sense) is based on the fact that fossils resembling cells have been found in geological strata dating as far back as 3.5 billion years but the conditions of primitive Earth appear to have been too violent for life to have survived prior to 3.8 billion years ago. How life originated and how the first cell came into being are matters of speculation, since these events cannot be reproduced in the laboratory. Nonetheless, several types of experiments provide important evidence bearing on some steps of the process.

It was first suggested in the 1920s that simple organic molecules could form and spontaneously polymerize into macromolecules under the conditions thought to exist in primitive Earth's atmosphere. At the time life arose, the atmosphere of Earth is thought to have contained little or no free oxygen, instead consisting principally of $CO_2$ and $N_2$ in addition to smaller amounts of gases such as $H_2$, $H_2S$, and CO. Such an atmosphere provides reducing conditions in which organic molecules, given a source of energy such as sunlight or electrical discharge, can form spontaneously.

*Table 1.1* Prokaryotic and Eukaryotic Cells

| Characteristic | Prokaryote | Eukaryote |
|---|---|---|
| Nucleus | Absent | Present |
| Diameter of a typical cell | $\approx 1\ \mu m$ | $10-100\ \mu m$ |
| Cytoskeleton | Absent | Present |
| Cytoplasmic organelles | Absent | Present |
| DNA content (base pairs) | $1 \times 10^6$ to $5 \times 10^6$ | $1.5 \times 10^7$ to $5 \times 10^9$ |
| Chromosomes | Single circular DNA molecule | Multiple linear DNA molecules |

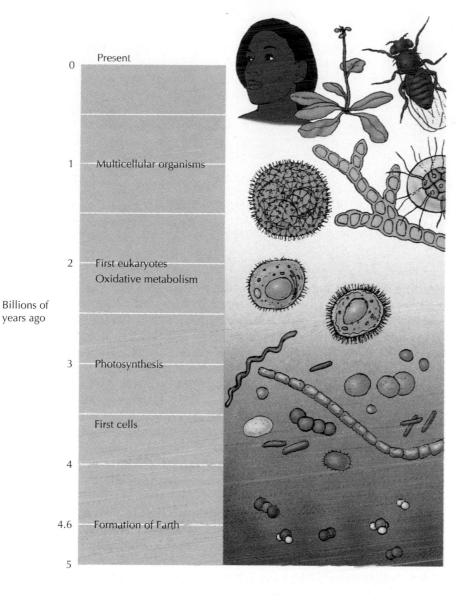

*Figure 1.1*
**Time scale of evolution** The scale indicates the approximate times at which some of the major events in the evolution of cells are thought to have occurred.

1   Multicellular organisms

2   First eukaryotes
Oxidative metabolism

Billions of
years ago

*Figure 1.2*
**Spontaneous formation of organic molecules** Water vapor was refluxed through an atmosphere consisting of $CH_4$, $NH_3$, and $H_2$, into which electric sparks were discharged. Analysis of the reaction products revealed the formation of a variety of organic molecules, including the amino acids alanine, aspartic acid, glutamic acid, and glycine.

3   Photosynthesis

First cells

4

4.6   Formation of Earth

5

The spontaneous formation of organic molecules was first demonstrated experimentally in the 1950s, when Stanley Miller (then a graduate student) showed that the discharge of electric sparks into a mixture of $H_2$, $CH_4$, and $NH_3$, in the presence of water, led to the formation of a variety of organic molecules, including several amino acids (Figure 1.2). Although Miller's experiments did not precisely reproduce the conditions of primitive Earth, they clearly demonstrated the plausibility of the spontaneous synthesis of organic molecules, providing the basic materials from which the first living organisms arose.

The next step in evolution was the formation of macromolecules. The monomeric building blocks of macromolecules have been demonstrated to polymerize spontaneously under plausible prebiotic conditions. Heating dry mixtures of amino acids, for example, results in their polymerization to form polypeptides. But the critical characteristic of the macromolecule from which life evolved must have been the ability to replicate itself. Only a macromolecule capable of directing the synthesis of new copies of itself would have been capable of reproduction and further evolution.

Of the two major classes of informational macromolecules in present-day cells (nucleic acids and proteins), only the nucleic acids are capable of

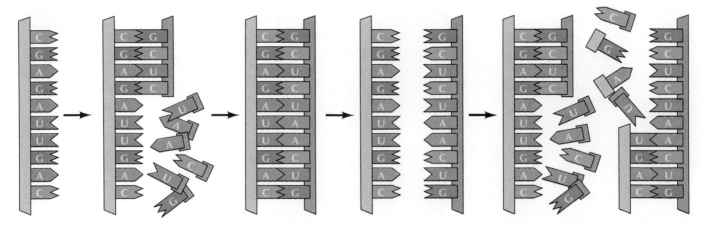

*Figure 1.3*
**Self-replication of RNA**    Complementary pairing between nucleotides (adenine [A] with uracil [U] and guanine [G] with cytosine [C]) allows one strand of RNA to serve as a template for the synthesis of a new strand with the complementary sequence.

directing their own self-replication. Nucleic acids can serve as templates for their own synthesis as a result of specific base pairing between complementary nucleotides (Figure 1.3). A critical step in understanding molecular evolution was thus reached in the early 1980s, when it was discovered in the laboratories of Sid Altman and Tom Cech that RNA is capable of catalyzing a number of chemical reactions, including the polymerization of nucleotides. RNA is thus uniquely able both to serve as a template for and to catalyze its own replication. Consequently, RNA is generally believed to have been the initial genetic system, and an early stage of chemical evolution is thought to have been based on self-replicating RNA molecules—a period of evolution known as the **RNA world**. Ordered interactions between RNA and proteins then evolved into the present-day genetic code, and DNA eventually replaced RNA as the genetic material.

The first cell is presumed to have arisen by the enclosure of self-replicating RNA in a membrane composed of **phospholipids** (Figure 1.4). As discussed in detail in the next chapter, phospholipids are the basic components of all present-day biological membranes, including the plasma membranes of both prokaryotic and eukaryotic cells. The key characteristic of the phospholipids that form membranes is that they are **amphipathic** molecules, meaning that one portion of the molecule is soluble in water and another portion is not. Phospholipids have long, water-insoluble

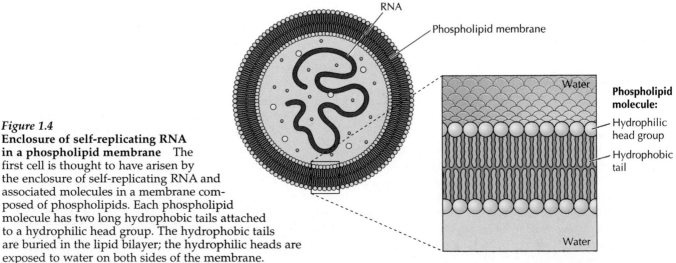

*Figure 1.4*
**Enclosure of self-replicating RNA in a phospholipid membrane**    The first cell is thought to have arisen by the enclosure of self-replicating RNA and associated molecules in a membrane composed of phospholipids. Each phospholipid molecule has two long hydrophobic tails attached to a hydrophilic head group. The hydrophobic tails are buried in the lipid bilayer; the hydrophilic heads are exposed to water on both sides of the membrane.

(**hydrophobic**) hydrocarbon chains joined to water-soluble (**hydrophilic**) head groups that contain phosphate. When placed in water, phospholipids spontaneously aggregate into a bilayer with their phosphate-containing head groups on the outside in contact with water and their hydrocarbon tails in the interior in contact with each other. Such a phospholipid bilayer forms a stable barrier between two aqueous compartments—for example, separating the interior of the cell from its external environment.

The enclosure of self-replicating RNA and associated molecules in a phospholipid membrane would thus have maintained them as a unit, capable of self-reproduction and further evolution. RNA-directed protein synthesis may already have evolved by this time, in which case the first cell would have consisted of self-replicating RNA and its encoded proteins.

## The Evolution of Metabolism

Because cells originated in a sea of organic molecules, they were able to obtain food and energy directly from their environment. But such a situation is self-limiting, so cells needed to evolve their own mechanisms for generating energy and synthesizing the molecules necessary for their replication. The generation and controlled utilization of metabolic energy is central to all cell activities, and the principal pathways of energy metabolism (discussed in detail in Chapter 2) are highly conserved in present-day cells. All cells use **adenosine 5′–triphosphate (ATP)** as their source of metabolic energy to drive the synthesis of cell constituents and carry out other energy-requiring activities, such as movement (e.g., muscle contraction). The mechanisms used by cells for the generation of ATP are thought to have evolved in three stages, corresponding to the evolution of glycolysis, photosynthesis, and oxidative metabolism (Figure 1.5). The development of these metabolic pathways changed Earth's atmosphere, thereby altering the course of further evolution.

In the initially anaerobic atmosphere of Earth, the first energy-generating reactions presumably involved the breakdown of organic molecules in the absence of oxygen. These reactions are likely to have been a form of present-day **glycolysis**—the anaerobic breakdown of glucose to lactic acid, with the net energy gain of two molecules of ATP. In addition to using ATP as their source of intracellular chemical energy, all present-day cells carry out glycolysis, consistent with the notion that these reactions arose very early in evolution.

Glycolysis provided a mechanism by which the energy in preformed

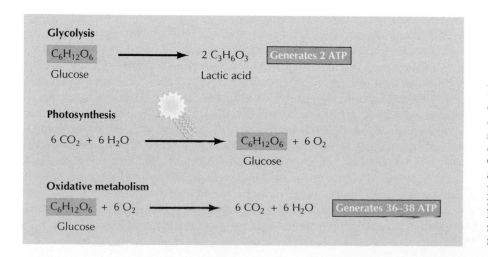

**Glycolysis**

$$C_6H_{12}O_6 \longrightarrow 2\ C_3H_6O_3 \quad \boxed{\text{Generates 2 ATP}}$$

Glucose $\qquad\qquad$ Lactic acid

**Photosynthesis**

$$6\ CO_2 + 6\ H_2O \longrightarrow C_6H_{12}O_6 + 6\ O_2$$

$\qquad\qquad\qquad\qquad\qquad$ Glucose

**Oxidative metabolism**

$$C_6H_{12}O_6 + 6\ O_2 \longrightarrow 6\ CO_2 + 6\ H_2O \quad \boxed{\text{Generates 36–38 ATP}}$$

Glucose

*Figure 1.5*
**Generation of metabolic energy** Glycolysis is the anaerobic breakdown of glucose to lactic acid. Photosynthesis utilizes energy from sunlight to drive the synthesis of glucose from $CO_2$ and $H_2O$, with the release of $O_2$ as a by-product. The $O_2$ released by photosynthesis is used in oxidative metabolism, in which glucose is broken down to $CO_2$ and $H_2O$, releasing much more energy than is obtained from glycolysis.

organic molecules (e.g., glucose) could be converted to ATP, which could then be used as a source of energy to drive other metabolic reactions. The development of **photosynthesis** is generally thought to have been the next major evolutionary step, which allowed the cell to harness energy from sunlight and provided independence from the utilization of preformed organic molecules. The first photosynthetic bacteria, which evolved more than 3 billion years ago, probably utilized $H_2S$ to convert $CO_2$ to organic molecules—a pathway of photosynthesis still used by some bacteria. The use of $H_2O$ as a donor of electrons and hydrogen for the conversion of $CO_2$ to organic compounds evolved later and had the important consequence of changing Earth's atmosphere. The use of $H_2O$ in photosynthetic reactions produces the by-product free $O_2$; this mechanism is thought to have been responsible for making $O_2$ abundant in Earth's atmosphere approximately 2 billion years ago.

The release of $O_2$ as a consequence of photosynthesis changed the environment in which cells evolved and is commonly thought to have led to the development of **oxidative metabolism**. Alternatively, oxidative metabolism may have evolved before photosynthesis, with the increase in atmospheric $O_2$ then providing a strong selective advantage for organisms capable of using $O_2$ in energy-producing reactions. In either case, $O_2$ is a highly reactive molecule, and oxidative metabolism, utilizing this reactivity, has provided a mechanism for generating energy from organic molecules that is much more efficient than anaerobic glycolysis. For example, the complete oxidative breakdown of glucose to $CO_2$ and $H_2O$ yields energy equivalent to that of 36 to 38 molecules of ATP, in contrast to the 2 ATP molecules formed by anaerobic glycolysis. With few exceptions, present-day cells use oxidative reactions as their principal source of energy.

### Present-Day Prokaryotes

Present-day prokaryotes, which include all the various types of bacteria, are divided into two groups—the **archaebacteria** and the **eubacteria**—which diverged early in evolution. Some archaebacteria live in extreme environments, which are unusual today but may have been prevalent in primitive Earth. For example, thermoacidophiles live in hot sulfur springs with temperatures as high as 80°C and pH values as low as 2. The eubacteria include the common forms of present-day bacteria—a large group of organisms that live in a wide range of environments, including soil, water, and other organisms (e.g., human pathogens).

Most bacterial cells are spherical, rod-shaped, or spiral, with diameters of 1 to 10 $\mu$m. Their DNA contents range from about 0.6 million to 5 million base pairs, an amount sufficient to encode about 5000 different proteins. The largest and most complex prokaryotes are the **cyanobacteria**, bacteria in which photosynthesis evolved.

The structure of a typical prokaryotic cell is illustrated by *Escherichia coli* (*E. coli*), a common inhabitant of the human intestinal tract (Figure 1.6). The cell is rod-shaped, about 1 $\mu$m in diameter and about 2 $\mu$m long. Like most other prokaryotes, *E. coli* is surrounded by a rigid **cell wall** composed of polysaccharides and peptides. Within the cell wall is the **plasma membrane**, which is a bilayer of phospholipids and associated proteins. Whereas the cell wall is porous and readily penetrated by a variety of molecules, the plasma membrane provides the functional separation between the inside of the cell and its external environment. The DNA of *E. coli* is a single circular molecule in the nucleoid, which, in contrast to the nucleus of eukaryotes, is not surrounded by a membrane separating it from the cytoplasm. The cytoplasm contains approximately 30,000 **ribosomes** (the sites of protein synthesis), which account for its granular appearance.

*Figure 1.6*
**Electron micrograph of *E. coli***
The cell is surrounded by a cell wall, within which is the plasma membrane. DNA is located in the nucleoid. (Biology Media/Photo Researchers, Inc.)

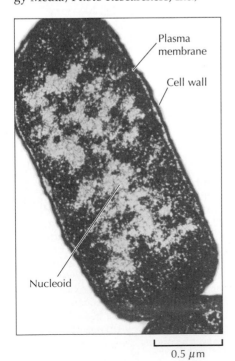

Plasma membrane

Cell wall

Nucleoid

0.5 $\mu$m

## Eukaryotic Cells

Like prokaryotic cells, all eukaryotic cells are surrounded by plasma membranes and contain ribosomes. However, eukaryotic cells are much more complex and contain a nucleus, a variety of cytoplasmic organelles, and a cytoskeleton (Figure 1.7). The largest and most prominent organelle of eukaryotic cells is the **nucleus**, with a diameter of approximately 5 $\mu$m. The nucleus contains the genetic information of the cell, which in eukaryotes is organized as linear rather than circular DNA molecules. The nucleus is the site of DNA replication and of RNA synthesis; the translation of RNA into proteins takes place on ribosomes in the cytoplasm.

In addition to a nucleus, eukaryotic cells contain a variety of membrane-enclosed organelles within their cytoplasm. These organelles provide compartments in which different metabolic activities are localized. Eukaryotic cells are generally much larger than prokaryotic cells, frequently having a cell volume at least a thousandfold greater. The compartmentalization provided by cytoplasmic organelles is what allows eukaryotic cells to function efficiently. Two of these organelles, **mitochondria** and **chloroplasts**, play critical roles in energy metabolism. Mitochondria, which are found in almost all eukaryotic cells, are the sites of oxidative metabolism and are thus responsible for generating most of the ATP derived from the breakdown of organic molecules. Chloroplasts are the sites of photosynthesis and are found only in the cells of plants and green algae. **Lysosomes** and **peroxisomes** also provide specialized metabolic compartments for the digestion of macromolecules and for various oxidative reactions, respectively. In addition, most plant cells contain large **vacuoles** that perform a variety of functions, including the digestion of macromolecules and the storage of both waste products and nutrients.

Because of the size and complexity of eukaryotic cells, the transport of proteins to their correct destinations within the cell is a formidable task. Two cytoplasmic organelles, the **endoplasmic reticulum** and the **Golgi apparatus**, are specifically devoted to the sorting and transport of proteins destined for secretion, incorporation into the plasma membrane, and incorporation into lysosomes. The endoplasmic reticulum is an extensive network of intracellular membranes, extending from the nuclear membrane throughout the cytoplasm. It functions not only in the processing and transport of proteins, but also in the synthesis of lipids. From the endoplasmic reticulum, proteins are transported within small membrane vesicles to the Golgi apparatus, where they are further processed and sorted for transport to their final destinations. In addition to this role in protein transport, the Golgi apparatus serves as a site of lipid synthesis and (in plant cells) as the site of synthesis of some of the polysaccharides that compose the cell wall.

Eukaryotic cells have another level of internal organization: the **cytoskeleton**, a network of protein filaments extending throughout the cytoplasm. The cytoskeleton provides the structural framework of the cell, determining cell shape and the general organization of the cytoplasm. In addition, the cytoskeleton is responsible for the movements of entire cells (e.g., the contraction of muscle cells) and for the intracellular transport and positioning of organelles and other structures, including the movements of chromosomes during cell division.

The eukaryotes developed approximately 2 billion years ago, following 1.5 billion years of prokaryotic evolution. Studies of their DNA sequences indicate that the archaebacteria and eubacteria are as different from each other as either is from present-day eukaryotes. Therefore, a very early event in evolution appears to have been the divergence of three lines of descent

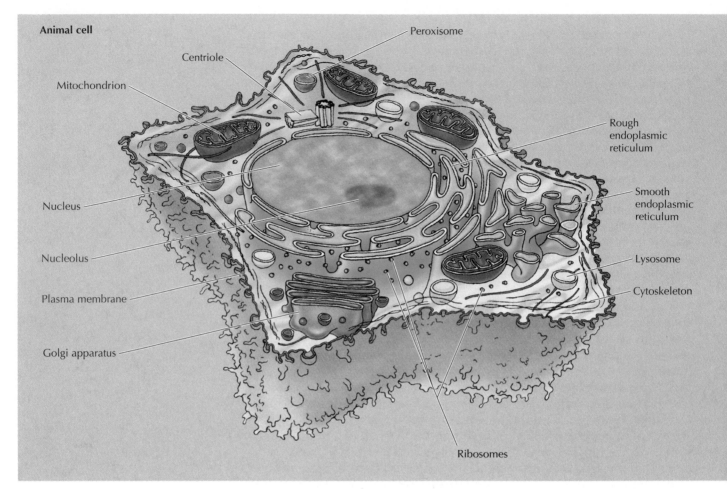

Animal cell

Peroxisome

Centriole

Mitochondrion

Rough endoplasmic reticulum

Smooth endoplasmic reticulum

Nucleus

Nucleolus

Lysosome

Plasma membrane

Cytoskeleton

Golgi apparatus

Ribosomes

*Figure 1.7*
**Structures of animal and plant cells**
Both animal and plant cells are surrounded by a plasma membrane and
contain a nucleus, a cytoskeleton, and
many cytoplasmic organelles in common. Plant cells are also surrounded
by a cell wall and contain chloroplasts
and large vacuoles.

from a common ancestor, giving rise to present-day archaebacteria, eubacteria, and eukaryotes (Figure 1.8). A critical step in the evolution of eukaryotic cells was the acquisition of membrane-enclosed subcellular organelles,
allowing the development of the complexity characteristic of these cells.
The organelles are thought to have been acquired as a result of the association of prokaryotic cells with the ancestor of eukaryotes.

The hypothesis that eukaryotic cells evolved from a symbiotic association of prokaryotes—**endosymbiosis**—is particularly well supported by
studies of mitochondria and chloroplasts, which are thought to have
evolved from bacteria living in large cells. Both mitochondria and chloroplasts are similar to bacteria in size, and like bacteria, they reproduce by
dividing in two. Most important, both mitochondria and chloroplasts contain their own DNA, which encodes some of their components. The mitochondrial and chloroplast DNAs are replicated each time the organelle
divides, and the genes they encode are transcribed within the organelle
and translated on organelle ribosomes. Mitochondria and chloroplasts thus
contain their own genetic systems, which are distinct from the nuclear
genome of the cell. Furthermore, the ribosomes and ribosomal RNAs of
these organelles are more closely related to those of bacteria than to those
encoded by the nuclear genomes of eukaryotes.

An endosymbiotic origin for these organelles is now generally accepted,
with mitochondria thought to have evolved from aerobic bacteria and
chloroplasts from photosynthetic bacteria, such as the cyanobacteria. The
acquisition of aerobic bacteria would have provided an anaerobic cell with

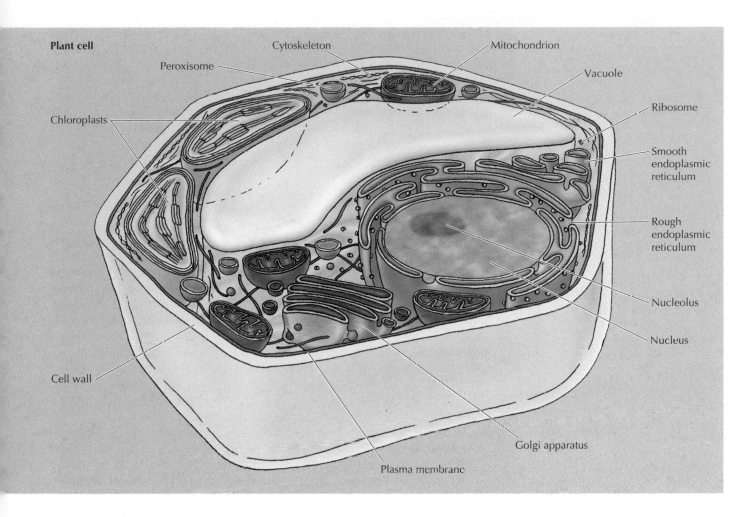

**Plant cell**

Peroxisome

Cytoskeleton

Mitochondrion

Vacuole

Ribosome

Chloroplasts

Smooth endoplasmic reticulum

Rough endoplasmic reticulum

Nucleolus

Nucleus

Cell wall

Golgi apparatus

Plasma membrane

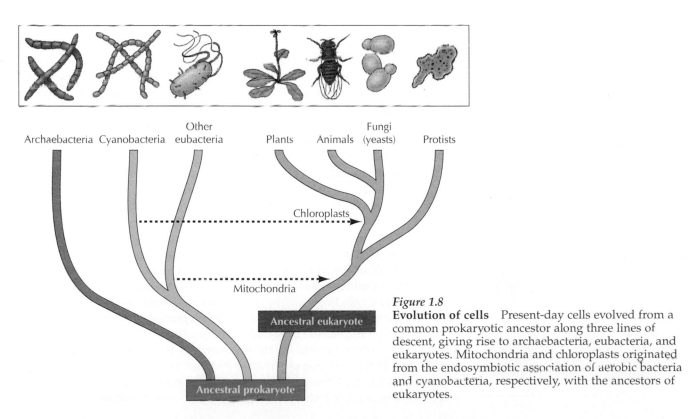

Archaebacteria   Cyanobacteria   Other eubacteria   Plants   Animals   Fungi (yeasts)   Protists

Chloroplasts

Mitochondria

Ancestral eukaryote

Ancestral prokaryote

*Figure 1.8*
**Evolution of cells**   Present-day cells evolved from a common prokaryotic ancestor along three lines of descent, giving rise to archaebacteria, eubacteria, and eukaryotes. Mitochondria and chloroplasts originated from the endosymbiotic association of aerobic bacteria and cyanobacteria, respectively, with the ancestors of eukaryotes.

*Table 1.2* DNA Content of Cells

| Organism | Haploid DNA content (millions of base pairs) |
|---|---|
| **Bacteria** | |
| *Mycoplasma* | 0.6 |
| *E. coli* | 4.7 |
| **Unicellular eukaryotes** | |
| *Saccharomyces cerevisiae* (yeast) | 14 |
| *Dictyostelium discoideum* | 70 |
| *Euglena* | 3000 |
| **Plants** | |
| *Arabidopsis thaliana* | 70 |
| *Zea mays* (corn) | 5000 |
| **Animals** | |
| *Caenorhabditis elegans* (nematode) | 100 |
| *Drosophila melanogaster* (fruit fly) | 165 |
| Chicken | 1200 |
| Mouse | 3000 |
| Human | 3000 |

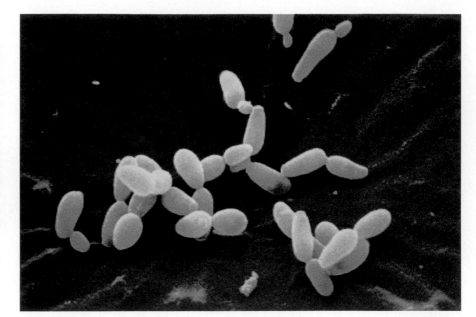

*Figure 1.9*
**Scanning electron micrograph of *Saccharomyces cerevisiae*** Artificial color has been added to the micrograph (Andrew Syed/Science Photo Library/ Photo Researchers, Inc.)

5 μm

the ability to carry out oxidative metabolism. The acquisition of photosynthetic bacteria would have provided the nutritional independence afforded by the ability to perform photosynthesis. Thus, these endosymbiotic associations were highly advantageous to their partners and were selected for in the course of evolution. Through time, most of the genes originally present in these bacteria apparently became incorporated into the nuclear genome of the cell, so only a few components of mitochondria and chloroplasts are still encoded by the organelle genomes.

## The Development of Multicellular Organisms

Many eukaryotes are unicellular organisms that, like bacteria, consist of only single cells capable of self-replication. The simplest eukaryotes are the **yeasts**. Yeasts are more complex than bacteria, but much smaller and simpler than the cells of animals or plants. For example, the commonly studied yeast *Saccharomyces cerevisiae* is about 6 μm in diameter and contains 14 million base pairs of DNA (Figure 1.9). Other unicellular eukaryotes, however, are far more complex cells, some containing as much DNA as human cells have (Table 1.2). They include organisms specialized to perform a variety of tasks, including photosynthesis, movement, and the capture and ingestion of other organisms as food. *Amoeba proteus*, for example, is a large, complex cell. Its volume is more than 100,000 times that of *E. coli*, and its length can exceed 1 mm when the cell is fully extended (Figure 1.10). Amoebas are highly mobile organisms that use cytoplasmic extensions, called **pseudopodia**, to move and to engulf other organisms, including bacteria and yeasts, as food. Other unicellular eukaryotes (the green algae) contain chloroplasts and are able to carry out photosynthesis.

Multicellular organisms evolved from unicellular eukaryotes about a billion years ago. Some unicellular eukaryotes form multicellular aggregates that appear to represent an evolutionary transition from single cells to multicellular organisms. For instance, the cells of many algae (e.g., the green

*Figure 1.10*
**Light micrograph of *Amoeba proteus***
(M. I. Walker/Photo Researchers, Inc.)

0.2 mm

alga *Volvox*) associate with each other to form multicellular colonies (Figure 1.11), which are thought to have been the evolutionary precursors of present-day plants. Increasing cell specialization then led to the transition from colonial aggregates to truly multicellular organisms. Continuing cell specialization and division of labor among the cells of an organism have led to the complexity and diversity observed in the many types of cells that make up present-day plants and animals, including human beings.

Plants are composed of fewer cell types than are animals, but each different kind of plant cell is specialized to perform specific tasks required by the organism as a whole (Figure 1.12). The cells of plants are organized into three main tissue systems: ground tissue, dermal tissue, and vascular tissue. The ground tissue contains **parenchyma cells**, which carry out most of the metabolic reactions of the plant, including photosynthesis. Ground tissue also contains two specialized cell types (**collenchyma cells** and **sclerenchyma cells**) that are characterized by thick cell walls and provide

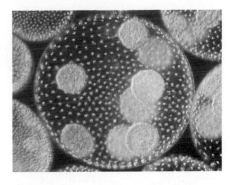

*Figure 1.11*
**Colonial green algae**   Individual cells of *Volvox* form colonies consisting of hollow balls in which hundreds or thousands of cells are embedded in a gelatinous matrix. (Cabisco/Visuals Unlimited.)

*Figure 1.12*
**Light micrographs of representative plant cells**   (A) Parenchyma cells, which are responsible for photosynthesis and other metabolic reactions. (B) Collenchyma cells, which are specialized for support and have thickened cell walls. (C) Epidermal cells on the surface of a leaf. Tiny pores (stomata) are flanked by specialized cells called guard cells. (D) Vessel elements and tracheids are elongated cells that are arranged end to end to form vessels of the xylem. (A, Jack M. Bastsack/Visuals Unlimited; B, A. J. Karpoff/Visuals Unlimited; C, Alfred Owczarzak/Biological Photo Service; D, Biophoto Associates/Science Source/Photo Researchers Inc.)

(A)

(B)

(C)

(D)

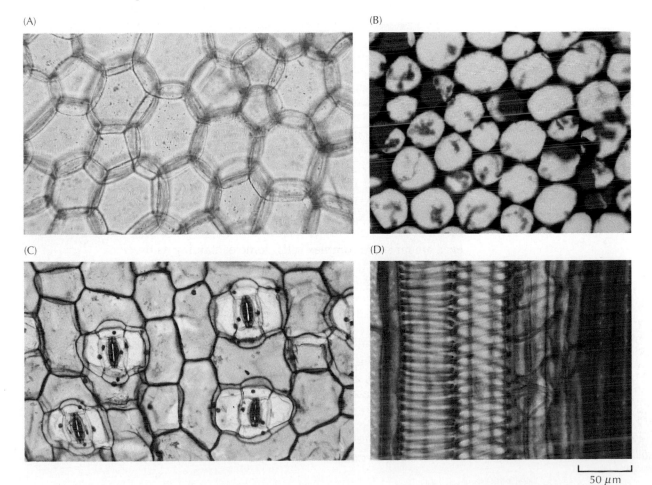

50 μm

structural support to the plant. Dermal tissue covers the surface of the plant and is composed of **epidermal cells**, which form a protective coat and allow the absorption of nutrients. Finally, several types of elongated cells form the vascular system (the xylem and phloem), which is responsible for the transport of water and nutrients throughout the plant.

The cells found in animals are considerably more diverse than those of plants. The human body, for example, is composed of more than 200 different kinds of cells, which are generally considered to be components of five main types of tissues: epithelial tissue, connective tissue, blood, nervous tissue, and muscle (Figure 1.13). **Epithelial cells** form sheets that cover the surface of the body and line the internal organs. There are many different types of epithelial cells, each specialized for a specific function,

(A)i  Mouth

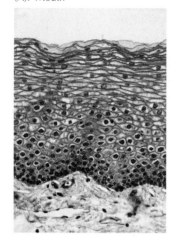

(A)ii  Bile duct

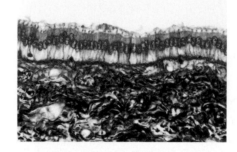

(A)iii  Intestine

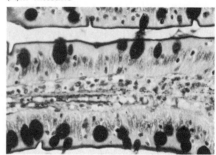

(B)

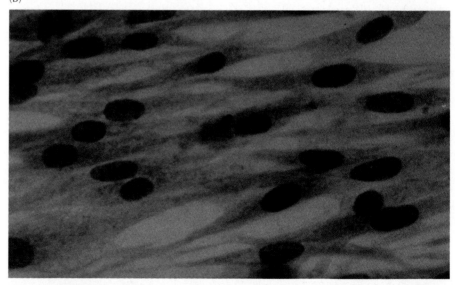

(C)

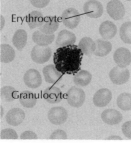

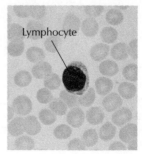

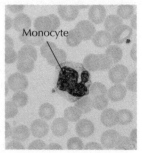

Erythrocyte

Granulocyte

Lymphocyte

Monocyte

*Figure 1.13*
**Light micrographs of representative animal cells**    (A) Epithelial cells of the mouth (a thick, multilayered sheet), bile duct, and intestine. (B) Fibroblasts are connective tissue cells characterized by their elongated spindle shape. (C) Erythrocytes, granulocytes, lymphocytes, and monocytes in human blood. [(A)i and (A)ii, G. W. Willis/Biological Photo Service; (A)iii, Biophoto Associates/Photo Researchers, Inc.; B, Don W. Fawcett/Visuals Unlimited; C, G. W. Willis/Biological Photo Service.]

including protection (the skin), absorption (e.g., the cells lining the small intestine), and secretion (e.g., cells of the salivary gland). Connective tissues include bone, cartilage, and adipose tissue, each of which is formed by different types of cells (osteoblasts, chondrocytes, and adipocytes, respectively). The loose connective tissue that underlies epithelial layers and fills the spaces between organs and tissues in the body is formed by another cell type, the **fibroblast**. Blood contains several different types of cells, which function in oxygen transport (red blood cells, or **erythrocytes**), inflammatory reactions (**granulocytes**, **monocytes**, and **macrophages**), and the immune response (**lymphocytes**). Nervous tissue is composed of nerve cells, or **neurons**, which are highly specialized to transmit signals throughout the body. Various types of sensory cells, such as cells of the eye and ear, are further specialized to receive external signals from the environment. Finally, several different types of muscle cells are responsible for the production of force and movement.

The evolution of animals clearly involved the development of considerable diversity and specialization at the cellular level. Understanding the mechanisms that control the growth and differentiation of such a complex array of specialized cells, starting from a single fertilized egg, is one of the major challenges facing contemporary cell and molecular biology.

## CELLS AS EXPERIMENTAL MODELS

The evolution of present-day cells from a common ancestor has important implications for cell and molecular biology as an experimental science. Because the fundamental properties of all cells have been conserved during evolution, the basic principles learned from experiments performed with one type of cell are generally applicable to other cells. On the other hand, because of the diversity of present-day cells, many kinds of experiments can be more readily undertaken with one type of cell than with another. Several different kinds of cells and organisms are commonly used as experimental models to study various aspects of cell and molecular biology. The features of some of these cells that make them particularly advantageous as experimental models are discussed in the sections that follow.

### E. coli

Because of their comparative simplicity, prokaryotic cells (bacteria) are ideal models for studying many fundamental aspects of biochemistry and molecular biology. The most thoroughly studied species of bacteria is *E. coli*, which has long been the favored organism for investigation of the basic mechanisms of molecular genetics. Most of our present concepts of molecular biology—including our understanding of DNA replication, the genetic code, gene expression, and protein synthesis—derive from studies of this humble bacterium.

*E. coli* has been especially useful to molecular biologists because of both its relative simplicity and the ease with which it can be propagated and studied in the laboratory. The genome of *E. coli*, for example, consists of approximately 4 million base pairs and encodes about 4000 different proteins. The human genome is nearly a thousand times more complex (approximately 3 billion base pairs) and encodes about 100,000 different proteins (see Table 1.2). The small size of the *E. coli* genome provides obvious advantages for genetic analysis, and the functions of nearly half of the total number of *E. coli* genes have been determined.

Molecular genetic experiments are further facilitated by the rapid growth of *E. coli* under well-defined laboratory conditions. Depending on the culture conditions, *E. coli* divide every 20 to 60 minutes. Moreover, a

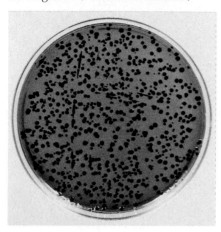

clonal population of *E. coli*, in which all cells are derived by division of a single cell of origin, can be readily isolated as a colony grown on semisolid agar-containing medium (Figure 1.14). Because bacterial colonies containing as many as $10^8$ cells can develop overnight, selecting genetic variants of an *E. coli* strain—for example, mutants that are resistant to an antibiotic, such as penicillin—is easy and rapid. The ease with which such mutants can be selected and analyzed was critical to the success of experiments that defined the basic principles of molecular genetics, discussed in Chapter 3.

The nutrient mixtures in which *E. coli* divide most rapidly include glucose, salts, and various organic compounds, such as amino acids, vitamins, and nucleic acid precursors. However, *E. coli* can also grow in much simpler media consisting only of salts, a source of nitrogen (such as ammonia), and a source of carbon and energy (such as glucose). In such a medium, the bacteria grow a little more slowly (with a division time of about 40 minutes) because they must synthesize all their own amino acids, nucleotides, and other organic compounds. The ability of *E. coli* to carry out these biosynthetic reactions in simple defined media has made them extremely useful in elucidating the biochemical pathways involved. Thus, the rapid growth and simple nutritional requirements of *E. coli* have greatly facilitated fundamental experiments in both molecular biology and biochemistry.

## Yeasts

Although bacteria have been an invaluable model for studies of many conserved properties of cells, they obviously cannot be used to study aspects of cell structure and function that are unique to eukaryotes. Yeasts, the simplest eukaryotes, have a number of experimental advantages similar to those of *E. coli*. Consequently, yeasts have provided a crucial model for studies of many fundamental aspects of eukaryotic cell biology.

The genome of the most frequently studied yeast, *Saccharomyces cerevisiae*, consists of 14 million base pairs of DNA—approximately three times more than the genome of *E. coli*, but far more manageable than the genomes of more complex eukaryotes, such as humans. Yet even in its simplicity, the yeast cell exhibits the typical features of eukaryotic cells (Figure 1.15): It contains a distinct nucleus surrounded by a nuclear membrane, its genomic DNA is organized as 16 linear chromosomes, and its cytoplasm contains a cytoskeleton and subcellular organelles.

Yeasts can be readily grown in the laboratory and can be studied by many of the same molecular genetic approaches that have proved so successful with *E. coli*. Although yeasts do not replicate as rapidly as bacteria, they still divide as frequently as every 2 hours and can easily be grown as colonies from a single cell. Consequently, yeasts can be used for a variety of genetic manipulations similar to those that can be performed using bacteria.

These features have made yeast cells the most approachable eukaryotic cells from the standpoint of molecular biology. Yeast mutants have been important in understanding many fundamental processes in eukaryotes, including DNA replication, transcription, RNA processing, protein sorting, and the regulation of cell division, as will be discussed in subsequent chapters. The unity of molecular cell biology is made abundantly clear by the fact that the general principles of cell structure and function revealed by studies of yeasts apply to all eukaryotic cells.

## Dictyostelium discoideum

*Dictyostelium discoideum* is a cellular slime mold, which, like yeast, is a comparatively simple unicellular eukaryote. The genome of *Dictyostelium* is approximately ten times larger than that of *E. coli*—more complex than the

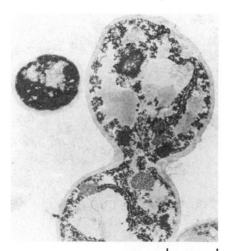

2 µm

yeast genome but considerably simpler than the genomes of higher eukaryotes. Moreover, *Dictyostelium* can be readily grown in the laboratory and is amenable to a variety of genetic manipulations.

Under conditions of plentiful food, *Dictyostelium* lives as a single-celled amoeba, feeding on bacteria and yeasts. It is a highly mobile cell, and this property has made *Dictyostelium* an important model for studying the molecular mechanisms responsible for animal cell movements (Figure 1.16). For example, introducing the appropriate mutations into *Dictyostelium* has revealed the roles of several genes in cell motility.

An additional interesting feature of *Dictyostelium* is the ability of single cells to aggregate into multicellular structures. If an adequate supply of food is not available, the cells associate to form wormlike structures called slugs, each consisting of up to 100,000 cells that function as a unit. *Dictyostelium* thus appears to straddle the border between unicellular and multicellular organisms, providing an important model for studies of cell signaling and cell–cell interactions.

**Figure 1.16** 10 μm
**Dictyostelium discoideum**
The amoeba is moving toward the pipette tip at the top of the micrograph. (Courtesy of Dr. Joel Swanson, Harvard Medical School.)

## Caenorhabditis elegans

The unicellular eukaryotes *Saccharomyces* and *Dictyostelium* are important models for studies of eukaryotic cells, but understanding the development of multicellular organisms requires the experimental analysis of plants and animals, organisms that are more complex. The nematode ***Caenorhabditis elegans*** (Figure 1.17) possesses several notable features that make it one of the most widely used models for studies of animal development and cell differentiation.

Although the genome of *C. elegans* (approximately 100 million base pairs) is larger than those of unicellular eukaryotes, it is simpler and more manageable than the genomes of most animals. *C. elegans* is a relatively simple multicellular organism: Adult worms consist of only 959 **somatic cells**, plus 1000 to 2000 **germ cells**. In addition, *C. elegans* can be easily grown and subjected to genetic manipulations in the laboratory.

The simplicity of *C. elegans* has enabled the course of its development to be studied in detail by microscopic observation. Such analyses have successfully traced the embryonic origin and lineage of all the cells in the adult worm. Genetic studies have also identified some of the mutations responsible for developmental abnormalities, leading to the isolation and characterization of critical genes that control nematode development and differentiation. Importantly, similar genes have also been found to function in complex animals (including humans), making *C. elegans* an important model for studies of animal development.

(a)

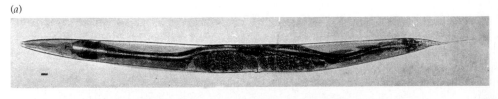

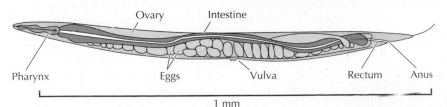

Pharynx  Ovary  Intestine  Eggs  Vulva  Rectum  Anus

1 mm

**Figure 1.17**
**Caenorhabditis elegans** (From J. E. Sulston and H. R. Horvitz, 1977, *Dev. Biol.* 56: 110.)

**Figure 1.18**
***Drosophila melanogaster*** (Darwin Dale/Photo Researchers, Inc.)

## Drosophila melanogaster

Like *C. elegans*, the fruit fly *Drosophila melanogaster* (Figure 1.18) has been a crucial model organism in developmental biology. The genome of *Drosophila* is similar in size to that of *C. elegans*, and *Drosophila* can be easily maintained and bred in the laboratory. Furthermore, the short reproductive cycle of *Drosophila* (about 2 weeks) makes it a very useful organism for genetic experiments. Many fundamental concepts of genetics—such as the relationship between genes and chromosomes—were derived from studies of *Drosophila* early in the twentieth century (see Chapter 3).

Extensive genetic analysis of *Drosophila* has uncovered many genes that control development and differentiation, and current methods of molecular biology have allowed the functions of these genes to be analyzed in detail. Consequently, studies of *Drosophila* have led to striking advances in understanding the molecular mechanisms that govern animal development, particularly with respect to formation of the body plan of complex multicellular organisms. As with *C. elegans*, similar genes and mechanisms exist in vertebrates, validating the use of *Drosophila* as a major experimental model in contemporary developmental biology.

## Arabidopsis thaliana

The study of plant molecular biology and development is an active and expanding field of considerable economic importance as well as intellectual interest. Since the genomes of plants cover a range of complexity comparable to that of animal genomes (see Table 1.2), an optimal model for studies of plant development would be a relatively simple organism with some of the advantageous properties of *C. elegans* and *Drosophila*. The small flowering plant ***Arabidopsis thaliana*** (Figure 1.19) meets these criteria and is therefore widely used as a model to study the molecular biology of plants.

*Arabidopsis* is notable for its genome of only about 70 million base pairs—a complexity only five times that of *Saccharomyces*, and similar to that of *C. elegans* and *Drosophila*. In addition, *Arabidopsis* is relatively easy to grow in the laboratory, and methods for molecular genetic manipulations of this plant have been developed. These studies have led to the identification of genes involved in various aspects of plant development, such as the development of flowers. Analysis of these genes points to clear similarities between the mechanisms that control the development of plants and animals, further emphasizing the fundamental unity of cell and molecular biology.

**Figure 1.19**
***Arabidopsis thaliana*** (Jeremy Burgess/Photo Researchers, Inc.)

## Vertebrates

The most complex animals are the vertebrates, including humans and other mammals. The human genome is approximately 3 billion base pairs—about 30 times larger than the genomes of *C. elegans*, *Drosophila*, or *Arabidopsis*. Moreover, the human body is composed of more than 200 different kinds of specialized cell types. This complexity makes the vertebrates difficult to study from the standpoint of cell and molecular biology, but much of the interest in biological sciences nonetheless stems from the desire to understand the human organism. Moreover, an understanding of many questions of immediate practical importance (e.g., in medicine) must be based directly on studies of human (or closely related) cell types.

One important approach to studying human and other mammalian cells is to grow isolated cells in culture, where they can be manipulated under controlled laboratory conditions. The use of cultured cells has allowed studies of many aspects of mammalian cell biology, including experiments that have elucidated the mechanisms of DNA replication, gene expression, protein synthesis and processing, and cell division. Moreover, the ability to culture cells in chemically defined media has allowed studies of the signaling mechanisms that normally control cell growth and differentiation within the intact organism.

The specialized properties of some highly differentiated cell types have made them important models for studies of particular aspects of cell biology. Muscle cells, for example, are highly specialized to undergo contraction, producing force and movement. Because of this specialization, muscle cells are a crucial model for studying cell movement at the molecular level. Another example is provided by nerve cells (neurons), which are specialized to conduct electrochemical signals over long distances. In humans, nerve cell axons may be more than a meter long, and some invertebrates, such as the squid, have giant neurons with axons as large as 1 mm in diameter. Because of their highly specialized structure and function, these giant neurons have provided important models for studies of ion transport across the plasma membrane, and of the role of the cytoskeleton in the transport of cytoplasmic organelles.

The frog *Xenopus laevis* is an important model for studies of early vertebrate development. *Xenopus* eggs are unusually large cells, with a diameter of approximately 1 mm (Figure 1.20). Because those eggs develop outside of the mother, all stages of development from egg to tadpole can be readily studied in the laboratory. In addition, *Xenopus* eggs can be obtained in large numbers, facilitating biochemical analysis. Because of these technical advantages, *Xenopus* has been widely used in studies of developmental biology and has provided important insights into the molecular mechanisms that control development, differentiation, and embryonic cell division.

Among mammals, the mouse is the most suitable for genetic analysis. Although the technical difficulties in studying mouse genetics (compared, for example, to the genetics of yeasts or *Drosophila*) are formidable, several mutations affecting mouse development have been identified. Most important, recent advances in molecular biology have enabled the production of **transgenic mice**, in which specific genes have been introduced into the mouse germ line, so that their effects on development or other aspects of cell function can be studied in the context of the whole animal. The suitability of the mouse as a model for human development is illustrated by the fact that mutations in homologous genes result in similar developmental defects in both species; piebaldism is a striking example (Figure 1.21).

*Figure 1.20*
**An egg of the frog *Xenopus laevis***
(Courtesy of Jeremy Green, Dana-Farber Cancer Institute.)

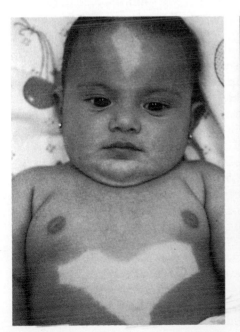

*Figure 1.21*
**The mouse as a model for human development** A child and a mouse show similar defects in pigmentation (piebaldism) as a result of mutations in a gene required for normal migration of melanocytes (the cells responsible for skin pigmentation) during embryonic development. (Courtesy of R. A. Fleischman, University of Texas Southwestern Medical Center.)

## TOOLS OF CELL BIOLOGY

As in all experimental sciences, research in cell biology depends on the laboratory methods that can be used to study cell structure and function. Many important advances in understanding cells have directly followed the development of new methods that have opened novel avenues of investigation. An appreciation of the experimental tools available to the cell biologist is thus critical to understanding both the current status and future directions of this rapidly moving area of science. Some of the important general methods of cell biology are described in the sections that follow. Other experimental approaches, including the methods of biochemistry and molecular biology, will be discussed in later chapters.

### Light Microscopy

Because most cells are too small to be seen by the naked eye, the study of cells has depended heavily on the use of microscopes. Indeed, the very discovery of cells arose from the development of the microscope: Robert Hooke first coined the term "cell" following his observations of a piece of cork with a simple light microscope in 1665 (Figure 1.22). Using a microscope that magnified objects up to about 300 times their actual size, Antony van Leeuwenhoek, in the 1670s, was able to observe a variety of different types of cells, including sperm, red blood cells, and bacteria. The proposal of the cell theory by Matthias Schleiden and Theodor Schwann in 1838 may be seen as the birth of contemporary cell biology. Microscopic studies of plant tissues by Schleiden and of animal tissues by Schwann led to the same conclusion: All organisms are composed of cells. Shortly thereafter, it was recognized that cells are not formed *de novo* but arise only from division of preexisting cells. Thus, the cell achieved its current recognition as the fundamental unit of all living organisms because of observations made with the light microscope.

The light microscope remains a basic tool of cell biologists, with technical improvements allowing the visualization of ever-increasing details of cell structure. Contemporary light microscopes are able to magnify objects up to about a thousand times. Since most cells are between 1 and 100 $\mu$m in diameter, they can be observed by light microscopy, as can some of the larger subcellular organelles, such as nuclei, chloroplasts, and mitochondria. However, the light microscope is not sufficiently powerful to reveal fine details of cell structure, for which **resolution**—the ability of a microscope to distinguish objects separated by small distances—is even more important than magnification. Images can be magnified as much as desired (for example, by projection onto a large screen) but such magnification does not increase the level of detail that can be observed.

The limit of resolution of the light microscope is approximately 0.2 $\mu$m; two objects separated by less than this distance appear as a single image, rather than being distinguished from one another. This theoretical limitation of light microscopy is determined by two factors—the wavelength ($\lambda$) of visible light and the light-gathering power of the microscope lens (numerical aperture, *NA*)—according to the following equation:

$$\text{Resolution} = \frac{0.61\lambda}{NA}$$

The wavelength of visible light is 0.4 to 0.7 $\mu$m, so the value of $\lambda$ is fixed at approximately 0.5 $\mu$m for the light microscope. The numerical aperture can be envisioned as the size of the cone of light that enters the microscope lens after passing through the specimen (Figure 1.23). It is given by the equation

*Figure 1.22*
**The cellular structure of cork**
A reproduction of Robert Hooke's drawing of a thin slice of cork examined with a light microscope. The "cells" that Hooke observed were actually only the cell walls remaining from cells that had long since died.

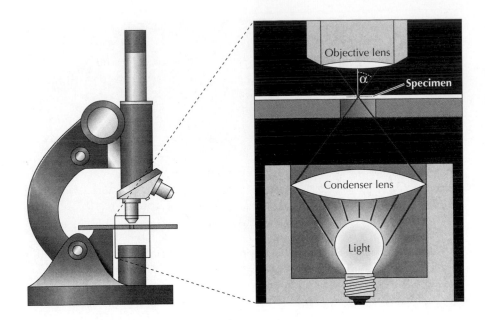

*Figure 1.23*
**Numerical aperture** Light is focused on the specimen by the condenser lens and then collected by the objective lens of the microscope. The numerical aperture is determined by the angle of the cone of light entering the objective lens ($\alpha$) and by the refractive index of the medium (usually air or oil) between the lens and the specimen.

$$NA = \eta \, \sin \alpha$$

where $\eta$ is the refractive index of the medium through which light travels between the specimen and the lens. The value of $\eta$ for air is 1.0, but it can be increased to a maximum of approximately 1.4 by using an oil-immersion lens to view the specimen through a drop of oil. The angle $\alpha$ corresponds to half the width of the cone of light collected by the lens. The maximum value of $\alpha$ is 90°, at which $\sin \alpha = 1$, so the highest possible value for the numerical aperture is 1.4.

The theoretical limit of resolution of the light microscope can therefore be calculated as follows:

$$\text{Resolution} = \frac{0.61 \times 0.5}{1.4} = 0.22 \; \mu\text{m}$$

Microscopes capable of achieving this level of resolution had been made already by the end of the nineteenth century; further improvements in this aspect of light microscopy cannot be expected.

Several different types of light microscopy are routinely used to study various aspects of cell structure. The simplest is **bright-field microscopy**, in which light passes directly through the cell and the ability to distinguish different parts of the cell depends on contrast resulting from the absorption of visible light by cell components. In many cases, cells are stained with dyes that react with proteins or nucleic acids in order to enhance the contrast between different parts of the cell. Prior to staining, specimens are usually treated with fixatives (such as alcohol, acetic acid, or formaldehyde) to stabilize and preserve their structures. The examination of fixed and stained tissues by bright-field microscopy is the standard approach for the analysis of tissue specimens in histology laboratories (Figure 1.24). Such staining procedures kill the cells, however, and therefore are not suitable for many experiments in which the observation of living cells is desired.

Without staining, the direct passage of light does not provide sufficient contrast to distinguish many parts of the cell, limiting the usefulness of bright-field microscopy. However, optical variations of the light microscope

*Figure 1.24*
**Bright-field micrograph of stained tissue**   Cross section of a hair follicle in human skin, stained with hematoxylin and eosin. (G. W. Willis/Biological Photo Service.)

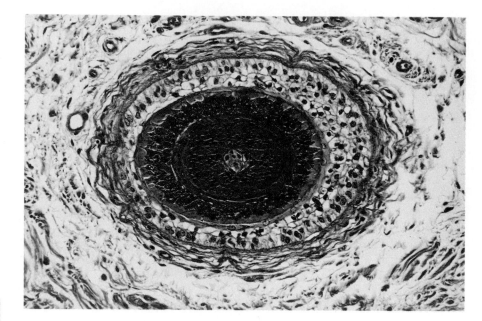

(A)

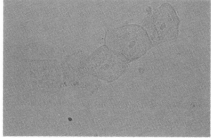

(B)

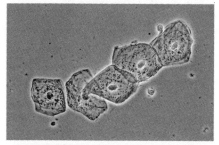

(C)

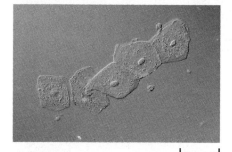

50 µm

*Figure 1.25*
**Microscopic observation of living cells**   Photomicrographs of human cheek cells obtained with (A) bright-field, (B) phase-contrast, and (C) differential interference-contrast microscopy. (Courtesy of Mort Abramowitz, Olympus America, Inc.)

can be used to enhance the contrast between light waves passing through regions of the cell with different densities. The two most common methods for visualizing living cells are **phase-contrast microscopy** and **differential interference-contrast microscopy** (Figure 1.25). Both kinds of microscopy use optical systems that convert variations in density or thickness between different parts of the cell to differences in contrast that can be seen in the final image. In bright-field microscopy, transparent structures (such as the nucleus) have little contrast because they absorb light poorly. However, light is slowed down as it passes through these structures so that its phase is altered compared to light that has passed through the surrounding cytoplasm. Phase-contrast and differential interference-contrast microscopy convert these differences in phase to differences in contrast, thereby yielding improved images of live, unstained cells.

The power of the light microscope has been considerably expanded by the use of video cameras and computers for image analysis and processing. Such electronic image-processing systems can substantially enhance the contrast of images obtained with the light microscope, allowing the visualization of small objects which otherwise could not be detected. For example, **video-enhanced differential interference-contrast microscopy** has allowed visualization of the movement of organelles along microtubules, which are cytoskeletal protein filaments with a diameter of only 0.025 µm (Figure 1.26). However, this enhancement does not overcome the theoretical limit of resolution of the light microscope, approximately 0.2 µm. Thus, although video enhancement allows the visualization of microtubules, the microtubules appear as blurred images at least 0.2 µm in diameter and an individual microtubule cannot be distinguished from a bundle of adjacent structures.

Light microscopy has been brought to the level of molecular analysis by methods for labeling specific molecules so that they can be visualized within cells. Specific genes or RNA transcripts can be detected by hybridization with nucleic acid probes of complementary sequence, and proteins can be detected using appropriate antibodies (see Chapter 3). Both nucleic acid probes and antibodies can be labeled with a variety of tags that allow their visualization in the light microscope, making it possible to determine the location of specific molecules within individual cells.

**Fluorescence microscopy** is a widely used and very sensitive method for studying the intracellular distribution of molecules (Figure 1.27). A fluorescent dye is used to label the molecule of interest within either fixed or living cells. The fluorescent dye is a molecule that absorbs light at one wavelength and emits light at a second wavelength. This fluorescence is detected by illuminating the specimen with a wavelength of light that excites the fluorescent dye and then using appropriate filters to detect the specific wavelength of light that the dye emits. Fluorescence microscopy can be used to study a variety of molecules within cells. One frequent application is to label antibodies directed against a specific protein with fluorescent dyes, so that the intracellular distribution of the protein can be determined.

**Confocal microscopy** combines fluorescence microscopy with electronic image analysis to obtain three-dimensional images. A small point of light, usually supplied by a laser, is focused on the specimen at a particular depth. The emitted fluorescent light is then collected using a detector, such as a video camera. Before the emitted light reaches the detector, however, it must pass through a pinhole aperture (called a confocal aperture) placed at precisely the point where light emitted from the chosen depth of the specimen comes to a focus (Figure 1.28). Consequently, only light emitted from the plane of focus is able to reach the detector. Scanning across the specimen generates a two-dimensional image of the plane of focus, a much sharper image than that obtained with standard fluorescence microscopy (Figure 1.29). Moreover, a series of images obtained at different depths can be used to reconstruct a three-dimensional image of the sample.

## Electron Microscopy

Because of the limited resolution of the light microscope, analysis of the details of cell structure has required the use of more powerful microscopic techniques—namely electron microscopy, which was developed in the 1930s and first applied to biological specimens by Albert Claude, Keith Porter, and George Palade in the 1940s and 1950s. The electron

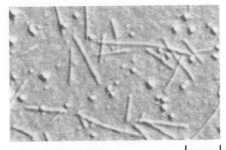

**Figure 1.26**
**Video-enhanced differential interference-contrast microscopy**
Electronic image processing allows the visualization of single microtubules. (Courtesy of E. D. Salmon, University of North Carolina, Chapel Hill.)

⊢————⊣ 2.5 µm

**Figure 1.27**
**Fluorescence microscopy** (A) Light passes through an excitation filter to select light of the wavelength (e.g., blue) that excites the fluorescent dye. A dichroic mirror then deflects the excitation light down to the specimen. The fluorescent light emitted by the specimen (e.g., green) then passes through the dichroic mirror and a second filter (the barrier filter) to select light of the wavelength emitted by the dye. (B) Fluorescence micrograph of a newt lung cell in which the DNA is stained blue and microtubules in the cytoplasm are stained green. (Conly S. Rieder/Biological Photo Service.)

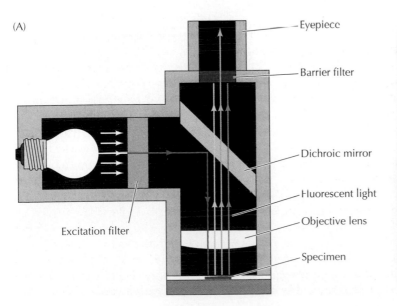

(A)

Eyepiece
Barrier filter
Dichroic mirror
Fluorescent light
Objective lens
Specimen
Excitation filter

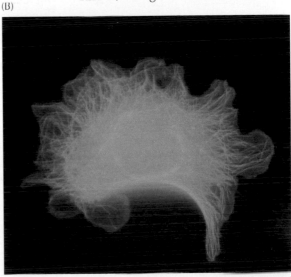

(B)

⊢————⊣ 10 µm

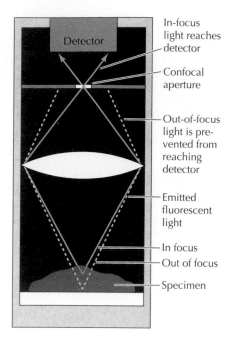

In-focus light reaches detector

Confocal aperture

Out-of-focus light is prevented from reaching detector

Emitted fluorescent light

In focus

Out of focus

Specimen

*Figure 1.28*
**Confocal microscopy**   A pinpoint of light is focused on the specimen at a particular depth, and emitted fluorescent light is collected by a detector. Before reaching the detector, the fluorescent light emitted by the specimen must pass through a confocal aperture placed at the point where light emitted from the chosen depth of the specimen comes into focus. As a result, only in-focus light emitted from the chosen depth of the specimen is detected.

microscope can achieve a much greater resolution than that obtained with the light microscope because the wavelength of electrons is shorter than that of light. The wavelength of electrons in an electron microscope can be as short as 0.004 nm—about 100,000 times shorter than the wavelength of visible light. Theoretically, this wavelength could yield a resolution of 0.002 nm, but such a resolution cannot be obtained in practice, because resolution is determined not only by wavelength, but also by the numerical aperture of the microscope lens. Numerical aperture is a limiting factor for electron microscopy because inherent properties of electromagnetic lenses limit their aperture angles to about 0.5 degrees, corresponding to numerical apertures of only about 0.01. Thus, under optimal conditions, the resolving power of the electron microscope is approximately 0.2 nm. Moreover, the resolution that can be obtained with biological specimens is further limited by their lack of inherent contrast. Consequently, for biological samples the practical limit of resolution of the electron microscope is 1 to 2 nm. Although this resolution is much less than that predicted simply from the wavelength of electrons, it represents more than a hundredfold improvement over the resolving power of the light microscope.

Two types of electron microscopy—transmission and scanning—are widely used to study cells. In principle, **transmission electron microscopy** is similar to the observation of stained cells with the bright-field light microscope. Specimens are fixed and stained with salts of heavy metals, which provide contrast by scattering electrons. A beam of electrons is then

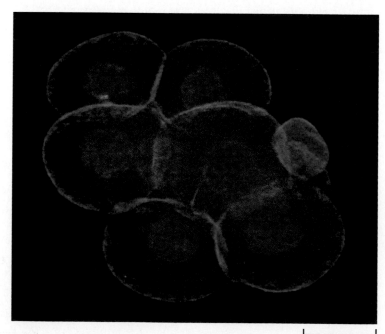

50 μm

*Figure 1.29*
**Confocal micrograph of mouse embryo cells**
Nuclei are stained red and actin filaments underlying the plasma membrane are stained green. (Courtesy of David Albertini, Tufts University School of Medicine.)

passed through the specimen and focused to form an image on a fluorescent screen. Electrons that encounter a heavy metal ion as they pass through the sample are deflected and do not contribute to the final image, so stained areas of the specimen appear dark.

Specimens to be examined by transmission electron microscopy can be prepared by either positive or negative staining. In positive staining, tissue specimens are cut into thin sections and stained with heavy metal salts (such as osmium tetroxide, uranyl acetate, and lead citrate) that react with lipids, proteins, and nucleic acids. These heavy metal ions bind to a variety of cell structures, which consequently appear dark in the final image (Figure 1.30). Alternative positive-staining procedures can also be used to identify specific macromolecules within cells. For example, antibodies labeled with electron-dense heavy metals (such as gold particles) are frequently used to determine the subcellular location of specific proteins in the electron microscope. This method is similar to the use of antibodies labeled with fluorescent dyes in fluorescence microscopy.

Negative staining is useful for the visualization of intact biological structures, such as bacteria, isolated subcellular organelles, and macromolecules (Figure 1.31). In this method, the biological specimen is deposited on a supporting film, and a heavy metal stain is allowed to dry around its surface. The unstained specimen is then surrounded by a film of electron-dense stain, producing an image in which the specimen appears light against a stained dark background.

**Metal shadowing** is another technique used to visualize the surface of isolated subcellular structures or macromolecules in the transmission electron microscope (Figure 1.32). The specimen is coated with a thin layer of evaporated metal, such as platinum. The metal is sprayed onto the specimen from an angle so that surfaces of the specimen that face the source of evaporated metal molecules are coated more heavily than others. This differential coating creates a shadow effect, giving the specimen a three-dimensional appearance in electron micrographs.

The preparation of samples by **freeze fracture**, in combination with metal shadowing, has been particularly important in studies of membrane structure. Specimens are frozen in liquid nitrogen (at –196°C) and then fractured with a knife blade. This process frequently splits the lipid bilayer, revealing the interior faces of a cell membrane (Figure 1.33). The specimen is then shadowed with platinum, and the biological material is dissolved with acid, producing a metal replica of the surface of the sample. Examination of such replicas in the electron microscope reveals many surface

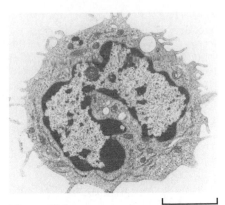

**Figure 1.30**     5 µm
**Positive staining**
Transmission electron micrograph of a positively stained white blood cell. (Don W. Fawcett/Visuals Unlimited.)

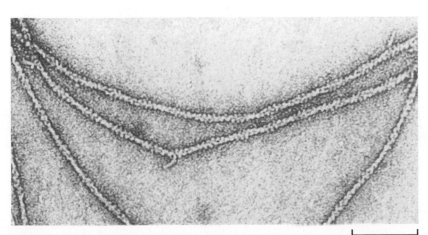

50 nm

**Figure 1.31**
**Negative staining**   Transmission electron micrograph of negatively stained actin filaments. (Courtesy of Roger Craig, University of Massachusetts Medical Center.)

*Figure 1.32*
**Metal shadowing**  Electron micrograph of actin/myosin filaments of the cytoskeleton prepared by metal shadowing. (Don Fawcett, J. Heuser/Photo Researchers, Inc.)

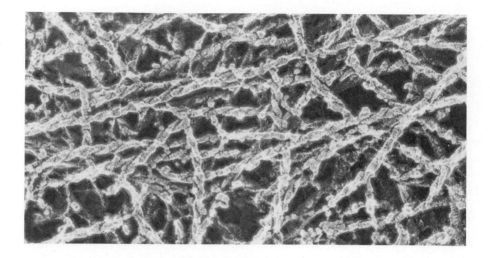

*Figure 1.33*
**Freeze fracture**  (A) Freeze fracture splits the lipid bilayer, leaving proteins embedded in the membrane associated with one of the two membrane halves. (B) Micrograph of freeze-fractured plasma membranes of two adjacent cells. Proteins that span the bilayer appear as intramembranous particles (arrow). (Don W. Fawcett/Photo Researchers, Inc.)

(A)

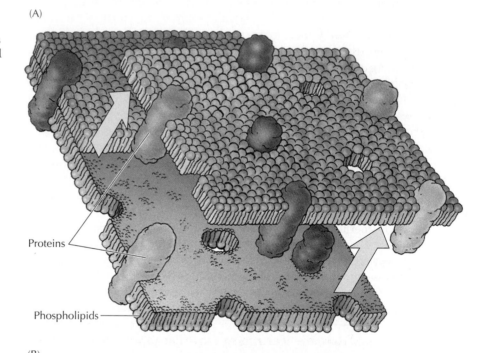

Proteins

Phospholipids

(B)

*Figure 1.34*
**Scanning electron microscopy**   Scanning electron micrograph of a macrophage.
(David Phillips/Visuals Unlimited.)

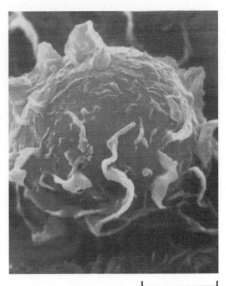

5 μm

bumps, corresponding to proteins that span the lipid bilayer. A variation of freeze fracture called **freeze etching** allows visualization of the external surfaces of cell membranes in addition to their interior faces.

The second type of electron microscopy, **scanning electron microscopy**, is used to provide a three-dimensional image of cells (Figure 1.34). In scanning electron microscopy the electron beam does not pass through the specimen. Instead, the surface of the cell is coated with a heavy metal, and a beam of electrons is used to scan across the specimen. Electrons that are scattered or emitted from the sample surface are collected to generate a three-dimensional image as the electron beam moves across the cell. Because the resolution of scanning electron microscopy is only about 10 nm, its use is generally restricted to studying whole cells rather than subcellular organelles or macromolecules.

## Subcellular Fractionation

Although the electron microscope has allowed detailed visualization of cell structure, microscopy alone is not sufficient to define the functions of the various components of eukaryotic cells. To address many questions concerning the function of subcellular organelles, it has proven necessary to isolate the organelles of eukaryotic cells in a form that can be used for biochemical studies. This is usually accomplished by **differential centrifugation**—a method developed largely by Albert Claude, Christian de Duve, and their colleagues in the 1940s and 1950s to separate the components of cells on the basis of their size and density.

The first step in subcellular fractionation is the disruption of the plasma membrane under conditions that do not destroy the internal components of the cell. Several methods are used, including sonication (exposure to high-frequency sound), grinding in a mechanical homogenizer, or treatment with a high-speed blender. All these procedures break the plasma membrane and the endoplasmic reticulum into small fragments while leaving other components of the cell (such as nuclei, lysosomes, peroxisomes, mitochondria, and chloroplasts) intact.

The suspension of broken cells (called a lysate or homogenate) is then fractionated into its components by a series of centrifugations in an **ultracentrifuge**, which rotates samples at very high speeds (up to 100,000 rpm) to produce forces up to 500,000 times greater than gravity. This force causes cell components to move toward the bottom of the centrifuge tube and form a pellet (a process called sedimentation) at a rate that depends on their size and density, with the largest and heaviest structures sedimenting most rapidly (Figure 1.35). Usually the cell homogenate is first centrifuged at a low speed, which sediments only unbroken cells and the largest subcellular structures—the nuclei. Thus, an enriched fraction of nuclei can be recovered from the pellet of such a low-speed centrifugation while the other cell components remain suspended in the supernatant (the remaining solution). The supernatant is then centrifuged at higher speed to sediment mitochondria, chloroplasts, lysosomes, and peroxisomes. Recentrifugation of the supernatant at still higher speed sediments fragments of the plasma membrane and the endoplasmic reticulum. A fourth centrifugation at still higher speed sediments ribosomes, leaving only the soluble portion of the cytoplasm (the cytosol) in the supernatant.

*Figure 1.35*
**Subcellular fractionation** Cells are lysed and subcellular components are separated by a series of centrifugations at increasing speeds. Following each centrifugation, the organelles that have sedimented to the bottom of the tube are recovered in the pellet. The supernatant is then recentrifuged at higher speed to sediment the next-largest organelles.

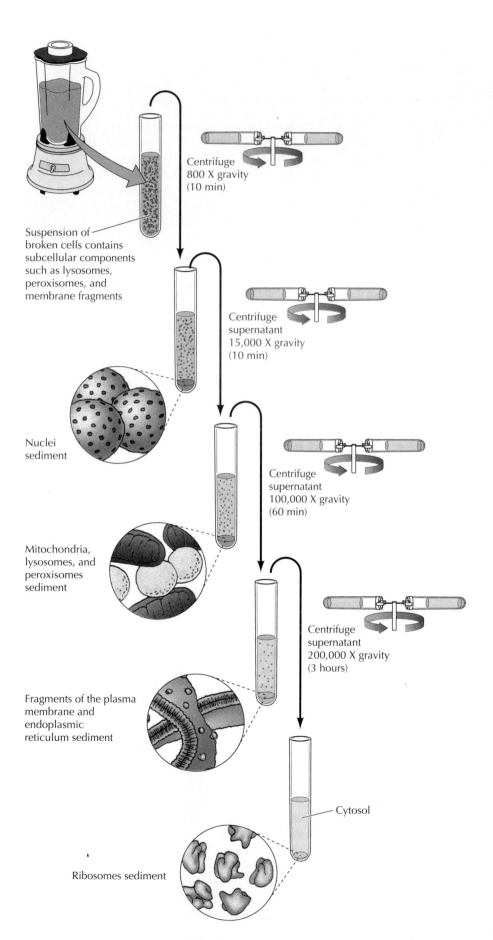

Suspension of broken cells contains subcellular components such as lysosomes, peroxisomes, and membrane fragments

Centrifuge 800 X gravity (10 min)

Centrifuge supernatant 15,000 X gravity (10 min)

Nuclei sediment

Centrifuge supernatant 100,000 X gravity (60 min)

Mitochondria, lysosomes, and peroxisomes sediment

Centrifuge supernatant 200,000 X gravity (3 hours)

Fragments of the plasma membrane and endoplasmic reticulum sediment

Cytosol

Ribosomes sediment

The fractions obtained from differential centrifugation correspond to enriched, but still not pure, organelle preparations. A greater degree of purification can be achieved by **density-gradient centrifugation**, in which organelles are separated by sedimentation through a gradient of a dense substance, such as sucrose. In **velocity centrifugation**, the starting material is layered on top of the sucrose gradient (Figure 1.36). Particles of different sizes sediment through the gradient at different rates, moving as discrete bands. Following centrifugation, the collection of individual fractions of the gradient provides sufficient resolution to separate organelles of similar size, such as mitochondria, lysosomes, and peroxisomes.

**Equilibrium centrifugation** in density gradients can be used to separate subcellular components on the basis of their buoyant density, independent of their size and shape. In this procedure, the sample is centrifuged in a gradient containing a high concentration of sucrose or cesium chloride. Rather than being separated on the basis of their sedimentation velocity, the sample particles are centrifuged until they reach an equilibrium posi-

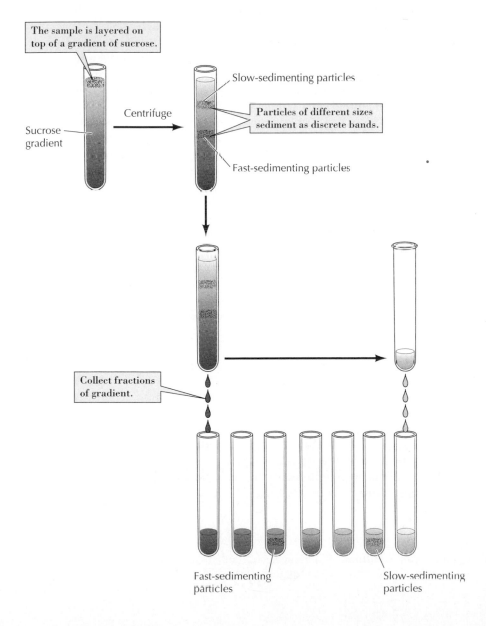

*Figure 1.36*
**Velocity centrifugation in a density gradient**   The sample is layered on top of a gradient of sucrose, and particles of different sizes sediment through the gradient as discrete bands. The separated particles can then be collected in individual fractions of the gradient, which can be obtained simply by puncturing the bottom of the centrifuge tube and collecting drops.

tion at which their buoyant density is equal to that of the surrounding sucrose or cesium chloride solution. Such equilibrium centrifugations are useful in separating different types of membranes from one another and are sufficiently sensitive to separate macromolecules that are labeled with different isotopes. A classic example, discussed in Chapter 3, is the analysis of DNA replication by separating DNA molecules containing heavy and light isotopes of nitrogen ($^{15}$N and $^{14}$N) by equilibrium centrifugation in cesium chloride gradients.

### Growth of Animal Cells in Culture

The ability to study cells depends largely on how readily they can be grown and manipulated in the laboratory. Although the process is technically far more difficult than the culture of bacteria or yeasts, a wide variety of animal and plant cells can be grown and manipulated in culture. Such *in vitro* cell culture systems have enabled scientists to study cell growth and differentiation, as well as to perform genetic manipulations required to understand gene structure and function.

Animal cell cultures are initiated by the dispersion of a piece of tissue into a suspension of its component cells, which is then added to a culture dish containing nutrient media. Most animal cell types, such as fibroblasts and epithelial cells, attach and grow on the plastic surface of dishes used for cell culture (Figure 1.37). Because they contain rapidly growing cells, embryos or tumors are frequently used as starting material. Embryo fibroblasts grow particularly well in culture and consequently are one of the most widely studied types of animal cells. Under appropriate conditions, however, some specialized cell types can also be grown in culture, allowing their differentiated properties to be studied in a controlled experimental environment.

The culture media required for the propagation of animal cells are much more complex than the minimal media sufficient to support the growth of bacteria and yeasts. Early studies of cell culture utilized media consisting of undefined components, such as plasma, serum, and embryo extracts. A major advance was thus made in 1955, when Harry Eagle described the first defined media that supported the growth of animal cells. In addition to salts and glucose, the media used for animal cell cultures contain various amino acids and vitamins, which the cells cannot make for themselves. The growth media for most animal cells in culture also include serum, which serves as a source of polypeptide growth factors that are required to stimulate cell division. Several such growth factors have been identified. They serve as critical regulators of cell growth and differentiation in multicellular organisms, providing signals by which different cells communicate with each other. For example, an important function of skin fibroblasts in the intact animal is to proliferate when needed to repair damage resulting from a cut or wound. Their division is triggered by a growth factor released from platelets during blood clotting, thereby stimulating proliferation of fibroblasts in the neighborhood of the damaged tissue. The identification of individual growth factors has made possible the culture of a variety of cells in serum-free media (media in which serum has been replaced by the specific growth factors required for proliferation of the cells in question).

10 μm

**Figure 1.37**
**Animal cells in culture** Scanning electron micrograph of human fibroblasts attached to the surface of a culture dish. (David M. Phillips/Visuals Unlimited.)

## KEY EXPERIMENT

# Animal Cell Culture

### Nutrition Needs of Mammalian Cells in Tissue Culture

Harry Eagle

National Institutes of Health, Bethesda, MD

*Science, Volume 122, 1955, pages 501–504*

Harry Eagle

## The Context

The earliest cell cultures involved the growth of cells from fragments of tissue that were embedded in clots of plasma—a culture system that was far from suitable for experimental analysis. In the late 1940s, a major advance was the establishment of cell lines that grew from isolated cells attached to the surface of culture dishes. But these cells were still grown in undefined media consisting of varying combinations of serum and embryo extracts. For example, a widely used human cancer cell line (called HeLa cells) was initially established in 1952 by growth in a medium consisting of chicken plasma, bovine embryo extract, and human placental cord serum. The use of such complex and undefined culture media made analysis of the specific growth requirements of animal cells impossible. Harry Eagle was the first to solve this problem, by carrying out a systematic analysis of the nutrients needed to support the growth of animal cells in culture.

## The Experiments

Eagle studied the growth of two established cell lines: HeLa cells and a mouse fibroblast line called L cells. He was able to grow these cells in a medium consisting of a mixture of salts, carbohydrates, amino acids, and vitamins, supplemented with serum protein. By systematically varying the components of this medium, Eagle was able to determine the specific nutrients required for cell growth. In addition to salts and glucose, these nutrients included 13 amino acids and several vitamins. A small amount of serum protein was also required. The basal medium developed by Eagle is described in the accompanying table, reprinted from his 1955 paper.

## The Impact

The medium developed by Eagle is still the basic medium used for animal cell culture. Its use has enabled scientists to grow a wide variety of cells under defined experimental conditions, which has been critical to studies of animal cell growth and differentiation, including identification of the growth factors present in serum—now known to include polypeptides that control the behavior of individual cells within intact animals.

Table 4. Basal media for cultivation of the HeLa cell and mouse fibroblast (*10*)

| L-Amino acids* (mM) | | | Vitamins‡ (mM) | | Miscellaneous | |
|---|---|---|---|---|---|---|
| Arginine | 0.1 | | Biotin | $10^{-3}$ | Glucose | 5mM§ |
| Cystine | 0.05 | (0.02)† | Choline | $10^{-3}$ | Penicillin | 0.005%# |
| Glutamine | 2.0 | (1.0)‖ | Folic acid | $10^{-3}$ | Streptomycin | 0.005%# |
| Histidine | 0.05 | (0.02)† | Nicotinamide | $10^{-3}$ | Phenol red | 0.0005%# |
| Isoleucine | 0.2 | | Pantothenic acid | $10^{-3}$ | | |
| Leucine | 0.2 | (0.1)† | Pyridoxal | $10^{-3}$ | | |
| Lysine | 0.2 | (0.1)† | Thiamine | $10^{-3}$ | For studies of cell nutrition | |
| Methionine | 0.05 | | Riboflavin | $10^{-4}$ | Dialyzed horse serum, 1%† | |
| Phenylalanine | 0.1 | (0.05)† | | | Dialyzed human serum, 5% | |
| Threonine | **0.2** | (0.1)† | Salts§ | | | |
| Tryptophan | 0.02 | (0.01)† | (mM) | | | |
| Tyrosine | 0.1 | | | | For stock cultures | |
| Valine | 0.2 | (0.1)† | NaCl | 100 | Whole horse serum, 5%† | |
| | | | KCl | 5 | Whole human serum, 10% | |
| | | | $NaH_2PO_4 \cdot H_2O$ | 1 | | |
| | | | $NaHCO_3$ | 20 | | |
| | | | $CaCl_2$ | 1 | | |
| | | | $MgCl_2$ | 0.5 | | |

\* Conveniently stored in the refrigerator as a single stock solution containing 20 times the indicated concentration of each amino acid.
† For mouse fibroblast.
‡ Conveniently stored as a single stock solution containing 100 or 1000 times the indicated concentration of each vitamin; kept frozen.
§ Conveniently stored in the refrigerator in two stock solutions, one containing NaCl, KCl, $NaH_2PO_4$, $NaHCO_3$, and glucose at 10 times the indicated concentration of each, and the second containing $CaCl_2$ and $MgCl_2$ at 20 times the indicated concentration.
‖ Conveniently stored as a 100mM stock solution; frozen when not in use.
# Conveniently stored as a single stock solution containing 100 times the indicated concentrations of penicillin, streptomycin, and phenol red.

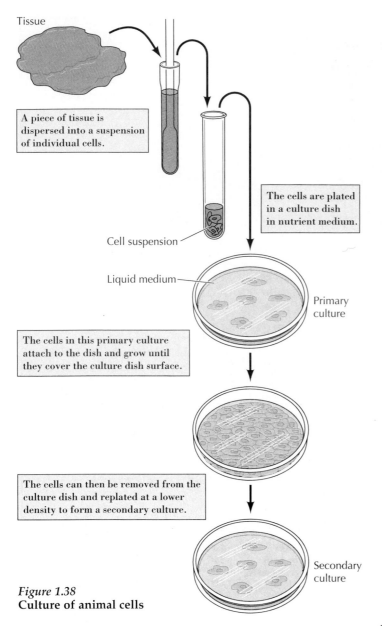

Tissue

A piece of tissue is dispersed into a suspension of individual cells.

The cells are plated in a culture dish in nutrient medium.

Cell suspension

Liquid medium

Primary culture

The cells in this primary culture attach to the dish and grow until they cover the culture dish surface.

The cells can then be removed from the culture dish and replated at a lower density to form a secondary culture.

Secondary culture

*Figure 1.38*
**Culture of animal cells**

The initial cell cultures established from a tissue are called **primary cultures** (Figure 1.38). The cells in a primary culture usually grow until they cover the culture dish surface. They can then be removed from the dish and replated at a lower density to form secondary cultures. This process can be repeated many times, but most normal cells cannot be grown in culture indefinitely. For example, normal human fibroblasts can usually be cultured for 50 to 100 population doublings, after which they stop growing and die. In contrast, cells derived from tumors frequently proliferate indefinitely in culture and are referred to as **immortal cell lines**. In addition, a number of immortalized rodent cell lines have been isolated from cultures of normal fibroblasts. Instead of dying as most of their counterparts do, a few cells in these cultures continue proliferating indefinitely, forming cell lines like those derived from tumors. Such permanent cell lines have been particularly useful for many types of experiments because they provide a continuous and uniform source of cells that can be manipulated, cloned, and indefinitely propagated in the laboratory.

Even under optimal conditions, the division time of most actively growing animal cells is on the order of 20 hours—ten times longer than the division time of yeasts. Consequently, experiments with cultured animal cells are more difficult and take much longer than those with bacteria or yeasts. For example, the growth of a visible colony of animal cells from a single cell takes a week or more, whereas colonies of *E. coli* or yeast develop from single cells overnight. Nonetheless, genetic manipulations of animal cells in culture have been indispensable to our understanding of cell structure and function.

### Culture of Plant Cells

Plant cells can also be cultured in nutrient media containing appropriate growth regulatory molecules. In contrast to the polypeptide growth factors that regulate the proliferation of most animal cells, the growth regulators of plant cells are small molecules that can pass through the plant cell wall. When provided with appropriate mixtures of these growth regulatory molecules, many types of plant cells proliferate in culture, producing a mass of undifferentiated cells called a callus (Figure 1.39).

A striking feature of plant cells that contrasts sharply to the behavior of animal cells is the phenomenon called **totipotency**. Differentiated animal cells, such as fibroblasts, cannot develop into other cell types, such as nerve cells. Many plant cells, however, are capable of forming any of the different cell types and tissues ultimately needed to regenerate an entire plant. Consequently, by appropriate manipulation of nutrients and growth regulatory molecules, undifferentiated plant cells in culture can be induced to form a variety of plant tissues, including roots, stems, and leaves. In many cases, even an entire plant can be regenerated from a single cultured cell. In addi-

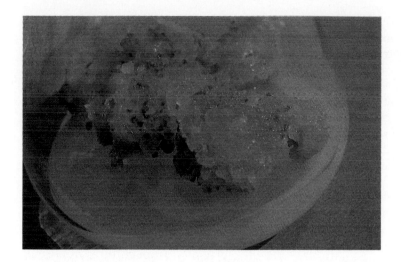

*Figure 1.39*
**Plant cells in culture**   An undifferentiated mass of plant cells (a callus) growing on a solid medium. (John N.A. Lott/Biological Photo Service.)

tion to its theoretical interest, the ability to produce a new plant from a single cell that has been manipulated in culture makes it easy to introduce genetic alterations into plants, opening important possibilities for agricultural genetic engineering.

## Viruses

Viruses are intracellular parasites that cannot replicate on their own. They reproduce by infecting host cells and usurping the cellular machinery to produce more virus particles. In their simplest forms, viruses consist only of genomic nucleic acid (either DNA or RNA) surrounded by a protein coat (Figure 1.40). Viruses are important in molecular and cellular biology because they provide simple systems that can be used to investigate the functions of cells. Because virus replication depends on the metabolism of the infected cells, studies of viruses have revealed many fundamental aspects of cell biology. Studies of bacterial viruses contributed substantially to our understanding of the basic mechanisms of molecular genetics, and experiments with a plant virus (tobacco mosaic virus) first demonstrated the genetic potential of RNA. Animal viruses have provided particularly sensitive probes for investigations of various activities of eukaryotic cells.

The rapid growth and small genome size of bacteria make them excellent subjects for experiments in molecular biology, and bacterial viruses (**bacteriophages**) have simplified the study of bacterial genetics even further. One of the most important bacteriophages is T4, which infects and replicates in *E. coli*. Infection with a single particle of T4 leads to the formation of approximately 200 progeny virus particles in 20 to 30 minutes. The initially infected cell then bursts (lyses), releasing progeny virus particles into the medium, where they can infect new cells. In a culture of bacteria growing on agar medium, the replication of T4 leads to the formation of a clear area of lysed cells (a plaque) in the lawn of bacteria (Figure 1.41). Just as infectious virus particles are easy to grow and assay, viral mutants—for example, viruses that will grow in one strain of *E. coli* but not another—are easy to isolate. Thus, T4 is manipulated even more readily than *E. coli* for studies of molecular genetics. Moreover, the genome of T4 is 20 times smaller than that of *E. coli*—approximately 0.2 million base pairs—further facilitating genetic analysis. Some other bacteriophages have even smaller genomes—the simplest consisting of RNA molecules of only about 3600 nucleotides.

*Figure 1.40*
**Structure of an animal virus**   (A) Papillomavirus particles contain a small circular DNA molecule enclosed in a protein coat (the capsid). (B) Electron micrograph of human papillomavirus particles. Artificial color has been added. (Alfred Pasieka/Science Photo Library/Photo Researchers, Inc.)

(A)

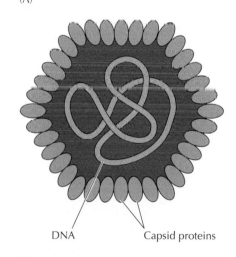

DNA                    Capsid proteins

(B)

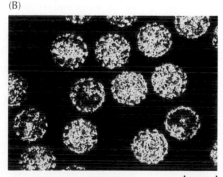

50 nm

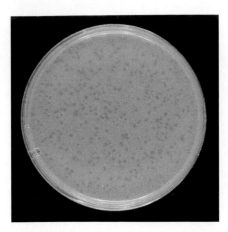

**Figure 1.41**
**Bacteriophage plaques**    T4 plaques are visible on a lawn of *E. coli*. Each plaque arises by the replication of a single virus particle. (E. C. S. Chen/Visuals Unlimited.)

Bacterial viruses have thus provided extremely facile experimental systems for molecular genetics. Studies of these viruses are largely what have led to the elucidation of many fundamental principles of molecular biology.

Because of the increased complexity of the animal cell genome, viruses have been even more important in studies of animal cells than in studies of bacteria. Many animal viruses replicate and can be assayed by plaque formation in cell cultures, much as bacteriophages can. Moreover, the genomes of animal viruses are similar in complexity to those of bacterial viruses (ranging from approximately 3000 to 300,000 base pairs), so animal viruses are far more manageable than are their host cells.

There are many diverse animal viruses, each containing either DNA or RNA as their genetic material (Table 1.3). One family of animal viruses—the **retroviruses**—contain RNA genomes in their virus particles but synthesize a DNA copy of their genome in infected cells. These viruses provide a good example of the importance of viruses as models, because studies of the retroviruses are what first demonstrated the synthesis of DNA from RNA templates—a fundamental mode of genetic information transfer now known to occur in both prokaryotic and eukaryotic cells. Other examples in which animal viruses have provided important models for investigations of their host cells include studies of DNA replication, transcription, RNA processing, and protein transport and secretion.

It is particularly noteworthy that infection by some animal viruses, rather than killing the host cell, converts a normal cell into a cancer cell. Studies of such cancer-causing viruses, first described by Peyton Rous in 1911, not only have provided the basis for our current understanding of cancer at the level of cell and molecular biology, but also have led to the elucidation of many of the molecular mechanisms that control animal cell growth and differentiation.

**Table 1.3** Examples of Animal Viruses

| Virus family | Representative member | Genome size (thousands of base pairs) |
|---|---|---|
| **RNA genomes** | | |
| Picornaviruses | Poliovirus | 7–8 |
| Togaviruses | Rubella virus | 12 |
| Flaviviruses | Yellow fever virus | 10 |
| Paramyxoviruses | Measles virus | 16–20 |
| Orthomyxoviruses | Influenza virus | 14 |
| Retroviruses | Human immunodeficiency virus | 9 |
| **DNA genomes** | | |
| Hepadnaviruses | Hepatitis B virus | 3.2 |
| Papovaviruses | Human papillomavirus | 5–8 |
| Adenoviruses | Adenovirus | 36 |
| Herpesviruses | Herpes simplex virus | 120–200 |
| Poxviruses | Vaccinia virus | 130–280 |

## MOLECULAR MEDICINE

# Viruses and Cancer

### The Disease

Cancer is a family of diseases characterized by uncontrolled cell proliferation. The growth of normal animal cells is carefully regulated to meet the needs of the complete organism. In contrast, cancer cells grow in an unregulated manner, ultimately invading and interfering with the function of normal tissues and organs. Cancer is the second most common cause of death (next to heart disease) in the United States. Approximately one out of every three Americans will develop cancer at some point in life and, in spite of major advances in treatment, nearly one out of every four Americans ultimately die of this disease. Understanding the causes of cancer and developing more effective methods of cancer treatment therefore represent major goals of medical research.

### Molecular and Cellular Basis

Cancer is now known to result from mutations in the genes that normally control cell proliferation. The major insights leading to identification of these genes came from studies of viruses that cause cancer in animals, the prototype of which was isolated by Peyton Rous in 1911. Rous found that sarcomas (a cancer of connective tissues) in chickens could be transmitted by a virus, now known as Rous sarcoma virus, or RSV. Because RSV is a retrovirus with a genome of only 10,000 base pairs, it can be subjected to molecular analysis much more readily than the complex genomes of chickens or other animal

cells can. Such studies eventually led to identification of a specific cancer-causing gene (oncogene) carried by the virus, and to the discovery of related genes in normal cells of all vertebrate species, including humans. Some cancers in humans are now known to be caused by viruses; others result from mutations in normal cell genes similar to the oncogene first identified in RSV.

### Prevention and Treatment

The human cancers that are caused by viruses include cervical and other anogenital cancers (papillomaviruses), liver cancer (hepatitis B virus), and some types of lymphomas (Epstein-Barr virus and human T-cell lymphotropic virus). Together, these virus-induced cancers account for about 20% of worldwide cancer incidence. In principle, these cancers could be prevented by vaccination against the

responsible viruses, and considerable progress in this area has been made by the development of an effective vaccine against hepatitis B virus.

Other human cancers are caused by mutations in normal cell genes, most of which occur during the lifetime of the individual rather than being inherited. Studies of cancer-causing viruses have led to the identification of many of the genes responsible for non-virus-induced cancers, and to an understanding of the molecular mechanisms responsible for cancer development. Major efforts are now under way to use these insights into the molecular and cellular biology of cancer to develop new approaches to cancer treatment.

### Reference

Rous, P. 1911. A sarcoma of the fowl transmissible by an agent separable from the tumor cells. *J. Exp. Med.* 13: 397–411.

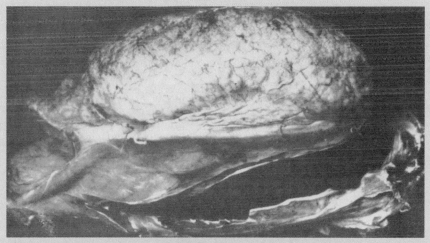

The transplantable tumor from which Rous sarcoma virus was isolated.

## KEY TERMS

prokaryotic cell, eukaryotic cell, RNA world, phospholipid, amphipathic, hydrophobic, hydrophilic

adenosine 5′-triphosphate (ATP), glycolysis, photosynthesis, oxidative metabolism

archaebacteria, eubacteria, cyanobacteria, *Escherichia coli* (*E. coli*), cell wall, plasma membrane, ribosome

nucleus, mitochondrion, chloroplast, lysosome, peroxisome, vacuole, endoplasmic reticulum, Golgi apparatus, cytoskeleton, endosymbiosis

yeast, *Saccharomyces cerevisiae*, pseudopodium, parenchyma cell, collenchyma cell, schlerenchyma cell, epidermal cell, epithelial cell, fibroblast, erythrocyte, granulocyte, monocyte, macrophage, lymphocyte, neuron

*Dictyostelium discoideum*

*Caenorhabditis elegans*, somatic cell, germ cell

*Drosophila melanogaster*

*Arabidopsis thaliana*

*Xenopus laevis*, transgenic mouse

## *Summary*

### THE ORIGIN AND EVOLUTION OF CELLS

*The First Cell:* All present-day cells, both prokaryotes and eukaryotes, are descended from a single ancestor. The first cell is thought to have arisen between 3.8 billion and 3.5 billion years ago as a result of enclosure of self-replicating RNA in a phospholipid membrane.

*The Evolution of Metabolism:* The earliest reactions for the generation of metabolic energy were a form of anaerobic glycolysis. Photosynthesis then evolved, followed by oxidative metabolism.

*Present-Day Prokaryotes:* Present-day prokaryotes are divided into two groups, the archaebacteria and the eubacteria, which diverged early in evolution.

*Eukaryotic Cells:* Eukaryotic cells, which are larger and more complex than prokaryotic cells, contain a nucleus, cytoplasmic organelles, and a cytoskeleton. They are thought to have evolved from symbiotic associations of prokaryotes.

*The Development of Multicellular Organisms:* The simplest eukaryotes are unicellular organisms, such as yeasts and amoebas. Multicellular organisms evolved from associations between such unicellular eukaryotes, and division of labor led to the development of the many kinds of specialized cells that make up present-day plants and animals.

### CELLS AS EXPERIMENTAL MODELS

*E. coli:* Because of their genetic simplicity and ease of study, bacteria such as *E. coli* are particularly useful for investigation of fundamental aspects of biochemistry and molecular biology.

*Yeasts:* As the simplest eukaryotic cells, yeasts are an important model for studying various aspects of eukaryotic cell biology.

*Dictyostelium discoideum:* The unicellular eukaryote *Dictyostelium* is widely used for experimental analysis of cell movement.

*Caenorhabditis elegans*: The nematode *C. elegans* is a simple multicellular organism that serves as an important model in developmental biology.

*Drosophila melanogaster:* Because of extensive genetic analysis, studies of the fruit fly *Drosophila* have led to major advances in understanding animal development.

*Arabidopsis thaliana:* The small flowering plant *Arabidopsis* is widely used as a model for studies of plant molecular biology and development.

*Vertebrates:* Many kinds of vertebrate cells can be grown in culture, where they can be studied under controlled laboratory conditions. Specialized cell types, such as neurons and muscle cells, provide useful models for investigating particular aspects of cell biology. The frog *Xenopus laevis* is an important model for studies of early vertebrate development, and the mouse is a mammalian species suitable for genetic analysis.

## TOOLS OF CELL BIOLOGY

*Light Microscopy:* A variety of methods are used to visualize cells and subcellular structures, and to determine the intracellular localization of specific molecules using the light microscope.

resolution, bright-field microscopy, phase-contrast microscopy, differential interference-contrast microscopy, video-enhanced differential interference-contrast microscopy, fluorescence microscopy, confocal microscopy

*Electron Microscopy:* Electron microscopy, with a resolution that is approximately a hundredfold greater than that of light microscopy, is used to analyze details of cell structure.

transmission electron microscopy, metal shadowing, freeze fracture, freeze etching, scanning electron microscopy

*Subcellular Fractionation:* The organelles of eukaryotic cells can be isolated for biochemical analysis by differential centrifugation.

differential centrifugation, ultracentrifuge, density-gradient centrifugation, velocity centrifugation, equilibrium centrifugation

*Growth of Animal Cells in Culture:* The propagation of animal cells in culture has allowed studies of the mechanisms that control cell growth and differentiation.

primary culture, immortal cell line

*Culture of Plant Cells:* Cultured plant cells can differentiate to form specialized cell types and, in some cases, can regenerate entire plants.

totipotency

*Viruses:* Viruses provide simple models for studies of cell function.

bacteriophage, retrovirus

## QUESTIONS

**1.** Why is the evolution of photosynthesis thought to have favored the subsequent evolution of oxidative metabolism?

**2.** Yeasts have been used as a model for the study of many aspects of the biology of eukaryotic cells. Why are they not a suitable model for analysis of animal cell movements?

**3.** What resolution can be obtained with a light microscope if the specimen is viewed through air rather than through oil? Assume that the wavelength of visible light is 0.5 $\mu$m.

**4.** You need to separate two organelles that have the same buoyant density but differ in size. Would you use velocity centrifugation or equilibrium centrifugation in density gradients?

**5.** One advantage of the T4 bacteriophage in the study of molecular genetics is the fact that its genome is 20 times smaller than that of *E. coli*. How much smaller is the genome of Rous sarcoma virus than that of its host?

## REFERENCES AND FURTHER READING

### The Origin and Evolution of Cells

Castresana, J. and M. Saraste. 1995. Evolution of energetic metabolism: The respiration-early hypothesis. *Trends Biochem. Sci.* 20: 443–448. [R]

Cech, T. R. 1986. A model for the RNA-catalyzed replication of RNA. *Proc. Natl. Acad. Sci. USA* 83: 4360–4363. [P]

Crick, F. H. C. 1968. The origin of the genetic code. *J. Mol. Biol.* 38: 367–379. [P]

Darnell, J. E. and W. F. Doolittle, 1986. Speculations on the early course of evolution. *Proc. Natl. Acad. Sci. USA* 83: 1271–1275. [P]

Fawcett, D. W. 1994. *Bloom and Fawcett: A Textbook of Histology.* 12th ed. New York: Chapman and Hall.

Gesteland, R. F. and J. F. Atkins (eds.). 1993. *The RNA World.* Plainview, NY: Cold Spring Harbor Laboratory Press.

Gilbert, W. 1986. The RNA world. *Nature* 319: 618. [R]

Joyce, G. F. 1989. RNA evolution and the origins of life. *Nature* 338: 217–224. [R]

Kasting, J. F. 1993. Earth's early atmosphere. *Science* 259: 920–926. [R]

Knoll, A. H. 1992. The early evolution of eukaryotes: A geological perspective. *Science* 256: 622–627. [R]

Lazcano, A. and S. L. Miller. 1996. The origin and early evolution of life: prebiotic chemistry, the pre-RNA world, and time. *Cell* 85: 793–798. [R]

Lake, J. A. and M. C. Rivera. 1994. Was the nucleus the first endosymbiont? *Proc. Natl. Acad. Sci. USA* 91: 2880–2881. [R]

Margulis, L. 1992. *Symbiosis in Cell Evolution.* 2nd ed. New York: W.H. Freeman.

Miller, S. L. 1953. A production of amino acids under possible primitive earth conditions. *Science* 117: 528–529. [P]

Orgel, L. E. 1992. Molecular replication. *Nature* 358: 203–209. [R]

Pace, N. R. 1992. New horizons for RNA catalysis. *Science* 256: 1402–1403. [R]

Raven, P. H., R. F. Evert and S. E. Eichhorn. 1992. *Biology of Plants*. 5th ed. New York: Worth.

Strickberger, M. W. 1995. *Evolution*. 2nd ed. Boston: Jones and Bartlett.

Woese, C. R., O. Kandler and M. L. Wheelis. 1990. Towards a natural system of organisms: Proposal for the domains Archae, Bacteria, and Eucarya. *Proc. Natl. Acad. Sci. USA* 87: 4576–4579. [P]

## Cells as Experimental Models

Ashburner, M. 1989. Drosophila: *A Laboratory Handbook*. Plainview, NY: Cold Spring Harbor Laboratory Press.

Broach, J. R., J. R. Pringle and E. W. Jones (eds.). 1991. *The Molecular and Cellular Biology of the Yeast* Saccharomyces. Plainview, NY: Cold Spring Harbor Laboratory Press.

Condeelis, J. 1993. Understanding the cortex of crawling cells: Insights from *Dictyostelium. Trends Cell Biol.* 3: 371–376. [R]

Hodgkin, J., R. H. A. Plasterk and R.H. Waterston. 1995. The nematode *Caenorhabditis elegans* and its genome. *Science* 270: 410–414. [R]

Hogan, B., R. Beddington, F. Costantini and E. Lacey. 1994. *Manipulating the Mouse Embryo*. 2nd ed. Plainview, NY: Cold Spring Harbor Laboratory Press.

Maliga, P., D. F. Klessig, A. R. Cashmore, W. Gruissem and J. E. Varner (eds.). 1994. *Methods in Plant Molecular Biology*. Plainview, NY: Cold Spring Harbor Laboratory Press.

Meyerowitz, E. M. 1989. *Arabidopsis*, a useful weed. *Cell* 56: 263–269. [R]

Meyerowitz, E. M. and C. R. Somerville (eds.). 1995. *Arabidopsis*. Plainview, NY: Cold Spring Harbor Laboratory Press.

Neidhardt, F. C., R. Curtiss III., J. L. Ingraham, E. C. C. Lin, K. B. Low Jr., B. Magasanik, W. Reznikoff, M. Riley, M. Schaechter and H. E. Umbarger (eds.). 1996. Escherichia coli *and* Salmonella: *Cellular and Molecular Biology*. 2nd ed. Washington, DC: ASM Press.

Neidhardt, F. C., J. L. Ingraham and M. Schaechter. 1990. *Physiology of the Bacterial Cell*. Sunderland, MA: Sinauer.

Slack, J. M. W. 1991. *From Egg to Embryo: Regional Specification in Early Development*. 2nd ed. New York: Cambridge University Press.

## Tools of Cell Biology

Bozzola, J. J. and L. D. Russell. 1992. *Electron Microscopy*. Boston: Jones and Bartlett.

Cairns, J., G. S. Stent and J. D. Watson (eds.). 1992. *Phage and the Origins of Molecular Biology*. Plainview, NY: Cold Spring Harbor Laboratory Press.

Claude, A. 1975. The coming of age of the cell. *Science* 189: 433–435. [R]

De Duve, C. 1975. Exploring cells with a centrifuge. *Science* 189: 186–194. [R]

Eagle, H. 1955. Nutrition needs of mammalian cells in tissue culture. *Science* 122: 501–504. [P]

Fields, B. N., D. M. Knipe, P. M. Howley, R. M. Chanock, J. L. Melnick, T. P. Monath, B. Roizman and S. E. Straus (eds.). 1996. *Fundamental Virology*. 3rd ed. New York: Lippincott-Raven.

Freshney, R. I. 1993. *Culture of Animal Cells*. 3rd ed. New York: Wiley.

Palade, G. 1975. Intracellular aspects of the process of protein synthesis. *Science* 189: 347–358. [R]

Porter, K. R., A. Claude and E. F. Fullam. 1945. A study of tissue culture cells by electron microscopy. *J. Exp. Med.* 81: 233–246. [P]

Rawlings, D. J. 1992. *Light Microscopy*. Oxford, England: BIOS Scientific.

Rickwood, D. (ed.). 1993. *Preparative Centrifugation: A Practical Approach*. Oxford, England: IRL Press.

Rous, P. 1911. A sarcoma of the fowl transmissible by an agent separable from the tumor cells. *J. Exp. Med.* 13: 397–411. [P]

Salmon, E. D. 1995. VE-DIC light microscopy and the discovery of kinesin. *Trends Cell Biol.* 5: 154–158. [R]

Slayter, E. M. and H. S. Slayter. 1992. *Light and Electron Microscopy*. Cambridge, England: Cambridge University Press.

Wang, Y. and D. L. Taylor (eds.). 1989. *Fluorescence Microscopy of Living Cells in Culture*. San Diego, CA: Academic Press.

# 2

# The Chemistry of Cells

CELLS ARE INCREDIBLY COMPLEX AND DIVERSE STRUCTURES, capable not only of self-replication—the very essence of life—but also of performing a wide range of specialized tasks in multicellular organisms. Yet cells obey the same laws of chemistry and physics that determine the behavior of nonliving systems. Consequently, modern cell biology seeks to understand cellular processes in terms of chemical and physical reactions.

This chapter considers the fundamental principles of biological chemistry that govern the lives of cells. It is intended neither to be a comprehensive discussion of biochemistry nor to chart all the metabolic reactions within cells. Rather, the chapter will focus on five major topics: the types of molecules within cells, the central role of proteins as biological catalysts, the generation and utilization of metabolic energy, the biosynthesis of major cell constituents, and the structure of biological membranes. An appreciation of these chemical foundations forms the basis for understanding the diverse aspects of cell structure and function that will be discussed throughout the rest of this text.

## THE MOLECULAR COMPOSITION OF CELLS

Cells are composed of water, inorganic ions, and carbon-containing (organic) molecules. Water is the most abundant molecule in cells, accounting for 70% or more of total cell mass. Consequently, the interactions between water and the other constituents of cells are of central importance in biological chemistry. The critical property of water in this respect is that it is a polar molecule, in which the hydrogen atoms have a slight positive charge and the oxygen has a slight negative charge (Figure 2.1). Because of their polar nature, water molecules can form hydrogen bonds with each other or with other polar molecules, as well as interacting with positively or negatively charged ions. As a result of these interactions, ions and polar molecules are readily soluble in water (hydrophilic). In contrast, nonpolar molecules, which cannot interact with water, are poorly soluble in an aqueous environment (hydrophobic). Consequently, nonpolar molecules tend to minimize their contact with water by associating closely with each other

(A)

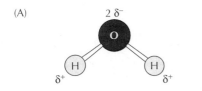

(B)

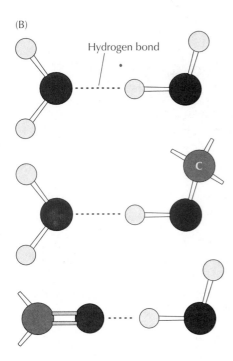

(C)

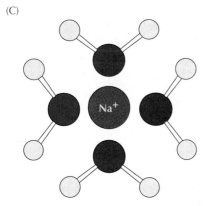

*Figure 2.1*
**Characteristics of water**   (A) Water is a polar molecule, with a slight negative charge ($\delta^-$) on the oxygen atom and a slight positive charge ($\delta^+$) on the hydrogen atoms. Because of this polarity, water molecules can form hydrogen bonds (dashed lines) either with each other or with other polar molecules (B), in addition to interacting with charged ions (C).

instead. As discussed later in this chapter, such interactions of polar and nonpolar molecules with water and with each other play crucial roles in the formation of biological structures, such as cell membranes.

The inorganic ions of the cell, including sodium ($Na^+$), potassium ($K^+$), magnesium ($Mg^{2+}$), calcium ($Ca^{2+}$), phosphate ($HPO_4^{2-}$), chloride ($Cl^-$), and bicarbonate ($HCO_3^-$), constitute 1% or less of the cell mass. These ions are involved in a number of aspects of cell metabolism, and thus play critical roles in cell function.

It is, however, the organic molecules that are the unique constituents of cells. Most of these organic compounds belong to one of four classes of molecules: carbohydrates, lipids, proteins, and nucleic acids. Proteins, nucleic acids, and most carbohydrates (the polysaccharides) are macromolecules formed by the joining (polymerization) of hundreds or thousands of low-molecular-weight precursors: amino acids, nucleotides, and simple sugars, respectively. Such macromolecules constitute 80 to 90% of the dry weight of most cells. Lipids are the other major constituent of cells. The remainder of the cell mass is composed of a variety of small organic molecules, including macromolecular precursors. The basic chemistry of cells can thus be understood in terms of the structures and functions of four major classes of organic molecules.

## Carbohydrates

The **carbohydrates** include simple sugars as well as polysaccharides. These simple sugars, such as glucose, are the major nutrients of cells. As discussed later in this chapter, their breakdown provides both a source of cellular energy and the starting material for the synthesis of other cell constituents. Polysaccharides are storage forms of sugars and form structural components of the cell. In addition, polysaccharides and shorter polymers of sugars act as markers for a variety of cell recognition processes, including the adhesion of cells to their neighbors and the transport of proteins to appropriate intracellular destinations.

The structures of representative simple sugars (**monosaccharides**) are illustrated in Figure 2.2. The basic formula for these molecules is $(CH_2O)_n$, from which the name carbohydrate is derived (C = "carbo" and $H_2O$ = "hydrate"). The six-carbon ($n = 6$) sugar glucose ($C_6H_{12}O_6$) is especially important in cells, since it provides the principal source of cellular energy. Other simple sugars have between three and seven carbons, with three- and five-carbon sugars being the most common. Sugars containing five or more carbons can cyclize to form ring structures, which are the predominant forms of these molecules within cells. As illustrated in Figure 2.2, the cyclized sugars exist in two alternative forms (called $\alpha$ or $\beta$), depending on the configuration of carbon 1.

Monosaccharides can be joined together by dehydration reactions, in which $H_2O$ is removed and the sugars are linked by a **glycosidic bond** between two of their carbons (Figure 2.3). If only a few sugars are joined together, the resulting polymer is called an **oligosaccharide**. If a large number (hundreds or thousands) of sugars are involved, the resulting polymers are macromolecules called **polysaccharides**.

Two common polysaccharides—**glycogen** and **starch**—are the storage forms of carbohydrates in animal and plant cells, respectively. Both glycogen and starch are composed entirely of glucose molecules in the $\alpha$ configuration (Figure 2.4). The principal linkage is between carbon 1 of one glucose and carbon 4 of a second. In addition, both glycogen and one form of starch (amylopectin) contain occasional $\alpha$ (1→6) linkages, in which carbon 1 of one glucose is joined to carbon 6 of a second. As illustrated in Figure 2.4, these linkages lead to the formation of branches resulting from the join-

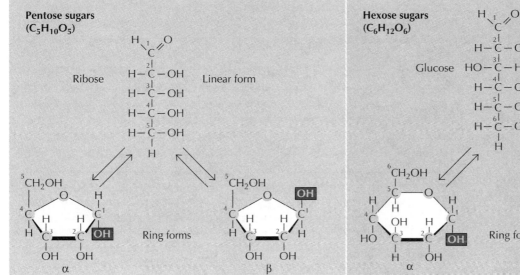

**Triose sugars (C$_3$H$_6$O$_3$)**

Glyceraldehyde

Dihydroxyacetone

*Figure 2.2*
**Structure of simple sugars**   Representative sugars containing three, five, and six carbons (triose, pentose, and hexose sugars, respectively) are illustrated. Sugars with five or more carbons can cyclize to form rings, which exist in two alternative forms ($\alpha$ and $\beta$) depending on the configuration of carbon 1.

**Pentose sugars (C$_5$H$_{10}$O$_5$)**

Ribose     Linear form

Ring forms

$\alpha$     $\beta$

**Hexose sugars (C$_6$H$_{12}$O$_6$)**

Glucose     Linear form

Ring forms

$\alpha$     $\beta$

ing of two separate $\alpha$ (1→4) linked chains. Such branches are present in glycogen and amylopectin, although another form of starch (amylose) is an unbranched molecule.

The structures of glycogen and starch are thus basically similar, as is their function: to store glucose. **Cellulose**, in contrast, has a quite distinct function as the principal structural component of the plant cell wall. Perhaps surprisingly, then, cellulose is also composed entirely of glucose molecules.

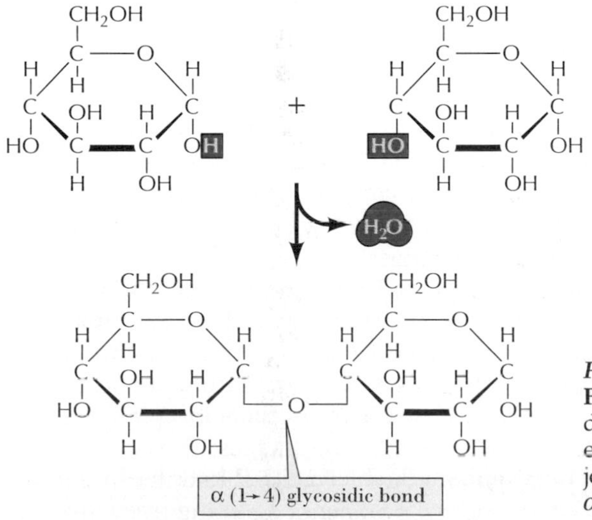

$\alpha$ (1→4) glycosidic bond

*Figure 2.3*
**Formation of a glycosidic bond**   Two simple sugars are joined by a dehydration reaction (a reaction in which water is removed).  In the example shown, two glucose molecules in the $\alpha$ configuration are joined by a bond between carbons 1 and 4, which is therefore called an $\alpha$ (1→4) glycosidic bond.

**Amylopectin (starch)**

**Glycogen**

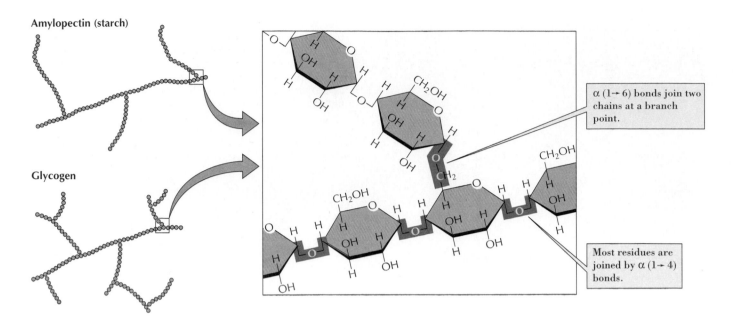

α (1 → 6) bonds join two chains at a branch point.

Most residues are joined by α (1 → 4) bonds.

**Cellulose**

*Figure 2.4*
**Structure of polysaccharides**
Polysaccharides are macromolecules consisting of hundreds or thousands of simple sugars. Glycogen, starch, and cellulose are all composed entirely of glucose residues, which are joined by $\alpha(1{\rightarrow}4)$ glycosidic bonds in glycogen and starch but by $\beta(1{\rightarrow}4)$ bonds in cellulose. Glycogen and one form of starch (amylopectin) also contain occasional $\alpha(1{\rightarrow}6)$ bonds, which serve as branch points by joining two separate $\alpha(1{\rightarrow}4)$ chains.

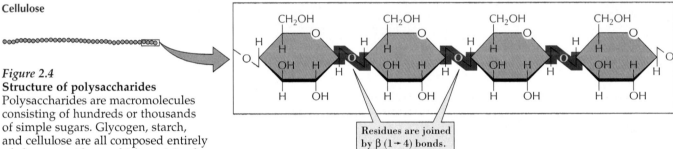

Residues are joined by β (1 → 4) bonds.

The glucose residues in cellulose, however, are in the $\beta$ rather than the $\alpha$ configuration, and cellulose is an unbranched polysaccharide (see Figure 2.4). The linkage of glucose residues by $\beta$ (1→4) rather than $\alpha$ (1→4) bonds causes cellulose to form long extended chains that pack side by side to form fibers of great mechanical strength.

In addition to their roles in energy storage and cell structure, oligosaccharides and polysaccharides are important in a variety of cell signaling processes. For example, oligosaccharides are frequently linked to proteins, where they serve as markers to target proteins for transport to the cell surface or incorporation into different subcellular organelles. Oligosaccharides and polysaccharides also serve as markers on the surface of cells, playing important roles in cell recognition and the interactions between cells in tissues of multicellular organisms.

## Lipids

**Lipids** have three major roles in cells. First, they provide an important form of energy storage. Second, and of great importance in cell biology, lipids are the major components of cell membranes. Third, lipids play important roles in cell signaling, both as steroid hormones (e.g., estrogen and testosterone) and as messenger molecules that convey signals from cell surface receptors to targets within the cell.

The simplest lipids are **fatty acids**, which consist of long hydrocarbon

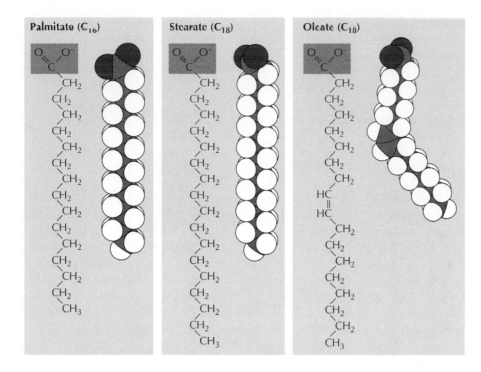

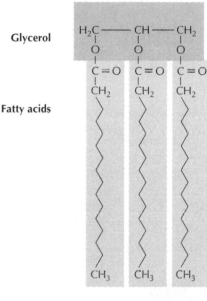

*Figure 2.5*
**Structure of fatty acids**  Fatty acids consist of long hydrocarbon chains terminating in a carboxyl group (COO⁻). Palmitate and stearate are saturated fatty acids consisting of 16 and 18 carbons, respectively. Oleate is an unsaturated 18-carbon fatty acid containing a double bond between carbons 9 and 10. Note that the double bond introduces a kink in the hydrocarbon chain.

*Figure 2.6*
**Structure of triacylglycerols**
Triacylglycerols (fats) contain three fatty acids joined to glycerol. In this example, all three fatty acids are palmitate, but triacylglycerols often contain a mixture of different fatty acids.

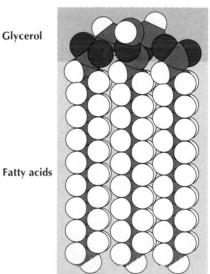

chains, most frequently containing 16 or 18 carbon atoms, with a carboxyl group (COO⁻) at one end (Figure 2.5). Unsaturated fatty acids contain one or more double bonds between carbon atoms; in saturated fatty acids all of the carbon atoms are bonded to the maximum number of hydrogen atoms. The long hydrocarbon chains of fatty acids contain only nonpolar C—H bonds, which are unable to interact with water. The hydrophobic nature of these fatty acid chains is responsible for much of the behavior of complex lipids, particularly in the formation of biological membranes.

Fatty acids are stored in the form of **triacylglycerols**, or **fats**, which consist of three fatty acids linked to a glycerol molecule (Figure 2.6). Triacylglycerols are insoluble in water and therefore accumulate as fat droplets in the cytoplasm. When required, they can be broken down for use in energy-yielding reactions discussed later in this chapter. It is noteworthy that fats are a more efficient form of energy storage than carbohydrates, yielding more than twice as much energy per weight of material broken down. Fats therefore allow energy to be stored in less than half the body weight that would be required to store the same amount of energy in carbohydrates—a particularly important consideration for animals because of their mobility.

**Phospholipids**, the principal components of cell membranes, consist of two fatty acids joined to a polar head group (Figure 2.7). In the **glycerol phospholipids**, the two fatty acids are bound to carbon atoms in glycerol, as in triacylglycerols. The third carbon of glycerol, however, is bound to a phosphate group, which is in turn frequently attached to another small polar molecule, such as choline, serine, inositol, or ethanolamine. **Sphingomyelin**, the only nonglycerol phospholipid in cell membranes, contains two hydrocarbon chains linked to a polar head group formed from serine rather than from glycerol. All phospholipids have hydrophobic tails, consisting of the two hydrocarbon chains, and hydrophilic head groups, consisting of the phosphate group and its polar attachments. Consequently, phospholipids are **amphipathic** molecules, part water-soluble and part water-insoluble. This property of phospholipids is the basis for the forma-

**Phosphatidic acid**

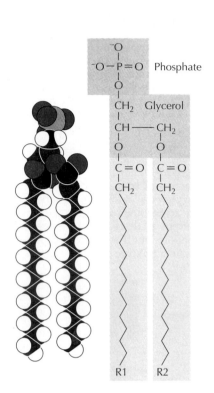

$^-$O
$^-$O—P=O    Phosphate
O
CH₂    Glycerol
CH——CH₂
O     O
C=O   C=O
CH₂    CH₂

R1    R2

**Phosphatidyl-
ethanolamine**

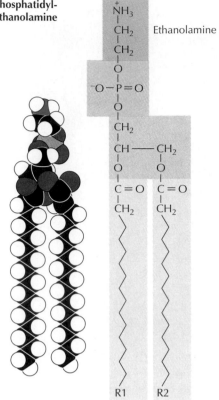

$^+$NH₃
CH₂    Ethanolamine
CH₂
O
$^-$O—P=O
O
CH₂
CH——CH₂
O     O
C=O   C=O
CH₂    CH₂

R1    R2

**Phosphatidyl-
choline**

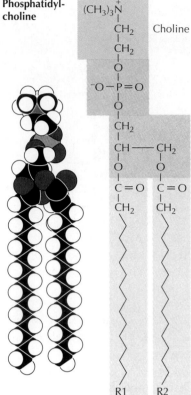

(CH₃)₃$^+$N
CH₂    Choline
CH₂
O
$^-$O—P=O
O
CH₂
CH——CH₂
O     O
C=O   C=O
CH₂    CH₂

R1    R2

**Phosphatidylserine**

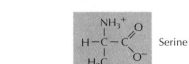

NH₃$^+$    O
H—C—C    Serine
H₂C     O$^-$
O
$^-$O—P=O
O
CH₂
CH——CH₂
O     O
C=O   C=O
CH₂    CH₂

R1    R2

**Phosphatidylinositol**

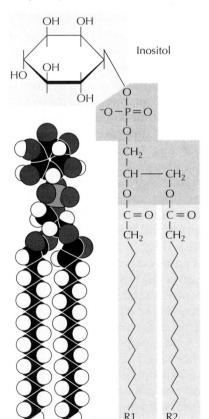

OH    OH

OH

HO

OH    Inositol

O
$^-$O—P=O
O
CH₂
CH——CH₂
O     O
C=O   C=O
CH₂    CH₂

R1    R2

**Sphingomyelin**

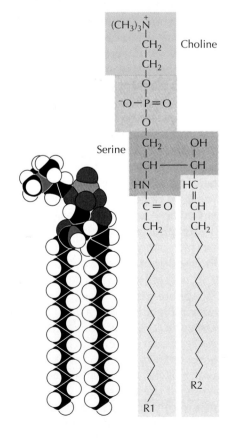

(CH₃)₃$^+$N
CH₂    Choline
CH₂
O
$^-$O—P=O
O
CH₂     OH
CH——CH
HN    HC
C=O    CH
CH₂    CH₂

Serine

R1    R2

◀ *Figure 2.7*
**Structure of phospholipids** Glycerol phospholipids contain two fatty acids joined to glycerol. The fatty acids may be different from each other and are designated R1 and R2. The third carbon of glycerol is joined to a phosphate group (forming phosphatidic acid), which in turn is frequently joined to another small polar molecule (forming phosphatidylethanolamine, phosphatidylcholine, phosphatidylserine, or phosphatidylinositol). In sphingomyelin, two hydrocarbon chains are bound to a polar head group formed from serine instead of glycerol.

tion of biological membranes, as discussed later in this chapter.

In addition to phospholipids, many cell membranes contain **glycolipids** and **cholesterol**. Glycolipids consist of two hydrocarbon chains linked to polar head groups that contain carbohydrates (Figure 2.8). They are thus similar to the phospholipids in their general organization as amphipathic molecules. Cholesterol, in contrast, consists of four hydrocarbon rings rather than linear hydrocarbon chains (Figure 2.9). The hydrocarbon rings are strongly hydrophobic, but the hydroxyl (OH) group attached to one end of cholesterol is weakly hydrophilic, so cholesterol is also amphipathic.

In addition to their roles as components of cell membranes, lipids function as signaling molecules, both within and between cells. The **steroid hormones** (such as estrogens and testosterone) are derivatives of cholesterol (see Figure 2.9). These hormones are a diverse group of chemical messengers, all of which contain four hydrocarbon rings to which distinct functional groups are attached. Derivatives of phospholipids also serve as messenger molecules within cells, acting to convey signals from cell surface receptors to intracellular targets (see Chapter 13).

## Nucleic Acids

The nucleic acids—DNA and RNA—are the principal informational molecules of the cell. **Deoxyribonucleic acid (DNA)** has a unique role as the genetic material, which in eukaryotic cells is located in the nucleus. Dif-

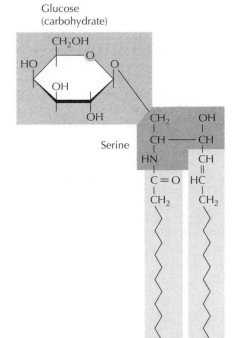

*Figure 2.8*
**Structure of glycolipids** Two hydrocarbon chains are joined to a polar head group formed from serine and containing carbohydrates (e.g., glucose).

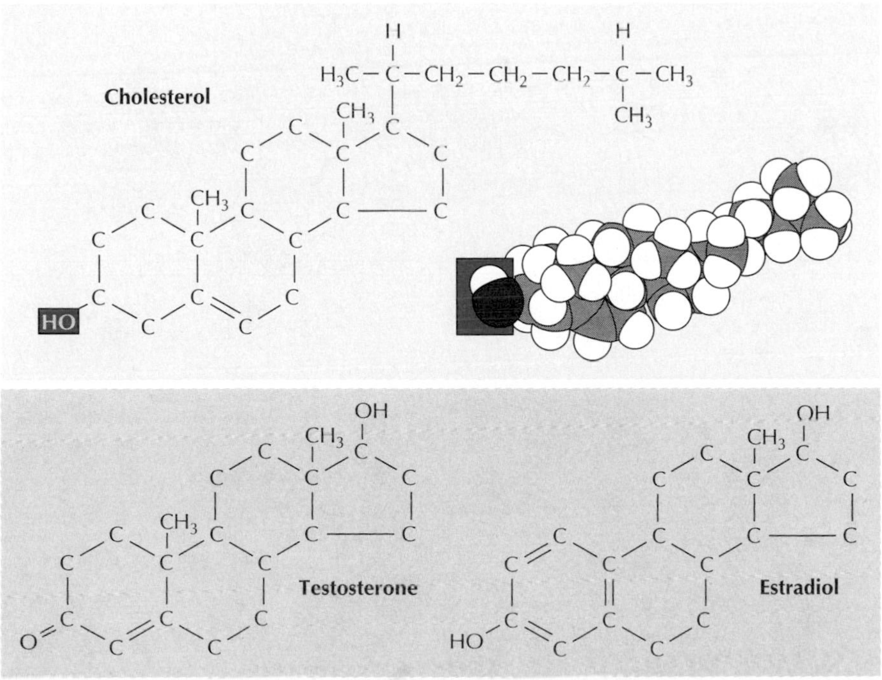

*Figure 2.9*
**Cholesterol and steroid hormones** Cholesterol, an important component of cell membranes, is an amphipathic molecule because of its polar hydroxyl group. Cholesterol is also a precursor to the steroid hormones, such as testosterone and estradiol (a form of estrogen). The hydrogen atoms bonded to the ring carbons are not shown in this figure.

ferent types of **ribonucleic acid (RNA)** participate in a number of cellular activities. **Messenger RNA (mRNA)** carries information from DNA to the ribosomes, where it serves as a template for protein synthesis. Two other types of RNA (**ribosomal RNA** and **transfer RNA**) are involved in protein synthesis. Still other kinds of RNAs are involved in the processing and transport of both RNAs and proteins. In addition to acting as an informational molecule, RNA is also capable of catalyzing a number of chemical reactions. In present-day cells, these include reactions involved in both protein synthesis and RNA processing.

DNA and RNA are polymers of **nucleotides**, which consist of **purine** and **pyrimidine** bases linked to phosphorylated sugars (Figure 2.10). DNA contains two purines (**adenine** and **guanine**) and two pyrimidines (**cytosine** and **thymine**). Adenine, guanine, and cytosine are also present in

*Figure 2.10*
**Components of nucleic acids**
Nucleic acids contain purine and pyrimidine bases linked to phosphorylated sugars. A nucleic acid base linked to a sugar alone is a nucleoside. Nucleotides additionally contain one or more phosphate groups.

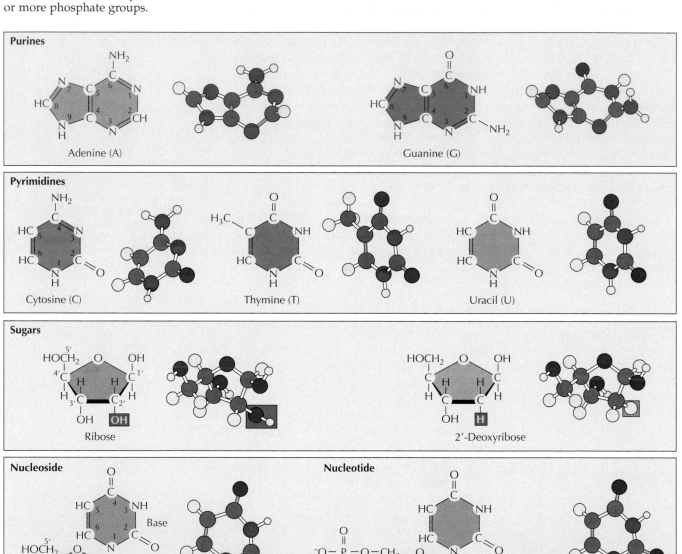

*Figure 2.11*
**Polymerization of nucleotides**    A phosphodiester bond is formed between the 3' hydroxyl group of one nucleotide and the 5' phosphate group of another. A polynucleotide chain has a sense of direction, one end terminating in a 5' phosphate group (the 5' end) and the other in a 3' hydroxyl group (the 3' end).

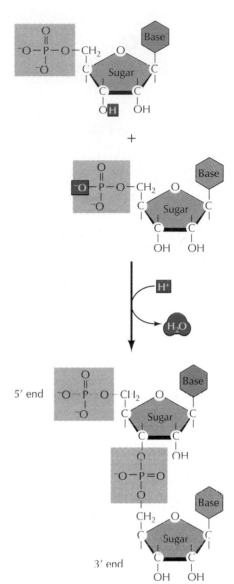

RNA, but RNA contains **uracil** in place of thymine. The bases are linked to sugars (**2'-deoxyribose** in DNA, or **ribose** in RNA) to form **nucleosides**. Nucleotides additionally contain one or more phosphate groups linked to the 5' carbon of nucleoside sugars.

The polymerization of nucleotides to form nucleic acids involves the formation of **phosphodiester bonds** between the 5' phosphate of one nucleotide and the 3' hydroxyl of another (Figure 2.11). **Oligonucleotides** are small polymers containing only a few nucleotides; the large **polynucleotides** that make up cellular RNA and DNA may contain thousands or millions of nucleotides, respectively. It is important to note that a polynucleotide chain has a sense of direction, with one end of the chain terminating in a 5' phosphate group and the other in a 3' hydroxyl group. Polynucleotides are always synthesized in the 5' to 3' direction, with a free nucleotide being added to the 3' OH group of a growing chain. By convention, the sequence of bases in DNA or RNA is also written in the 5' to 3' direction.

The information in DNA and RNA is conveyed by the order of the bases in the polynucleotide chain. DNA is a double-stranded molecule consisting of two polynucleotide chains running in opposite directions (see Chapter 3). The bases are on the inside of the molecule, and the two chains are joined by hydrogen bonds between complementary base pairs—adenine pairing with thymine and guanine with cytosine (Figure 2.12). The important consequence of such complementary base pairing is that one strand of DNA (or RNA) can act as a template to direct the synthesis of a complementary strand. Nucleic acids are thus uniquely capable of directing their own self-replication, allowing them to function as the fundamental informational molecules of the cell. The information carried by DNA and RNA directs the synthesis of specific proteins, which control most cellular activities.

Nucleotides are not only important as the building blocks of nucleic acids; they also play critical roles in other cell processes. Perhaps the most prominent example is adenosine 5'-triphosphate (ATP), which is the principal form of chemical energy within cells. Other nucleotides similarly function as carriers of either energy or reactive chemical groups in a wide variety of metabolic reactions. In addition, some nucleotides (e.g., cyclic AMP) are important signaling molecules within cells (see Chapter 13).

*Figure 2.12*
**Complementary pairing between nucleic acid bases**    The formation of hydrogen bonds between bases on opposite strands of DNA leads to the specific pairing of adenine (A) with thymine (T) and guanine (G) with cytosine (C).

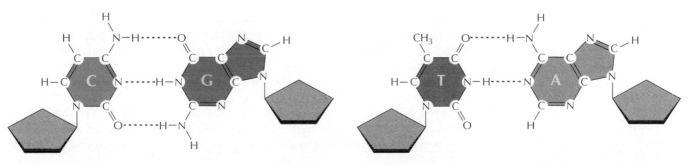

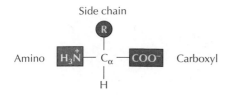

Side chain

Amino    Carboxyl

**Figure 2.13**
**Structure of amino acids** Each amino acid consists of a central carbon atom (the α carbon) bonded to a hydrogen atom, a carboxyl group, an amino group, and a specific side chain (designated R). At physiological pH, both the carboxyl and amino groups are ionized, as shown.

## Proteins

While nucleic acids carry the genetic information of the cell, the primary responsibility of **proteins** is to execute the tasks directed by that information. Proteins are the most diverse of all macromolecules, and each cell contains several thousand different proteins, which perform a wide variety of functions. The roles of proteins include serving as structural components of cells and tissues, acting in the transport and storage of small molecules (e.g., the transport of oxygen by hemoglobin), transmitting information between cells (e.g., protein hormones), and providing a defense against infection (e.g., antibodies). The most fundamental property of proteins, however, is their ability to act as enzymes, which, as discussed in the following section, catalyze nearly all the chemical reactions in biological systems. Thus, proteins direct virtually all activities of the cell. The central importance of proteins in biological chemistry is indicated by their name, which is derived from the Greek word *proteios*, meaning "of the first rank."

Proteins are polymers of 20 different **amino acids**. Each amino acid consists of a carbon atom (called the α carbon) bonded to a carboxyl group ($COO^-$), an amino group ($NH_3^+$), a hydrogen atom, and a distinctive side chain (Figure 2.13). The specific chemical properties of the different amino acid side chains determine the roles of each amino acid in protein structure and function.

The amino acids can be grouped into four broad categories according to the properties of their side chains (Figure 2.14). Ten amino acids have nonpolar side chains that do not interact with water. Glycine is the simplest amino acid, with a side chain consisting of only a hydrogen atom. Alanine, valine, leucine, and isoleucine have hydrocarbon side chains consisting of up to four carbon atoms. The side chains of these amino acids are hydrophobic and therefore tend to be located in the interior of proteins, where they are not in contact with water. Proline similarly has a hydrocarbon side chain, but it is unique in that its side chain is bonded to the nitrogen of the amino group as well as to the α carbon, forming a cyclic structure. The side chains of two amino acids, cysteine and methionine, contain sulfur atoms. Methionine is quite hydrophobic, but cysteine is less so because of its sulfhydryl (SH) group. As discussed later, the sulfhydryl group of cysteine plays an important role in protein structure because disulfide bonds can form between the side chains of different cysteine residues. Finally, two nonpolar amino acids, phenylalanine and tryptophan, have side chains containing very hydrophobic aromatic rings.

Five amino acids have uncharged but polar side chains. These include serine, threonine, and tyrosine, which have hydroxyl groups on their side chains, as well as asparagine and glutamine, which have polar amide ($O=C-NH_2$) groups. Because the polar side chains of these amino acids can form hydrogen bonds with water, these amino acids are hydrophilic and tend to be located on the outside of proteins.

The amino acids lysine, arginine, and histidine have side chains with charged basic groups. Lysine and arginine are very basic amino acids, and their side chains are positively charged in the cell. Consequently, they are very hydrophilic and are found in contact with water on the surface of proteins. Histidine can be either uncharged or positively charged at physiological pH, so it frequently plays an active role in enzymatic reactions involving the exchange of hydrogen ions, as illustrated in the example of enzymatic catalysis discussed in the following section.

Finally, two amino acids, aspartic acid and glutamic acid, have acidic side chains terminating in carboxyl groups. These amino acids are negatively charged within the cell and are therefore frequently referred to as aspartate

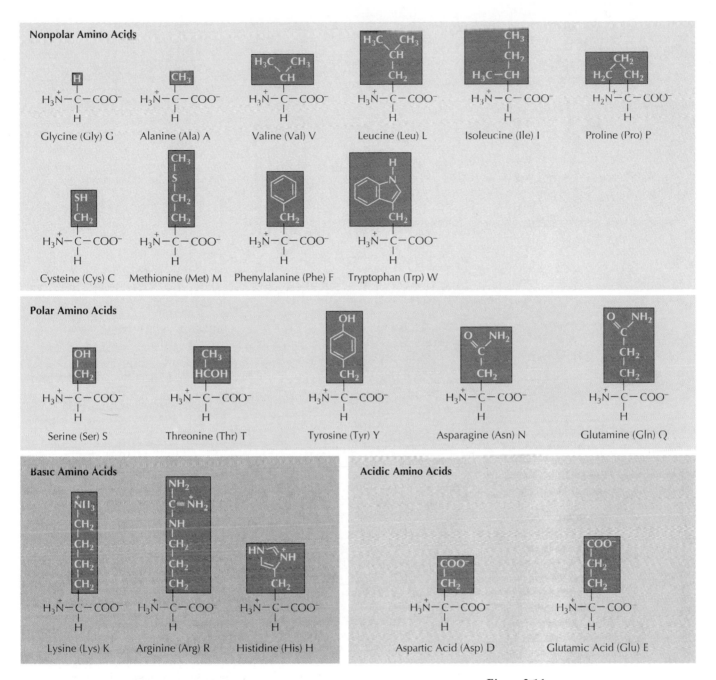

**Nonpolar Amino Acids**

Glycine (Gly) G    Alanine (Ala) A    Valine (Val) V    Leucine (Leu) L    Isoleucine (Ile) I    Proline (Pro) P

Cysteine (Cys) C    Methionine (Met) M    Phenylalanine (Phe) F    Tryptophan (Trp) W

**Polar Amino Acids**

Serine (Ser) S    Threonine (Thr) T    Tyrosine (Tyr) Y    Asparagine (Asn) N    Glutamine (Gln) Q

**Basic Amino Acids**

Lysine (Lys) K    Arginine (Arg) R    Histidine (His) H

**Acidic Amino Acids**

Aspartic Acid (Asp) D    Glutamic Acid (Glu) E

*Figure 2.14*
**The amino acids** The three-letter and one-letter abbreviations for each amino acid are indicated. The amino acids are grouped into four categories according to the properties of their side chains: nonpolar, polar, basic, and acidic.

and glutamate. Like the basic amino acids, these acidic amino acids are very hydrophilic and are usually located on the surface of proteins.

Amino acids are joined together by **peptide bonds** between the $\alpha$ amino group of one amino acid and the $\alpha$ carboxyl group of a second (Figure 2.15). **Polypeptides** are linear chains of amino acids, usually hundreds or thousands of amino acids in length. Each polypeptide chain has two distinct ends, one terminating in an $\alpha$ amino group (the amino, or N, terminus) and the other in an $\alpha$ carboxyl group (the carboxy, or C, terminus). Polypeptides are synthesized from the amino to the carboxy terminus, and the sequence of amino acids in a polypeptide is written (by convention) in the same order.

The defining characteristic of proteins is that they are polypeptides with specific amino acid sequences. In 1953 Frederick Sanger was the first to

# KEY EXPERIMENT

## The Folding of Polypeptide Chains

### Reductive Cleavage of Disulfide Bridges in Ribonuclease

Michael Sela, Frederick H. White, Jr., and Christian B. Anfinsen
National Institutes of Health, Bethesda, MD
*Science, Volume 125, 1957, pages 691–692.*

### The Context

Functional proteins are structurally far more complex than linear chains of amino acids. The formation of active enzymes or other proteins requires the folding of polypeptide chains into precise three-dimensional conformations. This difference between proteins and polypeptide chains raises critical questions with respect to understanding protein structure and function. How is the correct protein conformation chosen from the many possible conformations that could be adopted by a polypeptide, and what is the nature of the information that directs protein folding?

Classic experiments done by Christian Anfinsen and his colleagues provided the answer to these questions. Studying the enzyme ribonuclease, Anfinsen and his collaborators were able to demonstrate that denatured proteins can spontaneously refold into an active conformation. Therefore, all the information required to specify the correct three-dimensional conformation of a protein is contained in its primary amino acid sequence. A series of such experiments led Anfinsen to the conclusion that the native three-dimensional structure of a protein corresponds to its thermodynamically most stable conformation, as determined by interactions between its constituent amino acids. The original observations leading to the formulation of this critical principle were reported in this 1957 paper by Anfinsen, Michael Sela, and Fred White.

### The Experiments

The protein studied by Sela, White, and Anfinsen was bovine ribonuclease, a small protein of 124 amino acids that contains four disulfide (S—S) bonds between side chains of cysteine residues. The enzymatic activity of ribonuclease could be determined by its ability to degrade RNA to nucleotides, providing a ready assay for function of the native protein. This enzymatic activity was completely lost when ribonuclease was subjected to treatments that both disrupted noncovalent bonds (e.g., hydrogen bonds) and cleaved the disulfide bonds by reducing them to sulfhydryl (SH) groups. The denatured protein thus appeared to be in a random inactive conformation.

Sela, White, and Anfinsen made the crucial observation that enzymatic activity reappears if the denatured protein is incubated under conditions that allowed the polypeptide chain to refold and disulfide bonds to re-form. In these experiments the denaturing agents were removed, and the inactive enzyme was incubated at room temperature in a physiological buffer in the presence of $O_2$. This procedure led to oxidation of the sulfhydryl groups and the re-formation of disulfide bonds. During this process, the enzyme regained its catalytic activity, indicating that it had refolded to its native conformation. Because no

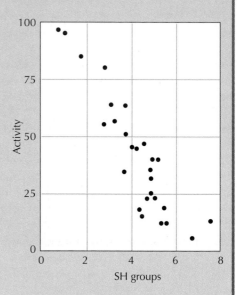

A summary of results of the renaturation experiments. The enzymatic activity of ribonuclease is plotted as a function of the number of sulfhydryl groups present after various treatments. Activity is expressed as percent activity of the native enzyme.

determine the complete amino acid sequence of a protein, the hormone insulin. Insulin was found to consist of two polypeptide chains, joined by disulfide bonds between cysteine residues (Figure 2.16). Most important, Sanger's experiment revealed that each protein consists of a specific amino acid sequence. Proteins are currently sequenced using automated methods, and the complete amino acid sequences of thousands of proteins are now known. Each consists of a unique sequence of amino acids, determined by the order of nucleotides in a gene (see Chapter 3).

## The Folding of Polypeptide Chains *(continued)*

other cellular components were present, all the information required for proper protein folding appeared to be contained in the primary amino acid sequence of the polypeptide chain.

### The Impact

Further experiments defined conditions under which denatured ribonuclease completely regains its native structure and enzymatic activity, establishing the "thermodynamic hypothesis" of protein folding—that is, the native three-dimensional structure of a protein corresponds to the thermodynamically most stable state under physiological conditions. Thermodynamic stability is governed by interactions of the constituent amino acids, so the three-dimension-

al conformations of proteins are determined directly by their primary amino acid sequences. Since the order of nucleotides in DNA specifies the amino acid sequence of a polypeptide, it follows that the nucleotide sequence of a gene contains all the information needed to determine the three-dimensional structure of its protein product.

Although Anfinsen's work established the thermodynamic basis of protein folding, understanding the mechanism of this process remains an active area of research. Protein folding is extremely complex, and we are still unable to deduce the three-dimensional structure of a protein directly from its amino acid sequence. It is also important to note that the

Christian Anfinsen

spontaneous folding of proteins *in vitro* is much slower than protein folding within the cell, which is assisted by enzymes (see Chapter 7). The protein folding problem thus remains a central challenge in biological chemistry.

**Figure 2.15**
**Formation of a peptide bond**   The carboxyl group of one amino acid is linked to the amino group of a second.

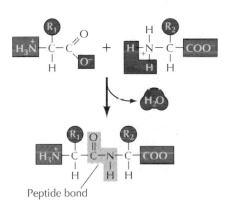

The amino acid sequence of a protein is only the first element of its structure. Rather than being extended chains of amino acids, proteins adopt distinct three-dimensional conformations that are critical to their function. These three-dimensional conformations of proteins are the result of interactions between their constituent amino acids, so the shapes of proteins are determined by their amino acid sequences. This was first demonstrated by experiments of Christian Anfinsen in which he disrupted the three-dimensional structures of proteins by treatments, such as heating, that break noncovalent bonds—a process called denaturation (Figure 2.17).

**Figure 2.16**
**Amino acid sequence of insulin**   Insulin consists of two polypeptide chains, one of 21 and the other of 30 amino acids (indicated here by their one-letter codes). The side chains of three pairs of cysteine residues are joined by disulfide bonds, two of which connect the polypeptide chains.

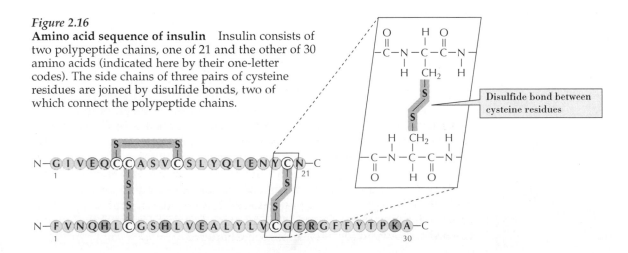

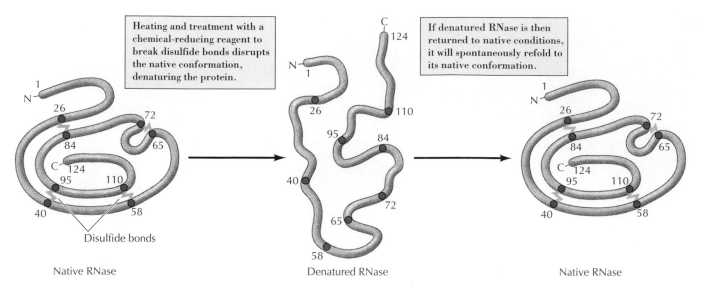

**Figure 2.17**
**Protein denaturation and refolding**
Ribonuclease (RNase) is a protein of 124 amino acids (indicated by numbers). The protein is normally folded into its native conformation, which contains four disulfide bonds (indicated as paired circles representing the cysteine residues).

**Figure 2.18**
**Three-dimensional structure of myoglobin**   Myoglobin is a protein of 153 amino acids that is involved in oxygen transport. The polypeptide chain is folded around a heme group that serves as the oxygen-binding site.

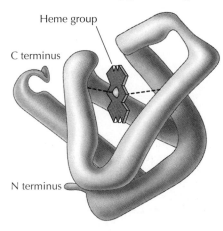

Following incubation under milder conditions, such denatured proteins often spontaneously returned to their native conformations, indicating that these conformations were directly determined by the amino acid sequence.

The three-dimensional structure of proteins is most frequently analyzed by **X-ray crystallography,** a high-resolution technique that can determine the arrangement of individual atoms within a molecule. A beam of X rays is directed at crystals of the protein to be analyzed, and the pattern of X rays that pass through the protein crystal is detected on X ray film. As the X rays strike the crystal, they are scattered in characteristic patterns determined by the arrangement of atoms in the molecule. The structure of the molecule can therefore be deduced from the pattern of scattered X rays (the diffraction pattern).

In 1958 John Kendrew was the first to determine the three-dimensional structure of a protein, myoglobin—a relatively simple protein of 153 amino acids (Figure 2.18). Since then, the three-dimensional structures of over 3000 proteins have been analyzed. Most, like myoglobin, are globular proteins with polypeptide chains folded into compact structures, although some (such as the structural proteins of connective tissues) are long fibrous molecules. Analysis of the three-dimensional structures of these proteins has revealed several basic principles that govern protein folding, although protein structure is so complex that predicting the three-dimensional structure of a protein directly from its amino acid sequence is impossible.

Protein structure is generally described as having four levels. The **primary structure** of a protein is the sequence of amino acids in its polypeptide chain. The **secondary structure** is the regular arrangement of amino acids within localized regions of the polypeptide. Two types of secondary structure, which were first proposed by Linus Pauling and Robert Corey in 1951, are particularly common: the $\alpha$ **helix** and the $\beta$ **sheet**. Both of these secondary structures are held together by hydrogen bonds between the CO and NH groups of peptide bonds. An $\alpha$ helix is formed when a region of a polypeptide chain coils around itself, with the CO group of one peptide bond forming a hydrogen bond with the NH group of a peptide bond located four residues downstream in the linear polypeptide chain (Figure 2.19). In contrast, a $\beta$ sheet is formed when two parts of a polypeptide chain lie side by side with hydrogen bonds between them. Such $\beta$ sheets can be formed between several polypeptide strands, which can be oriented

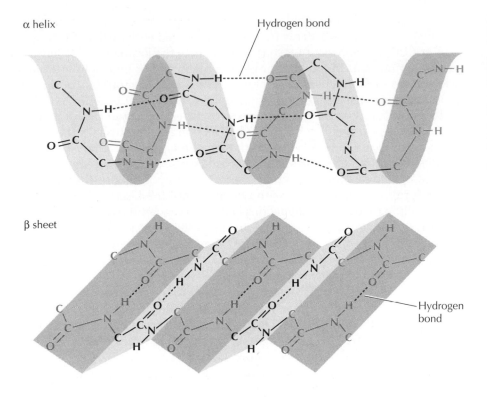

α helix

Hydrogen bond

β sheet

Hydrogen bond

**Figure 2.19**
**Secondary structure of proteins**
The most common types of secondary structure are the α helix and the β sheet. In an α helix, hydrogen bonds form between CO and NH groups of peptide bonds separated by four amino acid residues. In a β sheet, hydrogen bonds connect two parts of a polypeptide chain lying side by side. The amino acid side chains are not shown.

either parallel or antiparallel to each other.

**Tertiary structure** is the folding of the polypeptide chain as a result of interactions between the side chains of amino acids that lie in different regions of the primary sequence (Figure 2.20). In most proteins, combinations of α helices and β sheets, connected by loop regions of the polypeptide chain, fold into compact globular structures called **domains**, which are the basic units of tertiary structure. Small proteins, such as ribonuclease or myo-

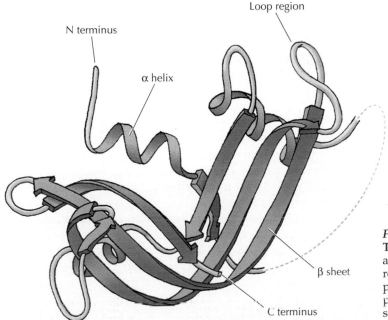

Loop region

N terminus

α helix

β sheet

C terminus

**Figure 2.20**
**Tertiary structure of ribonuclease** Regions of α-helix and β-sheet secondary structures, connected by loop regions, are folded into the native conformation of the protein. In this schematic representation of the polypeptide chain as a ribbon model, α helices are represented as spirals and β sheets as wide arrows.

Heme group    β chains

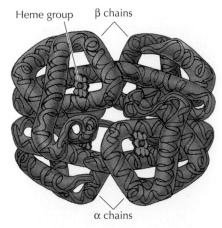

α chains

**Figure 2.21**
**Quaternary structure of hemoglobin**
Hemoglobin is composed of four poly-peptide chains, each of which is bound to a heme group. The two α chains and the two β chains are identical.

**Figure 2.22**
**Energy diagrams for catalyzed and uncatalyzed reactions** The reaction illustrated is the simple conversion of a substrate *S* to a product *P*. Because the final energy state of *P* is lower than that of *S*, the reaction proceeds from left to right. For the reaction to occur, however, *S* must first pass through a higher energy transition state. The energy required to reach this transition state (the activation energy) represents a barrier to the progress of the reaction and thereby determines the rate at which the reaction pro-ceeds. In the presence of a catalyst (e.g., an enzyme), the activation ener-gy is lowered and the reaction pro-ceeds at an accelerated rate.

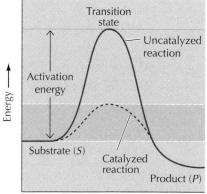

globin, contain only a single domain; larger proteins may contain a number of different domains, which are frequently associated with distinct functions.

A critical determinant of tertiary structure is the localization of hydro-phobic amino acids in the interior of the protein and of hydrophilic amino acids on the surface, where they interact with water. The interiors of folded proteins thus consist mainly of hydrophobic amino acids arranged in α helices and β sheets; these secondary structures are found in the hydro-phobic cores of proteins because hydrogen bonding neutralizes the polar character of the CO and NH groups of the polypeptide backbone. The loop regions connecting these elements of secondary structure are found on the surface of folded proteins, where the polar components of the peptide bonds form hydrogen bonds with water or with the polar side chains of hydrophilic amino acids. Interactions between polar amino acid side chains (hydrogen bonds and ionic bonds) on the protein surface are also impor-tant determinants of tertiary structure. In addition, the covalent disulfide bonds between the sulfhydryl groups of cysteine residues stabilize the folded structures of many cell-surface or secreted proteins.

The fourth level of protein structure, **quaternary structure**, consists of the interactions between different polypeptide chains in proteins composed of more than one polypeptide. Hemoglobin, for example, is composed of four polypeptide chains held together by the same types of interactions that maintain tertiary structure (Figure 2.21).

The distinct chemical characteristics of the 20 different amino acids thus lead to considerable variation in the three-dimensional conformations of folded proteins. Consequently, proteins constitute an extremely complex and diverse group of macromolecules, suited to the wide variety of tasks they perform in cell biology.

## THE CENTRAL ROLE OF ENZYMES AS BIOLOGICAL CATALYSTS

A fundamental task of proteins is to act as enzymes—catalysts that increase the rate of virtually all the chemical reactions within cells. Although RNAs are capable of catalyzing some reactions, most biological reactions are cat-alyzed by proteins. In the absence of enzymatic catalysis, most biochemical reactions are so slow that they would not occur under the mild conditions of temperature and pressure that are compatible with life. Enzymes accel-erate the rates of such reactions by well over a million-fold, so reactions that would take years in the absence of catalysis can occur in fractions of seconds if catalyzed by the appropriate enzyme. Cells contain thousands of different enzymes, and their activities determine which of the many possi-ble chemical reactions actually take place within the cell.

### The Catalytic Activity of Enzymes

Like all other catalysts, **enzymes** are characterized by two fundamental properties. First, they increase the rate of chemical reactions without them-selves being consumed or permanently altered by the reaction. Second, they increase reaction rates without altering the chemical equilibrium between reactants and products.

These principles of enzymatic catalysis are illustrated in the following example, in which a molecule acted upon by an enzyme (referred to as a **substrate** [*S*]) is converted to a **product** (*P*) as the result of the reaction. In the absence of the enzyme, the reaction can be written as follows:

$$S \rightleftharpoons P$$

The chemical equilibrium between *S* and *P* is determined by the laws of thermodynamics (as discussed further in the next section of this chapter) and is represented by the ratio of the forward and reverse reaction rates ($S \rightarrow P$ and $P \rightarrow S$, respectively). In the presence of the appropriate enzyme, the conversion of $S$ to $P$ is accelerated, but the equilibrium between $S$ and $P$ is unaltered. Therefore, the enzyme must accelerate both the forward and reverse reactions equally. The reaction can be written as follows:

$$S \overset{E}{\rightleftharpoons} P$$

Note that the enzyme ($E$) is not altered by the reaction, so the chemical equilibrium remains unchanged, determined solely by the thermodynamic properties of $S$ and $P$.

The effect of the enzyme on such a reaction is best illustrated by the energy changes that must occur during the conversion of $S$ to $P$ (Figure 2.22). The equilibrium of the reaction is determined by the final energy states of $S$ and $P$, which are unaffected by enzymatic catalysis. In order for the reaction to proceed, however, the substrate must first be converted to a higher energy state, called the **transition state**. The energy required to reach the transition state (the **activation energy**) constitutes a barrier to the progress of the reaction, limiting the rate of the reaction. Enzymes (and other catalysts) act by reducing the activation energy, thereby increasing the rate of reaction. The increased rate is the same in both the forward and reverse directions, since both must pass through the same transition state.

The catalytic activity of enzymes involves the binding of their substrates to form an enzyme-substrate complex ($ES$). The substrate binds to a specific region of the enzyme, called the **active site**. While bound to the active site, the substrate is converted into the product of the reaction, which is then released from the enzyme. The enzyme-catalyzed reaction can thus be written as follows:

$$S + E \rightleftharpoons ES \rightleftharpoons E + P$$

Note that $E$ appears unaltered on both sides of the equation, so the equilibrium is unaffected. However, the enzyme provides a surface upon which the reactions converting $S$ to $P$ can occur more readily. This is a result of interactions between the enzyme and substrate that lower the energy of activation and favor formation of the transition state.

## Mechanisms of Enzymatic Catalysis

The binding of a substrate to the active site of an enzyme is a very specific interaction. Active sites are clefts or grooves on the surface of an enzyme, usually composed of amino acids from different parts of the polypeptide chain that are brought together in the tertiary structure of the folded protein. Substrates initially bind to the active site by noncovalent interactions, including hydrogen bonds, ionic bonds, and hydrophobic interactions. Once a substrate is bound to the active site of an enzyme, multiple mechanisms can accelerate its conversion to the product of the reaction.

Although the simple example discussed in the previous section involved only a single substrate molecule, most biochemical reactions involve interactions between two or more different substrates. For example, the formation of a peptide bond involves the joining of two amino acids. For such reactions, the binding of two or more substrates to the active site in the proper position and orientation accelerates the reaction (Figure 2.23). The enzyme provides a template upon which the reactants are brought together and properly oriented to favor the formation of the transition state in which they interact.

Enzymes accelerate reactions also by altering the conformation of their substrates to approach that of the transition state. The simplest model of

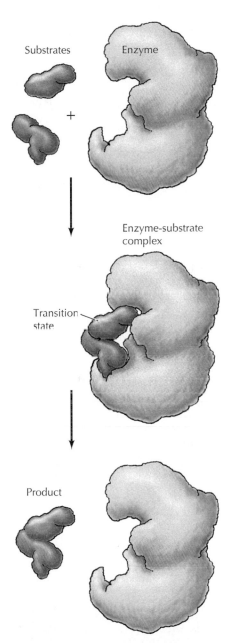

**Figure 2.23**
**Enzymatic catalysis of a reaction between two substrates**  The enzyme provides a template upon which the two substrates are brought together in the proper position and orientation to react with each other.

enzyme-substrate interaction is the **lock-and-key model**, in which the substrate fits precisely into the active site (Figure 2.24). In many cases, however, the configurations of both the enzyme and substrate are modified by substrate binding—a process called **induced fit**. In such cases the conformation of the substrate is altered so that it more closely resembles that of the transition state. The stress produced by such distortion of the substrate can further facilitate its conversion to the transition state by weakening critical bonds. Moreover, the transition state is stabilized by its tight binding to the enzyme, thereby lowering the required energy of activation.

In addition to bringing multiple substrates together and distorting the conformation of substrates to approach the transition state, many enzymes participate directly in the catalytic process. In such cases, specific amino acid side chains in the active site may react with the substrate and form bonds with reaction intermediates. The acidic and basic amino acids are often involved in these catalytic mechanisms, as illustrated in the following discussion of chymotrypsin as an example of enzymatic catalysis.

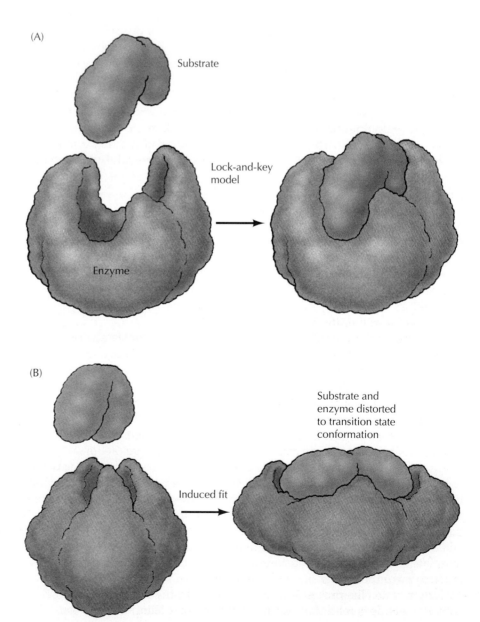

*Figure 2.24*
**Models of enzyme-substrate interaction**   (A) In the lock-and-key model, the substrate fits precisely into the active site of the enzyme. (B) In the induced-fit model, substrate binding distorts the conformations of both substrate and enzyme. This distortion brings the substrate closer to the conformation of the transition state, thereby accelerating the reaction.

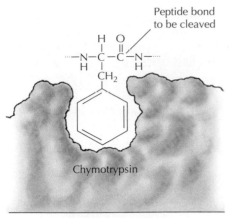

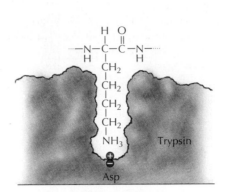

**Hydrophobic interaction**          **Ionic interaction**

*Figure 2.25*
**Substrate binding by serine proteases**   The amino acid adjacent to the peptide bond to be cleaved is inserted into a pocket at the active site of the enzyme. In chymotrypsin, the pocket binds hydrophobic amino acids; the binding pocket of trypsin contains a negatively charged aspartate residue that binds basic amino acids via an ionic interaction.

Chymotrypsin is a member of a family of enzymes (serine proteases) that digest proteins by catalyzing the hydrolysis of peptide bonds. The reaction can be written as follows:

$$\text{Protein} + H_2O \rightarrow \text{Peptide}_1 + \text{Peptide}_2$$

The different members of the serine protease family (including chymotrypsin, trypsin, elastase, and thrombin) have distinct substrate specificities; they preferentially cleave peptide bonds adjacent to different amino acids. For example, whereas chymotrypsin digests bonds adjacent to hydrophobic amino acids, such as tryptophan and phenylalanine, trypsin digests bonds next to basic amino acids, such as lysine and arginine. All the serine proteases, however, are similar in structure and use the same mechanism of catalysis. The active sites of these enzymes contain three critical amino acids— serine, histidine, and aspartate—that drive hydrolysis of the peptide bond. Indeed, these enzymes are called serine proteases because of the central role of the serine residue.

Substrates bind to the serine proteases by insertion of the amino acid adjacent to the cleavage site into a pocket at the active site of the enzyme (Figure 2.25). The nature of this pocket determines the substrate specificity of the different members of the serine protease family. For example, the binding pocket of chymotrypsin contains hydrophobic amino acids that interact with the hydrophobic side chains of its preferred substrates. In contrast, the binding pocket of trypsin contains a negatively charged acidic amino acid (aspartate), which is able to form an ionic bond with the lysine or arginine residues of its substrates.

Substrate binding positions the peptide bond to be cleaved adjacent to the active site serine (Figure 2.26). The proton of this serine is then transferred to the active site histidine. The conformation of the active site favors this proton transfer because the histidine interacts with the negatively charged aspartate residue. The serine reacts with the substrate, forming a tetrahedral transition state. The peptide bond is then cleaved, and the C-terminal portion of the substrate is released from the enzyme. However, the N-terminal peptide remains bound to serine. This situation is resolved when a water molecule (the second substrate) enters the active site and reverses the preceding reactions. The proton of the water molecule is transferred to histidine, and its hydroxyl group is transferred to the peptide, forming a second tetrahedral transition state. The proton is then transferred from histidine back to serine, and the peptide is released from the enzyme, completing the reaction.

*Figure 2.26*
**Catalytic mechanism of chymotrypsin**    Three amino acids at the active site (Ser-195, His-57, and Asp-102) play critical roles in catalysis.

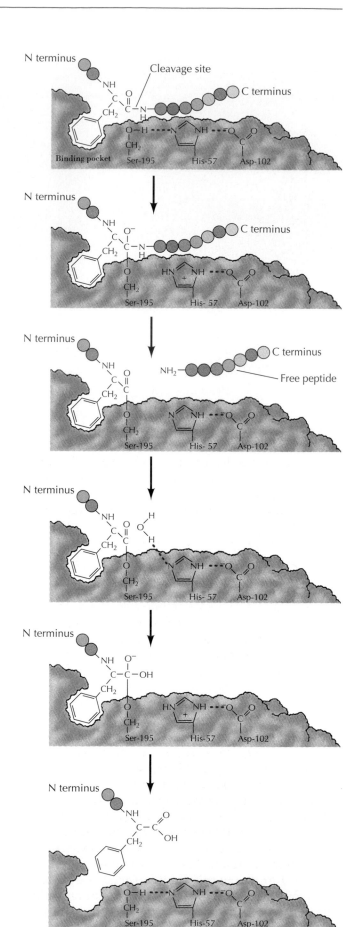

Ser-195 transfers H⁺ to His-57 and forms tetrahedral transition state with substrate. Asp-102 stabilizes the transfer of H⁺ to His-57 by an ionic interaction.

H⁺ transferred from His-57 to substrate; cleavage at peptide bond.

H₂O enters active site by forming hydrogen bond with His-57.

H₂O transfers H⁺ to His-57 and OH⁻ to substrate, forming second tetrahedral transition state.

H⁺ transferred from His-57 back to Ser-195; N-terminal peptide released from enzyme.

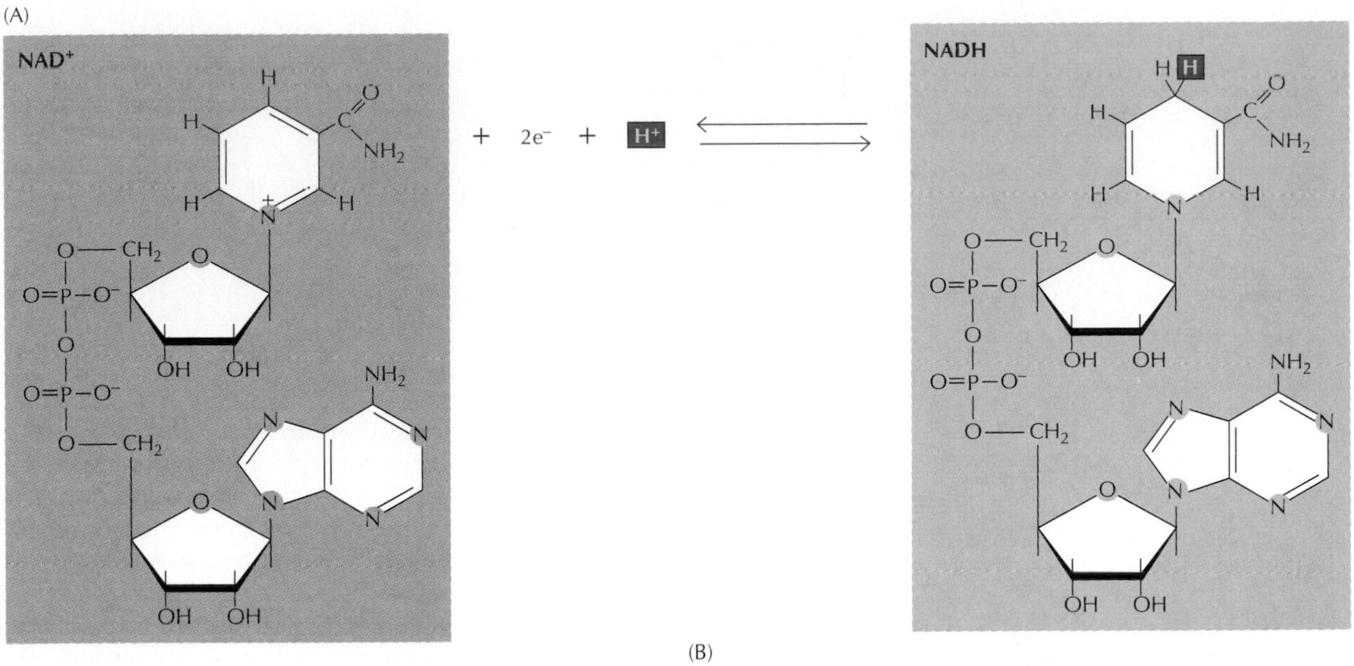

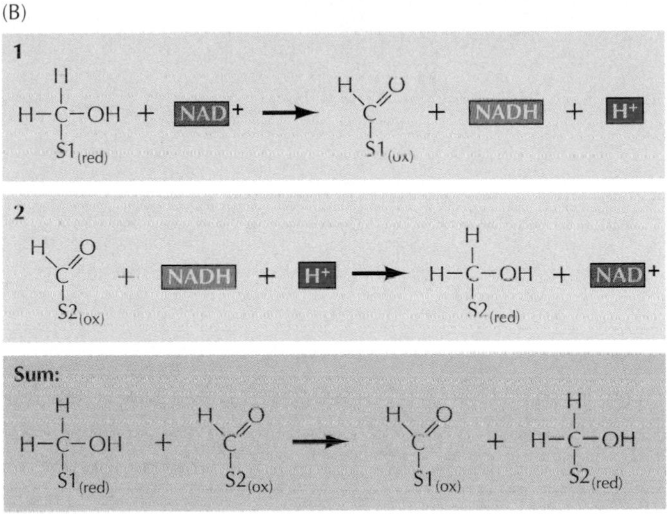

(A)

This example illustrates several features of enzymatic catalysis; the specificity of enzyme-substrate interactions, the positioning of different substrate molecules in the active site, and the involvement of active-site residues in the formation and stabilization of the transition state. Although the thousands of enzymes in cells catalyze many different types of chemical reactions, the same basic principles apply to their operation.

### Coenzymes

In addition to binding their substrates, the active sites of many enzymes bind other small molecules that participate in catalysis. **Prosthetic groups** are small molecules bound to proteins in which they play critical functional roles. For example, the oxygen carried by myoglobin and hemoglobin is bound to heme, a prosthetic group of these proteins. In many cases metal ions (such as zinc or iron) are bound to enzymes and play central roles in the catalytic process. In addition, various low-molecular-weight organic molecules participate in specific types of enzymatic reactions. These molecules are called **coenzymes** because they work together with enzymes to enhance reaction rates. In contrast to substrates, coenzymes are not irreversibly altered by the reactions in which they are involved. Rather, they are recycled and can participate in multiple enzymatic reactions.

Coenzymes serve as carriers of several types of chemical groups. A prominent example of a coenzyme is **nicotinamide adenine dinucleotide** (**NAD$^+$**), which functions as a carrier of electrons in oxidation–reduction reactions (Figure 2.27). NAD$^+$ can accept a hydrogen ion (H$^+$) and two electrons (e$^-$) from one substrate, forming NADH. NADH can then donate these electrons to a second substrate, re-forming NAD$^+$. Thus, NAD$^+$ transfers electrons from the first substrate (which becomes oxidized) to the second (which becomes reduced).

(B)

*Figure 2.27*
**Role of NAD$^+$ in oxidation-reduction reactions**   (A) Nicotinamide adenine dinucleotide (NAD$^+$) acts as a carrier of electrons in oxidation-reduction reactions by accepting electrons (e$^-$) to form NADH. (B) For example, NAD$^+$ can accept electrons from one substrate (S1), yielding oxidized S1 plus NADH. The NADH formed in this reaction can then transfer its electrons to a second substrate (S2), yielding reduced S2 and regenerating NAD$^+$. The net effect is the transfer of electrons (carried by NADH) from S1 (which becomes oxidized) to S2 (which becomes reduced).

*Table 2.1*    Examples of Coenzymes and Vitamins

| Coenzyme | Related vitamin | Chemical reaction |
|---|---|---|
| NAD+, NADP+ | Niacin | Oxidation–reduction |
| FAD | Riboflavin (B$_2$) | Oxidation–reduction |
| Thiamine pyrophosphate | Thiamine (B$_1$) | Aldehyde group transfer |
| Coenzyme A | Pantothenate | Acyl group transfer |
| Tetrahydrofolate | Folate | Transfer of one-carbon groups |
| Biotin | Biotin | Carboxylation |
| Pyridoxal phosphate | Pyridoxal (B$_6$) | Transamination |

Several other coenzymes also act as electron carriers, and still others are involved in the transfer of a variety of additional chemical groups (e.g., carboxyl groups and acyl groups; Table 2.1). The same coenzymes function together with a variety of different enzymes to catalyze the transfer of specific chemical groups between a wide range of substrates. Many coenzymes are closely related to vitamins, which contribute part or all of the structure of the coenzyme. Vitamins are not required by bacteria such as *E. coli* but are necessary components of the diets of human and other higher animals, which have lost the ability to synthesize these compounds.

### Regulation of Enzyme Activity

An important feature of most enzymes is that their activities are not constant but instead can be modulated. That is, the activities of enzymes can be regulated so that they function appropriately to meet the varied physiological needs that may arise during the life of the cell.

One common type of enzyme regulation is **feedback inhibition**, in which the product of a metabolic pathway inhibits the activity of an enzyme involved in its synthesis. For example, the amino acid isoleucine is synthesized by a series of reactions starting from the amino acid threonine (Figure 2.28). The first step in the pathway is catalyzed by the enzyme threonine deaminase, which is inhibited by isoleucine, the end product of the pathway. Thus, an adequate amount of isoleucine in the cell inhibits threonine deaminase, blocking further synthesis of isoleucine. If the concentration of isoleucine decreases, feedback inhibition is relieved, threonine deaminase is no longer inhibited, and additional isoleucine is synthesized. By so regulating the activity of threonine deaminase, the cell synthesizes the necessary amount of isoleucine but avoids wasting energy on the synthesis of more isoleucine than is needed.

Feedback inhibition is one example of **allosteric regulation**, in which enzyme activity is controlled by the binding of small molecules to regulatory sites on the enzyme (Figure 2.29). The term "allosteric regulation" derives from the fact that the regulatory molecules bind not to the catalytic site, but to a distinct site on the protein (*allo* = "other" and *steric* = "site"). Binding of the regulatory molecule changes the conformation of the protein, which in turn alters the shape of the active site and the catalytic activity of the enzyme. In the case of threonine deaminase, binding of the regulatory molecule (isoleucine) inhibits enzymatic activity. In other cases regulatory molecules serve as activators, stimulating rather than inhibiting their target enzymes.

The activities of enzymes can also be regulated by their interactions with other proteins and by covalent modifications, such as the addition of phosphate groups to serine, threonine, or tyrosine residues. **Phosphoryla-**

*Figure 2.28*
**Feedback inhibition**   The first step in the conversion of threonine to isoleucine is catalyzed by the enzyme threonine deaminase. The activity of this enzyme is inhibited by isoleucine, the end product of the pathway.

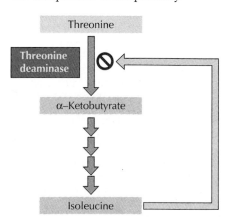

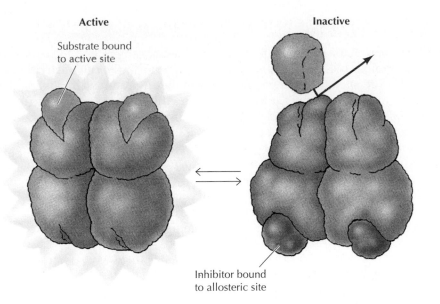

Active — Substrate bound to active site

Inactive — Inhibitor bound to allosteric site

*Figure 2.29*
**Allosteric regulation** In this example, enzyme activity is inhibited by the binding of a regulatory molecule to an allosteric site. In the absence of inhibitor, the substrate binds to the active site of the enzyme and the reaction proceeds. The binding of inhibitor to the allosteric site induces a conformational change in the enzyme and prevents substrate binding. Most allosteric enzymes consist of multiple subunits.

**tion** is a particularly common mechanism for regulating enzyme activity; the addition of phosphate groups either stimulates or inhibits the activities of many different enzymes (Figure 2.30). For example, muscle cells respond to epinephrine (adrenaline) by breaking down glycogen into glucose, thereby providing a source of energy for increased muscular activity. The breakdown of glycogen is catalyzed by the enzyme glycogen phosphorylase, which is activated by phosphorylation in response to the binding of epinephrine to a receptor on the surface of the muscle cell. Protein phosphorylation plays a central role in controlling not only metabolic reactions but also many other cellular functions, including cell growth and differentiation.

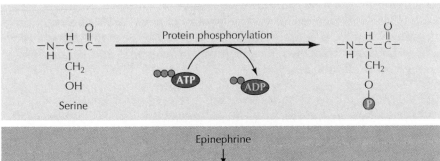
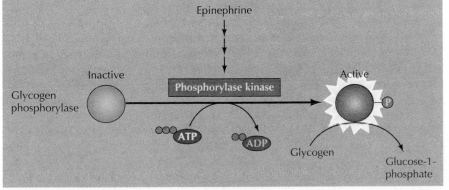

*Figure 2.30*
**Protein phosphorylation** Some enzymes are regulated by the addition of phosphate groups to the side-chain OH groups of serine (as shown here), threonine, or tyrosine residues. For example, the enzyme glycogen phosphorylase, which catalyzes the conversion of glycogen into glucose-1-phosphate, is activated by phosphorylation in response to the binding of epinephrine to muscle cells.

## METABOLIC ENERGY

Many tasks that a cell must perform, such as movement and the synthesis of macromolecules, require energy. A large portion of the cell's activities are therefore devoted to obtaining energy from the environment and using that energy to drive energy-requiring reactions. Although enzymes control the rates of virtually all chemical reactions within cells, the equilibrium position of chemical reactions is not affected by enzymatic catalysis. The laws of thermodynamics govern chemical equilibria and determine the energetically favorable direction of all chemical reactions. Many of the reactions that must take place within cells are energetically unfavorable, and are therefore able to proceed only at the cost of additional energy input. Consequently, cells must constantly expend energy derived from the environment. The generation and utilization of metabolic energy is thus fundamental to all of cell biology.

### Free Energy and ATP

The energetics of biochemical reactions are best described in terms of the thermodynamic function called **Gibbs free energy (G)**, named for Josiah Willard Gibbs. The change in free energy ($\Delta G$) of a reaction combines the effects of changes in enthalpy (the heat that is released or absorbed during a chemical reaction) and entropy (the degree of disorder resulting from a reaction) to predict whether or not a reaction is energetically favorable. All chemical reactions spontaneously proceed in the energetically favorable direction, accompanied by a decrease in free energy ($\Delta G < 0$). For example, consider a hypothetical reaction in which A is converted to B:

$$A \rightleftharpoons B$$

If $\Delta G < 0$, this reaction will proceed in the forward direction, as written. If $\Delta G > 0$, however, the reaction will proceed in the reverse direction and B will be converted to A.

The $\Delta G$ of a reaction is determined not only by the intrinsic properties of reactants and products, but also by their concentrations and other reaction conditions (e.g., temperature). It is thus useful to define the free-energy change of a reaction under standard conditions. (Standard conditions are considered to be a 1-$M$ concentration of all reactants and products, and 1 atm of pressure). The standard free-energy change ($\Delta G°$) of a reaction is directly related to its equilibrium position because the actual $\Delta G$ is a function of both $\Delta G°$ and the concentrations of reactants and products. For example, consider the reaction

$$A \rightleftharpoons B$$

The free-energy change can be written as follows:

$$\Delta G = \Delta G° + RT \ln [B]/[A]$$

where $R$ is the gas constant and $T$ is the absolute temperature.

At equilibrium, $\Delta G = 0$ and the reaction does not proceed in either direction. The equilibrium constant for the reaction ($K = [B]/[A]$ at equilibrium) is thus directly related to $\Delta G°$ by the above equation, which can be expressed as follows:

$$0 = \Delta G° + RT \ln K$$

or

$$\Delta G° = - RT \ln K$$

If the actual ratio [B]/[A] is greater than the equilibrium ratio ($K$), $\Delta G > 0$ and the reaction proceeds in the reverse direction (conversion of B to A). On the other hand, if the ratio [B]/[A] is less than the equilibrium ratio, $\Delta G < 0$ and A is converted to B.

The standard free-energy change ($\Delta G°$) of a reaction therefore determines its chemical equilibrium and predicts in which direction the reaction will proceed under any given set of conditions. For biochemical reactions, the standard free-energy change is usually expressed as $\Delta G°'$, which is the standard free-energy change of a reaction in aqueous solution at pH = 7, approximately the conditions within a cell.

Many biological reactions (such as the synthesis of macromolecules) are thermodynamically unfavorable ($\Delta G > 0$) under cellular conditions. In order for such reactions to proceed, an additional source of energy is required. For example, consider the reaction

$$A \rightleftharpoons B \qquad \Delta G = +10 \text{ kcal/mol}$$

The conversion of A to B is energetically unfavorable, so the reaction proceeds in the reverse rather than the forward direction. However, the reaction can be driven in the forward direction by coupling the conversion of A to B with an energetically favorable reaction, such as:

$$C \rightleftharpoons D \qquad \Delta G = -20 \text{ kcal/mol}$$

If these two reactions are combined, the coupled reaction can be written as follows:

$$A + C \rightleftharpoons B + D \qquad \Delta G = -10 \text{ kcal/mol}$$

The $\Delta G$ of the combined reaction is the sum of the free-energy changes of its individual components, so the coupled reaction is energetically favorable and will proceed as written. Thus, the energetically unfavorable conversion of A to B is driven by coupling it to a second reaction associated with a large decrease in free energy. Enzymes are responsible for carrying out such coupled reactions in a coordinated manner.

The cell uses this basic mechanism to drive the many energetically unfavorable reactions that must take place in biological systems. **Adenosine 5′-triphosphate (ATP)** plays a central role in this process by acting as a store of free energy within the cell (Figure 2.31). The bonds between the phosphates in ATP are known as **high-energy bonds** because their hydrolysis is accompanied by a relatively large decrease in free energy. There is nothing special about the chemical bonds themselves; they are called high-energy bonds only because a large amount of free energy is released when they are hydrolyzed within the cell. In the hydrolysis of ATP to ADP plus phosphate ($P_i$), $\Delta G°' = -7.3$ kcal/mol. Recall, however, that $\Delta G°'$ refers to "standard conditions," in which the concentrations of all products and reactants are 1 $M$. Actual intracellular concentrations of $P_i$ are approximately $10^{-2}$ $M$, and intracellular concentrations of ATP are higher than those of ADP. These differences between intracellular concentrations and those of the standard state favor ATP hydrolysis, so for ATP hydrolysis within a cell, $\Delta G$ is approximately $-12$ kcal/mol.

Alternatively, ATP can be hydrolyzed to AMP plus pyrophosphate ($PP_i$). This reaction yields about the same amount of free energy as the hydrolysis of ATP to ADP does. However, the pyrophosphate produced by this reaction is then itself rapidly hydrolyzed, with a $\Delta G$ similar to that of ATP hydrolysis. Consequently, the total free-energy change resulting from the hydrolysis of ATP to AMP is approximately twice that obtained by the hydrolysis of ATP to ADP. For comparison, the bond between the sugar

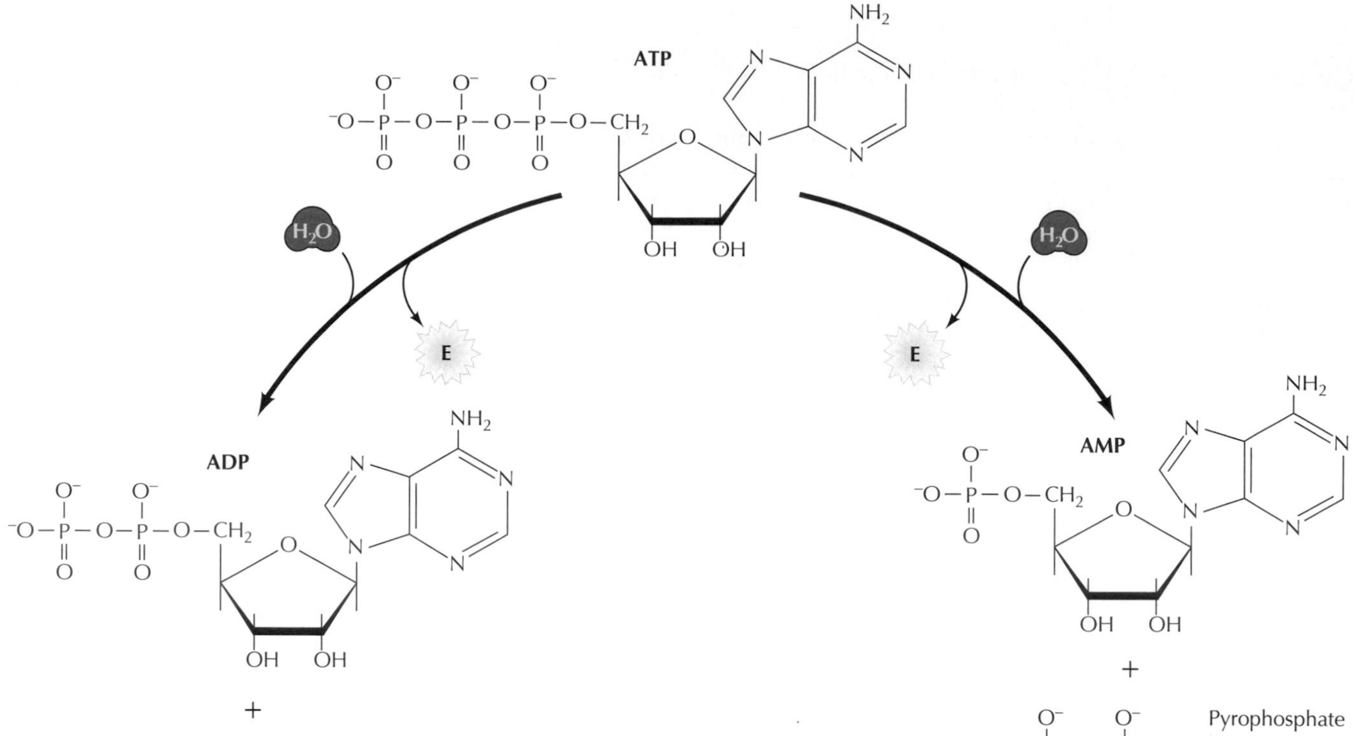

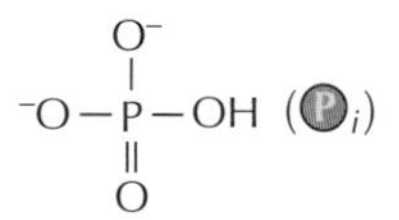

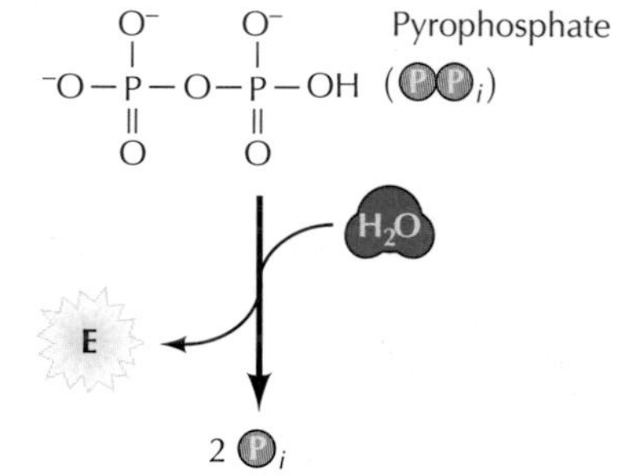

*Figure 2.31*
**ATP as a store of free energy** The bonds between the phosphate groups of ATP are called high-energy bonds because their hydrolysis results in a large decrease in free energy. ATP can be hydrolyzed either to ADP plus a phosphate group ($HPO_4^{2-}$) or to AMP plus pyrophosphate. In the latter case, pyrophosphate is itself rapidly hydrolyzed, releasing additional free energy.

and phosphate group of AMP, rather than having high energy, is typical of covalent bonds; for the hydrolysis of AMP, $\Delta G^{\circ\prime} = -3.3$ kcal/mol.

Because of the accompanying decrease in free energy, the hydrolysis of ATP can be used to drive other energy-requiring reactions within the cell. For example, the first reaction in glycolysis (discussed in the next section) is the conversion of glucose to glucose-6-phosphate. The reaction can be written as follows:

Glucose + Phosphate ($HPO_4^{2-}$) → Glucose-6-phosphate + $H_2O$

Because this reaction is energetically unfavorable as written ($\Delta G^{\circ\prime} = +3.3$ kcal/mol), it must be driven in the forward direction by being coupled to ATP hydrolysis ($\Delta G^{\circ\prime} = -7.3$ kcal/mol):

ATP + $H_2O$ → ADP + $HPO_4^{2-}$

The combined reaction can be written as follows:

Glucose + ATP → Glucose-6-phosphate + ADP

The free-energy change for this reaction is the sum of the free-energy changes for the individual reactions, so for the coupled reaction $\Delta G^{\circ\prime} = -4.0$ kcal/mol, favoring glucose-6-phosphate formation.

Other molecules, including other nucleoside triphosphates (e.g., GTP), also have high-energy bonds and can be used as ATP is to drive energy-requiring reactions. For most reactions, however, ATP provides the free energy. The energy-yielding reactions within the cell are therefore coupled to ATP synthesis, while the energy-requiring reactions are coupled to ATP hydrolysis. The high-energy bonds of ATP thus play a central role in cell metabolism by serving as a usable storage form of free energy.

## The Generation of ATP from Glucose

The breakdown of carbohydrates, particularly glucose, is a major source of cellular energy. The complete oxidative breakdown of glucose to $CO_2$ and $H_2O$ can be written as follows:

$$C_6H_{12}O_6 + 6\ O_2 \rightarrow 6\ CO_2 + 6\ H_2O$$

The reaction yields a large amount of free energy: $\Delta G^{\circ\prime} = -686$ kcal/mol. To harness this free energy in usable form, glucose is oxidized within cells in a series of steps coupled to the synthesis of ATP.

**Glycolysis,** the initial stage in the breakdown of glucose, is common to virtually all cells. Glycolysis occurs in the absence of oxygen and can provide all the metabolic energy of anaerobic organisms. In aerobic cells, however, glycolysis is only the first stage in glucose degradation.

The reactions of glycolysis result in the breakdown of glucose into pyruvate, with the net gain of two molecules of ATP (Figure 2.32). The initial reactions in the pathway actually consume energy, using ATP to phosphorylate glucose to glucose-6-phosphate and then fructose-6-phosphate to fructose-1,6-bisphosphate. The enzymes that catalyze these two reactions—hexokinase and phosphofructokinase respectively—are important regulatory points of the glycolytic pathway. The key control element is phosphofructokinase, which is inhibited by high levels of ATP. Inhibition of phosphofructokinase results in an accumulation of glucose-6-phosphate, which in turn inhibits hexokinase. Thus, when the cell has an adequate supply of metabolic energy available in the form of ATP, the breakdown of glucose is inhibited.

The reactions following the formation of fructose-1,6-bisphosphate constitute the energy-producing part of the glycolytic pathway. Cleavage of fructose-1,6-bisphosphate yields two molecules of the three-carbon sugar glyceraldehyde-3-phosphate, which is oxidized to 1,3-bisphosphoglycerate. The phosphate group of this compound has a very high free energy of hydrolysis ($\Delta G^{\circ\prime} = -11.5$ kcal/mol), so it is used in the next reaction of glycolysis to drive the synthesis of ATP from ADP. The product of this reaction, 3-phosphoglycerate, is then converted to phosphoenolpyruvate, the second high-energy intermediate in glycolysis. In the hydrolysis of the high-energy phosphate of phosphoenolpyruvate, $\Delta G^{\circ\prime} = -14.6$ kcal/mol, its conversion to pyruvate is coupled to the synthesis of ATP. Each molecule of glyceraldehyde-3-phosphate converted to pyruvate is thus coupled to the generation of two molecules of ATP; in total, four ATPs are synthesized from each starting molecule of glucose. Since two ATPs were required to prime the initial reactions, the net gain from glycolysis is two ATP molecules.

In addition to producing ATP, glycolysis converts two molecules of the coenzyme $NAD^+$ to NADH. In this reaction, $NAD^+$ acts as an oxidizing agent that accepts electrons from glyceraldehyde-3-phosphate. The NADH formed as a product must be recycled by serving as a donor of electrons for other oxidation–reduction reactions within the cell. In anaerobic conditions, the NADH formed during glycolysis is reoxidized to $NAD^+$ by the conversion of pyruvate to lactate or ethanol. In aerobic organisms, however, the NADH serves as an additional source of energy by donating its electrons to the electron transport chain, where they are ultimately used to reduce $O_2$ to $H_2O$, coupled to the generation of additional ATP.

In eukaryotic cells, glycolysis takes place in the cytosol. Pyruvate is then transported into mitochondria, where its complete oxidation to $CO_2$ and $H_2O$ yields most of the ATP derived from glucose breakdown. The next step in the metabolism of pyruvate is its oxidative decarboxylation in the presence of **coenzyme A (CoA)**, which serves as a carrier of acyl groups in

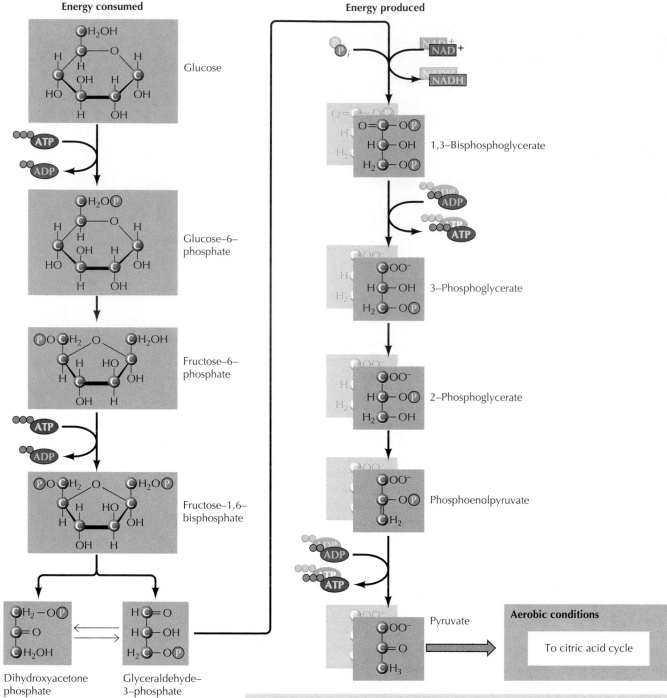

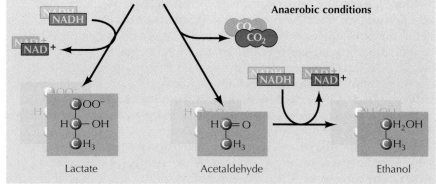

*Figure 2.32*
**Reactions of glycolysis**  Glucose is broken down to pyruvate, with the net formation of two molecules each of ATP and NADH. Under anaerobic conditions, the NADH is reoxidized by the conversion of pyruvate to ethanol or lactate. Under aerobic conditions, pyruvate is further metabolized by the citric acid cycle. Note that a single molecule of glucose yields two molecules each (shadow boxes) of the energy-producing 3-carbon derivatives.

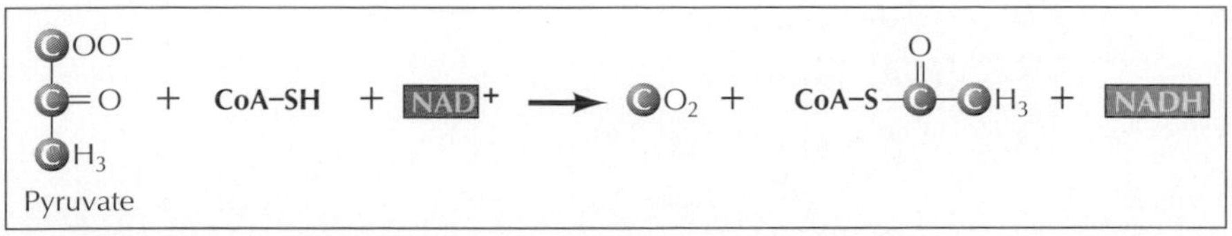

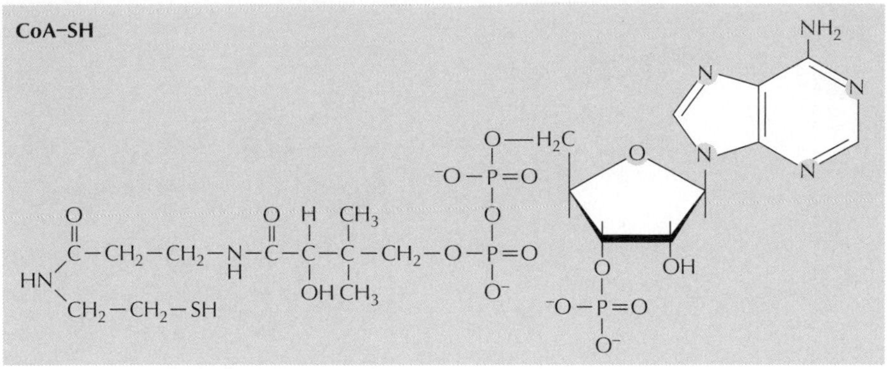

*Figure 2.33*
**Oxidative decarboxylation of pyruvate** Pyruvate is converted to $CO_2$ and acetyl CoA, and one molecule of NADH is produced in the process. Coenzyme A (CoA-SH) is a general carrier of activated acyl groups in a variety of reactions.

various metabolic reactions (Figure 2.33). One carbon of pyruvate is released as $CO_2$, and the remaining two carbons are added to CoA to form acetyl CoA. In the process, one molecule of $NAD^+$ is reduced to NADH.

The acetyl CoA formed by this reaction enters the **citric acid cycle** or **Krebs cycle** (Figure 2.34), which is the central pathway in oxidative metabolism. The two-carbon acetyl group combines with oxaloacetate (four carbons) to yield citrate (six carbons). Through eight further reactions, two carbons of citrate are completely oxidized to $CO_2$ and oxaloacetate is regenerated. During the cycle, one high-energy phosphate bond is formed in GTP, which is used directly to drive the synthesis of one ATP molecule. In addition, each turn of the cycle yields three molecules of NADH and one molecule of reduced **flavin adenine dinucleotide** (**FADH₂**), which is another carrier of electrons in oxidation–reduction reactions.

The citric acid cycle completes the oxidation of glucose to six molecules of $CO_2$. Four molecules of ATP are obtained directly from each glucose molecule—two from glycolysis and two from the citric acid cycle (one for each molecule of pyruvate). In addition, ten molecules of NADH (two from glycolysis, two from the conversion of pyruvate to acetyl CoA, and six from the citric acid cycle) and two molecules of FADH₂ are formed. The remaining energy derived from the breakdown of glucose comes from the reoxidation of NADH and FADH₂, with their electrons being transferred through the electron transport chain to (eventually) reduce $O_2$ to $H_2O$.

During **oxidative phosphorylation**, the electrons of NADH and FADH₂ combine with $O_2$, and the energy released from the process drives the synthesis of ATP from ADP. The transfer of electrons from NADH to $O_2$ releases a large amount of free energy: $\Delta G^{\circ\prime} = -52.5$ kcal/mol for each pair of electrons transferred. So that this energy can be harvested in usable form, the process takes place gradually by the passage of electrons through a series of carriers, which constitute the **electron transport chain** (Figure 2.35). The components of the electron transport chain are located in the inner mitochondrial membrane of eukaryotic cells, and oxidative phosphorylation is considered in more detail when mitochondria are discussed in Chapter 10. In aerobic bacteria, which use a comparable system, components of the electron transport chain are located in the plasma membrane. In either case, the transfer of electrons from NADH to $O_2$ yields sufficient energy to drive the

*Figure 2.34*
**The citric acid cycle**   A two-carbon acetyl group is transferred from acetyl CoA to oxaloacetate, forming citrate. Two carbons of citrate are then oxidized to $CO_2$ and oxaloacetate is regenerated. Each turn of the cycle yields one molecule of GTP, three of NADH, and one of $FADH_2$.

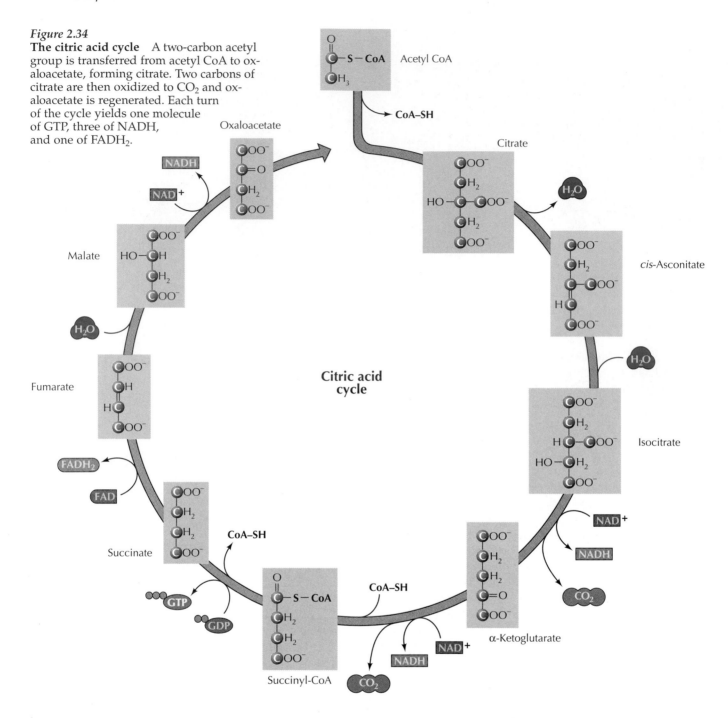

synthesis of approximately three molecules of ATP. Electrons from $FADH_2$ enter the electron transport chain at a lower energy level, so their transfer to $O_2$ yields less usable free energy, only two ATP molecules.

It is now possible to calculate the total yield of ATP from the oxidation of glucose. The net gain from glycolysis is two molecules of ATP and two molecules of NADH. The conversion of pyruvate to acetyl CoA and its metabolism via the citric acid cycle yields two additional molecules of ATP, eight of NADH, and two of $FADH_2$. Assuming that three molecules of ATP are derived from the oxidation of each NADH and two from each $FADH_2$, the total yield is 38 molecules of ATP per molecule of glucose. However, this yield is lower in some cells because the two molecules of NADH gen-

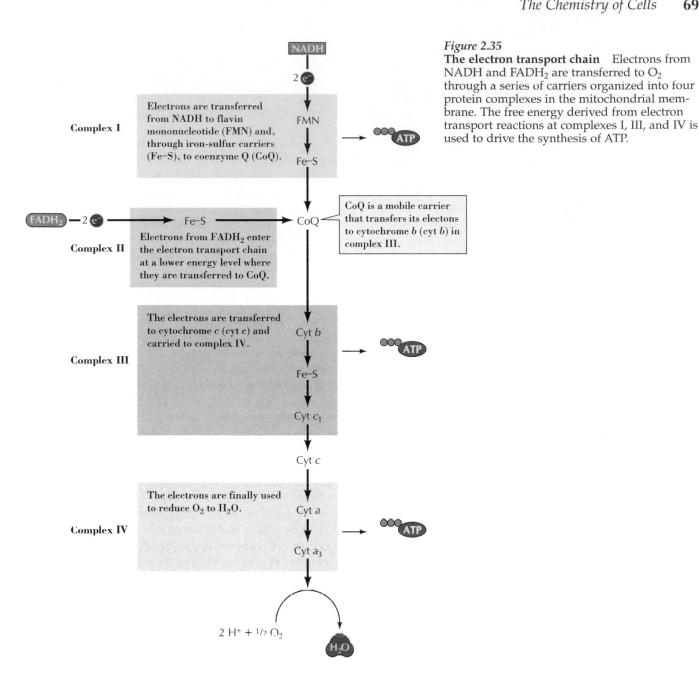

*Figure 2.35*
**The electron transport chain** Electrons from NADH and $FADH_2$ are transferred to $O_2$ through a series of carriers organized into four protein complexes in the mitochondrial membrane. The free energy derived from electron transport reactions at complexes I, III, and IV is used to drive the synthesis of ATP.

erated by glycolysis in the cytosol are unable to enter mitochondria directly. Instead, their electrons must be transferred into the mitochondrion via a shuttle system. Depending on the system used, this transfer may result in these electrons entering the electron transport chain at the level of $FADH_2$. In such cases, the two molecules of NADH derived from glycolysis give rise to two rather than three molecules of ATP, reducing the total yield to 36 rather than 38 ATPs per molecule of glucose.

### The Derivation of Energy from Other Organic Molecules

Energy in the form of ATP can be derived from the breakdown of other organic molecules, with the pathways involved in glucose degradation again playing a central role. Nucleotides, for example, can be broken down to sugars, which then enter the glycolytic pathway, and amino acids are degraded via the citric acid cycle. The two principal storage forms of energy within

CoA–SH   +   $\overset{O}{\underset{-O}{\parallel}}$ C$-(CH_2)_{14}-CH_3$

**Fatty acid**

ATP

AMP + PPi $\longrightarrow$ 2 Pi

$H_2O$

$\overset{O}{\parallel}$
CoA$-$S$-$C$-(CH_2)_{14}-CH_3$

**C$_{16}$ Acyl CoA**

FAD

FADH$_2$

$\overset{O}{\parallel}$   $\overset{H}{\underset{H}{|}}$
CoA$-$S$-$C$-$C$=$C$-(CH_2)_{12}-CH_3$

$H_2O$

$\overset{O}{\parallel}$    $\overset{OH}{\underset{H}{|}}$
CoA$-$S$-$C$-$CH$_2-$C$-(CH_2)_{12}-CH_3$

NAD$^+$

NADH + H$^+$

$\overset{O}{\parallel}$    $\overset{O}{\parallel}$
CoA$-$S$-$C$-$CH$_2-$C$-(CH_2)_{12}-CH_3$

CoA–SH

$\overset{O}{\parallel}$
CoA$-$S$-$C$-$CH$_3$

**Acetyl CoA**    +

$\overset{O}{\parallel}$
CoA$-$S$-$C$-(CH_2)_{12}-CH_3$

**C$_{14}$ Acyl CoA**

cells, polysaccharides and lipids, can also be broken down to produce ATP. Polysaccharides are broken down into free sugars, which are then metabolized as discussed in the previous section. Lipids, however, are an even more efficient energy storage molecule. Because lipids are more reduced than carbohydrates, consisting primarily of hydrocarbon chains, their oxidation yields substantially more energy per weight of starting material.

Fats (triacylglycerols) are the major storage form of lipids. The first step in their utilization is their breakdown to glycerol and free fatty acids. Each fatty acid is joined to coenzyme A, yielding a fatty acyl-CoA at the cost of one molecule of ATP (Figure 2.36). The fatty acids are then degraded in a stepwise oxidative process, two carbons at a time, yielding acetyl CoA plus a fatty acyl-CoA shorter by one two-carbon unit. Each round of oxidation also yields one molecule of NADH and one of FADH$_2$. The acetyl CoA then enters the citric acid cycle, and degradation of the remainder of the fatty acid continues in the same manner.

The breakdown of a 16-carbon fatty acid thus yields seven molecules of NADH, seven of FADH$_2$, and eight of acetyl CoA. In terms of ATP generation, this yield corresponds to 21 molecules of ATP derived from NADH ($3 \times 7$), 14 ATPs from FADH$_2$ ($2 \times 7$), and 96 from acetyl CoA ($8 \times 12$). Since one ATP was used to start the process, the net gain is 130 ATPs per molecule of a 16-carbon fatty acid. Compare this yield with the net gain of 38 ATPs per molecule of glucose. Since the molecular weight of a saturated 16-carbon fatty acid is 256 and that of glucose is 180, the yield of ATP is approximately 2.5 times greater per gram of the fatty acid—hence the advantage of lipids over polysaccharides as energy storage molecules.

## Photosynthesis

The generation of energy from oxidation of carbohydrates and lipids relies on the degradation of preformed organic compounds. The energy required for the synthesis of these compounds is ultimately derived from sunlight, which is harvested and used by plants and photosynthetic bacteria to drive the synthesis of carbohydrates. By converting the energy of sunlight to a usable form of chemical energy, photosynthesis is the source of virtually all metabolic energy in biological systems.

The overall equation of photosynthesis can be written as follows:

$$6\ CO_2 + 6\ H_2O \xrightarrow{\text{Light}} C_6H_{12}O_6 + 6\ O_2$$

The process is much more complex, however, and takes place in two distinct stages. In the first, called the **light reactions**, energy absorbed from sunlight drives the synthesis of ATP and NADPH (a coenzyme similar to NADH), coupled to the oxidation of $H_2O$ to $O_2$. The ATP and NADPH generated by the light reactions drive the synthesis of carbohydrates from $CO_2$

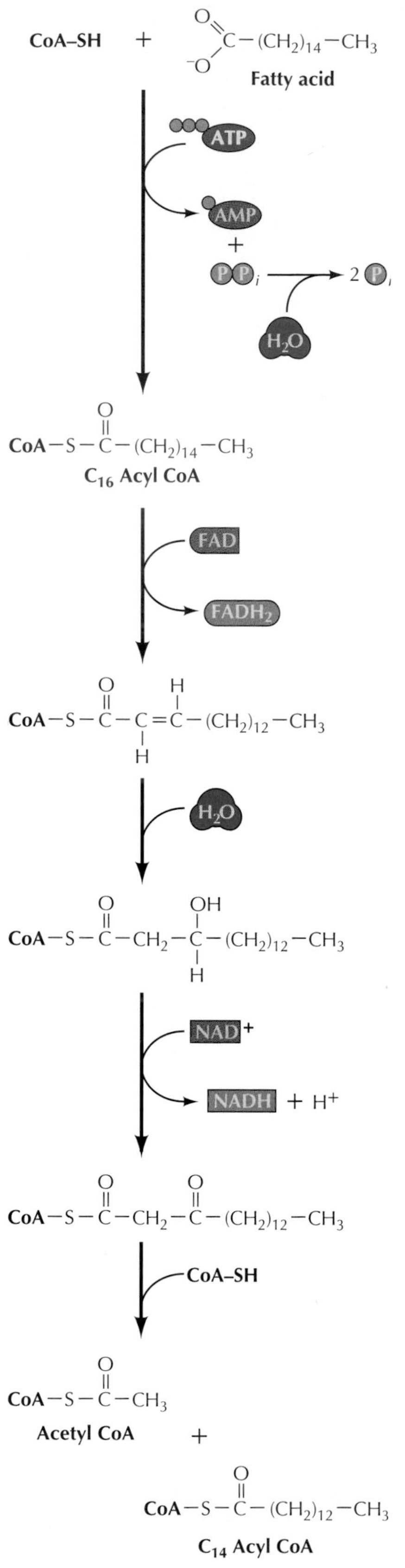

*Figure 2.36*
**Oxidation of fatty acids**   The fatty acid (e.g., the 16-carbon saturated fatty acid palmitate) is initially joined to coenzyme A at the cost of one molecule of ATP. Oxidation of the fatty acid then proceeds by stepwise removal of two-carbon units as acetyl CoA, coupled to the formation of one molecule each of NADH and FADH$_2$.

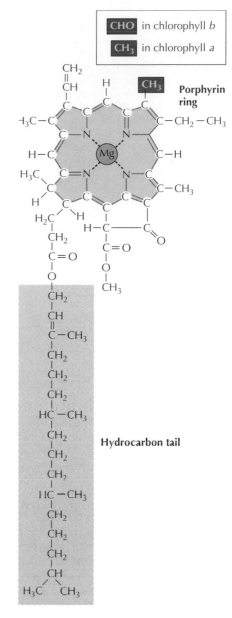

*Figure 2.37*
**The structure of chlorophyll**   Chlorophylls consist of porphyrin ring structures linked to hydrocarbon tails. Chlorophylls *a* and *b* differ by a single functional group in the porphyrin ring.

and $H_2O$ in a second set of reactions, called the **dark reactions** because they do not require sunlight. In eukaryotic cells, both the light and dark reactions occur in chloroplasts.

 **Photosynthetic pigments** capture energy from sunlight by absorbing photons. Absorption of light by these pigments causes an electron to move from its normal molecular orbital to one of higher energy, thus converting energy from sunlight into chemical energy. In plants the most abundant photosynthetic pigments are the **chlorophylls** (Figure 2.37), which together absorb visible light of all wavelengths other than green. Additional pigments absorb light of other wavelengths, so essentially the entire spectrum of visible light can be captured and utilized for photosynthesis.

 The energy captured by the absorption of light is used to convert $H_2O$ to $O_2$ (Figure 2.38). The high-energy electrons derived from this process then enter an electron transport chain, in which their transfer through a series of carriers is coupled to the synthesis of ATP. In addition, these high energy electrons reduce $NADP^+$ to NADPH.

 In the dark reactions, the ATP and NADPH produced from the light reactions drive the synthesis of carbohydrates from $CO_2$ and $H_2O$. One molecule of $CO_2$ at a time is added to a cycle of reactions—known as the **Calvin cycle** after its discoverer, Melvin Calvin—that leads to the formation of carbohydrates (Figure 2.39). Overall, the Calvin cycle consumes 18 molecules of ATP and 12 of NADPH for each molecule of glucose synthesized. Two electrons are needed to convert each molecule of $NADP^+$ to NADPH, so 24 electrons must pass through the electron transport chain to generate sufficient NADPH to synthesize one molecule of glucose. These electrons are obtained by the conversion of 12 molecules of $H_2O$ to six molecules of $O_2$, consistent with the formation of six molecules of $O_2$ for each molecule of glucose. It is not clear, however, whether the passage of the same 24 electrons through the electron transport chain is also sufficient to generate the 18 ATPs that are required by the Calvin cycle. Some of these ATP molecules may in-stead be generated by alternative electron transport chains that use the energy derived from sunlight to synthesize ATP without the synthesis of NADPH (see Chapter 10).

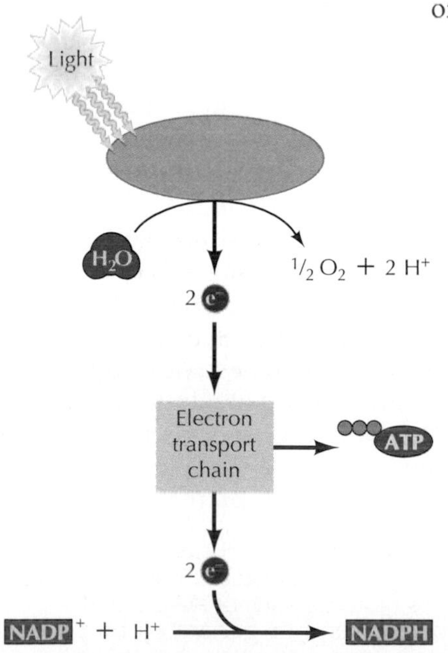

*Figure 2.38*
**The light reactions of photosynthesis**   Energy from sunlight is used to split $H_2O$ to $O_2$. The high-energy electrons derived from this process are then transported through a series of carriers and used to convert $NADP^+$ to NADPH. Energy derived from the electron transport reactions also drives the synthesis of ATP. The details of these reactions are discussed in Chapter 10.

*Figure 2.39*
**The Calvin cycle**   Shown here is the synthesis of one molecule of glucose from six molecules of $CO_2$. Each molecule of $CO_2$ is added to ribulose-1,5-bisphosphate to yield two molecules of 3-phosphoglycerate. Six molecules of $CO_2$ thus lead to the formation of 12 molecules of 3-phosphoglycerate, which are converted to 12 molecules of glyceraldehyde-3-phosphate at the cost of 12 molecules each of ATP and NADPH. Two molecules of glyceraldehyde-3-phosphate are then used for synthesis of glucose and ten molecules continue in the Calvin cycle to form six molecules of ribulose-5-phosphate. The cycle is then completed by the use of six additional ATP molecules for the synthesis of ribulose-1,5-bisphosphate.

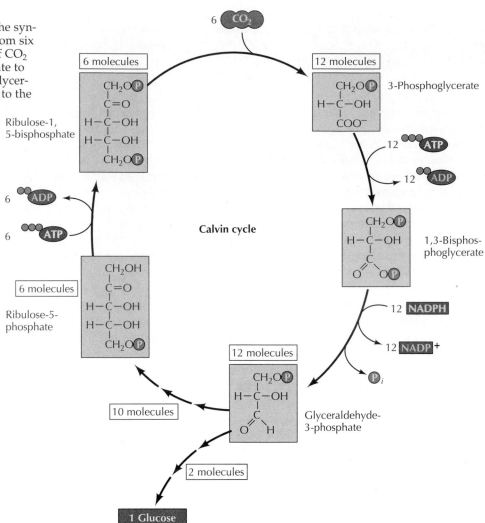

## THE BIOSYNTHESIS OF CELL CONSTITUENTS

The preceding section reviewed the major metabolic reactions by which the cell obtains and stores energy in the form of ATP. This metabolic energy is then used to accomplish various tasks, including the synthesis of macro-molecules and other cell constituents. Thus, energy derived from the breakdown of organic molecules (catabolism) is used to drive the synthesis of other required components of the cell. Most catabolic pathways involve the oxidation of organic molecules coupled to the generation of both energy (ATP) and reducing power (NADH). In contrast, biosynthetic (anabolic) pathways generally involve the use of both ATP and reducing power (usually in the form of NADPH) for the production of new organic compounds. One major biosynthetic pathway, the synthesis of carbohydrates from $CO_2$ and $H_2O$ during the dark reactions of photosynthesis, was discussed in the preceding section. Additional pathways leading to the biosynthesis of major cellular constituents (carbohydrates, lipids, proteins, and nucleic acids) are reviewed in the sections that follow.

### Carbohydrates

In addition to being obtained directly from food or generated by photosynthesis, glucose can be synthesized from other organic molecules. In animal cells, glucose synthesis (**gluconeogenesis**) usually starts with lactate (produced by anaerobic glycolysis), amino acids (derived from the breakdown of proteins), or glycerol (produced by the breakdown of lipids). Plants (but not animals) are also able to synthesize glucose from fatty acids—a process that is particularly important during the germination of seeds, when energy stored as fats must be converted to carbohydrates to support growth of the plant. In both animal and plant cells, simple sugars are polymerized and stored as polysaccharides.

Gluconeogenesis involves the conversion of pyruvate to glucose—essentially the reverse of glycolysis. However, as discussed earlier, the glycolytic conversion of glucose to pyruvate is an energy-yielding pathway, generating two molecules each of ATP and NADH. Although some reactions of glycolysis are readily reversible, others will proceed only in the direction of glucose breakdown, because they are associated with a large decrease in free energy. These energetically favorable reactions of glycolysis are bypassed during gluconeogenesis by other reactions (catalyzed by different enzymes) that are coupled to the expenditure of ATP and NADH in order to drive them in the direction of glucose synthesis. Overall, the generation of glucose from two molecules of pyruvate requires four molecules of ATP, two of GTP, and two of NADH. This process is considerably more costly than the simple reversal of glycolysis (which would require two molecules of ATP and two of NADH), illustrating the additional energy required to drive the pathway in the direction of biosynthesis.

In both plant and animal cells, glucose is stored in the form of polysaccharides (starch and glycogen, respectively). The synthesis of polysaccharides, like that of all other macromolecules, is an energy-requiring reaction. As noted earlier, the linkage of two sugars by a glycosidic bond can be written as a dehydration reaction, in which $H_2O$ is removed (see Figure 2.3). Such a reaction, however, is energetically unfavorable and therefore unable to proceed in the forward direction. Consequently, the formation of a glycosidic bond must be coupled to an energy-yielding reaction. This coupling is accomplished by the use of nucleotide sugars as intermediates in polysaccharide synthesis (Figure 2.40). Glucose is first phosphorylated in an ATP-driven reaction to glucose-6-phosphate, which is then converted to glucose-1-phosphate. Glucose-1-phosphate reacts with UTP (uridine triphosphate), yielding UDP-glucose plus pyrophosphate, which is hydrolyzed to phosphate with the release of additional free energy. UDP-glucose is an activated intermediate that then donates its glucose residue to a growing polysaccharide chain in an energetically favorable reaction. Thus, chemical energy in the form of ATP and UTP drives the synthesis of polysaccharides from simple sugars.

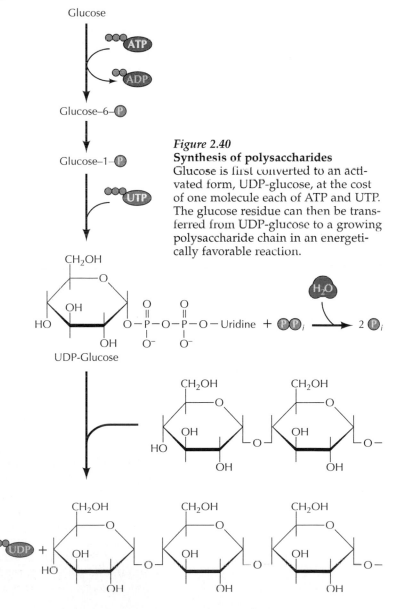

*Figure 2.40*
**Synthesis of polysaccharides**
Glucose is first converted to an activated form, UDP-glucose, at the cost of one molecule each of ATP and UTP. The glucose residue can then be transferred from UDP-glucose to a growing polysaccharide chain in an energetically favorable reaction.

# MOLECULAR MEDICINE

## *Phenylketonuria*

### The Disease

Phenylketonuria, or PKU, is an inborn error of amino acid metabolism with devastating effects. It afflicts approximately one in 10,000 newborn infants and, if untreated, results in severe mental retardation. Fortunately, understanding of the nature of the defect responsible for phenylketonuria has allowed early diagnosis and effective treatment.

### Molecular and Cellular Basis

Phenylketonuria is caused by a deficiency of the enzyme phenylalanine hydroxylase, which converts phenylalanine to tyrosine. This deficiency causes phenylalanine to accumulate to very high levels and undergo other reactions, such as conversion to phenylpyruvate. Phenylalanine, phenylpyruvate, and other abnormal metabolites accumulate in the blood and are secreted at high levels in the urine (the name of the disease derives from the high levels of phenylpyruvate, a phenylketone, found in the urine of affected children). Although the precise biochemical cause is unknown, mental retardation is a critical consequence of the accumulation of these abnormal phenylalanine metabolites.

### Prevention and Treatment

The enzyme deficiency causes no difficulties while the fetus is within the uterus, so children with phenylketonuria are normal at birth. If untreated, however, affected children become permanently and severely retarded within the first year of life. Fortunately, newborns with phenylketonuria can be readily identified by routine screening tests that detect elevated levels of phenylalanine in blood. Mental retardation can be prevented by feeding affected infants a synthetic diet that is low in phenylalanine. This dietary treatment eliminates the accumulation of toxic phenylalanine metabolites and effectively prevents the permanent mental deficiency that would otherwise ensue. Routine screening for phenylketonuria is thus a critical test for all newborn infants.

Abnormal metabolism of phenylalanine in patients with phenylketonuria.

## *Lipids*

Lipids are important energy storage molecules and the major constituent of cell membranes. They are synthesized from acetyl CoA, which is formed from the breakdown of carbohydrates, in a series of reactions that resemble the reverse of fatty acid oxidation. As with carbohydrate biosynthesis, however, the reactions leading to the synthesis of fatty acids differ from those involved in their degradation and are driven in the biosynthetic direction by being coupled to the expenditure of both energy in the form of ATP and reducing power in the form of NADPH. Fatty acids are synthesized by the stepwise addition of two-carbon units derived from acetyl CoA to a growing chain. The addition of each of these two-carbon units requires the expenditure of one molecule of ATP and two molecules of NADPH.

The major product of fatty acid biosynthesis, which occurs in the cytosol of eukaryotic cells, is the 16-carbon fatty acid palmitate. The principal constituents of cell membranes (phospholipids, sphingomyelin, and glycolipids) are then synthesized from free fatty acids in the endoplasmic reticulum and Golgi apparatus (see Chapter 9).

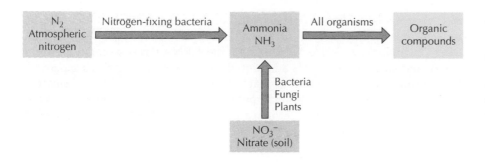

**Figure 2.41**
**Assimilation of nitrogen into organic compounds** Ammonia is incorporated into organic compounds by all organisms. Some bacteria are capable of converting atmospheric nitrogen to ammonia; and most bacteria, fungi, and plants can utilize nitrate from soil.

## Proteins

In contrast to carbohydrates and lipids, proteins (as well as nucleic acids) contain nitrogen in addition to carbon, hydrogen, and oxygen. Nitrogen is incorporated into organic compounds from different sources in different organisms (Figure 2.41). Some bacteria can use atmospheric $N_2$ by a process called **nitrogen fixation**, in which $N_2$ is reduced to $NH_3$ at the expense of energy in the form of ATP. Although relatively few species of bacteria are capable of nitrogen fixation, most bacteria, fungi, and plants can use nitrate ($NO_3^-$), which is a common constituent of soil, by reducing it to $NH_3$ via electrons derived from NADH or NADPH. Finally, all organisms are able to incorporate ammonia ($NH_3$) into organic compounds.

$NH_3$ is incorporated into organic molecules primarily during the synthesis of the amino acids glutamate and glutamine, which are derived from the citric acid cycle intermediate $\alpha$-ketoglutarate. These amino acids then serve as donors of amino groups during the synthesis of the other amino acids, which are also derived from central metabolic pathways, such as glycolysis and the citric acid cycle (Figure 2.42). The raw material for amino acid synthesis is thus obtained from glucose, and the amino acids are synthesized at the cost of both energy (ATP) and reducing power (NADPH). Many bacteria and plants can synthesize all 20 amino acids. Humans and other mammals, however, can synthesize only about half of the required amino acids; the remainder must be obtained from dietary sources (Table 2.2).

The polymerization of amino acids to form proteins also requires energy. Like the synthesis of polysaccharides, the formation of the peptide bond can be considered a dehydration reaction, which must be driven in the direction of protein synthesis by being coupled to another source of metabolic energy. In the biosynthesis of polysaccharides, this coupling is accomplished through the conversion of sugars to activated intermediates, such as UDP-glucose. Amino acids are similarly activated before being used for protein synthesis.

A critical difference between the synthesis of proteins and that of polysaccharides is that the amino acids are incorporated into proteins in a unique order, specified by a gene. The order of nucleotides in a gene specifies the amino acid sequence of a protein via translation, in which messenger RNA (mRNA) acts as a template for pro-

**Figure 2.42**
**Biosynthesis of amino acids**
The carbon skeletons of the amino acids are derived from intermediates in glycolysis and in the citric acid cycle.

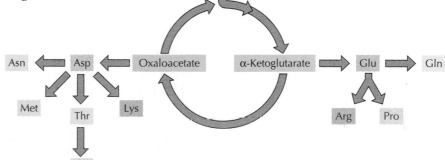

**Table 2.2** Dietary Requirements for Amino Acids in Humans

| Essential | Nonessential |
|---|---|
| Histidine | Alanine |
| Isoleucine | Arginine[a] |
| Leucine | Asparagine |
| Lysine | Aspartate |
| Methionine | Cysteine |
| Phenylalanine | Glutamate |
| Threonine | Glutamine |
| Tryptophan | Glycine |
| Valine | Proline |
| | Serine |
| | Tyrosine |

The essential amino acids must be obtained from dietary sources; the nonessential amino acids can be synthesized by human cells.

[a] Although arginine is classified as a nonessential amino acid, growing children must obtain additional arginine from their diet.

tein synthesis (see Chapter 3). Each amino acid is first attached to a specific transfer RNA (tRNA) molecule in a reaction coupled to ATP hydrolysis (Figure 2.43). The aminoacyl tRNAs then align on the mRNA template bound to ribosomes, and each amino acid is added to the C terminus of a growing peptide chain through a series of reactions that will be discussed in detail in Chapter 7. During the process, two additional molecules of GTP are hydrolyzed, so the incorporation of each amino acid into a protein is coupled to the hydrolysis of one ATP and two GTP molecules.

## Nucleic Acids

The precursors of nucleic acids, the nucleotides, are composed of phosphorylated five-carbon sugars joined to nucleic acid bases. Nucleotides can be synthesized from carbohydrates and amino acids; they can also be

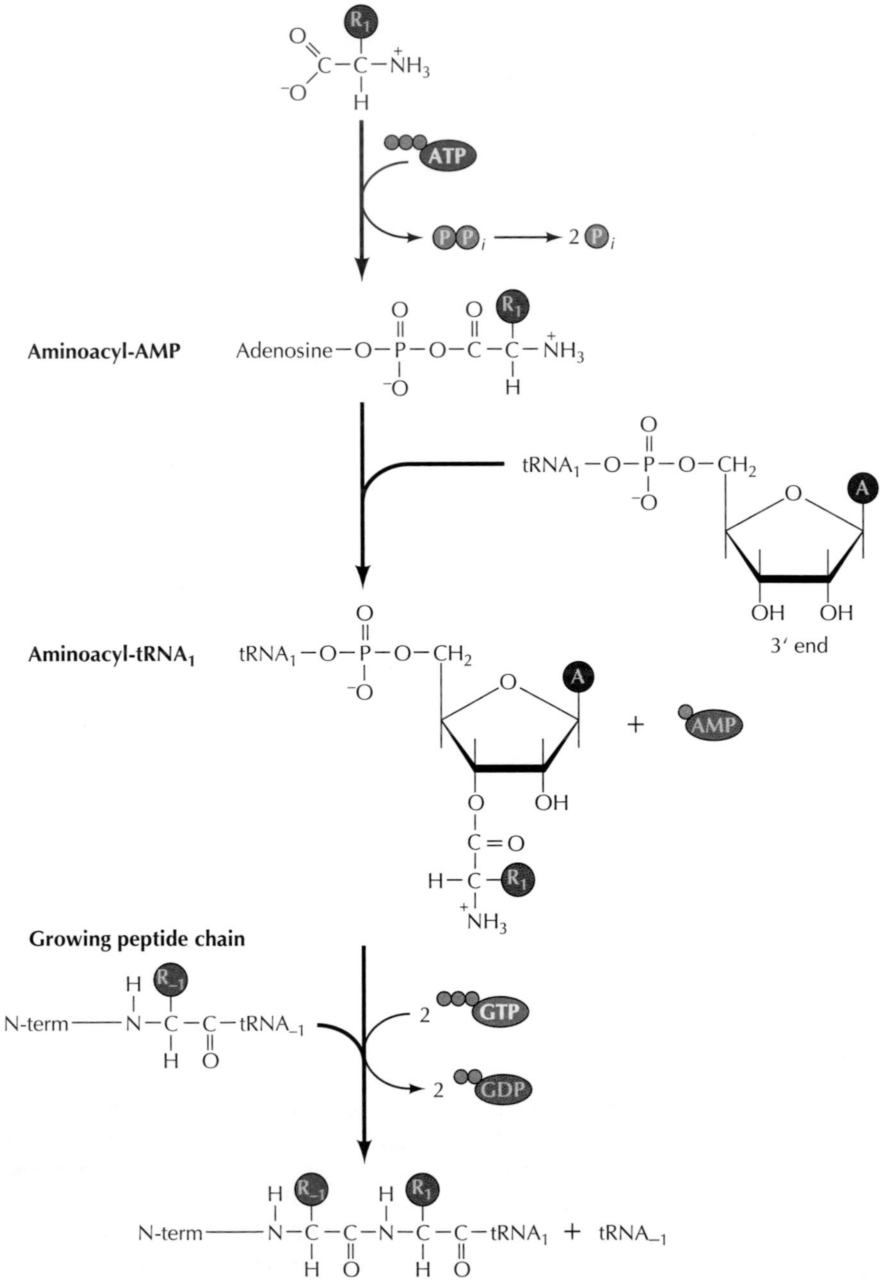

**Figure 2.43**
**Formation of the peptide bond** An amino acid is first activated by attachment to its tRNA in a two-step reaction involving the hydrolysis of ATP to AMP. The tRNAs serve as adaptors to align the amino acids on an mRNA template bound to ribosomes. Each amino acid is then transferred to the C terminus of the growing peptide chain at the cost of two additional molecules of GTP.

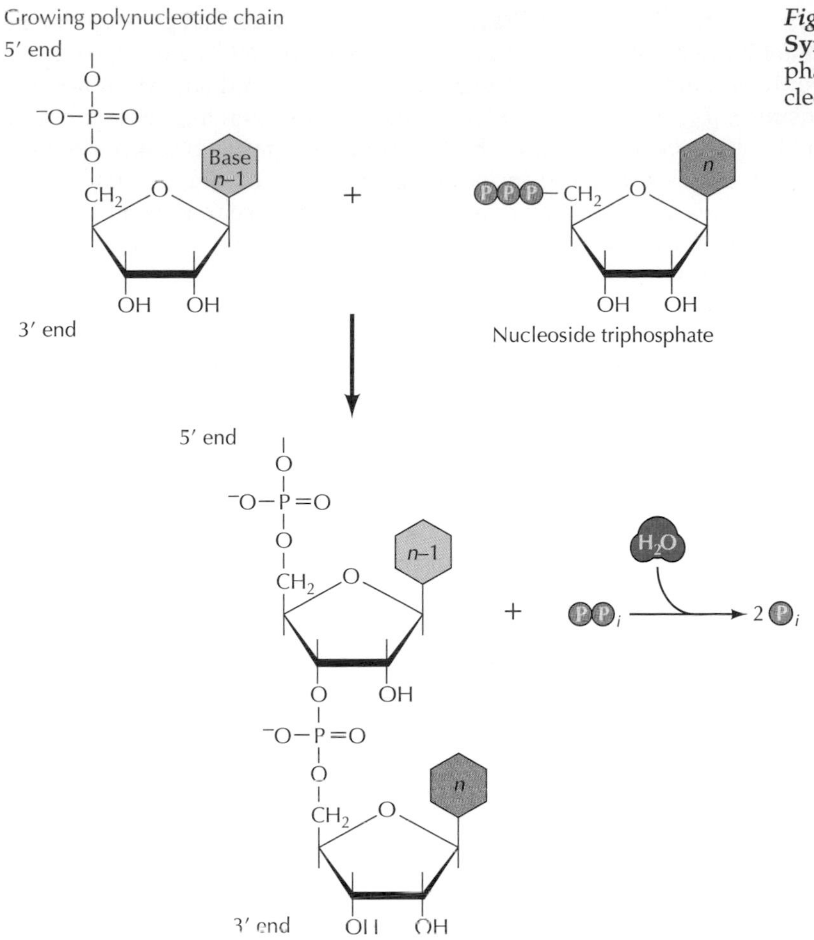

Growing polynucleotide chain

*Figure 2.44*
**Synthesis of polynucleotides**  Nucleoside triphosphates are joined to the 3′ end of a growing polynucleotide chain with the release of pyrophosphate.

obtained from dietary sources or reused following nucleic acid breakdown. The starting point for nucleotide biosynthesis is the phosphorylated sugar ribose-5-phosphate, which is derived from glucose-6-phosphate. Divergent pathways then lead to the synthesis of purine and pyrimidine ribonucleotides, which are the immediate precursors for RNA synthesis. These ribonucleotides are converted to deoxyribonucleotides, which serve as the monomeric building blocks of DNA.

RNA and DNA are polymers of nucleoside monophosphates. As for other macromolecules, however, direct polymerization of nucleoside monophosphates is energetically unfavorable, and the synthesis of polynucleotides instead uses nucleoside triphosphates as activated precursors (Figure 2.44). A nucleoside 5′-triphosphate is added to the 3′ hydroxyl group of a growing polynucleotide chain, with the release and subsequent hydrolysis of pyrophosphate serving to drive the reaction in the direction of polynucleotide synthesis.

## CELL MEMBRANES

The structure and function of cells are critically dependent on membranes, which not only separate the interior of the cell from its environment but also define the internal compartments of eukaryotic cells, including the nucleus and cytoplasmic organelles. The formation of biological membranes is based on the properties of lipids, and all cell membranes share a common struc-

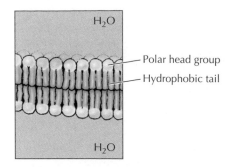

H₂O

— Polar head group
— Hydrophobic tail

H₂O

*Figure 2.45*
**A phospholipid bilayer**   Phospholipids spontaneously form stable bilayers, with their polar head groups exposed to water and their hydrophobic tails buried in the interior of the membrane.

tural organization: bilayers of phospholipids with associated proteins. These membrane proteins are responsible for many specialized functions; some act as receptors that allow the cell to respond to external signals, some are responsible for the selective transport of molecules across the membrane, and others participate in electron transport and oxidative phosphorylation. In addition, membrane proteins control the interactions between cells of multicellular organisms. The common structural organization of membranes thus underlies a variety of biological processes and specialized membrane functions, which will be discussed in detail in later chapters.

## Membrane Lipids

The fundamental building blocks of all cell membranes are phospholipids, which are amphipathic molecules, consisting of two hydrophobic fatty acid chains linked to a phosphate-containing hydrophilic head group (see Figure 2.7). Because their fatty acid tails are poorly soluble in water, phospholipids spontaneously form bilayers in aqueous solutions, with the hydrophobic tails buried in the interior of the membrane and the polar head groups exposed on both sides, in contact with water (Figure 2.45). Such **phospholipid bilayers** form a stable barrier between two aqueous compartments and represent the basic structure of all biological membranes.

Lipids constitute approximately 50% of the mass of most cell membranes, although this proportion varies depending on the type of membrane. Plasma membranes, for example, are approximately 50% lipid and 50% protein. The inner membrane of mitochondria, on the other hand, contains an unusually high fraction (about 75%) of protein, reflecting the abundance of protein complexes involved in electron transport and oxidative phosphorylation. The lipid composition of different cell membranes also varies (Table 2.3). The plasma membrane of *E. coli* consists predominantly of phosphatidylethanolamine, which constitutes 80% of total lipid. Mammalian plasma membranes are more complex, containing four major phospholipids—phosphatidylcholine, phosphatidylserine, phosphatidylethanolamine, and sphingomyelin—which together constitute 50 to 60% of total membrane lipid. In addition to the phospholipids, the plasma membranes of animal cells contain glycolipids and cholesterol, which generally correspond to about 40% of the total lipid molecules.

An important property of lipid bilayers is that they behave as two-dimensional fluids in which individual molecules (both lipids and proteins) are free to rotate and move in lateral directions (Figure 2.46). Such fluidity is a critical property of membranes and is determined by both tem-

*Table 2.3*   Lipid Composition of Cell Membranes

| Lipid | Plasma membrane E. coli | Plasma membrane Erythrocyte | Rough endoplasmic reticulum | Outer mitochondrial membranes |
|---|---|---|---|---|
| Phosphatidylcholine | 0 | 17 | 55 | 50 |
| Phosphatidylserine | 0 | 6 | 3 | 2 |
| Phosphatidylethanolamine | 80 | 16 | 16 | 23 |
| Sphingomyelin | 0 | 17 | 3 | 5 |
| Glycolipids | 0 | 2 | 0 | 0 |
| Cholesterol | 0 | 45 | 6 | <5 |

*Source:* Data from P. L. Yeagle, *The Membranes of Cells*, 2nd ed. (1993). San Diego, CA: Academic Press. Membrane compositions are indicated as the mole percentages of major lipid constituents.

perature and lipid composition. For example, the interactions between shorter fatty acid chains are weaker than those between longer chains, so membranes containing shorter fatty acid chains are less rigid and remain fluid at lower temperatures. Lipids containing unsaturated fatty acids similarly increase membrane fluidity because the presence of double bonds introduces kinks in the fatty acid chains, making them more difficult to pack together.

Because of its hydrocarbon ring structure (see Figure 2.9), cholesterol plays a distinct role in determining membrane fluidity. Cholesterol molecules insert into the bilayer with their polar hydroxyl groups close to the hydrophilic head groups of the phospholipids (Figure 2.47). The rigid hydrocarbon rings of cholesterol therefore interact with the regions of the fatty acid chains that are adjacent to the phospholipid head groups. This interaction decreases the mobility of the outer portions of the fatty acid chains, making this part of the membrane more rigid. On the other hand, insertion of cholesterol interferes with interactions between fatty acid chains, thereby maintaining membrane fluidity at lower temperatures.

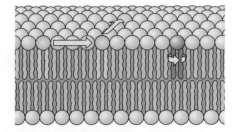

*Figure 2.46*
**Mobility of phospholipids in a membrane** Individual phospholipids can rotate and move laterally within a bilayer.

## Membrane Proteins

Proteins are the other major constituent of cell membranes, constituting 25 to 75% of the mass of the various membranes of the cell. The current model of membrane structure, proposed by Jonathan Singer and Garth Nicolson in 1972, views membranes as a **fluid mosaic** in which proteins are inserted into a lipid bilayer (Figure 2.48). While phospholipids provide the basic structural organization of membranes, membrane proteins carry out the specific functions of the different membranes of the cell. These proteins are divided into two general classes, based on the nature of their association with the membrane. **Integral membrane proteins** are embedded directly within the lipid bilayer. **Peripheral membrane proteins** are not inserted into the lipid bilayer but are associated with the membrane indirectly, generally by interactions with integral membrane proteins.

Many integral membrane proteins (called **transmembrane proteins**) span the lipid bilayer, with portions exposed on both sides of the membrane. The membrane-spanning portions of these proteins are usually $\alpha$-helical regions of 20 to 25 nonpolar amino acids. The hydrophobic side chains of these amino acids interact with the fatty acid chains of membrane lipids, and the formation of an $\alpha$ helix neutralizes the polar character of the peptide bonds, as discussed earlier in this chapter with respect to protein folding. Like the phospholipids, transmembrane proteins are amphipathic molecules, with their hydrophilic portions exposed to the aqueous environment on both sides of the membrane. Some transmembrane proteins span the membrane only once; others have multiple membrane-spanning regions. Most transmembrane proteins of eukaryotic plasma membranes have been modified by the addition of carbohydrates, which are exposed on the surface of the cell and may participate in cell-cell interactions.

Proteins can also be anchored in membranes by lipids that are covalently attached to the polypeptide chain (see Chapter 7). Distinct lipid modifications anchor proteins to the cytosolic and extracellular faces of the plasma membrane. Proteins can be anchored to the cytosolic face of the membrane either by the addition of a 14-carbon fatty acid (myristic acid) to their amino terminus or by the addition of either a 16-carbon fatty acid (palmitic acid) or 15- or 20-carbon prenyl groups to the side chains of cysteine residues. Alternatively, proteins are anchored to the extracellular face of the plasma membrane by the addition of glycolipids to their carboxy terminus.

*Figure 2.47*
**Insertion of cholesterol in a membrane** Cholesterol inserts into the membrane with its polar hydroxyl group close to the polar head groups of the phospholipids.

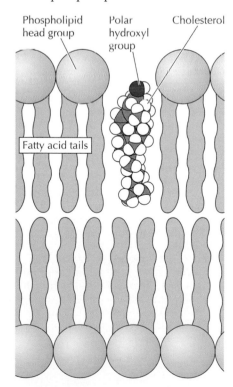

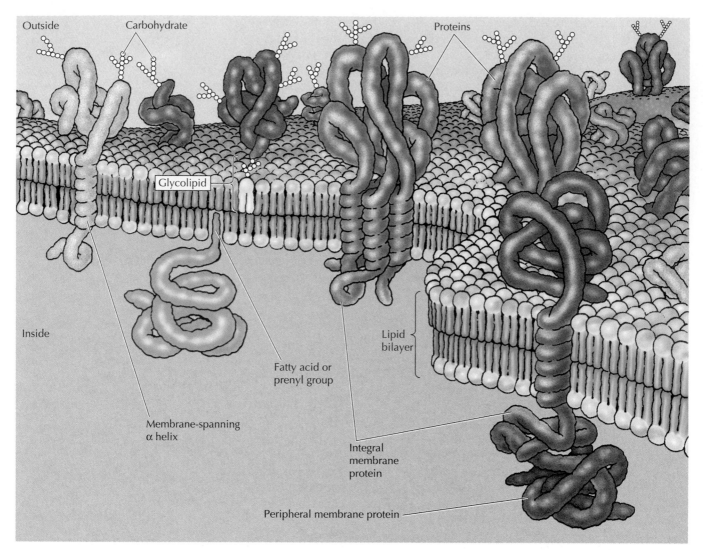

*Figure 2.48*
**Fluid mosaic model of membrane structure**    Biological membranes consist of proteins inserted into a lipid bilayer. Integral membrane proteins are embedded in the membrane, usually via $\alpha$-helical regions of 20 to 25 hydrophobic amino acids. Some transmembrane proteins span the membrane only once; others have multiple membrane-spanning regions. In addition, some proteins are anchored in the membrane by lipids that are covalently attached to the polypeptide chain. These proteins can be anchored to the extracellular face of the plasma membrane by glycolipids and to the cytosolic face by fatty acids or prenyl groups (see Chapter 7 for structures). Peripheral membrane proteins are not inserted in the membrane but are attached via interactions with integral membrane proteins.

## Transport across Cell Membranes

The selective permeability of biological membranes to small molecules allows the cell to control and maintain its internal composition. Only small uncharged molecules can diffuse freely through phospholipid bilayers (Figure 2.49). Small nonpolar molecules, such as $O_2$ and $CO_2$, are soluble in the lipid bilayer and therefore can readily cross cell membranes. Small uncharged polar molecules, such as $H_2O$, also can diffuse through membranes, but larger uncharged polar molecules, such as glucose, cannot. Charged molecules, such as ions, are unable to diffuse through a phospholipid bilayer regardless of size; even $H^+$ ions cannot cross a lipid bilayer by free diffusion.

Although ions and most polar molecules cannot diffuse across a lipid bilayer, many such molecules (such as glucose) are able to cross cell membranes. These molecules pass across membranes via the action of specific transmembrane proteins, which act as transporters. Such transport proteins determine the selective permeability of cell membranes and thus play a critical role in membrane function. They contain multiple membrane-spanning regions that form a passage through the lipid bilayer, allowing polar or charged molecules to cross the membrane through a protein pore without interacting with the hydrophobic fatty acid chains of the membrane phospholipids.

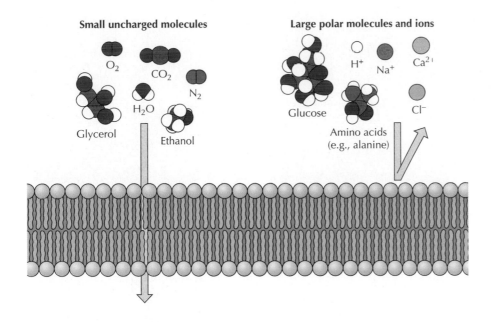

**Small uncharged molecules**

O$_2$

CO$_2$

N$_2$

H$_2$O

Glycerol

Ethanol

**Large polar molecules and ions**

H$^+$    Na$^+$    Ca$^{2+}$

Cl$^-$

Glucose

Amino acids
(e.g., alanine)

**Figure 2.49**
**Permeability of phospholipid
bilayers** Small uncharged molecules
can diffuse freely through a phospholipid bilayer. However, the bilayer is
impermeable to larger polar molecules
(such as glucose and amino acids) and
to ions.

As discussed in detail in Chapter 12, there are two general classes of membrane transport proteins (Figure 2.50). **Channel proteins** form open pores
through the membrane, allowing the free passage of any molecule of the
appropriate size. Ion channels, for example, allow the passage of inorganic
ions such as Na$^+$, K$^+$, Ca$^{2+}$, and Cl$^-$ across the plasma membrane. Once open,
channel proteins form small pores through which ions of the appropriate size
and charge can cross the membrane by free diffusion. The pores formed by
these channel proteins are not permanently open; rather, they can be selectively opened and closed in response to extracellular signals, allowing the
cell to control the movement of ions across the membrane. Such regulated
ion channels have been particularly well studied in nerve and muscle cells,
where they mediate the transmission of electrochemical signals.

In contrast to channel proteins, **carrier proteins** selectively bind and
transport specific small molecules, such as glucose. Rather than forming
open channels, carrier proteins act like enzymes to facilitate the passage of
specific molecules across membranes. In particular, carrier proteins bind
specific molecules and then undergo conformational changes that open

**Figure 2.50**
**Channel and carrier proteins**
(A) Channel proteins form open pores
through which molecules of the appropriate size (e.g., ions) can cross the
membrane. (B) Carrier proteins selectively bind the small molecule to be
transported and then undergo a conformational change to release the molecule on the other side of the membrane.

(A)

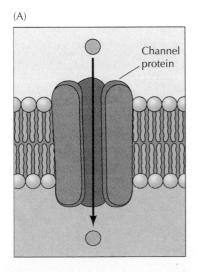

Channel
protein

(B)

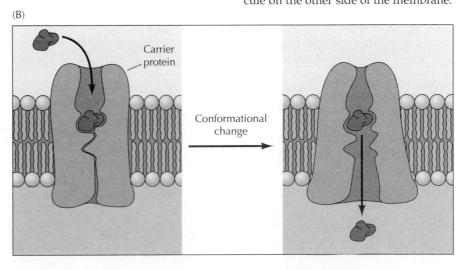

Carrier
protein

Conformational
change

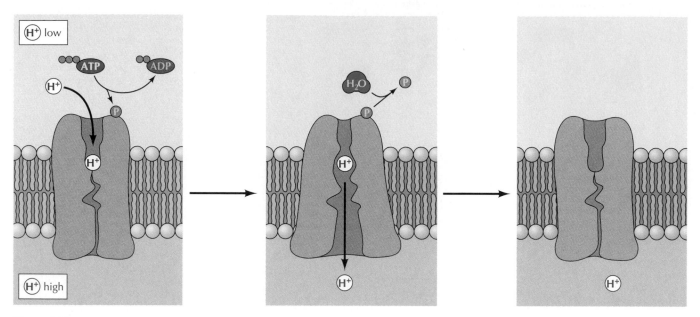

*Figure 2.51*
**Model of active transport** Energy derived from the hydrolysis of ATP is used to transport $H^+$ against the electrochemical gradient (from low to high $H^+$ concentration). Binding of $H^+$ is accompanied by phosphorylation of the carrier protein, which induces a conformational change that drives $H^+$ transport against the electrochemical gradient. Release of $H^+$ and hydrolysis of the bound phosphate group then restore the carrier to its original conformation.

channels through which the molecule to be transported can pass across the membrane and be released on the other side.

As described so far, molecules transported by either channel or carrier proteins cross membranes in the energetically favorable direction, as determined by concentration and electrochemical gradients—a process known as **passive transport**. However, carrier proteins also provide a mechanism through which the energy changes associated with transporting molecules across a membrane can be coupled to the use or production of other forms of metabolic energy, just as enzymatic reactions can be coupled to the hydrolysis or synthesis of ATP. For example, molecules can be transported in an energetically unfavorable direction across a membrane (e.g., against a concentration gradient) if their transport in that direction is coupled to ATP hydrolysis as a source of energy—a process called **active transport** (Figure 2.51). The free energy stored as ATP can thus be used to control the internal composition of the cell, as well as to drive the biosynthesis of cell constituents.

## Summary

### THE MOLECULAR COMPOSITION OF CELLS

*Carbohydrates:* Carbohydrates include simple sugars and polysaccharides. Polysaccharides serve as storage forms of sugars, structural components of cells, and markers for cell recognition processes.

*Lipids:* Lipids are the principal components of cell membranes, and they serve as energy storage and signaling molecules. Phospholipids consist of two hydrophobic fatty acid chains linked to a hydrophilic phosphate-containing head group.

*Nucleic Acids:* Nucleic acids are the principal informational molecules of the cell. Both DNA and RNA are polymers of purine and pyrimidine nucleotides. Hydrogen bonding between complementary base pairs allows nucleic acids to direct their self-replication.

deoxyribonucleic acid (DNA), ribonucleic acid (RNA), messenger RNA (mRNA), ribosomal RNA, transfer RNA, nucleotide, purine, pyrimidine, adenine, guanine, cytosine, thymine, uracil, 2′-deoxyribose, ribose, nucleoside, phosphodiester bond, oligo-nucleotide, polynucleotide

*Proteins:* Proteins are polymers of 20 different amino acids, each of which has a distinct side chain with specific chemical properties. Each protein has a unique amino acid sequence, which determines its three-dimensional structure. In most proteins, combinations of $\alpha$ helices and $\beta$ sheets fold into globular domains with hydrophobic amino acids in the interior and hydrophilic amino acids on the surface.

protein, amino acid, peptide bond, polypeptide, X-ray crystallography, primary structure, secondary structure, $\alpha$ helix, $\beta$ sheet, tertiary structure, domain, quaternary structure

## THE CENTRAL ROLE OF ENZYMES AS BIOLOGICAL CATALYSTS

*The Catalytic Activity of Enzymes:* Virtually all chemical reactions within cells are catalyzed by enzymes.

enzyme, substrate, product, transition state, activation energy, active site

*Mechanisms of Enzymatic Catalysis:* Enzymes increase reaction rates by binding substrates in the proper position, by altering the conformation of substrates to approach the transition state, and by participating directly in chemical reactions.

lock-and-key model, induced fit

*Coenzymes:* Coenzymes function in conjunction with enzymes to carry chemical groups between substrates.

prosthetic group, coenzyme, nicotinamide adenine dinucleotide (NAD+)

*Regulation of Enzyme Activity:* The activities of enzymes are regulated to meet the physiological needs of the cell. Enzyme activity can be controlled by the binding of small molecules, by interactions with other proteins, and by covalent modifications.

feedback inhibition, allosteric regulation, phosphorylation

## METABOLIC ENERGY

*Free Energy and ATP:* ATP serves as a store of free energy, which is used to drive energy-requiring reactions within cells.

Gibbs free energy (G), adenosine 5′-triphosphate (ATP), high-energy bond

*The Generation of ATP from Glucose:* The breakdown of glucose provides a major source of cellular energy. In aerobic cells, the complete oxidation of glucose yields 36 to 38 molecules of ATP. Most of this ATP is derived from electron transport reactions in which $O_2$ is reduced to $H_2O$.

glycolysis, coenzyme A (CoA), citric acid cycle, Krebs cycle, flavin adenine dinucleotide ($FADH_2$), oxidative phosphorylation, electron transport chain

*The Derivation of Energy from Other Organic Molecules:* ATP can also be derived from the breakdown of organic molecules other than glucose. Because fats are more reduced than carbohydrates, they provide a more efficient form of energy storage.

*Photosynthesis:* The energy required for the synthesis of organic molecules is ultimately derived from sunlight, which is harvested by plants and photosynthetic bacteria. In the first stage of photosynthesis, energy from sunlight is used to drive the synthesis of ATP and NADPH, coupled to the oxidation of $H_2O$ to $O_2$. The ATP and NADPH produced by these reactions are then used to synthesize glucose from $CO_2$ and $H_2O$.

light reactions, dark reactions, photosynthetic pigment, chlorophyll, Calvin cycle

**gluconeogenesis**

**nitrogen fixation**

**phospholipid bilayer**

**fluid mosaic, integral membrane protein, peripheral membrane protein, transmembrane protein**

**channel protein, carrier protein, passive transport, active transport**

## THE BIOSYNTHESIS OF CELL CONSTITUENTS

*Carbohydrates:* Glucose can be synthesized from other organic molecules, using energy and reducing power in the forms of ATP and NADH, respectively. Additional ATP is then needed to drive the synthesis of polysaccharides from simple sugars.

*Lipids:* Lipids are synthesized from acetyl CoA, which is formed from the breakdown of carbohydrates.

*Proteins:* The amino acids are synthesized from intermediates in glycolysis and the citric acid cycle. Their polymerization to form proteins requires additional energy in the form of ATP and GTP.

*Nucleic Acids:* Purine and pyrimidine nucleotides are synthesized from carbohydrates and amino acids. Their polymerization to DNA and RNA is driven by the use of nucleoside triphosphates as activated precursors.

## CELL MEMBRANES

*Membrane Lipids:* The basic structure of all cell membranes is a phospholipid bilayer. Membranes of animal cells also contain glycolipids and cholesterol.

*Membrane Proteins:* Proteins can either be inserted into the lipid bilayer or associated with the membrane indirectly, by protein-protein interactions. Some proteins span the lipid bilayer; others are anchored to one side of the membrane.

*Transport across Cell Membranes:* Lipid bilayers are permeable only to small uncharged molecules. Ions and most polar molecules are transported across cell membranes by specific transport proteins, the action of which can be coupled to the hydrolysis or synthesis of ATP.

## QUESTIONS

**1.** What key molecular characteristic of phospholipids makes them the basic components of cell membranes?

**2.** Where would you expect to find phenylalanine residues in a folded protein? What amino acid residues would you expect to find in the loop regions connecting different $\alpha$ helices?

**3.** The binding pocket of trypsin contains an aspartate residue. How do you think changing this amino acid to lysine would affect the enzyme's activity?

**4.** Consider the reaction A $\rightleftharpoons$ B + C, in which $\Delta G^{\circ\prime}$ = +3.5 kcal/mol. Calculate $\Delta G$ under intracellular conditions, given that the concentration of A is $10^{-2}$ $M$ and the concentrations of B and C are each $10^{-3}$ $M$. In which direction will the reaction proceed in the cell? For your calculation $R = 1.98 \times 10^{-3}$ kcal/mol/degree and $T = 298$ K (25°C). Note that $\ln(x) = 2.3 \log10(x)$.

**5.** Which process releases more free energy: the degradation of saturated fatty acids or the degradation of unsaturated fatty acids?

**6.** Why are different enzymes needed for gluconeogenesis and glycolysis?

**7.** Why are the membrane-spanning regions of transmembrane proteins frequently $\alpha$-helical?

## REFERENCES AND FURTHER READING

### General References

Abeles, R. H., P. A. Frey, and W. P. Jencks. 1992. *Biochemistry*. Boston: Jones and Bartlett.

Mathews, C. K., and K. E. van Holde. 1996. *Biochemistry*. 2nd ed. Redwood City, CA: Benjamin Cummings.

Stryer, L. 1995. *Biochemistry*. 4th ed. New York: W. H. Freeman.

### The Molecular Composition of Cells

Anfinsen, C. B. 1973. Principles that govern the folding of protein chains. *Science* 181: 223–230. [P]

Branden, C. and J. Tooze. 1991. *Introduction to Protein Structure*. New York: Garland.

Chothia, C. and A. V. Finkelstein. 1990. The classification and origins of protein folding patterns. *Ann. Rev. Biochem.* 59: 1007–1039. [R]

Hunkapiller, M. W. and L. E. Hood. 1983. Protein sequence analysis: Automated microsequencing. *Science* 219: 650–655. [R]

Kendrew, J. C. 1961. The three–dimensional structure of a protein molecule. *Sci. Am.* 205(6): 96–111. [R]

Pauling, L., R. B. Corey, and H. R. Branson. 1951. The structure of proteins: Two hydrogen bonded configurations of the polypeptide chain. *Proc. Natl. Acad. Sci. USA* 37: 205–211. [P]

Richardson, J. S. 1981. The anatomy and taxonomy of protein structure. *Adv. Protein Chem.* 34: 167–339. [R]

Rost, B., R. Schneider, and C. Sander. 1993. Progress in protein structure prediction? *Trends Biochem. Sci.* 18: 120–123. [R]

Sanger, F. 1988. Sequences, sequences, and sequences. *Ann. Rev. Biochem.* 57: 1–28. [R]

Tanford, C. 1978. The hydrophobic effect and the organization of living matter. *Science* 200: 1012–1018. [R]

### The Central Role of Enzymes as Biological Catalysts

Dressler, D. and H. Potter. 1991. *Discovering Enzymes*. New York: Scientific American Library.

Fersht, A. 1985. *Enzyme Structure and Mechanism*. 2nd ed. New York: W. H. Freeman.

Koshland, D. E. 1984. Control of enzyme activity and metabolic pathways. *Trends Biochem. Sci.* 9: 155–159. [R]

Lienhard, G. E. 1973. Enzymatic catalysis and transition-state theory. *Science* 180: 149–154. [R]

Lipscomb, W. N. 1983. Structure and catalysis of enzymes. *Ann. Rev. Biochem.* 52: 17–34. [R]

Lolis, E. and G. A. Pestko. 1990. Transition-state analogues in protein crystallography: probes of the structural source of enzyme catalysis. *Ann. Rev. Biochem.* 59: 597–630. [R]

Monod, J., J.-P. Changeux, and F. Jacob. 1963. Allosteric proteins and cellular control systems. *J. Mol. Biol.* 6: 306–329. [P]

Neurath, H. 1984. Evolution of proteolytic enzymes. *Science* 224: 350–357. [R]

Walsh, C. T. 1979. *Enzymatic Reaction Mechanisms*. New York: W. H. Freeman.

### Metabolic Energy

Bennett, J. 1979. The protein that harvests sunlight. *Trends Biochem. Sci.* 4: 268–271. [R]

Calvin, M. 1962. The path of carbon in photosynthesis. *Science* 135: 879–889. [R]

Deisenhofer, J. and H. Michel. 1991. Structures of bacterial photosynthetic reaction centers. *Ann. Rev. Cell Biol.* 7: 1–23. [R]

Klotz, I. 1967. *Energy Changes in Biochemical Reactions*. New York: Academic Press.

Krebs, H. A. 1970. The history of the tricarboxylic cycle. *Perspect. Biol. Med.* 14: 154–170. [R]

Kuhlbrandt, W., D. N. Wang and Y. Fujiyoshi. 1994. Atomic model of plant light-harvesting complex by electron crystallography. *Nature* 367: 614–621. [P]

Nicholls, D. G. and S. J. Ferguson. 1992. *Bioenergetics 2*. London: Academic Press.

### The Biosynthesis of Cell Constituents

Hers, H. G. and L. Hue. 1983. Gluconeogenesis and related aspects of glycolysis. *Ann. Rev. Biochem.* 52: 617–653. [R]

Jones, M. E. 1980. Pyrimidine nucleotide biosynthesis in animals: Genes, enzymes, and regulation of UMP biosynthesis. *Ann. Rev. Biochem.* 49: 253–279. [R]

Kornberg, A. and T. A. Baker. 1992. *DNA Replication*. 2nd ed. New York: W. H. Freeman.

Reichard, P. and A. Ehrenberg. 1983. Ribonucleotide reductase: A radical enzyme. *Science* 221: 514–519. [R]

Tolbert, N. E. 1981. Metabolic pathways in peroxisomes and glyoxysomes. *Ann. Rev. Biochem.* 50: 133–157. [R]

Umbarger, H. E. 1978. Amino acid biosynthesis and its regulation. *Ann. Rev. Biochem.* 47: 533–606. [R]

Van den Bosch, H., R. B. H. Schutgens, R. J. A. Wanders, and J.M. Tager. 1992. Biochemistry of peroxisomes. *Ann. Rev. Biochem.* 61: 157–197. [R]

Wakil, S. J., J. K. Stoops and V. C. Joshi. 1983. Fatty acid synthesis and its regulation. *Ann. Rev. Biochem.* 52: 537–579. [R]

### Cell Membranes

Bretscher, M. 1985. The molecules of the cell membrane. *Sci. Am.* 253(4): 100–108. [R]

Clarke, S. 1992. Protein isoprenylation and methylation at carboxy-terminal cysteine residues. *Ann. Rev. Biochem.* 61: 355–386. [R]

Cross, G. A. M. 1990. Glycolipid anchoring of plasma membrane proteins. *Ann. Rev. Cell Biol.* 6: 1–39. [R]

Englund, P. T. 1993. The structure and biosynthesis of glycosyl phosphatidyl-inositol protein anchors. *Ann. Rev. Biochem.* 62: 121–138. [R]

Petty, H. R. 1993. *Molecular biology of membranes: Structure and function*. New York: Plenum Press.

Singer, S. J. 1990. The structure and insertion of integral proteins in membranes. *Ann. Rev. Cell Biol.* 6: 247–296. [R]

Singer, S. J. and G. L. Nicolson. 1972. The fluid mosaic model of the structure of cell membranes. *Science* 175: 720–731. [P]

Towler, D. A., J. I. Gordon, S. P. Adams and L. Glaser. 1988. The biology and enzymology of eukaryotic protein acylation. *Ann. Rev. Biochem.* 57: 69–99. [R]

Yeagle, P. L. 1993. *The Membranes of Cells*. 2nd ed. San Diego, CA: Academic Press.

# 3

# *Fundamentals of Molecular Biology*

CONTEMPORARY MOLECULAR BIOLOGY is concerned principally with understanding the mechanisms responsible for transmission and expression of the genetic information that ultimately governs cell structure and function. As reviewed in Chapter 1, all cells share a number of basic properties, and this underlying unity of cell biology is particularly apparent at the molecular level. Such unity has allowed scientists to choose simple organisms (such as bacteria) as models for many fundamental experiments, with the expectation that similar molecular mechanisms are operative in organisms as diverse as *E. coli* and humans. Numerous experiments have established the validity of this assumption, and it is now clear that the molecular biology of cells provides a unifying theme to understanding diverse aspects of cell behavior.

Initial advances in molecular biology were made by taking advantage of the rapid growth and readily manipulable genetics of simple bacteria, such as *E. coli*, and their viruses. More recently, not only the fundamental principles but also many of the experimental approaches first developed in prokaryotes have been successfully applied to eukaryotic cells. The development of recombinant DNA has had a tremendous impact, allowing individual eukaryotic genes to be isolated and characterized in detail. Current advances in recombinant DNA technology have made even the determination of the complete sequence of the human genome a feasible project.

## HEREDITY, GENES, AND DNA

Perhaps the most fundamental property of all living things is the ability to reproduce. All organisms inherit the genetic information specifying their structure and function from their parents. Likewise, all cells arise from preexisting cells, so the genetic material must be replicated and passed from parent to progeny cell at each cell division. How genetic information is replicated and transmitted from cell to cell and organism to organism thus represents a question that is central to all of biology. Consequently, elucidation of the mechanisms of genetic transmission and identification of the genetic material as DNA were discoveries that formed the foundation of our current understanding of biology at the molecular level.

### Genes and Chromosomes

The classical principles of genetics were deduced by Gregor Mendel in 1865, on the basis of the results of breeding experiments with peas. Mendel studied the inheritance of a number of well-defined traits, such as seed color, and was able to deduce general rules for their transmission. In all cases, he could correctly interpret the observed patterns of inheritance by assuming that each trait is determined by a pair of inherited factors, which are now called **genes**. One gene copy (called an **allele**) specifying each trait is inherited from each parent. For example, breeding two strains of peas—one having yellow seeds, and the other green seeds—yields the following results (Figure 3.1). The parental strains each have two identical copies of the gene specifying yellow ($Y$) or green ($y$) seeds, respectively. The progeny plants are therefore hybrids, having inherited one gene for yellow seeds ($Y$) and one for green seeds ($y$). All these progeny plants (the first filial, or $F_1$, generation) have yellow seeds, so yellow ($Y$) is said to be **dominant** and green ($y$) **recessive**. The **genotype** (genetic composition) of the $F_1$ peas is thus $Yy$, and their **phenotype** (physical appearance) is yellow. If one $F_1$ offspring is bred with another, giving rise to $F_2$ progeny, the genes for yellow and green seeds segregate in a characteristic manner such that the ratio between $F_2$ plants with yellow seeds and those with green seeds is 3:1.

Mendel's findings, apparently ahead of their time, were largely ignored until 1900, when Mendel's laws were rediscovered and their importance recognized. Shortly thereafter, the role of **chromosomes** as the carriers of genes was proposed. It was realized that most cells of higher plants and animals are **diploid**—containing two copies of each chromosome. Formation of the germ cells (the sperm and egg), however, involves a unique type of cell division (**meiosis**) in which only one member of each chromosome

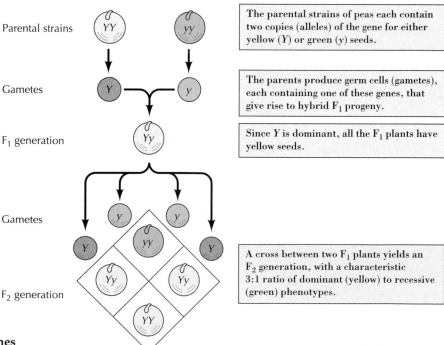

**Figure 3.1**
**Inheritance of dominant and recessive genes**

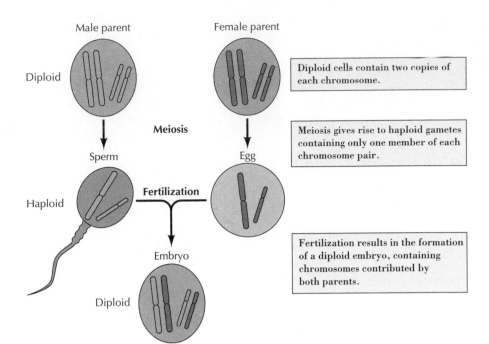

*Figure 3.2*
**Chromosomes at meiosis and fertilization**    Two chromosome pairs of a hypothetical organism are illustrated.

Male parent

Female parent

Diploid cells contain two copies of each chromosome.

Diploid

**Meiosis**

Sperm

Egg

Meiosis gives rise to haploid gametes containing only one member of each chromosome pair.

Haploid

**Fertilization**

Fertilization results in the formation of a diploid embryo, containing chromosomes contributed by both parents.

Embryo

Diploid

pair is transmitted to each progeny cell (Figure 3.2). Consequently, the sperm and egg are **haploid**, containing only one copy of each chromosome. The union of these two haploid cells at fertilization creates a new diploid organism, now containing one member of each chromosome pair derived from the male and one from the female parent. The behavior of chromosome pairs thus parallels that of genes, leading to the conclusion that genes are carried on chromosomes.

The fundamentals of mutation, genetic linkage, and the relationships between genes and chromosomes were largely established by experiments performed with the fruit fly, *Drosophila melanogaster*. *Drosophila* can be easily maintained in the laboratory, and they reproduce about every two weeks, which is a considerable advantage for genetic experiments. Indeed, these features continue to make *Drosophila* an organism of choice for genetic studies of animals, particularly the genetic analysis of development and differentiation.

In the early 1900s, a number of genetic alterations (**mutations**) were identified in *Drosophila*, usually affecting readily observable characteristics such as eye color or wing shape. Breeding experiments indicated that some of the genes governing these traits are inherited independently of each other, suggesting that these genes are located on different chromosomes that segregate independently during meiosis (Figure 3.3). Other genes, however, are frequently inherited together as paired characteristics. Such genes are said to be linked to each other by virtue of being located on the same chromosome. The number of groups of linked genes is the same as the number of chromosomes (four in *Drosophila*), supporting the idea that chromosomes are carriers of the genes.

Linkage between genes is not complete, however; chromosomes exchange material during meiosis, leading to **recombination** between linked genes (Figure 3.4). The frequency of recombination between two linked genes depends on the distance between them on the chromosome; genes that are close to each other recombine less frequently than do genes farther apart. Thus, the frequencies with which different genes recombine can be used to

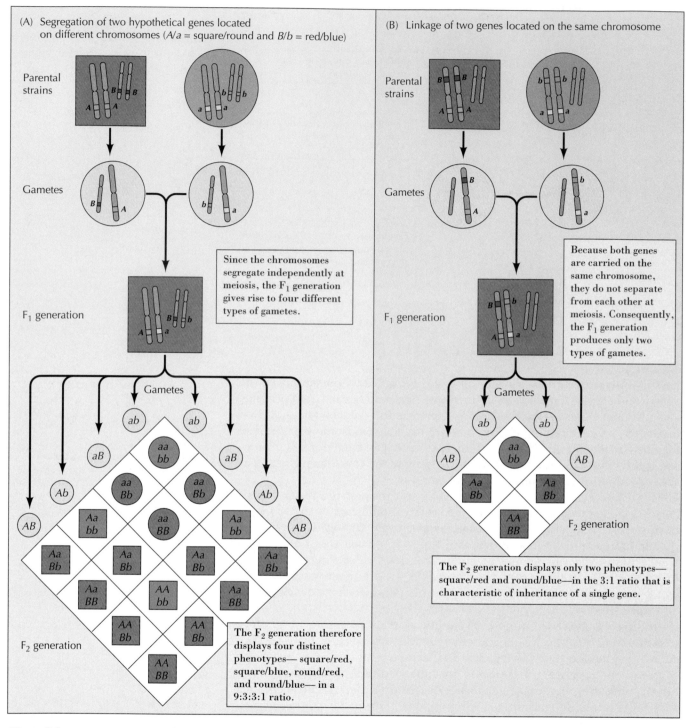

**Figure 3.3**
**Gene segregation and linkage**
(A) Segregation of two hypothetical genes for shape (*A/a* = square/round) and color (*B/b* = red/blue) located on different chromosomes. (B) Linkage of two genes located on the same chromosome.

determine their relative positions on the chromosome, allowing the construction of **genetic maps** (Figure 3.5). By 1915, nearly a hundred genes had been defined and mapped onto the four chromosomes of *Drosophila*, leading to general acceptance of the chromosomal basis of heredity.

## Genes and Enzymes

Early genetic studies focused on the identification and chromosomal localization of genes that control readily observable characteristics, such as the eye color of *Drosophila*. How these genes lead to the observed phenotypes, however, was unclear. The first insight into the relationship between genes and enzymes came in 1909, when it was realized that the inherited human disease phenylketonuria (see Molecular Medicine in Chapter 2) results from a genetic defect in metabolism of the amino acid phenylalanine. This defect was hypothesized to result from a deficiency in the enzyme needed to catalyze the relevant metabolic reaction, leading to the general suggestion that genes specify the synthesis of enzymes.

Clearer evidence linking genes with the synthesis of enzymes came from experiments of George Beadle and Edward Tatum, performed in 1941 with the fungus *Neurospora crassa*. In the laboratory, *Neurospora* can be grown on minimal or rich media similar to those discussed in Chapter 1 for the growth of *E. coli*. For *Neurospora*, minimal media consist only of salts, glucose, and biotin; rich media are supplemented with amino acids, vitamins, purines, and pyrimidines. Beadle and Tatum isolated mutants of *Neurospora* that grew normally on rich media but could not grow on minimal media. Each mutant was found to require a specific nutritional supplement, such as a particular amino acid, for growth. Furthermore, the requirement for a specific nutritional supplement correlated with the failure of the mutant to synthesize that particular compound. Thus, each mutation resulted in a deficiency in a specific metabolic pathway. Since such metabolic pathways were known to be governed by enzymes, the conclusion from these experiments was that each gene specified the structure of a single enzyme—the **one gene–one enzyme hypothesis**. Many enzymes are now known to consist of multiple polypeptides, so the currently accepted statement of this hypothesis is that each gene specifies the structure of a single polypeptide chain.

### Identification of DNA as the Genetic Material

Understanding the chromosomal basis of heredity and the relationship between genes and enzymes did not in itself provide a molecular explanation of the gene. Chromosomes contain proteins as well as DNA, and it was initially thought that genes were proteins. The first evidence leading to the identification of DNA as the genetic material came from studies in bacteria. These experiments represent a prototype for current approaches to defining the function of genes by introducing new DNA sequences into cells, as discussed later in this chapter.

The experiments that defined the role of DNA were derived from studies of the bacterium that causes pneumonia (*Pneumococcus*). Virulent strains of *Pneumococcus* are surrounded by a polysaccharide capsule that protects the bacteria from attack by the immune system of the host. Because the capsule gives bacterial colonies a smooth appearance in culture, encapsulated strains are denoted S. Mutant strains that have lost the ability to make a capsule (denoted R) form rough-edged colonies in culture and are no longer lethal when inoculated into mice. In 1928 it was observed that mice inoculated with nonencapsulated (R) bacteria plus heat-killed encapsulated (S) bacteria developed pneumonia and died. Importantly, the bacteria that were then isolated from these mice were of the S type. Subsequent experiments showed that a cell-free extract of S bacteria was similarly capable of converting (or transforming) R bacteria to the S state. Thus, a substance in the S extract (called the transforming principle) was responsible for inducing the genetic **transformation** of R to S bacteria.

In 1944 Oswald Avery, Colin MacLeod, and Maclyn McCarty established that the transforming principle was DNA, both by purifying it from bacte-

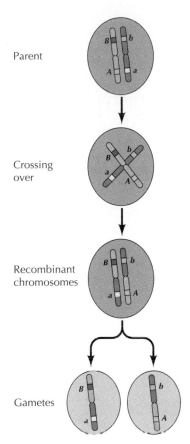

**Figure 3.4**
**Genetic recombination** During meiosis, members of chromosome pairs exchange material. The result is recombination between linked genes.

**Figure 3.5**
**A genetic map** Three genes are localized on a hypothetical chromosome based on frequencies of recombination between them (1% recombination between *a* and *b*; 3% between *b* and *c*; 4% between *a* and *c*). The frequencies of recombination are approximately proportional to the distances between genes on the chromosome.

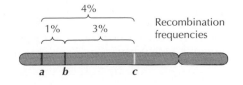

rial extracts and by demonstrating that the activity of the transforming principle is abolished by enzymatic digestion of DNA but not by digestion of proteins (Figure 3.6). Although these studies did not immediately lead to the acceptance of DNA as the genetic material, they were extended within a few years by experiments with bacterial viruses. In particular, it was shown that, when a bacterial virus infects a cell, the viral DNA rather than the viral protein must enter the cell in order for the virus to replicate. Moreover, the parental viral DNA (but not the protein) is transmitted to progeny virus particles. The concurrence of these results with continuing studies of the activity of DNA in bacterial transformation led to acceptance of the idea that DNA is the genetic material.

## The Structure of DNA

Our understanding of the three-dimensional structure of DNA, deduced in 1953 by James Watson and Francis Crick, has been the basis for present-day molecular biology. At the time of Watson and Crick's work, DNA was known to be a polymer composed of four nucleic acid bases—two purines (adenine [A] and guanine [G]) and two pyrimidines (cytosine [C] and thymine [T])—linked to phosphorylated sugars. Given the central role of DNA as the genetic material, elucidation of its three-dimensional structure appeared critical to understanding its function. Watson and Crick's consideration of the problem was heavily influenced by Linus Pauling's description of hydrogen bonding and the $\alpha$ helix, a common element of the secondary structure of proteins (see Chapter 2). Moreover, experimental data on the structure of DNA were available from X-ray crystallography studies by Maurice Wilkins and Rosalind Franklin. Analysis of these data revealed that the DNA molecule is a helix that turns every 3.4 nm. In addition, the data showed that the distance between adjacent bases is 0.34 nm, so there

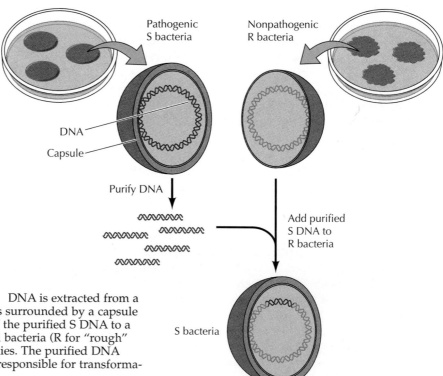

*Figure 3.6*
**Transfer of genetic information by DNA**    DNA is extracted from a pathogenic strain of *Pneumococcus*, which is surrounded by a capsule and forms smooth colonies (S). Addition of the purified S DNA to a culture of nonpathogenic, nonencapsulated bacteria (R for "rough" colonies) results in the formation of S colonies. The purified DNA therefore contains the genetic information responsible for transformation of R to S bacteria.

are ten bases per turn of the helix. An important finding was that the diameter of the helix is approximately 2 nm, suggesting that it is composed of not one but two DNA chains.

From these data, Watson and Crick built their model of DNA (Figure 3.7). The central features of the model are that DNA is a double helix with the sugar–phosphate backbones on the outside of the molecule. The bases are on the inside, oriented such that hydrogen bonds are formed between purines and pyrimidines on opposite chains. The base pairing is very specific: A always pairs with T and G with C. This specificity accounts for the earlier results of Erwin Chargaff, who had analyzed the base composition of various DNAs and found that the amount of adenine was always equal to that of thymine, and the amount of guanine to that of cytosine. Because of this specific base pairing, the two strands of a DNA molecule are complementary: Each strand contains all the information required to specify the sequences of bases on the other.

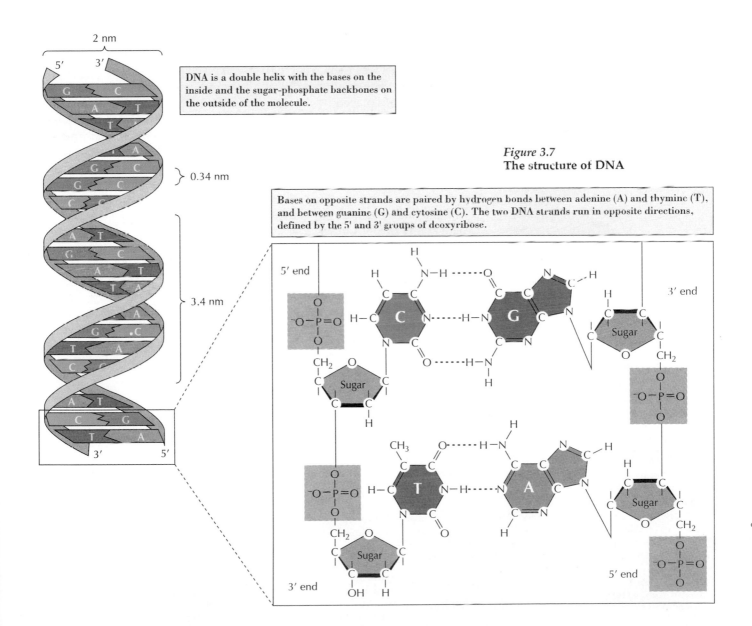

DNA is a double helix with the bases on the inside and the sugar-phosphate backbones on the outside of the molecule.

*Figure 3.7*
**The structure of DNA**

Bases on opposite strands are paired by hydrogen bonds between adenine (A) and thymine (T), and between guanine (G) and cytosine (C). The two DNA strands run in opposite directions, defined by the 5' and 3' groups of deoxyribose.

## Replication of DNA

The discovery of complementary base pairing between DNA strands immediately suggested a molecular solution to the question of how the genetic material could direct its own replication—a process that is required each time a cell divides. It was proposed that the two strands of a DNA molecule could separate and serve as templates for synthesis of new complementary strands, the sequence of which would be dictated by the specificity of base pairing (Figure 3.8). The process is called **semiconservative replication** because one strand of parental DNA is conserved in each progeny DNA molecule.

Direct support for semiconservative DNA replication was obtained in 1958 as a result of elegant experiments, performed by Matthew Meselson and Frank Stahl, in which DNA was labeled with isotopes that altered its density (Figure 3.9). *E. coli* were first grown in media containing the heavy isotope of nitrogen ($^{15}N$) in place of the normal light isotope ($^{14}N$). The DNA of these bacteria consequently contained $^{15}N$ and was heavier than that of bacteria grown in $^{14}N$. Such heavy DNA could be separated from DNA containing $^{14}N$ by equilibrium centrifugation in a density gradient of CsCl. This ability to separate heavy ($^{15}N$) DNA from light ($^{14}N$) DNA enabled the study of DNA synthesis. *E. coli* that had been grown in $^{15}N$ were transferred to media containing $^{14}N$ and allowed to replicate one more time. Their DNA was then extracted and analyzed by CsCl density gradient centrifugation. The results of this analysis indicated that all of the heavy DNA had been replaced by newly synthesized DNA with a density intermediate between that of heavy ($^{15}N$) and that of light ($^{14}N$) DNA molecules. The implication was that during replication, the two parental strands of heavy DNA separated and served as templates for newly synthesized progeny strands of light DNA, yielding double-stranded molecules of intermediate density. This experiment thus provided direct evidence for semiconservative DNA replication, clearly underscoring the importance of complementary base pairing between strands of the double helix.

The ability of DNA to serve as a template for its own replication was further established with the demonstration that an enzyme purified from *E. coli* (**DNA polymerase**) could catalyze DNA replication *in vitro*. In the presence of DNA to act as a template, DNA polymerase was able to direct the incorporation of nucleotides into a complementary DNA molecule.

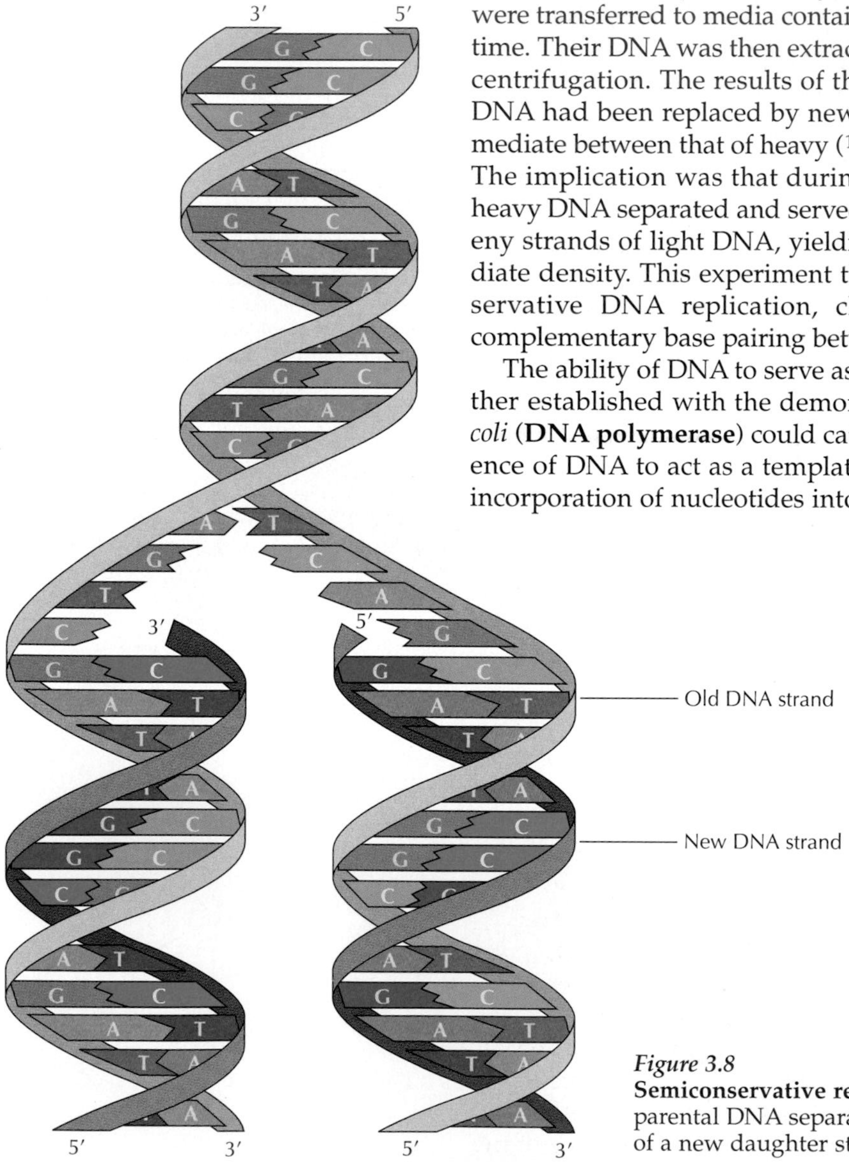

Old DNA strand

New DNA strand

**Figure 3.8**
**Semiconservative replication of DNA**   The two strands of parental DNA separate, and each serves as a template for synthesis of a new daughter strand by complementary base pairing.

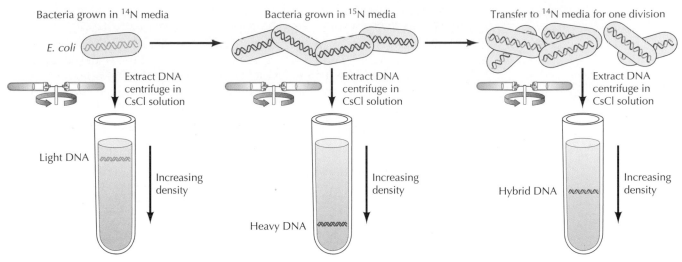

Bacteria grown in $^{14}$N media

*E. coli*

Extract DNA centrifuge in CsCl solution

Light DNA

Increasing density

Bacteria grown in $^{15}$N media

Extract DNA centrifuge in CsCl solution

Increasing density

Heavy DNA

Transfer to $^{14}$N media for one division

Extract DNA centrifuge in CsCl solution

Hybrid DNA

Increasing density

**Figure 3.9**
**Experimental demonstration of semi-conservative replication**   Bacteria grown in medium containing the normal isotope of nitrogen ($^{14}$N) are transferred into medium containing the heavy isotope ($^{15}$N) and grown in this medium for several generations. They are then transferred back to medium containing $^{14}$N and grown for one additional generation. DNA is extracted from these bacteria and analyzed by equilibrium ultracentrifugation in a CsCl solution. The CsCl sediments to form a density gradient, and the DNA molecules band at a position where their density is equal to that of the CsCl solution. DNA of the bacteria transferred from $^{15}$N to $^{14}$N medium for a single generation bands at a density intermediate between that of $^{15}$N DNA and that of $^{14}$N DNA, indicating that it represents a hybrid molecule with one heavy and one light strand.

## EXPRESSION OF GENETIC INFORMATION

Genes act by determining the structure of proteins, which are responsible for directing cell metabolism through their activity as enzymes. The identification of DNA as the genetic material and the elucidation of its structure revealed that genetic information must be specified by the order of the four bases (A, C, G, and T) that make up the DNA molecule. Proteins, in turn, are polymers of 20 amino acids, the sequence of which determines their structure and function. The first direct link between a genetic mutation and an alteration in the amino acid sequence of a protein was made in 1957, when it was found that patients with the inherited disease sickle-cell anemia had hemoglobin molecules that differed from normal ones by a single amino acid substitution. Deeper understanding of the molecular relationship between DNA and proteins came, however, from a series of experiments that took advantage of *E. coli* and its viruses as genetic models.

### Colinearity of Genes and Proteins
The simplest hypothesis to account for the relationship between genes and enzymes was that the order of nucleotides in DNA specified the order of amino acids in a protein. Mutations in a gene would correspond to alterations in the sequence of DNA, which might result from the substitution of one nucleotide for another or from the addition or deletion of nucleotides. These changes in the nucleotide sequence of DNA would then lead to corresponding changes in the amino acid sequence of the protein encoded by the gene in question. This hypothesis predicted that different mutations within a single gene could alter different amino acids in the encoded protein, and that the positions of mutations in a gene should reflect the positions of amino acid alterations in its protein product.

The rapid replication and the simplicity of the genetic system of *E. coli* were of major help in addressing these questions. A variety of mutants of *E. coli* could be isolated, including nutritional mutants that (like the *Neurospora* mutants discussed earlier) require particular amino acids for growth. Importantly, the rapid growth of *E. coli* made feasible the isolation and mapping of multiple mutants in a single gene, leading to the first demonstration of the linear relationship between genes and proteins. In these studies, Charles Yanofsky and his colleagues mapped a series of mutations in the gene that encodes an enzyme required for synthesis of the

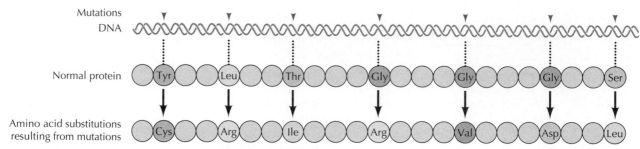

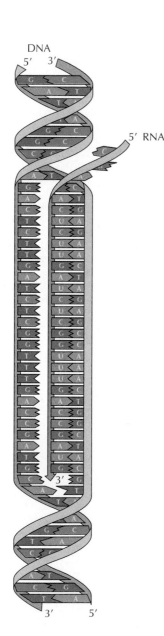

*Figure 3.10*
**Colinearity of genes and proteins**   A series of mutations (arrowheads) were mapped in the *E. coli* gene encoding tryptophan synthetase (top line). The amino acid substitutions resulting from each of the mutations was then determined by sequence analysis of the proteins of mutant bacteria (bottom line). These studies revealed that the order of mutations in DNA was the same as the order of amino acid substitutions in the encoded protein.

amino acid tryptophan. Analysis of the enzymes encoded by the mutant genes indicated that the relative positions of the amino acid alterations were the same as those of the corresponding mutations (Figure 3.10). Thus, the sequence of amino acids in the protein was colinear with that of mutations in the gene, as expected if the order of nucleotides in DNA specifies the order of amino acids in proteins.

## The Role of Messenger RNA

Although the sequence of nucleotides in DNA appeared to specify the order of amino acids in proteins, it did not necessarily follow that DNA itself directs protein synthesis. Indeed, this appeared not to be the case, since DNA is located in the nucleus of eukaryotic cells, whereas protein synthesis takes place in the cytoplasm. Some other molecule was therefore needed to convey genetic information from DNA to the sites of protein synthesis (the ribosomes).

RNA appeared a likely candidate for such an intermediate because the similarity of its structure to that of DNA suggested that RNA could be synthesized from a DNA template (Figure 3.11). RNA differs from DNA in that it is single-stranded rather than double-stranded, its sugar component is ribose instead of deoxyribose, and it contains the pyrimidine base uracil (U) instead of thymine (T) (see Figure 2.10). However, neither the change in sugar nor the substitution of U for T alters base pairing, so the synthesis of RNA can be readily directed by a DNA template. Moreover, since RNA is located primarily in the cytoplasm, it appeared a logical intermediate to convey information from DNA to the ribosomes. These characteristics of RNA suggested a pathway for the flow of genetic information that is known as the **central dogma** of molecular biology:

$$DNA \rightarrow RNA \rightarrow Protein$$

According to this concept, RNA molecules are synthesized from DNA templates (a process called **transcription**), and proteins are synthesized from RNA templates (a process called **translation**).

*Figure 3.11*
**Synthesis of RNA from DNA**   The two strands of DNA unwind, and one is used as a template for synthesis of a complementary strand of RNA.

Experimental evidence for the RNA intermediates postulated by the central dogma was obtained by Sidney Brenner, Francois Jacob, and Matthew Meselson in studies of *E. coli* infected with the bacteriophage T4. The synthesis of *E. coli* RNA stops following infection by T4, and the only new RNA synthesized in infected bacteria is transcribed from T4 DNA. This T4 RNA becomes associated with bacterial ribosomes, thus conveying the information from DNA to the site of protein synthesis. Because of their role as intermediates in the flow of genetic information, RNA molecules that serve as templates for protein synthesis are called **messenger RNAs (mRNAs)**. They are transcribed by an enzyme (**RNA polymerase**) that catalyzes the synthesis of RNA from a DNA template.

In addition to mRNA, two other types of RNA molecules are important in protein synthesis. **Ribosomal RNA (rRNA)** is a component of ribosomes, and **transfer RNAs (tRNAs)** serve as adaptor molecules that align amino acids along the mRNA template. The structures and functions of these molecules are discussed in the following section and in more detail in Chapters 6 and 7.

## The Genetic Code

How is the nucleotide sequence of mRNA translated into the amino acid sequence of a protein? In this step of gene expression genetic information is transferred between chemically unrelated types of macromolecules—nucleic acids and proteins—raising two new types of problems in understanding the action of genes.

First, since amino acids are structurally unrelated to the nucleic acid bases, direct complementary pairing between mRNA and amino acids during the incorporation of amino acids into proteins seemed impossible. How then could amino acids align on an mRNA template during protein synthesis? This question was solved by the discovery that tRNAs serve as adaptors between amino acids and mRNA during translation (Figure 3.12). Prior to its use in protein synthesis, each amino acid is attached by a specific enzyme to its appropriate tRNA. Base pairing between a recognition sequence on each tRNA and a complementary sequence on the mRNA then directs the attached amino acid to its correct position on the mRNA template.

The second problem in the translation of nucleotide sequence to amino acid sequence was determination of the **genetic code**. How could the information contained in the sequence of four different nucleotides be converted to the sequences of 20 different amino acids in proteins? Because 20 amino acids must be specified by only four nucleotides, at least three nucleotides must be used to encode each amino acid. Used singly, four nucleotides could encode only four amino acids and, used in pairs, four nucleotides could encode only sixteen ($4^2$) amino acids. Used as triplets, however, four nucleotides could encode 64 ($4^3$) different amino acids—more than enough to account for the 20 amino acids actually found in proteins.

Direct experimental evidence for the triplet code was obtained by studies of bacteriophage T4 bearing mutations in an extensively studied gene called *rII*. Phages with mutations in this gene form abnormally large plaques, which can be clearly distinguished from those formed by wild-type phages. Hence, isolating and mapping a number of *rII* mutants was easy and led to the establishment of a detailed genetic map of this locus. Study of recombinants between *rII* mutants that had arisen by additions or deletions of nucleotides revealed that phages containing additions or deletions of one or two nucleotides always exhibited the mutant phenotype.

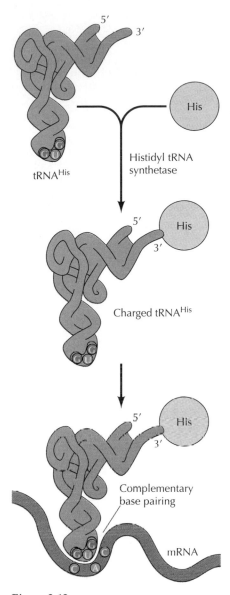

*Figure 3.12*
**Function of transfer RNA**    Transfer RNA serves as an adaptor during protein synthesis. Each amino acid (e.g., histidine) is attached to the 3′ end of a specific tRNA by an appropriate enzyme (an aminoacyl tRNA synthetase). The charged tRNAs then align on an mRNA template by complementary base pairing.

***Figure 3.13***
**Genetic evidence for a triplet code**
A series of mutations consisting of
additions of one, two, or three
nucleotides were studied in the *rII*
gene of bacteriophage T4. Additions of
one or two nucleotides alter the reading frame of the remainder of the
gene. Therefore, all the subsequent
amino acids are abnormal, and an
inactive protein is produced, giving
rise to mutant phage. Additions of
three nucleotides, however, alter only
a single amino acid. The reading frame
of the remainder of the gene is normal,
and an active protein giving rise to
wild-type (WT) phage is produced.

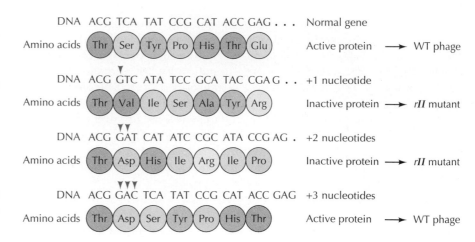

Phages containing additions or deletions of three nucleotides, however,
were frequently wild-type in function (Figure 3.13). These findings suggested that the gene is read in groups of three nucleotides, starting from a
fixed point. Additions or deletions of one or two nucleotides would then
alter the reading frame of the entire gene, leading to the coding of abnormal amino acids throughout the encoded protein. In contrast, additions or
deletions of three nucleotides would lead to the addition or deletion of
only a single amino acid; the rest of the amino acid sequence would remain
unaltered, frequently yielding an active protein.

Deciphering the genetic code thus became a problem of assigning
nucleotide triplets to their corresponding amino acids. This problem was
approached using *in vitro* systems that could carry out protein synthesis
(***in vitro*** **translation**). Cell extracts containing ribosomes, amino acids,
tRNAs, and the enzymes responsible for attaching amino acids to the
appropriate tRNAs (aminoacyl tRNA synthetases) were known to catalyze the incorporation of amino acids into proteins. However, such protein synthesis depends on the presence of mRNA bound to the ribosomes,
and can be greatly enhanced by the addition of purified mRNA. Since
added mRNA directs protein synthesis in such systems, the genetic code
could be deciphered by study of the translation of synthetic mRNAs of
known base sequence.

The first such experiment, performed by Marshall Nirenberg and Hein-
rich Matthaei, involved the *in vitro* translation of a synthetic RNA polymer
containing only uracil (Figure 3.14). This poly-U template was found to
direct the incorporation of only a single amino acid—phenylalanine—into
a polypeptide consisting of repeated phenylalanine residues. Therefore,
the triplet UUU encodes the amino acid phenylalanine. Similar experiments with RNA polymers containing only single nucleotides established
that AAA encodes lysine and CCC encodes proline. The remainder of the
code was deciphered using RNA polymers containing mixtures of
nucleotides, leading to the coding assignment of all 64 possible triplets
(called **codons**) (Table 3.1). Of the 64 codons, 61 specify an amino acid; the
remaining three (UAA, UAG, and UGA) are stop codons that signal the
termination of protein synthesis. The code is degenerate; that is, many
amino acids are specified by more than one codon. With few exceptions
(discussed in Chapter 10), all organisms utilize the same genetic code, providing strong support for the conclusion that all present-day cells evolved
from a common ancestor.

***Figure 3.14***
**The triplet UUU encodes phenylala-
nine**    *In vitro* translation of a synthetic RNA consisting of repeated uracils
(a poly-U template) results in the synthesis of a polypeptide containing only
phenylalanine.

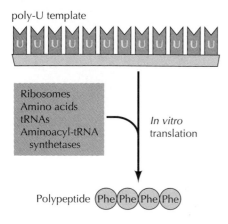

*Table 3.1* The Genetic Code

| First position | Second position | | | | Third position |
|---|---|---|---|---|---|
| | U | C | A | G | |
| U | Phe | Ser | Tyr | Cys | U |
| | Phe | Ser | Tyr | Cys | C |
| | Leu | Ser | STOP | STOP | A |
| | Leu | Ser | STOP | Trp | G |
| C | Leu | Pro | His | Arg | U |
| | Leu | Pro | His | Arg | C |
| | Leu | Pro | Gln | Arg | A |
| | Leu | Pro | Gln | Arg | G |
| A | Ile | Thr | Asn | Ser | U |
| | Ile | Thr | Asn | Ser | C |
| | Ile | Thr | Lys | Arg | A |
| | Met | Thr | Lys | Arg | G |
| G | Val | Ala | Asp | Gly | U |
| | Val | Ala | Asp | Gly | C |
| | Val | Ala | Glu | Gly | A |
| | Val | Ala | Glu | Gly | G |

## RNA Viruses and Reverse Transcription

With the elucidation of the genetic code, the fundamental principles of the molecular biology of cells appeared to have been established. According to the central dogma, the genetic material consists of DNA, which is capable of self-replication as well as being transcribed into mRNA, which serves in turn as the template for protein synthesis. However, as noted in Chapter 1, many viruses contain RNA rather than DNA as their genetic material, implying the use of other modes of information transfer.

RNA genomes were first discovered in plant viruses, many of which were found to be composed of only RNA and protein. Direct proof that RNA acts as the genetic material of these viruses was obtained in the 1950s by experiments demonstrating that RNA purified from tobacco mosaic virus could infect new host cells, giving rise to infectious progeny virus. The mode of replication of most viral RNA genomes was subsequently determined by studies of the RNA bacteriophages of *E. coli*. These viruses were found to encode a specific enzyme that could catalyze the synthesis of RNA from an RNA template (RNA-directed RNA synthesis), using the same mechanism of base pairing between complementary strands as is employed during DNA replication or transcription of RNA from DNA.

However, RNA-directed RNA synthesis did not appear to account for the replication of certain animal viruses (RNA tumor viruses), which were of particular interest because of their ability to cause cancer in infected animals. Although these viruses contain genomic RNA in their viral particles, experiments performed by Howard Temin in the early 1960s indicated that their replication requires DNA synthesis in infected cells, leading to the hypothesis that the RNA tumor viruses (now called **retroviruses**) replicate via synthesis of a DNA intermediate, called a DNA provirus (Figure 3.15). This hypothesis was initially met with widespread disbelief because it involves RNA-directed synthesis of DNA—a reversal of the central dogma. In 1970, however, Temin and David Baltimore inde-

## KEY EXPERIMENT

# The DNA Provirus Hypothesis

**Nature of the Provirus of Rous Sarcoma**

Howard M. Temin
McArdle Laboratory, University of Wisconsin, Madison, WI
*National Cancer Institute Monographs, Volume 17, 1964, pages 557–570.*

### The Context

Rous sarcoma virus (RSV), the first cancer-causing virus to be described, was of considerable interest as an experimental system for studying the molecular biology of cancer. Howard Temin began his research in this area when, as a graduate student in 1958, he developed the first assay for the transformation of normal cells to cancer cells in culture following infection with RSV. The availability of such a quantitative *in vitro* assay provided the tool needed for further studies of both cell transformation and virus replication. As Temin proceeded with these studies, he made a series of unexpected observations indicating that the replication of RSV was fundamentally different from that of other RNA viruses. These experiments led to Temin's proposal of the DNA provirus hypothesis, which stated that the viral RNA was copied into DNA in infected cells—a proposal that ran directly counter to the universally accepted central dogma of molecular biology.

### The Experiments

The DNA provirus hypothesis was based on several different types of experimental evidence. First, studies of cell transformation using mutants of RSV indicated that important characteristics of transformed cells were determined by genetic information of the virus. This information was regularly transmitted to daughter cells following cell division, even in the absence of virus replication. Temin therefore proposed that the viral genome was present in infected cells in a stably inherited form, which he called a provirus.

Evidence that the provirus was DNA was then derived from experiments with metabolic inhibitors. First, actinomycin D, which inhibits the synthesis of RNA from a DNA template, was found to inhibit virus production by RSV-infected cells (see figure). Second, inhibitors of DNA synthesis inhibited early stages of cell infection by RSV. Thus, DNA synthesis appeared to be required early in infection, and DNA-directed RNA synthesis appeared to be needed subsequently for the production of progeny viruses, leading to the proposal that the provirus was a DNA copy of the viral RNA genome. Temin sought further evidence for this proposal by using nucleic acid hybridization to detect viral sequences in infected cell DNA, but the sensitivity of the available techniques was limited and the data were unconvincing.

### The Impact

The DNA provirus hypothesis was thus proposed principally on the basis of genetic experiments and the effects of metabolic inhibitors. It was a radical proposal, which contradicted the accepted central dogma of molecular biology. In this setting, Temin's hypothesis that RSV replicated by the transfer of information from RNA to DNA not only failed to win the acceptance of the scientific community, but was met with gener-

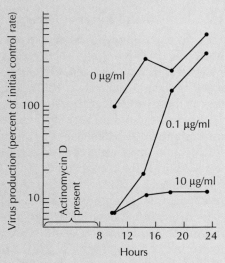

Effect of actinomycin D on RSV replication. RSV-infected cells were cultured with the indicated concentrations of actinomycin D for 8 hours. Actinomycin D was then removed and the amount of virus produced was determined.

pendently discovered that the RNA tumor viruses contain a novel enzyme that catalyzes the synthesis of DNA from an RNA template. In addition, clear-cut evidence for the existence of viral DNA sequences in infected cells was obtained. The synthesis of DNA from RNA, now called **reverse transcription**, was thus established as a mode of information transfer in biological systems.

Reverse transcription is important not only in the replication of retro-

## The DNA Provirus Hypothesis *(continued)*

al derision. Nonetheless, Temin persevered through the 1960s, continuing with experiments to test his hypothesis and providing increasingly convincing evidence in its support. These efforts culminated in 1970 with the discovery by Temin and Satoshi Mizutani, and at the same time by David Baltimore, of a viral enzyme, now known as reverse transcriptase, that synthesizes DNA from an RNA template—an unambiguous biochemical demonstration that the central dogma could be reversed.

Temin concluded his 1970 paper with the statement that the results

"constitute strong evidence that the DNA provirus hypothesis is correct and that RNA tumour viruses have a DNA genome when they are in cells and an RNA genome when they are in virions. This result would have strong implications for theories of viral carcinogenesis and, possibly, for theories of information transfer in other biological systems." As Temin predicted, the discovery of RNA-directed DNA synthesis has led to major advances in our understanding of cancer, human retroviruses, and gene rearrangements. Reverse transcriptase has further provided a critical tool for cDNA

Howard Temin

cloning, thereby impacting virtually all areas of contemporary cell and molecular biology.

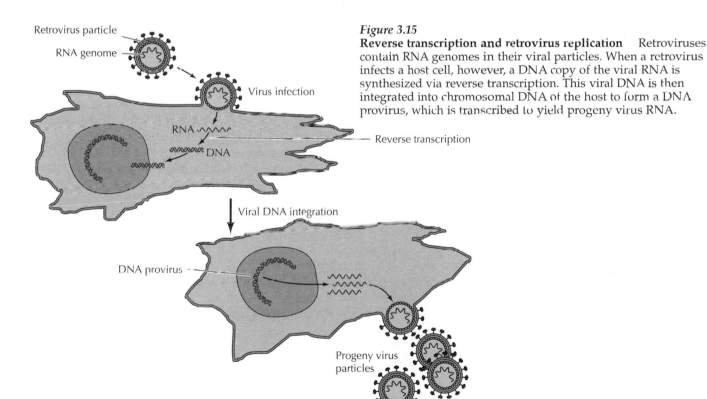

*Figure 3.15*
**Reverse transcription and retrovirus replication**    Retroviruses contain RNA genomes in their viral particles. When a retrovirus infects a host cell, however, a DNA copy of the viral RNA is synthesized via reverse transcription. This viral DNA is then integrated into chromosomal DNA of the host to form a DNA provirus, which is transcribed to yield progeny virus RNA.

viruses, but also in at least two other broad aspects of molecular and cellular biology. First, reverse transcription is not restricted to retroviruses; it also occurs in cells and, as discussed in Chapter 5, is frequently responsible for the transposition of DNA sequences from one chromosomal location to another. Second, enzymes that catalyze RNA-directed DNA synthesis (**reverse transcriptases**) can be used experimentally to generate DNA copies of any RNA molecule. The use of reverse transcriptase has thus allowed mRNAs of eukaryotic cells to be studied using the molecular approaches that are currently applied to the manipulation of DNA, as discussed in the following section.

## RECOMBINANT DNA

Classical experiments in molecular biology were strikingly successful in developing our fundamental concepts of the nature and expression of genes. Since these studies were based primarily on genetic analysis, their success depended largely on the choice of simple, rapidly replicating organisms (such as bacteria and viruses) as models. It was not clear, however, how these fundamental principles could be extended to provide a molecular understanding of the complexities of eukaryotic cells, since the genomes of most eukaryotes (e.g., the human genome) are up to a thousand times more complex than that of *E. coli*. In the early 1970s, the possibility of studying such genomes at the molecular level seemed daunting. In particular, there appeared to be no way in which individual genes could be isolated and studied.

This obstacle to the progress of molecular biology was overcome by the development of recombinant DNA technology, which provided scientists with the ability to isolate, sequence, and manipulate individual genes derived from any type of cell. The application of recombinant DNA has thus enabled detailed molecular studies of the structure and function of eukaryotic genes, thereby revolutionizing our understanding of cell biology.

### Restriction Endonucleases

The first step in the development of recombinant DNA technology was the characterization of **restriction endonucleases**—enzymes that cleave DNA at specific sequences. These enzymes were identified in bacteria, where they apparently provide a defense against the entry of foreign DNA (e.g., from a virus) into the cell. Bacteria have a variety of restriction endonucleases that cleave DNA at more than a hundred distinct recognition sites, each of which consists of a specific sequence of four to eight base pairs (examples are given in Table 3.2).

Since restriction endonucleases digest DNA at specific sequences, they can be used to cleave a DNA molecule at unique sites. For example, the restriction endonuclease *Eco*RI recognizes the six-base-pair sequence GAATTC. This sequence is present at five sites in DNA of the bacteriophage λ, so *Eco*RI digests λ DNA into six fragments ranging from 3.6 to 21.2 kilo-

*Table 3.2* Recognition Sites of Representative Restriction Endonucleases

| Enzyme[a] | Source | Recognition site[b] |
|---|---|---|
| *Bam*HI | *Bacillus amyloliquefaciens* H | GGATCC |
| *Eco*RI | *Escherichia coli* RY13 | GAATTC |
| *Hae*III | *Haemophilus aegyptius* | GGCC |
| *Hind*III | *Haemophilus influenzae* Rd | AAGCTT |
| *Hpa*I | *Haemophilus parainfluenzae* | GTTAAC |
| *Hpa*II | *Haemophilus parainfluenzae* | CCGG |
| *Mbo*I | *Moraxella bovis* | GATC |
| *Not*I | *Nocardia otitidis-caviarum* | GCGGCCGC |
| *Sfi*I | *Streptomyces fimbriatus* | GGCCNNNNNGGCC |
| *Taq*I | *Thermus aquaticus* | TCGA |

[a]Enzymes are named according to their species of isolation, followed by a number to distinguish different enzymes isolated from the same organism (e.g., *Hpa*I and *Hpa*II).
[b]Recognition sites show the sequence of only one strand of double-stranded DNA. "N" represents any base.

## MOLECULAR MEDICINE

# HIV and AIDS

### The Disease

Acquired immune deficiency syndrome (AIDS) is a new disease, first described in 1981. It has now become a worldwide pandemic, with more than 4 million people currently estimated to have developed AIDS. In the United States, more than 500,000 cases have been reported and AIDS is the leading cause of death among men between the ages of 25 and 44. The clinical manifestations of AIDS result principally from failure of the immune system to function normally. In the absence of normal immunity, AIDS patients are sensitive to opportunistic infections by agents (viruses, bacteria, fungi, and protozoans) against which a healthy individual would be resistant. Victims of AIDS also suffer a high frequency of some types of cancers, particularly lymphomas and Kaposi's sarcoma, although it is the opportunistic infections that are responsible for most deaths.

### Molecular and Cellular Basis

AIDS is caused by a retrovirus (human immunodeficiency virus or HIV) that was discovered by the research groups of Robert Gallo and Luc Montagnier in 1983. HIV infects principally a specific type of lymphocyte (the T4 lymphocyte) that is required for a normal immune response. In contrast to many other retroviruses, such as Rous sarcoma virus, HIV infection does not cause the cells it infects to become cancerous. Instead, HIV eventually kills the cells in which it replicates, ultimately resulting in depletion of the population of T4 lymphocytes and failure of the immune system in infected individuals. This failure of the immune system in turn leads to the opportunistic infections and cancers that represent the clinical manifestations of AIDS.

### Prevention and Treatment

At present, the only means of preventing AIDS is to avoid HIV infection. HIV is a fragile virus that quickly loses infectivity outside the body, so it cannot be transmitted by casual contact with an infected person. HIV can be transmitted in three ways: through sexual contact, through contaminated blood products, and from mother to child during pregnancy or breast-feeding. Following the isolation of HIV, screening tests were developed to ensure the safety of clotting factors and blood supplies used for transfusions. Prevention of HIV infection by other routes currently depends on individuals minimizing their personal risk of infection by adhering to safe sexual practices and avoiding sources of contaminated blood, such as shared needles used for intravenous drug injection.

Beyond modifying individual behavior to reduce the risk of infection, the identification of HIV as the cause of AIDS opens possibilities for

prevention and treatment. A vaccine to prevent HIV infection is being actively pursued, although several features of the biology of HIV pose difficulties to this approach. Alternatively, drugs that inhibit virus replication may provide a means to treat HIV-infected individuals. One such drug (azidothymidine, or AZT) has been in use since 1987. The effectiveness of AZT results from its specificity for HIV reverse transcriptase. In particular, HIV reverse transcriptase efficiently uses AZT in place of thymidine, resulting in incorporation of AZT and inactivation of HIV proviral DNA. Because cellular DNA polymerases utilize AZT much less efficiently than thymidine, AZT acts as a selective inhibitor of HIV replication. Unfortunately, while AZT treatment clearly benefits infected individuals, it is not sufficiently effective to be a cure. Major efforts are therefore under way to develop more effective drugs or drug combinations targeted against either reverse transcriptase or other HIV proteins.

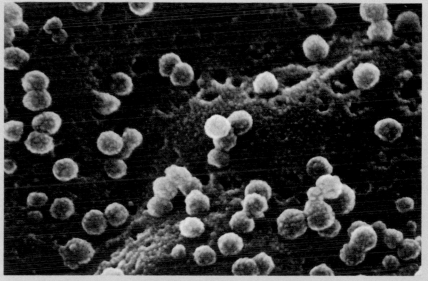

Scanning electron micrograph of HIV budding from T lymphocytes. (Cecil Fox/Photo Researchers, Inc.)

0.1 µm

**Figure 3.16**
**EcoRI digestion and gel electro-**
**phoresis of λ DNA**    *Eco*RI cleaves λ
DNA at five sites (arrows), yielding six
DNA fragments. These fragments are
then separated by electrophoresis in an
agarose gel. The DNA fragments
migrate toward the positive electrode,
with smaller fragments moving more
rapidly through the gel. Following
electrophoresis, the DNA is stained
with a fluorescent dye and photographed. The sizes of DNA fragments are indicated.

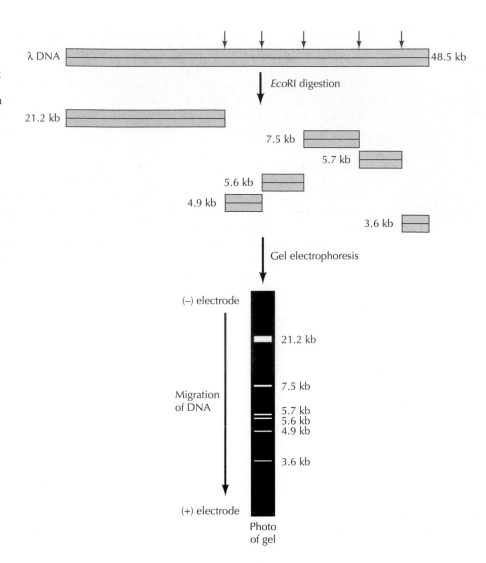

bases long (1 kilobase, or kb = 1000 base pairs) (Figure 3.16). These fragments can be separated according to size by **gel electrophoresis**—a common method in which molecules are separated based on the rates of their
migration in an electric field. A gel, usually formed from agarose or polyacrylamide, is placed between two buffer compartments containing electrodes. The sample (e.g., the mixture of DNA fragments to be analyzed) is
then pipetted into preformed slots in the gel, and the electric field is turned
on. Nucleic acids are negatively charged (because of their phosphate backbone), so they migrate toward the positive electrode. The gel acts like a
sieve, selectively retarding the movement of larger molecules. Smaller molecules therefore move through the gel more rapidly, allowing a mixture of
nucleic acids to be separated on the basis of size.

In addition to size, the order of restriction fragments can be determined by
a variety of methods, yielding (for example) a map of the *Eco*RI sites in λ DNA.
The locations of cleavage sites for multiple different restriction endonucleases
can be used to generate detailed **restriction maps** of DNA molecules, such as
viral genomes (Figure 3.17). In addition, individual DNA fragments produced
by restriction endonuclease digestion can be isolated following electrophoresis
for further study—including determination of their DNA sequence. The DNAs
of many viruses have been characterized by this approach.

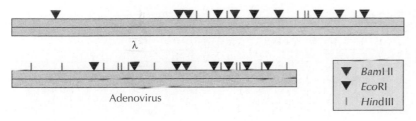

*Figure 3.17*
**Restriction maps of λ and adenovirus DNAs**
The locations of cleavage sites for *Bam*HI, *Eco*RI, and *Hind*III are shown in the DNAs of *E. coli* bacteriophage λ (48.5 kb) and human adenovirus-2 (35.9 kb).

Restriction endonuclease digestion alone, however, does not provide sufficient resolution for the analysis of larger DNA molecules, such as cellular genomes. A restriction endonuclease with a six-base-pair recognition site (such as *Eco*RI) cleaves DNA with a statistical frequency of once every 4096 base pairs ($1/4^6$). A molecule the size of λ DNA (48.5 kb) would therefore be expected to yield about ten *Eco*RI fragments, consistent with the results illustrated in Figure 3.16. However, restriction endonuclease digestion of larger genomes yields quite different results. For example, the human genome is approximately $3 \times 10^6$ kb long and is therefore expected to yield more than 500,000 *Eco*RI fragments. Such a large number of fragments cannot be separated from one another, so agarose gel electrophoresis of *Eco*RI-digested human DNA yields a continuous smear rather than a discrete pattern of DNA fragments. Because it is impossible to isolate single restriction fragments from such digests, restriction endonuclease digestion alone does not yield a source of homogeneous DNA suitable for further analysis. Quantities of such purified DNA fragments, however, can be obtained through molecular cloning.

### Generation of Recombinant DNA Molecules

The basic strategy in **molecular cloning** is to insert a DNA fragment of interest (e.g., a segment of human DNA) into a DNA molecule (called a **vector**) that is capable of independent replication in a host cell. The result is a **recombinant molecule** or **molecular clone**, composed of the DNA insert linked to vector DNA sequences. Large quantities of the inserted DNA can be obtained if the recombinant molecule is allowed to replicate in an appropriate host. For example, fragments of human DNA can be cloned in bacteriophage λ vectors (Figure 3.18). These recombinant molecules can then be introduced into *E. coli*, where they replicate efficiently to yield millions of progeny phages containing the human DNA insert. The DNA of these phages can then be isolated, yielding large quantities of recombinant molecules containing a single fragment of human DNA. Whereas this fragment might represent one part in 100,000 of human genomic DNA, it represents approximately one part in 10 after being cloned in the λ vector. Moreover, the fragment can be easily isolated from the rest of the vector DNA by restriction endonuclease digestion and gel electrophoresis, allowing a pure fragment of human DNA to be analyzed and further manipulated.

The DNA fragments used to create recombinant molecules are usually generated by digestion with restriction endonucleases. Many of these enzymes cleave their recognition sequences at staggered sites, leaving

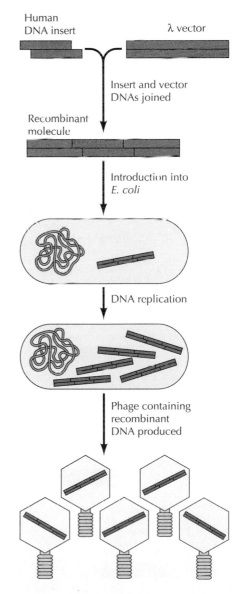

*Figure 3.18*
**Generation of a recombinant DNA molecule**    A fragment of human DNA is inserted into a λ DNA vector. The resulting recombinant molecule is then introduced into *E. coli*, where it replicates to yield recombinant progeny phage containing the human DNA insert.

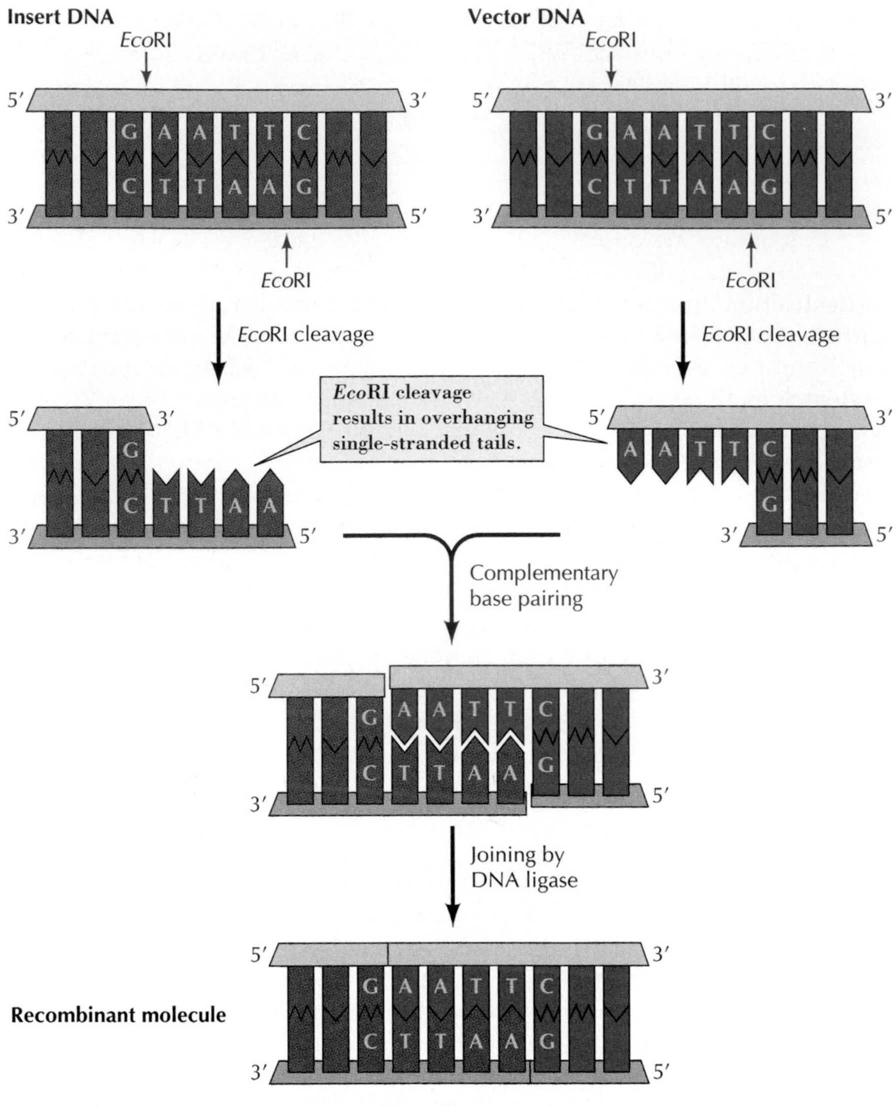

*Figure 3.19*
**Joining of DNA molecules** Vector and insert DNAs are digested with a restriction endonuclease (such as *Eco*RI) which cleaves at staggered sites leaving overhanging single-stranded tails. Vector and insert DNAs can then associate by complementary base pairing, and covalent joining of the DNA strands by DNA ligase yields a recombinant molecule.

**Insert DNA**

*Eco*RI

5′   G A A T T C   3′
C T T A A G
3′       5′

↑
*Eco*RI

↓ *Eco*RI cleavage

5′ G   3′
C T T A A
3′   5′

**Vector DNA**

*Eco*RI

5′   G A A T T C   3′
C T T A A G
3′       5′

↑
*Eco*RI

↓ *Eco*RI cleavage

5′ A A T T C   3′
G
3′   5′

*Eco*RI cleavage results in overhanging single-stranded tails.

↓ Complementary base pairing

5′   G A A T T C   3′
C T T A A G
3′     5′

↓ Joining by DNA ligase

**Recombinant molecule**

5′   G A A T T C   3′
C T T A A G
3′     5′

*Figure 3.20*
**cDNA cloning**

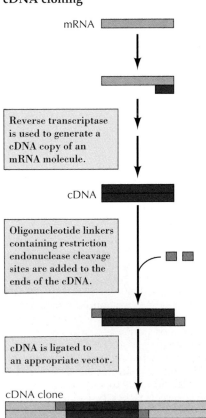

mRNA

Reverse transcriptase is used to generate a cDNA copy of an mRNA molecule.

cDNA

Oligonucleotide linkers containing restriction endonuclease cleavage sites are added to the ends of the cDNA.

cDNA is ligated to an appropriate vector.

cDNA clone

overhanging or cohesive single-stranded tails that can associate with each other by complementary base pairing (Figure 3.19). The association between such paired complementary ends can be established permanently by treatment with **DNA ligase**, an enzyme that seals breaks in DNA strands (see Chapter 5). Thus, two different fragments of DNA (e.g., a human DNA insert and a λ DNA vector) prepared by digestion with the same restriction endonuclease can be readily joined to create a recombinant DNA molecule.

The fragments of DNA that can be cloned are not limited to those that terminate in restriction endonuclease cleavage sites. Synthetic DNA "linkers" containing a variety of restriction endonuclease sites can be added to the blunt ends of any DNA fragment. Linkers are short oligonucleotides that can be readily obtained by chemical synthesis, allowing virtually any fragment of DNA to be prepared for ligation to a vector.

Not only DNA, but also RNA sequences can be cloned (Figure 3.20). The first step is to synthesize a DNA copy of the RNA using the enzyme reverse transcriptase. The DNA product (called a **cDNA** because it is *complementary* to the RNA used as a template) can then be ligated to vector DNA as already described. Since eukaryotic genes are usually interrupted by noncoding sequences (introns; see Chapter 4), which are removed from mRNA by splicing, the ability to clone cDNA as well as genomic DNA has been critical for understanding gene structure and function.

## Vectors for Recombinant DNA

Depending on the size of the insert DNA and the purpose of the experiment, many different types of cloning vectors can be used for the generation of recombinant molecules. The basic vector systems used for the isolation and propagation of cloned DNAs are reviewed here. Other vectors developed for the expression of cloned DNAs and the introduction of recombinant molecules into eukaryotic cells are discussed in subsequent sections.

Bacteriophage λ vectors are frequently used for the initial isolation of either genomic or cDNA clones from eukaryotic cells (Figure 3.21). In λ cloning vectors, sequences of the bacteriophage genome that are dispensable for virus replication have been removed and replaced with unique restriction sites for insertion of cloned DNA. DNA inserts can be as large as about 15 kb and still yield a recombinant genome that can be packaged into phage particles. To isolate genomic clones of human DNA, for example, random fragments of human DNA with an average size of about 15 kb are ligated to λ vector arms. These recombinant DNA molecules can then be efficiently packaged into phage particles by mixing DNA with λ proteins

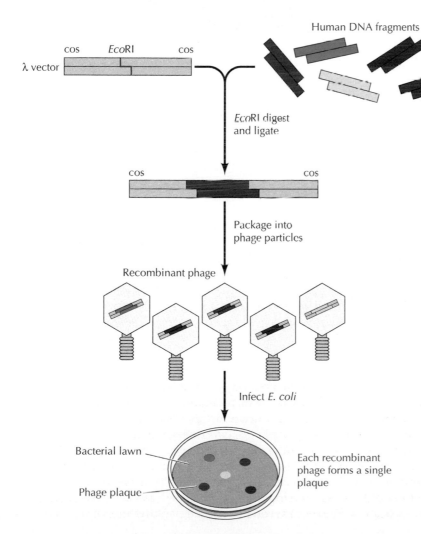

*Figure 3.21*
**Cloning in bacteriophage λ vectors**    The vector contains a restriction site (e.g., an *Eco*RI site) for insertion of cloned DNA. In addition, cos sites (cohesive ends), which are required for packaging DNA into phage particles, are present on both ends of the vector DNA. Insert DNA (e.g., human DNA) is ligated to the vector, and the recombinant molecules are packaged into phage particles by being mixed with phage proteins. The recombinant phages are then used to infect *E. coli*. Each recombinant phage, which carries a unique insert of cloned DNA, forms a single plaque in the infected bacterial culture. Progeny phage carrying unique DNA inserts can then be isolated from individual plaques and grown in large quantity.

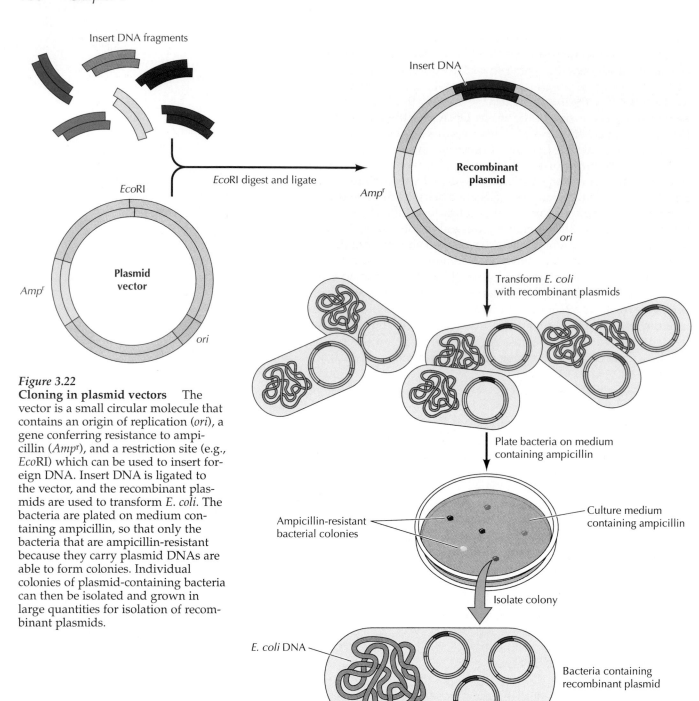

**Figure 3.22**
**Cloning in plasmid vectors** The vector is a small circular molecule that contains an origin of replication (*ori*), a gene conferring resistance to ampicillin (*Amp*^r), and a restriction site (e.g., *Eco*RI) which can be used to insert foreign DNA. Insert DNA is ligated to the vector, and the recombinant plasmids are used to transform *E. coli*. The bacteria are plated on medium containing ampicillin, so that only the bacteria that are ampicillin-resistant because they carry plasmid DNAs are able to form colonies. Individual colonies of plasmid-containing bacteria can then be isolated and grown in large quantities for isolation of recombinant plasmids.

(called packaging extracts) *in vitro*. The phage particles are then used to infect cultures of *E. coli*. Since each recombinant phage forms a single plaque, recombinants carrying unique inserts of human DNA can be isolated. In addition, recombinant phages containing particular genes of interest can be identified by nucleic acid hybridization or other screening methods, as discussed in the next section.

**Plasmid** vectors (Figure 3.22) allow easier manipulation of cloned DNA sequences than do phage vectors. Plasmids are small circular DNA molecules that can replicate independently—without being associated with chromosomal DNA—in bacteria. All that is required on the plasmid DNA is an **origin of replication**—the DNA sequence that signals the host cell DNA polymerase to replicate the DNA molecule. In addition, plasmid vectors carry genes that confer resistance to antibiotics (e.g., ampicillin resistance), so bacteria carrying the plasmids can be selected. Plasmid vectors usually consist of only 2 to 4 kb of DNA, in contrast to the 30 to 45 kb of phage DNA present in λ vectors, facilitating the analysis of an inserted DNA fragment.

To be cloned into a plasmid vector, a fragment of the insert DNA is ligated to an appropriate restriction site in the vector and the recombinant molecule is used to transform *E. coli*. Antibiotic-resistant colonies, which contain plasmid DNA, are selected. Such plasmid-containing bacteria can then be grown in large quantities and their DNA extracted. The small circular plasmid DNA molecules, of which there are often hundreds of copies per cell, can be separated from the bacterial chromosomal DNA; the result is purified plasmid DNA that is suitable for analysis of the cloned insert.

For some studies involving analysis of genomic DNA, it is desirable to clone even larger fragments of DNA than are accommodated by λ vectors. **Cosmid** and **yeast artificial chromosome (YAC)** vectors can be used for this purpose. Cosmid vectors (Figure 3.23) accommodate inserts of approximately 45 kb. These vectors contain bacteriophage λ sequences that allow efficient packaging of the cloned DNA into phage particles. In addition, cosmids contain origins of replication and the genes for antibiotic resistance that are characteristic of plasmids, so they are able to replicate as plasmids in bacterial cells. Even larger fragments of DNA (hundreds of kilobases) can be cloned in YAC vectors, which replicate as chromosomes in yeast cells. These vectors are particularly useful for chromosome mapping studies, as discussed in Chapter 4.

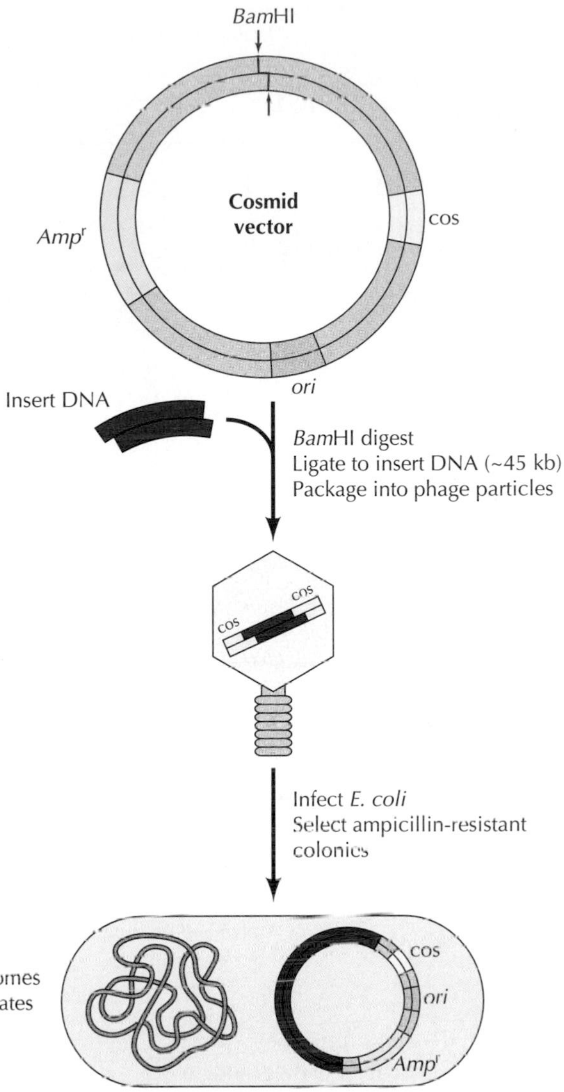

*Figure 3.23*
**Cloning in cosmid vectors**   A cosmid is a plasmid containing cos sites, which allow DNA to be packaged into λ phage particles. Large fragments of insert DNA (approximately 45 kb) are ligated to a cloning site (e.g., *Bam*HI) to yield molecules of approximately the size of the λ genome (48.5 kb), which is appropriate for incorporation into phage particles. The recombinant molecules are made linear (so that a cos site is present at each end), packaged into phage, and used to infect *E. coli*. The cosmid DNA becomes circular again within the infected bacteria, yielding molecules that replicate as plasmids and confer ampicillin resistance to infected cells.

## DNA Sequencing

Molecular cloning allows the isolation of individual fragments of DNA in quantities suitable for detailed characterization, including the determination of nucleotide sequence. Indeed, determination of the nucleotide sequences of many genes has elucidated not only the structure of their protein products, but also the properties of DNA sequences that regulate gene expression. Furthermore, the coding sequences of novel genes are frequently related to those of previously studied genes, and the functions of newly isolated genes can often be correctly deduced on the basis of such sequence similarities.

There are two main methods of DNA sequencing, both yielding rapid and accurate sequence information. Determining the sequence of several kilobases of DNA is a straightforward task for most molecular biology laboratories. Thus, it is now far easier to clone and sequence DNA than it is to determine the amino acid sequence of a protein. Since the nucleotide sequence of a gene can be readily translated into the amino acid sequence of its encoded protein, the easiest way of determining protein sequence has become the sequencing of a cloned gene.

**Maxam-Gilbert sequencing** is based on chemical degradation of DNA (Figure 3.24). The fragment of DNA to be sequenced is labeled at one end with a radioisotope. Samples of the DNA are then treated with chemicals that cleave the DNA molecule at specific bases, under controlled conditions

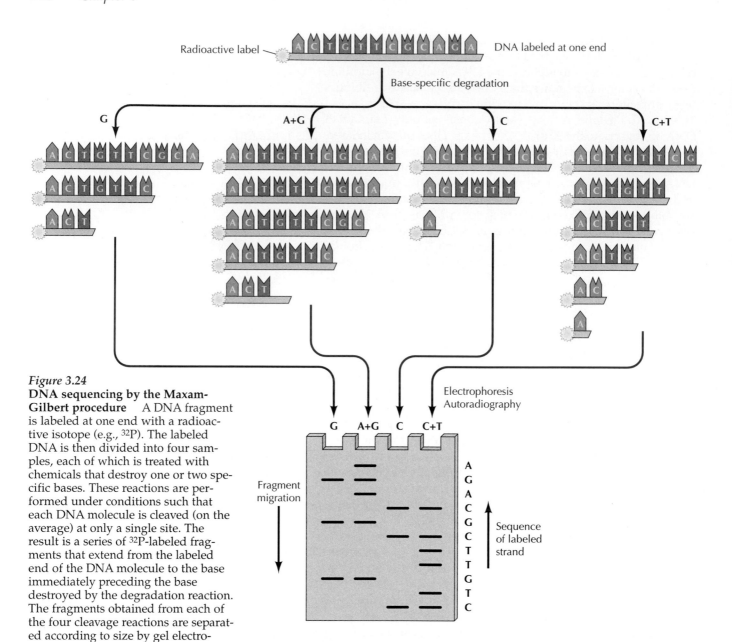

**Figure 3.24**
**DNA sequencing by the Maxam-Gilbert procedure** A DNA fragment is labeled at one end with a radioactive isotope (e.g., $^{32}$P). The labeled DNA is then divided into four samples, each of which is treated with chemicals that destroy one or two specific bases. These reactions are performed under conditions such that each DNA molecule is cleaved (on the average) at only a single site. The result is a series of $^{32}$P-labeled fragments that extend from the labeled end of the DNA molecule to the base immediately preceding the base destroyed by the degradation reaction. The fragments obtained from each of the four cleavage reactions are separated according to size by gel electrophoresis. Radioactive fragments are detected by autoradiography, and the DNA sequence can be read directly from the order of fragments on the autoradiogram. Note that the labeled terminal nucleotide (A in this example) cannot be identified, because it is destroyed in the chemical degradation reaction.

such that each DNA molecule is cleaved at only one or a few sites. Four separate reactions are performed, using reagents that cleave DNA at G, A + G, C, and C + T residues. The fragments of DNA produced in these cleavage reactions are then separated according to size by gel electrophoresis, and those that are radiolabeled are detected by exposure of the gel to X-ray film (**autoradiography**). Each radiolabeled fragment terminates just before the base specifically destroyed by the chemical with which the sample was treated, so the nucleotide sequence corresponds to the order of fragments read from the gel.

**Sanger sequencing**, which is currently the most common method, is based on premature termination of DNA synthesis resulting from the inclusion of chain-terminating **dideoxynucleotides** (which do not contain the deoxyribose 3′ hydroxyl group) in DNA polymerase reactions (Figure 3.25). DNA synthesis is initiated from a radiolabeled primer. Four separate reactions are run, each including one dideoxynucleotide (either A, C, G, or T) in addition to its normal counterpart. Incorporation of a dideoxynucleotide

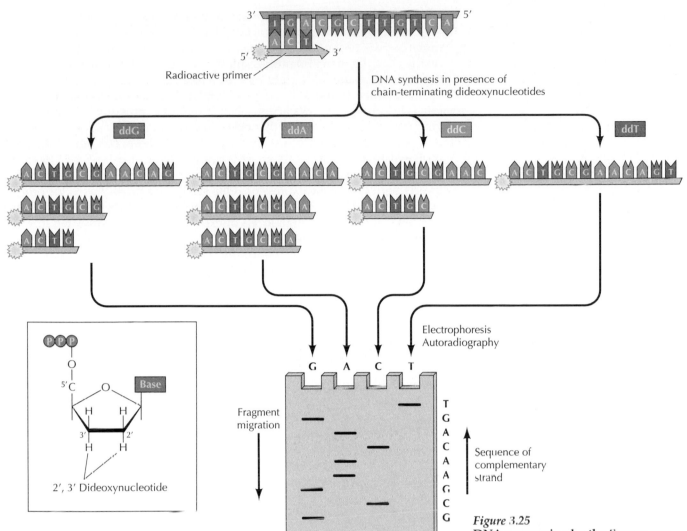

Radioactive primer

DNA synthesis in presence of chain-terminating dideoxynucleotides

ddG     ddA     ddC     ddT

2', 3' Dideoxynucleotide

Electrophoresis
Autoradiography

Fragment
migration

Sequence of
complementary
strand

*Figure 3.25*
**DNA sequencing by the Sanger procedure**   Dideoxynucleotides, which lack OH groups at the 3' as well as the 2' position of deoxyribose, are used to terminate DNA synthesis at specific bases. These molecules are incorporated normally into growing DNA strands. Because they lack a 3' OH, however, the next nucleotide cannot be added, so synthesis of that DNA strand terminates. DNA synthesis is initiated with a radioactive primer. Four separate reactions are carried out, each containing one dideoxynucleotide mixed with its normal counterpart as well as the three other normal deoxynucleotides. When the dideoxynucleotide is incorporated, DNA synthesis stops, so each reaction yields a series of products extending from the radioactive primer to the base substituted by a dideoxynucleotide. Products of the four reactions are separated by electrophoresis and analyzed by autoradiography to determine the DNA sequence.

stops further DNA synthesis because no 3' hydroxyl group is available for addition of the next nucleotide. Thus, a series of labeled DNA molecules is generated, each terminating at the base represented by the dideoxynucleotide in each reaction. As in Maxam-Gilbert sequencing, these products are separated by gel electrophoresis, and the DNA sequence is determined from the order of fragments.

### Expression of Cloned Genes

In addition to enabling determination of the nucleotide sequences of genes—and hence the amino acid sequences of their protein products— molecular cloning has provided new approaches to obtaining large amounts of proteins for structural and functional characterization. Many proteins of interest are present at only low levels in eukaryotic cells and therefore cannot be purified in significant amounts by conventional biochemical techniques. Given a cloned gene, however, this problem can be solved by the engineering of vectors that lead to high levels of gene expression in either bacteria or eukaryotic cells.

To express a eukaryotic gene in *E. coli*, the cDNA of interest is cloned into a plasmid or phage vector (called an **expression vector**) that contains sequences that drive transcription and translation of the inserted gene in bacterial cells (Figure 3.26). Inserted genes often can be expressed at levels high enough that the protein encoded by the cloned gene corresponds to as much as 10% of the total bacterial protein. Purifying the protein encoded by the cloned gene in quantities suitable for detailed biochemical or structural studies is then a straightforward matter.

It is frequently useful to express high levels of a cloned gene in eukaryotic cells, rather than in bacteria. This mode of expression may be important, for example, to ensure that posttranslational modifications of the protein (such as addition of carbohydrates or lipids) occur normally. Such protein expression in eukaryotic cells can be achieved, as in *E. coli*, by insertion of the cloned gene into a vector (usually derived from a virus) that directs high-level gene expression. One system frequently used for protein expression in eukaryotic cells is infection of insect cells by **baculovirus** vectors, which direct very high levels of expression of genes inserted in place of a viral structural protein.

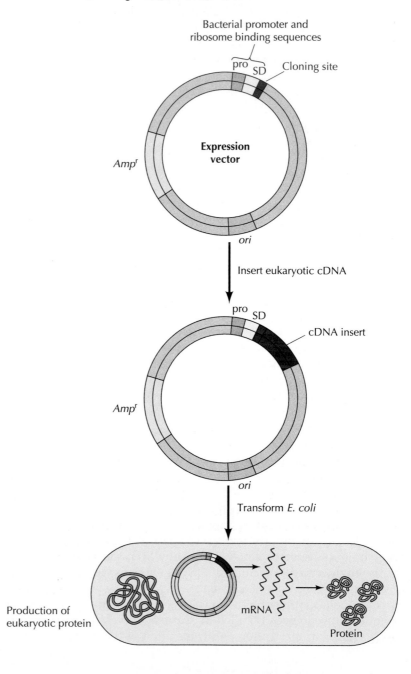

**Figure 3.26**
**Expression of cloned genes in bacteria**  Expression vectors contain promoter sequences (pro) that direct transcription of inserted DNA in bacteria, and sequences required for binding of mRNA to bacterial ribosomes (Shine-Delgarno [SD] sequences). A eukaryotic cDNA inserted adjacent to these sequences can be efficiently expressed in *E. coli*, resulting in production of eukaryotic proteins in transformed bacteria.

## Amplification of DNA by the Polymerase Chain Reaction

Molecular cloning allows individual DNA fragments to be propagated in bacteria and isolated in large amounts. An alternative method to isolating large amounts of a single DNA molecule is the **polymerase chain reaction (PCR)**, which was developed by Kary Mullis in 1988. Provided that some sequence of the DNA molecule is known, PCR can achieve a striking amplification of DNA via reactions carried out entirely *in vitro*. Essentially, DNA polymerase is used for repeated replication of a defined segment of DNA. The number of DNA molecules increases exponentially, doubling with each round of replication, so a substantial quantity of DNA can be obtained from a small number of initial template copies. For example, a single DNA molecule amplified through 30 cycles of replication would theoretically yield $2^{30}$ (approximately 1 billion) progeny molecules. Single DNA molecules can thus be amplified to yield readily detectable quantities of DNA that can be isolated by molecular cloning or further analyzed directly by restriction endonuclease digestion or nucleotide sequencing.

The general procedure for PCR amplification of DNA is illustrated in Figure 3.27. The starting material can be either a cloned DNA fragment or a mixture of DNA molecules—for example, total DNA from human cells. A specific region of DNA can be amplified from such a mixture, provided that the nucleotide sequence surrounding the region is known so that primers can be designed to initiate DNA synthesis at the desired point. Such primers are usually chemically synthesized oligonucleotides containing 15 to 20 bases of DNA. Two primers are used to initiate DNA synthesis in opposite directions from complementary DNA strands. The reaction is started by heating the template DNA to a high temperature (e.g., 95°C) so that the two strands separate. The temperature is then lowered to allow the primers to pair with their complementary sequences on the template strands. DNA polymerase then uses the primers to synthesize a new strand complementary to each template. Thus in one cycle of amplification, two new DNA molecules are synthesized from one template molecule. The process can be repeated multiple times, with a twofold increase in DNA molecules resulting from each round of replication.

The multiple cycles of heating and cooling involved in PCR are performed by programmable heating blocks called thermocyclers. The DNA polymerase (*Taq* polymerase) used in these reactions is a heat-stable enzyme from the bacterium *Thermus aquaticus*, which lives in hot springs at temperatures of about 75°C. The *Taq* polymerase is stable even at the high temperatures used to separate the strands of double-stranded DNA, so PCR amplification can be performed rapidly and automatically. RNA sequences can also be amplified by this method if reverse transcriptase is used to synthesize a cDNA copy prior to PCR amplification.

If enough of the sequence of a gene is known that primers can be specified, PCR amplification provides an extremely powerful method of obtaining readily detectable and manipulable amounts of DNA from starting material that may contain only a few molecules of the desired DNA sequence in a complex mixture of other molecules. For example, defined DNA sequences of up to several kilobases can be readily amplified from total genomic DNA, or a single cDNA can be amplified from total cell RNA. These amplified DNA segments can then be further manipulated or analyzed, for example, to detect mutations within a gene of interest. PCR is thus a powerful addition to the repertoire of recombinant DNA techniques. Its power is particularly apparent in applications such as the diagnosis of inherited diseases, studies of gene expression during development, and forensic medicine.

**Starting DNA**

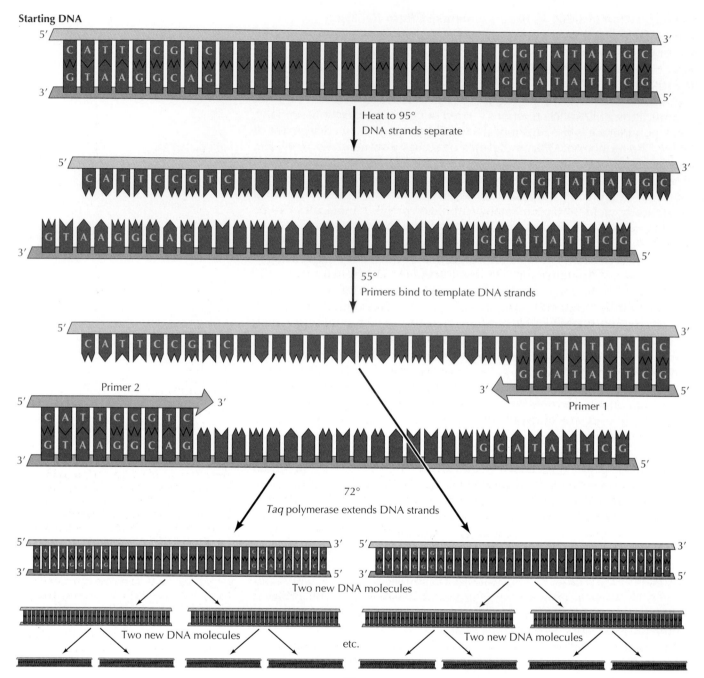

*Figure 3.27*
**Amplification of DNA by PCR**
The region of DNA to be amplified is flanked by two sequences used to prime DNA synthesis. The starting double-stranded DNA is heated to separate the strands and then cooled to allow primers (usually oligonucleotides of 15 to 20 bases) to bind to each strand of DNA. DNA polymerase from *Thermus aquaticus* (*Taq* polymerase) is used to synthesize new DNA strands starting from the primers, resulting in the formation of two new DNA molecules. The process can be repeated for multiple cycles, each resulting in a twofold amplification of DNA.

## DETECTION OF NUCLEIC ACIDS AND PROTEINS

The advent of molecular cloning has enabled the isolation and characterization of individual genes from eukaryotic cells. Understanding the role of genes within cells, however, requires analysis of the intracellular organization and expression of individual genes and their encoded proteins. In this section, the basic procedures currently available for detection of specific nucleic acids and proteins are discussed. These approaches are important for a wide variety of studies, including the mapping of genes to chromosomes, the analysis of gene expression, and the localization of proteins to subcellular organelles. The same general procedures are also used to isolate specific genes as molecular clones.

## Nucleic Acid Hybridization

The key to detection of specific nucleic acid sequences is base pairing between complementary strands of RNA or DNA. At high temperatures (e.g., 90 to 100°C), the complementary strands of DNA separate (denature), yielding single-stranded molecules. If such denatured DNA strands are then incubated under appropriate conditions (e.g., 65°), they will renature to form double-stranded molecules as dictated by complementary base pairing—a process called **nucleic acid hybridization**. Nucleic acid hybrids can be formed between two strands of DNA, two strands of RNA, or one strand of DNA and one of RNA.

Nucleic acid hybridization provides a means for detecting DNA or RNA sequences that are complementary to any isolated nucleic acid, such as a viral genome or a cloned DNA sequence (Figure 3.28). The cloned DNA is radiolabeled, usually by being synthesized in the presence of radioactive nucleotides. This radioactive DNA is then used as a **probe** for hybridization to complementary DNA or RNA sequences, which are detected by virtue of the radioactivity of the resulting double-stranded hybrids.

**Southern blotting** (a technique developed by E. M. Southern) is widely used for detection of specific genes in cellular DNA (Figure 3.29). The DNA to be analyzed is digested with a restriction endonuclease, and the digested DNA fragments are separated by gel electrophoresis. The gel is then overlaid with a nitrocellulose filter or nylon membrane, to which the DNA fragments are transferred (blotted) to yield a replica of the gel. The filter is then incubated with a radiolabeled probe, which hybridizes to the DNA fragments that contain the complementary sequence. These fragments are then visualized by exposure of the filter to X-ray film.

**Northern blotting** is a variation of the Southern blotting technique (hence its name) that is used for detection of RNA instead of DNA. In this method, total cellular RNAs are extracted and fractionated according to size by gel electrophoresis. As in Southern blotting, the RNAs are transferred to a filter and detected by hybridization with a radioactive probe. Northern blotting is frequently used in studies of gene expression—for example, to determine whether specific mRNAs are present in different types of cells.

Nucleic acid hybridization can be used to detect homologous DNA or RNA sequences not only in cell extracts, but also in chromosomes or intact cells—a procedure called *in situ* hybridization (Figure 3.30). In this case, the hybridization of radioactive or fluorescent probes to specific cells or subcellular structures is analyzed by microscopic examination. For example, labeled probes can be hybridized to intact chromosomes in order to identify the chromosomal regions that contain a gene of interest. *In situ* hybridization can also be used to detect specific mRNAs in different types of cells within a tissue.

## Detection of Small Amounts of DNA or RNA by PCR

Amplification of DNA by the polymerase chain reaction is a much more sensitive technique for detecting cellular DNA or RNA sequences than is Southern or Northern blotting. Approximately 100,000 copies of a DNA or RNA sequence are required for detection by blot hybridization. In contrast, PCR can amplify single copies of DNA (or RNA after reverse transcription) to readily detectable levels.

As discussed earlier, the specificity of amplification in PCR is provided by the use of oligonucleotide primers that hybridize to complementary sequences on the template molecule. Therefore, PCR can selectively amplify a specific DNA molecule from complex mixtures, such as total cell DNA or RNA. Consequently, PCR amplification can be used to detect specific DNA or RNA molecules in very small amounts of starting material,

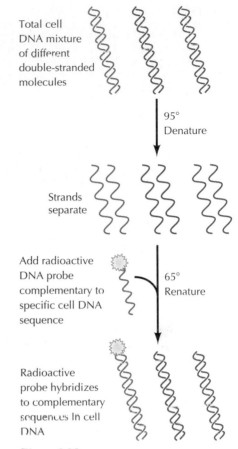

Total cell DNA mixture of different double-stranded molecules

95° Denature

Strands separate

Add radioactive DNA probe complementary to specific cell DNA sequence

65° Renature

Radioactive probe hybridizes to complementary sequences in cell DNA

**Figure 3.28**
**Detection of DNA by nucleic acid hybridization**    A specific sequence can be detected in total cell DNA by hybridization with a radiolabeled DNA probe. The DNA is denatured by heating to 95°C, yielding single-stranded molecules. The radiolabeled probe is then added and the temperature is lowered to 65°C, allowing complementary DNA strands to renature by pairing with each other. The radioactive probe hybridizes to complementary sequences in cell DNA, which can then be detected as radioactive double-stranded molecules.

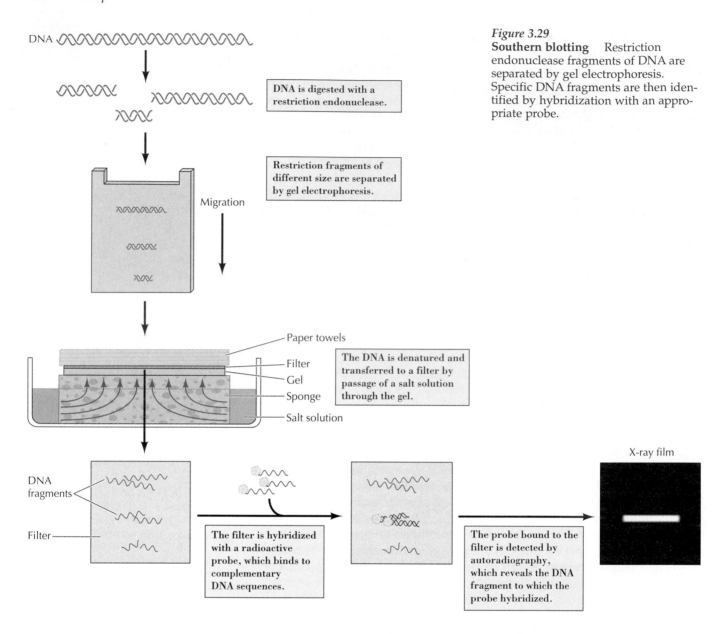

DNA

DNA is digested with a restriction endonuclease.

Restriction fragments of different size are separated by gel electrophoresis.

Migration

Paper towels
Filter
Gel
Sponge
Salt solution

The DNA is denatured and transferred to a filter by passage of a salt solution through the gel.

DNA fragments

Filter

The filter is hybridized with a radioactive probe, which binds to complementary DNA sequences.

The probe bound to the filter is detected by autoradiography, which reveals the DNA fragment to which the probe hybridized.

X-ray film

*Figure 3.29*
**Southern blotting** Restriction endonuclease fragments of DNA are separated by gel electrophoresis. Specific DNA fragments are then identified by hybridization with an appropriate probe.

*Figure 3.30*
*In situ* **hybridization** (A) Hybridization of human chromosomes with fluorescent probes specific for X (green) and Y (orange) chromosome DNA sequences. Chromosomes are stained blue with a fluorescent dye that binds DNA. (B) A radioactive probe for a gene expressed specifically in oocytes was hybridized to a section of mouse ovary. The tissue section was coated with a photographic emulsion and radioactivity detected as the formation of silver grains (arrow). (A courtesy of Ann A. Kiessling, Harvard Medical School; B courtesy of Debra Wolgemuth, Columbia University College of Physicians and Surgeons.)

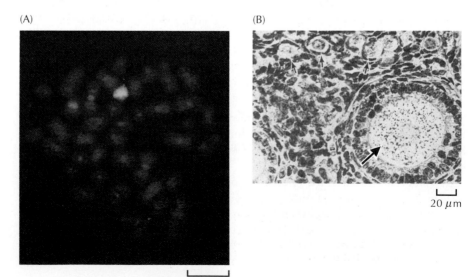

(A)

(B)

10 μm

20 μm

such as extracts of single cells. This extraordinary sensitivity has made PCR an important method for a variety of applications, including the analysis of gene expression in cells available in only limited quantities.

### Antibodies as Probes for Proteins

Studies of gene expression and function require the detection not only of DNA and RNA, but also of specific proteins. For these studies, **antibodies** take the place of nucleic acid probes as reagents that can selectively react with unique protein molecules. Antibodies are proteins produced by cells of the immune system (B lymphocytes) that react against molecules (**antigens**) that the host organism recognizes as foreign substances—for example, the protein coat of a virus. The immune systems of vertebrates are capable of producing millions of different antibodies, each of which specifically recognizes a unique antigen, which may be a protein, a carbohydrate, or a nonbiological molecule. An individual lymphocyte produces only a single type of antibody, but the antibody genes of different lymphocytes vary as a result of programmed gene rearrangements during development of the immune system (see Chapter 5). This variation gives rise to an array of lymphocytes with distinct antibody genes, programmed to respond to different antigens.

Antibodies can be generated by inoculation of an animal with any foreign protein. For example, antibodies against human proteins are frequently raised in rabbits. The sera of such immunized animals contain a mixture of antibodies (produced by different lymphocytes) that react against multiple sites on the immunizing antigen. However, single species of antibodies (**monoclonal antibodies**) can also be produced by the culturing of clonal lines of B lymphocytes from immunized animals (usually mice). Because each lymphocyte is programmed to produce only a single antibody, a clonal line of lymphocytes produces a monoclonal antibody that recognizes only a single antigenic determinant, thereby providing a highly specific immunological reagent.

Although antibodies can be raised against proteins purified from cells, other materials may also be used for immunization. For example, animals may be immunized with intact cells to raise antibodies against unknown proteins expressed by a specific cell type (e.g., a cancer cell). Such antibodies may then be used to identify proteins specifically expressed by the cell type used for immunization. In addition, antibodies are frequently raised against proteins expressed in bacteria as recombinant clones. In this way, molecular cloning allows the production of antibodies against proteins that may be difficult to isolate from eukaryotic cells. Moreover, antibodies can be raised against synthetic peptides that consist of only 10 to 15 amino acids, rather than against intact proteins. Therefore, once the sequence of a gene is known, antibodies against peptides synthesized from part of its predicted protein sequence can be produced. Because antibodies against these synthetic peptides frequently react with the intact protein as well, it is possible to produce antibodies against a protein starting with only the sequence of a cloned gene.

Antibodies can be used in a variety of ways to detect proteins in cell extracts. Two common methods are **immunoblotting** (also called **Western blotting**) and **immunoprecipitation**. Western blotting (Figure 3.31) is another variation of Southern blotting. Proteins in cell extracts are first separated according to size by gel electrophoresis. Because proteins have dif-

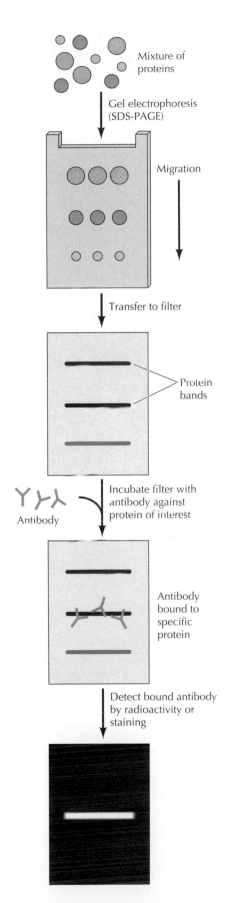

**Figure 3.31**
**Western blotting**   Proteins are separated according to size by SDS-polyacrylamide gel electrophoresis and transferred from the gel to a filter. The filter is incubated with an antibody directed against a protein of interest. The antibody bound to the filter can then be detected by reaction with various reagents, such as a radioactive probe that binds to the antibody.

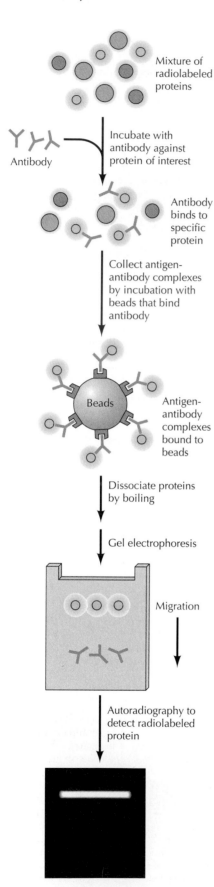

Antibody

Mixture of radiolabeled proteins

Incubate with antibody against protein of interest

Antibody binds to specific protein

Collect antigen-antibody complexes by incubation with beads that bind antibody

Beads

Antigen-antibody complexes bound to beads

Dissociate proteins by boiling

Gel electrophoresis

Migration

Autoradiography to detect radiolabeled protein

*Figure 3.32*
**Immunoprecipitation** Radiolabeled proteins are incubated with an antibody, which forms complexes with the protein against which it is directed (the antigen). These antigen-antibody complexes are collected on beads that bind the antibody. The beads are then boiled to dissociate the antigen-antibody complexes, and the recovered proteins are analyzed by SDS-polyacrylamide gel electrophoresis. The radioactive protein that was immunoprecipitated is detected by autoradiography.

ferent shapes and charges, however, this process requires a modification of the methods used for electrophoresis of nucleic acids. Proteins are separated by a method known as **SDS-polyacrylamide gel electrophoresis (SDS-PAGE)**, in which they are dissolved in a solution containing the negatively charged detergent sodium dodecyl sulfate (SDS). Each protein binds many detergent molecules, which denature the protein and give the protein an overall negative charge. Under these conditions, all proteins migrate toward the positive electrode—their rates of migration determined (like those of nucleic acids) only by size. Following electrophoresis, the proteins are transferred to a filter, which is then allowed to react with antibodies against the protein of interest. The antibody bound to the filter can be detected by various methods, thereby identifying the protein against which the antibody is targeted.

In immunoprecipitation, antibodies are used to isolate the proteins against which they are directed (Figure 3.32). Typically, cells are incubated with radioactive amino acids to label their proteins. Such a radiolabeled cell extract is then incubated with an antibody, which binds to its antigenic target protein. The resulting antigen-antibody complexes are isolated and subjected to electrophoresis, allowing detection of the radioactive antigen by autoradiography.

As discussed in Chapter 1, antibodies can also be used to visualize proteins within cells, as well as in cell lysates. For example, cells can be stained with antibodies labeled with fluorescent dyes, and the subcellular localization of the antigenic proteins can be visualized by fluorescence microscopy (see Figure 1.27). Antibodies can also be labeled with tags that are visible in the electron microscope, such as heavy metals, allowing visualization of antigenic proteins at the ultrastructural level.

## Probes for Screening Recombinant DNA Libraries

The same basic methods for detecting nucleic acids and proteins in cell extracts are used for identifying molecular clones that contain specific cellular DNA inserts. For example, nucleic acid hybridization can be used to identify genomic or cDNA clones that contain DNA sequences for which a probe is available. Cloned cDNAs in expression vectors can also be identified by the use of antibodies against their encoded proteins.

The first step in isolation of either genomic or cDNA clones is frequently the preparation of **recombinant DNA libraries**—collections of clones that contain all the genomic or mRNA sequences of a particular cell type (Figure 3.33). A **genomic library** of human DNA, for example, might be prepared by the cloning of random DNA fragments with average sizes of about 15 kb in a λ vector. Since the size of the human genome is about $3 \times 10^6$ kb, the DNA equivalent of one genome would be represented by 200,000 such λ clones. Because of statistical fluctuations in sampling, however, many genes will not be represented in a library of 200,000 recombinants, while other genes will be represented by multiple clones. Therefore, larger libraries, consisting of approximately 1 million recombinant phages are usually prepared to ensure a high likelihood that any gene of interest is represented in the collection.

Any gene for which a probe is available can readily be isolated from such a recombinant library. The recombinant phages are plated on *E. coli*, and each phage replicates to produce a plaque on the lawn of bacteria. The plaques are then blotted onto filters in a process similar to the transfer of DNA from a gel to a filter during Southern blotting, and the filters are hybridized with a radiolabeled probe to identify the phage plaques that contain the gene of interest. The appropriate plaque can then be isolated from the original plate in order to propagate the recombinant phage that carries the desired cell insert. Similar procedures can be used to screen bacterial colonies carrying plasmid DNA clones, so specific clones can be isolated by hybridization from either phage or plasmid libraries.

A variety of probes can be used for screening recombinant libraries. For example, a cDNA clone can be used as a probe to isolate the corresponding genomic clone, or a gene cloned from one species (e.g., mouse) can be used to isolate a related gene from a different species (e.g., human). In addition

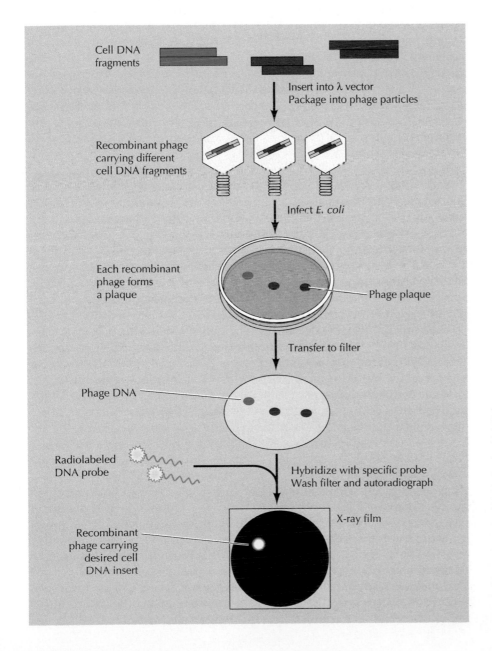

*Figure 3.33*
**Screening a recombinant library by hybridization**    Fragments of cell DNA are cloned in a bacteriophage λ vector and packaged into phage particles, yielding a collection of recombinant phage carrying different cell inserts. The phage are used to infect bacteria, and the culture is overlaid with a filter. Some of the phage in each plaque are transferred to the filter, which is then hybridized with a radiolabeled probe to identify the phage plaque containing the desired gene. The appropriate phage plaque can then be isolated from the original culture plate.

to isolated DNA fragments, synthetic oligonucleotides can be used as probes, enabling the isolation of genes on the basis of partial amino acid sequences of their encoded proteins. For example, oligonucleotides consisting of 15 to 20 bases can be synthesized on the basis of the partial amino acid sequence of a protein of interest. These oligonucleotides can then be used as probes to isolate cDNA clones, which (as already discussed) are much easier to sequence and to characterize than is the protein itself. It is thus possible to proceed experimentally from a partial peptide sequence of a protein to the isolation of a cloned gene.

An alternative approach to isolating a gene on the basis of its protein product is the use of antibodies as probes to screen expression libraries. In this case a **cDNA library** is generated in an expression vector that drives protein synthesis in *E. coli*. Phage plaques or bacterial colonies are then transferred to a filter as already described, but the filter is then reacted with an antibody (as in Western blotting) to identify clones that are producing the protein of interest.

These procedures for identifying molecular clones and detecting genes and gene products in cells illustrate the flexibility of recombinant DNA technology. Starting with any cloned gene, it is possible not only to determine the nucleotide sequence of that gene and use it as a probe for studies of gene organization and transcription, but also to express and raise antisera against its encoded protein. Conversely, genes can be cloned on the basis of limited characterization of a protein of interest, using either oligonucleotide or antibody probes. Thus, recombinant DNA has allowed experimental analyses to proceed either from DNA to protein or from protein to DNA, providing great versatility to the strategies currently employed for studies of eukaryotic genes and their encoded proteins.

## GENE FUNCTION IN EUKARYOTES

The recombinant DNA techniques discussed in the preceding sections provide powerful approaches to the isolation and detailed characterization of the genes of eukaryotic cells. Understanding the function of a gene, however, requires analysis of the gene within cells or intact organisms—not simply as a molecular clone in bacteria. In classical genetics, the function of genes has generally been revealed by the altered phenotypes of mutant organisms. The advent of recombinant DNA has added a new dimension to studies of gene function. Namely, it has become possible to investigate the function of a cloned gene directly by reintroducing the cloned DNA into eukaryotic cells. In simpler eukaryotes, such as yeasts, this technique has made possible the isolation of molecular clones corresponding to virtually any mutant gene. In addition, there are several methods by which cloned genes can be introduced into cultured animal and plant cells, as well as intact organisms, for functional analysis. These approaches can be coupled with the ability to introduce mutations in cloned DNA *in vitro*, extending the power of recombinant DNA to allow functional studies of the genes of more complex eukaryotes.

### Genetic Analysis in Yeasts

Yeasts are particularly advantageous for studies of eukaryotic molecular biology (see Chapter 1). The genome of *Saccharomyces cerevisiae*, which consists of approximately $1.4 \times 10^7$ base pairs, is 200 times smaller than the human genome. Moreover, yeasts can easily be grown in culture, reproducing with a division time of about 2 hours. Thus, yeasts offer the same basic advantages—a small genome and rapid reproduction—that are afforded by bacteria.

Mutations in yeasts can be identified as readily as in *E. coli*. For example, yeast mutants that require a particular amino acid or other nutrient for growth can easily be isolated. In addition, yeasts with defects in genes

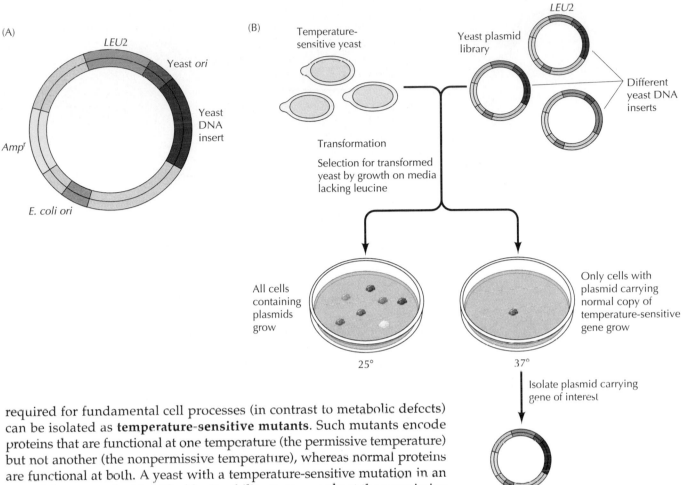

**Figure 3.34**
**Cloning of yeast genes** (A) A yeast vector. The vector contains a bacterial origin of replication (*ori*) and an ampicillin resistance gene (*Amp*^r), allowing it to be propagated as a plasmid in *E. coli*. In addition, the vector contains a yeast origin of replication and a marker gene (*LEU2*), allowing the selection of transformed yeast. The *LEU2* gene encodes an enzyme required for synthesis of the amino acid leucine, so transformation of yeast strains lacking this enzyme can be selected for by growth on medium lacking leucine. (B) Isolation of a yeast gene. A gene of interest is identified by a temperature-sensitive mutation, which allows yeast to grow at 25°C but not at 37°C. To isolate a clone of the gene, the temperature-sensitive yeasts are transformed with a plasmid library containing a collection of genes encompassing the entire yeast genome. All yeasts transformed by plasmid DNAs are able to grow on media lacking leucine at 25°C, but only those yeasts transformed by a plasmid carrying a normal copy of the gene of interest are able to grow at 37°C. The desired plasmid can be isolated from transformed yeasts that form colonies at the nonpermissive temperature.

required for fundamental cell processes (in contrast to metabolic defects) can be isolated as **temperature-sensitive mutants**. Such mutants encode proteins that are functional at one temperature (the permissive temperature) but not another (the nonpermissive temperature), whereas normal proteins are functional at both. A yeast with a temperature-sensitive mutation in an essential gene can be identified by its ability to grow only at the permissive temperature. The ability to isolate such temperature-sensitive mutants has allowed the identification of yeast genes controlling many fundamental cell processes, such as RNA synthesis and processing, progression through the cell cycle, and transport of proteins between cellular compartments.

The relatively simple genetics of yeast also enables a gene corresponding to any yeast mutation to be cloned, simply on the basis of its functional activity (Figure 3.34). First, a genomic library of normal yeast DNA is prepared in vectors that replicate as plasmids in yeasts as well as in *E. coli*. The small size of the yeast genome means that a complete library consists of only a few thousand plasmids. A mixture of such plasmids is then used to transform a temperature-sensitive yeast mutant, and transformants that are able to grow at the nonpermissive temperature are selected. Such transformants have acquired a normal copy of the gene of interest on plasmid DNA, which can then be easily isolated from the transformed yeast cells for further characterization.

Yeast genes encoding a wide variety of essential proteins have been identified in this manner. In many cases, such genes isolated from yeasts have also been useful in identifying and cloning related genes from mammalian cells. Thus, the simple genetics of yeast has not only provided an important model for eukaryotic cells, but has also led directly to the cloning of related genes from more complex eukaryotes.

## Gene Transfer in Plants and Animals

Although the cells of complex eukaryotes are not amenable to the simple genetic manipulations possible in yeasts, gene function can still be

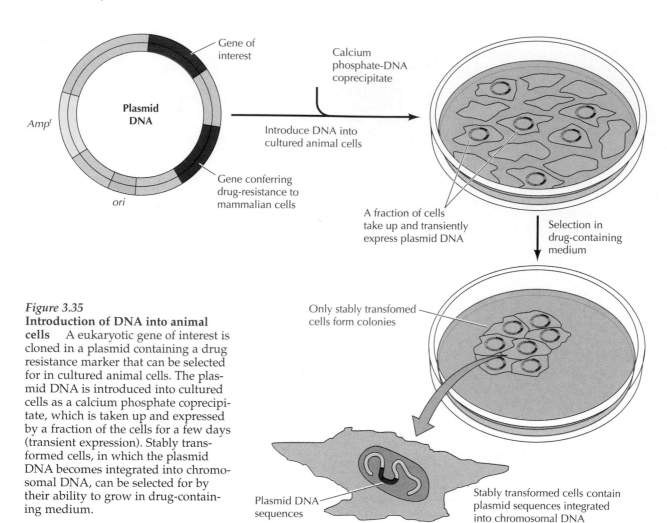

*Figure 3.35*
**Introduction of DNA into animal cells**   A eukaryotic gene of interest is cloned in a plasmid containing a drug resistance marker that can be selected for in cultured animal cells. The plasmid DNA is introduced into cultured cells as a calcium phosphate coprecipitate, which is taken up and expressed by a fraction of the cells for a few days (transient expression). Stably transformed cells, in which the plasmid DNA becomes integrated into chromosomal DNA, can be selected for by their ability to grow in drug-containing medium.

assayed by the introduction of cloned DNA into plant and animal cells. Such experiments (generally called **gene transfer**) have proven critical to addressing a wide variety of questions, including studies of the mechanisms that regulate gene expression and protein processing. In addition, as discussed later in the book, gene transfer has enabled the identification and characterization of genes that control animal cell growth and differentiation, including a variety of genes responsible for the abnormal growth of human cancer cells.

DNA is usually introduced into animal cells in culture as a coprecipitate with calcium phosphate (Figure 3.35). The methodology was initially developed for the introduction of infectious viral DNAs into animal cells and is therefore frequently called **transfection** (a word derived from *trans*formation + in*fection*). Cells exposed to such a DNA precipitate take up the DNA and transport it to the nucleus, where it can be transcribed for several days—a phenomenon called **transient expression**. In a smaller fraction of cells (0.1% or less), the foreign DNA becomes stably integrated into the cell genome and is transferred to progeny cells at cell division just as any other cell gene is. These stably transformed cells can be isolated if the transfected DNA contains a selectable marker, such as resistance to a drug that inhibits the growth of normal cells. Thus, any cloned gene can be introduced into mammalian cells by being transferred together with a drug resistance marker that can be used to isolate stable transformants. The effects of such cloned genes on cell behavior—for example, cell growth or differentiation—can then be analyzed.

**Figure 3.36**
**Retroviral vectors** The vector consists of retroviral sequences cloned in a plasmid that can be propagated in *E. coli*. Foreign DNA is inserted into the viral sequences, and recombinant plasmids are isolated in bacteria. Animal cells in culture are then transfected with the recombinant DNA. The DNA is taken up by a small fraction of the cells, which produce recombinant retroviruses carrying the inserted DNA. These recombinant retroviruses can be used to efficiently infect new cells, where the viral genome carrying the inserted gene integrates into chromosomal DNA as a provirus.

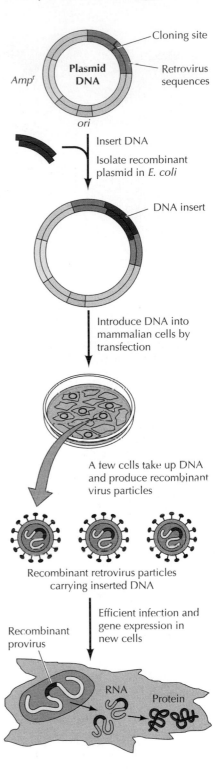

Various other methods, depending on the purpose of the experiment, can also be used to introduce DNA into mammalian cells. These methods include direct microinjection of DNA into the cell nucleus, incorporation of DNA into lipid vesicles (**liposomes**) that fuse with the plasma membrane, and exposure of cells to a brief electric pulse that transiently opens pores in the plasma membrane (**electroporation**). In addition, animal viruses can be used as vectors for more efficient introduction of cloned DNAs into cells. Retroviruses are particularly useful in this respect, since their life cycle involves the stable integration of viral DNA into the genome of infected cells (Figure 3.36). Consequently, retroviral vectors can be used to efficiently introduce cloned genes into a wide variety of cell types, making them an important vehicle for a broad range of applications, including gene therapy.

Cloned genes can also be introduced into the germ line of multicellular organisms, allowing them to be studied in the context of the intact animal rather than in cultured cells. Mice that carry such foreign genes (**transgenic mice**) are usually produced by microinjection of cloned DNA into the pronucleus of a fertilized egg (Figure 3.37). The injected eggs are then transferred to foster mothers and allowed to develop to term. In a fraction

**Figure 3.37**
**Production of transgenic mice** DNA is microinjected into a pronucleus of a fertilized mouse egg (fertilized eggs contain two pronuclei, one from the egg and one from the sperm). The microinjected eggs are then transferred to foster mothers and allowed to develop. Some of the offspring (transgenic) have incorporated the injected DNA into their genome.

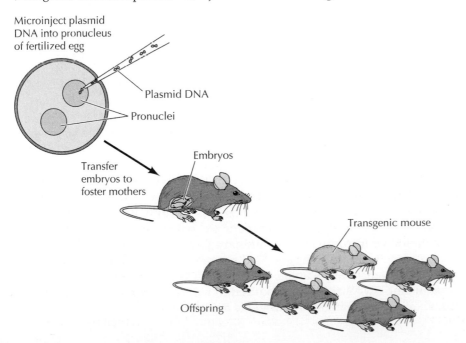

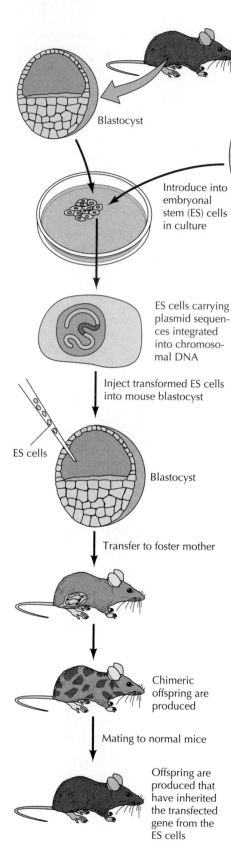

Blastocyst

Introduce into
embryonal
stem (ES) cells
in culture

Drug-resistance marker

Plasmid
DNA

Gene of interest

ES cells carrying
plasmid sequen-
ces integrated
into chromoso-
mal DNA

Inject transformed ES cells
into mouse blastocyst

ES cells

Blastocyst

Transfer to foster mother

Chimeric
offspring are
produced

Mating to normal mice

Offspring are
produced that
have inherited
the transfected
gene from the
ES cells

*Figure 3.38*
**Introduction of genes into mice via embryonal
stem cells**    Embryonal stem (ES) cells are cultured
cells derived from early mouse embryos (blasto-
cysts). DNA can be introduced into these cells in
culture, and stably transformed ES cells can be isolated. These transformed ES cells can then be injected into a recipient blastocyst, where they are able to
participate in normal development of the embryo.
Some of the progeny mice that develop after transfer
of injected embryos to foster mothers therefore contain cells derived from the transformed ES cells, as
well as from the normal cells of the blastocyst. Since
these mice are a mixture of two different cell types,
they are referred to as chimeric. Offspring carrying
the transfected gene can then be produced by the
breeding of chimeric mice in which descendants of
the transformed ES cells have been incorporated into
the germ line.

of the progeny mice (often about 10%), the foreign DNA will have integrated into the genome of the fertilized egg and is therefore present in all
cells of the animal. Since the foreign DNA is present in germ cells as well
as in somatic cells, it is transferred by breeding to new progeny mice just as
any other cell gene would be.

The properties of **embryonal stem (ES) cells** provide an alternative
means of introducing cloned genes into mice (Figure 3.38). ES cells can be
established in culture from early mouse embryos. They can also be reintro-
duced into early embryos, where they participate normally in development
and can give rise to cells in all tissues of the mouse—including germ cells.
It is thus possible to introduce cloned DNA into ES cells in culture, select
stably transformed cells, and then introduce these cells back into mouse
embryos. Such embryos give rise to chimeric offspring in which some cells
are derived from the normal embryo cells and some from the transfected
ES cells. In some such mice the transfected ES cells are incorporated into
the germ line. Breeding these mice therefore leads to the direct inheritance
of the transfected gene by their progeny.

Cloned DNAs can also be introduced into plant cells. One approach is to
remove the plant cell wall, forming protoplasts that are surrounded only by
a plasma membrane. DNA can then be introduced into such protoplasts by
electroporation, as was described for animal cells. Alternatively, DNA can
be introduced into intact plant cells by bombardment of the cells with
DNA-coated microprojectiles, such as small particles of tungsten. The DNA-
coated particles are shot directly into the plant cells; some of the cells are killed,
but others survive and become stably transformed.

Vectors for more efficient introduction of recombinant DNA into plant cells
have been developed from plant viruses. In addition, a plasmid from the bacterium *Agrobacterium tumefaciens* (the **Ti plasmid**) provides a novel vehicle
for the introduction of cloned DNAs into various plants (Figure 3.39). In
nature, *Agrobacterium* attaches to the leaves of plants and the Ti plasmid is
transferred into plant cells, where it becomes incorporated into chromosomal DNA. Vectors developed from the Ti plasmid therefore provide an efficient means of introducing recombinant DNA into sensitive plant cells.
Since many plants can be regenerated from single cultured cells (see Chapter 1), transgenic plants can be established directly from cells into which
recombinant DNA has been introduced in culture—a much simpler procedure than the production of transgenic animals.

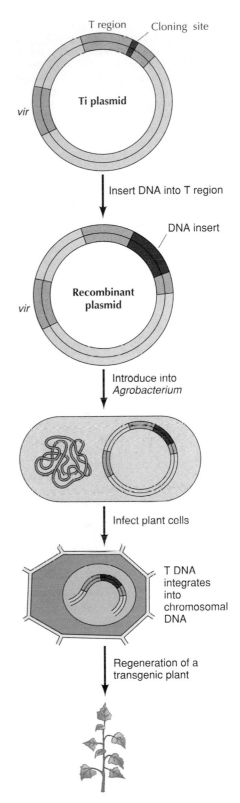

**Figure 3.39**
**Introduction of genes into plant cells via the Ti plasmid**   The Ti plasmid contains the T region, which is transferred to infected plant cells, and virulence (*vir*) genes, which function in T DNA transfer. In Ti plasmid vectors, foreign DNA is inserted into the T region. The recombinant plasmid is introduced into *Agrobacterium tumefaciens*, which is then used to infect cultured cells. The T region of the plasmid (carrying the inserted DNA) is transferred to the plant cells and integrates into chromosomal DNA. A transgenic plant can then be generated from the transformed cells.

## Mutagenesis of Cloned DNAs

In classical genetic studies (e.g., in bacteria or yeasts), mutants are the key to identifying genes and understanding their function by observing the altered phenotype of mutant organisms. In such studies, mutant genes are detected because they result in observable phenotypic changes—for example, temperature-sensitive growth or a specific nutritional requirement. The isolation of genes by recombinant DNA, however, has opened a different approach to mutagenesis. It is now possible to introduce any desired alteration into a cloned gene and to determine the effect of the mutation on gene function. Such procedures have been called **reverse genetics**, since a mutation is introduced into a gene first and its functional consequence is determined second. The ability to introduce specific mutations into cloned DNAs (*in vitro* **mutagenesis**) has proven to be a powerful tool in studying the expression and function of eukaryotic genes.

Cloned genes can be altered by many *in vitro* mutagenesis procedures, which can lead to the introduction of deletions, insertions, or single nucleotide alterations. Deletions, for example, are frequently generated by the digestion of a plasmid DNA with a restriction endonuclease that cleaves at a single site in the molecule (Figure 3.40). The linear DNA is then treated with an exonuclease, which digests DNA from the ends of the molecule. The digested DNA can be ligated again to yield a series of circular plasmids with deletions around the restriction site that was used initially to make the molecule linear. These deleted DNAs can be cloned and sequenced to determine the extent of the deletion in each of a series of individual plasmids. The functional consequence of the deletions can then be determined by introduction of the characterized plasmids into cultured cells.

Deletion mutagenesis enables a relatively crude assessment of the functional role of a region of a cloned gene. More detailed information can be provided by procedures that alter smaller regions of the DNA molecule. One example of such a procedure is the use of synthetic oligonucleotides to generate mutations of single bases in a DNA sequence (Figure 3.41). In this procedure a synthetic oligonucleotide bearing the mutant base is used as a primer for DNA synthesis. Newly synthesized DNA molecules containing the mutation can then be isolated and characterized. For example, specific amino acids of a protein can be altered in order to characterize their role in protein function.

These examples only begin to illustrate the wide variety of procedures for producing a desired alteration in a cloned gene. The effects of such mutations on gene expression and function can then be determined by introduction of the gene into an appropriate cell type. *In vitro* mutagenesis has thus allowed detailed characterization of the functional roles of both the regulatory and protein-coding sequences of cloned genes.

*Figure 3.40*
**Deletion mutagenesis of a cloned gene**    Plasmid DNA is made linear by digestion with a restriction endonuclease that cleaves at a site within the gene of interest. The linear DNA is then treated with an exonuclease, which digests DNA starting from the ends of each molecule. This process yields a collection of digested DNAs, in which different extents of the gene have been deleted. These molecules can then be recircularized by treatment with DNA ligase, yielding a set of plasmids with different deletions around the initial site of restriction endonuclease digestion.

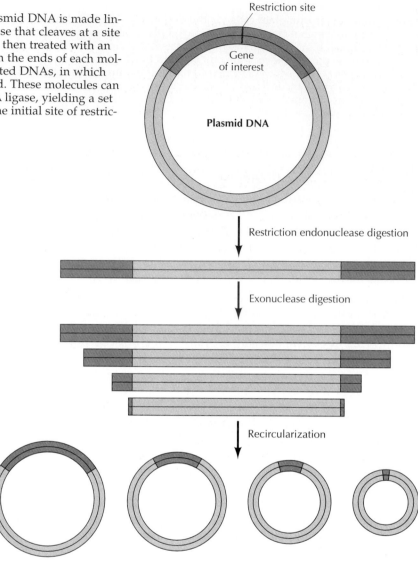

## Introducing Mutations into Cellular Genes

Although the transfer of cloned genes into cells (particularly in combination with *in vitro* mutagenesis) provides a powerful approach to studying gene structure and function, such experiments fall short of defining the role of an unknown gene in a cell or intact organism. The cells used as recipients for transfer of cloned genes usually already have normal copies of the gene in their chromosomal DNAs, and these normal copies continue to perform their roles in the cell. Determining the biological role of a cloned gene therefore requires that the activity of the normal cellular gene copies be eliminated. As discussed in the following section, this can be readily accomplished in yeasts. In mammalian cells, this task is considerably more difficult, although several approaches to either inactivating the chromosomal copies of a cloned gene or inhibiting normal gene function are now available.

Mutating the chromosomal genes in yeasts is relatively easy because DNA introduced into yeast cells frequently undergoes **homologous recombination** with its chromosomal copy (Figure 3.42). In homologous recombination, the cloned yeast gene replaces the normal allele, so mutations introduced

*Figure 3.41*
**Mutagenesis with synthetic oligonucleotides**    An oligonucleotide carrying the desired base alteration is used to prime DNA synthesis from plasmid DNA, and the DNA is circularized by incubation with DNA ligase. This process yields plasmids in which one strand contains the normal base and the other strand the mutant base. Replication of the plasmid DNAs after transformation of *E. coli* therefore yields a mixture of both starting and mutant plasmids.

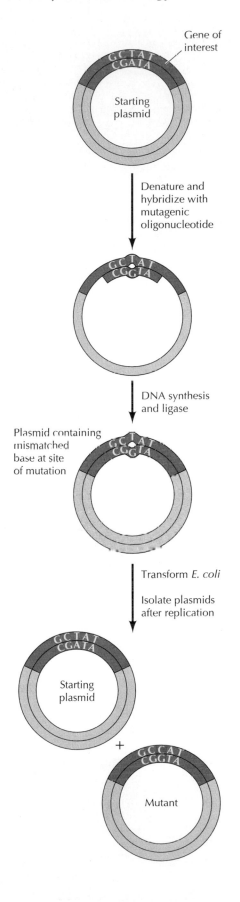

into the cloned gene *in vitro* become incorporated into the chromosomal copy of the yeast gene. In the simplest case, mutations that inactivate the cloned gene can be introduced in place of the normal gene copy in order to determine its role in cellular processes. Since yeasts have both haploid and diploid stages of their life cycle, even genes that are required for cell growth can be inactivated and studied. An inactive gene copy is introduced into diploid cells, which then contain one functional and one inactive copy of the target gene. The cells are induced to undergo meiosis, and the effect of gene inactivation on the progeny haploid cells can be observed.

Unfortunately, recombination between transferred DNA and the homologous chromosomal gene is a rare event in mammalian cells, so gene inactivation by this approach is much more difficult than it is in yeasts. Possibly because the genomes of mammalian cells are so much larger than that of

*Figure 3.42*
**Gene inactivation by homologous recombination**    A mutated copy of the cloned gene is introduced into cells. The cloned gene may then replace the normal gene copy by homologous recombination, yielding a cell carrying the desired mutation in its chromosomal DNA.

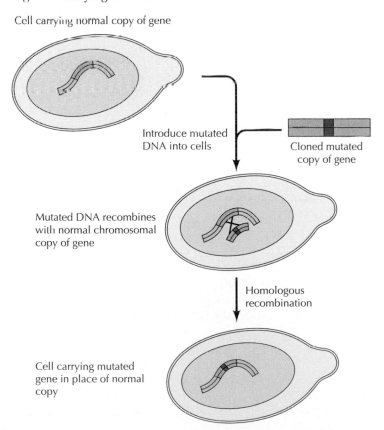

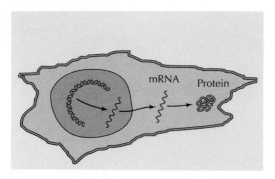

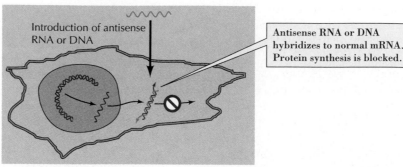

Antisense RNA or DNA hybridizes to normal mRNA. Protein synthesis is blocked.

Introduction of antisense RNA or DNA

*Figure 3.43*
**Inhibition of gene expression by antisense RNA or DNA**    Antisense RNA or single-stranded DNA is complementary to the mRNA of a gene of interest. Antisense nucleic acids therefore form hybrids with their target mRNAs, blocking the translation of mRNA into protein.

yeasts, most transfected DNA that integrates into the recipient cell genome does so at random sites by recombination with unrelated sequences. However, various procedures have been developed to isolate the transformed cells in which homologous recombination has occurred, so genes in mammalian cells can be inactivated by this approach. A particularly important application is the inactivation or disruption of genes in embryonal stem cells in culture. As already described, ES cells can be used to generate transgenic mice, so the effects of inactivation of a gene can be investigated in the context of the intact animal. Such studies have been particularly revealing for analysis of the function of genes in mouse development.

An alternative to gene inactivation by homologous recombination is the use of **antisense nucleic acids** to block gene expression (Figure 3.43). In this approach, RNA or single-stranded DNA complementary to the mRNA of the gene of interest (antisense) is introduced into a cell. The antisense RNA or DNA hybridizes with the mRNA and blocks its translation into protein. Moreover, the RNA-DNA hybrids resulting from the introduction of antisense DNA molecules are usually degraded within the cell. Anti-

*Figure 3.44*
**Direct inhibition of protein function**    Microinjected antibodies can bind to proteins within cells, thereby inhibiting their normal function. In addition, some mutant proteins interfere with the function of normal proteins—for example, by competing with the normal protein for interaction with target molecules.

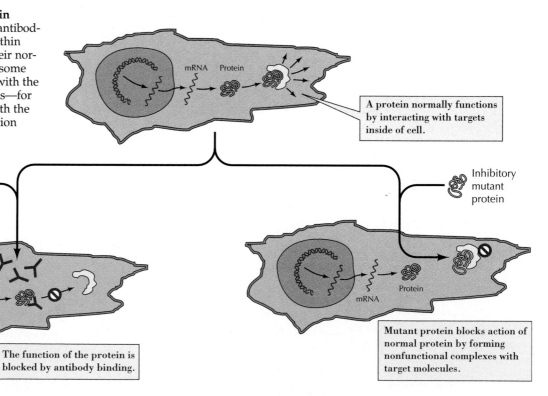

A protein normally functions by interacting with targets inside of cell.

Antibody against protein of interest

Inhibitory mutant protein

The function of the protein is blocked by antibody binding.

Mutant protein blocks action of normal protein by forming nonfunctional complexes with target molecules.

sense RNAs can be introduced into cells directly by microinjection. Alternatively, cells can be transfected with vectors that have been engineered to express antisense RNA. Antisense DNA is usually in the form of short oligonucleotides (about 20 bases long) that are microinjected into cells. Alternatively, because cells are able to take up such oligonucleotides from the culture medium, antisense oligonucleotides can simply be added to the cell culture.

In addition to inactivating a gene or blocking mRNA translation, it is sometimes possible to interfere directly with the function of proteins within cells (Figure 3.44). One approach is to microinject antibodies that block the activity of the protein against which they are directed. Alternatively, some mutant proteins interfere with the function of their normal counterparts when they are expressed within the same cell—for example, by competing with the normal protein for binding to its target molecule. Cloned DNAs encoding such mutant proteins (called **dominant inhibitory mutants**) can be introduced into cells by gene transfer and used to study the effects of blocking normal gene function.

## Summary

| KEY TERMS |
|---|

### HEREDITY, GENES, AND DNA

*Genes and Chromosomes*: Chromosomes are the carriers of genes.

gene, allele, dominant, recessive, genotype, phenotype, chromosome, diploid, meiosis, haploid, mutation, recombination, genetic map

*Genes and Enzymes*: A gene specifies the amino acid sequence of a polypeptide chain.

one gene–one enzyme hypothesis

*Identification of DNA as the Genetic Material*: DNA was identified as the genetic material by bacterial transformation experiments.

transformation

*The Structure of DNA*: DNA is a double helix in which hydrogen bonds form between purines and pyrimidines on opposite strands. Because of specific base pairing—A with T and G with C—the two strands of a DNA molecule are complementary in sequence.

*Replication of DNA*: DNA replicates by semiconservative replication, in which the two strands separate and each serves as a template for synthesis of a new progeny strand.

semiconservative replication, DNA polymerase

### EXPRESSION OF GENETIC INFORMATION

*Colinearity of Genes and Proteins*: The order of nucleotides in DNA specifies the order of amino acids in proteins.

*The Role of Messenger RNA*: Messenger RNA functions as an intermediate to convey information from DNA to the ribosomes, where it serves as a template for protein synthesis.

central dogma, transcription, translation, messenger RNA (mRNA), RNA polymerase, ribosomal RNA (rRNA), transfer RNA (tRNA)

*The Genetic Code*: Transfer RNAs serve as adaptors between amino acids and mRNA during translation. Each amino acid is specified by a codon consisting of three nucleotides.

genetic code, *in vitro* translation, codon

*RNA Viruses and Reverse Transcription*: DNA can be synthesized from RNA templates, as first discovered in retroviruses.

retrovirus, reverse transcription, reverse transcriptase

restriction endonuclease, gel electrophoresis, restriction map

molecular cloning, vector, recombinant molecule, molecular clone, DNA ligase, cDNA

plasmid, origin of replication, cosmid, yeast artificial chromosome (YAC)

Maxam-Gilbert sequencing, autoradiography, Sanger sequencing, dideoxynucleotide

expression vector, baculovirus

polymerase chain reaction (PCR)

nucleic acid hybridization, probe, Southern blotting, Northern blotting, *in situ* hybridization

antibody, antigen, monoclonal antibody, immunoblotting, Western blotting, immunoprecipitation, SDS-polyacrylamide gel electrophoresis (SDS-PAGE)

recombinant DNA library, genomic library, cDNA library

temperature-sensitive mutant

gene transfer, transfection, transient expression, liposome, electroporation, transgenic mouse, embryonal stem (ES) cell, Ti plasmid

reverse genetics, *in vitro* mutagenesis

homologous recombination, antisense nucleic acids, dominant inhibitory mutant

# RECOMBINANT DNA

*Restriction Endonucleases*: Restriction endonucleases cleave specific DNA sequences, yielding defined fragments of DNA molecules.

*Generation of Recombinant DNA Molecules*: Recombinant DNA molecules consist of a DNA fragment of interest ligated to a vector that is able to replicate independently in an appropriate host cell.

*Vectors for Recombinant DNA*: A variety of vectors are used to clone different sizes of DNA fragments.

*DNA Sequencing*: The nucleotide sequences of cloned DNA fragments can be readily determined.

*Expression of Cloned Genes*: The proteins encoded by cloned genes can be expressed at high levels in either bacteria or eukaryotic cells.

*Amplification of DNA by the Polymerase Chain Reaction*: PCR allows the amplification and isolation of specific fragments of DNA *in vitro*.

# DETECTION OF NUCLEIC ACIDS AND PROTEINS

*Nucleic Acid Hybridization*: Nucleic acid hybridization allows the detection of specific DNA or RNA sequences.

*Detection of Small Amounts of DNA or RNA by PCR*: PCR provides a sensitive method for detecting small amounts of specific DNA or RNA molecules.

*Antibodies as Probes for Proteins*: Antibodies are used to detect specific proteins in cells or cell extracts.

*Probes for Screening Recombinant DNA Libraries*: Specific DNA inserts can be detected in recombinant DNA libraries by the use of either nucleic acid hybridization or antibody probes.

# GENE FUNCTION IN EUKARYOTES

*Genetic Analysis in Yeasts*: The simple genetics and rapid replication of yeasts facilitate the molecular cloning of a gene corresponding to any yeast mutation.

*Gene Transfer in Plants and Animals*: Cloned genes can be introduced into complex eukaryotic cells and multicellular organisms for functional analysis.

*Mutagenesis of Cloned DNAs*: *In vitro* mutagenesis of cloned DNAs is used to study the effect of engineered mutations on gene function.

*Introducing Mutations into Cellular Genes*: Mutations can be introduced into chromosomal gene copies by homologous recombination with cloned DNA sequences. In addition, the expression or function of specific gene products can be blocked by antisense nucleic acids or dominant inhibitory mutants.

# QUESTIONS

**1.** What is the genetic map of a locus containing genes *x*, *y*, and *z*? The frequencies of recombination are 0.5% between *x* and *z*, 0.2% between *x* and *y*, and 0.7% between *y* and *z*.

*z ⟵ .5 ⟶ x ⟵ .7 ⟶ y* [handwritten]

**2.** You are studying an enzyme in which an active-site cysteine residue is encoded by the triplet UGU. How would mutating the third base to a C affect enzyme function? How about mutating it to an A?

*no effect* [handwritten]

*UGA stop codon* [handwritten]

**3.** Digestion of a 4-kb DNA molecule with *Eco*RI yields two fragments of 1 kb and 3 kb each. Digestion of the same molecule with *Hin*dIII yields fragments of 1.5 kb and 2.5 kb. Finally, digestion with *Eco*RI and *Hin*dIII in combination yields fragments of 0.5 kb, 1 kb, and 2.5 kb. Draw a restriction map indicating the positions of the *Eco*RI and *Hin*dIII cleavage sites.

**4.** Starting with DNA from a single sperm, how many copies of a specific gene sequence will be obtained after 10 cycles of PCR amplification? After 30 cycles?

**5.** You have cloned a cDNA of unknown function. How could you experimentally determine the subcellular localization of the protein it encodes?

**6.** You are interested in identifying the amino acid residues that are important for the catalytic activity of an enzyme. Assuming you have a cDNA clone available, what experimental strategies could you use?

# REFERENCES AND FURTHER READING

## General References

Freifelder, D. 1987. *Molecular Biology*. 2nd ed. Boston: Jones and Bartlett.

Watson, J. D., M. Gilman, J. Witkowski, and M. Zoller. 1992. *Recombinant DNA*. 2nd ed. New York: W. H. Freeman.

Watson, J. D., N. H. Hopkins, J. W. Roberts, J. A. Steitz, and A. M. Weiner. 1987. *Molecular Biology of the Gene*. 4th ed. Menlo Park, Calif: Benjamin Cummings.

## Heredity, Genes, and DNA

Avery, O. T., C. M. MacLeod, and M. McCarty. 1944. Studies on the chemical nature of the substance inducing transformation of pneumococcal types. *J. Exp. Med.* 79: 137–158. [P]

Franklin, R. E., and R. G. Gosling. 1953. Molecular configuration in sodium thymonucleate. *Nature* 171: 740–741. [P]

Kornberg, A. 1960. Biologic synthesis of deoxyribonucleic acid. *Science* 131: 1503–1508. [P]

Meselson, M. and F. W. Stahl. 1958. The replication of DNA in *Escherichia coli*. *Proc. Natl. Acad. Sci.* USA 44: 671–682. [P]

Watson, J.D. and F. H. C. Crick. 1953. Genetical implications of the structure of deoxyribonucleic acid. *Nature* 171: 964–967. [P]

Watson, J.D. and F. H. C. Crick. 1953. Molecular structure of nucleic acids: A structure for deoxyribose nucleic acid. *Nature* 171: 737–738. [P]

Wilkins, M. H. F., A. R. Stokes and H. R. Wilson. 1953. Molecular structure of deoxypentose nucleic acids. *Nature* 171: 738–740. [P]

## Expression of Genetic Information

Baltimore, D. 1970. RNA-dependent DNA polymerase in virions of RNA tumour viruses. *Nature* 226: 1209–1211. [P]

Brenner, S., F. Jacob and M. Meselson. 1961. An unstable intermediate carrying information from genes to ribosomes for protein synthesis. *Nature* 190: 576–581. [P]

Crick, F. H. C., L. Barnett, S. Brenner, and R. J. Watts-Tobin. 1961. General nature of the genetic code for proteins. *Nature* 192: 1227–1232. [P]

Ingram, V. M. 1957. Gene mutations in human hemoglobin: The chemical difference between normal and sickle cell hemoglobin. *Nature* 180: 326–328. [P]

Nirenberg, M. and P. Leder. 1964. RNA codewords and protein synthesis. *Science* 145: 1399–1407. [P]

Nirenberg, M. W. and J. H. Matthaei. 1961. The dependence of cell-free protein synthesis in *E. coli* upon naturally occurring or synthetic polyribonucleotides. *Proc. Natl. Acad. Sci.* USA 47: 1588–1602. [P]

Temin, H. M. and S. Mizutani. 1970. RNA-dependent DNA polymerase in virions of Rous sarcoma virus. *Nature* 226: 1211–1213. [P]

Yanofsky, C., B. C. Carlton, J. R. Guest, D. R. Helinski and U. Henning. 1964. On the colinearity of gene structure and protein structure. *Proc. Natl. Acad. Sci.* USA 51: 266–272. [P]

## Recombinant DNA

Arnheim, N. and H. Erlich. 1992. Polymerase chain reaction strategy. *Ann. Rev. Biochem.* 61: 131–156. [R]

Ausubel, F. M., R. Brent, R. E. Kingston, D. D. Moore, J. G. Seidman, J. A. Smith, and K. Struhl. eds. 1989. *Current Protocols in Molecular Biology*. New York: Greene Publishing and Wiley Interscience. [R]

Burke, D. T., G. F. Carle and M. V. Olson. 1987. Cloning of large segments of exogenous DNA into yeast by means of artificial chromosome vectors. *Science* 236: 806–812. [P]

Cohen, S. N., A. C. Y. Chang, H. W. Boyer and R. B. Helling. 1973. Construction of biologically functional bacterial plasmids in vitro. *Proc. Natl. Acad. Sci.* USA 70: 3240–3244. [P]

Glick, B. R. and J. J. Pasternak. 1994. *Molecular Biotechnology. Principles and Applications of Recombinant DNA*. Washington, D. C.: ASM Press.

Hames, B. D. and D. Glover. eds. 1995. *DNA Cloning. A Practical Approach*. Oxford, England: IRL Press.

Maxam, A. M. and W. Gilbert. 1977. A new method for sequencing DNA. *Proc. Natl. Acad. Sci.* USA 74: 560–564. [P]

McPherson, M. J., B. D. Hames, and G. Taylor. eds. 1995. *PCR. A Practical Approach*. Oxford, England: IRL Press.

Nathans, D. and H. O. Smith. 1975. Restriction endonucleases in the analysis and restructuring of DNA molecules. *Ann. Rev. Biochem.* 44: 273–293. [R]

Saiki, R. K., D. H. Gelfand, S. Stoffel, S. J. Scharf, R. Higuchi, G. T. Horn, K. B. Mullis and H. A. Erlich. 1988. Primer-directed enzymatic amplification of DNA with a thermostable DNA polymerase. *Science* 239: 487–491. [P]

Sambrook, J., E. F. Fritsch and T. Maniatis. 1989. *Molecular Cloning: A Laboratory Manual*, 2d ed. Plainview, N.Y.: Cold Spring Harbor Laboratory Press.

Sanger, F., S. Nicklen and A. R. Coulson. 1977. DNA sequencing with chain-terminating inhibitors. *Proc. Natl. Acad. Sci. USA* 74: 5463–5467. [P]

## Detection of Nucleic Acids and Proteins

Ausubel, F. M., R. Brent, R. E. Kingston, D. D. Moore, J. G. Seidman, J. A. Smith, and K. Struhl. eds. 1989. *Current Protocols in Molecular Biology.* New York: Greene Publishing and Wiley Interscience.

Benton, W. D. and R. W. Davis. 1977. Screening λgt recombinant clones by hybridization to single plaques *in situ*. *Science* 196: 180–182. [P]

Broome, S. and W. Gilbert. 1978. Immunological screening method to detect specific translation products. *Proc. Natl. Acad. Sci. USA* 75: 2746–2749. [P]

Caruthers, M. H. 1985. Gene synthesis machines: DNA chemistry and its uses. *Science* 230: 281–285. [R]

Grunstein, M. and D. S. Hogness, 1975. Colony hybridization: A method for the isolation of cloned DNAs that contain a specific gene. *Proc. Natl. Acad. Sci. USA* 72: 3961–3965. [P]

Harlow, E. and D. Lane. 1988. *Antibodies: A Laboratory Manual.* Plainview, N.Y.: Cold Spring Harbor Laboratory Press.

Kohler, G. and C. Milstein. 1975. Continuous cultures of fused cells secreting antibody of predefined specificity. *Nature* 256: 495–497. [P]

Laemmli, U. K. 1970. Cleavage of structural proteins during the assembly of the head of bacteriophage T4. *Nature* 227: 680–685. [P]

Maniatis, T., R. C. Hardison, E. Lacy, J. Lauer, C. O'Connell, D. Quon, G. K. Sim and A. Efstratiadis. 1978. The isolation of structural genes from libraries of eucaryotic DNA. *Cell* 15: 687–701. [P]

Sambrook, J., E. F. Fritsch and T. Maniatis. 1989. *Molecular Cloning: A Laboratory Manual*, 2nd ed. Plainview, N.Y.: Cold Spring Harbor Laboratory Press.

Schildkraut, C. L., J. Marmur and P. Doty. 1961. The formation of hybrid DNA molecules, and their use in studies of DNA homologies. *J. Mol. Biol.* 3: 595–617. [P]

Southern, E. M. 1975. Detection of specific sequences among DNA fragments separated by gel electrophoresis. *J. Mol. Biol.* 98: 503–517. [P]

## Gene Function in Eukaryotes

Botstein, D. and D. Shortle. 1985. Strategies and applications of *in vitro* mutagenesis. *Science* 229: 1193–1201. [R]

Bronson, S. K. and O. Smithies. 1994. Altering mice by homologous recombination using embryonic stem cells. *J. Biol. Chem.* 269: 27155–27158. [R]

Capecchi, M. R. 1989. Altering the genome by homologous recombination. *Science* 244: 1288–1292. [R]

Gasser, C. S. and R. T. Fraley. 1989. Genetically engineering plants for crop improvement. *Science* 244: 1293–1299. [R]

Gordon, J. W., G. A. Scangos, D. J. Plotkin, J. A. Barbosa, and F. H. Ruddle. 1980. Genetic transformation of mouse embryos by microinjection of purified DNA. *Proc. Natl. Acad. Sci. USA* 77: 7380–7384. [P]

Herskowitz, I. 1987. Functional inactivation of genes by dominant negative mutations. *Nature* 329: 219–222. [R]

Horsch, R. B., J. E. Fry, N. L. Hoffmann, D. Eichholtz, S. G. Rogers and R. T. Fraley. 1985. A simple and general method for transferring genes into plants. *Science* 227: 1229–1231. [P]

Izant, J. G. and H. Weintraub. 1984. Inhibition of thymidine kinase gene expression by antisense RNA: A molecular approach to genetic analysis. *Cell* 36: 1007–1015. [P]

Jaenisch, R. 1988. Transgenic animals. *Science* 240: 1468–1474. [R]

Joyner, A. L., ed. 1993. *Gene Targeting. A Practical Approach.* Oxford, England: IRL Press.

Kaiser, C., S. Michaelis and A. Mitchell. 1994. *Methods in Yeast Genetics: A Laboratory Course Manual.* Plainview, N.Y.: Cold Spring Harbor Laboratory Press.

Kuhn, R., F. Schwenk, M. Aguet and K. Rajewsky. 1995. Inducible gene targeting in mice. *Science* 269: 1427–1429. [P]

Maliga, P., D. F. Klessig, A. R. Cashmore, W. Gruissem and J. E. Varner. eds. 1994. *Methods in Plant Molecular Biology: A Laboratory Course Manual.* Plainview, N.Y.: Cold Spring Harbor Laboratory Press.

Palmiter, R. D. and R. L. Brinster. 1986. Germ-line transformation of mice. *Ann. Rev. Genet.* 20: 465–499. [R]

Robertson, E., A. Bradley, M. Kuehn and M. Evans. 1986. Germline transmission of genes introduced into cultured pluripotential cells by retroviral vector. *Nature* 323: 445–448. [P]

Smith, M. 1985. *In vitro* mutagenesis. *Ann. Rev. Genet.* 19: 423–462. [R]

Struhl, K. 1983. The new yeast genetics. *Nature* 305: 391–397. [R]

Thomas, K. R. and M. R. Capecchi, 1987. Site-directed mutagenesis by gene targeting in mouse embryo-derived stem cells. *Cell* 51: 503–512. [P]

Wagner, R. W. 1994. Gene inhibition using antisense oligonucleotides. *Nature* 372: 333–335. [R]

Wigler, M., R. Sweet, G. K. Sim, B. Wold, A. Pellicer, E. Lacy, T. Maniatis, S. Silverstein and R. Axel. 1979. Transformation of mammalian cells with genes from prokaryotes and eukaryotes. *Cell* 16: 777–785. [P]

*Water jar, Zuni, 1880*

# PART II The Flow of Genetic Information

# 4

# The Organization of Cellular Genomes

A S THE GENETIC MATERIAL, DNA PROVIDES A BLUEPRINT that directs all cellular activities and specifies the developmental plan of multicellular organisms. An understanding of gene structure and function is therefore fundamental to an appreciation of the molecular biology of cells. The development of gene cloning represented a major step toward this goal, enabling scientists to dissect complex eukaryotic genomes and probe the functions of eukaryotic genes. Continuing advances in recombinant DNA technology have now brought us to the exciting point of determining the sequences of entire genomes, providing a new approach to deciphering the genetic basis of cell behavior.

As reviewed in Chapter 3, the initial applications of recombinant DNA were directed toward the isolation and analysis of individual genes. Within the last few years, the more global approach of sequencing entire genomes has also become feasible—not only for bacteria and yeasts, but also for the larger genomes of complex plants and animals, including humans. Several genome sequencing projects are currently under way, and the first fruits of these efforts are already being realized. There is little doubt that the continued sequencing of cellular genomes will provide a rich harvest of information, including the identification of many hitherto unknown genes. The results of these genome sequencing projects can be expected to stimulate many years of future research in molecular and cellular biology, and to have a profound impact on our understanding of many inherited human diseases.

## THE COMPLEXITY OF EUKARYOTIC GENOMES

The genomes of most eukaryotes are larger and more complex than those of prokaryotes (Figure 4.1). This larger size of eukaryotic genomes is not inherently surprising, since one would expect to find more genes in organisms that are more complex. However, the genome size of many eukaryotes does not appear to be related to genetic complexity. For example, the genomes of salamanders and lilies contain more than ten times the amount of DNA that is in the human genome, yet these organisms are clearly not ten times more complex than humans.

**135**

*Figure 4.1*
**Genome size**   The range of sizes of the genomes of representative groups of organisms are shown on a logarithmic scale.

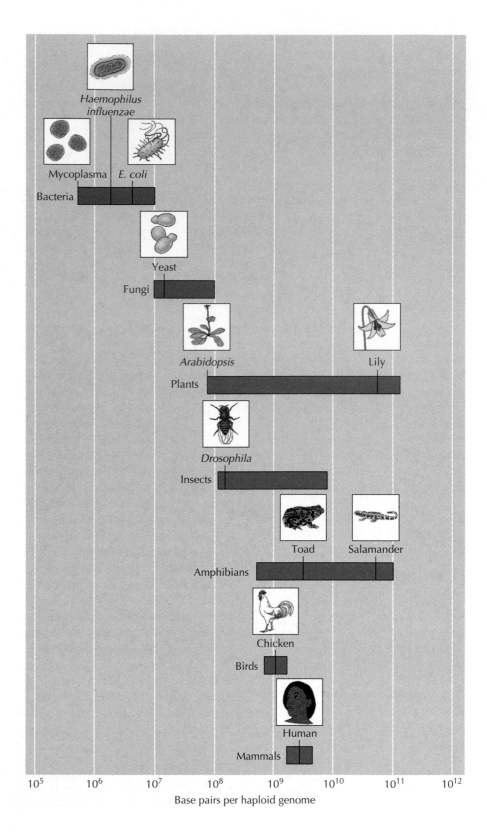

This apparent paradox was resolved by the discovery that the genomes of most eukaryotic cells contain not only functional genes but also large amounts of DNA sequences that do not code for proteins. The difference in the sizes of the salamander and human genomes thus reflects larger amounts of noncoding DNA, rather than more genes, in the genome of the salamander. The presence of large amounts of noncoding sequences is a general property of the genomes of complex eukaryotes. Thus, the thousandfold greater size of the human genome compared to that of *E. coli* is not due solely to a larger number of human genes. The human genome is thought to contain approximately 100,000 genes—only about 25 times more than *E. coli* has. Much of the complexity of eukaryotic genomes thus results from the abundance of several different types of noncoding sequences, which constitute most of the DNA of higher eukaryotic cells.

### Introns and Exons

In molecular terms, a **gene** can be defined as a segment of DNA that is expressed to yield a functional product, which may be either an RNA (e.g., ribosomal and transfer RNAs) or a polypeptide. Some of the noncoding DNA in eukaryotes is accounted for by long DNA sequences that lie between genes (**spacer sequences**). However, large amounts of noncoding DNA are also found within most eukaryotic genes. Such genes have a split structure in which segments of coding sequence (called **exons**) are separated by noncoding sequences (intervening sequences, or **introns**) (Figure 4.2). The entire gene is transcribed to yield a long RNA molecule and the introns are then removed by splicing, so only exons are included in the mRNA. Although most introns have no known function, they account for a substantial fraction of DNA in the genomes of higher eukaryotes.

Introns were first discovered in 1977, independently in the laboratories of Phillip Sharp and Richard Roberts, during studies of the replication of adenovirus in cultured human cells. Adenovirus is a useful model for studies of gene expression, both because the viral genome is only about $3.5 \times 10^4$ base pairs long and because adenovirus mRNAs are produced at high levels in infected cells. One approach used to characterize the adenovirus mRNAs was to determine the locations of the corresponding viral genes by examination of RNA-DNA hybrids in the electron microscope. Because RNA-DNA hybrids are distinguishable from single-stranded DNA, the positions of RNA transcripts on a DNA molecule can be determined. Surprisingly, such experiments revealed that adenovirus mRNAs do not hybridize to only a single region of viral DNA (Figure 4.3). Instead, a single mRNA molecule hybridizes to several separated regions of the viral genome. Thus, the adenovirus mRNA does not correspond to an uninterrupted transcript of the template DNA; rather the mRNA is assembled from several distinct blocks of sequence that originated from different parts of the viral DNA. This was subsequently shown to occur by **RNA splicing**, which will be discussed in detail in Chapter 6.

Soon after the discovery of introns in adenovirus, similar observations were made on cloned genes of eukaryotic cells. For example, electron

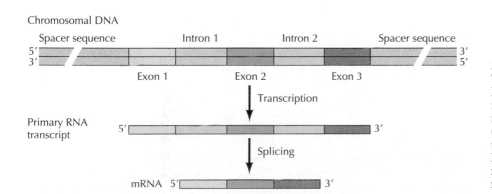

*Figure 4.2*
**The structure of eukaryotic genes**
Most eukaryotic genes contain segments of coding sequences (exons) interrupted by noncoding sequences (introns). Both exons and introns are transcribed to yield a long primary RNA transcript. The introns are then removed by splicing to form the mature mRNA.

# KEY EXPERIMENT

## *The Discovery of Introns*

### Spliced Segments at the 5′ Terminus of Adenovirus 2 Late mRNA

Susan M. Berget, Claire Moore, and Phillip A. Sharp
Massachusetts Institute of Technology, Cambridge, Massachusetts
*Proceedings of the National Academy of Sciences USA, Volume 74, 1977, pages 3171–3175*

### The Context

Prior to molecular cloning, little was known about mRNA synthesis in eukaryotic cells. However, it was clear that this process is more complex in eukaryotes than in bacteria. The synthesis of eukaryotic mRNAs appeared to require not only transcription, but also processing reactions that modify the structure of primary transcripts. Most notably, eukaryotic mRNAs appeared to be synthesized as long primary transcripts, found in the nucleus, which were then cleaved to yield much shorter mRNA molecules that were exported to the cytoplasm.

These processing steps were generally assumed to involve the removal of sequences from the 5′ and 3′ ends of the primary transcripts. In this model, mRNAs embedded within long primary transcripts would be encoded by uninterrupted DNA sequences. This view of eukaryotic mRNA was changed radically by the discovery of splicing, made independently by Berget, Moore, and Sharp, and by Louise Chow, Richard Gelinas, Tom Broker, and Richard Roberts (An amazing sequence arrangement at the 5′ ends of adenovirus 2 messenger RNA, *Cell* 12:1-8, 1977).

### The Experiments

Both of the research groups that discovered splicing used adenovirus 2 to investigate mRNA synthesis in human cells. The major advantage of the virus is that it provides a model that is much simpler than the host

cell. Viral DNA can be isolated directly from virus particles, and mRNAs encoding the viral structural proteins are present in such high amounts that they can be purified directly from infected cells. Berget, Moore, and Sharp focused their experiments on an abundant mRNA that encodes a viral structural polypeptide known as the hexon.

To map the hexon mRNA on the viral genome, purified mRNA was hybridized to adenovirus DNA and the hybrid molecules were examined by electron microscopy. As expected, the body of the hexon mRNA formed hybrids with restriction fragments of adenovirus DNA that had previously been shown to contain the hexon gene. Surprisingly, however, sequences at the 5′ end of hexon

mRNA failed to hybridize to DNA sequences adjacent to those encoding the body of the message, suggesting that the 5′ end of the mRNA had arisen from sequences located elsewhere in the viral genome.

This possibility was tested by hybridization of hexon mRNA to a restriction fragment extending upstream of the hexon gene. The mRNA-DNA hybrids formed in this experiment displayed a complex loop structure (see figure). The body of the mRNA formed a long hybrid region with the previously identified hexon DNA sequences. Strikingly, the 5′ end of the hexon mRNA hybridized to three short upstream regions of DNA, which were separated from each other and from the body of the message by large single-stranded DNA loops. The sequences at the 5′ end of hexon mRNA thus appeared to be transcribed from three separate regions of the viral genome, which were spliced to the body of the mRNA during the processing of a long primary transcript.

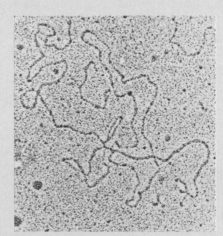 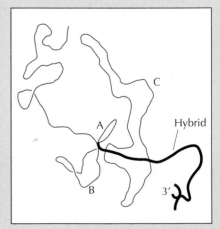

An electron micrograph and tracing of hexon mRNA hybridized to adenovirus DNA. The single-stranded loops designated A, B, and C correspond to introns.

## The Discovery of Introns (continued)

### The Impact

The discovery of splicing in adenovirus mRNA was quickly followed by similar experiments with cellular mRNAs, demonstrating that eukaryotic genes had a previously unexpected structure. Rather than being continuous, their coding sequences were interrupted by introns, which were removed from primary transcripts by splicing. Introns are now known to account for much of the DNA in eukaryotic genomes, and the roles of introns in the evolution and regulation of gene expression continue to be active areas of investigation. The discovery of splicing also stimulated intense interest in the mechanism of this unexpected RNA processing reaction. As discussed in Chapter 6, these studies have not only illuminated new mechanisms of regulating gene expression; they have also revealed novel catalytic activities of RNA and provided critical evidence supporting the hypothesis that early evolution was based on self-replicating RNA molecules. The unexpected structure of adenovirus mRNAs has thus had a major impact on diverse areas of cellular and molecular biology.

Phillip Sharp

Richard Roberts

(A)

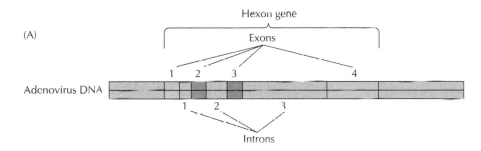

(B)

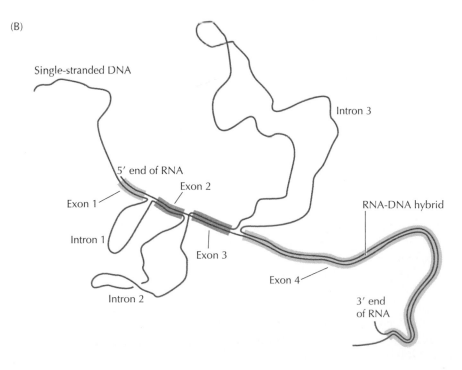

*Figure 4.3*
**Identification of introns in adenovirus mRNA**   (A) The gene encoding the adenovirus hexon (a major structural protein of the viral particle) consists of four exons, interrupted by three introns. (B) This tracing illustrates an electron micrograph of a hypothetical hybrid between hexon mRNA and a portion of adenovirus DNA. The exons are seen as regions of RNA-DNA hybrid, which are separated by single-stranded DNA loops corresponding to the introns.

*Figure 4.4*
**The mouse β-globin gene**   This gene contains two introns, which divide the coding region among three exons. Exon 1 encodes amino acids 1 to 30, exon 2 encodes amino acids 31 to 104, and exon 3 encodes amino acids 105 to 146. Exons 1 and 3 also contain untranslated regions (UTRs) at the 5′ and 3′ ends of the mRNA, respectively.

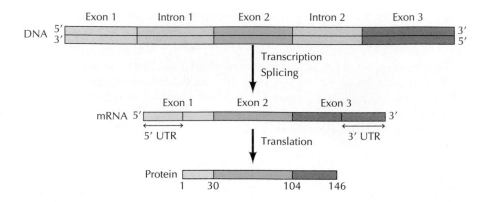

microscopic analysis of RNA-DNA hybrids and subsequent nucleotide sequencing of cloned genomic DNAs and cDNAs indicated that the coding region of the mouse β-globin gene (which encodes the β subunit of hemoglobin) is interrupted by two introns that are removed from the mRNA by splicing (Figure 4.4). The intron-exon structure of many eukaryotic genes is quite complicated, and the amount of DNA in the intron sequences is often greater than that in the exons. The chicken ovalbumin gene, for example, contains eight exons and seven introns distributed over approximately 7700 base pairs (7.7 **kilobases**, or **kb**) of genomic DNA. The exons total only about 1.9 kb, so approximately 75% of the gene consists of introns. An extreme example is the human gene that encodes the blood clotting protein factor VIII. This gene spans approximately 186 kb of DNA and is divided into 26 exons. The mRNA is only about 9 kb long, so the gene contains introns totaling more than 175 kb. On average, introns are estimated to account for about ten times more DNA than exons in the genes of higher eukaryotes.

Introns are present in most genes of complex eukaryotes, although they are not universal. Almost all histone genes, for example, lack introns, so introns are clearly not required for gene function in eukaryotic cells. In addition, introns are not found in most genes of simple eukaryotes, such as yeasts. Conversely, introns *are* present in rare genes of prokaryotes. The presence or absence of introns is therefore not an absolute distinction between prokaryotic and eukaryotic genes, although introns are much more prevalent in higher eukaryotes (both plants and animals), where they account for a substantial amount of total genomic DNA.

Most introns have no known cellular function, although a few have been found to encode functional RNAs or proteins. Introns are generally thought to represent remnants of sequences that were important earlier in evolution. In particular, introns may have helped accelerate evolution by facilitating recombination between protein-coding regions (exons) of different genes—a process known as exon shuffling. Exons frequently encode functionally distinct protein domains, so recombination between introns of different genes would result in new genes containing novel combinations of protein-coding sequences. As predicted by this hypothesis, DNA sequencing studies have demonstrated that some genes are chimeras of exons derived from several other genes, providing direct evidence that new genes can be formed by recombination between intron sequences.

Although exon shuffling provides a satisfactory explanation for the role of introns in evolution, the origin of introns remains an unanswered question. One theory (the "introns-early" view) proposes that introns were present early in evolution, prior to the divergence of prokaryotic and eukaryotic cells. According to this hypothesis, introns played an important role in the initial assembly of protein-coding sequences in the ancient ancestors of present-day cells. Introns were subsequently lost from most genes of prokaryotes and simpler eukaryotes (e.g., yeasts) in response to evolutionary selection for rapid replication, which led to streamlining the genomes of these organisms. However, since rapid cell division is not an advantage to higher eukaryotes, introns have been retained in their genomes. The alternative view ("introns-late") suggests that introns arose later in evolution as a result of the insertion of DNA sequences into genes that had already been formed as continuous protein-coding sequences. Exon shuffling would then have played an important role in the further evolution of genes in higher eukaryotes but would not account for the initial assembly of protein-coding sequences prior to the evolutionary divergence of prokaryotic and eukaryotic cells.

## Gene Families and Pseudogenes

Another factor contributing to the large size of eukaryotic genomes is that some genes are repeated many times. Whereas most prokaryotic genes are represented only once in the genome, many eukaryotic genes are present in multiple copies, called **gene families**. In some cases, multiple copies of genes are needed to produce RNAs or proteins required in large quantities, such as ribosomal RNAs or histones. In other cases, distinct members of a gene family may be transcribed in different tissues or at different stages of development. For example, the $\alpha$ and $\beta$ subunits of hemoglobin are both encoded by gene families in the human genome, with different members of these families being expressed in embryonic, fetal, and adult tissues (Figure 4.5). Members of many gene families (e.g., the globin genes) are clustered within a region of DNA; members of other gene families are dispersed to different chromosomes.

Gene families are thought to have arisen by duplication of an original ancestral gene, with different members of the family then diverging as a consequence of mutations during evolution. Such divergence can lead to the evolution of related proteins that are optimized to function in different tissues or at different stages of development. For example, fetal globins have a higher affinity for $O_2$ than do adult globins—a difference that allows the fetus to obtain $O_2$ from the maternal circulation.

As might be expected, however, not all mutations enhance gene function. Some gene copies have instead sustained mutations that result in their loss of ability to produce a functional gene product. For example, the human $\alpha$- and $\beta$-globin gene families each contain two genes that have been inacti-

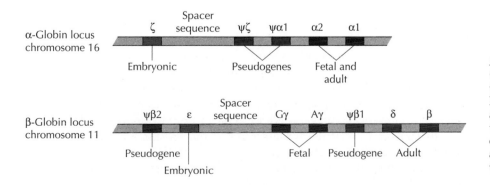

*Figure 4.5*
**Globin gene families** Members of the human $\alpha$- and $\beta$-globin gene families are clustered on chromosomes 16 and 11, respectively. Each family contains genes that are specifically expressed in embryonic, fetal, and adult tissues, in addition to nonfunctional gene copies (pseudogenes).

vated by mutations. Such nonfunctional gene copies (called **pseudogenes**) represent evolutionary relics that significantly increase the size of eukaryotic genomes without making a functional genetic contribution.

### Repetitive DNA Sequences

A substantial portion of eukaryotic genomes consists of highly repeated noncoding DNA sequences. These sequences, sometimes present in hundreds of thousands of copies per genome, were first demonstrated by Roy Britten and David Kohne during studies of the rates of reassociation of denatured fragments of cellular DNAs (Figure 4.6). Denatured strands of DNA hybridize to each other (reassociate), re-forming double-stranded molecules (see Figure 3.28). Since DNA reassociation is a bimolecular reaction (two separated strands of denatured DNA must collide with each other in order to hybridize), the rate of reassociation depends on the concentration of DNA strands. When fragments of *E. coli* DNA were denatured and allowed to hybridize with each other, all of the DNA reassociated at the same rate, as expected if each DNA sequence were represented once per genome. However, reassociation of fragments of DNA extracted from mammalian cells showed a very different pattern. Approximately 60% of the DNA fragments reassociated at the rate expected for sequences present once per genome, but the remainder reassociated much more rapidly than expected. The interpretation of these results was that some sequences were present in multiple copies and therefore reassociated more rapidly than those sequences that were represented only once per genome. In particular, these experiments indicated that approximately 40% of mammalian DNA consists of highly repetitive sequences, some of which are repeated $10^5$ to $10^6$ times.

Further analysis has identified several types of these highly repeated sequences. One class (called **simple-sequence DNA**) contains tandem arrays of thousands of copies of short sequences, ranging from 5 to 200 nucleotides. For example, one type of simple-sequence DNA in *Drosophila* consists of tandem repeats of the seven nucleotide unit ACAAACT. Because of their distinct base compositions, many simple-sequence DNAs can be separated from the rest of the genomic DNA by equilibrium centrifugation in CsCl density gradients. The density of DNA is determined by its base composition, with AT-rich sequences being less dense than GC-rich

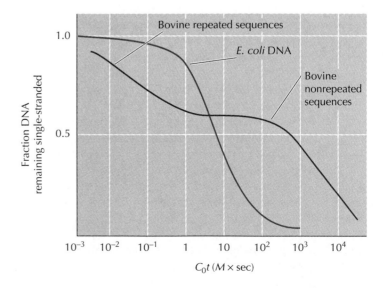

**Figure 4.6**
**Identification of repetitive sequences by DNA reassociation**   The kinetics of the reassociation of fragments of *E. coli* and bovine DNAs are illustrated as a function of $C_0t$, which is the initial concentration of DNA multiplied by the time of incubation. The *E. coli* DNA reassociates at a uniform rate, consistent with each fragment of DNA being represented once in a genome of $4.7 \times 10^6$ base pairs. In contrast, the bovine DNA fragments exhibit two distinct steps in their reassociation. About 60% of the DNA fragments (the nonrepeated sequences) reassociate more slowly than *E. coli* DNA, as expected for sequences represented as single copies in the larger bovine genome ($3 \times 10^9$ base pairs). However, the other 40% of the bovine DNA fragments (the repeated sequences) reassociate more rapidly than *E. coli* DNA, indicating that multiple copies of these sequences are present.

sequences. Therefore, an AT-rich simple-sequence DNA bands in CsCl gradients at a lower density than the bulk of *Drosophila* genomic DNA (Figure 4.7). Since such repeat-sequence DNAs band as "satellites" separate from the main band of DNA, they are frequently referred to as **satellite DNAs**. These sequences are repeated millions of times per genome, accounting for 10 to 20% of the DNA of most higher eukaryotes. Simple-sequence DNAs are not transcribed and do not convey functional genetic information. Some, however, may play important roles in chromosome structure.

Other repetitive DNA sequences are scattered throughout the genome rather than being clustered as tandem repeats. These sequences are classified as **SINEs** (short interspersed elements) or **LINEs** (long interspersed elements). The major SINEs in mammalian genomes are *Alu* **sequences**, so-called because they usually contain a single site for the restriction endonuclease *Alu*I. *Alu* sequences are approximately 300 base pairs long and about a million such sequences are dispersed throughout the genome, accounting for nearly 10% of the total cellular DNA. Although *Alu* sequences are transcribed into RNA, they do not encode proteins and their function is unknown. The major human LINEs (which belong to the LINE 1, or L1, family) are about 6000 base pairs long and repeat approximately 50,000 times in the genome. L1 sequences are transcribed and at least some encode proteins, but like *Alu* sequences, they have no known function in cell physiology.

Both *Alu* and L1 sequences are examples of transposable elements, which are capable of moving to different sites in genomic DNA (see Chapter 5). Some of these sequences may help regulate gene expression, but most *Alu* and L1 sequences appear not to make a useful contribution to the cell. They may, however, have played important evolutionary roles by contributing to the generation of genetic diversity.

### The Number of Genes in Eukaryotic Cells

Having discussed several kinds of noncoding DNA that contribute to the genomic complexity of higher eukaryotes, it is of interest to consider estimates of the total number of genes in eukaryotic genomes. Assuming that the average polypeptide is approximately 400 amino acids long, the average size of the coding sequence of a gene is 1200 base pairs. The genome of *E. coli* is $4.7 \times 10^6$ base pairs long, and current data suggest that *E. coli* may contain approximately 4000 genes, consistent with the use of most of the *E. coli* DNA as protein-coding sequence.

Most genes of yeasts, like those of *E. coli*, do not contain introns. If most of the yeast genome (consisting of $1.4 \times 10^7$ base pairs) were similarly used as protein-coding sequence, one might expect yeasts to contain 5000 to 10,000 genes. This estimate is roughly compatible with the number of yeast genes identified by the sequencing of yeast genomic DNA, as discussed later in this chapter.

In contrast to *E. coli* and yeasts, higher eukaryotes (including humans) have genomes that contain large amounts of noncoding DNA. Thus, only a small fraction of the total $3 \times 10^9$ base pairs of the human genome is expected to correspond to protein-coding sequence. Approximately one-third of the genome corresponds to highly repetitive sequences, leaving an estimated $2 \times 10^9$ base pairs for functional genes, pseudogenes, and non-repetitive spacer sequences. If the average gene spans 10,000–20,000 base pairs (including introns), one might expect the human genome to consist of about 100,000 genes, with protein-coding sequences corresponding to only about 3% of human DNA. Although this estimate is generally accepted as plausible, it remains to be verified or corrected by the anticipated results of human genome sequencing.

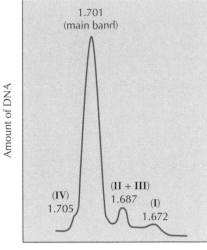

**Figure 4.7**
**Satellite DNA** Equilibrium centrifugation of *Drosophila* DNA in a CsCl gradient separates satellite DNAs (designated I-IV) with buoyant densities (in g/cm³) of 1.672, 1.687, and 1.705 from the main band of genomic DNA (buoyant density 1.701).

## CHROMOSOMES AND CHROMATIN

Not only are the genomes of most eukaryotes much more complex than those of prokaryotes, but the DNA of eukaryotic cells is also organized differently from that of prokaryotic cells. The genomes of prokaryotes are contained in single chromosomes, which are usually circular DNA molecules. In contrast, the genomes of eukaryotes are composed of multiple chromosomes, each containing a linear molecule of DNA. Although the numbers and sizes of chromosomes vary considerably between different species (Table 4.1), their basic structure is the same in all eukaryotes. The DNA of eukaryotic cells is tightly bound to small basic proteins (histones) that package the DNA in an orderly way in the cell nucleus. This task is substantial, given the DNA content of most eukaryotes. For example, the total extended length of DNA in a human cell is nearly 2 m, but this DNA must fit into a nucleus with a diameter of only 5 to 10 $\mu$m. Although DNA packaging is also a problem in bacteria, the mechanism by which prokaryotic DNAs are packaged in the cell appears distinct from that of eukaryotes and is not well understood.

### Chromatin

The complexes between eukaryotic DNA and proteins are called **chromatin**, which typically contains about twice as much protein as DNA. The major proteins of chromatin are the **histones**—small proteins containing a high proportion of basic amino acids (arginine and lysine) that facilitate binding to the negatively charged DNA molecule. There are five major types of histones—called H1, H2A, H2B, H3, and H4—which are very similar among different species of eukaryotes (Table 4.2). The histones are extremely abundant proteins in eukaryotic cells; together, their mass is approximately equal to that of the cell's DNA. In addition, chromatin contains an approximately equal mass of a wide variety of nonhistone chromosomal proteins. There are more than a thousand different types of these

*Table 4.1*   Chromosome Numbers of Eukaryotic Cells

| Organism | Genome size (Mb)[a] | Chromosome number[a] |
|---|---|---|
| Yeast (*Saccharomyces cerevisiae*) | 14 | 16 |
| Slime mold (*Dictyostelium*) | 70 | 7 |
| *Arabidopsis thaliana* | 70 | 5 |
| Corn | 5,000 | 10 |
| Onion | 15,000 | 8 |
| Lily | 50,000 | 12 |
| Nematode (*Caenorhabditis elegans*) | 100 | 6 |
| Fruit fly (*Drosophila*) | 165 | 4 |
| Toad (*Xenopus laevis*) | 3,000 | 18 |
| Lungfish | 50,000 | 17 |
| Chicken | 1,200 | 39 |
| Mouse | 3,000 | 20 |
| Cow | 3,000 | 30 |
| Dog | 3,000 | 39 |
| Human | 3,000 | 23 |

[a] Both genome size and chromosome number are for haploid cells.
Mb = millions of base pairs.

**Table 4.2** The Major Histone Proteins

| Histone[a] | Molecular weight | Number of amino acids | Percentage Lys + Arg |
|---|---|---|---|
| H1 | 22,500 | 244 | 30.8 |
| H2A | 13,960 | 129 | 20.2 |
| H2B | 13,774 | 125 | 22.4 |
| H3 | 15,273 | 135 | 22.9 |
| H4 | 11,236 | 102 | 24.5 |

[a] Data are for rabbit (H1) and bovine histones.

proteins, which are involved in a range of activities, including DNA replication and gene expression. The DNA of prokaryotes is similarly associated with proteins, some of which presumably function as histones do, packaging the DNA within the bacterial cell. Histones, however, are a unique feature of eukaryotic cells and are responsible for the distinct structural organization of eukaryotic chromatin.

The basic structural unit of chromatin, the **nucleosome**, was described by Roger Kornberg in 1974 (Figure 4.8). Two types of experiments led to Kornberg's proposal of the nucleosome model. First, partial digestion of chromatin with micrococcal nuclease (an enzyme that degrades DNA) was found to yield DNA fragments approximately 200 base pairs long. In contrast, a similar digestion of naked DNA (not associated with proteins) yielded a continuous smear of randomly sized fragments. These results suggested that the binding of proteins to DNA in chromatin protects

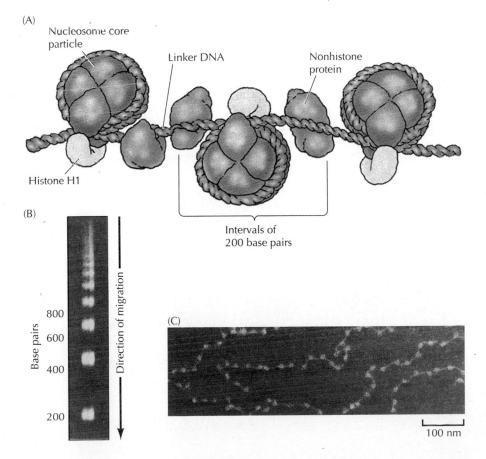

(A)
Nucleosome core particle
Linker DNA
Nonhistone protein
Histone H1

(B)
Base pairs
800
600
400
200
Direction of migration

Intervals of 200 base pairs

(C)

100 nm

*Figure 4.8*
**The organization of chromatin in nucleosomes** (A) The DNA is wrapped around histones in nucleosome core particles and sealed by histone H1. Nonhistone proteins bind to the linker DNA between nucleosome core particles. (B) Gel electrophoresis of DNA fragments obtained by partial digestion of chromatin with micrococcal nuclease. The linker DNA between the nucleosome core particles is preferentially sensitive, so limited digestion of chromatin yields fragments corresponding to multiples of 200 base pairs. (C) An electron micrograph of an extended chromatin fiber, illustrating its beaded appearance. (B courtesy of Roger Kornberg, Stanford University; C courtesy of Ada L. Olins and Donald E. Olins, Oak Ridge National Laboratory.)

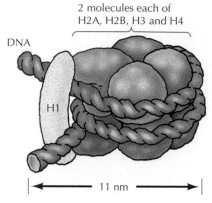

2 molecules each of
H2A, H2B, H3 and H4

DNA

H1

|← 11 nm →|

*Figure 4.9*
**Structure of a chromatosome** The nucleosome core particle consists of 146 base pairs of DNA wrapped 1.75 turns around a histone octamer consisting of two molecules each of H2A, H2B, H3, and H4. A chromatosome contains two full turns of DNA (166 base pairs) locked in place by one molecule of H1.

regions of the DNA from nuclease digestion, so that the enzyme can attack DNA only at sites separated by approximately 200 base pairs. Consistent with this notion, electron microscopy revealed that chromatin fibers have a beaded appearance, with the beads spaced at intervals of approximately 200 base pairs. Thus, both the nuclease digestion and the electron microscopic studies suggested that chromatin is composed of repeating 200-base-pair units, which were called nucleosomes.

More extensive digestion of chromatin with micrococcal nuclease was found to yield particles (called **nucleosome core particles**) that correspond to the beads visible by electron microscopy. Detailed analysis of these particles has shown that they contain 146 base pairs of DNA wrapped 1.75 times around a histone core consisting of two molecules each of H2A, H2B, H3, and H4 (the core histones) (Figure 4.9). One molecule of the fifth histone, H1, is bound to the DNA as it enters and exits each nucleosome core particle. This forms a chromatin subunit known as a **chromatosome**, which consists of 166 base pairs of DNA wrapped around the histone core and held in place by H1 (a linker histone). Although the core histones are conserved in all eukaryotes, the yeast *Saccharomyces cerevisiae* has no histone H1, indicating that linker histones are not essential for nucleosome formation.

The packaging of DNA with histones yields a chromatin fiber approximately 10 nm in diameter that is composed of chromatosomes separated by linker DNA segments averaging about 80 base pairs in length (Figure 4.10). In the electron microscope, this 10-nm fiber has the beaded appearance that suggested the nucleosome model. Packaging of DNA into such a 10-nm chromatin fiber shortens its length approximately sixfold. The chromatin can then be further condensed by coiling into 30-nm fibers, the structure of which still remains to be determined. Interactions between histone H1 molecules appear to play an important role in this stage of chromatin condensation, and it is noteworthy that the chromatin of *S. cerevisiae* (which lacks H1) remains in a more extended conformation.

*Figure 4.10*
**Chromatin fibers** The packaging of DNA into nucleosomes yields a chromatin fiber approximately 10 nm in diameter. The chromatin is further condensed by coiling into a 30-nm fiber, containing about six nucleosomes per turn. (Photographs courtesy of Ada L. Olins and Donald E. Olins, Oak Ridge National Laboratory.)

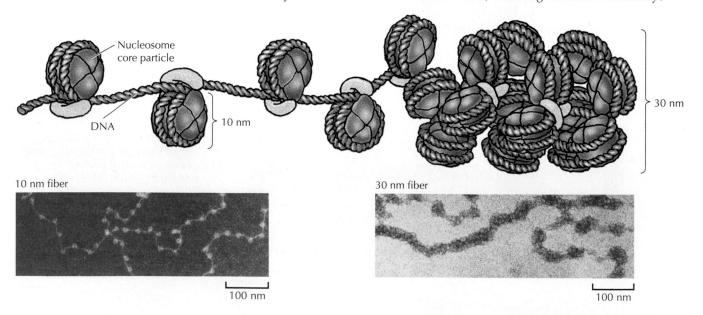

Nucleosome core particle

DNA

10 nm

30 nm

10 nm fiber

30 nm fiber

|— 100 nm

|— 100 nm

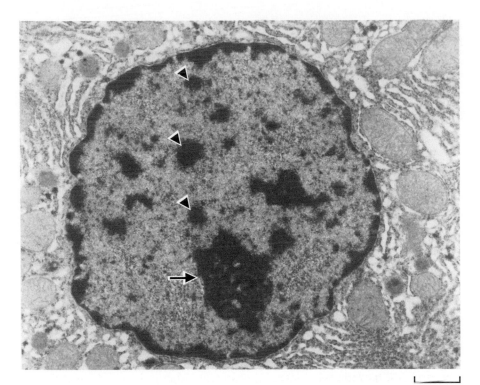

1 µm

*Figure 4.11*
**Interphase chromatin**   Electron
micrograph of an interphase nucleus.
The euchromatin is distributed
throughout the nucleus. The heterochromatin is indicated by arrowheads,
and the nucleolus by an arrow. (Courtesy of Ada L. Olins and Donald E.
Olins, Oak Ridge National Laboratory.)

The extent of chromatin condensation varies during the life cycle of the cell. In interphase (nondividing) cells, most of the chromatin (called **euchromatin**) is relatively decondensed and distributed throughout the nucleus (Figure 4.11). During this period of the cell cycle, genes are transcribed and the DNA is replicated in preparation for cell division. Most of the euchromatin in interphase nuclei appears to be in the form of 30-nm fibers, organized into large loops containing approximately 50 to 100 kb of DNA. About 10% of the euchromatin, containing the genes that are actively transcribed, is in a more decondensed state (the 10-nm conformation) that allows transcription. Chromatin structure is thus intimately linked to the control of gene expression in eukaryotes, as will be discussed in Chapter 6.

In contrast to euchromatin, about 10% of interphase chromatin (called **heterochromatin**) is in a very highly condensed state that resembles the chromatin of cells undergoing mitosis. Heterochromatin is transcriptionally inactive and contains highly repeated DNA sequences, such as those present at centromeres and telomeres.

As cells enter mitosis, their chromosomes become highly condensed so that they can be distributed to daughter cells. The loops of 30-nm chromatin fibers are thought to fold upon themselves further to form the compact metaphase chromosomes of mitotic cells, in which the DNA has been condensed nearly 10,000-fold (Figure 4.12). Such condensed chromatin can no longer be used as a template for RNA synthesis, so transcription ceases during mitosis. Electron micrographs indicate that the DNA in metaphase chromosomes is organized into large loops attached to a protein scaffold (Figure 4.13), but we currently understand neither the detailed structure of this highly condensed chromatin nor the mechanism of chromatin condensation.

Metaphase chromosomes are so highly condensed that their morphology can be studied using the light microscope (Figure 4.14). Several stain-

*Figure 4.12*
**Chromatin condensation during mito-
sis**   Scanning electron micrograph of
metaphase chromosomes. Artificial
color has been added. (Biophoto Asso-
ciates/Photo Researchers Inc.)

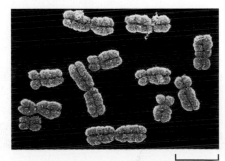

10 µm

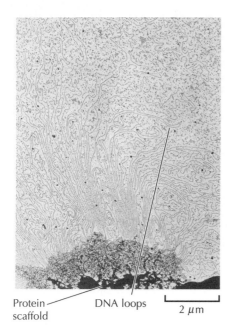

*Figure 4.13*
**Structure of metaphase chromosomes**   An electron micrograph of DNA loops attached to the protein scaffold of metaphase chromosomes that have been depleted of histones. (From J. R. Paulson and U. K. Laemmli, 1977. *Cell* 12: 817.)

Protein scaffold     DNA loops     |—— 2 μm ——|

ing techniques yield characteristic patterns of alternating light and dark chromosome bands, which result from the preferential binding of stains or fluorescent dyes to AT-rich versus GC-rich DNA sequences. These bands are specific for each chromosome and appear to represent distinct chromosome regions. Genes can be localized to specific chromosome bands by *in situ* hybridization, indicating that the packaging of DNA into metaphase chromosomes is a highly ordered and reproducible process.

## Centromeres

The **centromere** is a specialized region of the chromosome that plays a critical role in ensuring the correct distribution of duplicated chromosomes to daughter cells during mitosis (Figure 4.15). The cellular DNA is replicated during interphase, resulting in the formation of two copies of each chromosome prior to the beginning of mitosis. As the cell enters mitosis, chromatin condensation leads to the formation of metaphase chromosomes consisting of two identical sister chromatids. These sister chromatids are held together at the centromere, which is seen as a constricted chromosomal region. As mitosis proceeds, microtubules of the mitotic spindle attach to the centromere, and the two sister chromatids separate and move to opposite poles of the spindle. At the end of mitosis, nuclear membranes reform and the chromosomes decondense, resulting in the formation of daughter nuclei containing one copy of each parental chromosome.

*Figure 4.14*
**Human metaphase chromosomes**   (A) A light micrograph of human chromosomes spread from a metaphase cell. (B) Human chromosomes are arranged in pairs numbered from the largest (chromosome 1) to the smallest. The chromosomes shown are from a female, so there are 22 pairs of autosomes and two X chromosomes.  (Craig Holmes/Biological Photo Service.)

(A)

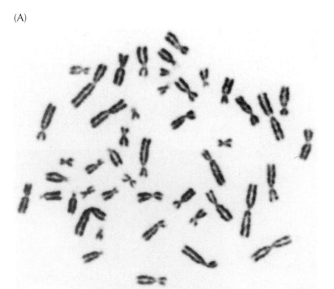

|—— 10 μm ——|

(B)

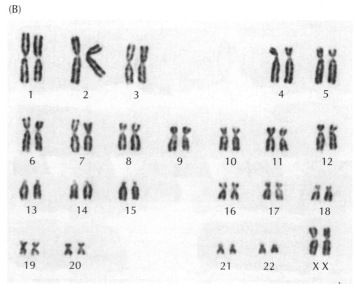

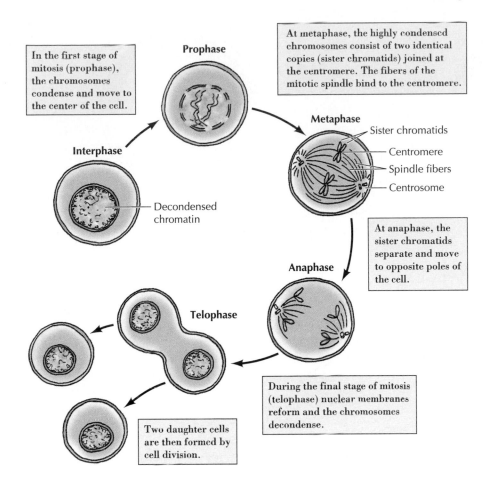

In the first stage of mitosis (prophase), the chromosomes condense and move to the center of the cell.

**Prophase**

At metaphase, the highly condensed chromosomes consist of two identical copies (sister chromatids) joined at the centromere. The fibers of the mitotic spindle bind to the centromere.

**Interphase**

**Metaphase**

Sister chromatids
Centromere
Spindle fibers
Centrosome

Decondensed chromatin

At anaphase, the sister chromatids separate and move to opposite poles of the cell.

**Anaphase**

**Telophase**

During the final stage of mitosis (telophase) nuclear membranes reform and the chromosomes decondense.

Two daughter cells are then formed by cell division.

*Figure 4.15*
**Chromosomes during mitosis** Since DNA replicates during interphase, the cell contains two identical duplicated copies of each chromosome prior to entering mitosis.

The centromeres thus serve both as the sites of association of sister chromatids and as the attachment sites for microtubules of the mitotic spindle. They consist of specific DNA sequences to which a number of centromere-associated proteins bind, forming a specialized structure called the **kinetochore** (Figure 4.16). The binding of microtubules to kinetochore proteins mediates the attachment of chromosomes to the mitotic spindle. Proteins associated with the kinetochore then act as "molecular motors" that drive the movement of chromosomes along the spindle fibers, segregating the chromosomes to daughter nuclei.

Centromeric DNA sequences have been defined best in yeasts, where their function can be assayed by following the segregation of plasmids at mitosis (Figure 4.17). Plasmids that contain functional centromeres segregate like chromosomes and are equally distributed to daughter cells following mitosis. In the absence of a functional centromere, however, the plasmid does not segregate properly, and many daughter cells fail to inherit plasmid DNA. Assays of this type have enabled determination of the sequences required for centromere function. Such experiments first showed that the centromere sequences of the well-studied yeast *Saccha-*

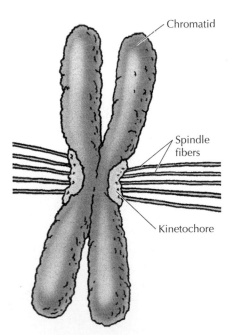

Chromatid

Spindle fibers

Kinetochore

*Figure 4.16*
**The centromere of a metaphase chromosome** The centromere is the region at which the two sister chromatids remain attached at metaphase. Specific proteins bind to centromeric DNA, forming the kinetochore, which is the site of spindle fiber attachment.

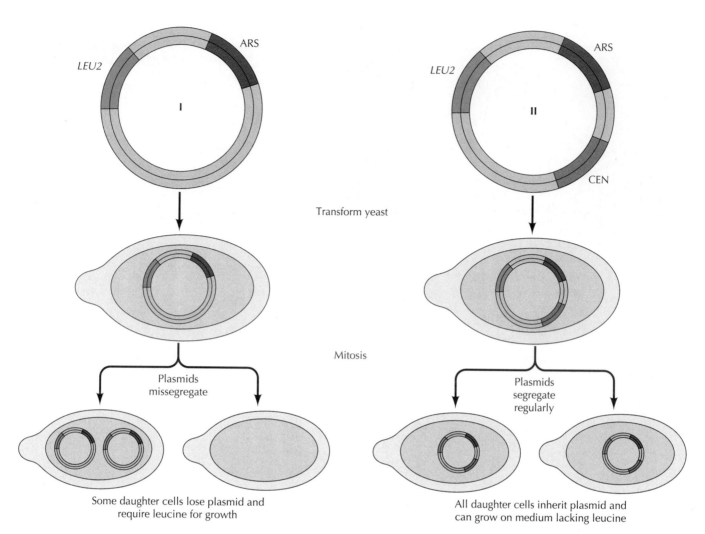

*Figure 4.17*
**Assay of a centromere in yeast**   Both plasmids shown contain a selectable marker (*LEU2*) and DNA sequences that serve as origins of replication in yeast (ARS, which stands for autonomously replicating sequence). However, plasmid I lacks a centromere and is therefore frequently lost as a result of missegregation during mitosis. In contrast, the presence of a centromere (CEN) in plasmid II ensures its regular transmission to daughter cells.

*romyces cerevisiae* are contained in approximately 125 base pairs consisting of three sequence elements: two short sequences of 8 and 25 base pairs separated by 78 to 86 base pairs of very AT-rich DNA (Figure 4.18).

The short centromere sequences defined in *S. cerevisiae*, however, do not appear to reflect the situation in other eukaryotes. More recent studies have defined the centromeres of the fission yeast *Schizosaccharomyces pombe* by a similar functional approach. Although *S. cerevisiae* and *S. pombe* are both yeasts, they appear to be as divergent from each other as either is from humans and are quite different in many aspects of their cell biology. These two yeast species thus provide complementary models for simple and easily studied eukaryotic cells. The centromeres of *S. pombe* span 50 to 100 kb of DNA; they are approximately a thousand times larger than those of *S. cerevisiae*. They consist of a central core of 7000 base pairs of single-copy DNA flanked by tandem repeats of three sets of repetitive sequences (Figure 4.19). Not only the central core but also the flanking repeated sequences are

**Figure 4.18**
**Centromere sequences of *S. cerevisiae***
The centromere (CEN) sequences consist of two short conserved sequences (CDEI and CDEIII) separated by 78 to 86 base pairs (bp) of AT-rich DNA (CDEII). The sequences shown are consensus sequences derived from analysis of the centromere sequences of individual yeast chromosomes. Pu = A or G; x = A or T; y = any base.

required for centromere function, so the centromeres of *S. pombe* appear to be considerably more complex than those of *S. cerevisiae*.

The centromeres of higher eukaryotes have not yet been defined by functional studies, but mammalian centromeres are characterized by extensive regions of heterochromatin consisting of highly repetitive satellite DNA sequences. In humans and other primates the primary centromeric sequence is α satellite DNA, which is a 171-base-pair sequence arranged in tandem repeats spanning up to millions of base pairs. The α satellite DNA appears to play a role in centromere structure and function, since it has been found to bind centromere-associated proteins. However, the precise role of α satellite DNA, as well as the potential activities of other repetitive sequences in mammalian centromeres, remains to be established. Consistent with their large size, mammalian centromeres form large kinetochores that bind 30 to 40 microtubules, whereas only single microtubules bind to the centromeres of *S. cerevisiae*.

## Telomeres

The sequences at the ends of eukaryotic chromosomes, called **telomeres**, play critical roles in chromosome replication and maintenance. Telomeres were initially recognized as distinct structures because broken chromosomes were highly unstable in eukaryotic cells, implying that specific sequences are required at normal chromosomal termini. This was subsequently demonstrated by experiments in which telomeres from the protozoan *Tetrahymena* were added to the ends of linear molecules of yeast plasmid DNA. The addition of these telomeric DNA sequences allowed these plasmids to replicate as linear chromosome-like molecules in yeasts, demonstrating directly that telomeres are required for the replication of linear DNA molecules.

The telomere DNA sequences of a variety of eukaryotes are similar, consisting of repeats of a simple-sequence DNA containing clusters of G residues on one strand (Table 4.3). For example, the sequence of telomere repeats in humans and other mammals is AGGGTT, and the telomere repeat in *Tetrahymena* is GGGGTT. These sequences are repeated hundreds or thousands of times, thus spanning up to several kilobases.

Telomeres play a critical role in replication of the ends of linear DNA molecules (see Chapter 5). DNA polymerase is able to extend a growing DNA chain but cannot initiate synthesis of a new chain at the terminus of a linear DNA molecule. Consequently, the ends of linear chromosomes cannot be replicated by the normal action of DNA polymerase. This problem has been solved by the evolution of a special mechanism, involving reverse transcriptase activity, to replicate telomeric DNA sequences.

**Figure 4.19**
**Centromeres of *S. pombe***   The arrangement of sequences at the centromere of chromosome II is illustrated. The centromere consists of a central core (CC) of unique-sequence DNA, flanked by tandem repeats of three repetitive sequence elements (B, K, and L).

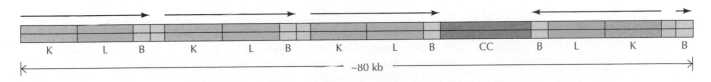

*Table 4.3*   Telomeric DNAs

| Organism | Telomeric repeat sequence |
|---|---|
| Yeasts | |
| *Saccharomyces cerevisiae* | $G_{1-3}T$ |
| *Schizosaccharomyces pombe* | $G_{2-5}TTAC$ |
| Protozoans | |
| *Tetrahymena* | GGGGTT |
| *Dictyostelium* | $G_{1-8}A$ |
| Plant | |
| (*Arabidopsis*) | AGGGTTT |
| Mammal | |
| (human) | AGGGTT |

## MAPPING AND SEQUENCING COMPLETE GENOMES

Some of the most exciting endeavors in molecular biology are the current efforts directed at obtaining the complete nucleotide sequences of both the human genome and the genomes of several model organisms, including *E. coli, Saccharomyces cerevisiae, Caenorhabditis elegans, Drosophila, Arabidopsis,* and the mouse. Recombinant DNA has had an enormous impact on our understanding of the molecular basis of cell biology by allowing the isolation and sequencing of a wide variety of important genes. Until the last few years, scientists had generally focused on the cloning, sequencing, and further characterization of specific genes. Now, however, recombinant DNA technology allows the alternative approach of cloning and sequencing entire genomes. In principle, this approach has the potential of identifying all the genes in an organism, which then become accessible for investigations of their structure and function. Genome sequencing is thus expected to provide scientists with a unique database, consisting of the nucleotide sequences of complete sets of genes. Since many of these genes will not have been previously identified, determination of their functions will form the basis of many future studies in cell biology.

### Prokaryotic Genomes

The first complete sequence of a cellular genome, reported in 1995 by a team of researchers led by J. Craig Venter, was that of the bacterium *Haemophilus influenzae*, a common inhabitant of the human respiratory tract. The genome of *H. influenzae* is approximately $1.8 \times 10^6$ base pairs (1.8 **megabases,** or **Mb**), slightly less than half the size of the *E. coli* genome. The complete nucleotide sequence indicated that the *H. influenzae* genome is a circular molecule containing 1,830,137 base pairs of DNA. The sequence was then analyzed to identify the genes encoding rRNAs, tRNAs, and proteins. Potential protein-coding regions were identified by computer analysis of the DNA sequence to detect **open-reading frames**—long stretches of nucleotide sequence that can encode polypeptides because they contain none of the three chain-terminating codons (UAA, UAG, and UGA). Since these chain-terminating codons occur randomly once in every 21 codons (3 chain-terminating codons out of 64 total), open-reading frames that extend for more than a hundred codons usually represent functional genes.

This analysis identified six copies of rRNA genes, 54 different tRNA genes, and 1743 potential protein-coding regions in the *H. influenzae* genome (Figure 4.20). More than a thousand of these could be assigned a biological role (e.g., an enzyme of the citric acid cycle) on the basis of their

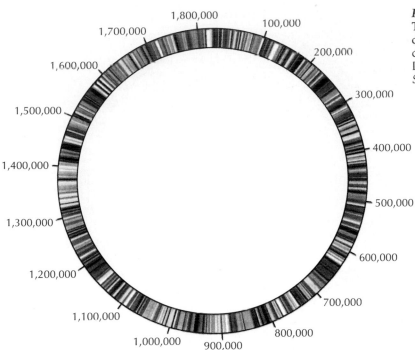

**Figure 4.20**
**The genome of** *Haemophilus influenzae* Predicted protein coding regions are designated by colored bars. Numbers indicate base pairs of DNA. (From R. D. Fleischmann et al., 1995. *Science* 269: 496.)

relationships to known protein sequences, but the others represent genes of currently unknown function. The predicted coding sequences have an average size of approximately 900 base pairs, so they cover about 1.6 Mb of DNA. Protein-coding sequences thus correspond to more than 85% of the genome of *H. influenzae*, consistent with previous estimates that protein-coding sequences account for most of the DNA in bacterial genomes.

The complete sequence of the genome of *Mycoplasma genitalium* has also been obtained. Mycoplasmas are of particular interest because they represent the simplest present-day bacteria and contain the smallest genomes of all known cells. The genome of *M. genitalium* is only 580 kb (0.58 Mb) long and may represent the minimal set of genes required to maintain a self-replicating organism. Analysis of its DNA sequence indicates that *M. genitalium* contains only 470 predicted protein-coding sequences, which correspond to approximately 88% of genomic DNA (Figure 4.21). Many of these sequences were identified as genes encoding proteins involved in DNA replication, transcription, translation, membrane transport, and energy metabolism. However, *M. genitalium* contains many fewer genes for metabolic enzymes than does *H. influenzae*, reflecting its more limited metabolism. For example, many genes known to encode components of biosynthetic pathways are lacking in the genome of *M. genitalium*, consistent with its need to obtain amino acids and nucleotide precursors from a host organism. Interestingly, the *Mycoplasma* genome also includes approximately 150 genes of currently unknown function. Thus, even in the simplest of cells, the biological roles of many genes remain to be determined.

Although the relative simplicity and facile genetics of *E. coli* have made it a favored organism of molecular biologists, the 4.7-Mb *E. coli* genome has not yet been completely sequenced. However, a great deal is known about the genome of *E. coli* from genetic analysis. More than 1400 genes have been identified in mutant strains of *E. coli*, and these genes have been positioned on a map of the *E. coli* chromosome. Indeed, the ability to isolate and map a large number of mutants first established that the genome of *E.*

*Figure 4.21*
**The genome of *Mycoplasma genitalium***
Predicted protein-coding sequences are designated by colored bars. Numbers indicate base pairs of DNA. (From C. M. Fraser et al., 1995. *Science* 270: 397.)

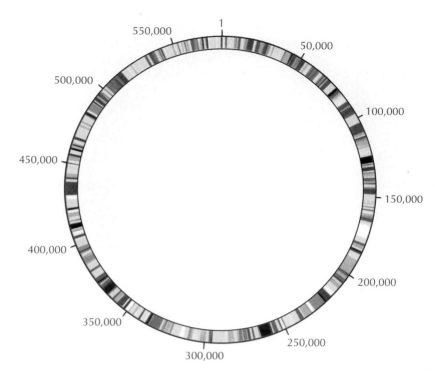

*Figure 4.22*
**Separation of yeast chromosomes**
The results of pulsed-field gel electrophoresis of the 16 chromosomes of *S. cerevisiae* are shown, with the estimated size of each chromosome indicated in kilobases (kb) of DNA. Note that the largest chromosome (chromosome XII) migrates anomalously under these conditions. (From M. V. Olson, 1991. In *The Molecular and Cellular Biology of the Yeast* Saccharomyces, ed. J. R. Broach et al., Cold Spring Harbor Laboratory Press.)

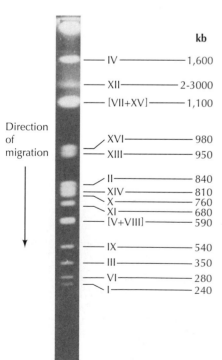

*coli* is circular, in contrast to the linear chromosomes of eukaryotes. The sequences of known genes, however, account for only about one-third of the genome, so it is reasonable to expect that *E. coli* may contain 2000 to 3000 additional genes that remain to be characterized. The portions of the *E. coli* genome sequenced up to now indicate that approximately 85% of *E. coli* DNA consists of protein-coding regions. This percentage is consistent with the *H. influenzae* and *M. genitalium* sequences and suggests that *E. coli* contains approximately 4000 genes. Thus, even for *E. coli*, complete sequencing of the genome can be expected to lead to the discovery of many new genes that have not been identified by extensive genetic analysis.

### Sequencing the Yeast Genome

As noted already, the simplest eukaryotic genome ($1.4 \times 10^7$ base pairs of DNA) is found in the yeast *Saccharomyces cerevisiae*. Moreover, yeasts grow rapidly and are subject to simple genetic manipulations. Thus, in many ways yeasts are model eukaryotic cells that can be studied much more readily than the cells of mammals or other higher eukaryotes. Consequently, determination of the complete nucleotide sequence of the yeast genome is viewed as a major first step toward understanding the genetic complexity of eukaryotes.

The small size of the yeast genome has greatly facilitated physical mapping, which has been based on the use of **pulsed-field gel electrophoresis (PFGE)** to separate extremely large DNA molecules according to size. The genome of *S. cerevisiae* contains 16 chromosomes ranging in size from approximately 240 to 2000 kb, and the resolving power of PFGE allows the intact yeast chromosomal DNAs to be separated from each other (Figure 4.22). Electrophoresis in agarose gels is routinely used to separate DNA fragments generated by restriction endonuclease digestion (see Figure 3.16). Under the usual conditions, the migration of DNA through a gel is driven by a constant electric field, and the separation of fragments is based on their rates of migration through pores in the gel matrix. This process provides good separation of DNA fragments up to about 30 kb long. Larger DNA molecules are not resolved by this technique, however, because they stretch out lengthwise as they migrate through the gel. In PFGE, the electric field applied to the gel is not constant, but periodically alternates in different directions. These periodic changes in the direction of the electric field force the DNA molecules to reorient in the direction of the field before they can continue to migrate through the gel. The time required for this reorientation depends on the size of the DNA fragments. Long DNA fragments reorient more slowly than shorter fragments, so their rate of migration through the gel is reduced. Repeated changes in the direction of the field continuously increase the separation between fragments of different sizes, allowing PFGE to separate DNA molecules ranging up to several thousand kilobases.

The separation of yeast chromosomes by PFGE allows cloned genes to be mapped to a chromosome simply by Southern blot hybridization. The ability to separate large DNA molecules by PFGE has also allowed restriction maps of the yeast chromosomes to be prepared using enzymes such as *Not*I and *Sfi*I (Figure 4.23). Since these restriction endonucleases recognize sequences of eight base pairs (e.g., GCGGCCGC for *Not*I), they cleave DNA at very rare sites, yielding fragments with average sizes of about 100 kb. More detailed physical maps have also been derived by the analysis of multiple overlapping clones of yeast DNA in bacteriophage λ and cosmid vectors.

A milestone in studies of eukaryotic genomes was reached in 1992 with the complete mapping and sequencing of the 315-kb yeast chromosome III—the first sequence analysis of an entire chromosome (Figure 4.24). This achievement was followed by the complete sequencing of chromosome VIII (563 kb) and chromosome XI (666 kb) and capped by the recent completion of the sequence of the entire yeast genome.

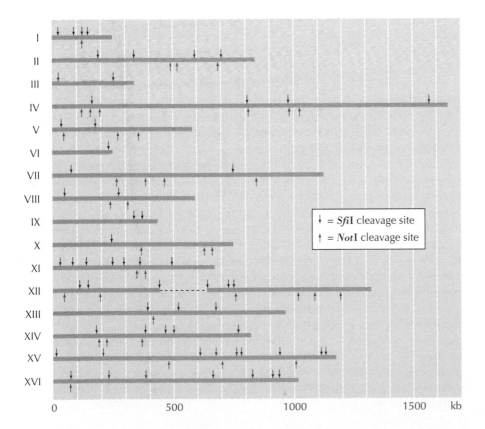

*Figure 4.23*
**Restriction maps of yeast chromosomes** The positions of cleavage sites for *Sfi*I and *Not*I are indicated in the 16 chromosomes of *S. cerevisiae*. The dashed portion of chromosome XII indicates the region of repeated genes encoding ribosomal RNA, which varies from 1000 to 2000 kb. (From M. V. Olson, 1991. In *The Molecular and Cellular Biology of the Yeast* Saccharomyces, ed. J. R. Broach et al., Cold Spring Harbor Laboratory Press.)

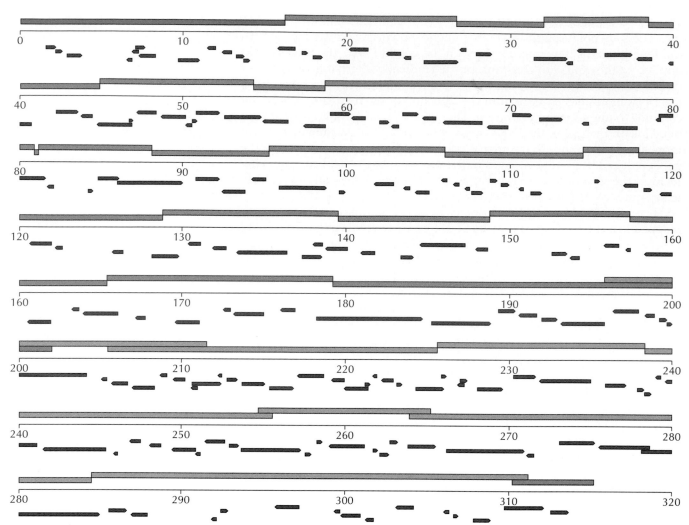

*Figure 4.24*
**Yeast chromosome III**   The upper
blue bars designate the clones used for
DNA sequencing. Open-reading
frames are indicated by arrows. (From
S. G. Oliver et al., 1992. *Nature* 357: 38.)

Analysis of the yeast genomic sequence has revealed that yeast have a
high density of open-reading frames, with an average length of approximately 1.5 kb. For example, 182 open-reading frames were identified in
chromosome III, 269 in chromosome VIII, and 331 in chromosome XI. On
average, the yeast genome contains one protein-coding sequence for every
two kb of DNA, so protein-coding sequences account for approximately
75% of the yeast genome. With a total size of 14 Mb, the yeast genome contains a total of about 6000 genes—slightly less than twice the number of
genes in *E. coli*.

Approximately half of the yeast genes that have been identified by
genomic sequencing can be assigned a function based on their relationship
to previously characterized genes. Thus, approximately 3000 previously
unrecognized genes have been discovered by sequencing the *S. cerevisiae*
genome in its entirety. Yeasts are particularly amenable to studies of the
functions of unknown genes because of the facility with which normal
chromosomal loci can be inactivated by homologous recombination with
cloned sequences (see Figure 3.42). Therefore, direct investigation of the
function of yeast genes that were initially identified only on the basis of
their nucleotide sequence will be possible. Sequencing of the yeast genome
will thus open the door to studies of many new areas of the biology of this
simple eukaryotic cell.

## The Genome of Caenorhabditis elegans

The genomes of *C. elegans*, *Drosophila*, and *Arabidopsis* are intermediate in size and complexity between those of yeasts and humans. Distinctive features of each of these organisms make them important models for genome analysis: *Arabidopsis* provides a simple model for a plant genome, *C. elegans* is widely used for studies of animal development, and *Drosophila* has been especially well analyzed genetically. The genomes of these organisms, however, are about tenfold larger than those of yeasts, introducing a new order of difficulty in genome mapping and sequencing.

The nematode *C. elegans* is accessible to genetic analysis, and more than 1400 genes have currently been identified. In addition, a physical map of the entire genome (approximately 100 Mb distributed among six chromosomes) is nearly complete. The initial phases of mapping the *C. elegans* genome used DNA fragments cloned in cosmids, which were then analyzed to detect overlapping DNA inserts. Cosmid clones could thus be arranged in overlapping clusters, called contigs (Figure 4.25). This approach, however, was unable to cover the complete genome, and final linking of the cosmid clones was accomplished by the cloning of much larger pieces of DNA in **yeast artificial chromosome (YAC)** vectors.

The unique feature of YACs is that they contain centromeres and telomeres, allowing them to replicate as linear chromosome-like molecules in yeasts (Figure 4.26). They can therefore be used to clone DNA fragments the size of yeast chromosomal DNAs; that is, YACs can accommodate inserts up to thousands of kilobases long, rather than the approximately 40-kb inserts carried by cosmids. The large DNA inserts that can be cloned in YACs are critically important for mapping complex genomes. In the case of *C. elegans*, their use allowed separate sets of cosmid contigs to be linked to each other, resulting in the preparation of a nearly complete physical map of the genome. The importance of YACs in genome analysis will be even more apparent in our following discussion of mapping genomes of still greater complexity, such as that of humans.

More than one-fifth of the genome of *C. elegans* has already been sequenced, and completion of the *C. elegans* sequence is anticipated by the end of 1998. The data obtained indicate that *C. elegans* contains about 13,000 genes. About half of the genes sequenced to date are unrelated to previously described sequences.

Analysis of the functions of these hitherto undiscovered genes of *C. elegans* is likely to be particularly exciting in terms of understanding animal

*Figure 4.25*
**Mapping of cosmid and YAC clones in *C. elegans*** In the example illustrated, 13 cosmid clones, containing inserts of approximately 40 kb each, could be arranged in three overlapping sets (contigs). Analysis of larger fragments of DNA (hundreds of kilobases) cloned in YACs then allowed the cosmid contigs to be linked to each other.

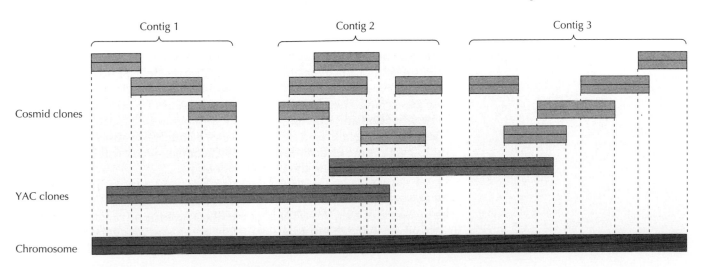

*Figure 4.26*
**Cloning large DNA fragments in YACs**    The vector contains a bacterial origin of replication (*ori*) and a gene conferring ampicillin resistance (*Amp*ʳ), allowing it to be propagated in *E. coli*. In addition, it contains a yeast origin of replication (ARS), a centomere (CEN), a selectable marker allowing growth in the absence of uracil (*URA3*), and two telomeres (TEL). The vector is digested with *Eco*RI and *Bam*HI, yielding two vector arms terminating with telomeres and *Eco*RI sites. Large fragments of cell DNA prepared by partial digestion with *Eco*RI are ligated to the vector arms. Yeast cells are then transformed with the recombinant YACs, which replicate as linear chromosome-like molecules.

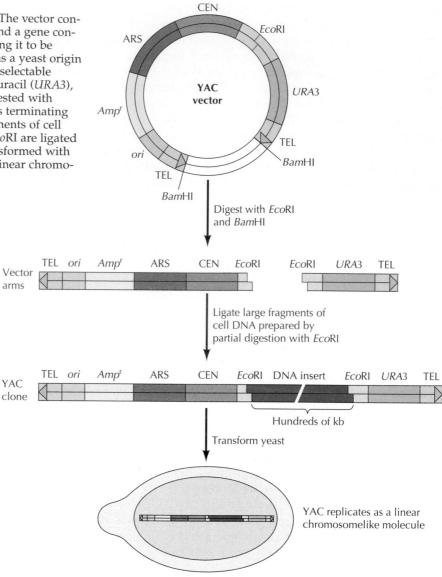

development. Although adult *C. elegans* are about 1 mm long and contain only 959 somatic cells in the entire body, they have all of the differentiated tissues of animals that are more complicated. Moreover, the complete pattern of cell divisions leading to *C. elegans* development has been described, including analysis of the connections made by all 302 neurons in the adult animal. Many of the genes involved in *C. elegans* development and differentiation have already been identified by classical genetic studies. Some of these genes have been found to be related to genes involved in controlling the proliferation and differentiation of mammalian cells, substantiating the validity of *C. elegans* as a model for more complex animals. With little doubt, many more critical developmental control genes will be uncovered as a result of sequencing the *C. elegans* genome.

### The Drosophila Genome

The mapping of the *Drosophila* genome illustrates the advantages provided by the extensive genetic analysis of this organism. The genome contains $1.65 \times 10^8$ base pairs distributed among four chromosomes. Nearly 4000 genes have been identified and mapped by genetic recombination, out of an estimated total of more than 15,000 genes in the *Drosophila* genome.

The advantages of *Drosophila* for genetic analysis include its relatively simple genome and the fact that it can be easily maintained and bred in the laboratory. In addition, a special tool for genetic analysis in *Drosophila* is provided by the giant **polytene chromosomes** that are found in some tissues, such as the salivary glands of larvae. These chromosomes arise in nondividing cells as a consequence of repeated replication of DNA strands that fail to separate from each other. Thus, each polytene chromosome contains hundreds of identical DNA molecules aligned in parallel. Because of their size, these polytene chromosomes are visible in the light microscope, and appropriate staining procedures reveal a distinct banding pattern (Figure 4.27). The banding of polytene chromosomes provides a much greater degree of resolution than that achieved with metaphase chromosomes (e.g., see Figure 4.14). The polytene chromosomes are decondensed interphase chromosomes that contain actively expressed genes. More than 5000 bands are visible, each corresponding to an average length of approximately 20 kb of DNA. In contrast, the bands identified in human metaphase chromosomes contain several megabases of DNA.

The banding pattern of polytene chromosomes thus provides a high-resolution physical map of the *Drosophila* genome. Moreover, this map of the chromosomes can be readily aligned with the genetic map. Gene deletions can often be correlated with the loss of a specific chromosomal band, thus defining the physical location of the gene on the chromosome (Figure 4.28).

Cloned DNAs can be similarly mapped by *in situ* hybridization to polytene chromosomes, often with sufficient resolution to localize cloned genes to specific bands (Figure 4.29). Thus, the map positions of cosmid or YAC clones (which span many bands) can readily be determined. Most of the *Drosophila* genome has been cloned in YACs and cosmids, and the ordering of these clones to provide a complete molecular map of the *Drosophila* genome is rapidly progressing, providing the raw material for genomic sequence analysis. Because of the power of *Drosophila* genetics, the identification and subsequent analysis of new genes in this organism will be important to many areas of molecular and cellular biology, including studies of development and differentiation.

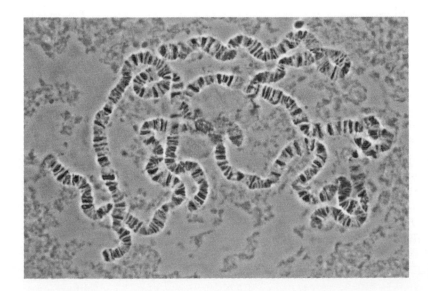

**Figure 4.27**
**Polytene chromosomes of *Drosophila***   A light micrograph of stained salivary gland chromosomes. The four chromosomes (X, 2, 3, and 4) are joined at their centromeres. (Peter J. Bryant/Biological Photo Service.)

*Figure 4.28*
**Drosophila gene mapping** Analysis of deletion mutants allows *Drosophila* genes to be mapped to bands of polytene chromosomes. The figure illustrates the localization of gene z to band 3A3 of the X chromosome by comparison of two deletions. Deletion 1 results in loss of the z gene; deletion 2 does not. Analysis of the polytene chromosome bands in flies bearing these deletions indicates that deletion 1 extends from band 3A3 to 3C2, whereas deletion 2 extends from 3A4 to 3C2. Therefore, the z gene must be located in band 3A3.

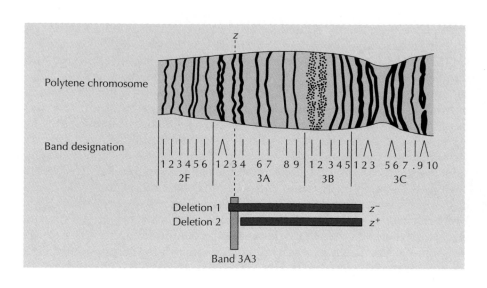

*Figure 4.29*
**In situ hybridization to a *Drosophila* polytene chromosome** Hybridization of a YAC clone to a polytene chromosome is illustrated. The region of hybridization is indicated by an arrow. (Courtesy of Daniel L. Hartl, Harvard University.)

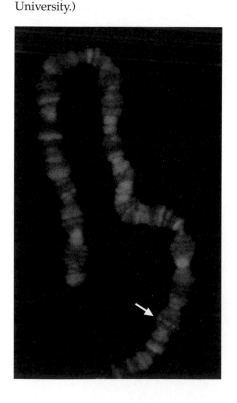

## THE HUMAN GENOME

The ultimate goal of genome analysis is to determine the complete nucleotide sequence of the human genome: $3 \times 10^9$ base pairs of DNA. To understand the magnitude of this undertaking (called the Human Genome Project), recall that the human genome is more than ten times larger than that of *Drosophila*; that the smallest human chromosome is several times larger than the entire yeast genome; and that the extended length of DNA that makes up the human genome is about 1 m long, whereas that of the fully sequenced bacterium *Haemophilus influenzae* is less than 1 mm long. Nonetheless, most scientists believe that completion of the sequence of human DNA within the next ten years is a reasonable goal. Identification of the estimated 100,000 human genes will not only open new horizons for research but will also have an immense impact on medicine by illuminating the genetic basis of many human diseases.

### Human Gene Mapping

The human genome is distributed among 24 chromosomes (22 autosomes and the 2 sex chromosomes), each containing between $5 \times 10^4$ and $26 \times 10^4$ kb of DNA (Figure 4.30). The first stage of human gene mapping is the assignment of known genes to a chromosomal locus, usually defined as a metaphase chromosome band. Of the estimated 100,000 human genes, approximately 6000 have been identified and about half of these have been mapped, in most cases using one of the three approaches described in this section.

The first general method developed for localizing human genes to individual chromosomes or subchromosomal regions was **somatic cell hybridization** (Figure 4.31). Cells in culture can be induced to fuse with each other by treatment with certain viruses or chemical agents, such as polyethylene glycol. The nuclei in such cells also fuse, yielding hybrid cells that contain chromosomes derived from both parents. Such hybrids can be made not only between cells of the same species but also between cells of different species, such as human and mouse. Importantly, the human chromosomes are unstable in such human-mouse hybrids and are progressively lost during growth of the hybrid cells. It is therefore possible to derive hybrid cell lines that contain only one or a few human chromosomes. In some cases, hybrid cell lines containing only fragments of human chromosomes can also be obtained. The chromosomal locus of a human gene can then be determined

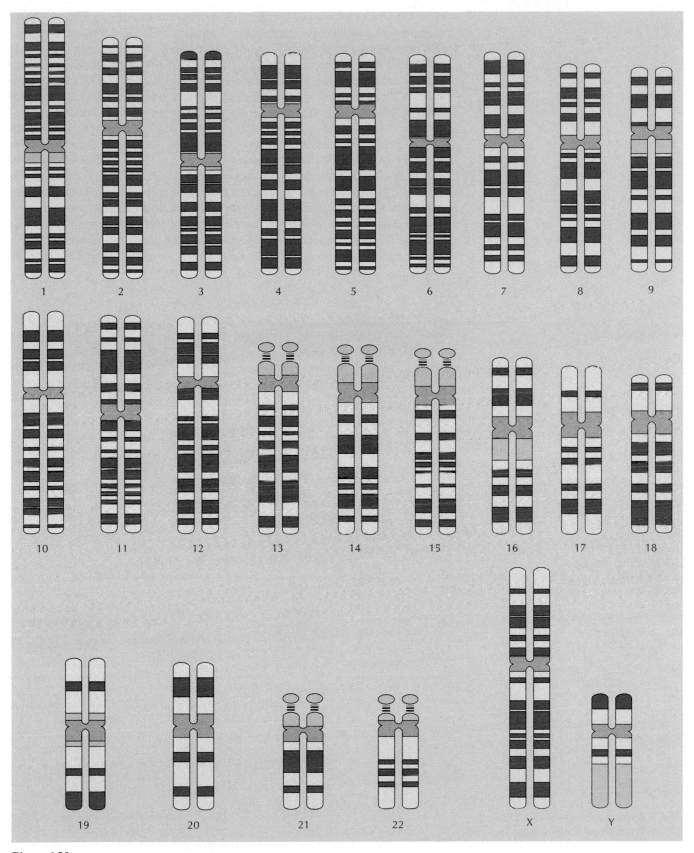

*Figure 4.30*
**The human chromosomes** A schematic of human metaphase chromosomes showing the banding pattern obtained after cytogenetic staining.

by screening of a panel of somatic cell hybrids containing different sets of human chromosomes. For example, hybridization of a cloned human gene to DNAs extracted from such a panel of somatic cell hybrids can be used to map the cloned gene to a human chromosome (see Figure 4.31).

Higher-resolution chromosomal localization of cloned genes is possible with *in situ* hybridization to chromosomes, usually using fluorescent probes—a method generally referred to as **fluorescence *in situ* hybridization,** or **FISH**. *In situ* hybridization to metaphase chromosomes allows the mapping of a cloned gene to a locus defined by a chromosome band (Figure 4.32). Although FISH provides greater resolution than analysis of somatic cell hybrids, each band of metaphase chromosomes still contains thousands of kilobases of DNA. *In situ* hybridization to metaphase chromosomes therefore does not provide the detailed mapping information obtained by hybridization to the polytene chromosomes of *Drosophila*, which allows the localization of genes to interphase chromosome bands containing only 10 to 20 kb of DNA. Higher resolution can be obtained,

***Figure 4.31***
**Somatic cell hybridization** Human and mouse cells in culture are fused by treatment with a virus or chemical agent, yielding a hybrid cell containing both human and mouse chromosomes. (In this example only three human and two mouse chromosomes are shown.) Human chromosomes are unstable in such hybrids and are gradually lost during the outgrowth of individual clones of the hybrid cells. Therefore, a panel of hybrid cell lines is obtained, with each line containing a different complement of human chromosomes. Hybridization of a cloned gene to DNAs extracted from such a panel of hybrid cell lines can be used to map the gene to a specific human chromosome. In this illustration, a human probe hybridizes to DNAs of cell lines A and D, but not B and C. Since only hybrid cell lines A and D have retained chromosome 1, these results map the cloned gene to this human chromosome.

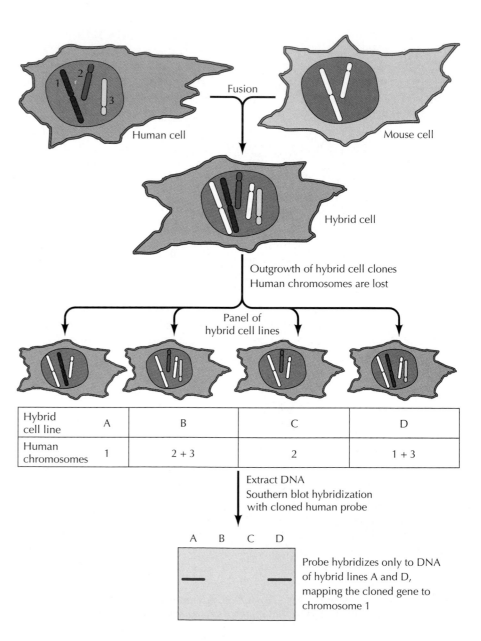

| Hybrid cell line | A | B | C | D |
|---|---|---|---|---|
| Human chromosomes | 1 | 2 + 3 | 2 | 1 + 3 |

*Figure 4.32*
**Fluorescence *in situ* hybridization**     A fluorescent probe for the gene encoding lamin B receptor is hybridized to stained human metaphase chromosomes (blue). Single gene hybridization signals are detected as red fluorescence. (Courtesy of K. L. Wyder and J. B. Lawrence, University of Massachusetts Medical Center.)

however, by hybridization to more extended human chromosomes from prometaphase or interphase cells, allowing the use of fluorescence *in situ* hybridization to map cloned genes to regions of about 100 kb.

Human genes can also be mapped by genetic linkage analysis, in which the distance between genes is estimated by recombination frequencies (see Figure 3.5). Genetic linkage was first used to map genes to the four *Drosophila* chromosomes and has subsequently been used to generate genetic maps of various other organisms, including *E. coli*, *S. cerevisiae*, *C. elegans*, and the mouse. However, construction of extensive genetic linkage maps generally requires the isolation of hundreds of single gene mutations, followed by extensive analysis of their inheritance in controlled breeding experiments. Although this approach applies to a variety of model organisms, it clearly cannot be applied directly to humans. This problem was overcome, however, by the realization that not only gene mutations but also any other detectable differences in DNA sequences between individuals can be used as genetic markers in human DNA. This approach greatly extended the range of available markers in the human genome and allowed analysis of recombination frequencies in small groups of families, leading to the development of a complete genetic map of the human genome.

The first sequences used as genetic markers were based on inherited differences in cleavage sites for restriction endonucleases. A single base change within a restriction endonuclease site is a readily detectable genetic marker because the mutated site is no longer cleaved by the enzyme in question. Two chromosomes that differ by such a mutation are then distinguishable on the basis of a **restriction fragment length polymorphism (RFLP)**, which arises because a particular cleavage site is present in only one of the two DNA molecules (Figure 4.33). A mutation that gives rise to an RFLP thus represents a genetic marker that can be detected by Southern blot hybridization of restriction endonuclease-digested DNA with an appropriate probe.

Many RFLPs have been identified simply by hybridization of cloned probes to Southern blots of restriction endonuclease-digested DNAs of a series of individuals. These RFLPs provide a large set of genetic markers that can be used to construct linkage maps by studies of a limited number of families. For example, the first complete linkage map of the human genome was generated in 1987 by studies of the inheritance of 393 RFLPs in 21 families, each of which consisted of grandparents, parents, and children.

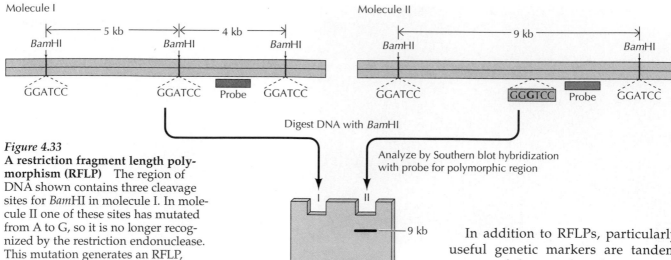

**Figure 4.33**
**A restriction fragment length polymorphism (RFLP)**  The region of DNA shown contains three cleavage sites for *Bam*HI in molecule I. In molecule II one of these sites has mutated from A to G, so it is no longer recognized by the restriction endonuclease. This mutation generates an RFLP, which can be detected by Southern blot hybridization of *Bam*HI-digested DNAs with a probe derived from the polymorphic region. In DNA I, the probe hybridizes to a 4-kb *Bam*HI fragment. In DNA II, however, the probe hybridizes to a 9-kb *Bam*HI fragment instead, because one of the *Bam*HI sites has been lost.

In addition to RFLPs, particularly useful genetic markers are tandem repeats of short nucleotide sequences (called **microsatellites**) distributed throughout the genome (Figure 4.34). For example, the human genome contains about 100,000 blocks of tandem repeats of the dinucleotide CA. Because the number of CA repeats at any given locus varies between individuals, differences in the number of tandem repeats can be used as genetic markers. Such variations can be easily scored by PCR using flanking-sequence primers, facilitating the use of microsatellites as genetic markers. The current linkage map of the human genome consists of more than 5000 loci (defined primarily by short tandem-repeat

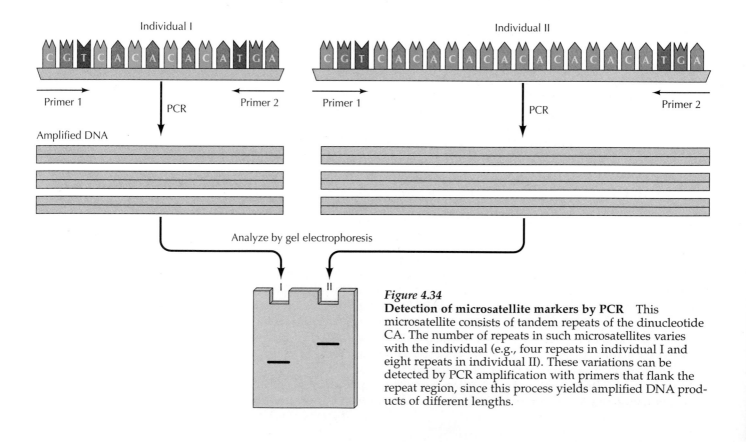

**Figure 4.34**
**Detection of microsatellite markers by PCR**  This microsatellite consists of tandem repeats of the dinucleotide CA. The number of repeats in such microsatellites varies with the individual (e.g., four repeats in individual I and eight repeats in individual II). These variations can be detected by PCR amplification with primers that flank the repeat region, since this process yields amplified DNA products of different lengths.

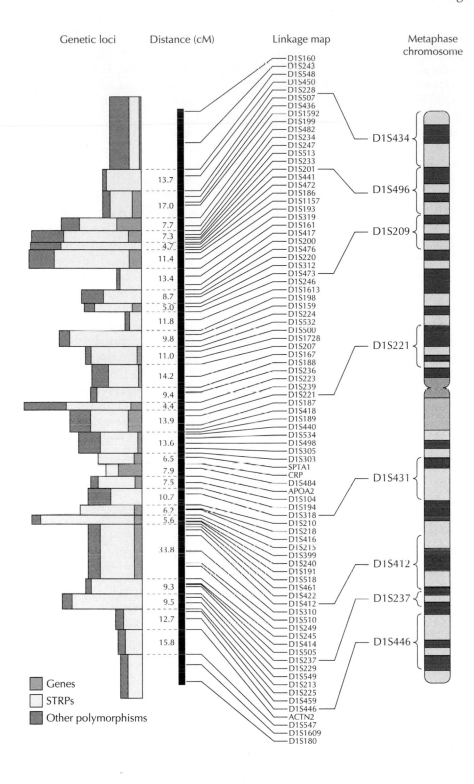

Genetic loci — Distance (cM) — Linkage map — Metaphase chromosome

*Figure 4.35*
**Linkage map of human chromosome 1**
The genetic linkage map is composed of loci containing short tandem-repeat polymorphisms (STRPs), with distances between markers indicated by frequencies of recombination (in centimorgans, or cM). One centimorgan corresponds to approximately 1 Mb of human DNA. Relationships between the genetic linkage map and metaphase chromosome bands are indicated to the right. The histogram to the left indicates the number of genes, STRPs, and other polymorphisms assigned to each map interval. (From Cooperative Human Linkage Center et al., 1994. Human Genetic Map: Genome Maps V. *Science* 265: 2055.)

polymorphisms), which are spaced at an average distance of less than 1 Mb of DNA (Figure 4.35).

Once markers such as RFLPs and short tandem-repeat polymorphisms (STRPs) have been mapped, they become available for use as reference markers for mapping other genes. Thus, the map position of any human gene can be determined by analysis of its linkage to these polymorphic DNA sequences. The most important application of this approach has been the genetic mapping and eventual cloning of genes responsible for inherited human diseases. Such isolation of human disease genes provides much of the impetus for mapping and sequencing the human genome. The basic approach to gene mapping is to analyze a series of polymorphic markers (e.g., RFLPs) in families carrying the disease gene of interest (Figure 4.36). Genetic linkage between an RFLP and the disease gene is then indicated by coinheritance of the disease and the RFLP marker in affected family members. Given that the RFLP has been mapped to a chromosomal locus, this linkage analysis determines the map position of the disease gene on a human chromosome.

Once the map position of a gene has been determined, that information can be used to isolate the gene as a molecular clone—a strategy known as **positional cloning**. This approach has been used to clone the genes for several human diseases; the gene for cystic fibrosis was one of the first examples. Cystic fibrosis is an inherited disease that affects approximately one in 2500 newborn Caucasians, although it is rare in other races. The disease is diagnosed in childhood and is inevitably fatal by about age 30. Most patients die from respiratory complications. The cystic fibrosis gene was first mapped by linkage to an RFLP, which was localized to chromosome 7 (Figure 4.37). Further mapping studies established that the cystic fibrosis gene is located between two RFLPs separated by approximately 1 Mb. Molecular clones from this region of DNA were isolated by the use of RFLPs as probes. These clones were then used as probes to isolate clones containing adjacent DNA sequences, leading to the isolation of a series of clones covering the region containing the cystic fibrosis gene. Analysis of these genomic clones, followed by cDNA cloning and sequencing, eventually identified a candidate gene that spans more than 250 kb of DNA and is

*Figure 4.36*
**Linkage of a human disease gene to an RFLP**   In this example, the allele leading to development of the disease (*D*) is dominant over the normal wild-type (*WT*) allele, so individuals with one copy of *D* are affected (pink symbols). The disease gene is linked to an RFLP defined by two alleles (*A1* and *A2*) that are distinguishable by Southern blot hybridization. In the family shown, the affected parent is heterozygous for the RFLP (*A1/A2*), while the normal parent is homozygous at this locus (*A2/A2*). Two children inherit the mutant gene; one inherits the wild-type allele. RFLP analysis of DNAs then indicates that both the affected children, but not the normal child, have also inherited *A1* from their affected parent, demonstrating genetic linkage between *A1* and *D*.

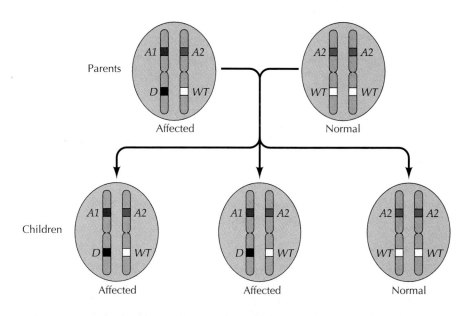

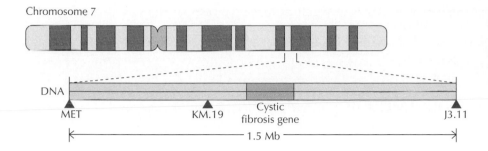

*Figure 4.37*
**The cystic fibrosis gene**   Linkage to the MET and J3.11 RFLPs mapped the gene to the indicated region of chromosome 7, which was further delineated by an additional RFLP designated KM.19. The cystic fibrosis gene, which spans approximately 250 kb, was then isolated by positional cloning.

transcribed to yield a 6-kb cDNA containing 24 exons. DNA sequencing confirmed that this gene is the cystic fibrosis gene by demonstrating that it is mutated in affected individuals.

The isolation of human disease genes has important implications not only for understanding the molecular and cellular basis of the disease, but also for prevention and treatment. An immediate benefit of cloning the gene responsible for an inherited disease is the potential for developing sensitive diagnostic tests for mutant alleles of the gene, such as the detection of mutations by PCR. These tests can identify individuals at risk for the disease and, at least for some diseases, enable appropriate preventive measures to be instituted before the disease develops. In addition, the ability to detect mutant alleles of disease genes opens the possibility of prenatal diagnosis to prevent transmission of the disease to future generations. Couples carrying mutant alleles of the gene can be identified, and their pregnancies monitored to determine if the mutant gene has been transmitted to the fetus. Because of the sensitivity of PCR analysis, mutations can be detected even in early embryos that have been derived by *in vitro* fertilization procedures prior to their return to the uterus.

In addition to these diagnostic measures, the cloning of disease genes suggests the potential use of **gene therapy** to correct the defect. The general strategy of gene therapy is to introduce a normal cDNA allele into the affected cells of patients. In some cases, genes can be introduced into target cells (e.g., lymphocytes) that can be maintained in culture and then returned to the patient. Alternatively, appropriate vectors can be used to introduce genes into some types of cells *in vivo*. For example, gene therapy trials for cystic fibrosis are evaluating the use of aerosols to deliver viral vectors to the affected epithelial cells lining the airways of the respiratory tract. Gene therapy is still in its early, experimental stages of development. However, the progress thus far clearly illustrates the general feasibility of the approach, suggesting that gene therapy will eventually become an important method of treating at least some human diseases.

### Physical Mapping of the Human Genome

A physical map of cloned DNA segments is a necessary intermediate between the genetic map and the nucleotide sequence of the human genome. Because of the large size of the genome, YACs provide the obvious vector for physical mapping of cloned human DNA. The most useful markers for analyzing these clones appear to be **sequence-tagged sites (STSs)**, which are readily detected by PCR analysis. STSs are short segments (usually 200 to 500 base pairs) of nonrepetitive DNA whose nucleotide sequences have been determined. Many STSs, each located at a unique site in the genome, can thus be defined. Since an STS corresponds to a segment of sequenced DNA, it is detectable in any sample of DNA by PCR

## MOLECULAR MEDICINE

# *Gene Therapy for Adenosine Deaminase Deficiency*

### The Disease

Adenosine deaminase deficiency is one of the causes of severe combined immunodeficiency (SCID), a group of hereditary diseases in which lymphocytes fail to develop normally. Affected individuals are deficient in all functions of the immune system. SCID is rare, with a frequency of about one in a million. Because of the profound defect in immune function, infants with this disease are susceptible to recurrent infections by various organisms, including viruses, bacteria, fungi, and protozoans. Unless treated, these infections are usually fatal within two years.

### Molecular and Cellular Basis

About 20% of SCID cases result from genetic deficiencies in the enzyme adenosine deaminase, which catalyzes the deamination of adenosine and deoxyadenosine. In the absence of adenosine deaminase, deoxyadenosine accumulates, resulting in corresponding increases in the concentration of deoxyadenosine triphosphate (dATP). High levels of dATP are toxic to proliferating cells because dATP inhibits the enzyme ribonucleotide reductase, which is required for synthesis of all four deoxyribonucleoside triphosphates. Consequently, increased concentrations of dATP block DNA synthesis.
Although all cells have adenosine deaminase, a deficiency in its activity is particularly toxic to lymphocytes, because they lack what other types of cells have: additional enzymes that degrade dAMP, thereby preventing dATP accumulation. Thus, a deficiency in adenosine deaminase specifically blocks lymphocyte repli-

Deoxyadenosine

Adenosine deaminase

Deoxyinosine

dAMP

dATP

Inhibition of ribonucleotide reductase and DNA synthesis

Metabolism of deoxyadenosine.

cation and prevents the development of a functional immune system.

### Prevention and Treatment

Some infants with adenosine deaminase deficiency can be cured by bone marrow transplantation, which is successful when compatible donor marrow can be obtained from an antigenically matched sibling. In other cases, the disease can be treated by enzyme replacement therapy, in which functional adenosine deaminase is administered to the patient by intravenous injection. Although this treatment is beneficial, it is not curative. Thus, adenosine deaminase deficiency represents a prime target for gene therapy.

In 1990, following preliminary studies in animals, the first clinical trial of gene therapy was initiated with the treatment of two children suffering from this disease. Lymphocytes obtained from the blood of these children were stimulated to proliferate in culture. A functional adenosine deaminase cDNA in a retroviral vector was then introduced into these lymphocytes, which were subsequently returned to the patients. These treatments were repeated 11 or 12 times over a two-year period. The results of this trial four years after its initiation demonstrate that the vector was successful-

*Gene Therapy for Adenosine Deaminase Deficiency* (continued)

ly transferred to the lymphocytes and has continued to be maintained for at least two years after the last treatment. Furthermore, gene therapy appears to have resulted in a substantial increase in both the numbers of blood lymphocytes and in immune function, as well as in clinical improvement in both children. The results of this initial trial thus support the efficacy of gene therapy as a treatment for this disease.

**Reference**

Blaese, R. M. and 18 others. 1995. T lymphocyte-directed gene therapy for ADA⁻ SCID: Initial results after 4 years. *Science* 270:475-480.

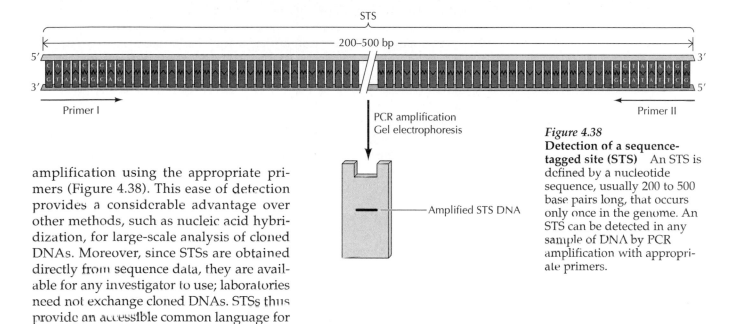

amplification using the appropriate primers (Figure 4.38). This ease of detection provides a considerable advantage over other methods, such as nucleic acid hybridization, for large-scale analysis of cloned DNAs. Moreover, since STSs are obtained directly from sequence data, they are available for any investigator to use; laboratories need not exchange cloned DNAs. STSs thus provide an accessible common language for genome mapping and sequencing.

**Figure 4.38**
**Detection of a sequence-tagged site (STS)** An STS is defined by a nucleotide sequence, usually 200 to 500 base pairs long, that occurs only once in the genome. An STS can be detected in any sample of DNA by PCR amplification with appropriate primers.

STS markers in YAC clones have been analyzed to construct physical maps both of individual human chromosomes and of the entire human genome. In the mapping of individual chromosomes, YAC libraries of human DNA are first screened so that clones carrying STS markers derived from the chromosome of interest can be identified. The mapping of chromosome 21, for example, used 198 STSs to screen more than 70,000 YAC clones, each containing inserts with an average size of 470 kb. This process identified 810 YACs containing inserts derived from chromosome 21. The STSs in these YACs were then aligned, yielding an array of overlapping YAC clones that spanned the chromosome (Figure 4.39).

Alternatively, STSs and other physical markers have been used to map collections of YACs covering the entire genome. Further screening of these clones with known genetic markers has allowed the construction of an integrated physical and genetic map covering all the human chromosomes, providing the raw material needed for genomic sequencing.

An alternative to analyzing the entire genome has been to focus on pro-

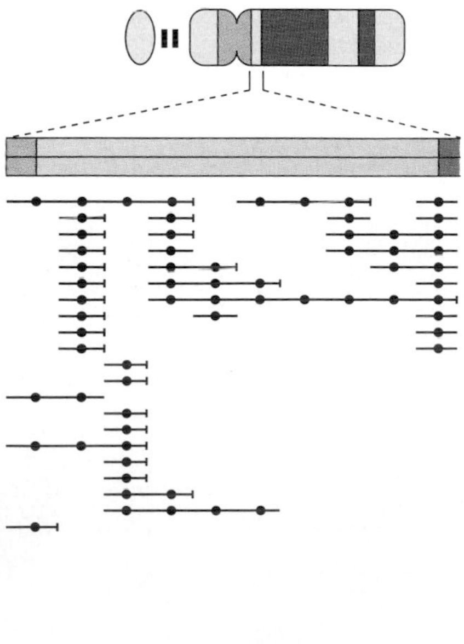

**Figure 4.39**
**Physical map of human chromosome 21** Overlapping YAC clones (lines) spanning the chromosome were aligned using STS markers (filled circles). Only a small portion of the map is shown. (From I. Chumakov et al., 1992. *Nature* 359: 380.)

tein-coding sequences by sequencing cDNAs. Assuming that only about 3% of the genome corresponds to protein-coding sequences, this task is obviously much more modest than sequencing the entire genome. The goal of sequencing protein-coding regions has been approached by the random sequencing of short regions (200 to 300 base pairs) of cDNAs. These short cDNA sequences are known as **expressed sequence tags (ESTs)** and can be used as markers for sequences in mRNAs. More than $5 \times 10^7$ base pairs of EST sequences have now been obtained, corresponding to a significant fraction of the total expected sequence present in human cDNA. The frequency of isolation of known sequences as ESTs confirms the general estimate of 100,000 genes in the human genome. The database of available ESTs has already proved a valuable resource—for example, by enabling the ready identification of new genes related to a cloned gene of interest. An important additional step will be the assignment of ESTs to positions on the genomic map. Delineation of the locations of transcribed sequences in the human genome may greatly facilitate the positional cloning of new genes of interest, including new human disease genes.

### Large-Scale DNA Sequencing

A cloned DNA map of the human genome will provide the starting material for nucleotide sequencing. Determining the sequence of the entire stretch of $3 \times 10^9$ base pairs of human DNA, however, is an immense task.

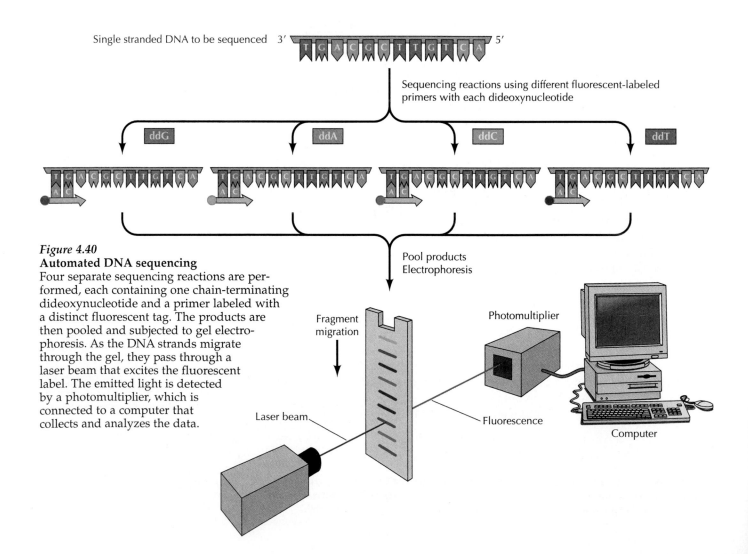

**Figure 4.40**
**Automated DNA sequencing**
Four separate sequencing reactions are performed, each containing one chain-terminating dideoxynucleotide and a primer labeled with a distinct fluorescent tag. The products are then pooled and subjected to gel electrophoresis. As the DNA strands migrate through the gel, they pass through a laser beam that excites the fluorescent label. The emitted light is detected by a photomultiplier, which is connected to a computer that collects and analyzes the data.

Consequently, an important goal of the Human Genome Project has been to develop improved methods of DNA sequencing, capable of completing the sequence of human DNA within a reasonable time and budget.

Although it was originally thought that this task would require the development of novel technologies, the problem now appears to have been solved as a result of continuing improvements in the currently available systems for automated DNA sequencing (Figure 4.40). The basic strategy for automated sequencing is to use fluorescence-labeled primers in dideoxynucleotide sequencing reactions (see Figure 3.25). As the newly synthesized DNA strands are electrophoresed through a gel, they pass through a laser beam that excites the fluorescent label. The resulting emitted light is then detected by a photomultiplier, and a computer collects and analyzes the data.

This type of automated DNA sequencing has now been used in a variety of genome sequencing projects, including the sequencing of bacterial, yeast, and *C. elegans* genomes. Improvements in sequencing strategy that have come from experience with these organisms suggest that it may be possible to sequence the entire human genome within 5 to 10 years. As the Human Genome Project enters this stage, we can anticipate a new database corresponding to the complete nucleotide sequence of human DNA—information that will have a profound impact on future research in human biology and medicine.

## Summary

**KEY TERMS**

### THE COMPLEXITY OF EUKARYOTIC GENOMES

*Introns and Exons:* Most eukaryotic genes have a split structure in which segments of coding sequence (exons) are interrupted by noncoding sequences (introns). In complex eukaryotes, introns account for about ten times more DNA than do exons.

gene, spacer sequence, exon, intron, RNA splicing, kilobase (kb)

*Gene Families and Pseudogenes:* Many eukaryotic genes are present in multiple copies, called gene families. Some members of gene families function in different tissues or at different stages of development. Other members of gene families (pseudogenes) have been inactivated by mutations and no longer represent functional genes.

gene family, pseudogene

*Repetitive DNA Sequences:* Approximately 40% of mammalian DNA consists of highly repetitive DNA sequences, some of which are present in $10^5$ to $10^6$ copies per genome.

simple-sequence DNA, satellite DNA, SINE, LINE, *Alu* sequence

*The Number of Genes in Eukaryotic Cells:* Only a small fraction of the genome in complex eukaryotes corresponds to protein-coding sequences. The human genome is estimated to contain about 100,000 genes.

### CHROMOSOMES AND CHROMATIN

*Chromatin:* The DNA of eukaryotic cells is wrapped around histones to form nucleosomes. Chromatin can be further compacted by the folding of nucleosomes into higher-order structures, including the highly condensed metaphase chromosomes of cells undergoing mitosis.

chromatin, histone, nucleosome, nucleosome core particle, chromatosome, euchromatin, heterochromatin

*Centromeres:* Centromeres are specialized regions of eukaryotic chromosomes that serve as the sites of association between sister chromatids and the sites of spindle fiber attachment during mitosis.

centromere, kinetochore

| | |
|---|---|
| **telomere** | *Telomeres:* Telomeres are specialized sequences required to maintain the ends of eukaryotic chromosomes. |

## MAPPING AND SEQUENCING COMPLETE GENOMES

**megabase (Mb), open-reading frame**

*Prokaryotic Genomes:* The genomes of *Haemophilus influenzae* and *Mycoplasma genitalium* have been completely sequenced. Most of these bacterial genomes consist of protein-coding sequences.

**pulsed-field gel electrophoresis (PFGE)**

*Sequencing the Yeast Genome:* Yeast chromosomes have been mapped using pulsed-field gel electrophoresis to separate large DNA molecules. Genome sequencing indicates that *S. cerevisiae* contains approximately 6000 genes, only half of which appear to be related to genes of known function.

**yeast artificial chromosome (YAC)**

*The Genome of Caenorhabditis elegans:* Yeast artificial chromosome (YAC) vectors have been used to map the *C. elegans* genome. Sequencing indicates that *C. elegans* contains approximately 13,000 genes, about half of which have not been previously identified.

**polytene chromosome**

*The Drosophila Genome:* Gene mapping in *Drosophila* is facilitated by genetic analysis and *in situ* hybridization of cloned genes to polytene chromosomes.

## THE HUMAN GENOME

**somatic cell hybridization, fluorescence *in situ* hybridization (FISH), restriction fragment length polymorphism (RFLP), microsatellite, positional cloning, gene therapy**

*Human Gene Mapping:* Human genes can be mapped by somatic cell hybridization, *in situ* hybridization, and genetic linkage analysis. The mapping of some human disease genes has allowed their isolation by positional cloning.

**sequence-tagged site (STS), expressed sequence tag (EST)**

*Physical Mapping of the Human Genome:* YAC clones have been used to construct maps of the human genome. The random sequencing of cDNA clones has provided markers for sequences present in mRNAs.

*Large-Scale DNA Sequencing:* Automated DNA sequencing is expected to enable the sequence of the human genome to be completed in 5 to 10 years.

## QUESTIONS

1. Repetitive DNA sequences were first identified by studies of rates of DNA reassociation. What relative rates of reassociation are expected for sequences repeated 1000 times in the genome compared to genes with only a single copy?

2. How many histone molecules are associated with the chromosomes of a human cell? Assuming that a typical cell contains about 1 ng of protein, what fraction of total cellular protein corresponds to histones?

3. Given current estimates of gene numbers, what fraction of DNA in *C. elegans* corresponds to protein-coding sequences?

4. Why are YAC vectors useful for analysis of complex genomes? What is the role of telomeres in these vectors?

5. You are interested in cloning a human disease gene and have established its position on the current linkage map of the genome. How much DNA will you have to clone in order to isolate your gene? Approximately how many cosmid clones would be required to encompass this region?

## REFERENCES AND FURTHER READING

### General References

Hartl, D. L. 1994. *Genetics.* 3rd ed. Boston: Jones and Bartlett.

Singer, M. and P. Berg. 1991. *Genes and Genomes.* Mill Valley, CA: University Science Books.

Watson, J. D., M. Gilman, J. Witkowski and M. Zoller. 1992. *Recombinant DNA.* 2nd ed. New York: Scientific American Books.

### The Complexity of Eukaryotic Genomes

Berget, S. M., C. Moore and P. A. Sharp. 1977. Spliced segments at the 5′ terminus of adenovirus 2 late mRNA. *Proc. Natl. Acad. Sci. USA* 74: 3171–3175. [P]

Breathnach, R., J. L. Mandel and P. Chambon. 1977. Ovalbumin gene is split in chicken DNA. *Nature* 270: 314–319. [P]

Britten, R. J. and D. E. Kohne. 1968. Repeated sequences in DNA. *Science* 161: 529–540. [P]

Charlesworth, B., P. Sniegowski and W. Stephan. 1994. The evolutionary dynamics of repetitive DNA in eukaryotes. *Nature* 371: 215–220. [R]

Chow, L. T., R. E. Gelinas, T. R. Broker and R. J. Roberts. 1977. An amazing sequence arrangement at the 5′ ends of adenovirus 2 messenger RNA. *Cell* 12: 1–8. [P]

Fritsch, E. F., R. M. Lawn and T. Maniatis. 1980. Molecular cloning and characterization of the human β-like globin gene cluster. *Cell* 19: 959–972. [P]

Gilbert, W. 1985. Genes-in-pieces revisited. *Science* 228: 823–824. [R]

Hentschel, C. C. and M. L. Birnsteil. 1981. The organization and expression of histone gene families. *Cell* 25: 301–313. [R]

Little, P. F. R. 1982. Globin pseudogenes. *Cell* 28: 683–684. [R]

Schmid, C. W. and W. R. Jelinek. 1982. The *Alu* family of dispersed repetitive sequences. *Science* 216: 1065–1070. [R]

Singer, M. F. 1982. SINEs and LINEs: Highly repeated short and long interspersed sequences in mammalian genomes. *Cell* 28: 433–434. [R]

Singer, M. F. and J. Skowronski. 1985. Making sense out of LINEs: Long interspersed repeat sequences in mammalian genomes. *Trends Biochem. Sci.* 10: 119–122. [R]

Stoltzfus, A., D. F. Spencer, M. Zuker, J. M. Logsdon, Jr. and W. F. Doolittle. 1994. Testing the exon theory of genes: The evidence from protein structure. *Science* 265: 202–207. [R]

Tilghman, S. M., P. J. Curtis, D. C. Tiemeier, P. Leder and C. Weissmann. 1978. The intervening sequence of a mouse β-globin gene is transcribed within the 15S β-globin mRNA precursor. *Proc. Natl. Acad. Sci. USA* 75: 1309–1313. [P]

### Chromosomes and Chromatin

Blackburn, E. H. 1991. Structure and function of telomeres. *Nature* 350: 569–573. [R]

Blackburn, E. H. 1994. Telomeres: No end in sight. *Cell* 77: 621–623. [R]

Bloom, K. 1993. The centromere frontier: Kinetochore components, microtubule-based motility and the CEN-value paradox. *Cell* 73: 621–624. [R]

Carbon, J. 1984. Yeast centromeres: Structure and function. *Cell* 37: 351–353. [R]

Clarke, L. 1990. Centromeres of budding and fission yeasts. *Trends Genet.* 6: 150–154. [R]

Clarke, L. and J. Carbon. 1980. Isolation of a yeast centromere and construction of functional small circular chromosomes. *Nature* 287: 504–509. [P]

Comings, D. E. 1978. Mechanisms of chromosome banding and implications for chromosome structure. *Ann. Rev. Genet.* 12: 25–46. [R]

Felsenfeld, G. 1992. Chromatin as an essential part of the transcriptional mechanism. *Nature* 355: 219–224. [R]

Grunstein, M. 1992. Histones as regulators of genes. *Sci. Am.* 267(4): 68–74B. [R]

Haaf, T., P. E. Warburton and H. F. Willard. 1992. Integration of human α-satellite DNA into simian chromosomes: Centromere protein binding and disruption of normal chromosome segregation. *Cell* 70: 681–696. [P]

Kornberg, R. D. 1974. Chromatin structure: A repeating unit of histones and DNA. *Science* 184: 868–871. [P]

Kornberg, R. D. and Y. Lorch. 1992. Chromatin structure and transcription. *Ann. Rev. Cell Biol.* 8: 563–587. [R]

Pardue, M. L. and J. G. Gall. 1970. Chromosomal localization of mouse satellite DNA. *Science* 168: 1356–1358. [P]

Paulson, J. R. and U. K. Laemmli. 1977. The structure of histone-depleted metaphase chromosomes. *Cell* 12: 817–828. [P]

Pluta, A. F., A. M. MacKay, A. M. Ainsztein, A. G. Goldberg and W. C. Earnshaw. 1995. The centromere: Hub of chromosomal activities. *Science* 270: 1591–1594. [R]

Richmond, T. J., J. T. Finch, B. Rushton, D. Rhodes and A. Klug. 1984. Structure of the nucleosome core particle at 7 Å resolution. *Nature* 311: 532–537. [R]

Saitoh, H., J. Tomkiel, C. A. Cooke, H. Ratrie III, M. Maurer, N. F. Rothfield and W. C. Earnshaw. 1992. CENP-C, an autoantigen in scleroderma, is a component of the human inner kinetochore plate. *Cell* 70: 115–125. [P]

Saitoh, Y. and U. K. Laemmli. 1994. Metaphase chromosome structure: Bands arise from a differential folding path of the highly AT-rich scaffold. *Cell* 76: 609–622. [P]

Schulman, I. and K. S. Bloom. 1991. Centromeres: An integrated protein/DNA complex required for chromosome movement. *Ann. Rev. Cell Biol.* 7: 311–336. [R]

Szostak, J. W. and E. H. Blackburn. 1982. Cloning yeast telomeres on linear plasmid vectors. *Cell* 29: 245–255. [P]

Van Holde, K. E. 1989. *Chromatin.* New York: Springer-Verlag.

Van Holde, K. E. and J. Zlatanovai. 1995. Chromatin higher order structure: Chasing a mirage. *J. Biol. Chem.* 270: 8373–8376. [R]

Willard, H. F. 1990. Centromeres of mammalian chromosomes. *Trends Genet.* 6: 410–416. [R]

Wolffe, A. 1995. *Chromatin: Structure and Function.* 2nd ed. New York: Academic Press.

Zakian, V. A. 1995 Telomeres: Beginning to understand the end. *Science* 270: 1601–1607. [R]

### Mapping and Sequencing Complete Genomes

Burke, D. T., G. F. Carle and M. V. Olson. 1987. Cloning of large segments of exogenous DNA into yeast by means of artificial chromosome vectors. *Science* 236: 806–812. [P]

Coulson, A., R. Waterston, J. Kiff, J. Sulston and Y. Kohara. 1988. Genome linking with yeast artificial chromosomes. *Nature* 335: 184–186. [P]

Daniels, D. L., G. Plunkett III, V. Burland and F. R. Blattner. 1992. Analysis of the *Escherichia coli* genome: DNA sequence of the region from 84.5 to 86.5 minutes. *Science* 257: 771–778. [P]

Dujon, B. and 107 others. 1994. Complete DNA sequence of yeast chromosome XI. *Nature* 369: 371–378. [P]

Fleischmann, R. D. and 39 others. 1995. Whole-genome random sequencing and assembly of *Haemophilus influenzae* Rd. *Science* 269: 496–512. [P]

Fraser, C. M. and 28 others. 1995. The minimal gene complement of *Mycoplasma genitalium.* *Science* 270: 397–403. [P]

Garza, D., J. W. Ajioka, D. T. Burke and D. L. Hartl. 1989. Mapping the *Drosophila* genome with yeast artificial chromosomes. *Science* 246: 641–646. [R]

Hodgkin, J., R. H. A. Plasterk and R. H. Waterston. 1995. The nematode *Caenorhabditis elegans* and its genome. *Science* 270: 410–414. [R]

Johnston, M. and 34 others. 1994. Complete nucleotide sequence of *Saccharomyces cerevisiae* chromosome VIII. *Science* 265: 2077–2082. [P]

Merriam, J., M. Ashburner, D. L. Hartl and F. C. Kafatos. 1991. Toward cloning and mapping the genome of *Drosophila*. *Science* 254: 221–225. [R]

Oliver, S. G. and 146 others. 1992. The complete DNA sequence of yeast chromosome III. *Nature* 357: 38–46. [P]

Olson, M. V. 1991. Genome structure and organization in *Saccharomyces cerevisiae*. In *The Molecular and Cellular Biology of the Yeast Saccharomyces: Genome Dynamics, Protein Synthesis, and Energetics* ed. J. R. Broach, J. R. Pringle and E. W. Jones, 1–39. Plainview, NY: Cold Spring Harbor Laboratory Press.

### The Human Genome

Adams, M. D. and 12 others. 1991. Complementary DNA sequencing: Expressed sequence tags and human genome project. *Science* 252: 1651–1656. [R]

Adams, M. D. and 85 others. 1995. Initial assessment of human gene diversity and expression patterns based upon 83 million nucleotides of cDNA sequence. *Nature* 377 Suppl.: 3–174. [P]

Boguski, M. S. 1995. The turning point in genome research. *Trends Biochem. Sci.* 20: 295–296. [R]

Botstein, D., R. L. White, M. Skolnick and R. W. Davis. 1980. Construction of a genetic linkage map in man using restriction fragment length polymorphism. *Am. J. Hum. Genet.* 32: 314–331. [P]

Chumakov, I. and 35 others. 1992. Continuum of overlapping clones spanning the entire human chromosome 21q. *Nature* 359: 380–387. [P]

Chumakov, I. M. and 62 others. 1995. A YAC contig map of the human genome. *Nature* 377 Suppl.: 175–298. [P]

Cohen, D., I. Chumakov and J. Weissenbach. 1993. A first-generation physical map of the human genome. *Nature* 366: 698–701. [P]

Collins, F. S. 1992. Cystic fibrosis: Molecular biology and therapeutic implications. *Science* 256: 774–779. [R]

Cooperative Human Linkage Center. 1994. A comprehensive human linkage map with centimorgan density. *Science* 265: 2049–2054. [P]

Crystal, R. G. 1995. Transfer of genes to humans: Early lessons and obstacles to success. *Science* 270: 404–410. [R]

Dib, C. and 13 others. 1996. A comprehensive genetic map of the human genome based on 5,264 microsatellites. *Nature* 380: 152–154. [P]

Donis-Keller, H. and 32 others. 1987. A genetic linkage map of the human genome. *Cell* 51: 319–337. [P]

Foote, S., D. Vollrath, A. Hilton and D. C. Page. 1992. The human Y chromosome: Overlapping DNA clones spanning the euchromatic region. *Science* 258: 60–66. [P]

Hudson, T. J. and 50 others. 1995. An STS-based map of the human genome. *Science* 270: 1945–1954. [P]

Hunkapiller, T., R. J. Kaiser, B. F. Koop and L. Hood. 1991. Large-scale and automated DNA sequence determination. *Science* 254: 59–67. [R]

McKusick, V. A. 1995. *Mendelian Inheritance in Man.* 11th ed. Baltimore: Johns Hopkins University Press.

Nakamura, Y., M. Leppert, P. O'Connell, R. Wolff, T. Holm, M. Culver, C. Martin, E. Fujimoto, M. Hoff, E. Kumlin and R. White. 1987. Variable number of tandem repeat (VNTR) markers for human gene mapping. *Science* 235: 1616–1622. [P]

Olson, M. V. 1995. A time to sequence. *Science* 270: 394–396. [R]

Olson, M., L. Hood, C. Cantor and D. Botstein. 1989. A common language for physical mapping of the human genome. *Science* 245: 1434–1435. [R]

Rommens and 14 others. 1989. Identification of the cystic fibrosis gene: Chromosome walking and jumping. *Science* 245: 1059–1065. [P]

Ruddle, F. H. 1981. A new era in mammalian gene mapping: Somatic cell genetics and recombinant DNA methodologies. *Nature* 294: 115–120. [R]

Weiss, M. C. and H. Green. 1967. Human-mouse hybrid cell lines containing partial complements of human chromosomes and functioning human genes. *Proc. Natl. Acad. Sci. USA* 58: 1104–1111. [P]

# 5

# Replication, Maintenance, and Rearrangements of Genomic DNA

THE FUNDAMENTAL BIOLOGICAL PROCESS OF REPRODUCTION requires the faithful transmission of genetic information from parent to offspring. Thus, the accurate replication of genomic DNA is essential to the lives of all cells and organisms. Each time a cell divides, its entire genome must be duplicated, and complex enzymatic machinery is required to copy the large DNA molecules that make up both prokaryotic and eukaryotic chromosomes. In addition, cells have evolved mechanisms to correct mistakes that sometimes occur during DNA replication and to repair DNA damage that can result from the action of environmental agents, such as radiation. Abnormalities in these processes result in a failure of accurate replication and maintenance of genomic DNA—a failure that can have disastrous consequences, such as the development of cancer.

Despite the importance of accurate DNA replication and maintenance, cell genomes are far from static. In order for species to evolve, mutations and gene rearrangements are needed to maintain genetic variation between individuals. Recombination between homologous chromosomes during meiosis plays an important role in this process by allowing parental genes to be rearranged into new combinations in the next generation. Rearrangements of DNA sequences within the genome are also thought to contribute to evolution by creating novel combinations of genetic information. In addition, some DNA rearrangements are programmed to regulate gene expression during the differentiation and development of individual cells and organisms. In humans, a prominent example is the rearrangement of antibody genes during development of the immune system. A careful balance between maintenance and variation of genetic information is thus critical to the development of individual organisms as well as to evolution of the species.

## DNA REPLICATION

As discussed in Chapter 3, DNA replication is a semiconservative process in which each parental strand serves as a template for the synthesis of a new complementary daughter strand. The central enzyme involved is DNA polymerase, which catalyzes the joining of deoxyribonucleoside

**175**

5'-triphosphates (dNTPs) to form the growing DNA chain. However, DNA replication is much more complex than a single enzymatic reaction. Other proteins are involved, and proofreading mechanisms are required to ensure that the accuracy of replication is compatible with the low frequency of errors that is needed for cell reproduction. Additional proteins and specific DNA sequences are also needed both to initiate replication and to copy the ends of eukaryotic chromosomes.

### DNA Polymerases

**DNA polymerase** was first identified in lysates of *E. coli* by Arthur Kornberg in 1956. The ability of this enzyme to accurately copy a DNA template provided a biochemical basis for the mode of DNA replication that was initially proposed by Watson and Crick, so its isolation represented a landmark discovery in molecular biology. Ironically, however, this first DNA polymerase to be identified (now called DNA polymerase I) is not the major enzyme responsible for *E. coli* DNA replication. Instead, it is now clear that both prokaryotic and eukaryotic cells contain several different DNA polymerases that play distinct roles in the replication and repair of DNA.

The multiplicity of DNA polymerases was first revealed by the isolation of a mutant strain of *E. coli* that was deficient in polymerase I (Figure 5.1). Cultures of *E. coli* were treated with a chemical (a **mutagen**) that induces a high frequency of mutations, and individual bacterial colonies were isolated and screened to identify a mutant strain lacking polymerase I. Analysis of a few thousand colonies led to the isolation of the desired mutant, which was almost totally defective in polymerase I activity. Surprisingly, the mutant bacteria grew normally, leading to the conclusion that polymerase I is not required for DNA replication. On the other hand, the mutant bacteria were extremely sensitive to agents that damage DNA (e.g., ultraviolet light), suggesting that polymerase I is involved primarily in the repair of DNA damage rather than in DNA replication *per se*.

The conclusion that polymerase I is not required for replication implied that *E. coli* must contain other DNA polymerases, and subsequent experiments led to the identification of two such enzymes, now called DNA polymerases II and III. The potential roles of these enzymes were investigated by the isolation of appropriate mutants. Strains of *E. coli* with mutations in polymerase II were found to grow and otherwise behave normally, so the role of this enzyme in the cell is unknown. Temperature-sensitive polymerase III mutants, however, were unable to replicate their DNA at high temperature, and subsequent studies have confirmed that polymerase III is the major replicative enzyme in *E. coli*.

It is now known that, in addition to polymerase III, polymerase I is also required for replication of *E. coli* DNA. The original polymerase I mutant was not completely defective in that enzyme, and later experiments showed that the residual polymerase I activity in this strain plays a key role in the replication process. The replication of *E. coli* DNA thus involves two distinct DNA polymerases, the specific roles of which are discussed below.

Eukaryotic cells contain five DNA polymerases: $\alpha$, $\beta$, $\gamma$, $\delta$, and $\varepsilon$. Polymerase $\gamma$ is located in mitochondria and is responsible for replication of mitochondrial DNA. The other four enzymes are located in the nucleus and are therefore candidates for involvement in nuclear DNA replication. Polymerases $\alpha$, $\delta$, and $\varepsilon$ are most active in dividing cells, suggesting that they function in replication. In contrast, polymerase $\beta$ is active in nondividing and dividing cells, suggesting that it may function primarily in the repair of DNA damage.

Two types of experiments have provided further evidence addressing

*Figure 5.1*
**Isolation of a mutant deficient in polymerase I**   A culture of *E. coli* was treated with a chemical mutagen, and individual bacterial colonies were isolated by growth on semisolid medium. Several thousand colonies were then cultured and screened to identify a mutant lacking polymerase I.

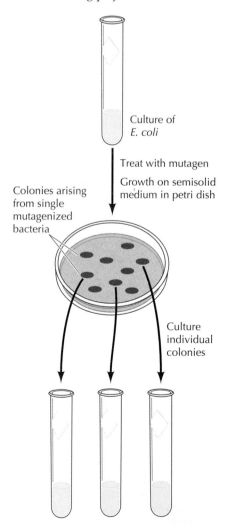

Culture of *E. coli*

Treat with mutagen

Growth on semisolid medium in petri dish

Colonies arising from single mutagenized bacteria

Culture individual colonies

Screen for DNA polymerase activity to identify mutant lacking polymerase I

the roles of polymerases $\alpha$, $\delta$, and $\varepsilon$ in DNA replication. First, replication of the DNAs of some animal viruses, such as SV40, can be studied in cell-free extracts. The ability to study replication *in vitro* has allowed direct identification of the enzymes involved, and analysis of such cell-free systems has shown that polymerases $\alpha$ and $\delta$ are required for SV40 DNA replication. Second, polymerases $\alpha$, $\delta$, and $\varepsilon$ are found in yeasts as well as in mammalian cells, enabling the use of the powerful approaches of yeast genetics (see Chapter 3) to test their biological roles directly. Such studies indicate that yeast mutants lacking any of these three DNA polymerases are unable to proliferate, implying a critical role for polymerase $\varepsilon$ as well as for $\alpha$ and $\delta$. However, since a biochemical role for polymerase $\varepsilon$ in DNA replication has not been demonstrated, polymerase $\varepsilon$ may function instead in a DNA repair pathway that is required for cell survival.

All known DNA polymerases share two fundamental properties that carry critical implications for DNA replication (Figure 5.2). First, all polymerases synthesize DNA only in the 5' to 3' direction, adding a dNTP to the 3' hydroxyl group of a growing chain. Second, DNA polymerases can add a new deoxyribonucleotide only to a preformed primer strand that is hydrogen-bonded to the template; they are not able to initiate DNA synthesis *de novo* by catalyzing the polymerization of free dNTPs. In this respect, DNA polymerases differ from RNA polymerases, which can initiate the synthesis of a new strand of RNA in the absence of a primer. As discussed later in this chapter, these properties of DNA polymerases appear critical for maintaining the high fidelity of DNA replication that is required for cell reproduction.

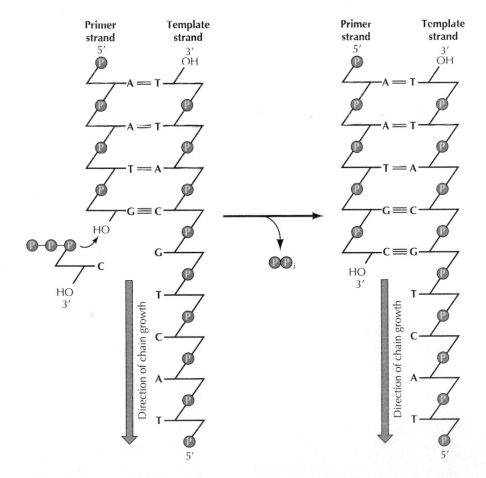

*Figure 5.2*
**The reaction catalyzed by DNA polymerase**  All known DNA polymerases add a deoxyribonucleoside 5'-triphosphate to the 3' hydroxyl group of a growing DNA chain (the primer strand).

## *The Replication Fork*

DNA molecules in the process of replication were first analyzed by John Cairns in experiments in which *E. coli* were grown in the presence of radioactive thymidine, which allowed subsequent visualization of newly replicated DNA by autoradiography (Figure 5.3). In some cases, complete circular molecules in the process of replicating could be observed. These DNA molecules contained two **replication forks**, representing the regions of active DNA synthesis. At each fork the parental strands of DNA separated and two new daughter strands were synthesized.

The synthesis of new DNA strands complementary to both strands of the parental molecule posed an important problem to understanding the biochemistry of DNA replication. Since the two strands of double-helical DNA run in opposite (antiparallel) directions, continuous synthesis of two new strands at the replication fork would require that one strand be synthesized in the 5′ to 3′ direction while the other is synthesized in the opposite (3′ to 5′) direction. But DNA polymerase catalyzes the polymerization of dNTPs only in the 5′ to 3′ direction. How, then, can the other progeny strand of DNA be synthesized?

This enigma was resolved by experiments showing that only one strand of DNA is synthesized in a continuous manner in the direction of overall DNA replication; the other is formed from small, discontinuous pieces of DNA that are synthesized backward with respect to the direction of movement of the replication fork (Figure 5.4). These small pieces of newly synthesized DNA (called **Okazaki fragments** after their discoverer) are joined by the action of **DNA ligase**, forming an intact new DNA strand. The continuously synthesized strand is called the **leading strand**, since its elongation in the direction of replication fork movement exposes the template used for the synthesis of Okazaki fragments (the **lagging strand**).

Although the discovery of discontinuous synthesis of the lagging strand provided a mechanism for the elongation of both strands of DNA at the replication fork, it raised another question: Since DNA polymerase requires a primer and cannot initiate synthesis *de novo*, how is the synthesis of Okazaki fragments initiated? The answer is that short fragments of RNA serve as primers for DNA replication (Figure 5.5). In contrast to DNA synthesis, the synthesis of RNA can initiate *de novo*, and an enzyme called **primase** synthesizes short fragments of RNA (e.g., three to ten nucleotides

**Figure 5.3**
**Replication of *E. coli* DNA** (A) An autoradiograph showing bacteria that were grown in [³H]thymidine for two generations to label the DNA, which was then extracted and visualized by exposure to photographic film. (B) This schematic illustrates the two replication forks shown in (A). (From J. Cairns, *Cold Spring Harbor Symp. Quant. Biol.*, 1963, 28: 43.)

(A)

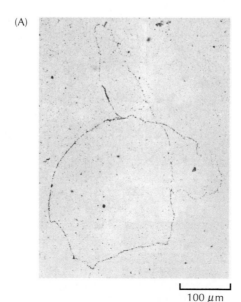

100 μm

(B)

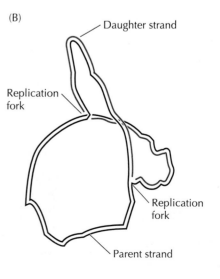

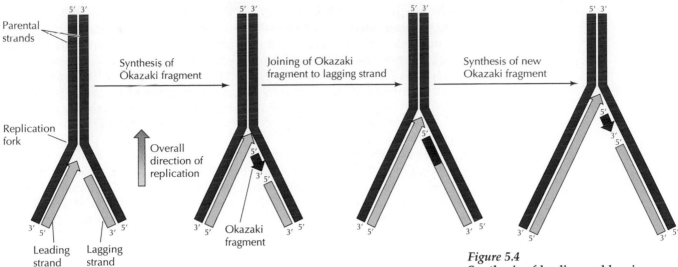

*Figure 5.4*
**Synthesis of leading and lagging strands of DNA**   The leading strand is synthesized continuously in the direction of replication fork movement. The lagging strand is synthesized in small pieces (Okazaki fragments) backward from the overall direction of replication. The Okazaki fragments are then joined by the action of DNA ligase.

long) complementary to the lagging strand template at the replication fork. Okazaki fragments are then synthesized via extension of these RNA primers by DNA polymerase. An important consequence of such RNA priming is that newly synthesized Okazaki fragments contain an RNA-DNA joint, the discovery of which provided critical evidence for the role of RNA primers in DNA replication.

To form a continuous lagging strand of DNA, the RNA primers must eventually be removed from the Okazaki fragments and replaced with DNA. In *E. coli*, RNA primers are removed by the combined action of **RNase H**, an enzyme that degrades the RNA strand of RNA-DNA hybrids, and polymerase I. This is the aspect of *E. coli* DNA replication in which polymerase I plays a critical role. In addition to its DNA poly-

*Figure 5.5*
**Initiation of Okazaki fragments with RNA primers**   Short fragments of RNA serve as primers that can be extended by DNA polymerase.

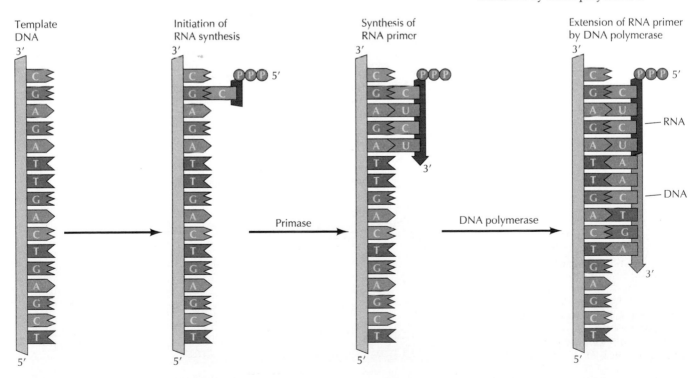

merase activity, polymerase I acts as an **exonuclease** that can hydrolyze DNA (or RNA) in either the 3′ to 5′ or 5′ to 3′ direction. The action of polymerase I as a 5′ to 3′ exonuclease removes ribonucleotides from the 5′ ends of Okazaki fragments, allowing them to be replaced with deoxyribonucleotides to yield fragments consisting entirely of DNA (Figure 5.6). In eukaryotic cells, other exonucleases take the place of *E. coli* polymerase I in removing primers, and the gaps between Okazaki fragments are filled by the action of polymerase δ. As in prokaryotes, these DNA fragments can then be joined by DNA ligase.

The different DNA polymerases thus play distinct roles at the replication fork (Figure 5.7). In prokaryotic cells, polymerase III is the major replicative polymerase, functioning in the synthesis both of the leading strand of DNA and of Okazaki fragments by the extension of RNA primers. Polymerase I then removes RNA primers and fills the gaps between Okazaki fragments. In eukaryotic cells, however, multiple DNA polymerases are required to do what in *E. coli* is accomplished by polymerase III alone. Polymerase α is found in a complex with primase, and it appears to function in conjunction with primase to synthesize short RNA-DNA fragments during lagging strand synthesis. Polymerase δ can then synthesize both the leading and lagging strands, acting to extend the RNA-DNA primers initially synthesized by the polymerase α-primase complex. In addition, polymerase δ can take the place of *E. coli* polymerase I in filling the gaps between Okazaki fragments following primer removal. Since the replication of SV40 DNA *in vitro* requires only polymerases α and δ, the role of ε is unclear; it may also function in cellular DNA replication, but its biochemical activities remain to be elucidated.

Not only polymerases and primase but also a number of other proteins act at the replication fork. These additional proteins have been identified

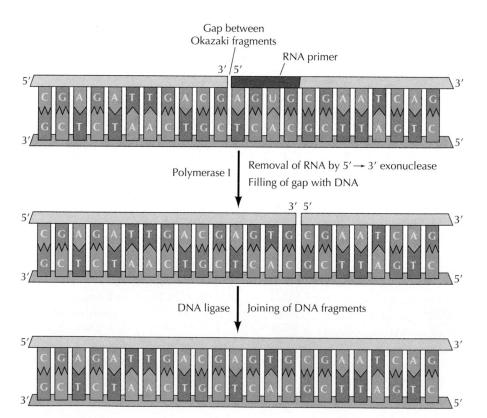

*Figure 5.6*
**Removal of RNA primers and joining of Okazaki fragments**   Because of its 5′ to 3′ exonuclease activity, DNA polymerase I removes RNA primers and fills the gaps between Okazaki fragments with DNA. The resultant DNA fragments can then be joined by DNA ligase.

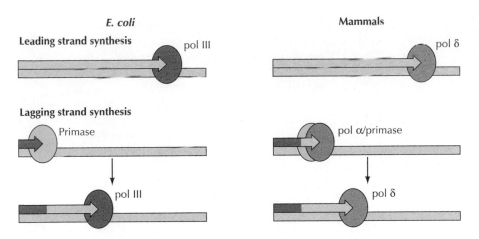

*Figure 5.7*
**Roles of DNA polymerases in *E. coli* and mammalian cells** The leading strand is synthesized by polymerase III (pol III) in *E. coli* and by polymerase δ (pol δ) in mammalian cells. In *E. coli*, lagging strand synthesis is initiated by primase, and RNA primers are extended by polymerase III. In mammalian cells, lagging strand synthesis is initiated by a complex of primase with polymerase α (pol α). The short RNA-DNA fragments synthesized by this complex are then extended by polymerase δ.

both by the analysis of *E. coli* mutants defective in DNA replication and by the purification of the mammalian proteins required for *in vitro* replication of SV40 DNA. One class of proteins required for replication binds to DNA polymerases, increasing the activity of the polymerases and causing them to remain bound to the template DNA so that they continue synthesis of a new DNA strand. Both *E. coli* polymerase III and eukaryotic polymerase δ are associated with two types of accessory proteins (sliding-clamp proteins and brace proteins) that load the polymerase onto the primer and maintain its stable association with the template (Figure 5.8). The brace proteins (called the γ complex in *E. coli* and replication factor C [RFC] in eukaryotes) specifically recognize and bind DNA at the junction between the primer and template. The sliding-clamp proteins (β protein in *E. coli* and proliferating cell nuclear antigen [PCNA] in eukaryotes) bind adjacent to the brace proteins, forming a ring around the template DNA. The brace and clamp proteins then load the DNA polymerase onto DNA at the primer-template junction. The ring formed by the sliding clamp maintains the association of the polymerase with its template as replication proceeds, allowing the uninterrupted synthesis of many thousands of nucleotides of DNA.

Other proteins unwind the template DNA and stabilize single-stranded regions (Figure 5.9). **Helicases** are enzymes that catalyze the unwinding of parental DNA, coupled to the hydrolysis of ATP, ahead of the replication fork. **Single-stranded DNA-binding proteins** (e.g., eukaryotic replication factor A [RFA]) then stabilize the unwound template DNA, keeping it in an extended single-stranded state so that it can be copied by the polymerase.

As the strands of parental DNA unwind, the DNA ahead of the replication fork is forced to rotate. Unchecked, this rotation would cause circular DNA molecules (such as SV40 DNA or the *E. coli* chromosome) to become twisted around themselves, eventually blocking replication (Figure 5.10). This problem is solved by **topoisomerases,** enzymes that catalyze the reversible breakage and rejoining of DNA strands. There are two types of these enzymes: Type I topoisomerases break just one strand of DNA; type II topoisomerases introduce simultaneous breaks in both strands. The breaks introduced by type I and type II topoisomerases serve as "swivels" that allow the two strands of template DNA to rotate freely around each other so that replication can proceed without twisting the DNA ahead of the fork (Figure 5.10). Although eukaryotic chromosomes are composed of linear rather than circular DNA molecules, their replication also requires topoisomerases: otherwise the complete chromosomes would have to rotate continually during DNA synthesis.

*Figure 5.8*
**Polymerase accessory proteins**
(A) The brace protein (RFC in mammalian cells) binds DNA at the junction between primer and template. The sliding-clamp protein (PCNA in mammalian cells) binds adjacent to the brace protein. DNA polymerase then binds to the brace-clamp complex. (B) Model of the *E. coli* sliding-clamp protein bound to DNA. The protein is a dimer, with one subunit shown in blue and the other in red. DNA is shown in orange and yellow. (Computer model by Dan Richardson.)

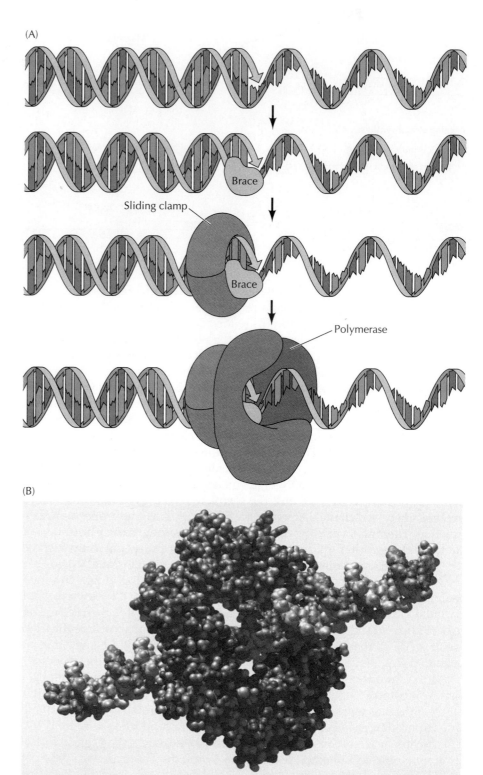

(B)

Type II topoisomerase is needed not only to unwind DNA but also to unravel newly replicated circular DNA molecules that become interwined with each other. In addition, topoisomerase II is a major component of the protein scaffold of eukaryotic chromosomes, to which large loops of DNA are attached (see Chapter 4). Studies of yeast mutants indicate that topoisomerase II is required for the separation of daughter chromatids at mito-

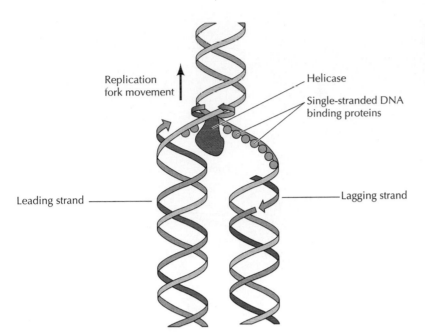

Replication fork movement

Helicase

Single-stranded DNA binding proteins

Leading strand

Lagging strand

*Figure 5.9*
**Action of helicases and single-stranded DNA-binding proteins**   Helicases unwind the two strands of parental DNA ahead of the replication fork. The unwound DNA strands are then stabilized by single-stranded DNA-binding proteins so that they can serve as templates for new DNA synthesis.

sis, suggesting that it also untangles newly replicated loops of DNA in the chromosomes of eukaryotes.

The enzymes involved in DNA replication act in a coordinated manner to synthesize both leading and lagging strands of DNA simultaneously at the replication fork (Figure 5.11). This task is accomplished by the formation of dimers of the replicative DNA polymerases (polymerase III in *E. coli* or polymerase δ in eukaryotes), each with its appropriate accessory proteins. One molecule of polymerase then acts in synthesis of the leading strand while the other acts in synthesis of the lagging strand. The lagging strand template is thought to form a loop at the replication fork so that the polymerase subunit engaged in lagging strand synthesis moves in the same overall direction as the other subunit, which is synthesizing the leading strand.

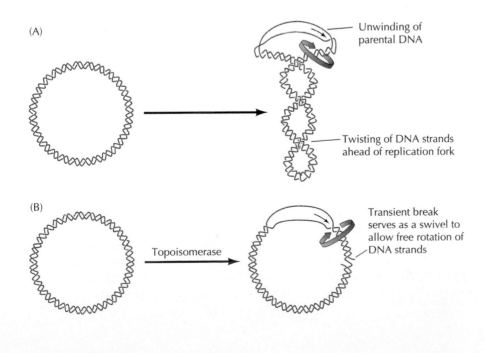

(A)

Unwinding of parental DNA

Twisting of DNA strands ahead of replication fork

(B)

Topoisomerase

Transient break serves as a swivel to allow free rotation of DNA strands

*Figure 5.10*
**Action of topoisomerases during DNA replication**   (A) As the two strands of template DNA unwind, the DNA ahead of the replication fork is forced to rotate in the opposite direction, causing circular molecules to become twisted around themselves. (B) This problem is solved by topoisomerases, which catalyze the reversible breakage and joining of DNA strands. The transient breaks introduced by these enzymes serve as swivels that allow the two strands of DNA to rotate freely around each other.

*Figure 5.11*
**Model of the *E. coli* replication fork**   Helicase, primase, and two molecules of DNA polymerase III carry out coordinated synthesis of both the leading and lagging strands of DNA. The lagging strand template is folded so that the polymerase responsible for lagging strand synthesis moves in the same direction as overall movement of the fork. Topoisomerase acts as a swivel ahead of the fork, and DNA polymerase I and ligase remove RNA primers and join Okazaki fragments behind the fork.

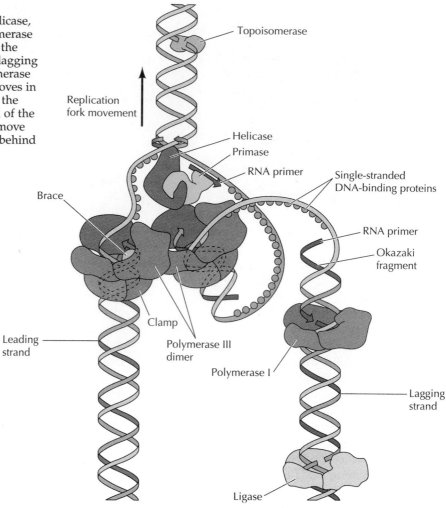

*Figure 5.12*
**Mismatching between rare configurations of nucleic acid bases**   In its normal configuration, guanine (G) specifically forms complementary hydrogen bonds (dashed lines) with cytosine (C). However, G occasionally assumes a rare configuration (tautomeric form) that instead forms hydrogen bonds with thymine (T).

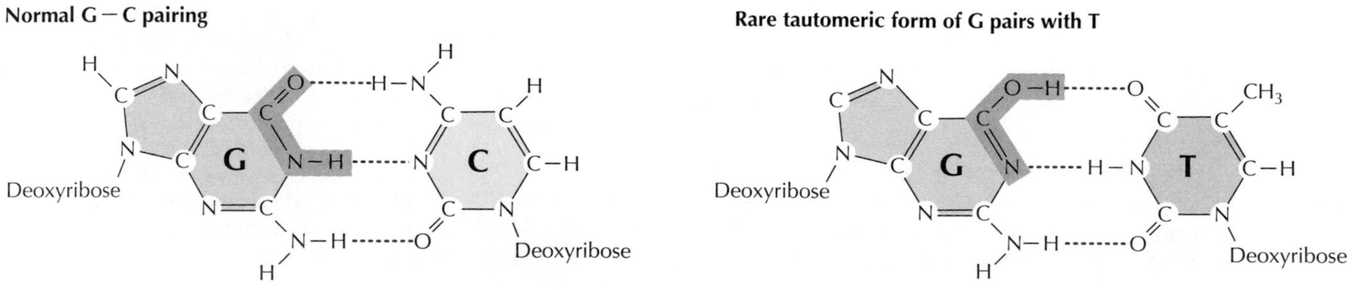

## The Fidelity of Replication

The accuracy of DNA replication is critical to cell reproduction, and estimates of mutation rates for a variety of genes indicate that the frequency of errors during replication corresponds to only one incorrect base per $10^9$ to $10^{10}$ nucleotides incorporated. This error frequency is much lower than would be predicted simply on the basis of complementary base pairing. In particular, the standard configurations of nucleic acid bases are in equilibrium with rare alternative conformations (tautomeric forms) that hydrogen-bond with the wrong partner (e.g., G with T) with a frequency of about one incorrect base per $10^4$ (Figure 5.12). The much higher degree of fidelity actually achieved results largely from the activities of DNA polymerase.

**Normal G — C pairing**

**Rare tautomeric form of G pairs with T**

One mechanism by which DNA polymerase increases the fidelity of replication is by helping to select the correct base for insertion into newly synthesized DNA. The polymerase does not simply catalyze incorporation of whatever nucleotide is hydrogen-bonded to the template strand. Instead, it actively discriminates against incorporation of a mismatched base, presumably by adapting to the conformation of a correct base pair. The molecular mechanisms responsible for the ability of DNA polymerases to select against incorrect bases are not yet entirely understood, but this selectivity appears to increase the accuracy of replication about a hundredfold, reducing the expected error frequency from $10^{-4}$ to approximately $10^{-6}$.

The other major mechanism responsible for the accuracy of DNA replication is the **proofreading** activity of DNA polymerase. As already noted, *E. coli* polymerase I has 3′ to 5′ as well as 5′ to 3′ exonuclease activity. The 5′ to 3′ exonuclease operates in the direction of DNA synthesis and helps remove RNA primers from Okazaki fragments. The 3′ to 5′ exonuclease operates in the reverse direction of DNA synthesis, and participates in proofreading newly synthesized DNA (Figure 5.13). Proofreading is effective because DNA polymerase requires a primer and is not able to initiate synthesis *de novo*. Primers that are hydrogen-bonded to the template are preferentially used, so when an incorrect base is incorporated, it is likely to be removed by the 3′ to 5′ exonuclease activity rather than being used to continue synthesis. Such 3′ to 5′ exonuclease activities are also associated with *E. coli* polymerase III and eukaryotic polymerases δ and ε. The 3′ to 5′ exonucleases of these polymerases selectively excise mismatched bases that have been incorporated at the end of a growing DNA chain, thereby increasing the accuracy of replication by a hundred- to a thousandfold.

The importance of proofreading may explain the fact that DNA polymerases require primers and catalyze the growth of DNA strands only in the 5′ to 3′ direction. When DNA is synthesized in the 5′ to 3′ direction, the energy required for polymerization is derived from hydrolysis of the 5′ triphosphate group of a free dNTP as it is added to the 3′ hydroxyl group of the growing chain (see Figure 5.2). If DNA were to be extended in the 3′ to 5′ direction, the energy of polymerization would instead have to be derived from hydrolysis of the 5′ triphosphate group of the terminal nucleotide already incorporated into DNA. This would eliminate the possibility of proofreading, because removal of a mismatched terminal nucleotide would also remove the 5′ triphosphate group needed as an energy source for further chain elongation. Thus, although the ability of DNA polymerase to extend a primer only in the 5′ to 3′ direction appears to make replication a complicated process, it is necessary for ensuring accurate duplication of the genetic material.

Combined with the ability to discriminate against the insertion of mismatched bases, the proofreading activity of DNA polymerases is sufficient to reduce the error frequency of replication to about one mismatched base per $10^9$. Additional mechanisms (discussed in the section "DNA Repair") act to remove mismatched bases that have been incorporated into newly synthesized DNA, further ensuring correct replication of the genetic information.

## Origins and the Initiation of Replication

The replication of both prokaryotic and eukaryotic DNAs starts at a unique sequence called the **origin of replication**, which serves as a specific binding site for proteins that initiate the replication process. The first origin to be defined was that of *E. coli*, in which genetic analysis indicated that replication always begins at a unique site on the bacterial chromosome. The *E. coli* origin has since been studied in detail and found to consist of 245 base pairs of DNA, elements of which serve as binding sites for proteins

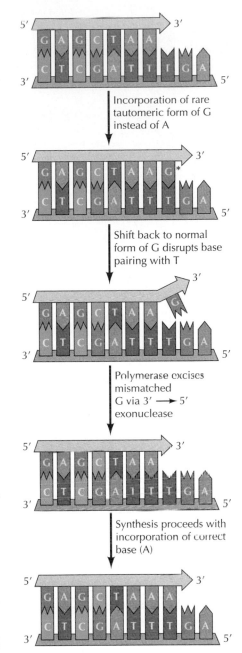

**Figure 5.13**
**Proofreading by DNA polymerase** A rare tautomeric form of G (G\*) is incorporated in place of A as a result of mispairing with T on the template strand. The subsequent shift of G back to its normal form disrupts this base pairing, so the 3′ terminal G is no longer hydrogen-bonded to the template strand. This mismatch at the 3′ terminus of the growing chain is recognized and excised by the 3′ to 5′ exonuclease activity of DNA polymerase, which requires a primer hydrogen-bonded to the template strand in order to continue synthesis. Following excision of the mismatched G, DNA synthesis can proceed with incorporation of the correct nucleotide (A).

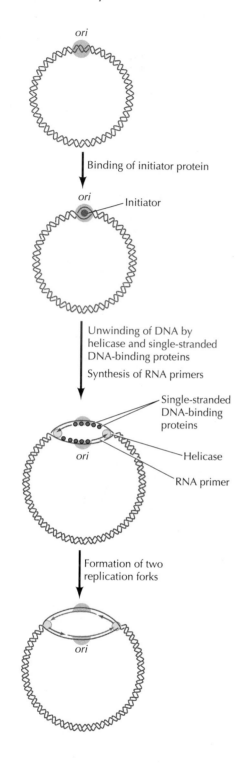

*Figure 5.15*
**Replication origins in eukaryotic chromosomes**   Replication initiates at multiple origins (*ori*), each of which produces two replication forks.

*Figure 5.14*
**Origin of replication in *E. coli***   Replication initiates at a unique site on the *E. coli* chromosome, designated the origin (*ori*). The first event is the binding of an initiator protein to *ori* DNA, which leads to partial unwinding of the template. The DNA continues to unwind by the actions of helicase and single-stranded DNA-binding proteins, and RNA primers are synthesized by primase. The two replication forks formed at the origin then move in opposite directions along the circular DNA molecule.

required to initiate DNA replication (Figure 5.14). The key step is the binding of an initiator protein to specific DNA sequences within the origin. The initiator protein begins to unwind the origin DNA and recruits the other proteins involved in DNA synthesis. Helicase and single-stranded DNA-binding proteins then act to continue unwinding and exposing the template DNA, and primase initiates the synthesis of leading strands. Two replication forks are formed and move in opposite directions along the circular *E. coli* chromosome.

The origins of replication of several animal viruses, such as SV40, have been studied as models for the initiation of DNA synthesis in eukaryotes. SV40 has a single origin of replication (consisting of 64 base pairs) that functions both in infected cells and in cell-free systems. Replication is initiated by a virus-encoded protein (called T antigen) that binds to the origin and also acts as a helicase. A single-stranded DNA-binding protein is required to stabilize the unwound template, and the DNA polymerase α-primase complex then initiates DNA synthesis.

Although single origins are sufficient to direct the replication of bacterial and viral genomes, multiple origins are needed to replicate the much larger genomes of eukaryotic cells within a reasonable period of time. For example, the entire genome of *E. coli* ($4 \times 10^6$ base pairs) is replicated from a single origin in approximately 30 minutes. If mammalian genomes ($3 \times 10^9$ base pairs) were replicated from a single origin at the same rate, DNA replication would require about 3 weeks (30,000 minutes). The problem is further exacerbated by the fact that the rate of DNA replication in mammalian cells is actually about tenfold lower than in *E. coli*, possibly as a result of the packaging of eukaryotic DNA in chromatin. Nonetheless, the genomes of mammalian cells are typically replicated within a few hours, necessitating the use of thousands of replication origins.

The presence of multiple replication origins in eukaryotic cells was first demonstrated by the exposure of cultured mammalian cells to radioactive

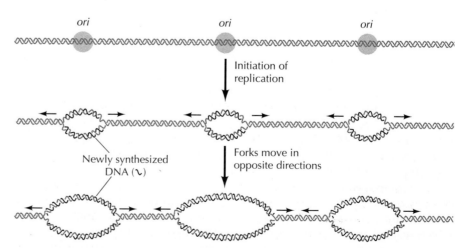

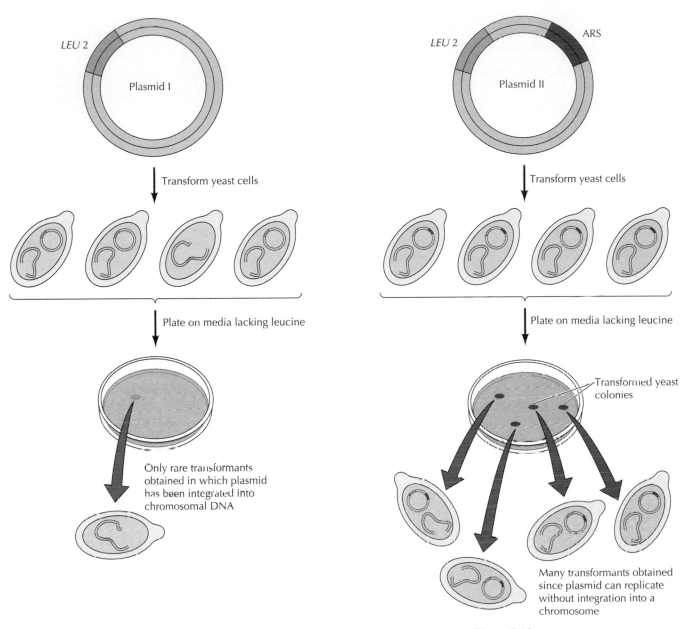

***Figure 5.16***
**Identification of origins of replication in yeast** Both plasmids I and II contain a selectable marker gene (*LEU* 2) that allows transformed cells to grow on medium lacking leucine. Only plasmid II, however, contains an origin of replication (ARS). The transformation of yeasts with plasmid I yields only rare transformants in which the plasmid has integrated into chromosomal DNA. Plasmid II, however, is able to replicate without integration into a yeast chromosome (autonomous replication), so many more transformants result from its introduction into yeast cells.

thymidine for different time intervals, followed by autoradiography to detect newly synthesized DNA. The results of such studies indicated that DNA synthesis is initiated at multiple sites, from which it then proceeds in both directions along the chromosome (Figure 5.15). The replication origins in mammalian cells are spaced at intervals of approximately 50 to 300 kb; thus the human genome has about 30,000 origins of replication. The genomes of simpler eukaryotes also have multiple origins; for example, replication in yeasts initiates at origins separated by intervals of approximately 40 kb.

The origins of replication of eukaryotic chromosomes have been studied best in yeasts, in which they have been identified as sequences that can support the replication of plasmids in transformed cells (Figure 5.16). This has provided a functional assay for these sequences, and several such elements (called **autonomously replicating sequences,** or **ARSs**) have been isolated. Their role as origins of replication has been verified by direct biochemical analysis, not only in plasmids but also in yeast chromosomal DNA.

Functional ARS elements span about 100 base pairs, including an 11-

*Figure 5.17*
**A yeast ARS element**   The element contains an 11-base-pair ARS consensus sequence (ACS), which is the specific binding site of the origin replication complex (ORC). Three additional elements (B1, B2, and B3) are individually not essential but together contribute to ARS function, possibly by being easily unwound to form single-stranded DNA.

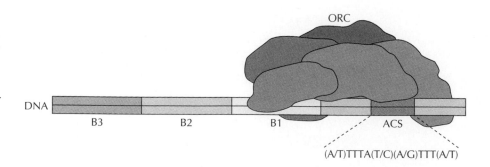

base-pair core sequence common to many different ARSs (Figure 5.17). This core sequence is essential for ARS function and has been found to be the binding site of a protein complex (called the **origin replication complex,** or **ORC**) that initiates DNA replication at yeast origins. The mechanism of initiation of DNA replication in yeasts thus appears similar to that in prokaryotes and eukaryotic viruses; that is, an initiator protein specifically binds to origin sequences.

In contrast to the well-defined ARS elements in yeasts, much less is known about the nature of replication origins in more complex eukaryotes. In particular, origins of replication in animal cells have not been characterized. However, proteins related to the yeast ORC proteins have recently been detected in a variety of eukaryotes, including *Drosophila, C. elegans, Arabidopsis,* and humans, suggesting that these proteins are general initiators of eukaryotic DNA replication. Further studies of ORC proteins may therefore facilitate the isolation and characterization of replication origins in more complex eukaryotic cells.

### Telomeres and Telomerase: Replicating the Ends of Chromosomes
Because DNA polymerases extend primers only in the 5′ to 3′ direction, they are unable to copy the extreme 5′ ends of linear DNA molecules. Consequently, special mechanisms are required to replicate the terminal sequences of the linear chromosomes of eukaryotic cells. These sequences (**telomeres**) consist of tandem repeats of simple-sequence DNA (see Chapter 4). They are replicated by the action of a unique enzyme called **telomerase**, which is able to maintain telomeres by catalyzing their synthesis in the absence of a DNA template.

Telomerase is a **reverse transcriptase**, one of a class of DNA polymerases, first discovered in retroviruses (see Chapter 3), that synthesize DNA from an RNA template. Importantly, telomerase carries its own template RNA, which is complementary to the telomere repeat sequences, as part of the enzyme complex. The use of this RNA as a template allows telomerase to generate multiple copies of the telomeric repeat sequences, thereby maintaining telomeres in the absence of a conventional DNA template to direct their synthesis.

The mechanism of telomerase action was initially elucidated in 1985 by Carol Greider and Elizabeth Blackburn during studies of the protozoan *Tetrahymena* (Figure 5.18). The *Tetrahymena* telomerase is complexed to a 159-nucleotide-long RNA that includes the sequence 3′–AACCCCAAC– 5′. This sequence is complementary to the *Tetrahymena* telomeric repeat (5′– TTGGGG–3′) and serves as the template for the synthesis of telomeric DNA. The use of this RNA as a template allows telomerase to extend the 3′ end of chromosomal DNA by one repeat unit beyond its original length. The complementary strand can then be synthesized by the polymerase

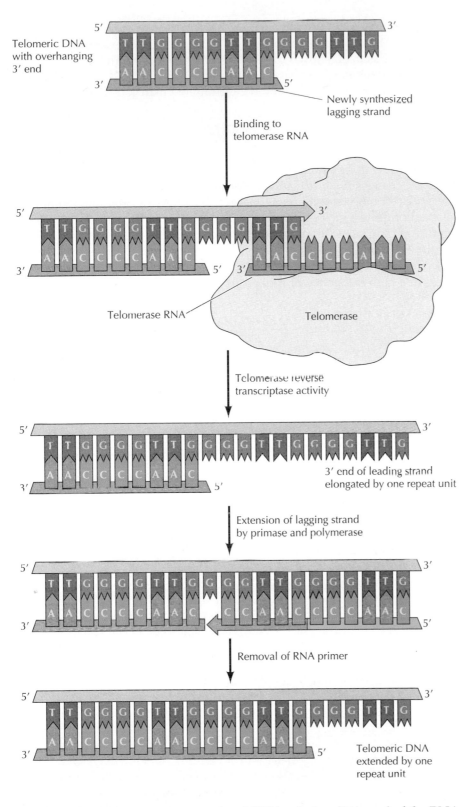

*Figure 5.18*
**Action of telomerase**   Telomeric DNA is a simple repeat sequence with an overhanging 3′ end on the newly synthesized leading strand. Telomerase carries its own RNA molecule, which is complementary to telomeric DNA, as part of the enzyme complex. The overhanging end of telomeric DNA binds to the telomerase RNA, which then serves as a template for extension of the leading strand by one repeat unit. The lagging strand of telomeric DNA can then be elongated by conventional RNA priming and DNA polymerase activity.

α-primase complex using conventional RNA priming. Removal of the RNA primer leaves an overhanging 3′ end of chromosomal DNA, which can be further extended by another round of telomerase action.

Telomerase has been identified in a variety of eukaryotes, and genes encoding telomerase RNAs have been cloned from *Tetrahymena*, yeasts, mice, and humans. In each case, the telomerase RNA contains sequences complementary to the telomeric repeat sequence of that organism (see Table 4.3). Moreover, the introduction of mutant telomerase RNA genes into yeasts has been shown to result in corresponding alterations of the chromosomal telomeric repeat sequences, directly demonstrating the function of telomerase in maintaining the termini of eukaryotic chromosomes.

## DNA REPAIR

DNA, like any other molecule, can undergo a variety of chemical reactions. Because DNA uniquely serves as a permanent copy of the cell genome, however, changes in its structure are of much greater consequence than are alterations in other cell components, such as RNAs or proteins. Mutations can result from the incorporation of incorrect bases during DNA replication. In addition, various chemical changes occur in DNA either spontaneously (Figure 5.19) or as a result of exposure to chemicals or radiation (Figure 5.20). Such damage to DNA can block replication or transcription, and can result in a high frequency of mutations—consequences that are unaccept-

(A) **Deamination**

Cytosine → Uracil

Adenine → Hypoxanthine

(B) **Depurination**

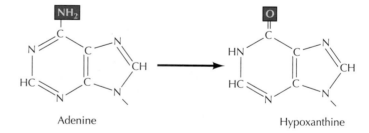

DNA chain          DNA chain

dGMP          AP site

*Figure 5.19*
**Spontaneous damage to DNA**   There are two major forms of spontaneous DNA damage: deamination of adenine, cytosine, and guanine (A), and depurination (loss of purine bases) resulting from cleavage of the bond between the purine bases and deoxyribose, leaving an apurinic (AP) site in DNA (B). dGMP = deoxyguanosine monophosphate.

(A) **Exposure to UV light**

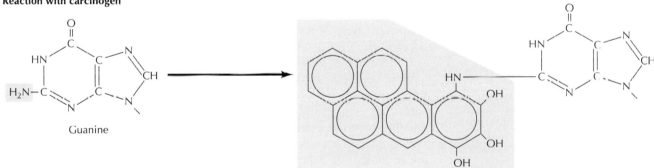

Adjacent thymines in DNA

Thymine dimer

Cyclobutane ring

(B) **Alkylation**

Guanine

$O^6$-methylguanine

(C) **Reaction with carcinogen**

Guanine

Bulky group addition

*Figure 5.20*
**Examples of DNA damage induced by radiation and chemicals**   (A) UV light induces the formation of pyrimidine dimers, in which two adjacent pyrimidines (e.g., thymines) are joined by a cyclobutane ring structure. (B) Alkylation is the addition of methyl or ethyl groups to various positions on the DNA bases. In this example, alkylation of the $O^6$ position of guanine results in formation of $O^6$-methylguanine. (C) Many carcinogens (e.g., benzo(a)pyrene) react with DNA bases, resulting in the addition of large bulky chemical groups to the DNA molecule.

able from the standpoint of cell reproduction. To maintain the integrity of their genomes, cells have therefore had to evolve mechanisms to repair damaged DNA. These mechanisms of DNA repair can be divided into two general classes: (1) direct reversal of the chemical reaction responsible for DNA damage, and (2) removal of the damaged bases followed by their replacement with newly synthesized DNA. Where DNA repair fails, additional mechanisms have evolved to enable cells to cope with the damage.

## Direct Reversal of DNA Damage

Most damage to DNA is repaired by removal of the damaged bases followed by resynthesis of the excised region. Some lesions in DNA, however, can be repaired by direct reversal of the damage, which may be a more effi-

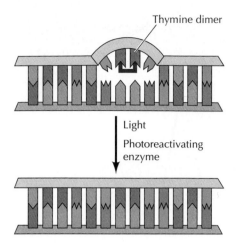

Thymine dimer

Light

Photoreactivating enzyme

**Figure 5.21**
**Direct repair of thymine dimers**
UV-induced thymine dimers can be repaired by photoreactivation, in which energy from visible light is used to split the bonds forming the cyclobutane ring.

cient way of dealing with specific types of DNA damage that occur frequently. Only a few types of DNA damage are repaired in this way, particularly pyrimidine dimers resulting from exposure to ultraviolet (UV) light and alkylated guanine residues that have been modified by the addition of methyl or ethyl groups at the $O^6$ position of the purine ring.

UV light is one of the major sources of damage to DNA and is also the most thoroughly studied form of DNA damage in terms of repair mechanisms. Its importance is illustrated by the fact that exposure to solar UV irradiation is the cause of almost all skin cancer in humans. The major type of damage induced by UV light is the formation of **pyrimidine dimers,** in which adjacent pyrimidines on the same strand of DNA are joined by the formation of a cyclobutane ring resulting from saturation of the double bonds between carbons 5 and 6 (see Figure 5.20A). The formation of such dimers distorts the structure of the DNA chain and blocks transcription or replication past the site of damage, so their repair is closely correlated with the ability of cells to survive UV irradiation. One mechanism of repairing UV-induced pyrimidine dimers is direct reversal of the dimerization reaction. The process is called **photoreactivation** because energy derived from visible light is utilized to break the cyclobutane ring structure (Figure 5.21). The original pyrimidine bases remain in DNA, now restored to their normal state. As might be expected from the fact that solar UV irradiation is a major source of DNA damage for diverse cell types, the repair of pyrimidine dimers by photoreactivation is common to a variety of prokaryotic and eukaryotic cells, including *E. coli*, yeasts, and some species of plants and animals. Curiously, however, photoreactivation is not universal; many species (including humans) lack this mechanism of DNA repair.

Another form of direct repair deals with damage resulting from the reaction between alkylating agents and DNA. Alkylating agents are reactive compounds that can transfer methyl or ethyl groups to a DNA base, thereby chemically modifying the base (see Figure 5.20B). A particularly important type of damage is methylation of the $O^6$ position of guanine, because the product, $O^6$-methylguanine, forms complementary base pairs with thymine instead of cytosine. This lesion can be repaired by an enzyme (called $O^6$-methylguanine methyltransferase) that transfers the methyl group from $O^6$-methylguanine to a cysteine residue in its active site (Figure 5.22). The potentially mutagenic chemical modification is thus removed, and the original guanine is restored. Enzymes that catalyze this direct repair reaction are widespread in both prokaryotes and eukaryotes, including humans.

### Excision Repair

Although direct repair is an efficient way of dealing with particular types of DNA damage, excision repair is a more general means of repairing a wide variety of chemical alterations to DNA. Consequently, the various types of excision repair are the most important DNA repair mechanisms in both prokaryotic and eukaryotic cells. In excision repair, the damaged DNA is recognized and removed, either as free bases or as nucleotides. The resulting gap is then filled in by synthesis of a new DNA strand, using the undamaged complementary strand as a template. Three types of excision repair—base-excision repair, nucleotide-excision repair, and mismatch repair—enable cells to cope with a variety of different kinds of DNA damage.

The repair of uracil-containing DNA is a good example of **base-excision repair**, in which single damaged bases are recognized and removed from the DNA molecule (Figure 5.23). Uracil can arise in DNA by two mechanisms: (1) Uracil (as dUTP [deoxyuridine triphosphate]) is occasionally

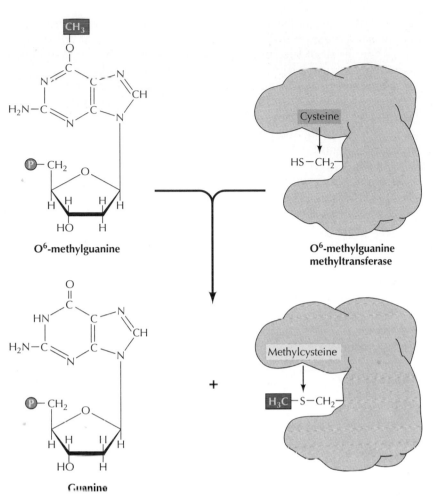

O⁶-methylguanine

Guanine

**Figure 5.22**
**Repair of O⁶-methylguanine**   O⁶-methylguanine methyltransferase transfers the methyl group from O⁶-methylguanine to a cysteine residue in the enzyme's active site.

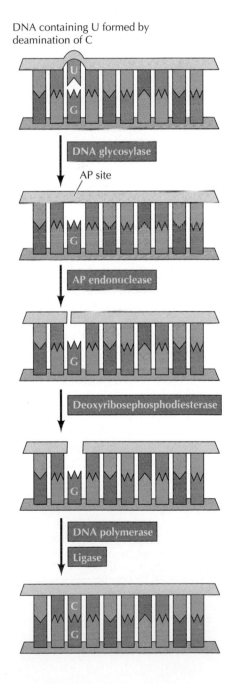

**Figure 5.23**
**Base-excision repair**   In this example, uracil (U) has been formed by deamination of cytosine (C) and is therefore opposite a guanine (G) in the complementary strand of DNA. The bond between uracil and the deoxyribose is cleaved by a DNA glycosylase, leaving a sugar with no base attached in the DNA (an AP site). This site is recognized by AP endonuclease, which cleaves the DNA chain. The remaining deoxyribose is removed by deoxyribosephosphodiesterase. The resulting gap is then filled by DNA polymerase and sealed by ligase, leading to incorporation of the correct base (C) opposite the G.

incorporated in place of thymine during DNA synthesis, and (2) uracil can be formed in DNA by the deamination of cytosine (see Figure 5.19A). The second mechanism is of much greater biological significance because it alters the normal pattern of complementary base pairing and thus represents a mutagenic event. The excision of uracil in DNA is catalyzed by **DNA glycosylase**, an enzyme that cleaves the bond linking the base (uracil) to the deoxyribose of the DNA backbone. This reaction yields free uracil and an apyrimidinic site—a sugar with no base attached. DNA glycosylases also recognize and remove other abnormal bases, including hypoxanthine formed by the deamination of adenine, pyrimidine dimers, alkylated purines other than O⁶-alkylguanine, and bases damaged by oxidation or ionizing radiation.

The result of DNA glycosylase action is the formation of an apyridiminic or apurinic site (generally called an AP site) in DNA. Similar AP sites are formed as the result of the spontaneous loss of purine bases (see Figure 5.19B), which occurs at a significant rate under normal cellular conditions. For example, each cell in the human body is estimated to lose several thousand purine bases daily. These sites are repaired by **AP endonuclease**, which cleaves adjacent to the AP site (see Figure 5.23). The remaining deoxyribose moiety is then removed, and the resulting single-base gap is filled by DNA polymerase and ligase.

Whereas DNA glycosylases recognize only specific forms of damaged bases, other excision repair systems recognize a wide variety of damaged bases that distort the DNA molecule, including UV-induced pyrimidine dimers and bulky groups added to DNA bases as a result of the reaction of many carcinogens with DNA (see Figure 5.20C). This widespread form of DNA repair is known as **nucleotide-excision repair**, because the damaged bases (e.g., a thymine dimer) are removed as part of an oligonucleotide containing the lesion (Figure 5.24).

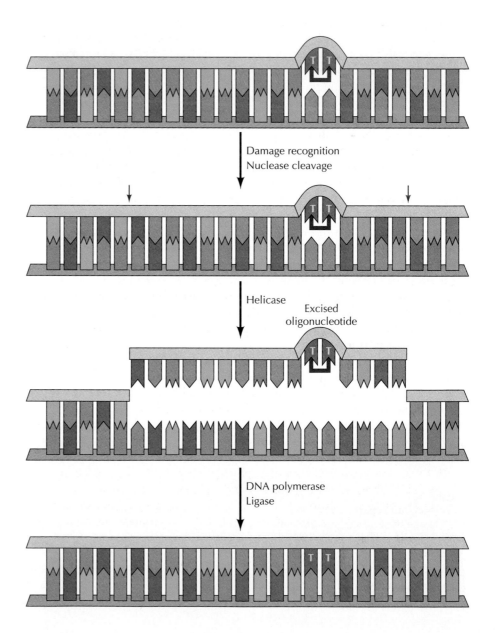

**Figure 5.24**
**Nucleotide-excision repair of thymine dimers** Damaged DNA is recognized and then cleaved on both sides of a thymine dimer by 3′ and 5′ nucleases. Unwinding by a helicase results in excision of an oligonucleotide containing the damaged bases. The resulting gap is then filled by DNA polymerase and sealed by ligase.

In *E. coli*, nucleotide-excision repair is catalyzed by the products of three genes (*uvrA, B,* and *C*) that were identified because mutations at these loci result in extreme sensitivity to UV light. The protein UvrA recognizes damaged DNA and recruits UvrB and UvrC to the site of the lesion. UvrB and UvrC then cleave on the 3′ and 5′ sides of the damaged site, respectively, thus excising an oligonucleotide consisting of 12 or 13 bases. The UvrABC complex is frequently called an **excinuclease**, a name that reflects its ability to directly *excise* an oligonucleotide. The action of a helicase is then required to remove the damage-containing oligonucleotide from the double-stranded DNA molecule, and the resulting gap is filled by DNA polymerase I and sealed by ligase.

Nucleotide-excision repair systems have also been studied extensively in eukaryotes, particularly in yeasts and in humans. In yeasts, as in *E. coli*, several genes involved in DNA repair (called *RAD* genes for *rad*iation sensitivity) have been identified by the isolation of mutants with increased sensitivity to UV light. In humans, DNA repair genes have been identified largely by studies of individuals suffering from inherited diseases resulting from deficiencies in the ability to repair DNA damage. The most extensively studied of these diseases is xeroderma pigmentosum (XP), a rare genetic disorder that affects approximately one in 250,000 people. Individuals with this disease are extremely sensitive to UV light and develop multiple skin cancers on the regions of their bodies that are exposed to sunlight. In 1968 James Cleaver made the key discovery that cultured cells from XP patients were deficient in the ability to carry out nucleotide-excision repair. This observation not only provided the first link between DNA repair and cancer, but also suggested the use of XP cells as an experimental system to identify human DNA repair genes. The identification of human DNA repair genes has been accomplished by studies not only of XP cells, but also of two other human diseases resulting from DNA repair defects (Cockayne's syndrome and trichothiodystrophy) and of UV-sensitive mutants of rodent cell lines. The availability of mammalian cells with defects in DNA repair has allowed the cloning of repair genes based on the ability of wild-type alleles to restore normal UV sensitivity to mutant cells in gene transfer assays, thereby opening the door to experimental analysis of nucleotide-excision repair in mammalian cells.

Molecular cloning has now identified seven different repair genes (designated *XPA* through *XPG*) that are mutated in cases of xeroderma pigmentosum, as well as in some cases of Cockayne's syndrome, trichothiodystrophy, and UV-sensitive mutants of rodent cells. Table 5.1 lists the enzymes encoded by these genes. Some UV-sensitive rodent cells have mutations in yet another repair gene, called *ERCC1* (for *e*xcision *r*epair *c*ross *c*omplementing), which has not been found to be mutated in known human diseases. It is notable that the proteins encoded by these human DNA repair genes are closely related to proteins encoded by yeast *RAD* genes, indicating that nucleotide-excision repair is highly conserved throughout eukaryotes.

With cloned yeast and human repair genes available, it has been possible to purify their encoded proteins and develop *in vitro* systems to study the repair process. Although some steps remain to be fully elucidated, these studies have led to the development of a basic model for nucleotide-excision repair in eukaryotic cells. In mammalian cells, the XPA protein initiates repair by recognizing damaged DNA and forming complexes with other proteins involved in the repair process. These include the XPB and XPD proteins, which act as helicases that unwind the damaged DNA. In addition, the binding of XPA to damaged DNA leads to the recruitment of

***Table 5.1*** Enzymes Involved in Nucleotide-Excision Repair

| Human | Yeast | Function |
|-------|-------|----------|
| XPA | RAD14 | Damage recognition |
| XPB | RAD25 | Helicase |
| XPC | RAD4 | DNA binding |
| XPD | RAD3 | Helicase |
| XPF | RAD1 | 5′ nuclease |
| XPG | RAD2 | 3′ nuclease |
| ERCC1 | RAD10 | Dimer with XPF |

XPF (as a heterodimer with ERCC1) and XPG to the repair complex. XPF and XPG are endonucleases, which cleave DNA on the 5′ and 3′ sides of the damaged site, respectively. This cleavage excises an oligonucleotide consisting of approximately 29 bases. The resulting gap then appears to be filled in by DNA polymerase δ or ε (in association with RFC and PCNA) and sealed by ligase.

An intriguing feature of nucleotide-excision repair is its relationship to transcription. A connection between transcription and repair was first suggested by experiments showing that transcribed strands of DNA are repaired more rapidly than nontranscribed strands in both *E. coli* and mammalian cells. Since DNA damage blocks transcription, this transcription-repair coupling is thought to be advantageous by allowing the cell to preferentially repair damage to actively expressed genes. Although the molecular mechanism of transcription-repair coupling is not yet known, it is noteworthy that the XPB and XPD helicases are also components of a multisubunit transcription factor (called TFIIH) that is required to initiate the transcription of eukaryotic genes (see Chapter 6). Thus, these helicases appear to be required for the unwinding of DNA during both transcription and nucleotide-excision repair, providing a direct biochemical link between these two processes. Note also that some cases of Cockayne's syndrome result from mutations in another helicase, called CSB. The consequence of mutational inactivation of CSB is a failure of the preferential repair of transcribed DNA strands, suggesting that CSB may function as a transcription-repair coupling factor.

A third excision repair system recognizes mismatched bases that are incorporated during DNA replication. Many such mismatched bases are removed by the proofreading activity of DNA polymerase. Those that are missed are subject to later correction by the **mismatch repair** system, which scans newly replicated DNA. If a mismatch is found, the enzymes of this repair system are able to identify and excise the mismatched base specifically from the newly replicated DNA strand, allowing the error to be corrected and the original sequence restored.

In *E. coli*, the ability of the mismatch repair system to distinguish between parental DNA and newly synthesized DNA is based on the fact that DNA of this bacterium is modified by the methylation of adenine residues within the sequence GATC to form 6-methyladenine (Figure 5.25). Since methylation occurs after replication, newly synthesized DNA strands are not methylated and thus can be specifically recognized by the mismatch repair enzymes. Mismatch repair is initiated by the protein MutS, which recognizes the mismatch and forms a complex with two other proteins called MutL and MutH. The MutH endonuclease then cleaves the unmethylated DNA strand at a GATC sequence. MutL and MutS then act together with an exonuclease and a helicase to excise the DNA between the strand break and the mismatch, with the resulting gap being filled by DNA polymerase and ligase.

Eukaryotes have a similar mismatch repair system, although the mechanism by which eukaryotic cells identify newly replicated DNA is not yet known. However, genes related to the *E. coli* genes *MutS* and *MutL* are required for mismatch repair in both yeast and mammalian cells. The importance of this repair system is dramatically illustrated by the fact that mutations in the human homologs of *MutS* and *MutL* are responsible for a common type of inherited colon cancer (hereditary nonpolyposis colorectal cancer, or HNPCC). HNPCC is one of the most common inherited diseases; it affects as many as one in 200 people and is responsible for about 15% of all colorectal cancers in this country. The relationship between HNPCC and defects in mismatch repair was discovered in 1993, when two

**Figure 5.25**
**Mismatch repair in *E. coli*** The mismatch repair system detects and excises mismatched bases in newly replicated DNA, which is distinguished from the parental strand because it has not yet been methylated. MutS binds to the mismatched base, followed by MutL. The binding of MutL activates MutH, which cleaves the unmodified strand opposite a site of methylation. MutS and MutL, together with a helicase and an exonuclease, then excise the portion of the unmodified strand that contains the mismatch. The gap is then filled by DNA polymerase and sealed by ligase.

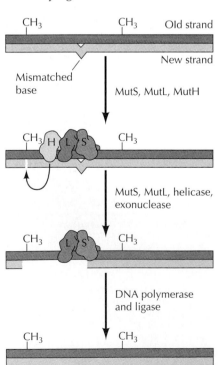

## MOLECULAR MEDICINE

# *Colon Cancer and DNA Repair*

### The Disease

Cancers of the colon and rectum (colorectal cancers) are one of the most common types of cancer in Western countries, accounting for nearly 140,000 cancer cases per year in the United States (approximately 10% of the total cancer incidence). Most colon cancers (like other types of cancer) are not inherited diseases; that is, they are not transmitted directly from parent to offspring. However, two inherited forms of colon cancer have been described. In both of these syndromes, the inheritance of a cancer susceptibility gene results in a very high likelihood of cancer development. One inherited form of colon cancer (familial adenomatous polyposis) is extremely rare, accounting for less than 1% of total colon cancer incidence. The second inherited form of colon cancer (hereditary nonpolyposis colorectal cancer, or HNPCC) is much more common and accounts for up to 15% of all colon cancer cases. Indeed, HNPCC is one of the most common inherited diseases, affecting as many as one in 200 people. Although colon cancers are the most common manifestation of this disease, affected individuals also suffer an increased incidence of other types of cancer, including cancers of the ovary and endometrium.

### Molecular and Cellular Basis

Like other cancers, colorectal cancer results from mutations in genes that regulate cell proliferation, leading to the uncontrolled growth of cancer cells. In most cases these mutations occur sporadically in somatic cells. In hereditary cancers, however, inher-

ited germ-line mutations predispose the individual to cancer development. A striking advance was made in 1993 with the discovery that a gene responsible for approximately 50% of HNPCC cases encodes an enzyme involved in mismatch repair of DNA; this gene is a human homolog of the *E. coli MutS* gene. Subsequent studies have shown that three other genes, responsible for most remaining cases of HNPCC, are homologs of *MutL* and thus are also involved in the mismatch repair pathway. Defects in these genes appear to result in a high frequency of mutations in other cell genes, with a correspondingly high likelihood that some of these mutations will eventually lead to the development of cancer by affecting genes that regulate cell proliferation.

### Prevention and Treatment

As with other inherited diseases, identification of the genes responsible for HNPCC allows individuals at risk for this inherited cancer to be identified by genetic testing. Moreover, prenatal genetic diagnosis may be of great importance to carriers of HNPCC mutations who are planning a family. However, the potential benefits of detecting these mutations are not limited to preventing the transmission of mutant genes to the next generation; their detection may also help prevent the development of cancer in affected individuals.

In terms of disease prevention, a key characteristic of colon cancer is that it develops gradually over several years. Early diagnosis of the disease substantially improves the

chances for patient survival. The initial stage of colon cancer development is the outgrowth of small benign polyps, which eventually become malignant and invade the surrounding connective tissue. Prior to the development of malignancy, however, polyps can be easily removed surgically, effectively preventing the outgrowth of a malignant tumor. Polyps and early stages of colon cancer can be detected by examination of the colon with a thin lighted tube (colonoscopy), so frequent colonoscopy of HNPCC patients may allow polyps to be removed before cancer develops. In addition, several drugs are being tested as potential inhibitors of colon cancer development, and these drugs may be of significant benefit to HNPCC patients. By allowing the timely application of such preventive measures, the identification of mutations responsible for HNPCC may make a significant contribution to disease prevention.

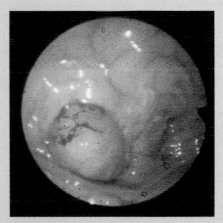

A colon polyp visualized by colonoscopy. (Custom Medical Stock Photo.)

groups of researchers cloned the human homolog of *MutS* and found that mutations in this gene were responsible for about half of all HNPCC cases. Subsequent studies have shown that most of the remaining cases of HNPCC are caused by mutations in one of three human genes that are homologs of *MutL*.

## Postreplication Repair

The direct reversal and excision repair systems act to correct DNA damage before replication, so that replicative DNA synthesis can proceed using an undamaged DNA strand as a template. Should these systems fail, however, the cell has alternative mechanisms for dealing with damaged DNA at the replication fork. Pyrimidine dimers and many other types of lesions cannot be copied by the normal action of DNA polymerases, so replication is blocked at the sites of such damage. Downstream of the damaged site, however, replication can be initiated again by the synthesis of an Okazaki fragment and can proceed along the damaged template strand (Figure 5.26). The result is a daughter strand that has a gap opposite the site of damage to the parental strand. One of two types of mechanisms may be used to repair such gaps in newly synthesized DNA: recombinational repair or error-prone repair.

**Recombinational repair** depends on the fact that one strand of the parental DNA was undamaged and therefore was copied during replication to yield a normal daughter molecule (see Figure 5.26). The undamaged parental strand can be used to fill the gap opposite the site of dam-

*Figure 5.26*
**Postreplication repair** The presence of a thymine dimer blocks replication, but DNA polymerase can bypass the lesion and reinitiate replication at a new site downstream of the dimer. The result is a gap opposite the dimer in the newly synthesized DNA strand. In recombinational repair, this gap is filled by recombination with the undamaged parental strand. Although this leaves a gap in the previously intact parental strand, the gap can be filled by the actions of polymerase and ligase, using the intact daughter strand as a template. Two intact DNA molecules are thus formed, and the remaining thymine dimer eventually can be removed by excision repair.

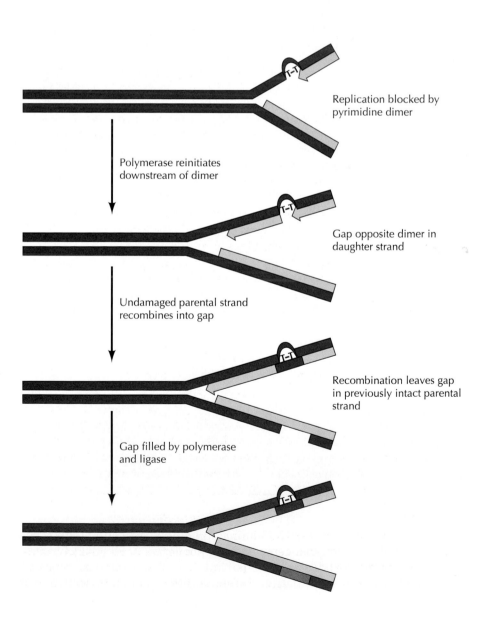

Replication blocked by pyrimidine dimer

Polymerase reinitiates downstream of dimer

Gap opposite dimer in daughter strand

Undamaged parental strand recombines into gap

Recombination leaves gap in previously intact parental strand

Gap filled by polymerase and ligase

age in the other daughter molecule by recombination between homologous DNA sequences (see the next section). Because the resulting gap in the previously intact parental strand is opposite an undamaged strand, it can be filled in by DNA polymerase. Although the other parent molecule still retains the original damage (e.g., a pyrimidine dimer), the damage now lies opposite a normal strand and can be dealt with later by excision repair.

In **error-prone repair**, a gap opposite a site of DNA damage is filled by newly synthesized DNA. Since the new DNA is synthesized from a damaged template strand, this form of DNA synthesis is very inaccurate and leads to frequent mutations. It is used only in bacteria that have been subjected to potentially lethal conditions, such as extensive UV irradiation. Such treatments induce the SOS response, which may be viewed as a mechanism for dealing with extreme environmental stress. The SOS response includes inhibition of cell division and induction of repair systems to cope with a high level of DNA damage. Under these conditions, error-prone repair mechanisms are used, presumably as a way of dealing with damage so extensive that cell death is the only alternative.

## RECOMBINATION BETWEEN HOMOLOGOUS DNA SEQUENCES

Accurate DNA replication and repair of DNA damage are essential to maintaining genetic information and ensuring its accurate transmission from parent to offspring. From the standpoint of evolution, however, it is also important to generate genetic diversity. Genetic differences between individuals provide the essential starting material of natural selection, which allows species to evolve and adapt to changing environmental conditions. Recombination plays an important role in this process by allowing genes to be reassorted into different combinations. For example, genetic recombination results in the exchange of genes between paired homologous chromosomes during meiosis (see Figure 3.4). However, increasing genetic diversity is not the only role of recombination. As discussed in the previous section, recombination is also an important mechanism for repairing damaged DNA. In addition, recombination is involved in rearrangements of specific DNA sequences that alter the expression and function of some genes during development and differentiation. Thus, recombination plays important roles in the lives of individual cells and organisms, as well as contributing to the genetic diversity of the species.

This section discusses the molecular mechanisms of recombination between DNA molecules that share extensive sequence homology. Examples include recombination between paired eukaryotic chromosomes during meiosis and recombination between bacterial chromosomes during mating. Since this type of recombination involves the exchange of genetic information between two homologous DNA molecules, it does not alter the overall arrangement of the genes on a chromosome. Other types of recombination, however, do not require extensive sequence homology and therefore can occur between unrelated DNAs. Recombination events of this type lead to gene rearrangements, which are discussed later in the chapter.

### DNA Molecules Recombine by Breaking and Rejoining

Genetic recombination was first defined by studies of *Drosophila*, on the basis of the observation that genes on different copies of homologous chromosomes can reassort during meiosis. With the subsequent discovery that genes consist of DNA, two alternative models to explain recombination at

**Copy choice**

Synthesis of new daughter DNAs                    Switching to copy other parental templates                    Recombinant daughter DNAs

**Breakage and rejoining**

Parental DNAs                    Breakage and crosswise rejoining                    Recombinant DNAs

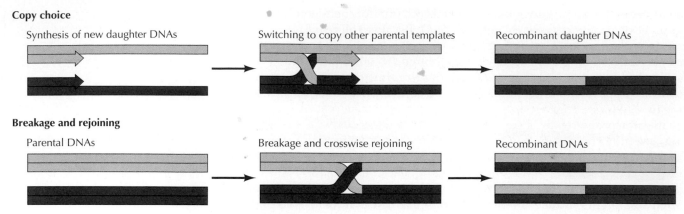

*Figure 5.27*
**Models of recombination**    In copy choice, recombination occurs during the synthesis of daughter DNA molecules. DNA replication starts with one parental DNA template and then switches to a second parental molecule, resulting in the synthesis of recombinant daughter DNAs containing sequences homologous to both parents. In breakage and rejoining, recombination occurs as a result of breakage and crosswise rejoining of parental DNA molecules.

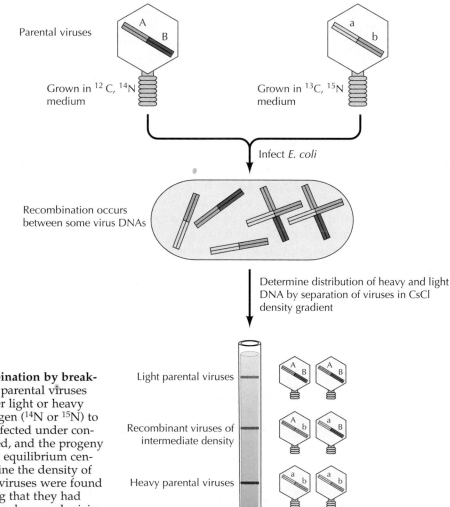

Parental viruses

Grown in $^{12}$C, $^{14}$N medium

Grown in $^{13}$C, $^{15}$N medium

Infect *E. coli*

Recombination occurs between some virus DNAs

Determine distribution of heavy and light DNA by separation of viruses in CsCl density gradient

*Figure 5.28*
**Experimental demonstration of recombination by breakage and rejoining**    Genetically distinct parental viruses were grown in medium containing either light or heavy isotopes of carbon ($^{12}$C or $^{13}$C) and nitrogen ($^{14}$N or $^{15}$N) to density-label their DNAs. *E. coli* were infected under conditions in which replication was inhibited, and the progeny viruses were harvested and analyzed by equilibrium centrifugation in a CsCl gradient to determine the density of genetic recombinants. The recombinant viruses were found to have intermediate densities, indicating that they had acquired DNA from both parents by a breakage and rejoining mechanism.

Light parental viruses

Recombinant viruses of intermediate density

Heavy parental viruses

the molecular level were considered (Figure 5.27). The "copy choice" model proposed that the recombinant molecule is generated during DNA synthesis, as a result of copying first one parental DNA and then switching to copy a different template. The alternative proposal was that recombination results from the breakage and rejoining of two parental DNA molecules rather than by synthesis of new DNA.

These alternatives were first distinguished in 1961 by studies of recombination between the genomes of bacterial viruses (Figure 5.28). Infection of *E. coli* with viruses carrying different genetic markers was known to yield recombinant progeny. To determine if this recombination involved breakage and rejoining of the parental DNAs, one of the parental viruses was grown in medium containing the heavy isotopes of carbon ($^{13}C$) and nitrogen ($^{15}N$), while the other was grown in medium containing the normal light isotopes ($^{12}C$ and $^{14}N$). The result was parental viruses having different densities, so they could be separated by equilibrium density centrifugation in a CsCl gradient. *E. coli* were then infected with these differentially labeled parental viruses under conditions in which replication was inhibited, and the progeny viruses produced were analyzed for both their density and their genetic characteristics. The important result was that genetic recombinant viruses were obtained that had intermediate densities, indicating that they had acquired DNA from both parents, as predicted by the breakage-and-rejoining, but not the copy choice, model.

### Models of Homologous Recombination

The finding that recombination occurs by breakage and rejoining raises a critical question: How can two parental DNA molecules be broken at precisely the same point, so that they can rejoin without mutations resulting from the gain or loss of nucleotides at the break point? During recombination between homologous DNA molecules (**general homologous recombination**), this alignment is provided, not surprisingly, by base pairing between complementary DNA strands (Figure 5.29). Overlapping single strands are exchanged between homologous DNA molecules, leading to the formation of a heteroduplex region, in which the two strands of the recombinant double helix are derived from different parents. If the heteroduplex region contains a genetic difference, the result is a single progeny DNA molecule that contains two genetic markers. In some cases, mispaired bases in a heteroduplex may be recognized and corrected by mismatch repair systems, as discussed in preceding sections of this chapter. Genetic evidence for the formation and repair of such heteroduplex regions, obtained in studies of recombination in fungi as well as in bacteria, led to the development of a molecular model for recombination in 1964. This model, known as the **Holliday model** (after Robin Holliday), has continued to provide the basis for current thinking about recombination mechanisms, although it has been modified as new data have been obtained.

The original version of the Holliday model proposed that recombination is initiated by the introduction of nicks at the same position on the two

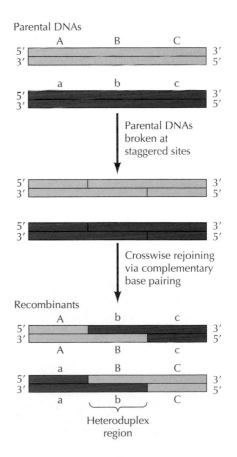

*Figure 5.29*
**Homologous recombination by complementary base pairing**  Parental DNAs are broken at staggered sites, and overlapping single-stranded regions are exchanged via base pairing with homologous sequences. The result is a heteroduplex region, in which the two DNA strands are derived from different parental molecules.

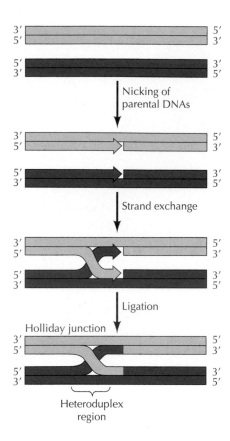

*Figure 5.30*
**The Holliday model for homologous recombination**   Single-strand nicks are introduced at the same position on both parental molecules. The nicked strands then exchange by complementary base pairing, and ligation produces a crossed-strand intermediate called a Holliday junction.

parental DNA molecules (Figure 5.30). The nicked DNA strands partially unwind, and each invades the other molecule by pairing with the complementary unbroken strand. Ligation of the broken strands then produces a crossed-strand intermediate, known as a **Holliday junction**, that is the central intermediate in recombination. The direct demonstration of Holliday junctions by electron microscopy has provided clear support for this model of recombination (Figure 5.31).

Once a Holliday junction is formed, it can be resolved by cutting and rejoining of the crossed strands to yield recombinant molecules (Figure 5.32). This can occur in two different ways, depending on the orientation of the Holliday junction, which can readily form two different isomers. In the isomer resulting from the initial strand exchange, the crossed strands are those that were nicked at the start of the recombination process. However, simple rotation of this structure yields a different isomer in which the unbroken parental strands are crossed. Resolution of these different isomers has distinct genetic consequences. In the first case, the progeny molecules have heteroduplex regions but are nonrecombinant for DNA that flanks these regions. If isomerization occurs, however, cutting and rejoining of the crossed strands results in progeny molecules that are recombinant for DNA that flanks the heteroduplex regions. The structure of the Holliday junction thus provides the possibility of generating both recombinant and nonrecombinant heteroduplexes, consistent with the genetic data upon which the Holliday model was based.

One modification of the Holliday model, proposed in 1975, eliminates a potential difficulty with the initial proposal—namely, how can both parental molecules be nicked simultaneously at the same position to initiate recombination? In this modified version, recombination is initiated by a nick in only one of the parental molecules (Figure 5.33). The nicked strand is then displaced, and the resulting single strand invades the other parental molecule by homologous base pairing. This process produces a displaced loop of DNA, which can then be cleaved and joined to the other parental molecule. The result is a crossed-strand Holliday junction, which can be resolved into recombinant or nonrecombinant heteroduplex molecules as already described.

Still another modification of the Holliday model suggests that recombination is initiated by a double-strand break, rather than a single-strand nick. Exonucleases at the site of the break then generate single-stranded tails, which can invade a homologous double-stranded molecule. Again the result is a Holliday junction, which can be resolved to yield either recombinant or nonrecombinant heteroduplexes. This double-strand break repair model appears to be particularly applicable to meiotic recombination in yeasts.

Multiple alternatives may thus account for the initial stages of recombination between two DNA molecules, and the details of recombination mechanisms, particularly in eukaryotic cells, have not been fully elucidated. But the crossed-strand Holliday junction, generated by strand exchange leading to the formation of a heteroduplex region, remains the central intermediate in consideration of the recombination process.

*Figure 5.31*
**Identification of Holliday junctions by electron microscopy**   Electron micrograph of a Holliday junction that was detected during recombination of plasmid DNAs in *E. coli*. An interpretive drawing of the structure is shown below. The molecule illustrates a Holliday junction in the open configuration resulting from rotation of the crossed-strand intermediate (see Figure 5.32). (Courtesy of Huntington Potter and David Dressler, Harvard Medical School.)

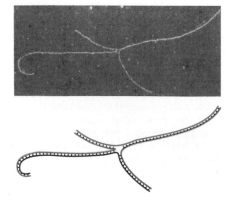

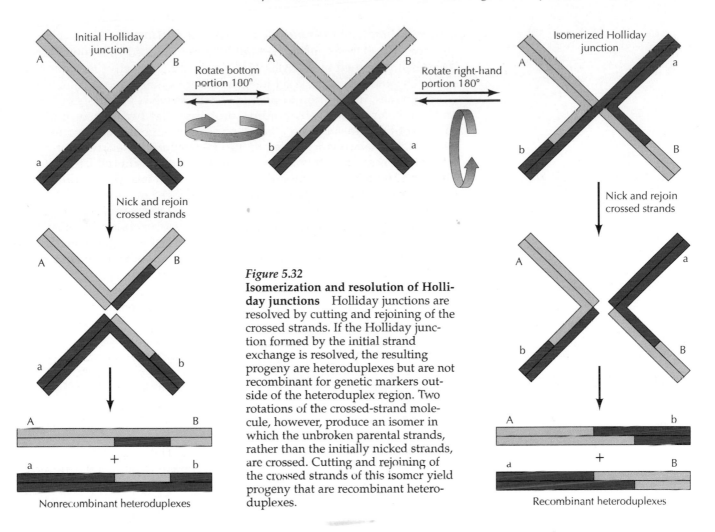

*Figure 5.32*
**Isomerization and resolution of Holliday junctions** Holliday junctions are resolved by cutting and rejoining of the crossed strands. If the Holliday junction formed by the initial strand exchange is resolved, the resulting progeny are heteroduplexes but are not recombinant for genetic markers outside of the heteroduplex region. Two rotations of the crossed-strand molecule, however, produce an isomer in which the unbroken parental strands, rather than the initially nicked strands, are crossed. Cutting and rejoining of the crossed strands of this isomer yield progeny that are recombinant heteroduplexes.

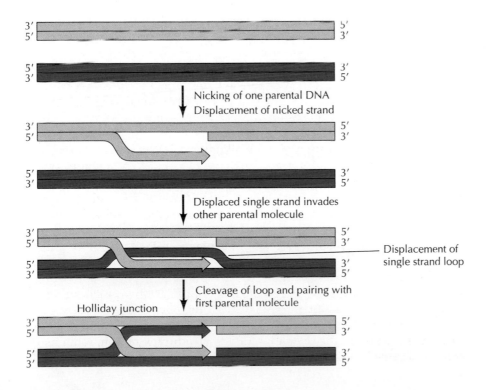

*Figure 5.33*
**Initiation of recombination by nicking only one parental molecule** The nicked DNA strand invades the other parental molecule by homologous base pairing, thereby displacing a single-stranded loop of DNA. This loop is then cleaved and pairs with the first parental molecule, yielding a crossed-strand Holliday junction.

## *Enzymes Involved in Homologous Recombination*

Most of the enzymes currently known to be involved in recombination have been identified by analysis of recombination-defective mutants of *E. coli*. Such genetic analysis has established that recombination requires specific enzymes, in addition to proteins (such as DNA polymerase, ligase, and single-stranded DNA-binding proteins) that function in multiple aspects of DNA metabolism. The identification of genes required for efficient recombination in *E. coli* led to the isolation of their encoded proteins, which have been characterized by biochemical analysis in cell-free systems. These studies have elucidated the action of several enzymes in catalyzing the formation and resolution of Holliday junctions.

The central protein involved in homologous recombination is **RecA**, which promotes the exchange of strands between homologous DNAs that causes heteroduplexes to form (Figure 5.34). The action of RecA can be considered in three stages. First, the RecA protein binds to single-stranded

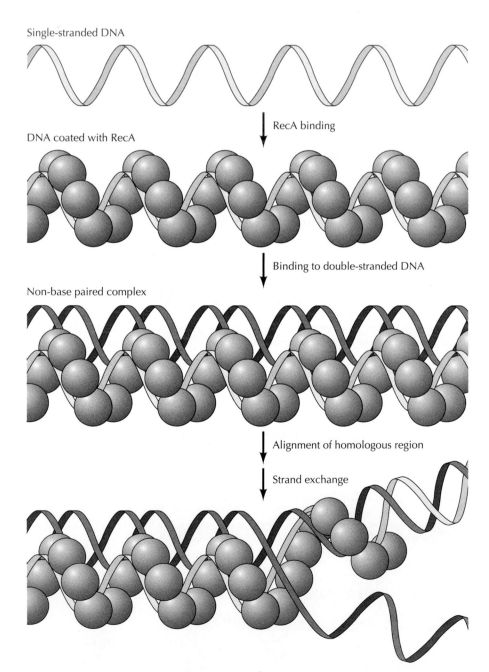

Single-stranded DNA

RecA binding

DNA coated with RecA

Binding to double-stranded DNA

Non-base paired complex

Alignment of homologous region

Strand exchange

**Figure 5.34**
**Function of the RecA protein**   RecA initially binds to single-stranded DNA to form a protein-DNA filament. The RecA protein that coats the single-stranded DNA then binds to a second, double-stranded DNA molecule to form a non-base-paired complex. Complementary base pairing and strand exchange follow, forming a heteroduplex region.

DNA, coating the DNA to form a protein-DNA filament. Because RecA has two DNA binding sites, the RecA protein bound to single-stranded DNA is able to bind a second, double-stranded DNA molecule, forming a complex between the two DNAs. This nonspecific RecA-mediated association is followed by specific base pairing between the single-stranded DNA and its complement. The RecA protein then catalyzes strand exchange, with the single strand originally coated with RecA displacing its homologous strand to form a heteroduplex. Thus, the RecA protein is capable of catalyzing, by itself, the strand exchange reactions that are central to the formation of Holliday junctions.

In yeast, a RecA-related protein, designated RAD51, is required for genetic recombination as well as for the repair of double-strand breaks. RAD51 is not only structurally similar to RecA; like RecA, it is also able to catalyze strand exchange reactions *in vitro*. Proteins related to RAD51 have been identified in complex eukaryotes, including humans, indicating that proteins related to RecA play key roles in homologous recombination in both prokaryotic and eukaryotic cells.

Most recombination events in *E. coli* also require the RecBCD enzyme, which is a complex of three proteins (RecB, C, and D). The properties of RecBCD are consistent with the hypothesis that it initiates recombination by providing the single-stranded DNA to which RecA binds. RecBCD accomplishes this task by unwinding and nicking double-stranded DNA (Figure 5.35). The RecBCD complex binds to the end of a DNA molecule and then acts as a helicase, transiently unwinding the DNA as it travels along the molecule. When it encounters a specific nucleotide sequence

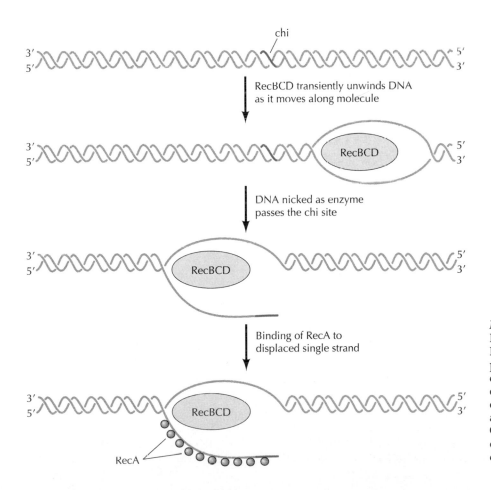

*Figure 5.35*
**Initiation of recombination by RecBCD** The *E. coli* RecBCD complex binds to the end of a DNA molecule and unwinds the DNA as it travels along the molecule. When it encounters a specific sequence (called a chi site), it nicks the DNA strand. Continued unwinding then forms a displaced single strand to which RecA can bind.

Holliday junction

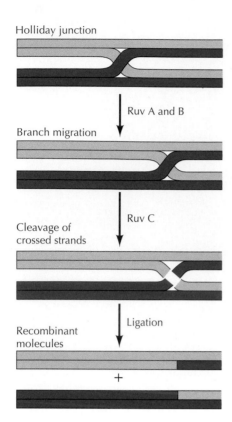

Branch migration

Cleavage of
crossed strands

Recombinant
molecules

+

*Figure 5.36*
**Branch migration and resolution of Holliday junctions** Two *E. coli* proteins (RuvA and RuvB) together catalyze the movement of the crossed-strand site in Holliday junctions (branch migration). RuvC resolves the Holliday junctions by cleaving the crossed strands, which are then joined by ligase.

(GCTGGTGG, called a chi site), RecBCD acts as a nuclease to introduce a single-strand nick. It then continues to unwind the double helix, forming a displaced single strand to which RecA can bind to initiate strand exchange.

Once a Holliday junction is formed, three other *E. coli* proteins (RuvA, B, and C) become involved in recombination (Figure 5.36). RuvA and RuvB act as a complex to drive the migration of the site at which the strands cross in the Holliday junction, thereby varying the extent of the heteroduplex region and the position at which the crossed strands will be cut and rejoined. RuvC then resolves Holliday junctions by cleaving the crossed DNA strands. Rejoining of the cleaved strands by ligation completes the process, yielding two recombinant molecules.

In yeasts, Holliday junctions are resolved by a complex of RAD1 and RAD10, with RAD1 cleaving single-stranded DNA at the crossover junction. RAD1 and RAD10 are homologs of the mammalian XPF and ERCC1 DNA repair proteins and also cleave damaged DNA during nucleotide-excision repair (see Table 5.1). RAD1 and RAD10 homologs thus may have a conserved function in recombination in eukaryotic cells.

## DNA REARRANGEMENTS

Homologous recombination results in the reassortment of genes between chromosome pairs without altering the arrangement of genes within the genome. In contrast, other types of recombinational events lead to rearrangements of genomic DNA. Some of these DNA rearrangements are important in controlling gene expression in specific cell types; others may play an evolutionary role by contributing to genetic diversity.

The discovery that genes can move to different chromosomal locations came from Barbara McClintock's studies of corn in the 1940s. Purely on the basis of genetic analysis, McClintock described novel genetic elements that could move to different locations in the genome and alter the expression of adjacent genes. Nearly three decades elapsed, however, before the physical basis of McClintock's work was elucidated by the discovery of transposable elements in bacteria and the notion of movable genetic elements became widely accepted by scientists. Several types of DNA rearrangements, including the transposition of elements initially described by McClintock, are now recognized in both prokaryotic and eukaryotic cells.

### Site-Specific Recombination

In contrast to general homologous recombination, which occurs at any extensive region of sequence homology, **site-specific recombination** occurs between specific DNA sequences, which are usually homologous over only a short stretch of DNA. The principal interaction in this process is mediated by proteins that recognize the specific DNA target sequences rather than by complementary base pairing.

The prototype of site-specific recombination has been provided by studies of the bacteriophage λ. When λ infects *E. coli*, it can either replicate to cause cell lysis or it can integrate into the bacterial chromosome, forming a prophage that is then maintained as part of the *E. coli* genome (a process called **lysogeny**) (Figure 5.37). Under appropriate conditions, DNA integration can be reversed, resulting in excision of the λ DNA and initiation of lytic viral replication. Both the integration and the excision of λ DNA involve site-specific recombination between viral and host cell DNA sequences.

E. coli DNA and λ DNA recombine at specific sites, called attachment (*att*) sites. Thus, integration of λ DNA involves recombination between *att*

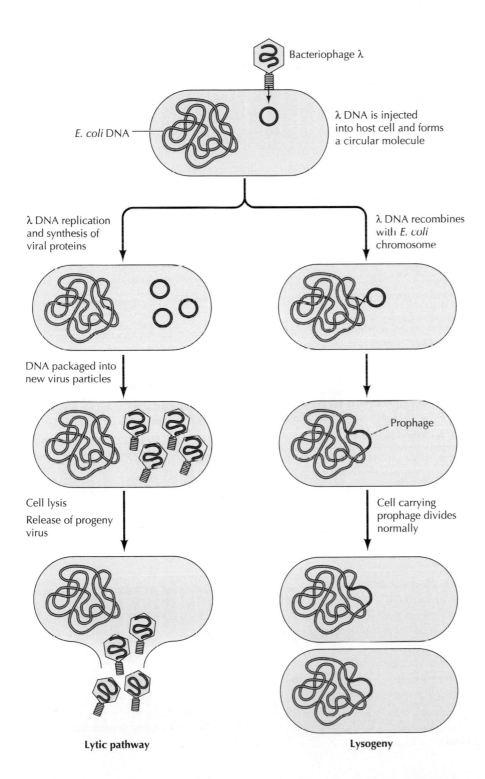

Bacteriophage λ

λ DNA is injected into host cell and forms a circular molecule

E. coli DNA

λ DNA replication and synthesis of viral proteins

λ DNA recombines with E. coli chromosome

DNA packaged into new virus particles

Prophage

Cell lysis
Release of progeny virus

Cell carrying prophage divides normally

**Lytic pathway**

**Lysogeny**

*Figure 5.37*
**Lytic and lysogenic pathways of bacteriophage** λ   Infection of *E. coli* is initiated by the injection of λ DNA, which then becomes circular within the host cell. In lytic infection, the λ DNA replicates and directs the synthesis of viral proteins. The viral DNA is then packaged into progeny virus particles, which are released upon cell lysis. In lysogenic infection, the λ DNA recombines with the host genome to form a prophage that is integrated into the *E. coli* chromosome. The integrated λ DNA does not direct the synthesis of progeny viruses, but is instead replicated along with the rest of the bacterial genome.

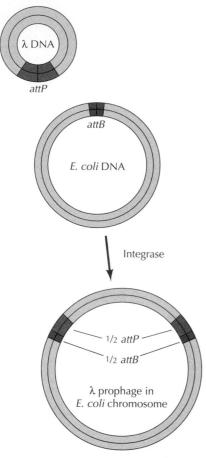

sites of the phage (*attP*) and the bacterium (*attB*), which are about 240 and 25 nucleotides long, respectively (Figure 5.38). The process is mediated by a λ protein called integrase (Int), which specifically binds to both *attP* and *attB* sequences. Int initially binds to *attP*, forming a complex in which the *attP* DNA is wrapped around multiple copies of the Int protein. The Int-*attP* complex binds to *attB*, aligning the phage and bacterial *att* sites. The phage and bacterium then exchange strands within a 15-nucleotide core sequence shared by *attB* and *attP* (Figure 5.39). The Int protein introduces staggered cuts within the core homology region of *attB* and *attP*, catalyzes

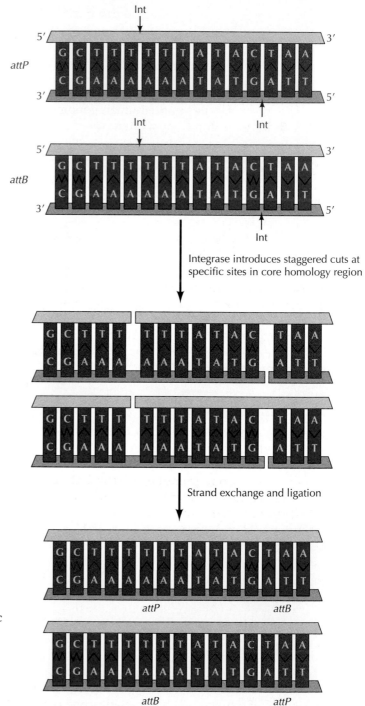

**Figure 5.38**
**Integration of λ DNA by site-specific recombination**   Integration results from recombination between specific sequences in the λ and *E. coli* genomes, called *attP* and *attB*, respectively. The process is catalyzed by a virus-encoded enzyme (integrase), which recognizes both *attP* and *attB* sequences.

**Figure 5.39**
**Mechanism of λ site-specific recombination**   Site-specific recombination occurs within a 15-nucleotide homologous core sequence shared by *attP* and *attB*. Integrase (Int) cleaves at specific sites within this sequence to generate staggered single-stranded DNA tails. It then catalyzes strand exchange and ligation, resulting in recombination between *attP* and *attB* and integration of λ DNA.

strand exchange, and then ligates the broken strands, integrating the λ DNA into the *E. coli* chromosome. The Int protein also acts in excision of the λ prophage, which is essentially the reverse of integration.

Site-specific recombination is important not only in the interaction of viruses such as λ with their host cells, but also in programmed gene rearrangements within cell genomes. In vertebrates, site-specific recombination is critical to the development of the immune system, which recognizes foreign substances (**antigens**) and provides protection against infectious agents. There are two major classes of immune responses, which are mediated by B and T lymphocytes. B lymphocytes secrete antibodies (**immunoglobulins**) that react with soluble antigens; T lymphocytes express cell surface proteins (called **T cell receptors**) that react with antigens expressed on the surfaces of other cells. The key feature of both immunoglobulins and T cell receptors is their enormous diversity, which enables different antibody or T cell receptor molecules to recognize a vast array of foreign antigens. For example, each individual is capable of producing more than $10^{11}$ different antibody molecules, which is far in excess of the total number of genes in the human genome (approximately $10^5$). Rather than being encoded in germ-line DNA, these diverse antibodies (and T cell receptors) are encoded by unique lymphocyte genes that are formed during development of the immune system as a result of site-specific recombination between distinct segments of immunoglobulin and T cell receptor genes.

The role of site-specific recombination in the formation of immunoglobulin genes was first demonstrated by Susumu Tonegawa in 1976. Immunoglobulins consist of pairs of identical heavy and light polypeptide chains (Figure 5.40). Both the heavy and light chains are composed of C-terminal constant regions and N-terminal variable regions. The variable regions, which have different amino acid sequences in different immunoglobulin molecules, are responsible for antigen binding, and it is the diversity of variable region amino acid sequences that allows different individual antibodies to recognize unique antigens. Although every individual is capable of producing a vast spectrum of different antibodies, each B lymphocyte produces only a single type of antibody. Tonegawa's key discovery was that each antibody is encoded by unique genes formed by site-specific recombination during B lymphocyte development. These gene rearrangements create different immunoglobulin genes in different individual B lymphocytes, so the population of approximately $10^{12}$ B lymphocytes in the human body includes cells capable of producing antibodies against a diverse array of foreign antigens.

The genes that encode immunoglobulin light chains consist of three regions: a V region that encodes the 95 to 96 N-terminal amino acids of the polypeptide variable region; a joining (J) region that encodes the 12 to 14 C-terminal amino acids of the polypeptide variable region; and a C region that encodes the polypeptide constant region (Figure 5.41). The major class of light-chain genes in the mouse are formed from combinations of approximately 250 V regions and four J regions with a single C region. Site-specific recombination during lymphocyte development leads to a gene rearrangement in which a single V region recombines with a single J region to generate a functional light-chain gene. Different V and J regions are rearranged in different B lymphocytes, so the possible combinations of 250 V regions with 4 J regions can generate approximately 1000 (4 × 250) unique light chains.

The heavy-chain genes include a fourth region (known as the diversity, or D, region), which encodes amino acids lying between V and J (Figure 5.42). Assembly of a functional heavy-chain gene requires two recombina-

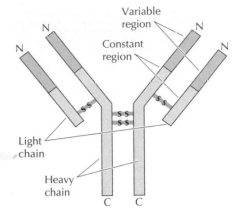

**Figure 5.40**
**Structure of an immunoglobulin**
Immunoglobulins are composed of two heavy chains and two light chains, joined by disulfide bonds. Both the heavy and the light chains consist of variable and constant regions.

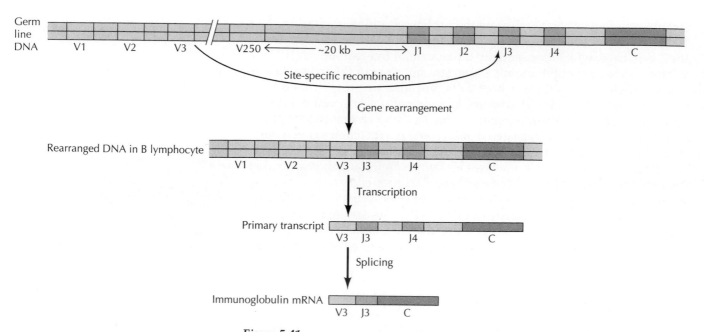

**Figure 5.41**
**Rearrangement of immunoglobulin light-chain genes**   Each light-chain gene
(mouse κ light chains are illustrated) consists of a constant region (C), a joining
region (J), and a variable region (V). There are approximately 250 different V
regions, which are separated from J and C by about 20 kb in germ-line DNA. During the development of B lymphocytes, site-specific recombination joins one of the
V regions to one of the four J regions. This rearrangement activates transcription,
resulting in the formation of a primary transcript containing the rearranged VJ
region together with the remaining J regions and C. The remaining unused J
regions and the introns between J and C are then removed by splicing, yielding a
functional mRNA.

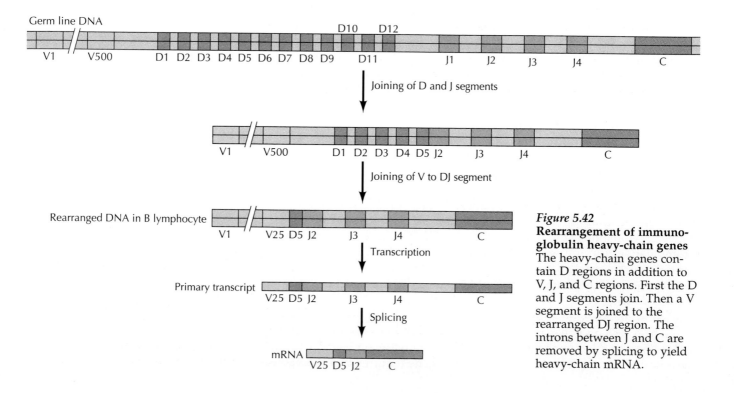

**Figure 5.42**
**Rearrangement of immuno-
globulin heavy-chain genes**
The heavy-chain genes contain D regions in addition to
V, J, and C regions. First the D
and J segments join. Then a V
segment is joined to the
rearranged DJ region. The
introns between J and C are
removed by splicing to yield
heavy-chain mRNA.

tion events: A D region first recombines with a J region, and a V region then recombines with the rearranged DJ segment. In the mouse, there are about 500 heavy-chain V regions, 12 D regions, and 4 J regions, so the total number of heavy chains that can be generated by the recombination events is 24,000 (500 × 12 × 4).

Combinations between the 1000 different light chains and 24,000 different heavy chains formed by site-specific recombination can generate approximately $2 \times 10^7$ different immunoglobulin molecules. This diversity is further increased because the joining of immunoglobulin gene segments is often imprecise, with one to several nucleotides frequently being lost or gained at the sites of joining. The mutations resulting from these deletions and insertions increase the diversity of immunoglobulin variable regions approximately a hundredfold, corresponding to the formation of about $10^5$ different light chains and $2 \times 10^6$ heavy chains, which can then combine to form more than $10^{11}$ distinct antibodies. Still further antibody diversity is generated after the formation of rearranged immunoglobulin genes by a process known as somatic hypermutation, which results in the introduction of frequent mutations into the variable regions of both heavy-chain and light-chain genes.

T cell receptors similarly consist of two chains (called α and β), each of which contains variable and constant regions (Figure 5.43). The genes encoding these polypeptides are generated by recombination between V and J segments (the α chain) or between V, D, and J segments (the β chain), analogous to the formation of immunoglobulin genes. Site-specific recombination between these distinct segments of DNA, in combination with mutations introduced during recombination, generates a degree of diversity in T cell receptors that is similar to that in immunoglobulins. However, T cell receptors differ from immunoglobulins in that they are not subject to the introduction of further diversity by somatic hypermutation.

## Transposition via DNA Intermediates

Site-specific recombination occurs between two specific sequences that contain at least a small core of homology. In contrast, **transposition** involves the movement of sequences throughout the genome and has no requirement for sequence homology. Elements that move by transposition, such as those first described by McClintock, are called **transposable elements,** or **transposons**. They are divided into two general classes, depending on whether they transpose via DNA intermediates or via RNA intermediates. The first class of transposable elements is discussed here; transposition via RNA intermediates is considered in the next section.

The best-studied transposons are those of bacteria, all of which move via DNA intermediates (Figure 5.44). The simplest of these elements are the

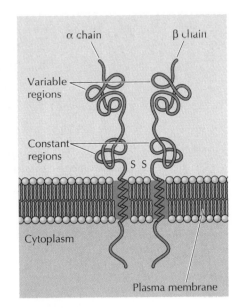

**Figure 5.43**
**Structure of a T cell receptor**   T cell receptors consist of two polypeptide chains (α and β) that span the plasma membrane and are joined by disulfide bonds. Both the α and β chains are composed of variable and constant regions.

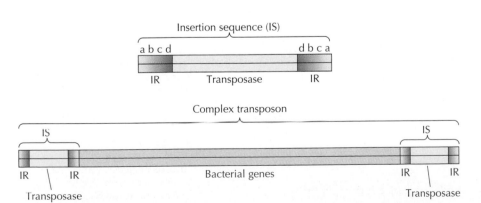

**Figure 5.44**
**Bacterial transposons**   Insertion sequences (IS) range from 800 to 2000 nucleotides and contain a gene for transposase flanked by inverted repeats (IR) of about 20 nucleotides. Complex transposons consist of two insertion sequences flanking other genes and are typically 5 to 20 kb long.

## KEY EXPERIMENT

# *Rearrangement of Immunoglobulin Genes*

**Evidence for Somatic Rearrangement of Immunoglobulin Genes Coding for Variable and Constant Regions**

Nobumichi Hozumi and Susumu Tonegawa
Basel Institute for Immunology, Basel, Switzerland
*Proceedings of the National Academy of Sciences, USA, Volume 73, 1976, pages 3628–3632*

### The Context

The ability of the vertebrate immune system to recognize a seemingly infinite variety of foreign molecules implies that lymphocytes can produce a correspondingly vast array of antibodies. Since this antibody diversity is key to immune recognition, understanding the mechanism by which an apparently unlimited number of distinct immunoglobulins are encoded in genomic DNA is a central issue in immunology.

Prior to the experiments of Hozumi and Tonegawa, protein sequencing of multiple immunoglobulins had demonstrated that both heavy and light chains consist of distinct variable and constant regions. Genetic studies further indicated that mice inherit only single copies of the constant-region genes. These observations first led to the proposal that immunoglobulins are encoded by multiple variable-region genes that can associate with a single constant-region gene. The discovery of immunoglobulin gene rearrangements by Hozumi and Tonegawa provided the first direct experimental support for this hypothesis and laid the groundwork for understanding the molecular basis of antibody diversity.

### The Experiments

Hozumi and Tonegawa tested the possibility that the genes encoding immunoglobulin variable and constant regions were joined at the DNA level during lymphocyte development. Their experimental approach was to use restriction endonuclease digestion to compare the organization of variable-region and constant-region sequences in DNAs extracted from mouse embryos and from cells of a mouse plasmacytoma (a B lymphocyte tumor that produces a single species of immunoglobulin).

Embryo and plasmacytoma DNAs were digested with the restriction endonuclease *Bam*HI, and DNA fragments of different sizes were separated by electrophoresis in agarose gels. The gel was then cut into slices, and DNA extracted from each slice was hybridized with radiolabeled probes that had been prepared from immunoglobulin mRNA isolated from the plasmacytoma cells. Two probes were used, corresponding either to the complete immunoglobulin mRNA or to the 3′ half of the mRNA, consisting only of constant-region sequences.

The critical result was that completely different patterns of variable-region and constant-region sequences were detected in embryo versus plasmacytoma DNAs (see figure). In embryo DNA, the complete probe hybridized to two *Bam*HI fragments of approximately 8.6 and 5.6 kb, respectively. Only the 8.6-kb fragment hybridized to the 3′ probe, suggest-

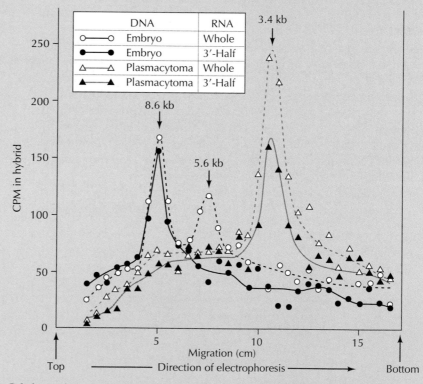

Gel electrophoresis of embryo and plasmacytoma DNAs digested with *Bam*HI and hybridized to probes corresponding to either the whole or the 3′ half of the plasmacytoma mRNA. Data are presented as the radioactivity detected in hybrid molecules with DNA from each gel slice.

## Rearrangement of Immunoglobulin Genes *(continued)*

ing that the 5.6-kb fragment contained constant-region sequences and the 8.6-kb fragment contained variable-region sequences. In striking contrast, both probes hybridized to only a single 3.4-kb fragment in plasmacytoma DNA. The interpretation of these results was that the variable- and constant-region sequences were separated in embryo DNA but rearranged to form a single immunoglobulin gene during lymphocyte development.

### The Impact

The initial results of Hozumi and Tonegawa, based on the relatively indirect approach of restriction endonuclease mapping, were confirmed and extended by the molecular cloning and sequencing of immunoglobulin genes. Such studies have now unambiguously established that these genes are generated by site-specific recombination between distinct segments of DNA in B lymphocytes. In T lymphocytes, similar DNA rearrangements are responsible for formation of the genes encoding T cell receptors. Thus, site-specific recombination and programmed gene rearrangements are central to the development of the immune system.

Further studies have shown that the variable regions of immunoglobulins and T cell receptors are generated by rearrangements of two or three distinct segments of DNA. The ability of these segments to recombine, together with a high frequency of mutations introduced at the recombination sites,

Susumu Tonegawa

is largely responsible for immunoglobulin and T cell receptor diversity. The discovery of immunoglobulin gene rearrangements thus provided the basis for understanding how the immune system can recognize and respond to a virtually unlimited range of foreign substances.

insertion sequences, ranging in size from about 800 to 2000 nucleotides. Insertion sequences consist only of a gene for the enzyme involved in transposition (transposase) flanked by short inverted repeats, which are the sites at which transposase acts. Complex transposons consist of two insertion sequences flanking other genes, which move as a unit.

Insertion sequences move from one chromosomal site to another without replicating their DNA (Figure 5.45). Transposase introduces a staggered break in the target DNA and cleaves at the ends of the transposon inverted-repeat sequences. Although transposase acts specifically at the transposon inverted repeats, it is usually nonspecific with respect to the sequence of the target DNA, so it catalyzes the movement of transposons throughout the genome. Following the cleavage of transposon and target site DNAs, transposase joins the overhanging ends of the target DNA to the transposable element. The resulting gap in the target-site DNA is repaired by DNA synthesis, followed by ligation to the other strand of the tranposon. The result of this process is a short direct repeat of the target-site DNA on both sides of the transposable element—a hallmark of transposon integration.

This transposition mechanism causes the transposon to move from one chromosomal site to another. Other types of transposons move by a more complex mechanism, in which the transposon is replicated in concert with its integration into a new target site. This mechanism results in the integration of one copy of the transposon into a new position in the genome, while another copy remains at its original location.

Transposons that move via DNA intermediates are present in eukaryotes as well as in bacteria. The original transposable elements described by McClintock in corn move by a nonreplicative mechanism, as do some transposable elements in *Drosophila* and other insects. Like most bacterial transposons, these elements move to nonspecific target sites throughout the genome. Although the movement of these transposons itself is not likely to be useful to the cells in which it occurs, the mutations induced by transposon movement may play an important role in evolution.

*Figure 5.45*
**Transposition of insertion sequences**
Simple transposition does not involve replication of the transposon DNA. Transposase cleaves at both ends of the transposon and introduces a staggered cut in the target DNA. The overhanging ends of target DNA are then joined to the transposon, and gaps resulting from the staggered cuts at the target site are repaired. The result is the formation of short direct repeats of target-site DNA (5 to 10 nucleotides long) flanking the integrated transposon.

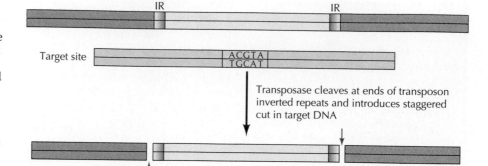

In yeasts and protozoans, however, transposition by a replicative mechanism is responsible for programmed DNA rearrangements that regulate gene expression. In these cases transposition is initiated by the action of a site-specific nuclease that cleaves a specific target site, at which a copy of the transposable element is then inserted. Transposable elements are thus capable not only of moving to nonspecific sites throughout the genome, but also of participating in specific gene rearrangements that result in programmed changes in gene expression.

### Transposition via RNA Intermediates

Most transposons in eukaryotic cells move via RNA intermediates rather than DNA intermediates. The mechanism of transposition of these elements is similar to the replication of retroviruses, which have provided the prototype system for studying this class of movable DNA sequences.

**Retroviruses** contain RNA genomes in their virus particles but replicate via the synthesis of a DNA provirus, which is integrated into the chromosomal DNA of infected cells (see Figure 3.15). A DNA copy of the viral RNA is synthesized by the viral enzyme **reverse transcriptase**. The mechanism by which this occurs results in the synthesis of a DNA molecule that contains direct repeats of several hundred nucleotides at both ends (Figure 5.46). These repeated sequences, called **long terminal repeats**, or **LTRs**, arise from duplication of the sites on viral RNA at which primers bind to initiate DNA synthesis. The LTR sequences thus play central roles in reverse transcription, in addition to being involved in the integration and subsequent transcription of proviral DNA.

Like all DNA polymerases, reverse transcriptase requires a primer, which

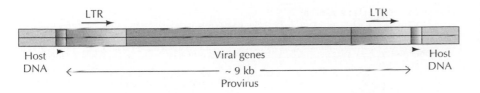

*Figure 5.46*
**The organization of retroviral DNA**
The integrated proviral DNA is flanked by long terminal repeats (LTRs), which are direct repeats of several hundred nucleotides. Viral genes, including genes for reverse transcriptase and for structural proteins of the virus particle, are located between the LTRs. The integrated provirus is flanked by short direct repeats of host DNA.

in the case of retroviruses is a tRNA molecule bound at a specific site (the primer-binding site) close to the 5′ terminus of the viral RNA (Figure 5.47). Since DNA synthesis proceeds in the 5′ to 3′ direction, only a short piece of DNA is synthesized before reverse transcriptase reaches the end of its template. Continuation of DNA synthesis then depends on the ability of reverse transcriptase to "jump" to the 3′ end of the template RNA molecule. This is accomplished via an RNase H activity of reverse transcriptase, which degrades the RNA strand of DNA-RNA hybrids. As a result, the newly synthesized DNA is converted to a single-stranded molecule, which can hybridize to a short repeated sequence present at both the 5′ and the 3′ ends of the viral RNA. DNA synthesis can then continue, yielding a single-stranded DNA complementary to viral RNA. Synthesis of the opposite strand of DNA is initiated by a fragment of viral RNA that acts as a primer, at a site near the 3′ end of the template DNA strand. Again the result is a short piece of DNA, which includes the primer-binding site copied from the tRNA used as the initial primer for reverse transcription. The primer-binding sequence of the tRNA is then degraded by RNase H, leaving an overhanging DNA strand that again "jumps" to pair with its complementary sequence at the other end of the template. DNA synthesis can then continue once more, finally yielding a linear DNA with LTRs at both ends.

The linear viral DNA integrates into the host cell chromosome by a process that resembles the integration of DNA transposable elements. Integration is catalyzed by a viral integration protein and occurs at nonspecific target sequences in cellular DNA. The integration protein cleaves two bases from the ends of viral DNA and introduces a staggered cut at the target site in cellular DNA. The overhanging ends of cellular DNA are then joined to the termini of viral DNA, and the gap is filled by DNA synthesis. The integrated provirus is therefore flanked by a direct repeat of cell sequences, similar to the repeats that flank DNA transposons.

The viral life cycle continues with transcription of the integrated provirus, which yields viral genomic RNA as well as mRNAs that direct the synthesis of viral proteins (including reverse transcriptase and integration protein). The genomic RNA is then packaged into viral particles, which are released from the host cell. These progeny viruses can infect a new cell, initiating another round of DNA synthesis and integration. The net effect can be viewed as the movement of the provirus from one chromosomal site to another, via the synthesis and reverse transcription of an RNA intermediate.

The retroviruses can be considered one type of **retrotransposon**—an element that moves via RNA intermediates. Other retrotransposons differ from retroviruses in that they are not packaged into infectious particles and therefore cannot spread from one cell to another. However, these retrotransposons can move to new chromosomal sites within the same cell, via mechanisms fundamentally similar to those involved in retrovirus replication.

Some retrotransposons (called class I retrotransposons) are structurally similar to retroviruses (Figure 5.48). Retrotransposons of this type have been studied in yeasts, *Drosophila*, mice, and plants. They have LTR sequences at

*Figure 5.47*
**Generation of LTRs during reverse transcription** LTRs consist of three sequence elements: a short repeat sequence (R) of about 20 nucleotides that is present at both ends of the viral RNA; a sequence unique to the 5′ end of viral RNA (U5); and a sequence unique to the 3′ end of viral RNA (U3). Repeats of these sequences are generated during DNA synthesis as reverse transcriptase jumps twice between the ends of its template. Synthesis is initiated using a tRNA primer bound to a primer-binding site (PBS) adjacent to U5 at the 5′ end of the viral RNA. The polymerase copies R, and the RNA strand of the RNA-DNA hybrid is then degraded by RNase H. The polymerase then jumps to the 3′ end of the viral RNA in order to synthesize a complete DNA strand complementary to the RNA template. The polymerase jumps again during synthesis of the second strand of DNA, which is also initiated by a primer bound close to the 5′ end of its template. The result of these jumps is the formation of LTRs that contain U3-R-U5 sequences.

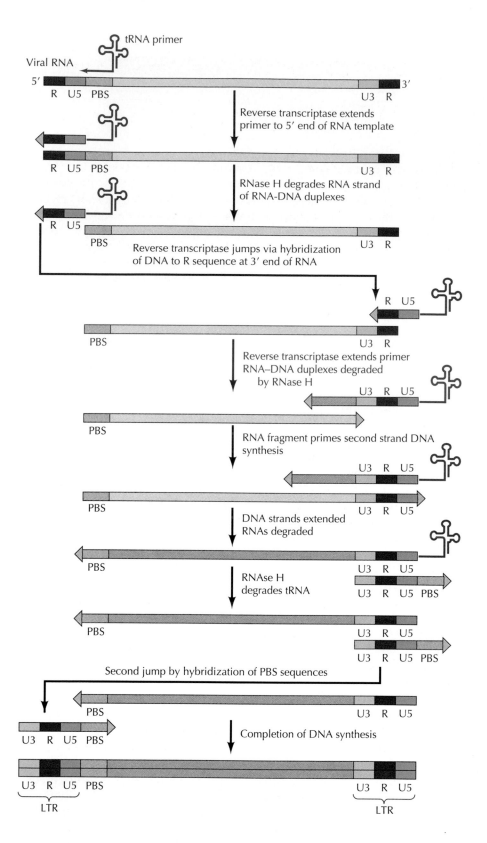

both ends; they encode reverse transcriptase and integration proteins; and they transpose (like retroviruses) via transcription into RNA, synthesis of a new DNA copy by reverse transcriptase, and integration into cellular DNA.

Class II retrotransposons differ from retroviruses in that they do not contain LTR sequences, although they do encode their own reverse transcrip-

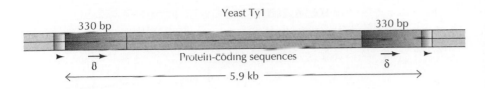

*Figure 5.48*
**Structure of a class I retrotransposon**
The yeast Ty1 transposable element displays the same organization as a retrovirus. Protein-coding sequences, including genes for reverse transcriptase and integration proteins, are flanked by LTRs (called δ elements) of about 330 base pairs (bp). The integrated transposon is flanked by short direct repeats of target-site DNA.

tase. Retrotransposons of this type are found in mammals, insects (including *Drosophila*), plants, and trypanosomes. In mammals, the major class of these retrotransposons consists of the highly repetitive long interspersed elements (**LINEs**), which are repeated approximately 50,000 times in the genome and account for about 5% of genomic DNA (see Chapter 4). A full-length LINE element is 6 to 7 kb long, although many members of the family are truncated at their 5′ end (Figure 5.49). At their 3′ end, LINEs have tracts of A-rich sequences thought to be derived by reverse transcription of the poly-A tails that are added to mRNAs following transcription (see Chapter 6). Like other transposable elements, LINEs are flanked by short direct repeats of the target-site DNA, indicating that integration involves staggered cuts and repair synthesis.

Since LINEs do not contain LTR sequences, the mechanism of their reverse transcription and subsequent integration into chromosomal DNA must differ from that of retroviruses and class I (LTR-containing) retrotransposons. These mechanisms are not yet understood, although one possibility is that reverse transcription is primed by a broken end of chromosomal DNA at the integration target site (Figure 5.50). This process might involve nicking of the target DNA by a nuclease encoded by the retrotransposon, but integration may simply occur at sites of nonspecific DNA damage. Reverse transcription would then initiate within the poly-A tract at the 3′ end of the transposon RNA and continue along the molecule. The opposite strand of DNA would be synthesized using the other broken end of target-site DNA as primer, resulting in simultaneous synthesis and integration of the retrotransposon DNA.

Other sequence elements, which do not encode their own reverse transcriptase, also transpose via RNA intermediates. These elements include the highly repetitive short interspersed elements (**SINEs**), of which there are nearly a million copies in mammalian genomes (see Chapter 4). The major family of these elements consists of the *Alu* sequences, which are about 300 bases long. These sequences have A-rich tracts at their 3′ end and are flanked by short duplications of target-site DNA sequences, a structure similar to that of retrotransposons that lack LTR sequences (e.g., LINEs). SINEs arose by reverse transcription of small RNAs, including tRNAs and small cytoplasmic RNAs involved in protein transport. Since SINEs no longer encode functional RNA products, they represent pseudogenes that arose via RNA-mediated transposition. Pseudogenes of many protein-coding genes (called **processed pseudogenes**) have similarly arisen by reverse transcription of mRNAs (Figure 5.51). Such processed pseudogenes are readily recognized not only because they terminate in an A-rich tract but also because the introns present in the corresponding normal gene have

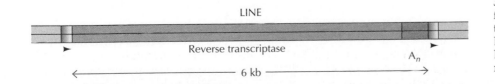

*Figure 5.50*
**Model for reverse transcription and integration of non-LTR-containing retrotransposons**   Reverse transcription, primed by a broken end of DNA at the target site, initiates within the poly-A tail at the 3′ end of retrotransposon RNA. Synthesis of the opposite strand of retrotransposon DNA is similarly primed by the other strand of DNA at the target site.

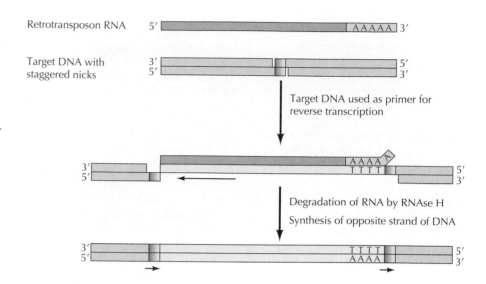

been removed during mRNA processing. Since these elements do not include a gene for reverse transcriptase, their transposition presumably involves a reverse transcriptase that is encoded elsewhere in the genome—possibly by class I or II retrotransposons, such as LINEs. The mechanism of reverse transcription and integration of SINEs and of other processed pseudogenes is unknown, although events similar to those responsible for the transposition of LINEs may be involved.

Although the highly repetitive SINEs and LINEs account for a significant fraction of genomic DNA, their transpositions to random sites in the genome are not likely to be useful for the cell in which they are located. These transposons induce mutations when they integrate at a new target site, and like mutations induced by other agents, most mutations resulting from transposon integration are expected to be harmful to the cell. Indeed, mutations resulting from the transposition of both LINEs and *Alu* elements have been associated with some cases of hemophilia, muscular dystrophy, and colon cancer. On the other hand, some mutations resulting from the movement of transposable elements may be beneficial, contributing in a positive way to evolution of the species. For example, some retrotransposons in mammalian genomes have been found to contain regulatory

*Figure 5.51*
**Formation of a processed pseudogene** The gene illustrated contains three exons, separated by two introns. The introns are removed from the primary transcript by splicing, and a poly-A tail is added to the 3′ end of the mRNA. Reverse transcription and integration then yield a processed pseudogene, which does not contain introns and has an A-rich tract at its 3′ end. The processed pseudogene is flanked by short direct repeats of target-site DNA that were generated during its integration.

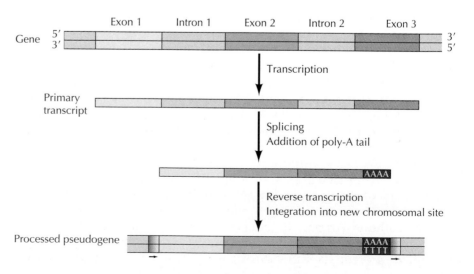

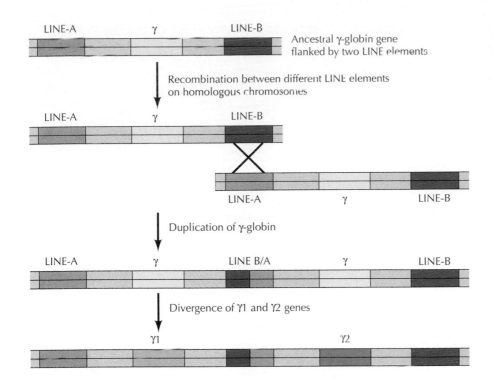

LINE-A     γ     LINE-B     Ancestral γ-globin gene flanked by two LINE elements

Recombination between different LINE elements on homologous chromosomes

LINE-A     γ     LINE-B

LINE-A     γ     LINE-B

Duplication of γ-globin

LINE-A     γ     LINE B/A     γ     LINE-B

Divergence of γ1 and γ2 genes

γ1            γ2

*Figure 5.52*
**Duplication of the γ-globin gene by recombination between LINEs** The γ-globin gene was duplicated during the evolution of primates. The ancestral gene was flanked by two LINEs (LINE-A and LINE-B). Duplication of the gene resulted from recombination between different LINEs on homologous chromosomes. The duplicated γ-globin gene (γ2) then evolved to a pattern of fetal rather than embryonic expression.

sequences that control the expression of adjacent genes.

In addition to their role as mutagens, retrotranspons may contribute to genetic diversity by stimulating rearrangements of other genes. For example, recombination between homologous transposable elements integrated at different sites in the genome can lead to rearrangements or duplications of linked genes. One such event has been documented during evolution of the globin gene family in primates, where recombination between two LINEs resulted in duplication of the γ-globin gene (Figure 5.52). Further evolution of the duplicated gene then led to a shift in γ-globin expression from embryonic to fetal tissues. Such gene duplications are a basic mechanism of evolution, since (as in this example) they allow duplicated genes to diverge in function. The involvement of retrotransposons in this process thus provides a good example of their potential evolutionary roles.

## Gene Amplification

The DNA rearrangements that have been discussed so far alter the position of a DNA sequence within the genome. **Gene amplification** may be viewed as a different type of alteration in genome structure; it increases the number of copies of a gene within a cell. Gene amplification results from repeated rounds of DNA replication, yielding multiple copies of a particular region (Figure 5.53). The amplified DNA sequences can be found either as free extrachromosomal molecules or as tandem arrays of sequences within a chromosome. In either case, the result is increased expression of the amplified gene, simply because more copies of the gene are available to be transcribed.

In some cases, gene amplification is responsible for developmentally programmed increases in gene expression. The prototypical example is amplification of the ribosomal RNA genes in amphibian oocytes (eggs). Eggs are extremely large cells, with correspondingly high requirements for protein synthesis. Amphibian oocytes in particular are about a million times larger in volume than typical somatic cells and must support large amounts of protein synthesis during early development. This requires

*Figure 5.53*
**DNA amplification** Repeated rounds of DNA replication yield multiple copies of a particular chromosomal region.

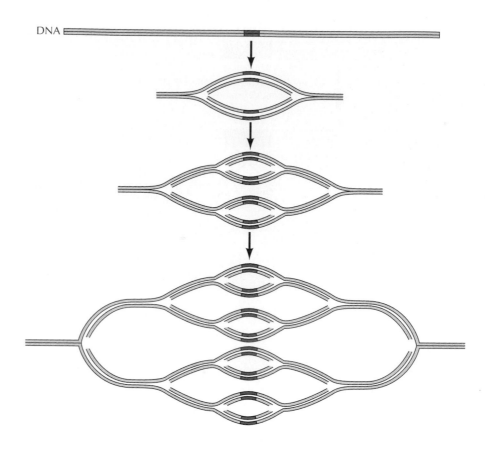

DNA

increased synthesis of ribosomal RNAs, which is accomplished in part by amplification of the ribosomal RNA genes. As discussed in Chapter 4, there are already several hundred copies of ribosomal RNA genes per genome, so that enough ribosomal RNA can be produced to meet the needs of somatic cells. In amphibian eggs, these genes are amplified an additional 2000-fold, to approximately 1 million copies per oocyte. Another example of programmed gene amplification occurs in *Drosophila*, where the genes that encode eggshell proteins (chorion genes) are amplified in ovarian cells to support the requirement for large amounts of these proteins. Like other programmed gene rearrangements, however, gene amplification is a relatively infrequent event that occurs in highly specialized cell types; it is not a common mechanism of gene regulation.

Gene amplification also occurs as an abnormal event in cancer cells, where it results in the increased expression of genes that contribute to uncontrolled cell growth. Such gene amplification was first recognized in cancer cells that had become resistant to methotrexate, a drug commonly used in cancer chemotherapy. Methotrexate inhibits the enzyme dihydrofolate reductase, which is involved in the synthesis of dNTPs and is therefore required for DNA synthesis. Resistance to methotrexate frequently develops by amplification of the dihydrofolate reductase gene, leading to increased production of the enzyme and consequently the loss of effective inhibition by methotrexate. In addition, gene amplification in cancer cells frequently results in the increased expression of genes that drive cell proliferation (oncogenes) and thereby directly contributes to tumor development (see Chapter 15). For example, amplification of the oncogene *erb*B-2 is frequently involved in human breast cancers. Thus, as with other types of DNA rearrangements, gene amplification can have either beneficial or deleterious consequences for the cell or organism in which it occurs.

## Summary

### DNA REPLICATION

*DNA Polymerases:* Different DNA polymerases play distinct roles in DNA replication and repair in both prokaryotic and eukaryotic cells. All known DNA polymerases synthesize DNA only in the 5′ to 3′ direction by the addition of dNTPs to a preformed primer strand of DNA.

*The Replication Fork:* Parental strands of DNA separate and serve as templates for the synthesis of two new strands at the replication fork. One new DNA strand (the leading strand) is synthesized in a continuous manner; the other strand (the lagging strand) is formed by the joining of small fragments of DNA that are synthesized backward with respect to the overall direction of replication. DNA polymerases and various other proteins act in a coordinated manner to synthesize both leading and lagging strands of DNA.

*The Fidelity of Replication:* DNA polymerases increase the accuracy of replication both by selecting the correct base for insertion and by proofreading newly synthesized DNA to eliminate mismatched bases.

*Origins and the Initiation of Replication:* DNA replication starts at specific origins of replication, which contain binding sites for proteins that initiate the process.

*Telomeres and Telomerase: Replicating the Ends of Chromosomes:* Telomeric repeat sequences at the ends of chromosomes are replicated by the action of a reverse transcriptase (telomerase) that carries its own template RNA.

### DNA REPAIR

*Direct Reversal of DNA Damage:* A few types of common DNA lesions, such as pyrimidine dimers and alkylated guanine residues, are repaired by direct reversal of the damage.

*Excision Repair:* Most types of DNA damage are repaired by excision of the damaged DNA. The resulting gap is filled by newly synthesized DNA, using the undamaged complementary strand as a template. In base-excision repair, specific types of single damaged bases are removed from the DNA molecule. In contrast, nucleotide excision repair systems recognize a wide variety of lesions that distort the structure of DNA and remove the damaged bases as part of an oligonucleotide. A third excision repair system specifically removes mismatched bases from newly synthesized DNA strands.

*Postreplication Repair:* Replication of damaged DNA can result in the synthesis of a daughter strand containing a gap opposite the site of damage in the parental template strand. Such gaps in newly synthesized DNA can be filled by recombination with the undamaged parental strand or by error-prone repair synthesis using the damaged parental strand as a template.

### RECOMBINATION BETWEEN HOMOLOGOUS DNA SEQUENCES

*DNA Molecules Recombine by Breaking and Rejoining:* The molecular mechanism of recombination involves the breaking and rejoining of parental DNA molecules.

DNA polymerase, mutagen

replication fork, Okazaki fragment, DNA ligase, leading strand, lagging strand, primase, RNase H, exonuclease, helicase, single-stranded DNA-binding protein, topoisomerase

proofreading

origin of replication, autonomously replicating sequence (ARS), origin replication complex (ORC)

telomere, telomerase, reverse transcriptase

pyrimidine dimer, photoreactivation

base-excision repair, DNA glycosylase, AP endonuclease, nucleotide-excision repair, excinuclease, mismatch repair

recombinational repair, error-prone repair

**general homologous recombination, Holliday model, Holliday junction**

**RecA**

**site-specific recombination, lysogeny, antigen, immunoglobulin, T cell receptor**

**transposition, transposable element, transposon**

**retrovirus, reverse transcriptase, long terminal repeat (LTR), retrotransposon, LINE, SINE, processed pseudogene**

**gene amplification**

*Models of Homologous Recombination:* Alignment between homologous DNA molecules is provided by complementary base pairing. Nicked strands of parental DNA invade the other parental molecule, yielding a crossed-strand intermediate known as a Holliday junction. Recombinant molecules are then formed by cleavage and rejoining of the crossed strands.

*Enzymes Involved in Homologous Recombination:* The central enzyme of homologous recombination is RecA, which catalyzes the exchange of strands between homologous DNAs. Other enzymes nick and unwind parental DNAs and resolve Holliday junctions.

## DNA REARRANGEMENTS

*Site-Specific Recombination:* Site-specific recombination takes place between specific DNA sequences that are recognized by proteins that mediate the process. In vertebrates, site-specific recombination plays a critical role in generating immunoglobulin and T cell receptor genes during development of the immune system.

*Transposition via DNA Intermediates:* Most DNA transposons move throughout the genome with no requirement for specific DNA sequences at their sites of insertion. In yeasts and protozoans, however, the transposition of some DNA sequences to specific target sites results in programmed DNA rearrangements that regulate gene expression.

*Transposition via RNA Intermediates:* Most transposons in eukaryotic cells move by reverse transcription of RNA intermediates, similar to the replication of retroviruses. These retrotransposons include the highly repeated LINE and SINE sequences of mammalian genomes.

*Gene Amplification:* Gene amplification results from repeated replication of a chromosomal region. In some cases, gene amplification provides a mechanism for increasing gene expression during development. Gene amplification also frequently occurs in cancer cells, where it can result in the elevated expression of genes that contribute to uncontrolled cell proliferation.

## QUESTIONS

**1.** How is the fidelity of DNA replication affected by the fact that DNA polymerase adds nucleotides only to a primer strand that is hydrogen-bonded to the template?

**2.** Would you expect the RNA fragments synthesized by primase to be accurate copies of the template DNA? How does this affect the overall accuracy of DNA replication?

**3.** Patients with xeroderma pigmentosum suffer an extremely high incidence of skin cancer but have not been found to have correspondingly high incidences of cancers of internal organs (e.g., colon cancer). What might this suggest about the kinds of DNA damage responsible for most internal cancers?

**4.** *RecA* mutants of *E. coli* are sensitive to UV irradiation in addition to being recombination-deficient. Why?

**5.** What phenotype would you predict for a mutant mouse lacking one of the genes required for site-specific recombination in lymphocytes?

**6.** Many of the drugs in clinical use and under evaluation for the treatment of AIDS are inhibitors of the HIV reverse transcriptase. What reverse transcriptases in human cells might also be inhibited by these drugs? What would be the consequences of inhibiting these enzymes?

## REFERENCES AND FURTHER READING

### General References

Berg, D. E. and M. M. Howe, eds. 1989. *Mobile DNA*. Washington, DC: American Society of Microbiology.

Friedberg, E. C., G. C. Walker and W. Siede. 1995. *DNA Repair and Mutagenesis*. Washington, DC: ASM Press.

Kornberg, A. and T. A. Baker. 1992. *DNA Replication*. 2nd ed. New York: W. H. Freeman.

Kucherlapati, R. and G. R. Smith. eds. 1988. *Genetic Recombination*. Washington, DC: American Society of Microbiology.

### DNA Replication

Blackburn, E. H. 1992. Telomerases. *Ann. Rev. Biochem.* 61: 113–129. [R]

Cairns, J. 1963. The chromosome of *Escherichia coli*. *Cold Spring Harbor Symp. Quant. Biol.* 28: 43–46. [P]

Challberg, M. D. and T. J. Kelly. 1989. Animal virus DNA replication. *Ann. Rev. Biochem.* 58: 671–717. [R]

Diller, J. D. and M. K. Raghuraman. 1994. Eukaryotic replication origins: Control in space and time. *Trends Biochem. Sci.* 19: 320–325. [R]

Echols, H. and M. F. Goodman. 1991. Fidelity mechanisms in DNA replication. *Ann. Rev. Biochem.* 60: 477–511. [R]

Fangman, W. L. and B. J. Brewer. 1991. Activation of replication origins within yeast chromosomes. *Ann. Rev. Cell Biol.* 7: 375–402. [R]

Gavin, K. A., M. Hidaka and B. Stillman. 1995. Conserved initiator proteins in eukaryotes. *Science* 270: 1667–1671. [P]

Heintz, N. H., L. Dailey, P. Held and N. Heintz. 1992. Eukaryotic replication origins as promoters of bidirectional DNA synthesis. *Trends Genet.* 8: 376–381. [R]

Huberman, J. A. and A. D. Riggs. 1968. On the mechanism of DNA replication in mammalian chromosomes. *J. Mol. Biol.* 32: 327–341. [P]

Hubscher, U. and P. Thommes. 1992. DNA polymerase epsilon: In search of a function. *Trends Biochem. Sci.* 17: 55–58. [R]

Kelman, Z. and M. O'Donnell. 1995. DNA polymerase III holoenzyme: Structure and function of a chromosomal replicating machine. *Ann. Rev. Biochem.* 64: 171–200. [R]

Kornberg, A., I. R. Lehman, M. J. Bessman and E. S. Simms. 1956. Enzymic synthesis of deoxyribonucleic acid. *Biochim. Biophys. Acta* 21: 197–198. [P]

Kunkel, T. A. 1992. DNA replication fidelity. *J. Biol. Chem.* 267: 18251–18254. [R]

Marians, K. J. 1992. Prokaryotic DNA replication. *Ann. Rev. Biochem.* 61: 673–719. [R]

Ogawa, T. and T. Okazaki. 1980. Discontinuous DNA replication. *Ann. Rev. Biochem.* 49: 421–457. [R]

Roca, J. 1995. The mechanisms of DNA topoisomerases. *Trends Biochem. Sci.* 20: 156–160. [R]

Stillman, B. 1994. Initiation of chromosomal DNA replication in eukaryotes. *J. Biol. Chem.* 269: 7047–7050. [R]

Stillman, B. 1994. Smart machines at the DNA replication fork. *Cell* 78: 725–728. [R]

Stinthcomb, D. T., K. Struhl and R. W. Davis. 1979. Isolation and characterization of a yeast chromosomal replicator. *Nature* 282: 39–43. [P]

Toyn, J. H., W. M. Toone, B. A. Morgan and L. H. Johnston. 1995. The activation of DNA replication in yeast. *Trends Biochem. Sci.* 20: 70–73. [R]

Waga, S. and B. Stillman. 1994. Anatomy of a DNA replication fork revealed by reconstitution of SV40 DNA replication *in vitro*. *Nature* 369: 207–212. [P]

Wang, J. C. 1991. DNA topoisomerases: Why so many? *J. Biol. Chem.* 266: 6659–6662. [R]

Wang, T. S.-F. 1991. Eukaryotic DNA polymerases. *Ann. Rev. Biochem.* 60: 513–552. [R]

Zakian, V. A. 1995. Telomeres: Beginning to understand the end. *Science* 270: 1601–1607. [R]

### DNA Repair

Cleaver, J. E. 1968. Defective repair replication of DNA in xeroderma pigmentosum. *Nature* 218: 652–656. [P]

Cleaver, J. E. 1994. It was a very good year for DNA repair. *Cell* 76: 1–4. [R]

Fishel, R., M. K. Lescoe, M. R. S. Rao, N. G. Copeland, N. A. Jenkins, J. Garber, M. Kane and R. Kolodner. 1993. The human mutator gene homolog *MSH2* and its association with hereditary nonpolyposis colon cancer. *Cell* 75: 1027–1038. [P]

Grossman, L., P. R. Caron, S. J. Mazur and E. Y. Oh. 1988. Repair of DNA-containing pyrimidine dimers. *FASEB J.* 2: 2696–2701. [R]

Hanawalt, P. C. 1994. Transcription-coupled repair and human disease. *Science* 266: 1957–1958. [R]

Kolodner, R. D. 1995. Mismatch repair: Mechanisms and relationship to cancer susceptibility. *Trends Biochem. Sci.* 20: 397–401. [R]

Lahue, R. S., K. G. Au and P. Modrich. 1989. DNA mismatch correction in a defined system. *Science* 245: 160–164. [R]

Leach, F. S. and 34 others. 1993. Mutations of a *mutS* homolog in hereditary nonpolyposis colorectal cancer. *Cell* 75: 1215–1225. [P]

Lehmann, A. R. 1995. Nucleotide excision repair and the link with transcription. *Trends Biochem. Sci.* 20: 402–405. [R]

Modrich, P. 1994. Mismatch repair, genetic stability, and cancer. *Science* 266: 1959–1960. [R]

Sancar, A. 1994. Mechanisms of DNA excision repair. *Science* 266: 1954–1956. [R]

Sancar, A. 1995. Excision repair in mammalian cells. *J. Biol. Chem.* 270: 15915–15918. [R]

Sancar, A., and Rupp, W.D. 1983. A novel repair enzyme: UVRABC excision nuclease of *Escherichia coli* cuts a DNA strand on both sides of the damaged region. *Cell* 33: 249–260. [P]

Sancar, A. and G. B. Sancar. 1988. DNA repair enzymes. *Ann. Rev. Biochem.* 57: 29–67. [R]

Seeberg, E., L. Eide and M. Bjoras. 1995. The base excision repair pathway. *Trends Biochem. Sci.* 20: 391–397. [R]

Tanaka, K. and R. D. Wood. 1994. Xeroderma pigmentosum and nucleotide excision repair of DNA. *Trends Biochem. Sci.* 19: 83–86. [R]

Van Houten, B. 1990. Nucleotide excision repair in *Escherichia coli*. *Microbiol. Rev.* 54: 18–51. [R]

### Recombination between Homologous DNA Sequences

DasGupta, C., A. M. Wu, R. Kahn, R. P. Cunningham and C. M. Radding. 1981. Concerted strand exchange and formation of Holliday structures by *E. coli* RecA protein. *Cell* 25: 507–516. [P]

Holliday, R. 1964. A mechanism for gene conversion in fungi. *Genet. Res.* 5: 282–304. [P]

Kowalczykowski, S. C. and A. K. Eggleston. 1994. Homologous pairing and DNA strand-exchange proteins. *Ann. Rev. Biochem.* 63: 991–1043. [R]

Meselson, M. and C. M. Radding. 1975. A general model for genetic recombination. *Proc. Natl. Acad. Sci. USA* 72: 358–361. [P]

Meselson, M. and J. J. Weigle. 1961. Chromosome breakage accompanying genetic recombination in bacteriophage. *Proc. Natl. Acad. Sci. USA* 47: 857–868. [P]

Potter, H. and D. Dressler. 1976. On the mechanism of genetic recombination: Electron microscopic observation of recombination intermediates. *Proc. Natl. Acad. Sci. USA* 73: 3000–3004. [P]

Radding, C. M. 1991. Helical interactions in homologous pairing and strand exchange driven by RecA protein. *J. Biol. Chem.* 266: 5355–5358. [R]

Shinohara, A. and T. Ogawa. 1995. Homologous recombination and the roles of double-strand breaks. *Trends Biochem. Sci.* 20: 387–391. [R]

Szostak, J. W., T. L. Orr-Weaver, R. J. Rothstein and F. W. Stahl. 1983. The double-strand-break repair model for recombination. *Cell* 33: 25–35. [P]

Taylor, A. F. 1992. Movement and resolution of Holliday junctions by enzymes from *E. coli*. *Cell* 69: 1063–1065. [R]

West, S. C. 1992. Enzymes and molecular mechanisms of genetic recombination. *Ann. Rev. Biochem.* 61: 603–640. [R]

West, S. C. 1994. The processing of recombination intermediates: Mechanistic insights from studies of bacterial proteins. *Cell* 76: 9–15. [R]

## DNA Rearrangements

Blackwell, T. K. and F. W. Alt. 1989. Molecular characterization of the lymphoid V(D)J recombination activity. *J. Biol. Chem.* 264: 10327–10330. [R]

Boeke, J. D., D. J. Garfinkel, C. A. Styles and G. R. Fink. 1985. Ty elements transpose through an RNA intermediate. *Cell* 40: 491–500. [P]

Borst, P. and D. R. Greaves. 1987. Programmed gene rearrangements altering gene expression. *Science* 235: 658–667. [R]

Craig, N. L. 1995. Unity in transposition reactions. *Science* 270: 253–254. [R]

Davis, M. M. 1990. T cell receptor gene diversity and selection. *Ann. Rev. Biochem.* 59: 475–496. [R]

Dombroski, B. A., S. L. Mathias, E. Nanthakumar, A. F. Scott and H. H. Kazazian Jr. 1991. Isolation of an active human transposable element. *Science* 254: 1805–1808. [P]

Fedoroff, N. V. 1989. About maize transposable elements and development. *Cell* 56: 181–191. [R]

Fedoroff, N. and D. Botstein. 1992. *The Dynamic Genome: Barbara McClintock's Ideas in the Century of Genetics*. Plainview, NY: Cold Spring Harbor Laboratory Press.

Finnegan, D. J. 1989. Eukaryotic transposable elements and genome evolution. *Trends Genet.* 5: 103–107. [R]

Fitch, D. H. A., W. J. Bailey, D. A. Table, M. Goodman, L. Sieu, and J. L. Slightom. 1991. Duplication of the γ-globin gene mediated by L1 long interspersed repetitive elements in an early ancestor of simian primates. *Proc. Natl. Acad. Sci. USA* 88: 7396–7400. [P]

Gilboa, E., S. W. Mitra, S. Goff and D. Baltimore. 1979. A detailed model of reverse transcription and tests of crucial aspects. *Cell* 18: 93–100. [P]

Grandbastien, M.-A. 1992. Retroelements in higher plants. *Trends Genet.* 8: 103–108. [R]

Hozumi, N. and S. Tonegawa. 1976. Evidence for somatic rearrangement of immunoglobulin genes coding for variable and constant regions. *Proc. Natl. Acad. Sci. USA* 73: 3628–3632. [P]

Janeway, C. A., Jr. and P. Travers. 1994. *Immunobiology: The Immune System in Health and Disease*. London/New York: Current Biology Ltd./Garland Publishing.

Landy, A. 1989. Dynamic, structural, and regulatory aspects of λ site-specific recombination. *Ann. Rev. Biochem.* 58: 913–949. [R]

Lewis, S. and M. Gellert. 1989. The mechanism of antigen receptor gene assembly. *Cell* 59: 585–588. [R]

Lieber, M. R. 1992. The mechanism of V(D)J recombination: A balance of diversity, specificity, and stability. *Cell* 70: 873–876. [R]

Luan, D. D., M. H. Korman, J. L. Jakubczak and T. H. Eickbush. 1993. Reverse transcription of R2Bm RNA is primed by a nick at the chromosomal target site: A mechanism for non-LTR retrotransposition. *Cell* 72: 595–605. [P]

McClintock, B. 1956. Controlling elements and the gene. *Cold Spring Harbor Symp. Quant. Biol.* 21: 197–216. [P]

Singer, M. F. 1995. Unusual reverse transcriptases. *J. Biol. Chem.* 270: 24623–24626. [R]

Stark, G. R., M. Debatisse, E. Giulotto and G. M. Wahl. 1989. Recent progress in understanding mechanisms of mammalian DNA amplification. *Cell* 57: 901–908. [R]

Stark, G. R. and G. M. Wahl. 1984. Gene amplification. *Ann. Rev. Biochem.* 53: 447–491. [R]

Temin, H. M. 1989. Retrons in bacteria. *Nature* 339: 254–255. [R]

Tonegawa, S. 1983. Somatic generation of antibody diversity. *Nature* 302: 575–581. [R]

Weiner, A. M., P. L. Deininger and A. Efstratiadis. 1986. Nonviral retroposons: Genes, pseudogenes, and transposable elements generated by the reverse flow of genetic information. *Ann. Rev. Biochem.* 55: 631–661. [R]

Whitcomb, J. M. and S. H. Hughes. 1992. Retroviral reverse transcription and integration: Progress and problems. *Ann. Rev. Cell Biol.* 8: 275–306. [R]

# 6

# RNA Synthesis and Processing

CHAPTERS 4 AND 5 DISCUSSED THE ORGANIZATION and maintenance of genomic DNA, which can be viewed as the set of genetic instructions governing all cellular activities. These instructions are implemented via the synthesis of RNAs and proteins. Importantly, the behavior of a cell is determined not only by what genes it inherits, but also by which of those genes are expressed at any given time. Regulation of gene expression allows cells to adapt to changes in their environments and is responsible for the distinct activities of the multiple differentiated cell types that make up complex plants and animals. Muscle cells and liver cells, for example, contain the same genes; the functions of these cells are determined not by differences in their genomes, but by regulated patterns of gene expression that govern development and differentiation.

The first step in expression of a gene, the transcription of DNA into RNA, is the primary level at which gene expression is regulated in both prokaryotic and eukaryotic cells. RNAs in eukaryotic cells are then modified in various ways—for example, introns are removed by splicing—to convert the primary transcript into its functional form. Different types of RNA play distinct roles in cells: Messenger RNAs (mRNAs) serve as templates for protein synthesis; ribosomal RNAs (rRNAs) and transfer RNAs (tRNAs) function in mRNA translation. Still other small RNAs function in splicing and protein sorting in eukaryotes. Transcription and RNA processing are discussed in this chapter. The final step in gene expression, the translation of mRNA to protein, is the subject of Chapter 7.

## TRANSCRIPTION IN PROKARYOTES

As in most areas of molecular biology, studies of *E. coli* have provided the model for subsequent investigations of transcription in eukaryotic cells. As reviewed in Chapter 3, mRNA was discovered first in *E. coli*. *E. coli* was also the first organism from which RNA polymerase was purified and studied. The basic mechanisms by which transcription is regulated were likewise elucidated by pioneering experiments in *E. coli*, in which regulated gene expression allows the cell to respond to variations in the environment, such

225

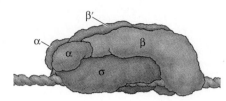

**Figure 6.1**
***E. coli* RNA polymerase** The complete enzyme consists of five subunits: two $\alpha$, one $\beta$, one $\beta'$, and one $\sigma$. The $\sigma$ subunit is relatively weakly bound and can be dissociated from the other four subunits, which constitute the core polymerase.

as changes in the availability of nutrients. An understanding of transcription in *E. coli* has thus provided the foundation for studies of the far more complex mechanisms that regulate gene expression in eukaryotic cells.

### RNA Polymerase and Transcription

The principal enzyme responsible for RNA synthesis is **RNA polymerase**, which catalyzes the polymerization of ribonucleoside 5′-triphosphates (NTPs) as directed by a DNA template. The synthesis of RNA is similar to that of DNA, and like DNA polymerase, RNA polymerase catalyzes the growth of RNA chains always in the 5′ to 3′ direction. Unlike DNA polymerase, however, RNA polymerase does not require a preformed primer to initiate the synthesis of RNA. Instead, transcription initiates *de novo* at specific sites at the beginning of genes. The initiation process is particularly important because this is the primary step at which transcription is regulated.

*E. coli* RNA polymerase, like DNA polymerase, is a complex enzyme made up of multiple polypeptide chains. The intact enzyme consists of four different types of subunits, called $\alpha$, $\beta$, $\beta'$, and $\sigma$ (Figure 6.1). The $\sigma$ subunit is relatively weakly bound and can be separated from the other subunits, yielding a core polymerase consisting of two $\alpha$, one $\beta$, and one $\beta'$ subunits. The core polymerase is fully capable of catalyzing the polymerization of NTPs into RNA, indicating that $\sigma$ is not required for the basic catalytic activity of the enzyme. However, the core polymerase does not bind specifically to the DNA sequences that signal the normal initiation of transcription; therefore, the $\sigma$ subunit is required to identify the correct sites for transcription initiation. The selection of these sites is a critical element of transcription because synthesis of a functional RNA must start at the beginning of a gene.

The DNA sequence to which RNA polymerase binds to initiate transcription of a gene is called the **promoter**. The DNA sequences involved in promoter function were first identified by comparisons of the nucleotide sequences of a series of different genes isolated from *E. coli*. These comparisons revealed that the region upstream of the transcription initiation site contains two sets of sequences that are similar in a variety of genes. These common sequences encompass six nucleotides each, and are located approximately 10 and 35 base pairs upstream of the transcription start site (Figure 6.2). They are called the –10 and –35 elements, denoting their position relative to the transcription initiation site, which is defined as the +1 position. The sequences at the –10 and –35 positions in different promoters are not identical, but they are all similar enough to establish consensus sequences—the bases most frequently found at each position.

Several types of experimental evidence support the functional importance of the –10 and –35 promoter elements. First, genes with promoters that differ from the consensus sequences are transcribed less efficiently than genes whose promoters match the consensus sequences more closely. Second, mutations introduced in either the –35 or –10 consensus sequences have strong effects on promoter function. Third, the sites at which RNA polymerase binds to promoters have been directly identified by **footprinting** experiments, which are widely used to determine the sites at which

**Figure 6.2**
**Sequences of *E. coli* promoters** *E. coli* promoters are characterized by two sets of sequences located 10 and 35 base pairs upstream of the transcription start site (+1). The consensus sequences shown correspond to the bases most frequently found in different promoters.

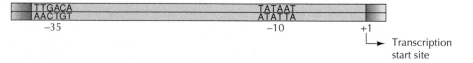

proteins bind to DNA (Figure 6.3). In experiments of this type, a DNA fragment is radiolabeled at one end, as for sequencing by the Maxam-Gilbert procedure (see Figure 3.24). The labeled DNA is incubated with the protein of interest (e.g., RNA polymerase) and then subjected to partial digestion with DNase. The principle of the method is that the regions of DNA to which the protein binds are protected from DNase digestion. These regions can therefore be identified by comparison of the digestion products of the protein-bound DNA with those resulting from identical DNase treatment of a parallel sample of DNA that was not incubated with protein. Variations of this basic method, which employ chemical reagents to modify and cleave DNA at particular nucleotides, can be used to identify the specific DNA bases that are in contact with protein. Such footprinting analysis has

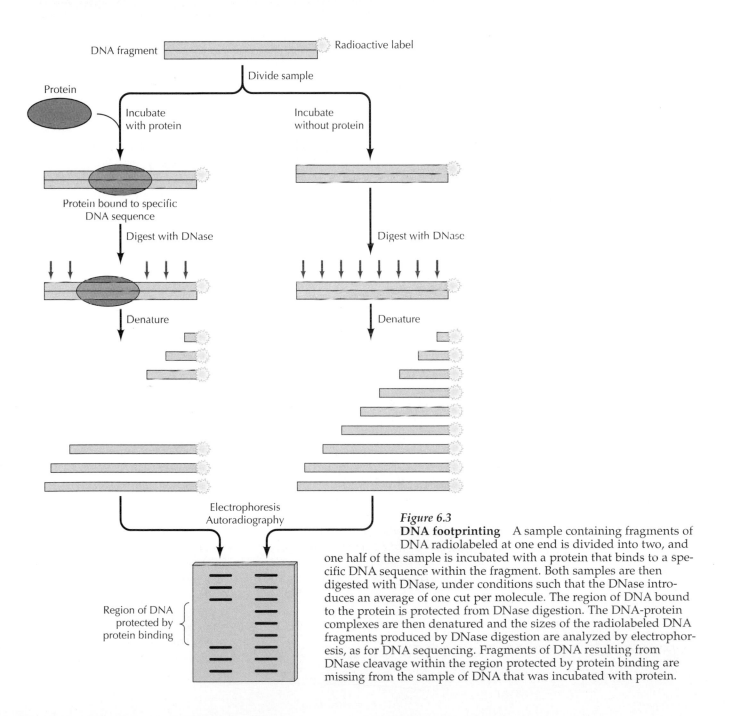

*Figure 6.3*
**DNA footprinting** A sample containing fragments of DNA radiolabeled at one end is divided into two, and one half of the sample is incubated with a protein that binds to a specific DNA sequence within the fragment. Both samples are then digested with DNase, under conditions such that the DNase introduces an average of one cut per molecule. The region of DNA bound to the protein is protected from DNase digestion. The DNA-protein complexes are then denatured and the sizes of the radiolabeled DNA fragments produced by DNase digestion are analyzed by electrophoresis, as for DNA sequencing. Fragments of DNA resulting from DNase cleavage within the region protected by protein binding are missing from the sample of DNA that was incubated with protein.

*Figure 6.4*
**Transcription by *E. coli* RNA polymerase**  The polymerase initially binds nonspecifically to DNA and migrates along the molecule until the σ subunit binds to the –35 and –10 promoter elements, forming a closed-promoter complex. The polymerase then unwinds DNA around the initiation site, and transcription is initiated by the polymerization of free NTPs. The σ subunit then dissociates from the core polymerase, which migrates along the DNA and elongates the growing RNA chain.

shown that RNA polymerase generally binds to promoters over approximately a 60-base-pair region, extending from –40 to +20 (i.e., from 40 nucleotides upstream to 20 nucleotides downstream of the transcription start site). The σ subunit binds specifically to sequences in both the –35 and –10 promoter regions, substantiating the importance of these sequences in promoter function. In addition, some *E. coli* promoters have a third sequence, located upstream of the –35 region, that serves as a specific binding site for the RNA polymerase α subunit.

In the absence of σ, RNA polymerase binds nonspecifically to DNA with low affinity. The role of σ is to direct the polymerase to promoters by binding specifically to both the –35 and –10 sequences, leading to the initiation of transcription at the beginning of a gene (Figure 6.4). The initial binding between the polymerase and a promoter is referred to as a closed-promoter complex because the DNA is not unwound. The polymerase then unwinds

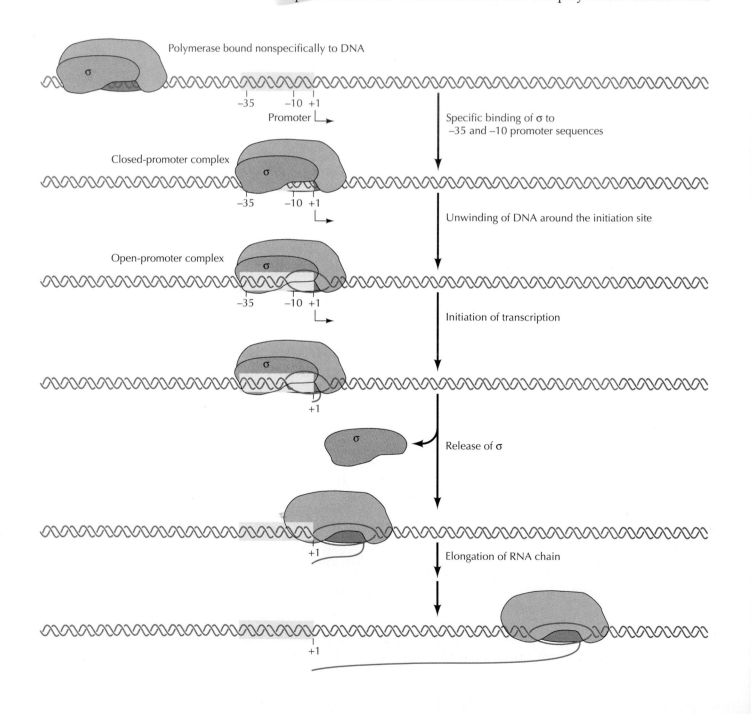

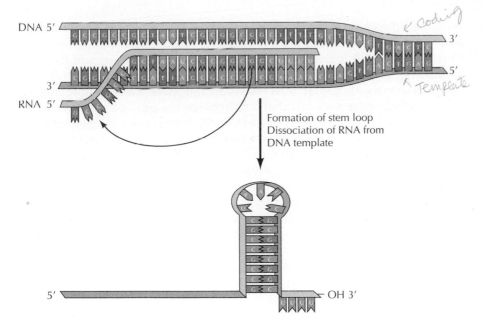

*Figure 6.5*
**Transcription termination** The termination of transcription is signaled by a GC-rich inverted repeat followed by four A residues. The inverted repeat forms a stable stem-loop structure in the RNA, causing the RNA to dissociate from the DNA template.

Formation of stem loop
Dissociation of RNA from
DNA template

approximately 15 bases of DNA around the initiation site to form an open-promoter complex in which single-stranded DNA is available as a template for transcription. Transcription is initiated by the joining of two free NTPs. After addition of about the first 10 nucleotides, σ is released from the polymerase, which then leaves the promoter and moves along the template DNA to continue elongation of the growing RNA chain. As it travels, the polymerase unwinds the template DNA ahead of it and rewinds the DNA behind it, maintaining an unwound region of about 17 base pairs in the region of transcription.

RNA synthesis continues until the polymerase encounters a termination signal, at which point transcription stops, the RNA is released from the polymerase, and the enzyme dissociates from its DNA template. The simplest and most common type of termination signal in *E. coli* consists of a symmetrical inverted repeat of a GC-rich sequence followed by four or more A residues (Figure 6.5). Transcription of the GC-rich inverted repeat results in the formation of a segment of RNA that can form a stable stem-loop structure by complementary base pairing. The formation of such a self-complementary structure in the RNA disrupts its association with the DNA template and terminates transcription. Because hydrogen bonding between A and U is weaker than that between G and C, the presence of A residues downstream of the inverted repeat sequences is thought to facilitate the dissociation of the RNA from its template.

## Repressors and Negative Control of Transcription

The pioneering studies of gene regulation in *E. coli* were carried out by François Jacob and Jacques Monod in the 1950s. These investigators and their colleagues analyzed the expression of enzymes involved in the metabolism of lactose, which can be used as a source of carbon and energy via cleavage to glucose and galactose (Figure 6.6). The enzyme that catalyzes the cleavage

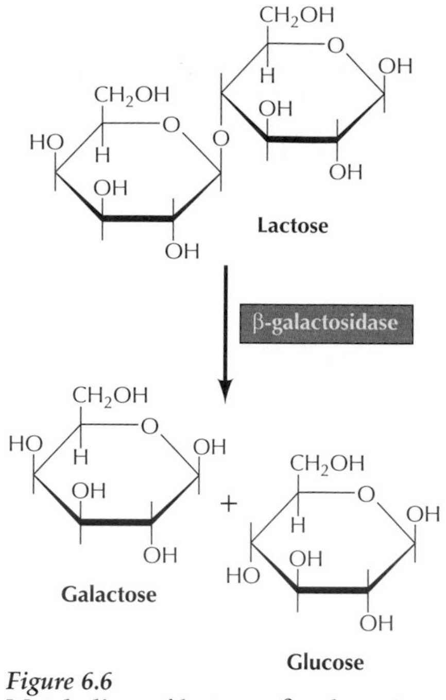

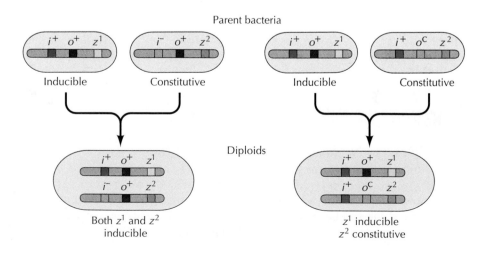

CH₂OH

Lactose

β-galactosidase

Galactose

Glucose

*Figure 6.6*
**Metabolism of lactose** β-galactosidase catalyzes the hydrolysis of lactose to glucose and galactose.

of lactose (β-galactosidase) and other enzymes involved in lactose metabolism are expressed only when lactose is available for use by the bacteria. Otherwise, the cell is able to economize by not investing energy in the synthesis of unnecessary RNAs and proteins. Thus, lactose induces the synthesis of enzymes involved in its own metabolism. In addition to requiring β-galactosidase, lactose metabolism involves the products of two other closely linked genes: lactose permease, which transports lactose into the cell, and a transacetylase, whose function in lactose metabolism is still unknown. On the basis of purely genetic experiments, Jacob and Monod deduced the mechanism by which the expression of these genes was regulated, thereby formulating a model that remains fundamental to our understanding of transcriptional regulation.

The starting point in this analysis was the isolation of mutants that were defective in regulation of the genes involved in lactose utilization. These mutants were of two types: constitutive mutants, which expressed all three genes even when lactose was not available, and noninducible mutants, which failed to express the genes even in the presence of lactose. Genetic mapping localized these regulatory mutants to two distinct loci, called $o$ and $i$, with $o$ located immediately upstream of the structural gene for β-galactosidase. Mutations affecting $o$ resulted in constitutive expression; mutants of $i$ were either constitutive or noninducible.

The function of these regulatory genes was probed by experiments in which two strains of bacteria were mated, resulting in diploid cells containing genes derived from both parents (Figure 6.7). Analysis of gene expression in such diploid bacteria provided critical insights by defining which alleles of these regulatory genes are dominant and which recessive. For example, when bacteria containing a normal $i$ gene ($i^+$) were mated with bacteria carrying an $i$ gene mutation resulting in constitutive expression (an $i^-$ mutation), the resulting diploid bacteria displayed normal inducibility; therefore, the normal $i^+$ gene was dominant over the $i^-$ mutant. In contrast, matings between normal bacteria and bacteria with an $o^c$ mutation (constitutive expression) yielded diploids with the constitutive expression phenotype, indicating that $o^c$ is dominant over $o^+$. Additional experiments in which mutations in $o$ and $i$ were combined with different mutations in the structural genes showed that $o$ affects the expression of only the genes to which it is physically linked, whereas $i$ affects the expression of genes on both chromosome copies in diploid bacteria. Thus, in an $o^c/o^+$ cell, only the structural genes that are linked to $o^c$

*Figure 6.7*
**Regulation of β-galactosidase in diploid *E. coli*** The mating of two bacterial strains results in diploid cells that contain genes from both parents. In these examples, it is assumed that the genes encoding β-galactosidase (the $z$ genes) can be distinguished on the basis of structural gene mutations, designated $z^1$ and $z^2$. In an $i^+/i^-$ diploid (left), both structural genes are inducible; therefore, $i^+$ is dominant over $i^-$ and affects expression of $z$ genes on both chromosomes. In contrast, in an $o^c/o^+$ diploid (right), the $z$ gene linked to $o^c$ is constitutively expressed, whereas that linked to $o^+$ is inducible. Therefore, $o$ affects expression of only the adjacent $z$ gene on the same chromosome.

Parent bacteria

$i^+$ $o^+$ $z^1$     $i^-$ $o^+$ $z^2$          $i^+$ $o^+$ $z^1$     $i^+$ $o^c$ $z^2$

Inducible     Constitutive          Inducible     Constitutive

Diploids

$i^+$ $o^+$ $z^1$          $i^+$ $o^+$ $z^1$
$i^-$ $o^+$ $z^2$          $i^+$ $o^c$ $z^2$

Both $z^1$ and $z^2$          $z^1$ inducible
inducible          $z^2$ constitutive

are constitutively expressed. In contrast, in an $i^+/i^-$ cell, structural genes on both chromosomes are regulated normally. These results led to the conclusion that $o$ represents a region of DNA that controls the transcription of adjacent genes, whereas the $i$ gene encodes a regulatory factor (e.g., a protein) that can diffuse throughout the cell and control genes on both chromosomes.

The model of gene regulation developed on the basis of these experiments is illustrated in Figure 6.8. The genes encoding $\beta$-galactosidase, permease, and transacetylase are expressed as a single unit, called an **operon**. Transcription of the operon is controlled by $o$ (the **operator**), which is adjacent to the transcription initiation site. The $i$ gene encodes a protein that regulates transcription by binding to the operator. Since $i^-$ mutants (which result in constitutive gene expression) are recessive, it was concluded that these mutants failed to make a functional gene product. This result implies that the normal $i$ gene product is a **repressor**, which blocks transcription when bound to $o$. The addition of lactose leads to induction of the operon because lactose binds to the repressor, thereby preventing it from binding to the operator DNA. In noninducible $i$ mutants (which are dominant over $i^+$), the repressor fails to bind lactose, so expression of the operon cannot be induced.

The model neatly fits the results of the genetic experiments from which it was derived. In $i^-$ cells, the repressor is not made, so the *lac* operon is constitutively expressed. Diploid $i^+/i^-$ cells are normally inducible, since functional repressor is encoded by the $i^+$ allele. Finally, in $o^c$ mutants a functional operator has been lost and repressor cannot be bound. Consequently, $o^c$ mutants are dominant but affect the expression only of linked structural genes.

Confirmation of this basic model has since come from a variety of experiments, including Walter Gilbert's isolation, in the 1960s, of the *lac* repressor and analysis of its binding to operator DNA. Molecular analysis has

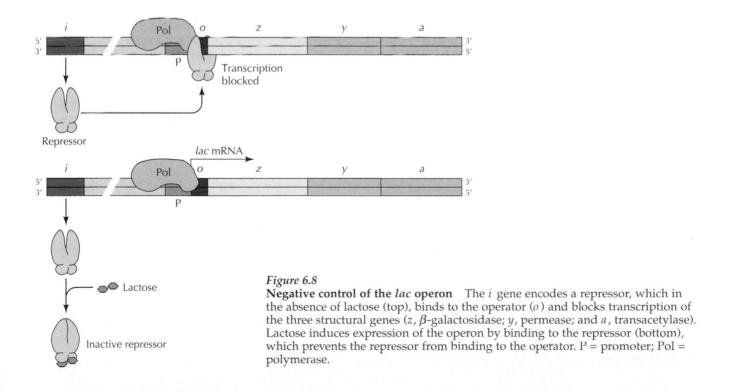

**Figure 6.8**
**Negative control of the *lac* operon** The $i$ gene encodes a repressor, which in the absence of lactose (top), binds to the operator ($o$) and blocks transcription of the three structural genes ($z$, $\beta$-galactosidase; $y$, permease; and $a$, transacetylase). Lactose induces expression of the operon by binding to the repressor (bottom), which prevents the repressor from binding to the operator. P = promoter; Pol = polymerase.

defined the operator as approximately 30 base pairs of DNA, starting a few bases before the transcription initiation site. Footprinting analysis has identified this region as the site to which the repressor binds, blocking transcription. As predicted, lactose binds to the repressor, which then no longer binds to operator DNA. Also as predicted, $o^c$ mutations alter sequences within the operator, thereby preventing repressor binding and resulting in constitutive gene expression.

The central principle of gene regulation exemplified by the lactose operon is that control of transcription is mediated by the interaction of regulatory proteins with specific DNA sequences. This general mode of regulation is broadly applicable to both prokaryotic and eukaryotic cells. Regulatory sequences like the operator are called ***cis*-acting control elements**, because they affect the expression of only linked genes on the same DNA molecule. On the other hand, proteins like the repressor are called ***trans*-acting factors** because they can affect the expression of genes located on other chromosomes within the cell. The *lac* operon is an example of negative control because binding of the repressor blocks transcription. This, however, is not always the case; many *trans*-acting factors are activators rather than inhibitors of transcription.

### Positive Control of Transcription

The first demonstration that repressors are not the only type of transcriptional regulatory proteins came from studies of the *E. coli* enzymes involved in the utilization of arabinose—a five-carbon sugar that, like lactose, can be used in place of glucose as a source of carbon and energy. The genes encoding three enzymes involved in arabinose metabolism are organized into an operon, which can be induced by arabinose (Figure 6.9). The operon is regulated by the product of a gene called *araC*, which, like the *lac i* gene, encodes a *trans*-acting protein. However, binding of the AraC protein to the operator activates rather than represses transcription. This effect was indicated initially by the results of genetic analyses, which showed that loss of a functional AraC protein led to noninducibility rather than constitutive expression of the *ara* operon. This result is opposite to the effect of *lac i⁻* mutants and indicates that the AraC protein activates transcription in the presence of arabinose, rather than repressing transcription in its absence. This conclusion has since been confirmed by biochemical isolation of the AraC protein, which (when bound to arabinose) binds to a site just upstream from the promoter and stimulates transcription.

The best-studied example of positive control in *E. coli* is the effect of glucose on the expression of genes that encode enzymes involved in the breakdown (catabolism) of other sugars (including lactose and arabinose) that provide alternative sources of carbon and energy. Glucose is preferentially utilized, so as long as glucose is available, enzymes involved in catabolism of alternative energy sources are not expressed. For example, if *E. coli* are grown in me-

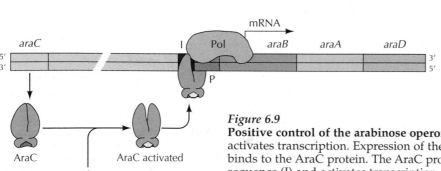

*Figure 6.9*
**Positive control of the arabinose operon** The *araC* gene encodes a protein that activates transcription. Expression of the operon is induced by arabinose, which binds to the AraC protein. The AraC protein then binds to a regulatory DNA sequence (I) and activates transcription. The operon contains three structural genes (*araB*, *araA*, and *araD*). P = promoter; Pol = polymerase.

**Figure 6.10**
**Regulation of gene expression by glucose**   Low levels of glucose activate adenylyl cyclase, which converts ATP to cyclic AMP (cAMP). Cyclic AMP then binds to the catabolite activator protein (CAP) and stimulates its binding to regulatory sequences of various operons concerned with the metabolism of alternative sugars, such as lactose and arabinose.

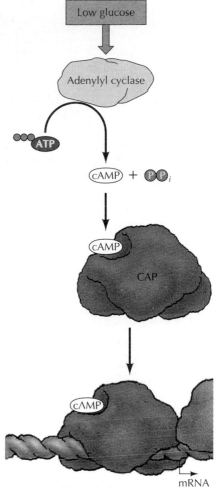

Transcription activation

dium containing both glucose and lactose, the *lac* operon is not induced and only glucose is used by the bacteria. Thus, glucose represses the *lac* operon even in the presence of the normal inducer (lactose).

Glucose repression (generally called catabolite repression) is now known to be mediated by a positive control system, which is coupled to levels of cyclic AMP (cAMP) (Figure 6.10). In bacteria, the enzyme adenylyl cyclase, which converts ATP to cAMP, is regulated such that levels of cAMP increase when glucose levels drop. cAMP then binds to a transcriptional regulatory protein called catabolite activator protein (CAP). The binding of cAMP stimulates the binding of CAP to its target DNA sequences, which in the *lac* operon are located approximately 50 to 70 bases upstream of the transcription start site. CAP then activates transcription, at least in part by interacting with the $\alpha$ subunit of RNA polymerase and facilitating the binding of polymerase to the promoter.

### Transcriptional Attenuation

Both the positive and negative control mechanisms that we have discussed act at the level of initiation of transcription. An additional mechanism, **transcriptional attenuation**, regulates the expression of some genes by controlling the ability of RNA polymerase to continue elongation past specific sites. This mode of regulation has been described best in the *E. coli trp* operon, which encodes five enzymes involved in biosynthesis of the amino acid tryptophan. These genes are expressed only when tryptophan is not available to the cell in its environment, since otherwise the synthesis of additional tryptophan is unnecessary.

The *trp* operon is regulated in part by a repressor that, when bound to tryptophan, blocks transcription (Figure 6.11). However, transcriptional attenuation provides an additional level of control that results in more stringent regulation than could be achieved by repression of initiation alone. The site of attenuation is located 162 nucleotides downstream of the transcription start site. If tryptophan is abundant, most transcription terminates at this site; only if tryptophan is scarce does transcription continue to yield functional Trp mRNA.

The mechanism of attenuation depends on the fact that translation in bacteria is coupled with transcription, so ribosomes begin translating the 5′ end of an mRNA while it is still being synthesized. Thus, the rate of translation can affect the structure of the growing RNA chain, which in turn determines whether further transcription can continue. Transcription termination is signaled by a stem-loop structure that forms by complementary base pairing between two specific sequences of the growing Trp mRNA chain (Figure 6.12). This structure forms if translation of the growing chain is proceeding at a normal rate, as it does when tryptophan is present in adequate supply. If tryptophan is scarce, however, protein synthesis stalls at a critical region of the message. If this occurs, the ribosomes bound to the mRNA block formation of the transcription-terminating stem loop, allowing Trp mRNA synthesis to continue.

The critical region of Trp mRNA contains two adjacent tryptophan codons,

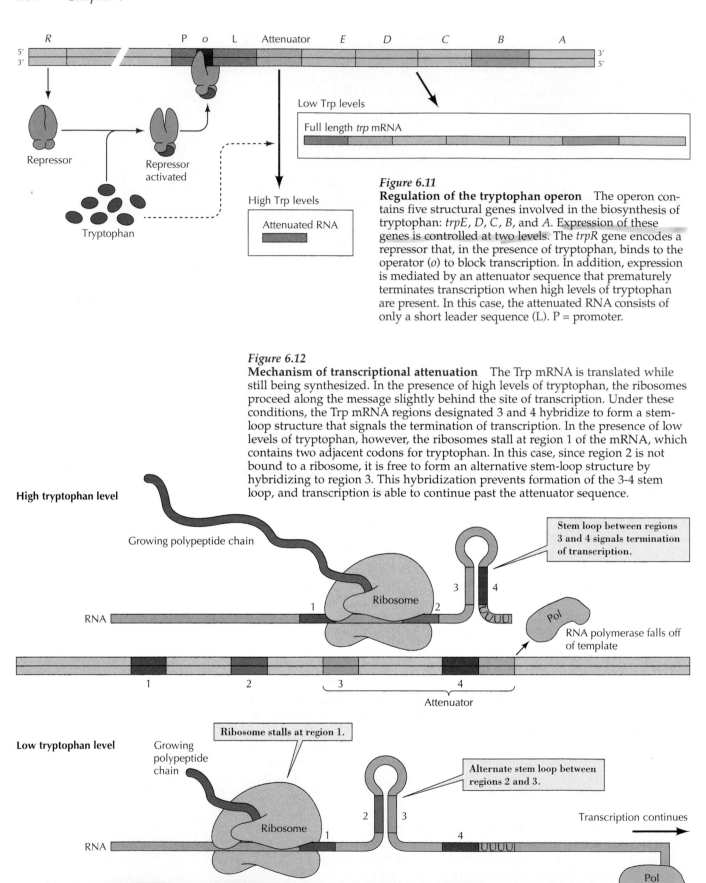

**Figure 6.11**
**Regulation of the tryptophan operon**   The operon contains five structural genes involved in the biosynthesis of tryptophan: *trpE, D, C, B,* and *A.* Expression of these genes is controlled at two levels. The *trpR* gene encodes a repressor that, in the presence of tryptophan, binds to the operator (*o*) to block transcription. In addition, expression is mediated by an attenuator sequence that prematurely terminates transcription when high levels of tryptophan are present. In this case, the attenuated RNA consists of only a short leader sequence (L). P = promoter.

**Figure 6.12**
**Mechanism of transcriptional attenuation**   The Trp mRNA is translated while still being synthesized. In the presence of high levels of tryptophan, the ribosomes proceed along the message slightly behind the site of transcription. Under these conditions, the Trp mRNA regions designated 3 and 4 hybridize to form a stem-loop structure that signals the termination of transcription. In the presence of low levels of tryptophan, however, the ribosomes stall at region 1 of the mRNA, which contains two adjacent codons for tryptophan. In this case, since region 2 is not bound to a ribosome, it is free to form an alternative stem-loop structure by hybridizing to region 3. This hybridization prevents formation of the 3-4 stem loop, and transcription is able to continue past the attenuator sequence.

so the rate of translation is highly dependent on tryptophan levels; this is the link between transcriptional attenuation and the availability of tryptophan. If tryptophan levels in the cell are low, the ribosome stalls at this point and transcription of Trp mRNA continues. If tryptophan is abundant, translation continues and transcription is terminated.

## EUKARYOTIC RNA POLYMERASES AND BASAL TRANSCRIPTION FACTORS

Although transcription proceeds by the same fundamental mechanisms in all cells, it is considerably more complex in eukaryotic cells than in bacteria. This is reflected in two distinct differences between the prokaryotic and eukaryotic systems. First, whereas all genes are transcribed by a single RNA polymerase in bacteria, eukaryotic cells contain multiple different RNA polymerases that transcribe distinct classes of genes. Second, rather than binding directly to promoter sequences, eukaryotic RNA polymerases need to interact with a variety of additional proteins to specifically initiate transcription. This increased complexity of eukaryotic transcription presumably facilitates the sophisticated regulation of gene expression needed to direct the activities of the many different cell types of multicellular organisms.

### Eukaryotic RNA Polymerases

Eukaryotic cells contain three distinct nuclear RNA polymerases that transcribe different classes of genes (Table 6.1). Protein-coding genes are transcribed by RNA polymerase II to yield mRNAs; ribosomal RNAs (rRNAs) and transfer RNAs (tRNAs) are transcribed by RNA polymerases I and III. RNA polymerase I is specifically devoted to transcription of the three largest species of rRNAs, which are designated 28S, 18S, and 5.8S according to their rates of sedimentation during velocity centrifugation. RNA polymerase III transcribes the genes for tRNAs and for the smallest species of ribosomal RNA (5S rRNA). Some of the small RNAs involved in splicing and protein transport (snRNAs and scRNAs) are also transcribed by RNA polymerase III, while others are polymerase II transcripts. In addition, separate RNA polymerases (which are similar to bacterial RNA polymerases) are found in chloroplasts and mitochondria, where they specifically transcribe the DNAs of those organelles.

All three of the nuclear RNA polymerases are complex enzymes, consisting of 8 to 14 different subunits each. Although they recognize different promoters and transcribe distinct classes of genes, they share several common features. The two largest subunits of all three eukaryotic RNA polymerases are related to the $\beta$ and $\beta'$ subunits of the single *E. coli* RNA polymerase. In addition, five subunits of the eukaryotic RNA polymerases are common to all three different enzymes. Consistent with these structural similarities, the different eukaryotic polymerases share several functional properties, including the need to interact with other proteins to appropriately initiate transcription.

### Basal Transcription Factors and Initiation of Transcription by RNA Polymerase II

Because RNA polymerase II is responsible for the synthesis of mRNA from protein-coding genes, it has been the focus of most studies of transcription in eukaryotes. Early attempts at studying this enzyme indicated that its activity is fundamentally different from that of prokaryotic RNA polymerase. The accurate transcription of bacterial genes that can be accomplished *in vitro* simply by the addition of purified RNA polymerase to DNA

**Table 6.1** Classes of Genes Transcribed by Eukaryotic RNA Polymerases

| Type of RNA synthesized | RNA polymerase |
|---|---|
| Nuclear genes | |
|   mRNA | II |
|   tRNA | III |
|   rRNA | |
|     5.8S, 18S, 28S | I |
|     5S | III |
|   snRNA and scRNA | II and III[a] |
| Mitochondrial genes | Mitochondrial[b] |
| Chloroplast genes | Chloroplast[b] |

[a] Some small nuclear (sn) and small cytoplasmic (sc) RNAs are transcribed by polymerase II and others by polymerase III.

[b] The mitochondrial and chloroplast RNA polymerases are similar to bacterial enzymes.

containing a promoter is not possible in eukaryotic systems. The basis of this difference was elucidated in 1979, when Robert Roeder and his colleagues discovered that RNA polymerase II is able to initiate transcription only if additional proteins are added to the reaction. Thus, transcription in the eukaryotic system appeared to require distinct initiation factors that (in contrast to bacterial $\sigma$ factors) were not associated with the polymerase.

Biochemical fractionation of nuclear extracts has now led to the identification of specific proteins (called **transcription factors**) that are required for RNA polymerase II to initiate transcription. Indeed, the identification and characterization of these factors represents a major part of ongoing efforts to understand transcription in eukaryotic cells. Two general types of transcription factors have been defined. **Basal transcription factors** are involved in transcription from all polymerase II promoters and therefore constitute part of the general transcription machinery. Additional transcription factors (discussed later in the chapter) bind to DNA sequences that control the expression of individual genes and are thus responsible for regulating gene expression.

At least five basal transcription factors are required for initiation of transcription by RNA polymerase II in reconstituted *in vitro* systems (Figure 6.13). The promoters of many genes transcribed by polymerase II contain a sequence similar to TATAA 25 to 30 nucleotides upstream of the transcription start site. This sequence (called the **TATA box**) resembles the −10 sequence element of bacterial promoters, and the results of introducing mutations into TATAA sequences have demonstrated their role in the initiation of transcription. The first step in formation of a transcription complex is the binding of a basal transcription factor called TFIID to the TATA box (*TF* indicates *tran*scription *f*actor; *II* indicates polymerase *II*). TFIID is itself composed of multiple subunits, including the **TATA-binding protein (TBP)**, which binds specifically to the TATAA consensus sequence, and other polypeptides, called **TBP-associated factors (TAFs)**. TBP then binds a second basal transcription factor (TFIIB) forming a TBP-TFIIB complex at the promoter (Figure 6.14). TFIIB in turn serves as a bridge to RNA polymerase, which binds to the TBP-TFIIB complex in association with a third factor, TFIIF.

Following recruitment of RNA polymerase II to the promoter, the binding of two additional factors (TFIIE and TFIIH) is required for initiation of transcription. TFIIH is a multisubunit factor that appears to play at least two important roles. First, two subunits of TFIIH are helicases, which may unwind DNA around the initiation site. (These subunits of TFIIH are also required for nucleotide excision repair, as discussed in Chapter 5.) Another subunit of TFIIH is a protein kinase that phosphorylates repeated sequences present in the C-terminal domain of the largest subunit of RNA polymerase II. Phosphorylation of these sequences is thought to release the polymerase from its association with the initiation complex, allowing it to proceed along the template as it elongates the growing RNA chain.

In addition to a TATA box, the promoters of many genes transcribed by RNA polymerase II contain a second important sequence element (an initiator, or Inr, sequence) that spans the transcription start site. Moreover, some RNA polymerase II promoters contain only an Inr element, with no TATA box. Initiation at these promoters still requires TFIID (and TBP), even though TBP obviously does not recognize these promoters by binding directly to the TATA sequence. Instead, other subunits of TFIID (TAFs) appear to bind to the Inr sequences. This binding recruits TBP to the promoter, and TFIIB, polymerase II, and additional transcription factors then assemble as already described. TBP thus plays a central role in initiating polymerase II transcription, even on promoters that lack a TATA box.

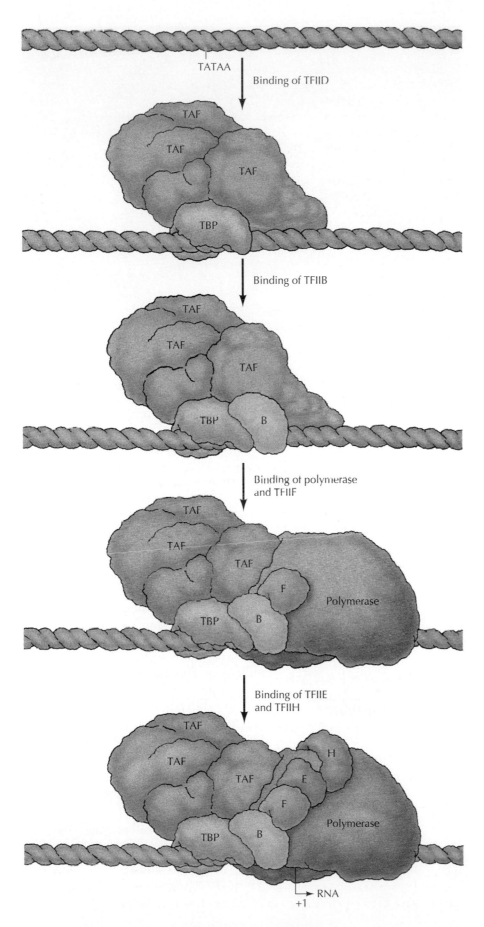

*Figure 6.13*
**Formation of a polymerase II tran-
scription complex**   Many polymerase
II promoters have a TATA box (consensus sequence TATAA) 25 to 30
nucleotides upstream of the transcription start site. This sequence is recognized by transcription factor TFIID,
which consists of the TATA-binding
protein (TBP) and TBP-associated factors (TAFs). TFIIB(B) then binds to
TBP, followed by binding of the polymerase in association with TFIIF(F).
Finally, TFIIE(E) and TFIIH(H) associate with the complex.

*Figure 6.14*
**Model of the TBP-TFIIB complex bound to DNA**   The DNA is shown as a stick figure consisting of yellow and green strands, with the site of transcription initiation designated +1. TBP consists of two repeats, colored light blue and dark blue. TFIIB repeats are colored orange and magenta. Note that TBP binding bends the DNA by approximately 110°. (From D. B. Nikolov et al., 1995, *Nature* 377: 119.)

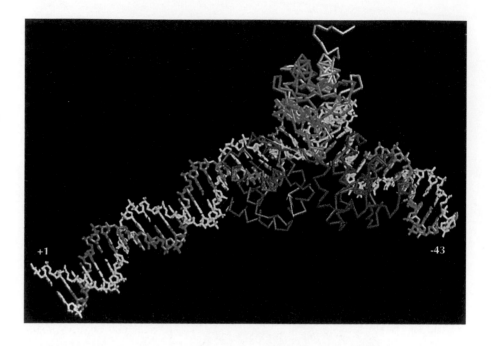

Despite the development of *in vitro* systems and the characterization of several basal transcription factors, much remains to be learned concerning the mechanism of polymerase II transcription in eukaryotic cells. The sequential recruitment of transcription factors described here represents the minimal system required for transcription *in vitro*; additional factors may be needed within the cell. Furthermore, RNA polymerase II appears to be able to associate with some transcription factors *in vivo* prior to the assembly of a transcription complex on DNA. In particular, preformed complexes of RNA polymerase II with TFIIB, TFIIF, and TFIIH have been detected in both yeast and mammalian cells. These large complexes (called polymerase II holoenzymes) can be recruited to a promoter via direct interaction with TFIID (Figure 6.15). The relative contributions of stepwise assembly of individual factors versus recruitment of the RNA polymerase II holoenzyme to promoters within the cell thus remain to be determined. Finally, the functions of many of the basal transcription factors are still unknown. The many unanswered questions concerning the critical process of mRNA synthesis in eukaryotes make this one of the most intensively studied and rapidly progressing areas in contemporary cell and molecular biology.

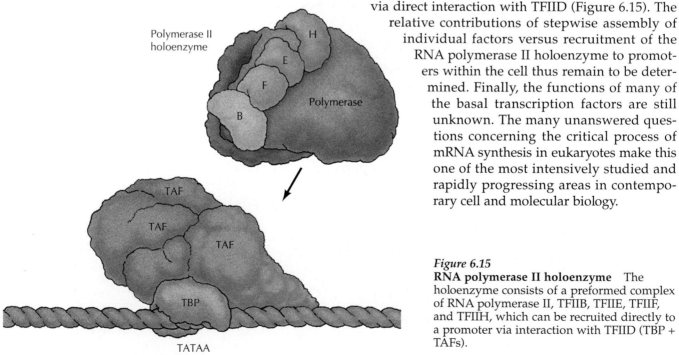

*Figure 6.15*
**RNA polymerase II holoenzyme**   The holoenzyme consists of a preformed complex of RNA polymerase II, TFIIB, TFIIE, TFIIF, and TFIIH, which can be recruited directly to a promoter via interaction with TFIID (TBP + TAFs).

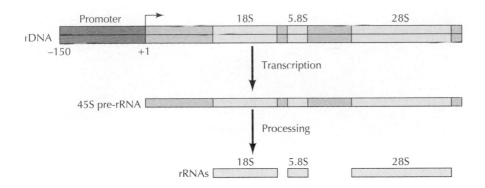

*Figure 6.16*
**The ribosomal RNA gene**   The ribosomal DNA (rDNA) is transcribed to yield a large RNA molecule (45S pre-rRNA), which is then cleaved into 28S, 18S, and 5.8S rRNAs.

## Transcription by RNA Polymerases I and III

As previously discussed, distinct RNA polymerases are responsible for the transcription of genes encoding ribosomal and transfer RNAs in eukaryotic cells. All three RNA polymerases, however, require additional transcription factors to associate with appropriate promoter sequences. Furthermore, although the three different polymerases in eukaryotic cells recognize distinct types of promoters, a common transcription factor—the TATA-binding protein (TBP)—appears to be required for initiation of transcription by all three enzymes.

RNA polymerase I is devoted solely to the transcription of ribosomal RNA genes, which are present in tandem repeats. Transcription of these genes yields a large 45S pre-rRNA, which is then processed to yield the 28S, 18S, and 5.8S rRNAs (Figure 6.16). The promoter of ribosomal RNA genes spans about 150 base pairs just upstream of the transcription initiation site. These promoter sequences are recognized by two transcription factors, UBF (upstream binding factor) and SL1 (selectivity factor 1), which bind cooperatively to the promoter and then recruit polymerase I to form an initiation complex (Figure 6.17). The SL1 transcription factor is composed of four protein subunits, one of which, surprisingly, is TBP. The role of TBP has been demonstrated directly by the finding that yeasts carrying mutations in TBP are defective not only for transcription by polymerase II, but also for transcription by polymerases I and III. Thus, TBP is a common transcription factor required by all three classes of eukaryotic RNA polymerases. Since the promoter for ribosomal RNA genes does not contain a TATA box, TBP does not bind to specific promoter sequences. Instead, the association of TBP with ribosomal RNA genes is mediated by the binding of other proteins in the SL1 complex to the promoter, a situation similar to the association of TBP with the Inr sequences of polymerase II genes that lack TATA boxes.

The genes for tRNAs, 5S rRNA, and some of the small RNAs involved in splicing and protein transport are transcribed by polymerase III. These

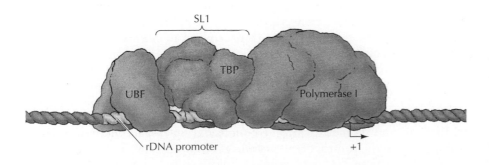

*Figure 6.17*
**Initiation of rDNA transcription**
Two transcription factors, UBF and SL1, bind cooperatively to the rDNA promoter and recruit RNA polymerase I to form an initiation complex. One subunit of SL1 is the TATA-binding protein (TBP).

*Figure 6.18*
**Transcription of polymerase III genes**
The promoters of 5S rRNA and tRNA genes are downstream of the transcription initiation site. Transcription of the 5S rRNA gene is initiated by the binding of TFIIIA, followed by the binding of TFIIIC, TFIIIB, and the polymerase. The tRNA promoters do not contain a binding site for TFIIIA, and TFIIIA is not required for their transcription. Instead, TFIIIC initiates the transcription of tRNA genes by binding to promoter sequences, followed by the association of TFIIIB and polymerase. The TATA-binding protein (TBP) is a subunit of TFIIIB.

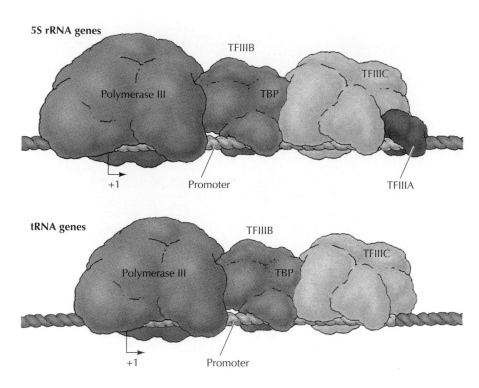

*Figure 6.19*
**Identification of eukaryotic regulatory sequences**    The regulatory sequence of a cloned eukaryotic gene is ligated to a reporter gene that encodes an easily detectable enzyme. The resulting plasmid is then introduced into cultured recipient cells by transfection. An active regulatory sequence directs transcription of the reporter gene, expression of which is then detected in the transfected cells.

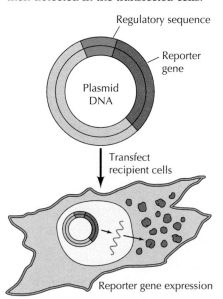

genes are characterized by promoters that lie within, rather than upstream of, the transcribed sequence (Figure 6.18). The most thoroughly studied of the genes transcribed by polymerase III are the 5S rRNA genes of *Xenopus*. TFIIIA (which is the first transcription factor to have been purified) initiates assembly of a transcription complex by binding to specific DNA sequences in the 5S rRNA promoter. This binding is followed by the sequential binding of TFIIIC, TFIIIB, and the polymerase. The promoters for the tRNA genes differ from the 5S rRNA promoter in that they do not contain the DNA sequence recognized by TFIIIA. Instead, TFIIIC binds directly to the promoter of tRNA genes, serving to recruit TFIIIB and polymerase to form a transcription complex. TFIIIB is composed of multiple subunits, one of which (once again) is the TATA-binding protein, TBP. As already noted, the requirement of TBP for transcription of polymerase III genes has been demonstrated directly by the isolation of yeast mutants. Thus, although the three RNA polymerases of eukaryotic cells recognize different promoters, TBP appears to be a common element that links promoter recognition with polymerase recruitment to the transcription complex.

## REGULATION OF TRANSCRIPTION IN EUKARYOTES

Although the control of gene expression is far more complex in eukaryotes than in bacteria, the same basic principles apply. The expression of eukaryotic genes is controlled primarily at the level of initiation of transcription, although in some cases transcription may be attenuated and regulated at subsequent steps. As in bacteria, transcription in eukaryotic cells is controlled by proteins that bind to specific regulatory sequences and modulate the activity of RNA polymerase. The intricate task of regulating gene expression in the many differentiated cell types of multicellular organisms is accomplished primarily by the combined actions of multiple different transcriptional regulatory proteins. In addition, the packaging of DNA into chromatin and its modification by methylation impart further levels of complexity to the control of eukaryotic gene expression.

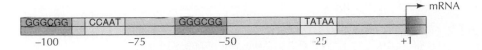

*Figure 6.20*
**A eukaryotic promoter** The promoter of the thymidine kinase gene of herpes simplex virus contains three sequence elements upstream of the TATA box that are required for efficient transcription: a CCAAT box and two GC boxes (consensus sequence GGGCGG).

## cis-*Acting Regulatory Sequences: Promoters and Enhancers*

As already discussed, transcription in bacteria is regulated by the binding of proteins to *cis*-acting sequences (e.g., the *lac* operator) that control the transcription of adjacent genes. Similar *cis*-acting sequences regulate the expression of eukaryotic genes. These sequences have been identified in mammalian cells largely by the use of gene transfer assays to study the activity of suspected regulatory regions of cloned genes (Figure 6.19). The eukaryotic regulatory sequences are usually ligated to a reporter gene that encodes an easily detectable enzyme. The expression of the reporter gene following its transfer into cultured cells then provides a sensitive assay for the ability of the cloned regulatory sequences to direct transcription. Biologically active regulatory regions can thus be identified, and *in vitro* mutagenesis can be used to determine the roles of specific sequences within the region.

Genes transcribed by RNA polymerase II have two core promoter elements, the TATA box and the Inr sequence, that serve as specific binding sites for basal transcription factors. Other *cis*-acting sequences serve as binding sites for a wide variety of regulatory factors that control the expression of individual genes. These *cis*-acting regulatory sequences are frequently, though not always, located upstream of the TATA box. For example, two regulatory sequences that are found in many eukaryotic genes were identified by studies of the promoter of the herpes simplex virus gene that encodes thymidine kinase (Figure 6.20). Both of these sequences are located within 100 base pairs upstream of the TATA box: their consensus sequences are CCAAT and GGGCGG (called a GC box). Specific proteins that bind to these sequences and stimulate transcription have since been identified.

In contrast to the relatively simple organization of CCAAT and GC boxes in the herpes thymidine kinase promoter, many genes in mammalian cells are controlled by regulatory sequences located farther away (sometimes more than 10 kilobases) from the transcription start site. These sequences, called **enhancers**, were first identified by Walter Schaffner in 1981 during studies of the promoter of another virus, SV40 (Figure 6.21). In addition to a TATA box and a set of six GC boxes, two 72-base-pair repeats located farther upstream are required for efficient transcription from this promoter. These sequences were found to stimulate transcription from other promoters as well as from that of SV40, and, surprisingly, their activity depended on neither their distance nor their orientation with respect to the transcription initiation site (Figure 6.22). They could stimulate transcription when placed either upstream or downstream of the promoter, in either a forward or backward orientation.

The ability of enhancers to function even when separated by long distances from transcription initiation sites at first suggested that they work by mechanisms different from those of promoters. However, this has

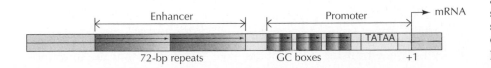

*Figure 6.21*
**The SV40 enhancer** The SV40 promoter for early gene expression contains a TATA box and six GC boxes arranged in three sets of repeated sequences. In addition, efficient transcription requires an upstream enhancer consisting of two 72-base-pair (bp) repeats.

*Figure 6.22*
**Action of enhancers**   Without an enhancer, the gene is transcribed at a low basal level (A). Addition of an enhancer (E)—for example, the SV40 72-base-pair repeats—stimulates transcription. The enhancer is active not only when placed just upstream of the promoter (B), but also when inserted up to several kilobases either upstream or downstream from the transcription start site (C and D). In addition, enhancers are active in either the forward or backward orientation (E).

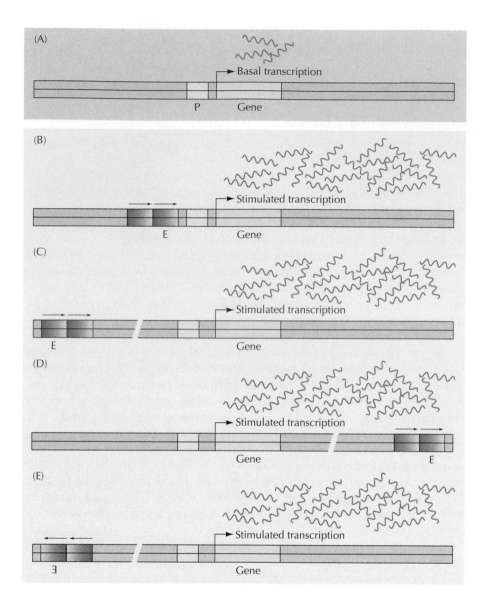

turned out not to be the case: Enhancers, like promoters, function by binding transcription factors that then regulate RNA polymerase. This is possible because of DNA looping, which allows a transcription factor bound to a distant enhancer to interact with RNA polymerase or basal transcription factors at the promoter (Figure 6.23). Transcription factors bound to distant enhancers can thus work by the same mechanisms as those bound adjacent to promoters, so there is no fundamental difference between the actions of enhancers and those of *cis*-acting regulatory sequences adjacent to transcription start sites. Interestingly, although enhancers were first identified in mammalian cells, they have subsequently been found in bacteria—an unusual instance in which studies of eukaryotes served as a model for the simpler prokaryotic systems.

The binding of specific transcriptional regulatory proteins to enhancers is responsible for the control of gene expression during development and differentiation, as well as during the response of cells to hormones and growth factors. One of the most thoroughly studied mammalian enhancers controls the transcription of immunoglobulin genes in B lymphocytes. Gene transfer experiments have established that the immuno-

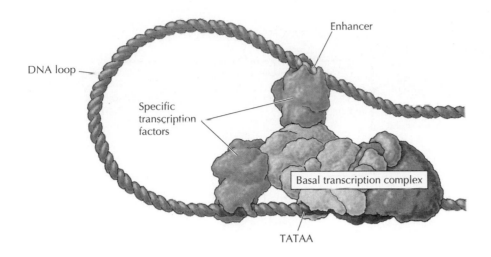

*Figure 6.23*
**DNA looping**  Transcription factors bound at distant enhancers are able to interact with basal transcription factors at the promoter because the intervening DNA can form loops. There is therefore no fundamental difference between the action of transcription factors bound to DNA just upstream of the promoter and to distant enhancers.

globulin enhancer is active in lymphocytes, but not in other types of cells. Thus, this regulatory sequence is at least partly responsible for tissue-specific expression of the immunoglobulin genes in the appropriate differentiated cell type.

An important aspect of enhancers is that they usually contain multiple functional sequence elements that bind different transcriptional regulatory proteins. These proteins work together to regulate gene expression. The immunoglobulin heavy-chain enhancer, for example, spans approximately 200 base pairs and contains at least nine distinct sequence elements that serve as protein-binding sites (Figure 6.24). Mutation of any one of these sequences reduces but does not abolish enhancer activity, indicating that the functions of individual proteins that bind to the enhancer are at least partly redundant. Many of the individual sequence elements of the immunoglobulin enhancer by themselves stimulate transcription in non-lymphoid cells. The restricted activity of the intact enhancer in B lymphocytes therefore does not result from the tissue-specific function of each of its components. Instead, tissue-specific expression results from the combination of the individual sequence elements that make up the complete enhancer. These elements include some *cis*-acting regulatory sequences that bind transcriptional activators that are expressed specifically in B lymphocytes, as well as other regulatory sequences that bind repressors in non-lymphoid cells. Thus, the immunoglobulin enhancer contains negative regulatory elements that inhibit transcription in inappropriate cell types, as well as positive regulatory elements that activate transcription in B lymphocytes. The overall activity of the enhancer is greater than the sum of its parts, reflecting the combined action of the proteins associated with each of its individual sequence elements.

## Transcriptional Regulatory Proteins

The isolation of a variety of transcriptional regulatory proteins has been based on their specific binding to promoter or enhancer sequences. Protein binding to these DNA sequences is commonly analyzed by two types of

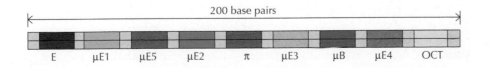

*Figure 6.24*
**The immunoglobulin enhancer**
The immunoglobulin heavy-chain enhancer spans about 200 bases and contains nine functional sequence elements (E, $\mu$E1–5, $\pi$, $\mu$B, and OCT), which together stimulate transcription in B lymphocytes.

*Figure 6.25*
**Electrophoretic-mobility shift assay**
A sample containing radiolabeled fragments of DNA is divided into two, and one half of the sample is incubated with a protein that binds to a specific DNA sequence. Samples are then analyzed by electrophoresis in a nondenaturing gel so that the protein remains bound to DNA. Protein binding is detected by the slower migration of DNA-protein complexes compared to that of free DNA. Only a fraction of the DNA in the sample is actually bound to protein, so both DNA-protein complexes and free DNA are detected following incubation of the DNA with protein.

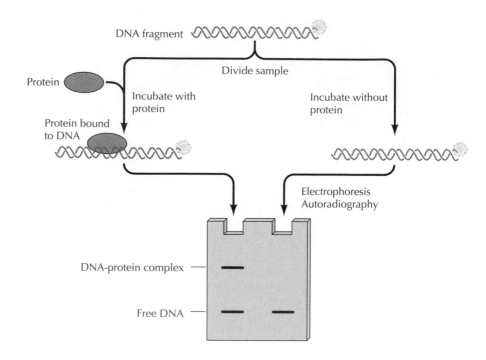

experiments. The first, footprinting, was described earlier in connection with the binding of RNA polymerase to prokaryotic promoters (see Figure 6.3). The second approach is the **electrophoretic-mobility shift assay**, in which a radiolabeled DNA fragment is incubated with a protein preparation and then subjected to electrophoresis through a nondenaturing gel (Figure 6.25). Protein binding is detected as a decrease in the electrophoretic mobility of the DNA fragment, since its migration through the gel is slowed by the bound protein. The combined use of footprinting and electrophoretic-mobility shift assays has led to the correlation of protein-binding sites with the regulatory elements of enhancers and promoters, indicating that these sequences generally constitute the recognition sites of specific DNA-binding proteins.

One of the prototypes of eukaryotic transcription factors was initially identified by Robert Tjian and his colleagues during studies of the transcription of SV40 DNA. This factor (called Sp1, for specificity protein 1) was found to stimulate transcription from the SV40 promoter, but not from several other promoters, in cell-free extracts. Then, stimulation of transcription by Sp1 was found to depend on the presence of the GC boxes in the SV40 promoter: If these sequences were deleted, stimulation by Sp1 was abol-

*Table 6.2* Examples of Transcription Factors and Their DNA-Binding Sites

| Transcription factor | Consensus binding site |
|---|---|
| Specificity protein 1 (Sp1) | GGGCGG |
| CCAAT/enhancer binding protein (C/EBP) | CCAAT |
| Activator protein 1 (AP1) | TGACTCA |
| Octamer binding proteins (OCT-1 and OCT-2) | ATGCAAAT |
| E-box binding proteins (E12, E47, E2-2) | CANNTG[a] |

[a] N stands for any nucleotide.

ished. Moreover, footprinting experiments established that Sp1 binds specifically to the GC box sequences. Taken together, these results indicate that the GC box represents a specific binding site for a transcriptional activator—Sp1. Similar experiments have established that many other transcriptional regulatory sequences, including the CCAAT sequence and the various sequence elements of the immunoglobulin enhancer, also represent recognition sites for sequence-specific DNA-binding proteins (Table 6.2).

The specific binding of Sp1 to the GC box not only established the action of Sp1 as a sequence-specific transcription factor; it also suggested a general approach to the purification of transcription factors. The isolation of these proteins initially presented a formidable challenge because they are present in very small quantities (e.g., only 0.001% of total cell protein) that are difficult to purify by conventional biochemical techniques. This problem was overcome in the purification of Sp1 by **DNA-affinity chromatography** (Figure 6.26). Multiple copies of oligonucleotides corresponding to the GC box sequence were bound to a solid support, and cell extracts were passed through the oligonucleotide column. Because Sp1 bound to the GC box with high affinity, it was specifically retained on the column while other proteins were not. Highly purified Sp1 could thus be obtained and used for further studies, including partial determination of its amino acid sequence, which in turn led to cloning of the gene for Sp1.

The general approach of DNA-affinity chromatography, first optimized for the purification of Sp1, has been used successfully to isolate a wide variety of sequence-specific DNA-binding proteins from eukaryotic cells.

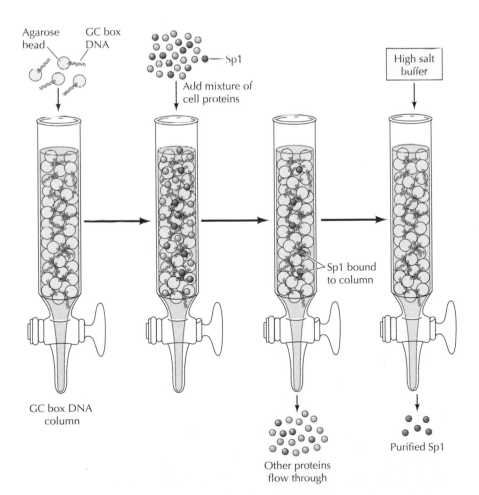

*Figure 6.26*
**Purification of Sp1 by DNA-affinity chromatography**   A double-stranded oligonucleotide containing repeated GC box sequences is bound to agarose beads, which are poured into a column. A mixture of cell proteins containing Sp1 is then applied to the column; because Sp1 specifically binds to the GC box oligonucleotide, it is retained on the column while other proteins flow through. Washing the column with high salt buffer then dissociates Sp1 from the GC box DNA, yielding purified Sp1.

## KEY EXPERIMENT

# Isolation of a Eukaryotic Transcription Factor

### Affinity Purification of Sequence-Specific DNA Binding Proteins

James T. Kadonaga and Robert Tjian

University of California, Berkeley

*Proceedings of the National Academy of Sciences, USA, 1986, Volume 83, pages 5889–5893*

### The Context

Starting with studies of the *lac* operon by Jacob and Monod, it became clear that transcription is regulated by proteins that bind to specific DNA sequences. One of the prototype systems for studies of gene expression in eukaryotic cells was the monkey virus SV40, in which several regulatory DNA sequences were identified in the early 1980s. In 1983 William Dynan and Robert Tjian first demonstrated that one of these sequence elements (the GC box) is the specific binding site of a protein detectable in nuclear extracts of human cells. This protein (called Sp1 for specificity protein 1) not only binds to the GC box sequence; it also stimulates transcription *in vitro*, demonstrating that it is a sequence-specific transcriptional activator.

To study the mechanism of Sp1 action, it then became necessary to obtain the transcription factor in pure form and eventually to clone the *Sp1* gene. The isolation of pure Sp1 thus became a high priority, but it also posed a daunting technical challenge. Sp1 and other transcription factors appeared to represent only about 0.001% of total cell protein, so they could not be purified by conventional biochemical techniques. James Kadonaga and Robert Tjian solved this problem by developing a method of DNA-affinity chromatography that led to the purification not only of Sp1 but also of many other eukaryotic transcription factors, thereby opening the door to molecular analysis of transcriptional regulation in eukaryotic cells.

### The Experiments

The DNA-affinity chromatography method developed by Kadonaga and Tjian exploited the specific high-affinity binding of Sp1 to the GC box sequence, GGGCGG. Synthetic oligonucleotides containing multiple copies of this sequence were coupled to solid beads, and a crude nuclear extract was passed through a column consisting of beads linked to GC box DNA. The beads were then washed to remove proteins that had failed to bind specifically to the oligonucleotides. Finally, the beads were washed with a high salt buffer (0.5 *M* KCl), which disrupted the binding of Sp1 to DNA, thereby releasing Sp1 from the column.

Gel electrophoresis demonstrated that the crude nuclear extract initial-

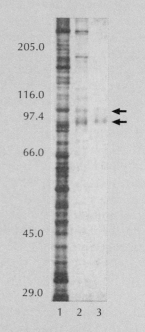

Purification of Sp1. Gel electrophoresis of proteins initially present in the crude nuclear extract (lane 1) and of proteins obtained after either one or two sequential cycles of DNA-affinity chromatography (lanes 2 and 3, respectively). The sizes of marker proteins (in kilodaltons) are indicated to the left of the gel, and the Sp1 polypeptides are indicated by arrows.

ly applied to the column was a complex mixture of proteins (see figure). In contrast, approximately 90% of the protein recovered after two cycles of DNA-affinity chromatography corresponded to only two polypeptides, which were identified as

Protein purification has been followed by gene cloning and nucleotide sequencing, leading to the accumulation of a great deal of information on the structure and function of these critical regulatory proteins.

### Structure and Function of Transcriptional Activators

Because transcription factors are central to the regulation of gene expression, understanding the mechanisms of their action is a major area of ongo-

---

### Isolation of a Eukaryotic Transcription Factor (continued)

Sp1 by DNA binding and by their activity in *in vitro* transcription assays. Thus, Sp1 had been successfully purified by DNA-affinity chromatography.

**The Impact**

In their 1986 paper, Kadonaga and Tjian stated that the DNA-affinity chromatography technique "should be generally applicable for the purification of other sequence-specific DNA binding proteins." This prediction has been amply verified; many eukaryotic transcription factors have been purified by this method. The genes that encode still other transcription factors have been isolated by an alternative approach (developed independently in 1988 in the laboratories of Phillip Sharp and Steven McKnight) in which cDNA expression libraries are screened with oligonucleotide probes to detect recombinant proteins that bind specifically to the desired DNA sequences. The ability to isolate sequence-specific DNA-binding proteins by these methods has led to detailed characterization of the structure and function of a wide variety of transcriptional regulatory proteins,

Robert Tjian

providing the basis for our current understanding of gene expression in eukaryotic cells.

---

ing research in cell and molecular biology. The most thoroughly studied of these proteins are **transcriptional activators**, which, like Sp1, bind to regulatory DNA sequences and stimulate transcription. In general, these factors have been found to consist of two domains: One region of the protein specifically binds DNA; the other activates transcription by interacting with other components of the transcriptional machinery (Figure 6.27). Transcriptional activators appear to be modular proteins, in the sense that the DNA binding and activation domains of different factors can frequently be interchanged using recombinant DNA techniques. Such manipulations result in hybrid transcription factors, which activate transcription by binding to promoter or enhancer sequences determined by the specificity of their DNA-binding domains. It therefore appears that the basic function of the DNA-binding domain is to anchor the transcription factor to the proper site on DNA; the activation domain then independently stimulates transcription by interacting with other proteins.

Many different transcription factors have now been identified in eukaryotic cells, as might be expected, given the intricacies of tissue-specific and

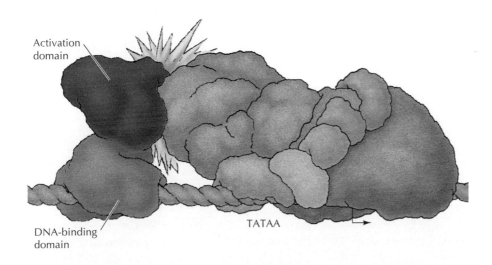

Activation domain

DNA-binding domain

TATAA

*Figure 6.27*
**Structure of transcriptional activators** Transcriptional activators consist of two independent domains. The DNA-binding domain recognizes a specific DNA sequence, and the activation domain interacts with other components of the transcriptional machinery.

*Figure 6.28*
**Families of DNA-binding domains**
(A) Zinc finger domains consist of loops in which an α helix and a β sheet coordinately bind a zinc ion. (B) Helix-turn-helix domains consist of three (or in some cases four) helical regions. One helix (helix 3) makes most of the contacts with DNA, while helices 1 and 2 lie on top and stabilize the interaction. (C) The DNA-binding domains of leucine zipper proteins are formed from two distinct polypeptide chains. Interactions between the hydrophobic side chains of leucine residues exposed on one side of a helical region (the leucine zipper) are responsible for dimerization. Immediately following the leucine zipper is a DNA-binding helix, which is rich in basic amino acids. (D) Helix-loop-helix domains are similar to leucine zippers, except that the dimerization domains of these proteins each consist of two helical regions separated by a loop.

inducible gene expression in complex multicellular organisms. Molecular characterization has revealed that the DNA-binding domains of many of these proteins are related to one another (Figure 6.28). **Zinc finger domains** contain repeats of cysteine and histidine residues that bind zinc ions and fold into looped structures ("fingers") that bind DNA. These domains were initially identified in the polymerase III transcription factor TFIIIA but are also common among transcription factors that regulate polymerase II promoters, including Sp1. Other examples of transcription factors that contain zinc finger domains are the **steroid hormone receptors**, which regulate gene transcription in response to hormones such as estrogen and testosterone.

The **helix-turn-helix** motif was first recognized in prokaryotic DNA-binding proteins, including the *E. coli* catabolite activator protein (CAP). In these proteins, one helix makes most of the contacts with DNA, while the other helices lie across the complex to stabilize the interaction. In eukaryotic cells, helix-turn-helix proteins include the **homeodomain** proteins, which play critical roles in the regulation of gene expression during embryonic development. The genes encoding these proteins were first discovered as developmental mutants in *Drosophila*. Some of the earliest recognized *Drosophila* mutants (termed homeotic mutants in 1894) resulted in the development of flies in which one body part was transformed into another. For example, in the homeotic mutant called *Antennapedia*, legs rather than antennae grow out of the head of the fly (Figure 6.29). Genetic analysis of these mutants, pioneered by Ed Lewis in the 1940s, has shown that *Drosophila* contains nine homeotic genes, each of which specifies the

(A) Zinc fingers

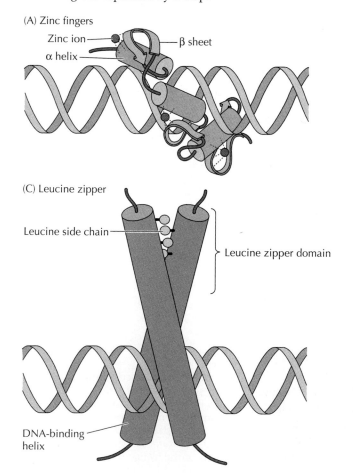

(C) Leucine zipper

(B) Helix-turn-helix

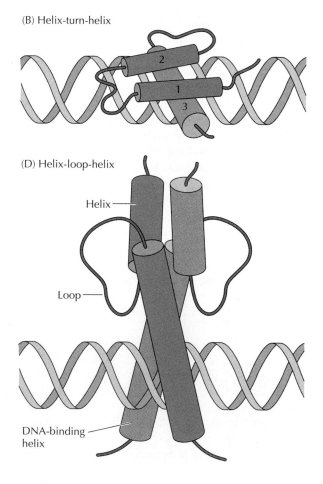

(D) Helix-loop-helix

(A)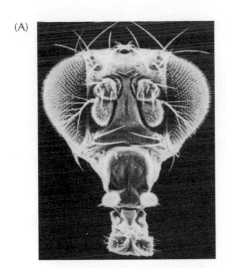

(B)

*Figure 6.29*
**The *Antennapedia* mutation** *Antennapedia* mutant flies have legs growing out of their heads in place of antennae. (A) Head of a normal fly. (B) Head of an *Antennapedia* mutant. (From Kaufman et al., 1990, *Adv. Genet.* 27: 309, courtesy of T. C. Kaufman)

identity of a different body segment. Molecular cloning and analysis of these genes then indicated that they contain conserved sequences of 180 base pairs (called **homeoboxes**) that encode the DNA-binding domains (homeodomains) of transcription factors. A wide variety of additional homeodomain proteins have since been identified in fungi, plants, and other animals, including humans. Vertebrate homeobox genes are strikingly similar to their *Drosophila* counterparts in both structure and function, demonstrating the highly conserved roles of these transcription factors in animal development.

Two other families of DNA-binding proteins, **leucine zipper** and **helix-loop-helix** proteins, contain DNA-binding domains formed by dimerization of two polypeptide chains. The leucine zipper contains four or five leucine residues spaced at intervals of seven amino acids, resulting in their hydrophobic side chains being exposed at one side of a helical region. This region serves as the dimerization domain for the two protein subunits, which are held together by hydrophobic interactions between the leucine side chains. Immediately following the leucine zipper is a region rich in positively charged amino acids (lysine and arginine) that binds DNA. The helix-loop-helix proteins are similar in structure, except that their dimerization domains are each formed by two helical regions separated by a loop. An important feature of both leucine zipper and helix-loop-helix transcription factors is that different members of these families can dimerize with each other. Thus, the combination of distinct protein subunits can form an expanded array of factors that can differ both in DNA sequence recognition and in transcription-stimulating activities. Both leucine zipper and helix-loop-helix proteins play important roles in regulating tissue-specific and inducible gene expression, and the formation of dimers between different members of these families is a critical aspect of the control of their function.

The activation domains of transcription factors are not as well characterized as their DNA-binding domains. Some, called acidic activation domains, are rich in negatively charged residues (aspartate and glutamate); others are rich in proline or glutamine residues. These activation domains are thought to stimulate transcription by interacting with basal transcription factors, such as TFIIB or TFIID, thereby facilitating the assembly of a transcription complex on the promoter. For example, the activation domains of several transcription factors (including Sp1) have been shown

# MOLECULAR MEDICINE

## *The Pit-1 Transcription Factor and Growth Hormone Deficiency*

### The Disease

Growth hormone is a protein produced by the pituitary gland that is required for normal growth and development. An estimated one out of every 5,000 to 10,000 newborn infants is deficient in growth hormone production. These infants develop into children with short stature—pituitary dwarfs. Such cases of growth hormone deficiency can arise from a variety of causes, including inherited mutations in genes affecting growth hormone production. In some cases, affected children are deficient not only in growth hormone but also in one or more of the other hormones produced by the pituitary gland, such as thyroid-stimulating hormone (TSH), adrenocorticotrophic hormone (ACTH), gonadotrophins, and prolactin. A well-characterized subset of these patients are deficient in production of growth hormone, thyroid-stimulating hormone, and prolactin but produce the other pituitary hormones at normal levels.

### Molecular and Cellular Basis

The combined human pituitary hormone deficiencies involving growth hormone, thyroid-stimulating hormone, and prolactin closely resemble the phenotypes of certain dwarf strains of mice. These dwarf mice were known to lack the specific pituitary cell types responsible for growth hor-

mone, thyroid-stimulating hormone, and prolactin secretion (called somatotrophs, thyrotrophs, and lactotrophs, respectively). In 1990 these defects were found to result from mutations in the gene encoding a transcription factor called Pit-1. Two groups of researchers extended these findings to humans with the identification of *Pit-1* mutations in patients with combined deficiencies in growth hormone, thyroid-stimulating hormone, and prolactin. Molecular analysis has demonstrated that Pit-1 is a homeodomain-containing protein required for transcription of the growth hormone and prolactin genes, as well as for the establishment of somatotrophs, thyrotrophs, and lactotrophs during pituitary development.

### Prevention and Treatment

Children with growth hormone deficiency can be treated effectively by human growth hormone injections. Until recently, this hormone was available only from human brains and consequently was in extremely short supply. In 1979, however, human growth hormone cDNA was successfully expressed in *E. coli*. This recombinant hormone was approved for clinical use in 1985, providing a readily available source of human growth hormone for the treatment of pituitary dwarfs. When growth hormone deficiency results from inherited *Pit-1* mutations, genetic screening

provides the further potential for presymptomatic diagnosis and treatment of affected children.

### Reference

Parks, J. S, H. Abdul-Latif, E. Kinoshita, L. R. Meacham., R. W. Pfaffle and M. R. Brown. 1993. Genetics of growth hormone gene expression. *Hormone Res.* 40: 54–61.

An individual of normal height (circus producer P. T. Barnum) and a pituitary dwarf who appeared in the circus as General Tom Thumb. (Bettmann Archive, Inc.)

to interact with TFIID by binding to TBP-associated factors (TAFs) (Figure 6.30). An important feature of these interactions is that different activators can bind to different basal transcription factors or TAFs, providing a mechanism by which the combined action of multiple factors can synergistically stimulate transcription—a key feature of transcriptional regulation in eukaryotic cells.

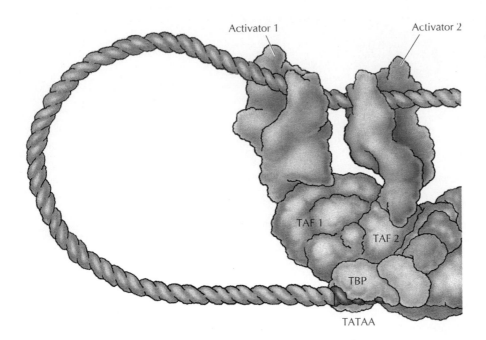

*Figure 6.30*
**Synergistic action of transcriptional activators** Different transcriptional activators can interact with basal transcription factor TFIID by binding to different TAFs.

## Eukaryotic Repressors

Gene expression in eukaryotic cells is regulated by repressors as well as by transcriptional activators. Like their prokaryotic counterparts, eukaryotic repressors bind to specific DNA sequences and inhibit transcription. In some cases, eukaryotic repressors simply interfere with the binding of other transcription factors to DNA (Figure 6.31A). For example, the binding of a repressor near the transcription start site can block the interaction of RNA polymerase or basal transcription factors with the promoter, which is similar to the action of repressors in bacteria. Other repressors compete with activators for binding to specific regulatory sequences. Some such repressors contain the same DNA-binding domain as the activator but lack its activation domain. As a result, their binding to a promoter or enhancer blocks the binding of the activator, thereby inhibiting transcription.

In contrast to repressors that simply interfere with activator binding, many repressors (called active repressors) contain specific functional domains that inhibit transcription via protein-protein interactions (Figure 6.31B). The first such active repressor was described in 1990 during studies of a gene called *Krüppel*, which is involved in embryonic development in *Drosophila*. Molecular analysis of the Krüppel protein demonstrated that it contains a discrete repression domain, which is linked to a zinc finger DNA-binding domain. The Krüppel repression domain could be interchanged with distinct DNA-binding domains of other transcription factors. These hybrid molecules also repressed transcription, indicating that the Krüppel repression domain inhibits transcription via protein-protein interactions, irrespective of its site of binding to DNA.

Many active repressors have since been found to play key roles in the regulation of transcription in animal cells, in many cases serving as critical regulators of cell growth and differentiation. As with transcriptional activators, several distinct types of repression domains have been identified. For example, the repression domain of Krüppel is rich in alanine residues, whereas other repression domains are rich in proline or acidic residues. The functional targets of repressors are also diverse. Some repressors inhibit transcription by interacting with basal transcription factors, such as TFIID; others are thought to interact with specific activator proteins.

*Figure 6.31*
**Action of eukaryotic repressors**
(A) Some repressors block the binding of activators to regulatory sequences. (B) Other repressors have active repression domains that inhibit transcription by interactions with basal transcription factors.

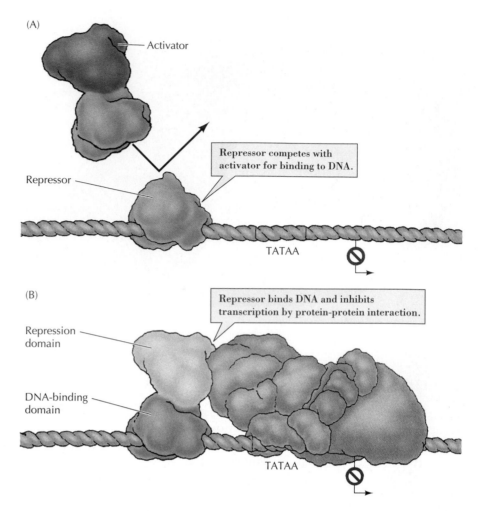

(A)

Activator

Repressor

> Repressor competes with activator for binding to DNA.

TATAA

(B)

> Repressor binds DNA and inhibits transcription by protein-protein interaction.

Repression domain

DNA-binding domain

TATAA

The regulation of transcription by repressors as well as by activators considerably extends the range of mechanisms that control the expression of eukaryotic genes. One important role of repressors may be to inhibit the expression of tissue-specific genes in inappropriate cell types. For example, as noted earlier, a repressor-binding site in the immunoglobulin enhancer is thought to contribute to its tissue-specific expression by suppressing transcription in nonlymphoid cell types. Other repressors play key roles in the control of cell proliferation and differentiation in response to hormones and growth factors (see Chapters 13 and 14).

### Relationship of Chromatin Structure to Transcription

In the preceding discussion, the transcription of eukaryotic genes was considered as if they were present within the nucleus as naked DNA. However, this is not the case. The DNA of all eukaryotic cells is tightly bound to histones, forming chromatin. The basic structural unit of chromatin is the nucleosome, which consists of 146 base pairs of DNA wrapped around two molecules each of histones H2A, H2B, H3, and H4, with one molecule of histone H1 bound to the DNA as it enters and exits the nucleosome core particle (see Figure 4.9). The chromatin is then further condensed by being coiled into higher-order structures organized into large loops of DNA. This packaging of eukaryotic DNA in chromatin clearly has important consequences in terms of its availability as a template for transcription, so chromatin structure must be considered a critical aspect of gene expression in eukaryotic cells.

The relationship between chromatin structure and transcription is evident at several levels. First, actively transcribed genes are found in decon-

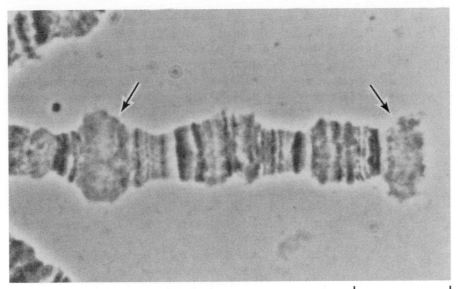

*Figure 6.32*
**Decondensed chromosome regions in** *Drosophila*    A light micrograph showing decondensed regions of polytene chromosomes (arrows), which are active in RNA synthesis. (Courtesy of Joseph Gall, Carnegie Institute.)

10 μm

densed chromatin, corresponding to the extended 10-nm chromatin fibers discussed in Chapter 4 (see Figure 4.10). For example, microscopic visualization of the polytene chromosomes of *Drosophila* indicates that regions of the genome that are actively engaged in RNA synthesis correspond to decondensed chromosome regions (Figure 6.32). In addition, treatment of intact nuclei of vertebrate cells with DNase indicates that about 10% of the DNA, corresponding to the genes that are actively transcribed in any given cell type, is preferentially sensitive to nuclease degradation. For example, the globin genes are sensitive to DNase digestion in reticulocytes (the precursors of red blood cells) but not in other types of differentiated cells. Actively transcribed genes thus appear to be present in a decondensed fraction of chromatin that is more accessible to transcription factors than is the rest of the genome. Such decondensed chromatin is characterized by modifications of histones (for example, acetylation) and by the binding of two nonhistone chromosomal proteins (called **HMG-14** and **HMG-17**) to nucleosomes. (HMG stands for high-mobility group proteins; these proteins migrate rapidly during gel electrophoresis.) The functional significance of these modifications is just beginning to be understood.

Even in decondensed chromatin, actively transcribed genes remain bound to histones and packaged in nucleosomes, so transcription factors and RNA polymerase are still faced with the problem of interacting with chromatin rather than with naked DNA. The tight winding of DNA around the nucleosome core particle is a major obstacle to transcription, affecting both the ability of transcription factors to bind DNA and the ability of RNA polymerase to transcribe through a chromatin template.

A variety of experiments have shown that nucleosomes bound to the promoter inhibit the initiation of transcription by blocking the binding of transcription factors and RNA polymerase. This inhibitory effect of nucleosomes is relieved by **nucleosome remodeling factors**, which facilitate the binding of transcription factors by disrupting chromatin structure, thereby allowing transcription factors to bind nucleosome DNA and direct the assembly of a transcription complex at the promoter (Figure 6.33). As a result of this chromatin remodeling, the enhancer and promoter regions of actively transcribed genes contain disrupted or fewer nucleosomes and are therefore even more sensitive to nuclease digestion than is the body of the gene. Con-

*Figure 6.33*
**Transcription of chromatin** The DNA in chromatin is tightly wound around nucleosome cores, which blocks the binding of transcription factors and RNA polymerase. Nucleosome remodeling factors facilitate the binding of transcription factors, thereby displacing nucleosomes from enhancer and promoter regions. RNA polymerase is then able to transcribe through a nucleosome template.

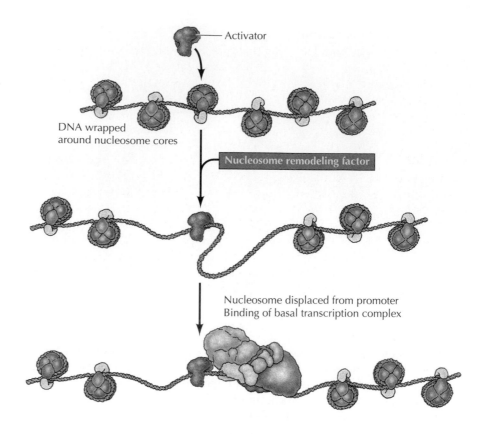

sequently, the regulatory regions of actively transcribed genes usually coincide with chromatin regions characterized by extreme sensitivity to DNase digestion, called **DNase-hypersensitive sites**. Eukaryotic transcriptional activators thus appear to play dual roles in controlling gene expression. In addition to stimulating transcription by interacting with basal transcription factors, they induce alterations of chromatin structure that alleviate repression by histones. The mechanism by which nucleosome remodeling factors are targeted to actively transcribed genes is not yet known; given the significance of these factors in the regulation of gene expression in eukaryotic cells, this question represents an important area of current research.

In contrast to promoter and enhancer sequences, the body of transcribed genes remains packaged in nucleosomes. Perhaps surprisingly, this packaging is not an impassable barrier to RNA polymerase, which is able to transcribe through a nucleosome core by disrupting histone-DNA contacts. The ability of RNA polymerase to transcribe chromatin templates is facilitated by acetylation of histones and by the association of the nonhistone chromosomal proteins HMG-14 and HMG-17 with the nucleosomes of actively transcribed genes. The binding sites of these proteins on nucleosomes overlap the binding site of histone H1, and HMG-14 and HMG-17 appear to stimulate transcription by disrupting the interaction of histone H1 with nucleosomes. Whereas histone H1 induces chromatin condensation and represses transcription, HMG-14 and HMG-17 maintain a decondensed chromatin structure that facilitates transcription through a nucleosome template. As with nucleosome remodeling factors, the signals that target HMG-14 and HMG-17 to actively transcribed genes remain to be elucidated by future research.

*Figure 6.34*
**DNA methylation** A methyl group is added to the 5-carbon position of cytosine residues in DNA.

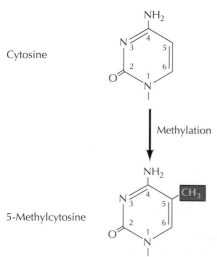

### DNA Methylation

The methylation of DNA is another general mechanism that may be involved in the control of transcription in vertebrates. Cytosine residues in vertebrate DNA can be modified by the addition of methyl groups at the 5-carbon position (Figure 6.34). DNA is methylated specifically at the C's that precede G's in the DNA chain (CpG dinucleotides). This methylation appears to be correlated with reduced transcriptional activity of at least some genes. Distinct patterns of methylation are found in different tissues, with the DNA of inactive genes being more heavily methylated than the DNA of genes that are actively transcribed. In addition, some genes contain high frequencies of CpG dinucleotides in the vicinity of their promoters. Methylation represses transcription of these genes via the action of a protein that specifically binds to methylated DNA and inhibits transcription.

Although DNA methylation is capable of inhibiting transcription, its general significance in gene regulation is unclear. In many cases, methylation of inactive genes is thought to be a consequence, rather than the primary cause, of their lack of transcriptional activity. However, an important regulatory role of DNA methylation has been established in the phenomenon known as **genomic imprinting**, which controls the expression of some genes involved in the development of mammalian embryos. In most cases, both the paternal and maternal alleles of a gene are expressed in diploid cells. However, there are a few imprinted genes (16 have been described in both mice and humans) whose expression depends on whether they are inherited from the mother or from the father. In some cases, only the paternal allele of an imprinted gene is expressed, and the maternal allele is transcriptionally inactive. For other imprinted genes, the maternal allele is expressed and the paternal allele is inactive.

Although the biological role of genomic imprinting is uncertain, DNA methylation appears to distinguish between the paternal and maternal alleles of imprinted genes. A good example is the gene *H19*, which is transcribed only from the maternal copy (Figure 6.35). The *H19* gene is specifically methylated during the development of male, but not female, germ cells. The union of sperm and egg at fertilization therefore yields an embryo containing a methylated paternal allele and an unmethylated maternal allele of the gene. These differences in methylation are maintained following DNA replication by an enzyme that specifically methylates CpG sequences of a daughter strand that is hydrogen-bonded to a methylated parental strand (Figure 6.36). The paternal *H19* allele therefore remains methylated, and transcriptionally inactive, in embryonic cells and somatic tissues. However, the paternal *H19* allele becomes demethylated in

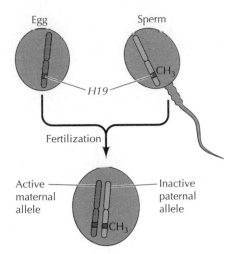

**Figure 6.35**
**Genomic imprinting**   The *H19* gene is specifically methylated during development of male germ cells. Therefore, sperm contain a methylated *H19* allele and eggs contain an unmethylated allele. Following fertilization, the methylated paternal allele remains transcriptionally inactive, and only the unmethylated maternal allele is expressed in the embryo.

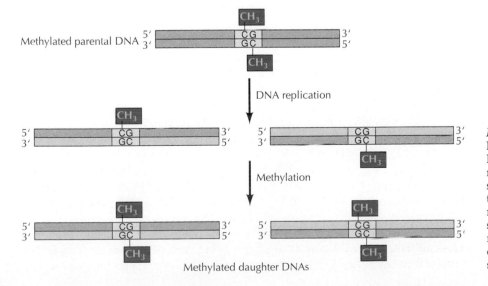

Methylated daughter DNAs

**Figure 6.36**
**Maintenance of methylation patterns**
In parental DNA, both strands are methylated at complementary CpG sequences. Following replication, only the parental strand of each daughter molecule is methylated. The newly synthesized daughter strands are then methylated by an enzyme that specifically recognizes CpG sequences opposite a methylation site.

the germ line, allowing a new pattern of methylation to be established for transmittal to the next generation.

## RNA PROCESSING AND TURNOVER

Although transcription is the first and most highly regulated step in gene expression, it is usually only the beginning of the series of events required to produce a functional RNA. Most newly synthesized RNAs must be modified in various ways to be converted to their functional forms. Bacterial mRNAs are an exception; as discussed earlier in this chapter, they are used immediately as templates for protein synthesis while still being transcribed. However, the primary transcripts of both rRNAs and tRNAs must undergo a series of processing steps in prokaryotic as well as eukaryotic cells. Primary transcripts of eukaryotic mRNAs similarly undergo extensive modifications, including the removal of introns by splicing, before they are transported from the nucleus to the cytoplasm to serve as templates for protein synthesis. Regulation of these processing steps provides an additional level of control of gene expression, as does regulation of the rates at which different mRNAs are subsequently degraded within the cell.

### Processing of Ribosomal and Transfer RNAs

The basic processing of ribosomal and transfer RNAs in prokaryotic and eukaryotic cells is similar, as might be expected given the fundamental roles of these RNAs in protein synthesis. As discussed previously, eukaryotes have four species of ribosomal RNAs (see Table 6.1), three of which (the 28S, 18S, and 5.8S rRNAs) are derived by cleavage of a single long precursor transcript, called a **pre-rRNA** (Figure 6.37). Prokaryotes have three ribosomal RNAs (23S, 16S, and 5S), which are equivalent to the 28S, 18S, and 5S rRNAs of eukaryotic cells and are also formed by the processing of a single pre-rRNA transcript. The only rRNA that is not processed extensively is the 5S rRNA in eukaryotes, which is transcribed from a separate gene.

*Figure 6.37*
**Processing of ribosomal RNAs** Prokaryotic cells contain three rRNAs (16S, 23S, and 5S), which are formed by cleavage of a pre-rRNA transcript. Eukaryotic cells (e.g., human cells) contain four rRNAs. One of these (5S rRNA) is transcribed from a separate gene; the other three (18S, 28S, and 5.8S) are derived from a common pre-rRNA. Following cleavage, the 5.8S rRNA (which is unique to eukaryotes) becomes hydrogen-bonded to 28S rRNA.

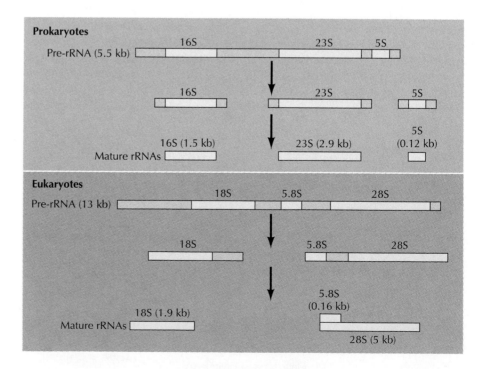

Prokaryotic and eukaryotic pre-rRNAs are processed in several steps. Initial cleavages of bacterial pre-rRNA yield separate precursors for the three individual rRNAs; these are then further processed by secondary cleavages to the final products. In eukaryotic cells, pre-rRNA is first cleaved at a site adjacent to the 5.8S rRNA on its 5' side, yielding two separate precursors that contain the 18S and the 28S + 5.8S rRNAs, respectively. Further cleavages then convert these to their final products, with the 5.8S rRNA becoming hydrogen-bonded to the 28S molecule. In addition to these cleavages, rRNA processing involves the addition of methyl groups to the bases and sugar moieties of specific nucleotides, although the function of these modifications is not known.

Like rRNAs, tRNAs in both bacteria and eukaryotes are synthesized as longer precursor molecules (**pre-tRNAs**), some of which contain several individual tRNA sequences (Figure 6.38). In bacteria, some tRNAs are included in the pre-rRNA transcripts. The processing of the 5' end of pre-tRNAs involves cleavage by an enzyme called **RNase P**, which is of special interest because it is a prototypical model of a reaction catalyzed by an RNA enzyme. RNase P consists of RNA and protein molecules, both of which are required for maximal activity. In 1983 Sidney Altman and his colleagues demonstrated that the isolated RNA component of RNase P is itself capable of catalyzing pre-tRNA cleavage. These experiments established that RNase P is a **ribozyme**—an enzyme in which RNA rather than protein is responsible for catalytic activity.

The 3' end of tRNAs is generated by the action of a conventional protein RNase, but the processing of this end of the tRNA molecule also involves an

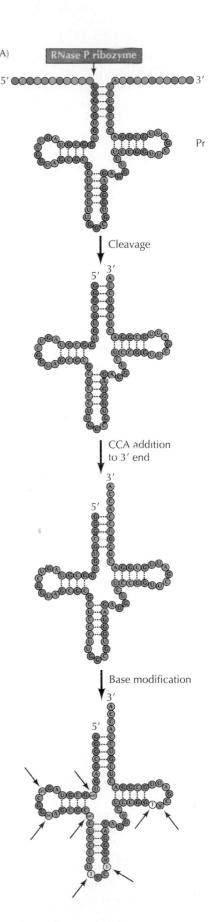

**Figure 6.38**
**Processing of transfer RNAs** (A) Transfer RNAs are derived from pre-tRNAs, some of which contain several individual tRNA molecules. Cleavage at the 5' end of the tRNA is catalyzed by the RNase P ribozyme; cleavage at the 3' end is catalyzed by a conventional protein RNase. A CCA terminus is then added to the 3' end of many tRNAs in a posttranscriptional processing step. Finally, some bases are modified at characteristic positions in the tRNA molecule. In this example, these modified nucleosides include dihydrouridine (DHU), methylguanosine (mG), inosine (I), ribothymidine (T), and pseudouridine ($\psi$). (B) Structure of modified bases. Ribothymidine, dihydrouridine, and pseudouridine are formed by modification of uridines in tRNA. Inosine and methylguanosine are formed by the modification of guanosines.

(B) Modified bases

| Dihydrouridine (DHU) | Ribothymidine (T) | Pseudouridine ($\psi$) |
| --- | --- | --- |
| Ribose | Ribose | Ribose |

| Inosine (I) | $2N$-Methylguanosine (mG) |
| --- | --- |
| Ribose | Ribose |

unusual activity: the addition of a CCA terminus. All tRNAs have the sequence CCA at their 3' ends. This sequence is the site of amino acid attachment, so it is required for tRNA function during protein synthesis. The CCA terminus is encoded in the DNA of some tRNA genes, but in others it is not, instead being added as an RNA processing step by an enzyme that recognizes and adds CCA to the 3' end of all tRNAs that lack this sequence.

Another unusual aspect of tRNA processing is the extensive modification of bases in tRNA molecules. Approximately 10% of the bases in tRNAs are altered to yield a variety of modified nucleotides at specific positions in tRNA molecules (see Figure 6.38). The functions of most of these modified bases are unknown, but some play important roles in protein synthesis by altering the base-pairing properties of the tRNA molecule (see Chapter 7).

Some pre-tRNAs, as well as pre-rRNAs in a few organisms, contain introns that are removed by splicing. These processing steps are discussed in the next section, together with other splicing reactions.

### Processing of mRNA in Eukaryotes

In contrast to the processing of ribosomal and transfer RNAs, the processing of messenger RNAs represents a major difference between prokaryotic and eukaryotic cells. In bacteria, ribosomes have immediate access to mRNA and translation begins on the nascent mRNA chain while transcription is still in progress. In eukaryotes, mRNA synthesized in the nucleus must first be transported to the cytoplasm before it can be used as a template for protein synthesis. Moreover, the initial products of transcription in eukaryotic cells (**pre-mRNAs**) are extensively modified before export from the nucleus. The processing of pre-mRNA includes modification of both ends of the molecule, as well as the removal of introns from its middle (Figure 6.39).

The 5' end of pre-mRNAs is modified soon after its synthesis by the addition of a structure called a **7–methylguanosine cap**. Capping is initiated by the addition of a GTP in reverse orientation to the 5' terminal nucleotide of the pre-mRNA. Then methyl groups are added to this G residue and to the ribose moieties of one or two 5' nucleotides of the RNA chain. The 5' cap aligns eukaryotic mRNAs on the ribosome during translation (see Chapter 7).

The 3' end of most eukaryotic mRNAs is defined not by termination of transcription, but by cleavage of the primary transcript and addition of a **poly-A tail**— a processing reaction called **polyadenylation** (Figure 6.40). In animal cells, the signal for polyadenylation is the hexanucleotide AAUAAA. This sequence is recognized by a complex of proteins, including an endonuclease that cleaves the RNA chain 10 to 30 nucleotides farther downstream. A separate poly-A polymerase then adds a poly-A tail of about 200 nucleotides to the transcript. The act of polyadenylation signals the termination of transcription by RNA polymerase, which usually occurs several hundred nucleotides downstream of the site of poly-A addition.

Poly-A tails have been shown to regulate both the stability and translation of mRNAs, although how they do so is not yet clear. In addition, some eukaryotic mRNAs (e.g., histone mRNAs) are not polyadenylated, so a poly-A tail is not an absolute requirement for mRNA function. On the other hand, polyadenylation plays an important regulatory role in early development, where changes in the length of poly-A tails control mRNA translation. For example, many mRNAs are stored in unfertilized eggs in an untranslated form with short poly-A tails (approximately 20 nucleotides long). Fertilization stimulates the lengthening of the poly-A tails of these stored mRNAs, which in turn activates their translation and the synthesis of proteins required for early embryonic development.

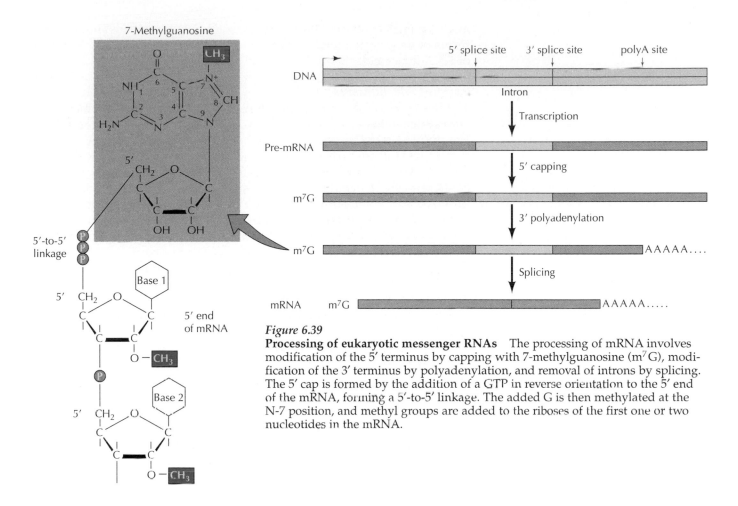

**Figure 6.39**

**Processing of eukaryotic messenger RNAs** The processing of mRNA involves modification of the 5' terminus by capping with 7-methylguanosine (m⁷G), modification of the 3' terminus by polyadenylation, and removal of introns by splicing. The 5' cap is formed by the addition of a GTP in reverse orientation to the 5' end of the mRNA, forming a 5'-to-5' linkage. The added G is then methylated at the N-7 position, and methyl groups are added to the riboses of the first one or two nucleotides in the mRNA.

The most striking modification of pre-mRNAs is the removal of introns by splicing. As discussed in Chapter 4, the coding sequences of most eukaryotic genes are interrupted by noncoding sequences (introns) that are precisely excised from the mature mRNA. Most genes contain multiple introns, which typically account for about ten times more pre-mRNA sequences than the exons do. The unexpected discovery of introns in 1977 generated an active research effort directed toward understanding the mechanism of splicing, which had to be highly specific to yield functional

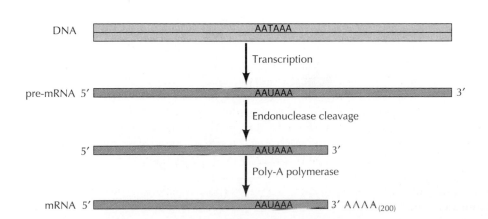

**Figure 6.40**

**Formation of the 3' ends of eukaryotic mRNAs** An endonuclease cleaves the pre-mRNA 10 to 30 nucleotides downstream of a polyadenylation signal (AAUAAA in animal cells). Poly-A polymerase then adds a poly-A tail consisting of about 200 A's to the 3' end of the RNA.

mRNAs. Further studies of splicing have not only illuminated new mechanisms of gene regulation; they have also revealed novel catalytic activities of RNA molecules.

### Splicing Mechanisms

The key to understanding pre-mRNA splicing was the development of *in vitro* systems that efficiently carried out the splicing reaction (Figure 6.41). Pre-mRNAs were synthesized *in vitro* by the cloning of structural genes (with their introns) adjacent to promoters for bacteriophage RNA polymerases, which could readily be isolated in large quantities. Transcription of these plasmids could then be used to prepare large amounts of pre-mRNAs that, when added to nuclear extracts of mammalian cells, were found to be correctly spliced. As with transcription, the use of such *in vitro* systems has allowed splicing to be analyzed in much greater detail than would have been possible in intact cells.

Analysis of the reaction products and intermediates formed *in vitro* revealed that pre-mRNA splicing proceeds in two steps (Figure 6.42). First, the pre-mRNA is cleaved at the 5′ splice site, and the 5′ end of the intron is joined to an adenine nucleotide within the intron (near its

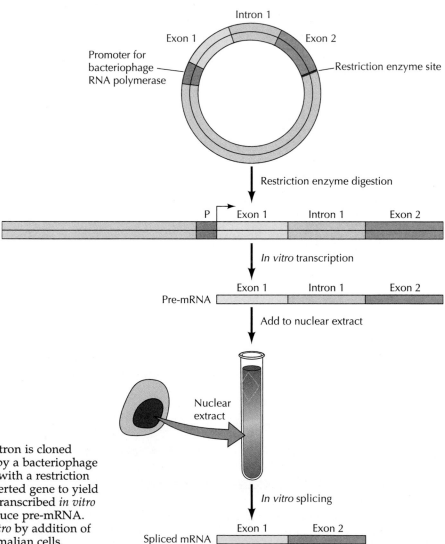

**Figure 6.41**
***In vitro* splicing** A gene containing an intron is cloned downstream of a promoter (P) recognized by a bacteriophage RNA polymerase. The plasmid is digested with a restriction enzyme that cleaves at the 3′ end of the inserted gene to yield a linear DNA molecule. This DNA is then transcribed *in vitro* with the bacteriophage polymerase to produce pre-mRNA. Splicing reactions can then be studied *in vitro* by addition of this pre-mRNA to nuclear extracts of mammalian cells.

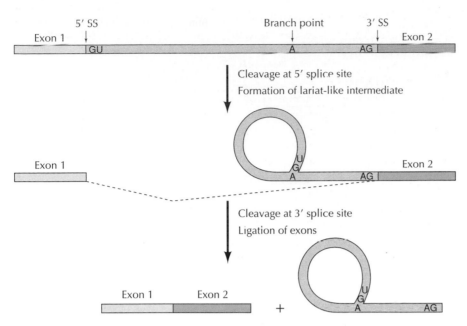

**Figure 6.42**
**Splicing of pre-mRNA**    The splicing reaction proceeds in two steps. The first step involves cleavage at the 5' splice site (SS) and joining of the 5' end of the intron to an A within the intron (the branch point). This reaction yields a lariat-like intermediate, in which the intron forms a loop. The second step is cleavage at the 3' splice site and simultaneous ligation of the exons, resulting in excision of the intron as a lariat-like structure.

3' end). In this step an unusual bond forms between the 5' end of the intron and the 2' hydroxyl group of the adenine nucleotide. The resulting intermediate is a lariat-like structure, in which the intron forms a loop. The second step in splicing then proceeds with simultaneous cleavage at the 3' splice site and ligation of the two exons. The intron is thus excised as a lariat-like structure, which is then linearized and degraded within the nucleus of intact cells.

These reactions define three critical sequence elements of pre-mRNAs: sequences at the 5' splice site, sequences at the 3' splice site, and sequences within the intron at the branch point (the point at which the 5' end of the intron becomes ligated to form the lariat-like structure) (see Figure 6.42). Pre-mRNAs contain similar consensus sequences at each of these positions, allowing the splicing apparatus to recognize pre-mRNAs and carry out the cleavage and ligation reactions involved in the splicing process.

Biochemical analysis of nuclear extracts has revealed that splicing takes place in large complexes, called **spliceosomes**, composed of proteins and RNAs. The RNA components of the spliceosome are five types of **small nuclear RNAs (snRNAs)** called U1, U2, U4, U5, and U6. These snRNAs, which range in size from approximately 50 to nearly 200 nucleotides, are complexed with six to ten protein molecules to form small nuclear ribonucleoprotein particles (**snRNPs**), which play central roles in the splicing process. The U1, U2, and U5 snRNPs each contain a single snRNA molecule, whereas U4 and U6 snRNAs are complexed to each other in a single snRNP.

The first step in spliceosome assembly is the binding of U1 snRNP to the 5' splice site of pre-mRNA (Figure 6.43). This recognition of 5' splice sites involves base pairing between the 5' splice site consensus sequence and a

*Figure 6.43*
**Assembly of the spliceosome** The first step in spliceosome assembly is the binding of U1 snRNP to the 5′ splice site (SS), followed by the binding of U2 snRNP to the branch point. A preformed complex consisting of U4/U6 and U5 snRNPs then enters the spliceosome, with U5 binding to both the 5′ and 3′ splice sites.

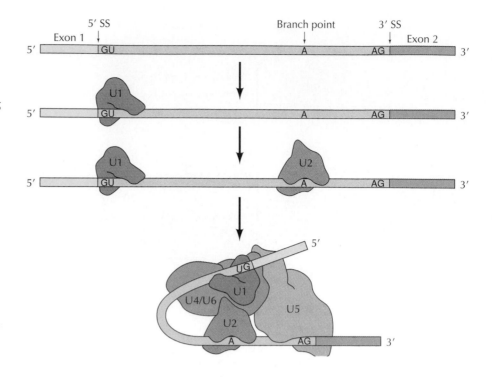

complementary sequence at the 5′ end of U1 snRNA (Figure 6.44). U2 snRNP then binds to the branch point, by similar complementary base pairing between U2 snRNA and branch point sequences. A preformed complex consisting of U4/U6 and U5 snRNPs is then incorporated into the spliceosome, with U5 binding to sequences at both the 5′ and 3′ splice sites (see Figure 6.43).

Not only do the snRNAs recognize consensus sequences at the branch points and splice sites of pre-mRNAs; they also appear to catalyze the splicing reaction directly. The catalytic role of RNAs in splicing was demonstrated by the discovery that some RNAs are capable of **self-splicing**; that is, they can catalyze the removal of their own introns in the absence of other protein or RNA factors. Self-splicing was first described by Tom Cech and his colleagues during studies of the 28S rRNA of the protozoan *Tetrahymena*. This RNA contains an intron of approximately 400 bases that is precisely removed following incubation of the pre-rRNA in the absence of added proteins. Further studies have revealed that splicing is catalyzed by the intron, which acts as a ribozyme to direct its own excision from the pre-rRNA molecule. The discovery of self-splicing of *Tetrahymena* rRNA, together with the studies of RNase P already discussed, provided the first demonstrations of the catalytic activity of RNA.

Additional studies have revealed self-splicing RNAs in mitochondria, chloroplasts, and bacteria. These self-splicing RNAs are divided into two classes on the basis of their reaction mechanisms (Figure 6.45). The first step in splicing for group I introns (e.g., *Tetrahymena* pre-rRNA) is cleavage at the 5′ splice site mediated by a guanosine cofactor. The 3′ end of the free exon then reacts with the 3′ splice site to excise the intron as a linear RNA. In contrast, the self-splicing reactions of group II introns (e.g., some mitochondrial pre-mRNAs) closely resemble those characteristic of nuclear pre-mRNA splicing, in which cleavage of the 5′ splice site results from attack by an adenosine nucleotide in the intron. As with pre-mRNA splicing, the result is a lariat-like intermediate, which is then excised.

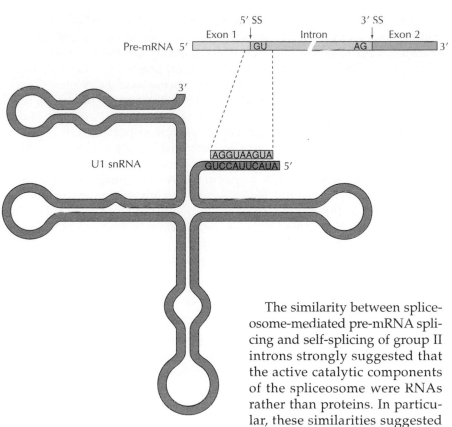

*Figure 6.44*
**Binding of U1 snRNA to the 5' splice site**   The 5' terminus of U1 snRNA binds to consensus sequences at 5' splice sites by complementary base pairing.

The similarity between spliceosome-mediated pre-mRNA splicing and self-splicing of group II introns strongly suggested that the active catalytic components of the spliceosome were RNAs rather than proteins. In particular, these similarities suggested that pre-mRNA splicing was catalyzed by the snRNAs of the spliceosome. Continuing studies of pre-mRNA splicing have provided clear support for this view; U2, U5, and U6 snRNAs have been identified as catalytic components of the spliceosome. Pre-mRNA splicing is thus considered to be an RNA-based reaction, catalyzed by spliceosome snRNAs acting analogously to group II self-splicing introns. Spliceosome proteins are also required, however, and the precise roles of both the snRNA and protein components of the spliceosome still

*Figure 6.45*
**Self-splicing introns**   Group I and group II self-splicing introns are distinguished by their reaction mechanisms. In group I introns, the first step in splicing is cleavage of the 5' splice site by reaction with a guanosine cofactor. The result is a linear intermediate with a G added to the 5' end of the intron. In group II introns (as in pre-mRNA splicing), the first step is cleavage of the 5' splice site by reaction with an A within the intron, forming a lariat-like intermediate. In both cases, the second step is simultaneous cleavage of the 3' splice site and ligation of the exons.

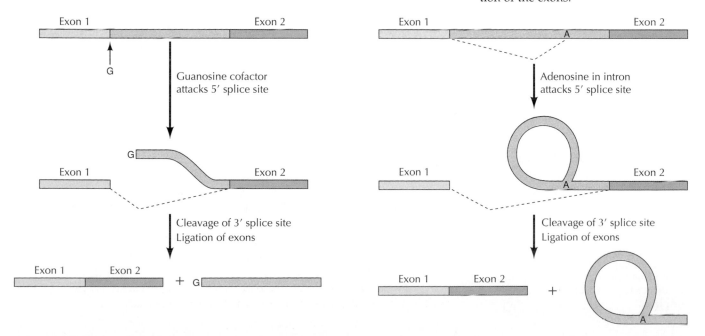

remain to be understood fully. One important role of protein factors is to identify and select the splice sites that are recognized by the snRNAs. For example, additional protein factors are required for the binding of both U1 and U2 snRNPs to the appropriate sites on pre-mRNA. Because many pre-mRNAs contain multiple introns, the splicing machinery must be able to identify and join the appropriate 5′ and 3′ splice sites to produce a functional mRNA. The mechanism of such splice site selection is not understood, but protein splicing factors clearly play an important role.

Finally, it should be noted that the mechanism of pre-tRNA splicing differs from both types of self-splicing reactions and from the splicing of nuclear pre-mRNA. Pre-tRNA splicing involves conventional protein enzymes with no known role for RNA catalysis. Instead, a protein endonuclease cleaves at the splice sites, excising the intron as a linear RNA. This step is followed by ligation of the exons, yielding the mature tRNA molecule. Although the introns of tRNAs may also participate in splicing, tRNA splicing appears to be fundamentally different from the other RNA-based splicing reactions.

### Variations of Pre-mRNA Splicing

The central role of splicing in the processing of pre-mRNA opens the possibility of regulation of gene expression by control of the activity of the splicing machinery. Moreover, since most pre-mRNAs contain multiple introns, different mRNAs can be produced from the same gene by different combinations of 5′ and 3′ splice sites. The possibility of joining exons in varied combinations provides a novel means of controlling gene expression by generating multiple mRNAs (and therefore multiple proteins) from the same pre-mRNA. This process, called **alternative splicing**, occurs frequently in genes of complex eukaryotes and provides an important mechanism for tissue-specific and developmental regulation of gene expression.

One interesting example of alternative splicing is provided by some genes that encode transcriptional regulatory proteins. In several cases, alternative splicing of these pre-mRNAs yields products with dramatically different functions—namely, the ability to act as either activators or repressors of transcription (Figure 6.46). As discussed earlier, transcriptional activators consist of two distinct domains: a DNA-binding domain and an activation domain. These domains are generally encoded in separate exons, so alternative splicing allows them to be reassorted into different combinations, thereby enabling the production of activators and repressors from the same gene. Splicing that yields an mRNA that contains exons encoding both DNA-binding and activation domains will result in synthesis of an activator

*Figure 6.46*
**Alternative splicing of pre-mRNA that encodes a transcription factor**
In this example, a transcriptional regulatory protein is encoded by four exons: the first encodes the DNA-binding domain; the third encodes the activation domain. The pre-mRNA is subject to two alternative-splicing pathways. The first results in ligation of all four exons, yielding an mRNA that encodes a transcriptional activator. In the second splicing pathway, however, exon 2 is joined directly to exon 4, yielding an mRNA that does not include the third exon. This mRNA is then translated to yield a protein lacking the activation domain, which therefore functions as a repressor rather than an activator of transcription.

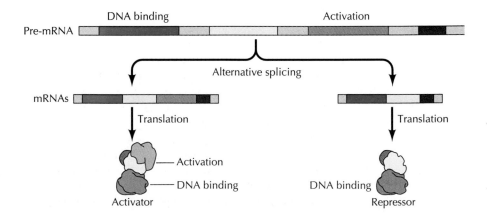

protein. Alternative splicing, however, may result in the formation of an mRNA encoding a DNA-binding domain but lacking an activation domain. Translation of such an alternatively spliced mRNA will result in synthesis of a repressor, which will suppress gene expression by competing with the activator for binding to target DNA sequences (see Figure 6.31).

Because patterns of alternative splicing can vary in different tissues, regulation of splicing provides an important means of regulating tissue-specific gene expression. Although the mechanism by which the correct splice sites are selected in a pre-mRNA is not known, several protein factors have been identified that contribute to splice site selection and can affect the use of alternative splice sites in a pre-mRNA molecule. Variations in the expression of such splicing factors in different cell types may result in tissue-specific patterns of alternative splicing, thereby contributing to the regulation of gene expression during development and differentiation. A particularly notable example is provided by sex determination in *Drosophila*, where alternative splicing of the same pre-mRNA determines whether a fly is male or female.

All of the splicing reactions already discussed have resulted in the joining of two exons that were originally present on the same pre-mRNA molecule. An intriguing variation on this theme is ***trans*-splicing**, in which exons originating from two separate transcripts are ligated together. This form of splicing was discovered in trypanosomes, in which all mRNAs contain an identical spliced leader sequence of 35 nucleotides at their 5′ end. The spliced leader is transcribed separately as the 5′ portion of a 137-nucleotide RNA and is then joined to 3′ splice sites of pre-mRNAs by a *trans*-splicing reaction (Figure 6.47). Nematodes (e.g., *C. elegans*) and *Euglena* also exhibit *trans*-splicing. In addition, mammalian cells are capable of carrying out *trans*-splicing reactions with the nematode spliced leader RNAs, indicating that the machinery required for *trans*-splicing, which includes at least some of the snRNPs required for the normal *cis*-splicing of nuclear pre-mRNAs, also exists in mammals. However, the potential biological significance of *trans*-splicing in higher eukaryotes, including mammals, remains to be determined.

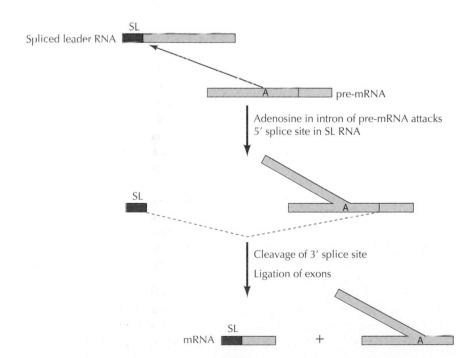

*Figure 6.47*
***trans*-splicing**   The reactions that are involved in *trans*-splicing are similar to those of conventional pre-mRNA splicing, except that in *trans*-splicing, exons originating from two separate RNA molecules are ligated. In the first step, the splice site of the spliced leader (SL) RNA is cleaved by reaction with an adenosine within the intron of the pre-mRNA. Then the 3′ splice site is cleaved and the SL and pre-mRNA exons are ligated.

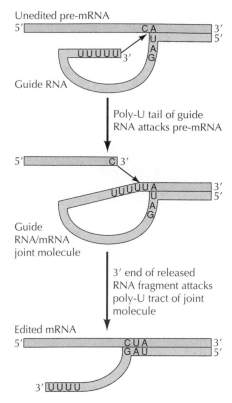

Unedited pre-mRNA

Guide RNA

Poly-U tail of guide
RNA attacks pre-mRNA

Guide
RNA/mRNA
joint molecule

3′ end of released
RNA fragment attacks
poly-U tract of joint
molecule

Edited mRNA

*Figure 6.48*
**Model of mitochondrial mRNA editing in trypanosomes**   U's are inserted at multiple sites along the pre-mRNA molecule. The information for editing is provided by a guide RNA, which is complementary to the edited portion of the mRNA and has a poly-U tail at its 3′ end. In the editing reaction shown, a single U is inserted between C and A residues of the pre-mRNA. This insertion is thought to be accomplished by cleavage between these residues and formation of a joint molecule in which the 3′ end of the guide RNA is joined to the A residue of the mRNA. The 3′ end of the released RNA fragment then attacks the poly-U tract of the joint molecule, forming an edited mRNA that contains one of the U residues from the poly-U tail of the guide RNA. *

* A recent study (M. L. Kable, S. D. Seiwert, S. Heidmann, and K. Stuart, 1996, *Science* 273: 1189–1195) indicates that inserted U's are derived from free UTP rather than from the guide RNA.

## RNA Editing

**RNA editing** refers to RNA processing events (other than splicing) that alter the protein-coding sequences of some mRNAs. This unexpected form of RNA processing was first discovered in mitochondrial mRNAs of trypanosomes but has more recently also been described in mitochondrial mRNAs of other organisms, chloroplast mRNAs of higher plants, and nuclear mRNAs of some mammalian genes.

The editing of mitochondrial mRNAs in trypanosomes and related protozoans involves the addition and deletion of U residues at multiple sites along the molecule (Figure 6.48). The information required for editing is encoded in "guide" RNAs, which are complementary to edited portions of the mature mRNA. The guide RNAs contain poly-U tails, which donate the U's added during editing via a series of reactions that appear similar to RNA splicing. In some cases, editing is so extensive that it accounts for more than half of the nucleotides in the mature mRNA.

Other forms of editing result in subtler changes in mRNA sequence—in particular, single base alterations. Examples of these types of editing are found in the mitochondrial and chloroplast RNAs of higher plants, and in nuclear mRNAs of mammalian cells. The best-studied example in mammals is editing of the mRNA for apolipoprotein B, which transports lipids in the blood. In this case, tissue-specific RNA editing results in two different forms of apolipoprotein B (Figure 6.49). In humans, Apo-B100 (4536 amino acids) is synthesized in the liver by translation of the unedited mRNA. However, a shorter protein (Apo-B48, 2152 amino acids) is synthesized in the intestine as a result of translation of an edited mRNA in which a C has been changed to a U. This substitution changes the codon for glutamine (CAA) in the unedited mRNA to a translation termination codon (UAA) in the edited mRNA, resulting in synthesis of the shorter Apo-B protein. The editing reaction involves direct enzymatic conversion of C to U by removal of the cytosine amino group. This type of editing thus resembles the base modifications involved in tRNA processing more closely than it resembles the editing of trypanosome mitochondrial RNAs. Tissue-specific editing of Apo-B results in the expression of structurally and functionally different proteins in liver and intestine. The full-length Apo-B100 produced by the liver transports lipids in the circulation; Apo-B48 functions in the absorption of dietary lipids by the intestine.

## RNA Degradation

The processing steps discussed in the previous section result in the formation of mature mRNAs, which then direct protein synthesis. However, what may be considered the final aspect of the processing of an RNA molecule is its eventual degradation within the cell. Since the intracellular level of any RNA is determined by a balance between synthesis and degradation, the rate at which individual RNAs are degraded is another level at which gene expression can be controlled. Both ribosomal and transfer RNAs are very stable, and this stability largely accounts for the high levels of these RNAs (greater than 90% of all RNA) in both prokaryotic and eukaryotic cells. In contrast, bacterial mRNAs are rapidly degraded, usually having half-lives of only 2 to 3 minutes. This rapid turnover of bacterial mRNAs allows the cell to respond quickly to alterations in its environment, such as changes in the availability of nutrients required for growth. In eukaryotic cells, however, different mRNAs are degraded at different rates, providing an additional parameter to the regulation of eukaryotic gene expression.

The degradation of most eukaryotic mRNAs is initiated by shortening of

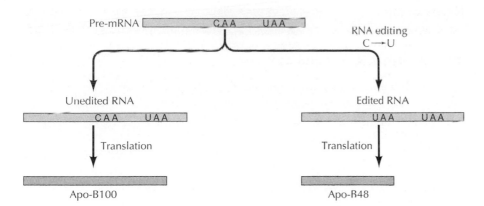

*Figure 6.49*
**Editing of apolipoprotein B mRNA**
In human liver, unedited mRNA is translated to yield a 4536-amino-acid protein called Apo-B100. In human intestine, however, the mRNA is edited by a base modification that changes a specific C to a U. This modification changes the codon for glutamine (CAA) to a termination codon (UAA), resulting in synthesis of a shorter protein (Apo-B48, consisting of only 2152 amino acids).

their poly-A tails. Then follows removal of the 5′ cap and degradation of the RNA by nucleases acting from both ends. The half-lives of mRNAs in mammalian cells vary from less than 30 minutes to approximately 20 hours. The unstable mRNAs frequently code for regulatory proteins, including certain transcription factors, whose levels within the cell vary rapidly in response to environmental stimuli. These mRNAs often contain specific AU-rich sequences near their 3′ ends that appear to signal rapid degradation by promoting deadenylation.

The stability of some mRNAs can also be regulated in response to extracellular signals. A good example is provided by the mRNA that encodes transferrin receptor—a cell surface protein involved in the uptake of iron by mammalian cells. The amount of transferrin receptor within cells is regulated by the availability of iron, largely as a result of modulation of the stability of its mRNA (Figure 6.50). In the presence of adequate amounts of iron, transferrin receptor mRNA is rapidly degraded as a result of specific nuclease cleavage at a sequence near its 3′ end. If an adequate supply of iron is not available, however, the mRNA is stabilized, resulting in increased synthesis of transferrin receptor and more iron uptake by the cell. This regulation is mediated by a protein that binds to specific sequences (called the iron response element, or IRE) near the 3′ end of transferrin receptor mRNA and protects the mRNA from cleavage. The binding of this regulatory protein to the IRE is in turn controlled by the levels of iron within the cell: If iron is scarce, the protein binds to the IRE and protects transferrin receptor mRNA from degradation. Similar changes in the stability of other mRNAs are involved in the regulation of gene expression by certain hormones. Thus, although transcription remains the primary level at which gene expression is regulated, variations in the rate of mRNA degradation also play an important role in controlling steady-state levels of mRNAs within the cell.

*Figure 6.50*
**Regulation of transferrin receptor mRNA stability** The levels of transferrin receptor mRNA are regulated by the availability of iron. If the supply of iron is adequate, the mRNA is rapidly degraded as a result of nuclease cleavage near the 3′ end. If iron is scarce, however, a regulatory protein (called the iron response element–binding protein, or IRE-BP) binds to a sequence near the 3′ end of the mRNA (the iron response element, or IRE), protecting the mRNA from nuclease cleavage.

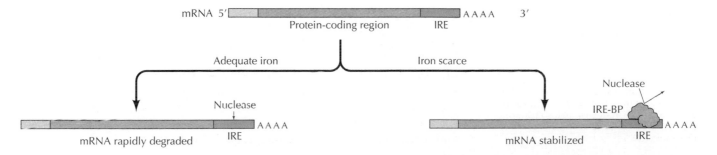

# Summary

## TRANSCRIPTION IN PROKARYOTES

*RNA Polymerase and Transcription:* E. coli RNA polymerase consists of $\alpha$, $\beta$, $\beta'$, and $\sigma$ subunits. Transcription is initiated by the binding of $\sigma$ to promoter sequences. After synthesis of the first few nucleotides of RNA, the core polymerase dissociates from $\sigma$ and travels along the template DNA as it elongates the RNA chain. Transcription then continues until the polymerase encounters a termination signal.

*Repressors and Negative Control of Transcription:* The prototype model for gene regulation in bacteria is the *lac* operon, which is regulated by the binding of a repressor to specific DNA sequences near the promoter.

*Positive Control of Transcription:* Some bacterial genes are regulated by transcriptional activators rather than repressors.

*Transcriptional Attenuation:* Transcription can also be regulated by control of the ability of RNA polymerase to elongate past specific sites in the template DNA.

## EUKARYOTIC RNA POLYMERASES AND BASAL TRANSCRIPTION FACTORS

*Eukaryotic RNA Polymerases:* Eukaryotic cells contain three distinct nuclear RNA polymerases that transcribe genes encoding mRNAs (polymerase II), rRNAs (polymerase I), and tRNAs (polymerase III).

*Basal Transcription Factors and Initiation of Transcription by RNA Polymerase II:* Eukaryotic RNA polymerases do not bind directly to promoter sequences; they require additional proteins (basal transcription factors) to initiate transcription. The promoter sequences of many polymerase II genes are recognized by the TATA-binding protein, which recruits additional transcription factors and RNA polymerase to the promoter.

*Transcription by RNA Polymerases I and III:* RNA polymerases I and III also require additional transcription factors to bind to the promoters of rRNA and tRNA genes.

## REGULATION OF TRANSCRIPTION IN EUKARYOTES

cis-*Acting Regulatory Sequences: Promoters and Enhancers:* Transcription of eukaryotic genes is controlled by proteins that bind to regulatory sequences, which can be located up to several kilobases away from the transcription start site. Enhancers typically contain binding sites for multiple proteins that work together to regulate gene expression.

*Transcriptional Regulatory Proteins:* Many eukaryotic transcription factors have been isolated on the basis of their binding to specific DNA sequences.

*Structure and Function of Transcriptional Activators:* Transcriptional activators are modular proteins, consisting of distinct DNA-binding and activation domains. DNA-binding domains mediate association with specific regulatory sequences; activation domains stimulate transcription by interacting with basal transcription factors.

*Eukaryotic Repressors:* Gene expression in eukaryotic cells is regulated by repressors as well as by activators. Some repressors interfere with the binding of activators or basal transcription factors to DNA. Other repressors contain discrete repression domains that inhibit transcription by interacting with either basal transcription factors or transcriptional activators.

*Relationship of Chromatin Structure to Transcription:* The packaging of DNA in nucleosomes presents an impediment to transcription in eukaryotic cells. Nucleosome remodeling factors disrupt chromatin structure and allow transcription factors to bind DNA sequences in enhancers and promoters. RNA polymerase is then able to transcribe through nucleosomes by disrupting histone-DNA contacts. This transcription is facilitated by the nonhistone chromosomal proteins HMG-14 and HMG-17, which are specifically associated with actively transcribed genes.

**HMG-14, HMG-17, nucleosome remodeling factor, DNase-hypersensitive site**

*DNA Methylation:* Methylation of cytosine residues can inhibit the transcription of vertebrate genes. Regulation of gene expression by methylation plays an important role in genomic imprinting, which controls the transcription of some genes involved in mammalian development.

**genomic imprinting**

## RNA PROCESSING AND TURNOVER

*Processing of Ribosomal and Transfer RNAs:* Ribosomal and transfer RNAs are derived by cleavage of long primary transcripts in both prokaryotic and eukaryotic cells. Methyl groups are added to rRNAs, and various bases are modified in tRNAs.

**pre-rRNA, pre-tRNA, RNase P, ribozyme**

*Processing of mRNA in Eukaryotes:* Eukaryotic pre-mRNAs are modified by the addition of 7–methylguanosine caps and 3′ poly-A tails, in addition to the removal of introns by splicing.

**pre-mRNA, 7-methylguanosine cap, poly-A tail, polyadenylation**

*Splicing Mechanisms:* Splicing of nuclear pre-mRNAs takes place in large complexes, called spliceosomes, composed of proteins and small nuclear RNAs (snRNAs). The snRNAs recognize sequences at the splice sites of pre-mRNAs and catalyze the splicing reaction. Some mitochondrial, chloroplast, and bacterial RNAs undergo self-splicing, in which the splicing reaction is catalyzed by intron sequences.

**spliceosome, small nuclear RNA (snRNA), snRNP, self-splicing**

*Variations of Pre-mRNA Splicing:* Exons can be joined in various combinations as a result of alternative splicing, which provides an important mechanism for tissue-specific control of gene expression in complex eukaryotes. In some simpler eukaryotes, exons from two different primary transcripts can be joined together by *trans*-splicing.

**alternative splicing, *trans*-splicing**

*RNA Editing:* Some mRNAs are modified by processing events that alter their protein-coding sequences. Editing of mitochondrial mRNAs in some protozoans involves the addition and deletion of U residues at multiple sites in the molecule. Other forms of RNA editing in plant and mammalian cells involve the modification of specific bases.

**RNA editing**

*RNA Degradation:* mRNAs in eukaryotic cells are degraded at different rates, providing an additional mechanism for control of gene expression. In some cases, rates of mRNA degradation are regulated by extracellular signals.

## QUESTIONS

**1.** The consensus sequence of the *E. coli* –10 promoter element is TATAAT. You are comparing two promoters that have –10 element sequences of TATGAT and CATGAT, respectively. Which would you expect to be transcribed more efficiently?

**2.** You are working with two strains of *E. coli*. One contains a wild-type β-galactosidase gene and an *i*⁻ mutation; the other contains a temperature-sensitive β-galactosidase gene and an *o^c* mutation. After mating these strains, you assay for the production of β-galactosi-

dase at both permissive and nonpermissive temperatures in the absence of lactose. What do you expect to find?

**3.** You are comparing the requirements for *in vitro* basal transcription of two polymerase II genes, one containing a TATA box and the other containing only an Inr sequence. Does transcription from these promoters require TBP or TFIID?

**4.** You are studying the enhancer of a gene that normally is expressed only in neurons. Constructs in which this en-

hancer is linked to a reporter gene are expressed in neuronal cells but not in fibroblasts. However, if you mutate a specific sequence element within the enhancer, you find expression in both fibroblasts and neuronal cells. What type of regulatory protein would you expect to bind to that enhancer element?

**5.** A transcription factor is found to activate transcription by binding to different DNA sequences in muscle cells and liver cells. How might alternative splicing be involved in determining this tissue specificity of activator function?

## REFERENCES AND FURTHER READING

### General References

McKnight, S. L. and K. R Yamamoto, eds. 1992. *Transcriptional Regulation*. Plainview, NY: Cold Spring Harbor Laboratory Press.

Ptashne, M. 1992. *A Genetic Switch*. 2nd ed. Cambridge, MA.: Cell Press and Blackwell Scientific Publications.

Tjian, R. 1995. Molecular machines that control genes. *Sci. Amer.* 272(2): 54–61.

### Transcription in Prokaryotes

Burgess, R. R., A. A. Travers, J. J. Dunn and E. K. F. Bautz. 1969. Factor stimulating transcription by RNA polymerase. *Nature* 221: 43–44. [P]

Busby, S. and R. H. Ebright. 1994. Promoter structure, promoter recognition, and transcription activation in prokaryotes. *Cell* 79: 743–746. [R]

Dombroski, A. J., W. A. Walter, M. T. Record Jr., D. A. Siegele and C. A Gross. 1992. Polypeptides containing highly conserved regions of transcription initiation factor σ⁷⁰ exhibit specificity of binding to promoter DNA. *Cell* 70: 501–512. [P]

Englesberg, E., J. Irr, J. Power and N. Lee. 1965. Positive control of enzyme synthesis by gene C in the L-arabinose system. *J. Bacteriol.* 90: 946–957. [P]

Gilbert, W. and B. Muller-Hill. 1966. Isolation of the *lac* repressor. *Proc. Natl. Acad. Sci. USA* 56: 1891–1899. [P]

Jacob, F. and J. Monod. 1961. Genetic and regulatory mechanisms in the synthesis of proteins. *J. Mol. Biol.* 3: 318–356. [P]

Kolb, A., S. Busby, H. Buc, S. Garges and S. Adhya. 1993. Transcriptional regulation by cAMP and its receptor protein. *Ann. Rev. Biochem.* 62: 749–795. [R]

Lewis, M., G. Chang, N. C. Horton, M. A. Kercher, H. C. Pace, M. A. Schumacher, R. G. Brennan and P. Lu. 1996. Crystal structure of the lactose operon repressor and its complexes with DNA and inducer. *Science* 271: 1247–1254. [P]

Pardee, A. B., F. Jacob and J. Monod. 1959. The genetic control and cytoplasmic expression of "inducibility" in the synthesis of β-galactosidase by *E. coli*. *J. Mol. Biol.* 1: 165–178. [P]

Yanofsky, C. 1981. Attenuation in the control of expression of bacterial operons. *Nature* 289: 751–758. [R]

### Eukaryotic RNA Polymerases and Basal Transcription Factors

Buratowski, S. 1994. The basics of basal transcription by RNA polymerase II. *Cell* 77: 1–3. [R]

Conaway, R. C. and J. C. Conaway. 1993. General initiation factors for RNA polymerase II. *Ann. Rev. Biochem.* 62: 161–190. [R]

Drapkin, R. and D. Reinberg. 1994. The multifunctional TFIIH complex and transcriptional control. *Trends Biochem. Sci.* 19: 504–508. [R]

Geiduschek, E. P. and G. P. Tocchini-Valentini. 1988. Transcription by RNA polymerase III. *Ann. Rev. Biochem.* 57: 873–914. [R]

Hernandez, N. 1993. TBP, a universal eukaryotic transcription factor? *Genes Dev.* 7: 1291–1308. [R]

Koleske, A. J. and R. A. Young. 1995. The RNA polymerase II holoenzyme and its implications for gene regulation. *Trends Biochem. Sci.* 20: 113–116. [R]

Matsui, T., J. Segall, P. A. Weil and R. G. Roeder. 1980. Multiple factors are required for accurate initiation of transcription by purified RNA polymerase II. *J. Biol. Chem.* 255: 11992–11996. [P]

Nikolov, D. B., H. Chen, E. D. Halay, A. A. Usheva, K. Hisatake, D. K. Lee, R. G. Roeder and S. K. Burley. 1995. Crystal structure of a TFIIB-TBP-TATA-element ternary complex. *Nature* 377: 119–128. [P]

Pugh, B. F. and R. Tjian. 1992. Diverse transcriptional functions of the multisubunit eukaryotic TFIID complex. *J. Biol. Chem.* 267: 679–682. [R]

Sharp, P. A. 1992. TATA-binding protein is a classless factor. *Cell* 68: 819–821. [R]

Weil, P. A., D. S. Luse, J. Segall and R. G. Roeder. 1979. Selective and accurate transcription at the Ad2 major late promoter in a soluble system dependent on purified RNA polymerase II and DNA. *Cell* 18: 469–484. [P]

Young, R. A. 1991. RNA polymerase II. *Ann. Rev. Biochem.* 60: 689–715. [R]

Zawel, L. and D. Reinberg. 1995. Common themes in assembly and function of eukaryotic transcription complexes. *Ann. Rev. Biochem.* 64: 533–561. [R]

### Regulation of Transcription in Eukaryotes

Adams, C. C. and J. L. Workman. 1993. Nucleosome displacement in transcription. *Cell* 72: 305–308. [R]

Atchison, M. L. 1988. Enhancers: mechanisms of action and cell specificity. *Ann. Rev. Cell. Biol.* 4: 127–153. [R]

Banerji, J., S. Rusconi and W. Schaffner. 1981. Expression of a β-globin gene is enhanced by remote SV40 DNA sequences. *Cell* 27: 299–308. [P]

Barlow, D. P. 1995. DNA methylation and genomic imprinting. *Cell* 77: 473–476. [R]

Beato, M., P. Herrlich and G. Schutz. 1995. Steroid hormone receptors: many actors in search of a plot. *Cell* 83: 851–857. [R]

Bird, A. 1992. The essentials of DNA methylation. *Cell* 70: 5–8. [R]

Brent, R. and M. Ptashne. 1985. A eukaryotic transcriptional activator bearing the DNA specificity of a prokaryotic repressor. *Cell* 43: 729–736. [P]

Buratowski, S. 1995. Mechanisms of gene activation. *Science* 270: 1773–1774. [R]

Choy, B. and M. R. Green. 1993. Eukaryotic activators function during multiple steps of preinitiation complex assembly. *Nature* 366: 531–536. [P]

Dynan, W. S. and R. Tjian. 1983. The promoter-specific transcription factor Sp1 binds to upstream sequences in the SV40 early promoter. *Cell* 35: 79–87. [P]

Felsenfeld, G. 1992. Chromatin as an essential part of the transcriptional mechanism. *Nature* 355: 219–224. [R]

Gehring, W. J., Y. Q. Qian, M. Billeter, K. Furukubo-Tokunaga, A. F. Schier, D. Resendez-Perez, M. Affolter, G. Otting and K. Wuthrich. 1994. Homeodomain-DNA recognition. *Cell* 78: 211–223. [R]

Goodrich, J. A., G. Cutler, and R. Tjian. 1996. Contacts in context: Promoter specificity and macromolecular interactions in transcription. *Cell* 84: 825–830. [R]

Grunstein, M. 1992. Histones as regulators of genes. *Sci. Amer.* 267(4). 68–74B. [R]

Hanna-Rose, W. and U. Hansen. 1996. Active repression mechanisms of eukaryotic transcription repressors. *Trends Genet.* 12: 229–234. [R]

Johnson, A. D. 1995. The price of repression *Cell* 81: 655–658. [R]

Johnson, P. F. and S. L. McKnight. 1989. Eukaryotic transcriptional regulatory proteins. *Ann. Rev. Biochem.* 58: 799–839. [R]

Kadonaga, J. T. and R. Tjian. 1986. Affinity purification of sequence-specific DNA binding proteins. *Proc. Natl. Acad. Sci. USA* 83: 5889–5893. [P]

Kornberg, R. D. and Y. Lorch. 1992. Chromatin structure and transcription. *Ann. Rev. Cell Biol.* 8: 563–587. [R]

Levine, M. and J. L. Manley. 1989. Transcriptional repression of eukaryotic promoters. *Cell* 59: 405-408. [R]

Licht, J. D., M. J. Grossel, J. Figge and U. M. Hansen. 1990. *Drosophila Krüppel* protein is a transcriptional repressor. *Nature* 346: 76–79. [P]

Maniatis, T., S. Goodbourn and J. A. Fischer. 1987. Regulation of inducible and tissue-specific gene expression. *Science* 236: 1237–1244. [R]

McGinnis, W. and R. Krumlauf. 1992. Homeobox genes and axial patterning. *Cell* 68: 283–302. [R]

McKnight, S. L. and R. Kingsbury. 1982. Transcriptional control signals of a eukaryotic protein-coding gene. *Science* 217: 316–324. [P]

Mitchell, P. and R. Tjian. 1989. Transcriptional regulation in mammalian cells by sequence-specific DNA binding proteins. *Science* 245: 371–378. [R]

Pabo, C. O. and R. T. Sauer. 1992. Transcription factors: structural families and principles of DNA recognition. *Ann. Rev. Biochem.* 61: 1053–1095. [R]

Paranjape, S. M., R. T. Kamakaka and J. T. Kadonaga. 1994. Role of chromatin structure in the regulation of transcription by RNA polymerase II. *Ann. Rev. Biochem.* 63: 265–297. [R]

Peterson, C. L. and J. W. Tamkun. 1995. The SWI-SNF complex: A chromatin remodeling machine? *Trends Biochem. Sci.* 20: 143–146. [R]

Ptashne, M. 1988. How eukaryotic transcriptional activators work. *Nature* 335: 683–689. [R]

Razin, A. and H. Cedar. 1994. DNA methylation and genomic imprinting. *Cell* 77: 473–476. [R]

Roberts, S. G. E. and M. R. Green. 1994. Activator-induced conformational change in general transcription factor TFIIB. *Nature* 371: 717–720. [P]

Sauer, F., S. K. Hansen and R. Tjian. 1995. Multiple TAF$_{II}$s directing synergistic activation of transcription. *Science* 270: 1783–1788. [P]

Staudt, L. M. and M. J. Lenardo. 1991. Immunoglobulin gene transcription. *Ann. Rev. Immunol.* 9: 373–398. [R]

Tjian, R. and T. Maniatis. 1994. Transcriptional activation: a complex puzzle with few easy pieces. *Cell* 77: 5–8. [R]

Wolffe, A. P. 1994. Transcription: In tune with the histones. *Cell* 77: 13–16. [R]

## RNA Processing and Turnover

Beelman, C. A. and R. Parker. 1995. Degradation of mRNA in eukaryotes. *Cell* 81: 179–183. [R]

Berget, S. M. 1995. Exon recognition in vertebrate splicing. *J. Biol. Chem.* 270: 2411–2414. [R]

Cech, T. R. 1990. Self-splicing of group I introns. *Ann. Rev. Biochem.* 59: 543–568. [R]

Curtis, D., R. Lehmann and P. D. Zamore. 1995. Translational regulation in development. *Cell* 81: 171–178. [R]

Foulkes, N. S. and P. Sassone-Corsi. 1992. More is better: activators and repressors from the same gene. *Cell* 68: 411–414. [R]

Green, M. R. 1991. Biochemical mechanisms of constitutive and regulated pre-mRNA splicing. *Ann. Rev. Cell Biol.* 7: 559–599. [R]

Guerrier-Takada, C., K. Gardiner, T. Marsh, N. Pace and S. Altman. 1983. The RNA moiety of ribonuclease P is the catalytic subunit of the enzyme. *Cell* 35: 849–857. [P]

Guthrie, C. 1991. Messenger RNA splicing in yeast: clues to why the spliceosome is a ribonucleoprotein. *Science* 253: 157–163. [R]

Hopper, A. K. and N. C. Martin. 1992. Processing of yeast cytoplasmic and mitochondrial precursor tRNAs. In *The Molecular and Cellular Biology of the Yeast Saccharomyces: Gene Expression.* E. W. Jones, J. R. Pringle and J. R. Broach, eds. pp. 99–141. Plainview, NY: Cold Spring Harbor Laboratory Press. [R]

Keller, W. 1995. No end yet to messenger RNA 3′ processing! *Cell* 81: 829–832. [R]

Klausner, R. D., T. A. Rouault and J. B. Harford. 1993. Regulating the fate of mRNA: The control of cellular iron metabolism. *Cell* 72: 19–28. [R]

Kruger, K., P. J. Grabowski, A. Zaug, A. J. Sands, D. E. Gottschling and T. R. Cech. 1982. Self-splicing RNA: Autoexcision and autocyclization of the ribosomal RNA intervening sequence of *Tetrahymena*. *Cell* 31: 147–157. [P]

Maniatis, T. 1991. Mechanisms of alternative pre-mRNA splicing. *Science* 251: 33–34. [R]

McKeown, M. 1992. Alternative mRNA splicing. *Ann. Rev. Cell Biol.* 8: 133–155. [R]

Nilsen, T. W. 1994. RNA–RNA interactions in the spliceosome: Unraveling the ties that bind. *Cell* 78: 1–4. [R]

Pace, N. R. and A. B. Burgin. 1990. Processing and evolution of the rRNAs. In *The Ribosome.* W. E. Hill, A. Dahlberg, R. A. Garrett, P. B. Moore, D. Schlessinger and J. R. Warner, eds. pp. 417–425. Washington, DC: American Society for Microbiology. [R]

Padgett, R. A., M. M. Konarska, P. J. Grabowski, S. F. Hardy and P. A. Sharp. 1984. Lariat RNAs as intermediates and products in the splicing of messenger RNA precursors. *Science* 225: 898–903. [P]

Padgett, R. A., S. M. Mount, J. A. Steitz and P. A. Sharp. 1983. Splicing of messenger RNA precursors is inhibited by antisera to small nuclear ribonucleoprotein. *Cell* 35: 101–107. [P]

Ruskin, B., A. R. Krainer, T. Maniatis and M. R. Green. 1984. Excision of an intact intron as a novel lariat structure during pre-mRNA splicing *in vitro*. *Cell* 38: 317–331. [P]

Sachs, A. B. 1993. Messenger RNA degradation in eukaryotes. *Cell* 74: 413–421. [R]

Scott, J. 1995. A place in the world for RNA editing. *Cell* 81: 833–836. [R]

Sharp, P. A. 1994. Split genes and RNA splicing. *Cell* 77: 805–815. [R]

Simpson, L. and O. H. Thiemann. 1995. Sense from nonsense: RNA editing in mitochondria of kinetoplastid protozoa and slime molds. *Cell* 81: 837–840. [R]

Sontheimer, E. J. and J. A. Steitz. 1993. The U5 and U6 small nuclear RNAs as active site components of the spliceosome. *Science* 262: 1989–1996. [P]

Wickens, M. 1990. In the beginning is the end: regulation of poly(A) addition and removal during early development. *Trends Biochem. Sci.* 15: 320–324. [R]

# 7

# Protein Synthesis, Processing, and Regulation

TRANSCRIPTION AND RNA PROCESSING ARE FOLLOWED BY TRANSLATION, the synthesis of proteins as directed by mRNA templates. Proteins are the active players in most cell processes, implementing the myriad tasks that are directed by the information encoded in genomic DNA. Protein synthesis is thus the final stage of gene expression. However, the translation of mRNA is only the first step in the formation of a functional protein. The polypeptide chain must then fold into the appropriate three-dimensional conformation and, frequently, undergo various processing steps before being converted to its active form. These processing steps, particularly in eukaryotes, are intimately related to the sorting and transport of different proteins to their appropriate destinations within the cell.

Although the expression of most genes is regulated primarily at the level of transcription (see Chapter 6), gene expression can also be controlled at the level of translation, and this control is an important element of gene regulation in both prokaryotic and eukaryotic cells. Of even broader significance, however, are the mechanisms that control the activities of proteins within cells. Once synthesized, most proteins can be regulated in response to extracellular signals by either covalent modifications or by association with other molecules. In addition, the levels of proteins within cells can be controlled by differential rates of protein degradation. These multiple controls of both the amounts and activities of intracellular proteins ultimately regulate all aspects of cell behavior.

## TRANSLATION OF mRNA

Proteins are synthesized from mRNA templates by a process that has been highly conserved throughout evolution (reviewed in Chapter 3). All mRNAs are read in the 5′ to 3′ direction, and polypeptide chains are synthesized from the amino to the carboxy terminus. Each amino acid is specified by three bases (a codon) in the mRNA, according to a nearly universal genetic code. The basic mechanics of protein synthesis are also the same in all cells: translation is carried out on ribosomes, with tRNAs serving as adaptors between the mRNA template and the amino acids being incorpo-

273

rated into protein. Protein synthesis thus involves interactions between three types of RNA molecules (mRNA templates, tRNAs, and rRNAs), as well as various proteins that are required for translation.

### Transfer RNAs

During translation, each of the 20 amino acids must be aligned with their corresponding codons on the mRNA template. All cells contain a variety of **tRNAs** that serve as adaptors for this process. As might be expected, given their common function in protein synthesis, different tRNAs share similar overall structures. However, they also possess unique identifying sequences that allow the correct amino acid to be attached and aligned with the appropriate codon in mRNA.

Transfer RNAs are approximately 70 to 80 nucleotides long and have characteristic cloverleaf structures that result from complementary base pairing between different regions of the molecule (Figure 7.1). X-ray crystallography studies have further shown that all tRNAs fold into similar compact L shapes, which are likely required for the tRNAs to fit onto ribosomes during the translation process. The adaptor function of the tRNAs involves two separated regions of the molecule. All tRNAs have the sequence CCA at their 3′ terminus, and amino acids are covalently attached to the ribose of the terminal adenosine. The mRNA template is then recognized by the **anticodon** loop, located at the other end of the folded tRNA, which binds to the appropriate codon by complementary base pairing.

The incorporation of the correctly encoded amino acids into proteins depends on the attachment of each amino acid to an appropriate tRNA, as well as on the specificity of codon-anticodon base pairing. The attachment of amino acids to specific tRNAs is mediated by a group of enzymes called

**Figure 7.1**
**Structure of tRNAs**   The structure of yeast phenylalanyl tRNA is illustrated in open "cloverleaf" form (A) to show complementary base pairing. Modified bases are indicated as mG, methylguanosine; mC, methylcytosine; DHU, dihydrouridine; T, ribothymidine; Y, a modified purine (usually adenosine); and ψ, pseudouridine. The folded form of the molecule is shown in (B) and a space-filling model in (C). (C courtesy of Dan Richardson.)

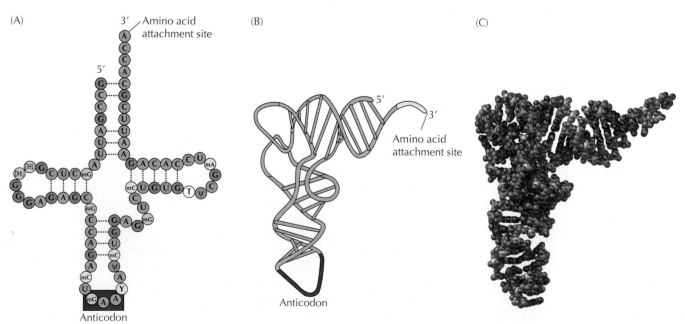

*Figure 7.2*
**Attachment of amino acids to tRNAs**   In the first reaction step, the amino acid is joined to AMP, forming an aminoacyl AMP intermediate. In the second step, the amino acid is transferred to the 3′ CCA terminus of the acceptor tRNA and AMP is released. Both steps of the reaction are catalyzed by aminoacyl tRNA synthetases.

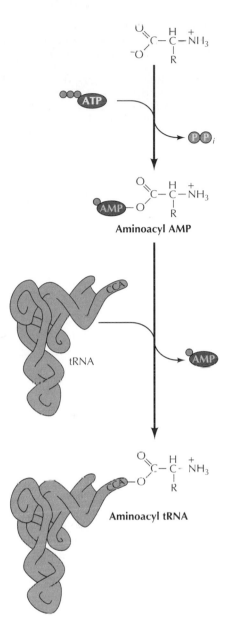

Aminoacyl AMP

tRNA

Aminoacyl tRNA

**aminoacyl tRNA synthetases**, which were discovered by Paul Zamecnik and Mahlon Hoagland in 1957. Each of these enzymes recognizes a single amino acid, as well as the correct tRNA (or tRNAs) to which that amino acid should be attached. The reaction proceeds in two steps (Figure 7.2). First, the amino acid is activated by reaction with ATP to form an aminoacyl AMP synthetase intermediate. The activated amino acid is then joined to the 3′ terminus of the tRNA. The aminoacyl tRNA synthetases must be highly selective enzymes that recognize both individual amino acids and specific base sequences that identify the correct acceptor tRNAs. In some cases, the high fidelity of amino acid recognition results in part from a proofreading function by which incorrect aminoacyl AMPs are hydrolyzed rather than being joined to tRNA during the second step of the reaction. Recognition of the correct tRNA by the aminoacyl tRNA synthetase is also highly selective; the synthetase recognizes specific nucleotide sequences (in most cases including the anticodon) that uniquely identify each species of tRNA.

After being attached to tRNA, an amino acid is aligned on the mRNA template by complementary base pairing between the mRNA codon and the anticodon of the tRNA. Codon-anticodon base pairing is somewhat less stringent than the standard A-U and G-C base pairing discussed in preceding chapters. The significance of this unusual base pairing in codon-anticodon recognition relates to the redundancy of the genetic code. Of the 64 possible codons, three are stop codons that signal the termination of translation; the other 61 encode amino acids (see Table 3.1). Thus, most of the amino acids are specified by more than one codon. In part, this redundancy results from the attachment of many amino acids to more than one species of tRNA. *E. coli*, for example, contain about 40 different tRNAs that serve as acceptors for the 20 different amino acids. In addition, some tRNAs are able to recognize more than one codon in mRNA, as a result of nonstandard base pairing (called wobble) between the tRNA anticodon and the third position of some complementary codons (Figure 7.3). Relaxed base pairing at this position results partly from the formation of G-U base pairs and partly from the modification of guanosine to inosine in the anticodons of several tRNAs during processing (see Figure 6.38). Inosine can base-pair with either C, U, or A in the third position, so its inclusion in the anticodon allows a single tRNA to recognize three different codons in mRNA templates.

## The Ribosome

**Ribosomes** are the sites of protein synthesis in both prokaryotic and eukaryotic cells. First characterized as particles detected by ultracentrifugation of cell lysates, ribosomes are usually designated according to their rates of sedimentation: 70S for bacterial ribosomes and 80S for the somewhat larger ribosomes of eukaryotic cells. Both prokaryotic and eukaryotic ribosomes are composed of two distinct subunits, each containing characteristic proteins and **rRNAs**. The fact that cells typically contain many ribosomes reflects the central importance of protein synthesis in cell metabolism. *E. coli*, for example, contain about 20,000 ribosomes, which account for approximately 25% of the dry weight of the cell, and rapidly growing mammalian cells contain about 10 million ribosomes.

*Figure 7.3*
**Nonstandard codon-anticodon base pairing** Base pairing at the third codon position is relaxed, allowing G to pair with U, and inosine (I) in the anticodon to pair with U, C, or A. Two examples of abnormal base pairing, allowing phenylalanyl (Phe) tRNA to recognize either UUC or UUU codons and alanyl (Ala) tRNA to recognize GCU, GCC, or GCA, are illustrated.

**Phenylalanyl tRNA pairing**

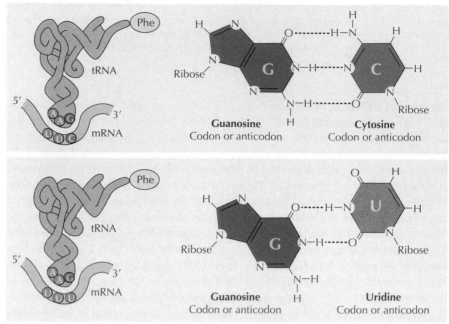

**Guanosine**
Codon or anticodon

**Cytosine**
Codon or anticodon

**Guanosine**
Codon or anticodon

**Uridine**
Codon or anticodon

**Alanyl tRNA pairing**

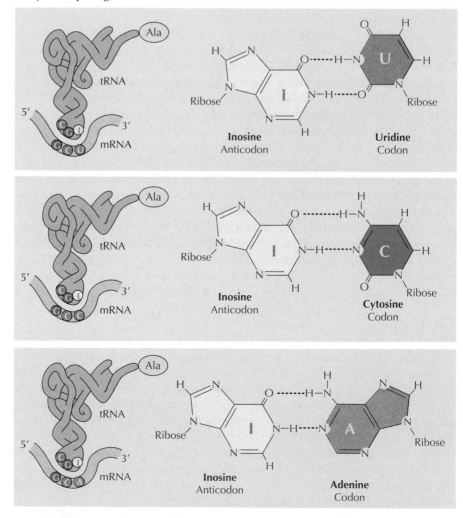

**Inosine**
Anticodon

**Uridine**
Codon

**Inosine**
Anticodon

**Cytosine**
Codon

**Inosine**
Anticodon

**Adenine**
Codon

The general structures of prokaryotic and eukaryotic ribosomes are similar, although they differ in some details (Figure 7.4). The small subunit (designated 30S) of *E. coli* ribosomes consists of the 16S rRNA and 21 proteins; the large subunit (50S) is composed of the 23S and 5S rRNAs and 34 proteins. Each ribosome contains one copy of the rRNAs and one copy of each of the ribosomal proteins, with one exception: one protein of the 50S subunit is present in four copies. The subunits of eukaryotic ribosomes are larger and contain more proteins than their prokaryotic counterparts have. The small subunit (40S) of eukaryotic ribosomes is composed of the 18S rRNA and approximately 30 proteins; the large subunit (60S) contains the 28S, 5.8S, and 5S rRNAs and about 45 proteins.

A noteworthy feature of ribosomes is that they can be formed *in vitro* by self-assembly of their RNA and protein constituents. As first described in 1968 by Masayasu Nomura, purified ribosomal proteins and rRNAs can be mixed together and, under appropriate conditions, will re-form a functional ribosome. Although ribosome assembly *in vivo* (particularly in

(A)

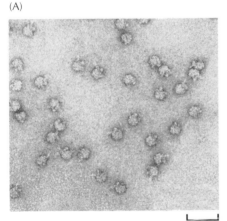

50 nm

(B)
70S ribosome

50S subunit

30S subunit

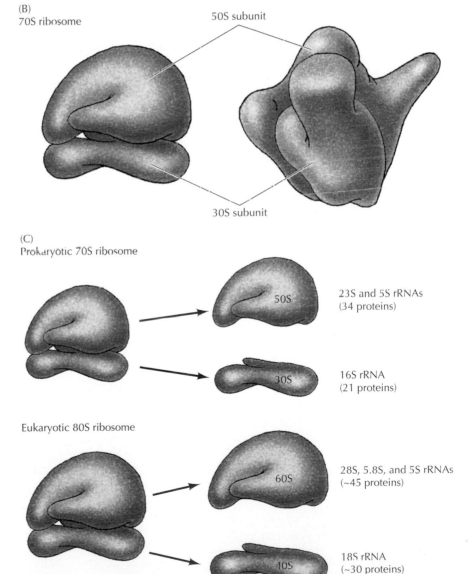

(C)
Prokaryotic 70S ribosome

50S — 23S and 5S rRNAs (34 proteins)

30S — 16S rRNA (21 proteins)

Eukaryotic 80S ribosome

60S — 28S, 5.8S, and 5S rRNAs (~45 proteins)

40S — 18S rRNA (~30 proteins)

**Figure 7.4**
**Ribosome structure** (A) Electron micrograph of *E. coli* 50S ribosomal subunits. (B) Model of ribosome structure. (C) Components of prokaryotic and eukaryotic ribosomes. Intact prokaryotic and eukaryotic ribosomes are designated 70S and 80S, respectively, on the basis of their sedimentation rates in ultracentrifugation. They consist of large and small subunits, which contain both ribosomal proteins and rRNAs. (A from M. Boublik et al., 1990, *The Ribosome*, p. 115. Courtesy of American Society for Microbiology.)

eukaryotic cells) is considerably more complicated, the ability of ribosomes to self-assemble *in vitro* has provided an important experimental tool, allowing analysis of the roles of individual proteins and rRNAs.

Like tRNAs, rRNAs form characteristic secondary structures by complementary base pairing (Figure 7.5). In association with ribosomal proteins the rRNAs fold further, into distinct three-dimensional structures. Initially, rRNAs were thought to play a structural role, providing a scaffold upon which ribosomal proteins assemble. However, with the discovery of the catalytic activity of other RNA molecules (e.g., RNase P and the self-splicing introns discussed in Chapter 6), the possible catalytic role of rRNA became widely considered. Consistent with this hypothesis, rRNAs were found to be absolutely required for the *in vitro* assembly of functional ribosomes. On the other hand, the omission of many ribosomal proteins resulted in a decrease, but not a complete loss, of ribosome activity.

Direct evidence for the catalytic activity of rRNA came from experiments

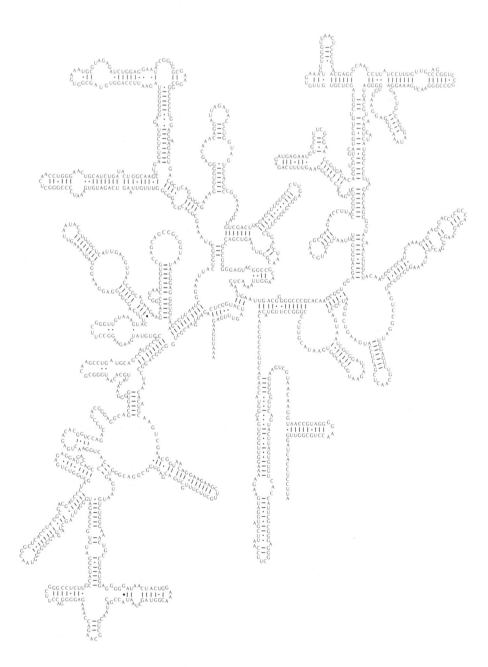

*Figure 7.5*
**The structure of 16S rRNA**
Complementary base pairing results in the formation of a distinct secondary structure. (From S. Stern, T. Powers, L.-I. Changchien, and H. F. Noller, 1989, *Science* 244: 783.)

of Harry Noller and his colleagues in 1992. These investigators demonstrated that the large ribosomal subunit is able to catalyze the formation of peptide bonds (the peptidyl transferase reaction) even after approximately 95% of the ribosomal proteins have been removed by standard protein extraction procedures. In contrast, treatment with RNase completely abolishes peptide bond formation, providing strong support for the hypothesis that the formation of a peptide bond is an RNA-catalyzed reaction. Further studies have shown that the 23S rRNA interacts with the 3′ CCA terminus of tRNA and is directly involved in the peptidyl transferase reaction. Rather than being the primary catalytic constituents of ribosomes, ribosomal proteins are now thought to facilitate proper folding of the rRNA and to enhance ribosome function by properly positioning the tRNAs.

The direct involvement of rRNA in the peptidyl transferase reaction has important evolutionary implications. RNAs are thought to have been the first self-replicating macromolecules (see Chapter 1). This notion is strongly supported by the fact that ribozymes, such as RNase P and self-splicing introns, can catalyze reactions that involve RNA substrates. The role of rRNA in the formation of peptide bonds extends the catalytic activities of RNA beyond self-replication to direct involvement in protein synthesis. Additional studies indicate that the *Tetrahymena* rRNA ribozyme can catalyze the attachment of amino acids to RNA, lending credence to the possibility that the original aminoacyl tRNA synthetases were RNAs rather than proteins. The ability of RNA molecules to catalyze the reactions required for protein synthesis as well as for self-replication may provide an important link for understanding the early evolution of cells.

## *The Organization of mRNAs and the Initiation of Translation*

Although the mechanisms of protein synthesis in prokaryotic and eukaryotic cells are similar, there are also differences, particularly in the signals that determine the positions at which synthesis of a polypeptide chain is initiated on an mRNA template (Figure 7.6). Translation does not simply begin at the 5′ end of the mRNA; it starts at specific initiation sites. The 5′ terminal portions of both prokaryotic and eukaryotic mRNAs are therefore noncoding sequences, referred to as **5′ untranslated regions**. Eukaryotic mRNAs usually encode only a single polypeptide chain, but many prokaryotic mRNAs encode multiple polypeptides that are synthesized

**Prokaryotic mRNA**

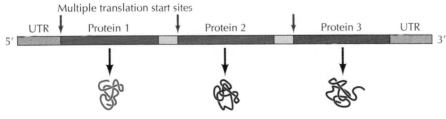

**Eukaryotic mRNA**

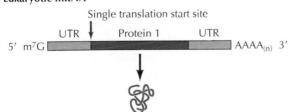

*Figure 7.6*
**Prokaryotic and eukaryotic mRNAs**  Both prokaryotic and eukaryotic mRNAs contain untranslated regions (UTRs) at their 5′ and 3′ ends. Eukaryotic mRNAs also contain 5′ 7-methylguanosine (m⁷G) caps and 3′ poly-A tails. Prokaryotic mRNAs are frequently polycistronic: They encode multiple proteins, each of which is translated from an independent start site. Eukaryotic mRNAs are usually monocistronic, encoding only a single protein.

# KEY EXPERIMENT

## *Catalytic Role of Ribosomal RNA*

### Unusual Resistance of Peptidyl Transferase to Protein Extraction Procedures

Harry F. Noller, Vernita Hoffarth, and Ludwika Zimniak
University of California at Santa Cruz
*Science, Volume 256, 1992, pages 1416–1419*

### The Context

The role of ribosomes in protein synthesis was elucidated in the 1960s. During this period, ribosomes were characterized as particles consisting of both proteins and RNAs, and the reconstitution of functional ribosomes from purified components was accomplished. At that time peptide bond formation (the peptidyl transferase reaction) was generally assumed to be catalyzed by ribosomal proteins, with the rRNAs delegated to a supporting role in ribosome structure. As early as the 1970s, however, evidence began to suggest that the rRNAs might play a more direct role in protein synthesis. For example, many ribosomal proteins were found to be nonessential for ribosome function. Conversely, the sequences of some parts of rRNA were shown to be extremely well conserved in evolution, suggesting a critical functional role for these portions of the rRNA molecule.

In the early 1980s, the catalytic activity of RNA molecules was established by Tom Cech's studies of the *Tetrahymena* ribozyme and Sidney Altman's studies of RNase P. These discoveries of RNA catalysis provided a precedent for the hypothesis that rRNA is directly involved in catalyzing peptide bond formation. Compelling evidence in support of such a role for rRNA was provided in this 1992 paper by Harry Noller and his colleagues.

### The Experiments

To study the catalytic activity of rRNA, Noller and colleagues used a simplified model reaction to assay peptidyl transferase activity. This reaction measures the transfer of radioactively labeled *N*-formylmethionine from a fragment of tRNA to the amino group of puromycin, an antibiotic that resembles an aminoacyl tRNA and can form peptide bonds with a growing peptide chain. The advantage of this peptidyl transferase model reaction is that it can be carried out by isolated 50S ribosomal subunits; small ribosomal subunits, other protein factors, and mRNA are not required.

The investigators then tested the role of rRNA by assaying the peptidyl transferase activity of 50S subunits from which the ribosomal proteins had been removed by standard protein extraction procedures. An important aspect of the experiments was the use of ribosomes from the bacterium *Thermus aquaticus*. Because these bacteria live at high temperatures, the structure of their rRNA was thought to be more stable than that of *E. coli* rRNA. The critical result was that the peptidyl transferase activity of *T. aquaticus* ribosomes was completely resistant to a vigorous extraction procedure involving treatment with detergents, proteases, and phenol (see figure). Most strikingly, full peptidyl transferase activity was retained even after repeated extractions that removed 95% of the ribosomal protein. In contrast, peptidyl transferase of either intact or extracted ribosomes was extremely sensitive to even a brief treatment with RNase. Although these experiments could not exclude a possible role for the

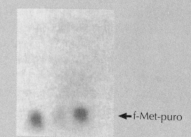

← f-Met-puro

| Protein extraction | − | − | + | + |
| RNase | − | + | − | + |

The peptidyl transferase reaction is assayed by the formation of radioactive *N*-formylmethionine-puromycin (f-Met-puro), which is detected by electrophoresis and autoradiography. Ribosomes from *T. aquaticus* were subjected to protein extraction or treatment with RNase, as indicated.

independently from distinct initiation sites. For example, the *E. coli lac* operon consists of three genes that are translated from the same mRNA (see Figure 6.8). Messenger RNAs that encode multiple polypeptides are called **polycistronic**, whereas **monocistronic** mRNAs encode a single polypeptide chain. Finally, both prokaryotic and eukaryotic mRNAs end in noncoding **3′ untranslated regions**.

## Catalytic Role of Ribosomal RNA *(continued)*

remaining ribosomal proteins, they provided strong support for the direct participation of 23S rRNA in the peptidyl transferase reaction.

### The Impact

The results of Noller's experiments, which support a catalytic role for rRNA in peptide bond formation, not only had a striking impact on our understanding of ribosome function but also profoundly extended the previously described catalytic activities of RNA molecules. Moreover, these findings were complemented by a simultaneous paper from Tom Cech's laboratory (Piccirilli et al., *Science*

256:1420–1424, 1992) that corroborated the ability of RNA to catalyze reactions involving amino acids and showed that RNA could act as an aminoacyl tRNA synthctase. These complementary findings provided substantial new support for the hypothesis that an early period of evolution was an RNA world populated by self-replicating RNA molecules. This hypothesis had previously been based on the ability of RNA molecules to catalyze the reactions required for their own replication. The discovery that RNA could catalyze reactions involved in protein synthesis provided a clear link between the RNA world

Harry F. Noller

and the flow of genetic information in present-day cells, with rRNA continuing to function in the key reaction of peptide bond formation.

---

In both prokaryotic and eukaryotic cells, translation always initiates with the amino acid methionine, usually encoded by AUG. Alternative initiation codons, such as GUG, are used occasionally in bacteria, but when they occur at the beginning of a polypeptide chain, these codons direct the incorporation of methionine rather than of the amino acid they normally encode (GUG normally encodes valine). In most bacteria, protein synthesis is initiated with a modified methioninc residue (*N*-formylmethionine), whereas unmodified methionines initiate protein synthesis in eukaryotes (except in mitochondria and chloroplasts, whose ribosomes resemble those of bacteria).

The signals that identify initiation codons are different in prokaryotic and eukaryotic cells, consistent with the distinct functions of polycistronic and monocistronic mRNAs (Figure 7.7). Initiation codons in bacterial mRNAs are preceded by a specific sequence (called a **Shine-Delgarno sequence**, after its discoverers) that aligns the mRNA on the ribosome for

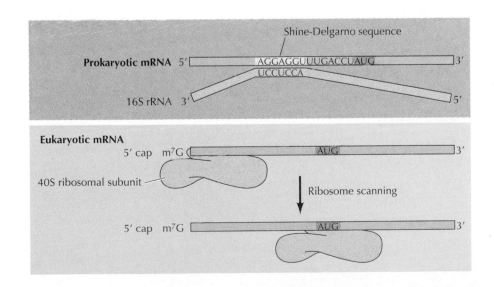

*Figure 7.7*
**Signals for translation initiation** Initiation sites in prokaryotic mRNAs are characterized by a Shine-Delgarno sequence that precedes the AUG initiation codon. Base pairing between the Shine-Delgarno sequence and a complementary sequence near the 3' terminus of 16S rRNA aligns the mRNA on the ribosome. In contrast, eukaryotic mRNAs are bound to the 40S ribosomal subunit by their 5' 7-methylguanosine caps. The ribosome then scans along the mRNA until it encounters an AUG initiation codon.

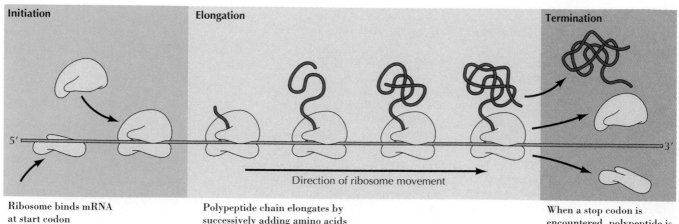

| Initiation | Elongation | Termination |
|---|---|---|
| 5′ | Direction of ribosome movement | 3′ |

Ribosome binds mRNA at start codon

Polypeptide chain elongates by successively adding amino acids

When a stop codon is encountered, polypeptide is released and ribosome dissociates

*Figure 7.8*
**Overview of translation**

translation by base-pairing with a complementary sequence near the 3′ terminus of 16S rRNA. This base-pairing interaction enables bacterial ribosomes to initiate translation not only at the 5′ end of an mRNA but also at the internal initiation sites of polycistronic messages. In contrast, ribosomes recognize eukaryotic mRNAs by binding to the 7-methylguanosine cap at their 5′ terminus (see Figure 6.39). The ribosomes then scan downstream of the 5′ cap until they encounter an AUG initiation codon. Sequences that surround AUGs affect the efficiency of initiation, so in many cases the first AUG in the mRNA is bypassed and translation initiates at an AUG farther downstream. However, eukaryotic mRNAs have no sequence equivalent to the Shine-Delgarno sequence of prokaryotic mRNAs. Translation of eukaryotic mRNAs is always initiated at a site determined by scanning from the 5′ terminus, consistent with their functions as monocistronic messages that encode only single polypeptides.

### The Process of Translation

Translation is generally divided into three stages: initiation, elongation, and termination (Figure 7.8). In both prokaryotes and eukaryotes the first step of the initiation stage is the binding of a specific initiator methionyl tRNA and the mRNA to the small ribosomal subunit. The large ribosomal subunit then joins the complex, forming a functional ribosome on which elongation of the polypeptide chain proceeds. A number of specific nonribosomal proteins are also required for the various stages of the translation process (Table 7.1).

*Table 7.1* Translation Factors

| | Translation factors | |
|---|---|---|
| **Role** | **Prokaryotes** | **Eukaryotes** |
| Initiation | IF-1, IF-2, IF-3 | eIF-1, eIF-2, eIF-2B, eIF-3, eIF-4A, eIF-4B, eIF-4C, eIF-4F, eIF-5, eIF-6 |
| Elongation | EF-Tu, EF-Ts, EF-G | eEF-1α, eEF-1βγ, eEF-2 |
| Termination | RF-1, RF-2 | eRF |

# MOLECULAR MEDICINE

## Antibiotics and Protein Synthesis

### The Disease

Bacteria are responsible for a wide variety of potentially lethal infectious diseases, including tuberculosis, bacterial pneumonia, childhood meningitis, infections of wounds and burns, syphilis, and gonorrhea. Prior to the 1940s, physicians had no effective treatments for these bacterial infections. At that time, the first antibiotics became available for clinical use. Previously untreatable infections became curable, and a significant increase in average life span has been attributed to the introduction of antibiotics into clinical practice. More than a hundred different antibiotics are now in use, providing effective treatment for bacterial infections.

### Molecular and Cellular Basis

To be effective clinically, an antibiotic must kill or inhibit the growth of bac-

teria without being toxic to humans. Thus, most clinically useful antibiotics are directed against targets that are present in bacteria but not in human cells. Penicillin, for example, inhibits synthesis of the bacterial cell wall (see Chapter 12). Many other commonly used antibiotics, however, inhibit different steps in protein synthesis (see table). Some of these antibiotics—including streptomycin, tetracycline, chloramphenicol, and erythromycin—are specific for prokaryotic ribosomes and therefore are effective agents for the treatment of bacterial infections. Other antibiotic inhibitors of protein synthesis act against both prokaryotes and eukaryotes (e.g., puromycin) or against eukaryotes only (e.g., cycloheximide). Although these antibiotics are obviously not useful clinically, they have provided important experimental tools for studies of protein

synthesis in both prokaryotic and eukaryotic cells.

### Prevention and Treatment

The use of antibiotics has had a major impact on modern medicine by enabling physicians to cure otherwise life-threatening bacterial infections. Unfortunately, however, mutations can lead to the development of antibiotic-resistant strains of bacteria. Many strains of bacteria are now resistant to one or more antibiotics, so physicians sometimes have to try several different antibiotics before they find one that is effective. Moreover, the spread of resistant bacterial strains has made some antibiotics obsolete. Most seriously, strains of bacteria that are resistant to multiple antibiotics have emerged, and a few strains of bacteria, frequently found in hospitals, are now resistant to all but one or a few known antibiotics. The emergence of these strains with multiple resistance raises the specter of untreatable infections caused by the spread of antibiotic-resistant bacteria—a scenario that would herald a return to the pre-antibiotic era of infectious diseases. Controlling antibiotic-resistant bacteria, both by optimizing the use of currently available antibiotics and by developing new antibiotics, thus represents an important concern in current medical practice.

### Antibiotic Inhibitors of Protein Synthesis

| Antibiotic | Target cells | Effect |
| --- | --- | --- |
| Streptomycin | Prokaryotic | Inhibits initiation and causes misreading |
| Tetracycline | Prokaryotic | Inhibits binding of aminoacyl tRNAs |
| Chloramphenicol | Prokaryotic | Inhibits peptidyl transferase activity |
| Erythromycin | Prokaryotic | Inhibits translocation |
| Puromycin | Prokaryotic and eukaryotic | Causes premature chain termination |
| Cycloheximide | Eukaryotic | Inhibits peptidyl transferase activity |

The first translation step in bacteria is the binding of three **initiation factors** (IF-1, IF-2, and IF-3) to the 30S ribosomal subunit (Figure 7.9). The mRNA and initiator *N*-formylmethionyl tRNA then join the complex, with IF-2 (which is bound to GTP) specifically recognizing the initiator tRNA. IF-3 is then released, allowing a 50S ribosomal subunit to associate with the complex. This association triggers the hydrolysis of GTP bound to IF-2, which leads to the release of IF-1 and IF-2 (bound to GDP). The result is the formation of a 70S initiation complex (with mRNA and initiator tRNA bound to the ribosome) that is ready to begin peptide bond formation during the elongation stage of translation.

*Figure 7.9*
**Initiation of translation in bacteria**   Three initiation factors (IF-1, IF-2, and IF-3) first bind to the 30S ribosomal subunit. This step is followed by binding of the mRNA and the initiator *N*-formylmethionyl (fMet) tRNA, which is recognized by IF-2 bound to GTP. IF-3 is then released, and a 50S subunit binds to the complex, triggering the hydrolysis of bound GTP, followed by the release of IF-1 and IF-2 bound to GDP.

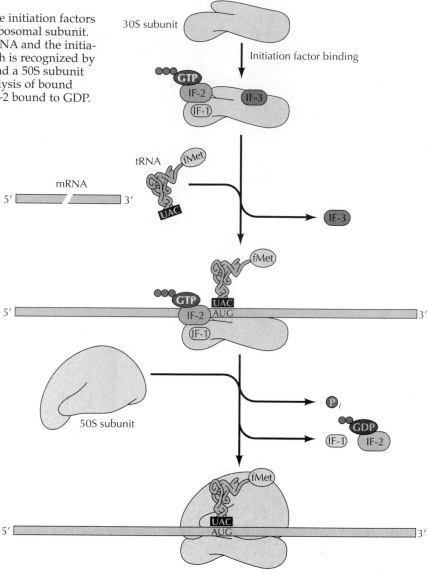

Initiation in eukaryotes is more complicated and requires at least ten proteins (each consisting of multiple polypeptide chains), which are designated eIFs (*e*ukaryotic *i*nitiation *f*actors; see Table 7.1). The factors eIF-1A and eIF-3 bind to the 40S ribosomal subunit, and eIF-2 (in a complex with GTP) associates with the initiator methionyl tRNA, bringing it to the 40S subunit (Figure 7.10). The 5′ cap of the mRNA is recognized by eIF-4E (the cap-binding subunit of eIF-4F), which, in association with eIF-4A and eIF-4B, brings the mRNA to the ribosome. The 40S ribosomal subunit, in association with the bound methionyl tRNA and eIFs, then scans the mRNA to identify the AUG initiation codon. When the AUG codon is reached, eIF-5 triggers the hydrolysis of GTP bound to eIF-2. Initiation factors (including eIF-2 bound to GDP) are then released, and a 60S subunit binds to the 40S subunit to form the 80S initiation complex of eukaryotic cells.

After the initiation complex has formed, translation proceeds by elongation of the polypeptide chain. The mechanism of elongation in prokary-

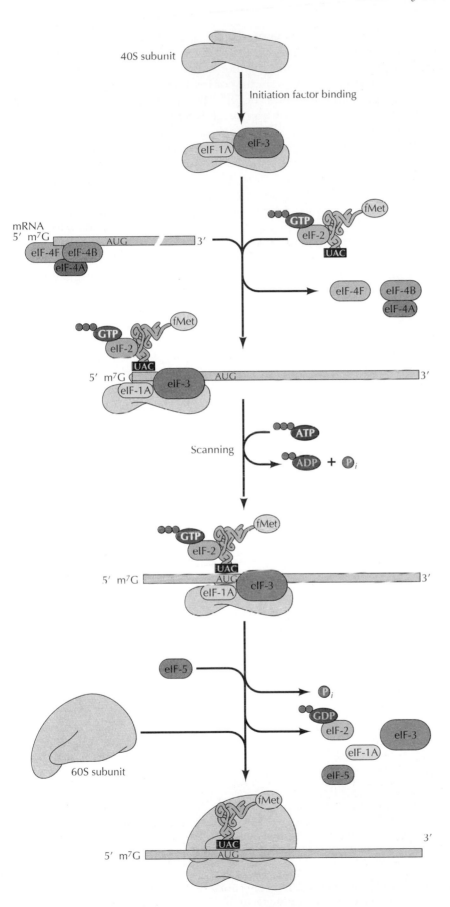

*Figure 7.10*
**Initiation of translation in eukaryotic cells**   Initiation factors eIF-3 and eIF-1A bind to the 40S ribosomal subunit. The initiator methionyl tRNA is brought to the ribosome by eIF-2 (complexed to GTP), and the mRNA by eIF-4F (which binds to the 5′ cap), eIF-4A, and eIF-4B. The ribosome then scans down the mRNA to identify the first AUG initiation codon. Scanning requires energy and is accompanied by ATP hydrolysis. When the initiating AUG is identified, eIF-5 triggers the hydrolysis of GTP bound to eIF-2, followed by the release of eIF-2 (complexed to GDP) and other initiation factors. The 60S ribosomal subunit then joins the 40S complex.

40S subunit

Initiation factor binding

eIF 1A  eIF-3

mRNA
5′ m7G    AUG    3′
eIF-4F  eIF-4B
eIF-4A

GTP  fMet
eIF-2
UAC

eIF-4F   eIF-4B
eIF-4A

GTP  fMet
eIF-2
UAC
5′ m7G    eIF-3    AUG    3′
eIF-1A

Scanning

ATP
ADP  +  P_i

GTP  fMet
eIF-2
UAC
5′ m7G    AUG    eIF-3    3′
eIF-1A

eIF-5

P_i

60S subunit

GDP
eIF-2    eIF-3

eIF-1A

eIF-5

fMet
UAC
5′ m7G    AUG    3′

otic and eukaryotic cells is very similar (Figure 7.11). The ribosome has three sites for tRNA binding, designated the P (peptidyl), A (aminoacyl), and E (exit) sites. The initiator methionyl tRNA is bound at the P site. The first step in elongation is the binding of the next aminoacyl tRNA to the A site by pairing with the second codon of the mRNA. The aminoacyl tRNA is escorted to the ribosome by an **elongation factor** (EF-Tu in prokaryotes, eEF-1α in eukaryotes), which is complexed to GTP. The GTP is hydrolyzed to GDP as the correct aminoacyl tRNA is inserted into the A site of the ribosome and the elongation factor bound to GDP is released. The requirement for hydrolysis of GTP before EF-Tu or eEF-1α is released from the ribosome is the rate-limiting step in elongation and provides a time interval during which an incorrect aminoacyl tRNA, which would bind less strongly to the mRNA codon, can dissociate from the ribosome rather than being used for protein synthesis. Thus, the expenditure of a high-energy GTP at this step is an important contribution to accurate protein synthesis; it allows time for proofreading of the codon-anticodon pairing before the peptide bond forms.

Once EF-Tu (or eEF-1α) has left the ribosome, a peptide bond can be formed between the initiator methionyl tRNA at the P site and the second aminoacyl tRNA at the A site. This reaction is catalyzed by the large ribosomal subunit, with the rRNA playing a critical role (as already discussed). The result is the transfer of methionine to the aminoacyl tRNA at the A site of the ribosome, forming a peptidyl tRNA at this position and leaving the uncharged initiator tRNA at the P site. The next step in elongation is translocation, which requires another elongation factor (EF-G in prokaryotes, eEF-2 in eukaryotes) and is again coupled to GTP hydrolysis. During translocation, the ribosome moves three nucleotides along the mRNA, positioning the next codon in an empty A site. This step translocates the peptidyl tRNA from the A site to the P site, and the uncharged tRNA from the P site to the E site. The ribosome is then left with a peptidyl tRNA bound at the P site, and an empty A site. The binding of a new aminoacyl tRNA to the A site then induces the release of the uncharged tRNA from the E site, leaving the ribosome ready for insertion of the next amino acid in the growing polypeptide chain.

As elongation continues, the EF-Tu (or eEF-1α) that is released from the ribosome bound to GDP must be reconverted to its GTP form (Figure 7.12). This conversion requires a third elongation factor, EF-Ts (eEF-1βγ in eukaryotes), which binds to the EF-Tu/GDP complex and promotes the exchange of bound GDP for GTP. This exchange results in the regeneration of EF-Tu/GTP, which is now ready to escort a new aminoacyl tRNA to the A site of the ribosome, beginning a new cycle of elongation. The regulation

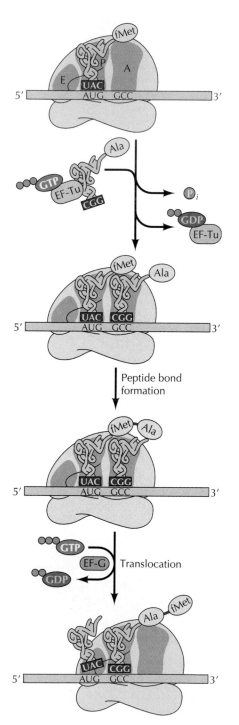

**Figure 7.11**
**Elongation stage of translation**    The ribosome has three tRNA-binding sites, designated P (peptidyl), A (aminoacyl), and E (exit). The initiating *N*-formylmethionyl tRNA is positioned in the P site, leaving an empty A site. The second aminoacyl tRNA (e.g., alanyl tRNA) is then brought to the A site by EF-Tu (complexed with GTP). Following GTP hydrolysis, EF-Tu (complexed with GDP) leaves the ribosome, with alanyl tRNA inserted into the A site. A peptide bond is then formed, resulting in the transfer of methionine to the aminoacyl tRNA at the A site. The ribosome then moves three nucleotides along the mRNA. This movement translocates the peptidyl (Met-Ala) tRNA to the P site and the uncharged tRNA to the E site, leaving an empty A site ready for addition of the next amino acid. Translocation is mediated by EF-G, coupled to GTP hydrolysis. The process, illustrated here for prokaryotic cells, is very similar in eukaryotes. (Table 7.1 gives the names of the eukaryotic elongation factors.)

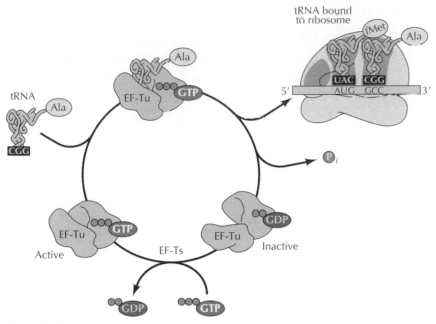

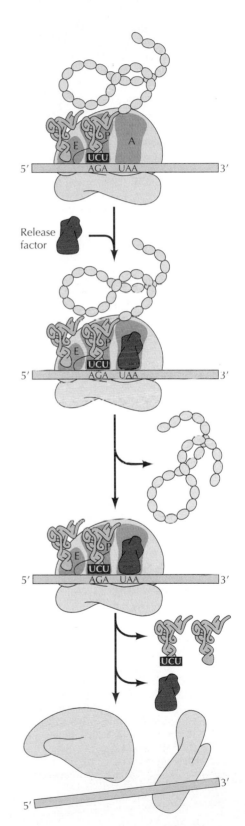

*Figure 7.12*
**Regeneration of EF-Tu/GTP**   EF-Tu complexed to GTP escorts the aminoacyl tRNA to the ribosome. The bound GTP is hydrolyzed as the correct tRNA is inserted, so EF-Tu complexed to GDP is released. The EF-Tu/GDP complex is inactive and unable to bind another tRNA. In order for translation to continue, the active EF-Tu/GTP complex must be regenerated by another factor, EF-Ts, which stimulates the exchange of the bound GDP for free GTP.

of EF-Tu by GTP binding and hydrolysis illustrates a common means of the regulation of protein activities. As will be discussed in later chapters, similar mechanisms control the activities of a wide variety of proteins involved in the regulation of cell growth and differentiation, as well as in protein transport and secretion.

Elongation of the polypeptide chain continues until a stop codon (UAA, UAG, or UGA) is translocated into the A site of the ribosome. Cells do not contain tRNAs with anticodons complementary to these termination signals; instead, they have **release factors** that recognize the signals and terminate protein synthesis (Figure 7.13). Prokaryotic cells contain two release factors: RF-1 recognizes UAA or UAG, and RF-2 recognizes UAA or UGA (see Table 7.1). Eukaryotic cells contain only one release factor (eRF),which recognizes all three termination codons. The release factors bind to a termination codon at the A site and stimulate hydrolysis of the bond between the tRNA and the polypeptide chain at the P site, resulting in release of the completed polypeptide from the ribosome. The tRNA is then released, and the ribosomal subunits and the mRNA template dissociate.

Messenger RNAs can be translated simultaneously by several ribosomes in both prokaryotic and eukaryotic cells. Once one ribosome has moved

*Figure 7.13*
**Termination of translation**   A termination codon (e.g., UAA) at the A site is recognized by a release factor rather than by a tRNA. The result is the release of the completed polypeptide chain, followed by the dissociation of tRNA and mRNA from the ribosome.

(A)

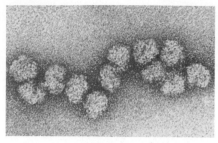

*Figure 7.14*
**Polysomes** Messenger RNAs are translated by a series of multiple ribosomes (a polysome). (A) Electron micrograph of a eukaryotic polysome. (B) Schematic of a generalized polysome. Note that the ribosomes closer to the 3′ end of the mRNA have longer polypeptide chains. (A from M. Boublik et al., 1990, *The Ribosome*, p. 117. Courtesy of American Society for Microbiology.)

(B)

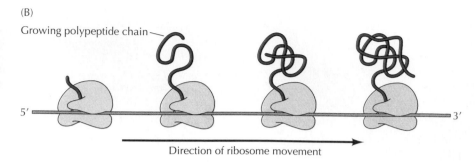

away from the initiation site, another can bind to the mRNA and begin synthesis of a new polypeptide chain. Thus, mRNAs are usually translated by a series of ribosomes, spaced at intervals of about 100 to 200 nucleotides (Figure 7.14). The group of ribosomes bound to an mRNA molecule is called a polyribosome, or **polysome**. Each ribosome within the group functions independently to synthesize a separate polypeptide chain.

## Regulation of Translation

Although transcription is the primary level at which gene expression is controlled, the translation of mRNAs is also regulated in both prokaryotic and eukaryotic cells. One mechanism of translational regulation is the binding of repressor proteins, which block translation, to specific mRNA sequences. The best understood example of this mechanism in eukaryotic cells is regulation of the synthesis of ferritin, a protein that stores iron within the cell. The translation of ferritin mRNA is regulated by the supply of iron: more ferritin is synthesized if iron is abundant (Figure 7.15). This regulation is mediated by a protein which (in the absence of iron) binds to a sequence (the iron response element, or IRE) in the 5′ untranslated region of ferritin mRNA, blocking its translation. In the presence of iron, the repressor no longer binds to the IRE and ferritin translation is able to proceed.

It is interesting to note that the regulation of translation of ferritin mRNA by iron is similar to the regulation of transferrin receptor mRNA stability, which was discussed in the previous chapter (see Figure 6.50). Namely, the stability of transferrin receptor mRNA is regulated by protein binding to an IRE in its 3′ untranslated region. The same protein binds to the IREs of both ferritin and transferrin receptor mRNAs. However, the consequences of protein binding to the two IREs are quite different. Protein bound to the transferrin receptor IRE protects the mRNA

*Figure 7.15*
**Translational regulation of ferritin** The mRNA contains an iron response element (IRE) near its 5′ cap. In the presence of adequate supplies of iron, translation of the mRNA proceeds normally. If iron is scarce, however, a protein (called the iron response element–binding protein, or IRE-BP) binds to the IRE, blocking translation of the mRNA.

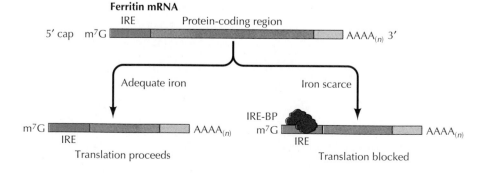

from degradation rather than inhibiting its translation. These distinct effects presumably result from the different locations of the IRE in the two mRNAs. To function as a repressor-binding site, the IRE must be located within 70 nucleotides of the 5' cap of ferritin mRNA, suggesting that protein binding to the IRE blocks translation by interfering with cap recognition and binding of the 40S ribosomal subunit. Rather than inhibiting translation, protein binding to the same sequence in the 3' untranslated region of transferrin receptor mRNA protects the mRNA from nuclease degradation. Binding of the same regulatory protein to different sites on mRNA molecules can thus have distinct effects on gene expression, in one case inhibiting translation and in the other stabilizing the mRNA to increase protein synthesis.

Another mechanism of translational regulation in eukaryotic cells, resulting in global effects on overall translational activity rather than on the translation of specific mRNAs, involves modulation of the activity of initiation factors, particularly eIF-2. As already discussed, eIF-2 (complexed with GTP) binds to the initiator methionyl tRNA, bringing it to the ribosome. The subsequent release of eIF-2 is accompanied by GTP hydrolysis, leaving eIF-2 as an inactive GDP complex. To participate in another cycle of initiation, the eIF-2/GTP complex must be regenerated by the exchange of bound GDP for GTP. This exchange is mediated by another factor, eIF-2B. The control of eIF-2 activity by GTP binding and hydrolysis is thus similar to that of EF-Tu (see Figure 7.12). However, the regulation of eIF-2 provides a critical control point in a variety of eukaryotic cells. In particular, eIF-2 can be phosphorylated by regulatory protein kinases. This phosphorylation blocks the exchange of bound GDP for GTP, thereby inhibiting initiation of translation. One type of cell in which such phosphorylation occurs is the reticulocyte, which is devoted to the synthesis of hemoglobin (Figure 7.16). The translation of globin mRNA is controlled by the availability of heme: the mRNA is translated only if adequate heme is available to form functional hemoglobin molecules. In the absence of heme, a protein kinase that phosphorylates eIF-2 is activated, and further translation is inhibited. Similar mechanisms have been found to control protein synthesis in other cell types, particularly virus-infected cells in which viral protein synthesis is inhibited by interferon.

Other studies have implicated eIF-4E, which binds to the 5' cap of mRNAs, as a translational regulatory protein. For example, the hormone insulin stimulates protein synthesis in adipocytes and muscle cells. This effect of insulin is mediated, at least in part, by phosphorylation of proteins associated with eIF-4E, resulting in stimulation of eIF-4E activity and increased rates of translational initiation.

Translational regulation is particularly important during early development. As discussed in Chapter 6, a variety of mRNAs are stored in oocytes in an untranslated form; the translation of these stored mRNAs is activated at fertilization or later stages of development. One mechanism of such translational regulation is the controlled polyadenylation of oocyte mRNAs. Many untranslated mRNAs are stored in oocytes with short poly-A tails (approximately 20 nucleotides). These stored mRNAs are subsequently recruited for translation at the appropriate stage of development by the lengthening of their poly-A tails to several hundred nucleotides. In addition, the translation of some mRNAs during development appears to be regulated by repressor proteins that bind to specific sequences in their 3' untranslated regions. The mechanism by which these proteins inhibit translation is not yet understood.

*Figure 7.16*
**Regulation of translation by phosphorylation of eIF-2** Translation in reticulocytes (which is devoted to synthesis of hemoglobin) is controlled by the supply of heme, which regulates the activity of eIF-2. The active form of eIF-2 (complexed with GTP) escorts initiator methionyl-tRNA to the ribosome (see Figure 7.10). The eIF-2 is then released from the ribosome in an inactive GDP-bound form, which must be reactivated by exchange of GTP for the bound GDP. If adequate heme is available, this exchange occurs and translation is able to proceed. If heme supplies are inadequate, however, a protein kinase that phosphorylates eIF-2 is activated. Phosphorylation of eIF-2 blocks the exchange of GTP for GDP, so eIF-2/GTP cannot be regenerated and translation is inhibited.

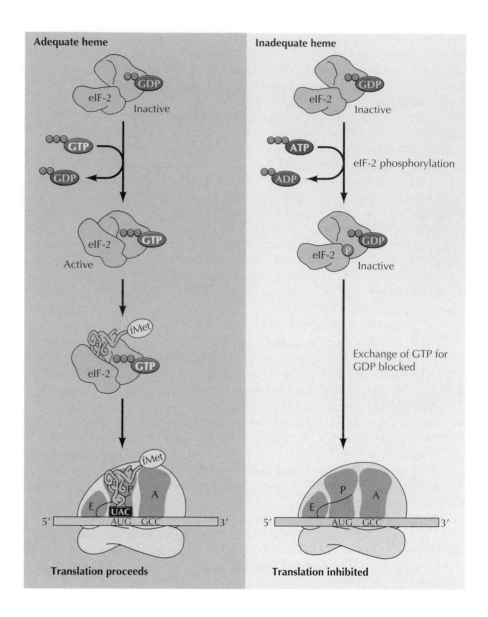

## PROTEIN FOLDING AND PROCESSING

Translation completes the flow of genetic information within the cell. The sequence of nucleotides in DNA has now been converted to the sequence of amino acids in a polypeptide chain. The synthesis of a polypeptide, however, is not equivalent to the production of a functional protein. To be useful, polypeptides must fold into distinct three-dimensional conformations, and in many cases multiple polypeptide chains must assemble into a functional complex. In addition, many proteins undergo further modifications, including cleavage and the covalent attachment of carbohydrates and lipids, that are critical for the function and correct localization of proteins within the cell.

### Chaperones and Protein Folding

The three-dimensional conformations of proteins result from interactions between the side chains of their constituent amino acids, as reviewed in Chapter 2. The classic principle of protein folding is that all the information required for a protein to adopt the correct three-dimensional conformation is provided by its amino acid sequence. This was initially established by Christian Anfinsen's experiments demonstrating that denatured RNase can spontaneously refold *in vitro* to its active conformation (see Figure 2.17). Protein folding thus appeared to be a self-assembly process that did not require additional cellular factors. More recent studies, however, have shown that this is not an adequate description of protein folding within the cell. The proper folding of proteins within cells is mediated by the activities of other proteins.

Proteins that facilitate the folding of other proteins are called molecular **chaperones**. The term "chaperone" was first used by Ron Laskey and his colleagues to describe a protein (nucleoplasmin) that is required for the assembly of nucleosomes from histones and DNA. Nucleoplasmin binds to histones and mediates their assembly into nucleosomes, but nucleoplasmin itself is not incorporated into the final nucleosome structure. Chaperones thus act as catalysts that facilitate assembly without being part of the assembled complex. Subsequent studies have extended the concept to include proteins that mediate a variety of other assembly processes, particularly protein folding.

It is important to note that chaperones do not convey additional information required for the folding of polypeptides into their correct three-dimensional conformations; the folded conformation of a protein is determined solely by its amino acid sequence. Rather, chaperones catalyze protein folding by assisting the self-assembly process. They appear to function by binding to and stabilizing partially folded polypeptides that are intermediates along the pathway leading to the final correctly folded state. In the absence of chaperones, unfolded or partially folded polypeptide chains would be unstable within the cell, frequently folding incorrectly or aggregating into insoluble complexes. The binding of chaperones stabilizes these unfolded polypeptides, thereby preventing incorrect folding or aggregation and allowing the polypeptide chain to fold into its correct conformation.

A good example is provided by chaperones that bind to nascent polypeptide chains that are still being translated on ribosomes, thereby preventing incorrect folding or aggregation of the amino-terminal portion of the polypeptide before synthesis of the chain is finished (Figure 7.17). Presumably, this interaction is particularly important for proteins in which the carboxy terminus (the last to be synthesized) is required for correct folding of the amino terminus. In such cases, chaperone binding stabilizes the amino-terminal portion in an unfolded conformation until the rest of the

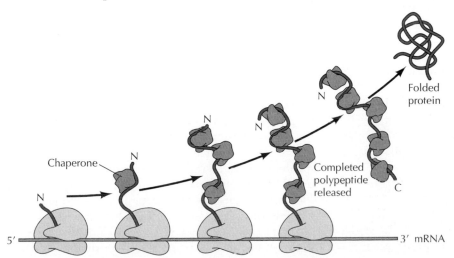

*Figure 7.17*
**Action of chaperones during translation**   Chaperones bind to the amino (N) terminus of the growing polypeptide chain, stabilizing it in an unfolded configuration until synthesis of the polypeptide is completed. The completed protein is then released from the ribosome and is able to fold into its correct three-dimensional conformation.

*Figure 7.18*
**Action of chaperones during protein transport**   A partially unfolded polypeptide is transported from the cytosol to a mitochondrion. Cytosolic chaperones stabilize the unfolded configuration. Mitochondrial chaperones facilitate transport and subsequent folding of the polypeptide chain within the organelle.

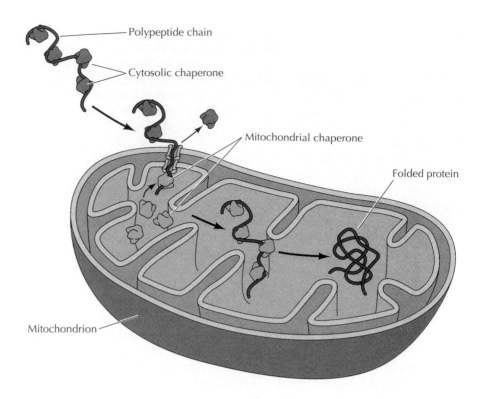

Polypeptide chain

Cytosolic chaperone

Mitochondrial chaperone

Folded protein

Mitochondrion

polypeptide chain is synthesized and the completed protein can fold correctly. Chaperones also stabilize unfolded polypeptide chains during their transport into subcellular organelles—for example, during the transfer of proteins into mitochondria from the cytosol (Figure 7.18). Proteins are transported across the mitochondrial membrane in partially unfolded conformations that are stabilized by chaperones in the cytosol. Chaperones within the mitochondrion then facilitate transfer of the polypeptide chain across the membrane and its subsequent folding within the organelle. In addition, chaperones are involved in the assembly of proteins that consist of multiple polypeptide chains, in the assembly of macromolecular structures (e.g., nucleoplasmin), and (as discussed later in this chapter) in the regulation of protein degradation.

Many of the proteins now known to function as molecular chaperones (Table 7.2) were initially identified as **heat-shock proteins**, a group of proteins expressed in cells that have been subjected to elevated temperatures or other forms of environmental stress. The heat-shock proteins (abbreviated Hsp), which are highly conserved in both prokaryotic and eukaryotic cells, are thought to stabilize and facilitate the refolding of proteins that have been partially denatured as a result of exposure to elevated temperature. However, many members of the heat-shock protein family are expressed and have essential cellular functions under normal growth conditions. These proteins serve as molecular chaperones, which are needed for polypeptide folding and transport under normal conditions as well as in cells subjected to environmental stress.

The Hsp70 and Hsp60 families of heat-shock proteins appear to be particularly important in the gen-

*Table 7.2* Molecular Chaperones

| Protein family | Chaperone proteins | |
| | Prokaryotes | Eukaryotes |
| --- | --- | --- |
| Hsp70 | DnaK | Hsc73 (cytosol) |
| | | BiP (endoplasmic reticulum) |
| | | mHsp70 (mitochondria) |
| | | ctHsp70 (chloroplasts) |
| Hsp60 | GroEL | Hsp60 (mitochondria) |
| | | Cpn60 (chloroplasts) |
| Hsp90 | HtpG | Hsp90 (cytosol) |
| | | Grp94 (endoplasmic reticulum) |
| TRiC | TF55 | TRiC (cytosol) |

eral pathways of protein folding in both prokaryotic and eukaryotic cells. The proteins of both families function by binding to unfolded regions of polypeptide chains. Members of the Hsp70 family stabilize unfolded polypeptide chains during translation (see, for example, Figure 7.17) as well as during the transport of polypeptides into a variety of subcellular compartments, such as mitochondria and the endoplasmic reticulum. These proteins bind to short segments (seven or eight amino acid residues) of unfolded polypeptides, maintaining the polypeptide chain in an unfolded configuration and preventing aggregation.

Members of the Hsp60 family (also called **chaperonins**) facilitate the folding of proteins into their native conformations. Each chaperonin consists of 14 subunits of approximately 60 kilodaltons (kd) each, arranged in two stacked rings to form a "double doughnut" structure (Figure 7.19). Unfolded polypeptide chains are shielded from the cytosol by being bound within the central cavity of the chaperonin cylinder. In this isolated environment protein folding can proceed while aggregation of unfolded segments of the polypeptide chain is prevented by their binding to the chaperonin. The binding of unfolded polypeptides to the chaperonin is a reversible reaction that is coupled to the hydrolysis of ATP as a source of energy. ATP hydrolysis thus drives multiple rounds of release and rebinding of unfolded regions of the polypeptide chain to the chaperonin, allowing the polypeptide to fold gradually into the correct conformation.

In some cases, members of the Hsp70 and Hsp60 families have been found to act together in a sequential fashion. For example, Hsp70 and Hsp60 family members act sequentially during the transport of proteins into mitochondria and during the folding of newly synthesized proteins in *E. coli* (Figure 7.20). First, an Hsp70 chaperone stabilizes nascent polypeptide chains until protein synthesis is completed. The unfolded polypeptide chain is then transferred to an Hsp60 chaperonin, within which protein folding takes place, yielding a protein correctly folded into its functional three-dimensional conformation. Members of the Hsp70 family are found in the cytosol and in subcellular organelles (e.g., mitochondria) of eukaryotic cells, as well as in bacteria (see Table 7.2). In contrast, members of the Hsp60 family are restricted to bacteria and related eukaryotic organelles (mitochondria and chloroplasts). However, the cytosol of eukaryotic cells contains chaperones that belong to the TRiC family of proteins, which form double ring structures and are thought to function analogously to the Hsp60 chaperonins. The principle of sequential Hsp70 and Hsp60 action may therefore represent a general pathway of protein folding.

## Enzymes and Protein Folding

In addition to chaperones, which facilitate protein folding by binding to and stabilizing partially folded intermediates, cells contain at least two types of enzymes that catalyze protein folding by breaking and re-forming covalent bonds. The formation of disulfide bonds between cysteine residues is important in stabilizing the folded structures of many proteins (see Figure 2.16). **Protein disulfide isomerase**, which was discovered by Christian Anfinsen in 1963, catalyzes the break-

*Figure 7.19*
**Structure of a chaperonin**   GroEL, a member of the Hsp60 family, is a porous cylinder composed of two stacked rings. Each ring consists of seven subunits. (Courtesy of Paul B. Sigler, Yale University.)

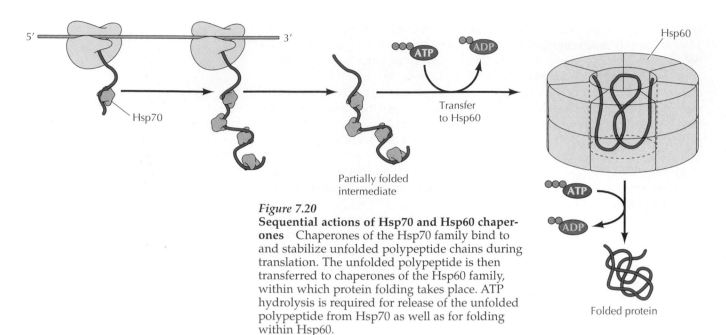

*Figure 7.20*
**Sequential actions of Hsp70 and Hsp60 chaperones**   Chaperones of the Hsp70 family bind to and stabilize unfolded polypeptide chains during translation. The unfolded polypeptide is then transferred to chaperones of the Hsp60 family, within which protein folding takes place. ATP hydrolysis is required for release of the unfolded polypeptide from Hsp70 as well as for folding within Hsp60.

age and re-formation of these bonds (Figure 7.21). For proteins that contain multiple cysteine residues, protein disulfide isomerase (PDI) plays an important role by promoting rapid exchanges between paired disulfides, thereby allowing the protein to attain the pattern of disulfide bonds that is compatible with its stably folded conformation. Disulfide bonds are generally restricted to secreted proteins and some membrane proteins because the cytosol contains reducing agents that maintain cysteine residues in their reduced (—SH form), thereby preventing the formation of disulfide (S—S) linkages. In eukaryotic cells, disulfide bonds form in the endoplasmic reticulum, in which an oxidizing environment is maintained. Consistent with the role of disulfide bonds in stabilizing secreted proteins, the activity of PDI in the endoplasmic reticulum is correlated with the level of protein secretion in different types of cells.

The second enzyme that plays a role in protein folding catalyzes the isomerization of peptide bonds that involve proline residues (Figure 7.22). Proline is an unusual amino acid in that the equilibrium between the *cis* and *trans* conformations of peptide bonds that precede proline residues is only slightly in favor of the *trans* form. In contrast, peptide bonds between other amino acids are almost always in the *trans* form. Isomerization between the *cis* and *trans* configurations of prolyl peptide bonds, which could otherwise represent a rate-limiting step in protein folding, is catalyzed by the enzyme **peptidyl prolyl isomerase**. This enzyme is widely

*Figure 7.21*
**The action of protein disulfide isomerase**   Protein disulfide isomerase (PDI) catalyzes the breakage and rejoining of disulfide bonds, resulting in exchanges between paired disulfides in a polypeptide chain. The enzyme forms a disulfide bond with a cysteine residue of the polypeptide and then exchanges its paired disulfide with another cysteine residue. In this example, PDI catalyzes the conversion of two incorrect disulfide bonds (1-2 and 3-4) to the correct pairing (1-3 and 2-4).

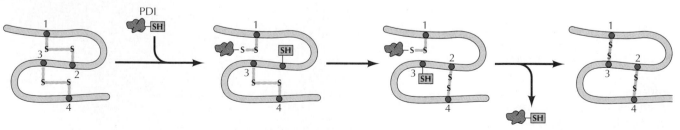

Incorrect disulfide bonds                                                                Correct disulfide bonds

distributed in both prokaryotic and eukaryotic cells and can catalyze the refolding of at least some proteins. However, its physiologically important substrates and role within cells have not yet been determined.

## Protein Cleavage

Cleavage of the polypeptide chain (**proteolysis**) is an important step in the maturation of many proteins. A simple example is removal of the initiator methionine from the amino terminus of many polypeptides, which occurs soon after the amino terminus of the growing polypeptide chain emerges from the ribosome. Additional chemical groups, such as acetyl groups or fatty acid chains (discussed shortly), are then frequently added to the amino-terminal residues.

Proteolytic modifications of the amino terminus also play a part in the translocation of many proteins across membranes, including secreted proteins in both bacteria and eukaryotes as well as proteins destined for incorporation into the plasma membrane, lysosomes, mitochondria, and chloroplasts of eukaryotic cells. These proteins are targeted for transport to their destinations by amino-terminal sequences that are removed by proteolytic cleavage as the protein crosses the membrane. For example, amino-terminal **signal sequences**, usually about 20 amino acids long, target secreted proteins to the plasma membrane of bacteria or to the endoplasmic reticulum of eukaryotic cells while translation is still in progress (Figure 7.23). The signal sequence, which consists predominantly of hydrophobic amino acids, is inserted into the membrane as it emerges from the ribosome. The remainder of the polypeptide chain passes through a channel in the membrane as translation proceeds. The signal sequence is then cleaved by a specific membrane protease (**signal peptidase**), and the mature protein is released. In eukaryotic cells, the translocation of growing polypeptide chains into the endoplasmic reticulum is the first step in targeting proteins for secretion, incorporation into the plasma membrane, or incorporation into lysosomes. The mechanisms that direct the transport of proteins to these destinations, as well as the role of other targeting sequences in directing the import of proteins into mitochondria and chloroplasts, will be discussed in detail in Chapters 9 and 10.

In other important instances of proteolytic processing, active enzymes or hormones form via cleavage of larger precursors. Insulin, which is synthesized as a longer precursor polypeptide, is a good example. Insulin forms by two cleavages. The initial precursor (preproinsulin) contains an amino-

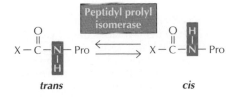

**Figure 7.22**
**The action of peptidyl prolyl isomerase** Peptidyl prolyl isomerase catalyzes the isomerization of peptide bonds that involve proline between the *cis* and *trans* conformations.

**Figure 7.23**
**The role of signal sequences in membrane translocation** Signal sequences target the translocation of polypeptide chains across the plasma membrane of bacteria or into the endoplasmic reticulum of eukaryotic cells (shown here). The signal sequence, a stretch of hydrophobic amino acids at the amino terminus of the polypeptide chain, inserts into a membrane channel as it emerges from the ribosome. The rest of the polypeptide is then translocated through the channel and the signal sequence is cleaved by the action of signal peptidase, releasing the mature translocated protein.

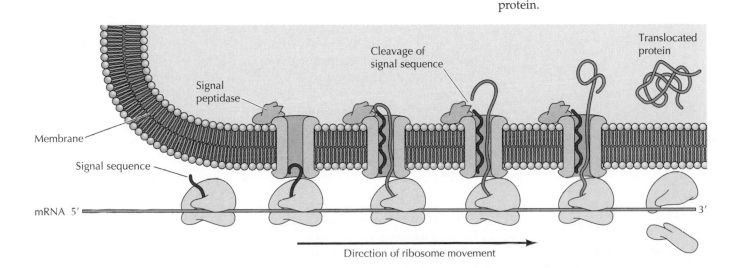

*Figure 7.24*
**Proteolytic processing of insulin**
The mature insulin molecule consists of two polypeptide chains (A and B) joined by disulfide bonds. It is synthesized as a precursor polypeptide (preproinsulin) containing an amino-terminal signal sequence that is cleaved during transfer of the growing polypeptide chain to the endoplasmic reticulum. This cleavage yields a second precursor (proinsulin), which is converted to insulin by further proteolysis, removing the internal connecting polypeptide.

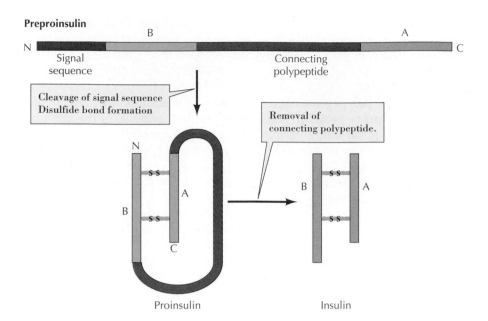

terminal signal sequence that targets the polypeptide chain to the endoplasmic reticulum (Figure 7.24). Removal of the signal sequence during transfer to the endoplasmic reticulum yields a second precursor, called proinsulin. This precursor is then converted to insulin, which consists of two chains held together by disulfide bonds, by proteolytic removal of an internal peptide. Other proteins activated by similar cleavage processes include digestive enzymes and the proteins involved in blood clotting.

It is interesting to note that the proteins of many animal viruses are derived from the cleavage of larger precursors. One particularly important example of the role of proteolysis in virus replication is provided by HIV. In the replication of HIV, a virus-encoded protease cleaves precursor polypeptides to form the viral structural proteins. Because of its central role in virus replication, the HIV protease (in addition to reverse transcriptase) is an important target for the development of drugs used for treating AIDS.

## Glycosylation

Many proteins, particularly in eukaryotic cells, are modified by the addition of carbohydrates, a process called **glycosylation**. The proteins to which carbohydrate chains have been added (called **glycoproteins**) are usually secreted or localized to the cell surface, although some nuclear and cytosolic proteins are also glycosylated. The carbohydrate moieties of glycoproteins are particularly important as recognition sites in cell-cell interactions and in the targeting of proteins for delivery to the appropriate intracellular compartments.

Glycoproteins are classified as either *N*-linked or *O*-linked, depending on the site of attachment of the carbohydrate side chain (Figure 7.25). In *N*-linked glycoproteins, the carbohydrate is attached to the nitrogen atom in the side chain of asparagine. In *O*-linked glycoproteins, the oxygen atom in the side chain of serine or threonine is the site of carbohydrate attachment. The sugars directly attached to these positions are usually either *N*-acetylglucosamine or *N*-acetylgalactosamine, respectively.

Most glycoproteins in eukaryotic cells are destined either for secretion or for incorporation into the plasma membrane. These proteins are transferred into the endoplasmic reticulum (with the cleavage of a signal sequence)

**Figure 7.25**

**Linkage of carbohydrate side chains to glycoproteins**   The carbohydrate chains of *N*-linked glycoproteins are attached to asparagine; those of *O*-linked glycoproteins are attached to either serine (shown) or threonine. The sugars joined to the amino acids are usually either *N*-acetylglucosamine (*N*-linked) or *N*-acetylgalactosamine (*O*-linked).

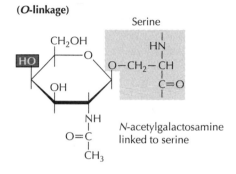

while their translation is still in progress. Glycosylation is also initiated in the endoplasmic reticulum before translation is complete. The first step is the transfer of a common oligosaccharide consisting of 14 sugar residues (2 *N*-acetylglucosamine, 3 glucose, and 9 mannose) to an asparagine residue of the growing polypeptide chain (Figure 7.26). The oligosaccharide is assembled within the endoplasmic reticulum on a lipid carrier (**dolichol phosphate**). It is then transferred as an intact unit to an acceptor asparagine (Asn) residue within the sequence Asn-X-Ser or Asn-X-Thr (where X is any amino acid other than proline).

In further processing, the common *N*-linked oligosaccharide is modified. Three glucose residues and one mannose are removed while the glycoprotein is in the endoplasmic reticulum. The oligosaccharide is then further modified in the Golgi apparatus, to which glycoproteins are transferred

**Figure 7.26**

**Synthesis of *N*-linked glycoproteins**   The first step in glycosylation is the addition of an oligosaccharide consisting of 14 sugar residues to a growing polypeptide chain in the endoplasmic reticulum (ER). The oligosaccharide (which consists of two *N*-acetylglucosamine, nine mannose, and three glucose residues) is assembled on a lipid carrier (dolichol phosphate) in the ER membrane. It is then transferred as a unit to an acceptor asparagine residue of the polypeptide.

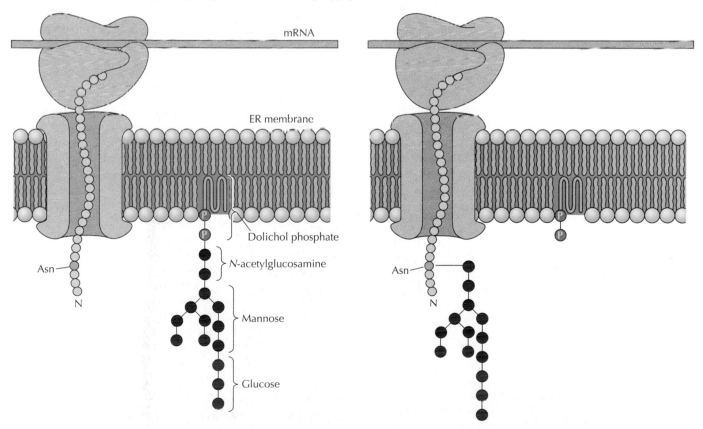

*Figure 7.27*
**Examples of *N*-linked oligosaccharides**  Various oligosaccharides form from further modifications of the common 14-sugar unit initially added in the endoplasmic reticulum (see Figure 7.26). In high-mannose oligosaccharides, the glucose residues and some mannose residues are removed, but no other sugars are added. In the synthesis of complex oligosaccharides, more mannose residues are removed and other sugars are added. Hybrid oligosaccharides are intermediate between high-mannose and complex oligosaccharides. The structures shown are representative examples.

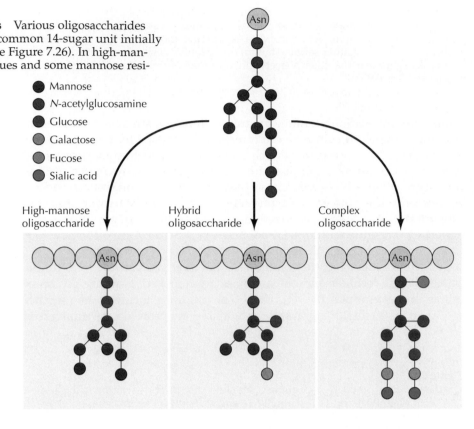

*Figure 7.28*
**Examples of *O*-linked oligosaccharides**  *O*-linked oligosaccharides usually consist of only a few carbohydrate residues, which are added one sugar at a time.

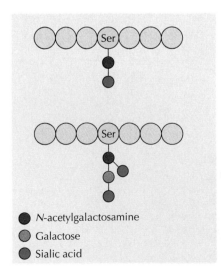

from the endoplasmic reticulum. These modifications (which will be discussed in Chapter 9) include both the removal and addition of carbohydrate residues as the glycoprotein is transported through the compartments of the Golgi (Figure 7.27). The *N*-linked oligosaccharides of different glycoproteins are processed to different extents, depending on both the enzymes present in different cells and on the accessibility of the oligosaccharide to the enzymes that catalyze its modification. Glycoproteins with inaccessible oligosaccharides do not have new sugars added to them in the Golgi. The relatively simple oligosaccharides of these glycoproteins are called high-mannose oligosaccharides because they contain a high proportion of mannose residues, similar to the common oligosaccharide originally added in the endoplasmic reticulum. In contrast, glycoproteins with accessible oligosaccharides are processed more extensively, resulting in the formation of a variety of complex oligosaccharides.

*O*-linked oligosaccharides are also added within the Golgi apparatus. In contrast to the *N*-linked oligosaccharides, *O*-linked oligosaccharides are formed by the addition of one sugar at a time and usually consist of only a few residues (Figure 7.28). Some cytosolic and nuclear proteins are also modified by the addition of *O*-linked carbohydrates, catalyzed by a different enzyme system. Many transcription factors, for example, are modified by the addition of single *O*-linked *N*-acetylglucosamine residues. Neither the sites of carbohydrate addition nor the roles of carbohydrates in the function of cytosolic and nuclear glycoproteins are yet understood.

*Attachment of Lipids*
Some proteins in eukaryotic cells are modified by the attachment of lipids to the polypeptide chain. Such modifications frequently target and anchor

these proteins to the plasma membrane, with which the hydrophobic lipid is able to interact (see Figure 2.48). Three general types of lipid additions— N-myristoylation, prenylation, and palmitoylation—are common in eukaryotic proteins associated with the cytosolic face of the plasma membrane. A fourth type of modification, the addition of glycolipids, plays an important role in anchoring some cell surface proteins to the extracellular face of the plasma membrane.

In some proteins, a fatty acid is attached to the amino terminus of the growing polypeptide chain during translation. In this process, called **N-myristoylation**, myristic acid (a 14-carbon fatty acid) is attached to an N-terminal glycine residue (Figure 7.29). The glycine is usually the second amino acid incorporated into the polypeptide chain; the initiator methionine is removed by proteolysis before fatty acid addition. Many proteins that are modified by N-myristoylation are associated with the inner face of the plasma membrane, and the role of the fatty acid in this association has been clearly demonstrated by analysis of mutant proteins in which the N-terminal glycine is changed to an alanine. This substitution prevents myristoylation and blocks the function of the mutant proteins by inhibiting their membrane association.

Lipids can also be attached to the side chains of cysteine, serine, and threonine residues. One important example of this type of modification is **prenylation**, in which specific types of lipids (prenyl groups) are attached to the sulfur atoms in the side chains of cysteine residues located near the C terminus of the polypeptide chain (Figure 7.30). Many plasma membrane–associated proteins involved in the control of cell growth and differentiation are modified in this way, including the Ras oncogene proteins, which are responsible for the uncontrolled growth of many human cancers (see Chapter 15). Prenylation of these proteins proceeds by three steps. First, the prenyl group is added to a cysteine located three amino acids from the carboxy terminus of the polypeptide chain. The prenyl groups added in this reaction are either farnesyl (15 carbons, as shown in Figure 7.30) or geranylgeranyl (20 carbons). The amino acids following the cysteine residue are then removed, leaving cysteine at the carboxy terminus. Finally, a methyl group is added to the carboxyl group of the C-terminal cysteine residue.

The biological significance of prenylation is indicated by the fact that mutations of the critical cysteine block the membrane association and function of Ras proteins. Because farnesylation is a relatively rare modification of cellular proteins, interest in this reaction has been stimulated by the possibility that inhibitors of the key enzyme (farnesyl transferase) might prove useful as drugs for the treatment of cancers that involve Ras proteins. Such inhibitors of Ras farnesylation have been found to interfere with the growth of cancer cells in experi-

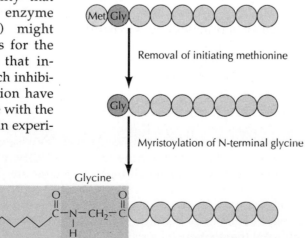

Removal of initiating methionine

Myristoylation of N-terminal glycine

Myristate    Glycine

*Figure 7.29*
**Addition of a fatty acid by N-myristoylation**   The initiating methionine is removed, leaving glycine at the N terminus of the polypeptide chain. Myristic acid (a 14-carbon fatty acid) is then added.

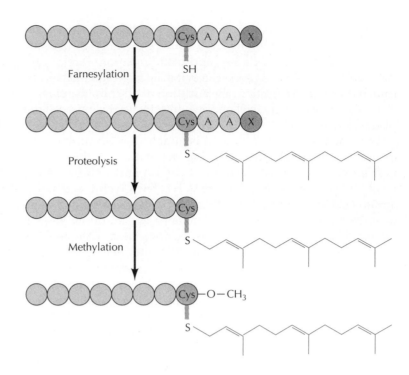

*Figure 7.30*
**Prenylation of a C-terminal cysteine residue**   The type of prenylation shown affects Ras proteins and proteins of the nuclear envelope (nuclear lamins). These proteins terminate with a cysteine residue (Cys) followed by two aliphatic amino acids (A) and any other amino acid (X) at the C terminus. The first step in their modification is addition of the 15-carbon farnesyl group to the side chain of cysteine (farnesylation). This step is followed by proteolytic removal of the three C-terminal amino acids and methylation of the cysteine, which is now at the C terminus.

*Figure 7.31*
**Palmitoylation**   Palmitate (a 16-carbon fatty acid) is added to the side chain of an internal cysteine residue.

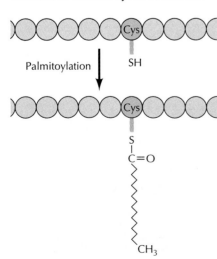

mental models and are awaiting evaluation of their efficacy against human tumors in clinical trials.

In the third type of fatty acid modification, **palmitoylation**, palmitic acid (a 16-carbon fatty acid) is added to sulfur atoms of the side chains of internal cysteine residues (Figure 7.31). Like N-myristoylation and prenylation, palmitoylation plays an important role in the association of some proteins with the cytosolic face of the plasma membrane.

Finally, lipids linked to oligosaccharides (**glycolipids**) are added to the C-terminal carboxyl groups of some proteins, where they serve as anchors that attach the proteins to the external face of the plasma membrane. Because the glycolipids attached to these proteins contain phosphatidylinositol, they are usually called **glycosylphosphatidylinositol**, or **GPI, anchors** (Figure 7.32). The oligosaccharide portions of GPI anchors are attached to the terminal carboxyl group of polypeptide chains. The inositol head group of phosphatidylinositol is in turn attached to the oligosaccharide, so the carbohydrate serves as a bridge between the protein and the fatty acid chains of the phospholipid. The GPI anchors are synthesized and added to proteins as a preassembled unit within the endoplasmic reticulum. Their addition is accompanied by cleavage of a peptide consisting of about 20 amino acids from the C terminus of the polypeptide chain. The modified protein is then transported to the cell surface, where the fatty acid chains of the GPI anchor mediate its attachment to the plasma membrane.

## REGULATION OF PROTEIN FUNCTION

A critical function of proteins is their activity as enzymes, which are needed to catalyze almost all biological reactions. Regulation of enzyme activity thus plays a key role in governing cell behavior. This is accomplished in part at the level of gene expression, which determines the amount of any enzyme (protein) synthesized by the cell. A further level of control is then obtained by regulation of protein function, which allows the

**Figure 7.32**
**Structure of a GPI anchor** The GPI anchor, attached to the C terminus, anchors the protein in the plasma membrane. The anchor is joined to the C-terminal amino acid by an ethanolamine, which is linked to an oligosaccharide that consists of mannose, *N*-acetylgalactosamine, and glucosamine residues. The oligosaccharide is in turn joined to the inositol head group of phosphatidylinositol. The two fatty acid chains of the lipid are embedded in the plasma membrane. The GPI anchor shown here is that of a rat protein, Thy-1.

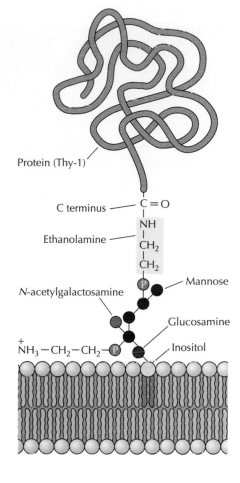

cell to regulate not only the amounts but also the activities of its protein constituents. Regulation of the activities of some of the proteins involved in transcription and translation has already been discussed in this and the preceding chapter, and many further examples of regulated protein function in the control of cell behavior will be evident throughout the remainder of this book. This section discusses the three general mechanisms by which the activities of cellular proteins are controlled.

### Regulation by Small Molecules

Most enzymes are controlled by changes in their conformation, which in turn alter catalytic activity. In many cases such conformational changes result from the binding of small molecules, such as amino acids or nucleotides, that regulate enzyme activity. This type of regulation commonly is responsible for controlling metabolic pathways by feedback inhibition. For example, the end products of many biosynthetic pathways (e.g., amino acids) inhibit the enzymes that catalyze the first step in their synthesis, thus ensuring an adequate supply of the product while preventing the synthesis of excess amounts (Figure 7.33).

Feedback inhibition is an example of **allosteric regulation**, in which a regulatory molecule binds to a site on an enzyme that is distinct from the catalytic site (*allo* = "other"; *steric* = "site"). The binding of such a regulatory molecule alters the conformation of the protein, thereby changing the shape of the catalytic site and affecting catalytic activity (see Figure 2.29). One of the best-studied allosteric enzymes is aspartate transcarbamylase, which catalyzes the first step in the synthesis of pyrimidine nucleotides and is regulated by feedback inhibition by cytidine triphosphate (CTP). Aspartate transcarbamylase consists of 12 distinct polypeptide chains: six catalytic subunits and six regulatory subunits. The binding of CTP to the

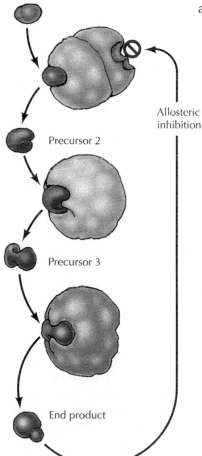

**Figure 7.33**
**Feedback inhibition** The end product of a biochemical pathway acts as an allosteric inhibitor of the enzyme that catalyzes the first step in its synthesis.

Active conformation

Inactive conformation

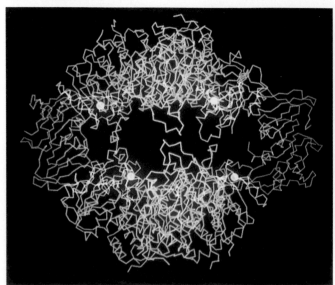

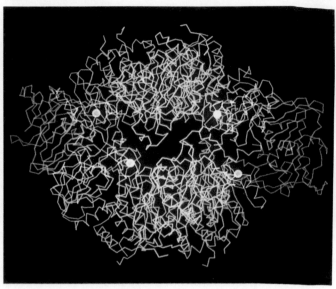

*Figure 7.34*
**Allosteric regulation of aspartate transcarbamylase**   The arrangement of two catalytic (green) and two regulatory (orange) subunits of aspartate transcarbamylase is shown in the active and inactive conformations. The complete enzyme consists of six catalytic and six regulatory subunits. (Courtesy of Irving Geis. Copyright © I. Geis and E. Gouaux, from coordinates by W. N. Lipscomb.)

*Figure 7.35*
**Conformational differences between active and inactive Ras proteins**   The Ras proteins alternate between active GTP-bound and inactive GDP-bound forms. The major effect of GTP binding versus GDP binding is alteration of the conformation of two regions of the molecule, designated the switch I and II regions. The backbone of the GTP complex is shown here in white; the backbone of the GDP complex of switch I and II are in blue and yellow, respectively. The guanine nucleotide is in red and $Mg^{2+}$ is yellow. (Courtesy of Sung-Hou Kim, University of California, Berkeley.)

regulatory subunits induces a major rearrangement of subunit positions, thereby inhibiting enzymatic activity (Figure 7.34).

Many transcription factors (discussed in Chapter 6) are also regulated by the binding of small molecules. For example, the binding of lactose or a metabolite to the *E. coli lac* repressor induces a conformational change that prevents the repressor from binding DNA. The opposite effect is illustrated in regulation of the arabinose operon: Binding of arabinose to the AraC protein induces a conformational change that activates rather than inhibits DNA binding. In eukaryotic cells, steroid hormones similarly control gene expression by binding to transcriptional regulatory proteins.

The regulation of translation factors such as EF-Tu by GTP binding (see Figure 7.12) illustrates another common mechanism by which the activities of intracellular proteins are controlled. In this case, the GTP-bound form of the protein is its active conformation, while the GDP-bound form is inactive. Many cellular proteins are similarly regulated by GTP or GDP binding. These proteins include the Ras oncogene proteins, which have been studied intensively because of their roles in the control of cell proliferation and in human cancers. X-ray crystallography analysis of these proteins has been particularly interesting, revealing subtle but functionally very important conformational differences between the inactive GDP-bound and active GTP-bound forms (Figure 7.35). This small difference in protein conformation determines whether Ras (in the active GTP-bound form) can interact with its target molecule, which signals the cell to divide. The importance of such subtle differences in protein conformation is dramatically illustrated by the fact that mutations in *ras* genes contribute to the development of about 15% of human cancers. Such mutations alter the structure of the Ras proteins so that they are locked in the active GTP-bound conformation and continually signal cell division, thereby driving the uncontrolled growth of cancer cells. In contrast, normal Ras proteins alternate between the GTP- and GDP-bound conformations, such that they are active only following stimulation by the hormones and growth factors that normally control cell proliferation in multicellular organisms.

## Protein Phosphorylation

The examples discussed in the previous section involve noncovalent associations of proteins with small-molecule inhibitors or activators. Since no covalent bonds form, the binding of these regulatory molecules to the protein is readily reversible, allowing the cell to respond rapidly to environmental changes. The activity of many proteins, however, is also regulated by covalent modifications. One example of this type of regulation is the activation of some enzymes by proteolytic cleavage of inactive precursors. As noted previously in this chapter, digestive enzymes and proteins involved in blood clotting are regulated by this mechanism. Since proteolysis is irreversible, however, it provides a means of controlling enzyme activation rather than of turning proteins on and off in response to changes in the environment. In contrast, other covalent modifications—particularly phosphorylation—are readily reversible within the cell and function, as allosteric regulation does, to reversibly activate or inhibit a wide variety of cellular proteins in response to environmental signals.

Protein phosphorylation is catalyzed by **protein kinases**, most of which transfer phosphate groups from ATP to the hydroxyl groups of the side chains of serine, threonine, or tyrosine residues (Figure 7.36). Most protein kinases phosphorylate either serine and threonine or tyrosine residues: these enzymes are called **protein-serine/threonine kinases** or **protein-tyrosine kinases**, respectively. Protein phosphorylation is reversed by **protein phosphatases**, which catalyze the hydrolysis of phosphorylated amino acid residues. Like protein kinases, most protein phosphatases are specific either for serine and threonine or for tyrosine residues, although some pro-

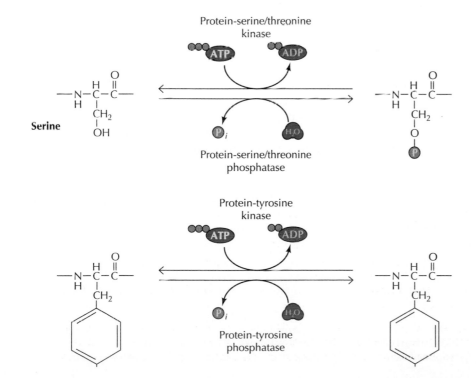

*Figure 7.36*
**Protein kinases and phosphatases**
Protein kinases catalyze the transfer of a phosphate group from ATP to the side chains of serine and threonine (protein-serine/threonine kinases) or tyrosine (protein-tyrosine kinases) residues. Protein phosphatases catalyze the removal of phosphate groups from the same amino acids by hydrolysis.

tein phosphatases recognize all three phosphoamino acids.

The combined action of protein kinases and protein phosphatases mediates the reversible phosphorylation of many cellular proteins. Frequently, protein kinases function as components of signal transduction pathways in which one kinase activates a second kinase, which may act on yet another kinase. The sequential action of a series of protein kinases can transmit a signal received at the cell surface to target proteins within the cell, resulting in changes in cell behavior in response to environmental stimuli.

The prototype of the action of protein kinases came from studies of glycogen metabolism by Ed Fischer and Ed Krebs in 1955. In muscle cells the hormone epinephrine (adrenaline) signals the breakdown of glycogen to glucose-1-phosphate, providing an available source of energy for increased muscular activity. Glycogen breakdown is catalyzed by the enzyme glycogen phosphorylase, which is regulated by a protein kinase (Figure 7.37). Epinephrine binds to a cell surface receptor that triggers the conversion of ATP to cyclic AMP (cAMP), which then binds to and activates a protein kinase, called cAMP-dependent protein kinase. This kinase phosphorylates and activates a second protein kinase, called phosphorylase kinase. Phosphorylase kinase in turn phosphorylates and activates glycogen phosphorylase, leading to glucose production. The activating phosphorylations of both phosphorylase kinase and glycogen phosphorylase can be reversed by specific phosphatases, so removal of the initial stimulus (epinephrine) inhibits further glycogen breakdown.

The signaling pathway that leads to activation of glycogen phosphorylase is initiated by the binding of small molecules at the cell surface—epinephrine binding to its receptor and cAMP binding to cAMP-dependent protein kinase. The signal is then transmitted to its intracellular target by the sequential action of protein kinases. Similar signaling pathways, in which protein kinases and phosphatases play central roles, are involved in regulating almost all aspects of the behavior of eukaryotic cells (see Chapters 13 and 14). Aberrations in these pathways, frequently involving abnormalities of protein kinases, are also responsible for many diseases associated with improper regulation of cell growth and differentiation, particularly the development of cancer.

## Protein-Protein Interactions

Many proteins consist of multiple subunits, each of which is an independent polypeptide chain. In some proteins the subunits are identical; other proteins are composed of two or more distinct polypeptides. In either case, interactions between the polypeptide chains are important in regulation of protein activity. The importance of these interactions is evident in many allosteric enzymes, such as aspartate transcarbamylase, in which the binding of a regulatory molecule alters protein conformation by changing the interactions between subunits.

Many other enzymes are similarly regulated by protein-protein interactions. A good example is cAMP-dependent protein kinase, which is composed of two regulatory and two catalytic subunits (Figure 7.38). In this

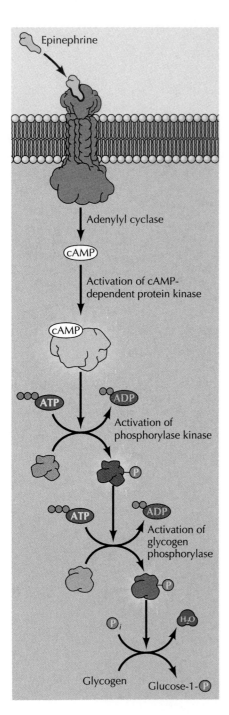

*Figure 7.37*
**Regulation of glycogen breakdown by protein phosphorylation**   The binding of epinephrine (adrenaline) to its cell surface receptor triggers the production of cyclic AMP (cAMP), which activates cAMP-dependent protein kinase. cAMP-dependent protein kinase phosphorylates and activates phosphorylase kinase, which in turn phosphorylates and activates glycogen phosphorylase. Glycogen phosphorylase then catalyzes the breakdown of glycogen to glucose-1-phosphate.

state, the enzyme is inactive; the regulatory subunits inhibit the enzymatic activity of the catalytic subunits. The enzyme is activated by cAMP, which binds to the regulatory subunits and induces a conformational change leading to dissociation of the complex; the free catalytic subunits are then enzymatically active protein kinases. Cyclic AMP thus acts as an allosteric regulator by altering protein-protein interactions.

The transcriptional regulatory proteins discussed in Chapter 6 provide another important example of protein-protein interactions. Many eukaryotic transcription factors function as activators or repressors via protein-protein interactions with components of the basal transcription machinery. As discussed in later chapters, similar protein-protein interactions, which can themselves be regulated by the binding of small molecules and phosphorylation, play critical roles in the control of many different aspects of cell behavior.

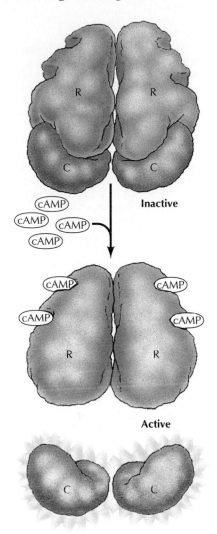

**Figure 7.38**
**Regulation of cAMP-dependent protein kinase**   In the inactive state, the enzyme consists of two regulatory (R) and two catalytic (C) subunits. Cyclic AMP binds to the regulatory subunits, inducing a conformational change that leads to their dissociation from the catalytic subunits. The free catalytic subunits are enzymatically active.

## PROTEIN DEGRADATION

The levels of proteins within cells are determined not only by rates of synthesis, but also by rates of degradation. The half-lives of proteins within cells vary widely, from minutes to several days, and differential rates of protein degradation are an important aspect of cell regulation. Many rapidly degraded proteins function as regulatory molecules, such as transcription factors. The rapid turnover of these proteins is necessary to allow their levels to change quickly in response to external stimuli. Other proteins are rapidly degraded in response to specific signals, providing another mechanism for the regulation of intracellular enzyme activity. In addition, faulty or damaged proteins are recognized and rapidly degraded within cells, thereby eliminating the consequences of mistakes made during protein synthesis. In eukaryotic cells, two major pathways—the ubiquitin-proteasome pathway and lysosomal proteolysis—mediate protein degradation.

### The Ubiquitin-Proteasome Pathway

The major pathway of selective protein degradation in eukaryotic cells uses **ubiquitin** as a marker that targets cytosolic and nuclear proteins for rapid proteolysis (Figure 7.39). Ubiquitin is a 76-amino acid polypeptide that is highly conserved in all eukaryotes (yeasts, animals, and plants). Proteins are marked for degradation by the attachment of ubiquitin to the amino group of the side chain of a lysine residue. Additional ubiquitins are then added to form a multiubiquitin chain. Such polyubiquinated proteins are recognized and degraded by a large, multisubunit protease complex, called the **proteasome**. Ubiquitin is released in the process, so it can be reused in another cycle. It is noteworthy that both the attachment of ubiquitin and the degradation of marked proteins require energy in the form of ATP.

Since the attachment of ubiquitin marks proteins for rapid degradation, the stability of many proteins is determined by whether they become ubiquitinated. Ubiquitination is a multistep process. First, ubiquitin is activated by being attached to the ubiquitin-activating enzyme, E1. The ubiquitin is then transferred to a second enzyme, called ubiquitin-conjugating enzyme (E2). In some cases, the ubiquitin is transferred from E2 to the target protein. In most cases, however, the ubiquitin is first transferred from E2 to a third enzyme (ubiquitin ligase, or E3) and then to the target protein. Most cells contain a single E1, but both the E2 and E3 enzymes are members of large families of proteins. Different members of the E2 and E3 families recognize different substrate proteins, and the specificity of these enzymes is what selectively targets cellular proteins for degradation by the ubiquitin-protea-

*Figure 7.39*
**The ubiquitin-proteasome pathway**
Proteins are marked for rapid degradation by the covalent attachment of several molecules of ubiquitin. Ubiquitin is first activated by the enzyme E1. Activated ubiquitin is then transferred to one of several different ubiquitin-conjugating enzymes (E2). In most cases, the ubiquitin is then transferred to a ubiquitin ligase (E3) and then to a specific target protein. Multiple ubiquitins are then added, and the polyubiquinated proteins are degraded by a protease complex (the proteasome).

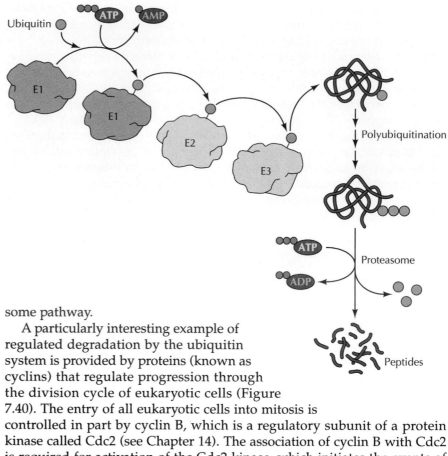

some pathway.

A particularly interesting example of regulated degradation by the ubiquitin system is provided by proteins (known as cyclins) that regulate progression through the division cycle of eukaryotic cells (Figure 7.40). The entry of all eukaryotic cells into mitosis is controlled in part by cyclin B, which is a regulatory subunit of a protein kinase called Cdc2 (see Chapter 14). The association of cyclin B with Cdc2 is required for activation of the Cdc2 kinase, which initiates the events of mitosis (including chromosome condensation and nuclear envelope breakdown) by phosphorylating various cellular proteins. Cdc2 also activates a ubiquitin-mediated proteolysis system that degrades cyclin B toward the end of mitosis. This degradation of cyclin B inactivates Cdc2, allowing the cell to exit mitosis and progress to interphase of the next cell cycle. The ubiquitination of cyclin B is a highly selective reaction, targeted by a 9-amino acid cyclin B sequence called the destruction box. Mutations of this sequence prevent cyclin B proteolysis and lead to the arrest of dividing cells in mitosis, demonstrating the importance of regulated protein degradation in controlling the fundamental process of cell division.

## Lysosomal Proteolysis

The other major pathway of protein degradation in eukaryotic cells involves the uptake of proteins by **lysosomes**. Lysosomes are membrane-enclosed organelles that contain an array of digestive enzymes, including several proteases (see Chapter 9). They have several roles in cell metabolism, including the digestion of extracellular proteins taken up by endocytosis as well as the gradual turnover of cytoplasmic organelles and cytosolic proteins.

The containment of proteases and other digestive enzymes within lysosomes prevents uncontrolled degradation of the contents of the cell. Therefore, in order to be degraded by lysosomal proteolysis, cellular proteins must first be taken up by lysosomes. One pathway for this uptake of cellular proteins, **autophagy**, involves the formation of vesicles (autophagosomes) in which small areas of cytoplasm or cytoplasmic organelles are

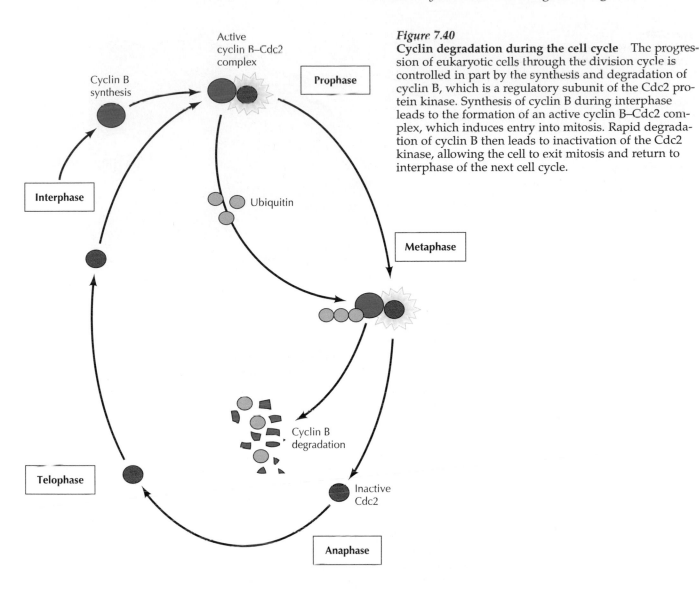

**Figure 7.40**
**Cyclin degradation during the cell cycle**   The progression of eukaryotic cells through the division cycle is controlled in part by the synthesis and degradation of cyclin B, which is a regulatory subunit of the Cdc2 protein kinase. Synthesis of cyclin B during interphase leads to the formation of an active cyclin B–Cdc2 complex, which induces entry into mitosis. Rapid degradation of cyclin B then leads to inactivation of the Cdc2 kinase, allowing the cell to exit mitosis and return to interphase of the next cell cycle.

enclosed in membranes derived from the endoplasmic reticulum (Figure 7.41). These vesicles then fuse with lysosomes, and the degradative lysosomal enzymes digest their contents. The uptake of proteins into autophagosomes appears to be nonselective, so it results in the eventual slow degradation of long-lived cytoplasmic proteins.

However, not all protein uptake by lysosomes is nonselective. For example, lysosomes are able to take up and degrade certain cytosolic proteins in a selective manner as a response to cellular starvation. The proteins degraded by lysosomal proteases under these conditions contain amino acid sequences similar to the broad consensus sequence Lys-Phe-Glu-Arg-Gln, which presumably targets them to lysosomes. A member of the Hsp70 family of molecular chaperones is also required for the lysosomal degradation of these proteins, presumably acting to unfold the polypeptide chains during their transport across the lysosomal membrane. The proteins susceptible to degradation by this pathway are thought to be normally long-lived but dispensable proteins. Under starvation conditions, these proteins are sacrificed to provide amino acids and energy, allowing some basic metabolic processes to continue.

*Figure 7.41*
**The lysosome system** Lysosomes contain various digestive enzymes, including proteases. Lysosomes take up cellular proteins by fusion with autophagosomes, which are formed by the enclosure of areas of cytoplasm or organelles (e.g., a mitochondrion) in fragments of the endoplasmic reticulum. This fusion yields a phagolysosome, which digests the contents of the autophagosome.

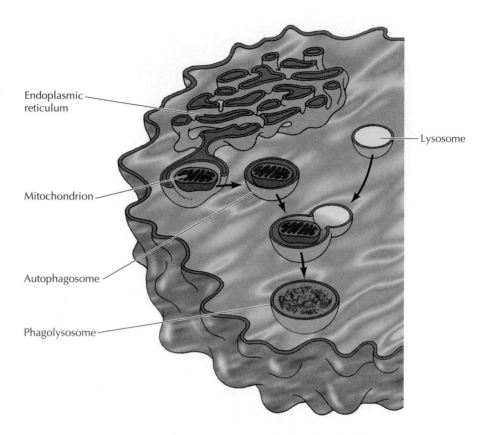

Endoplasmic reticulum

Lysosome

Mitochondrion

Autophagosome

Phagolysosome

## KEY TERMS

tRNA, anticodon, aminoacyl tRNA synthetase

ribosome, rRNA

5′ untranslated region, polycistronic, monocistronic, 3′ untranslated region, Shine-Delgarno sequence

initiation factor, elongation factor, release factor, polysome

## Summary

### TRANSLATION OF mRNA

*Transfer RNAs:* Transfer RNAs serve as adaptors that align amino acids on the mRNA template. Aminoacyl tRNA synthetases attach amino acids to the appropriate tRNAs, which then bind to mRNA codons by complementary base pairing.

*The Ribosome:* Ribosomes consist of two subunits, which are composed of proteins and ribosomal RNAs. The 23S rRNA is the primary catalyst of peptide bond formation.

*The Organization of mRNAs and the Initiation of Translation:* Translation of both prokaryotic and eukaryotic mRNAs initiates with a methionine residue. In bacteria, initiation codons are preceded by a sequence that aligns the mRNA on the ribosome by base pairing with 16S rRNA. In eukaryotes, initiation codons are identified by scanning from the 5′ end of the mRNA, which is recognized by its 7-methylguanosine cap.

*The Process of Translation:* Translation is initiated by the binding of methionyl tRNA and mRNA to the small ribosomal subunit. The large ribosomal subunit then joins the complex, and the polypeptide chain elongates until the ribosome reaches a termination codon in the mRNA. A variety of nonribosomal factors are required for initiation, elongation, and termination of translation in both prokaryotic and eukaryotic cells.

*Regulation of Translation:* Translation of specific mRNAs can be regulated by the binding of repressor proteins. General translational activity can be regulated by modification of initiation factors. Polyadenylation of mRNA is also an important mechanism for the regulation of translation during early development.

## PROTEIN FOLDING AND PROCESSING

*Chaperones and Protein Folding:* Molecular chaperones facilitate the intracellular folding of polypeptide chains into their correct three-dimensional conformations by binding to and stabilizing unfolded or partially folded polypeptide chains.

**chaperone, heat-shock protein, chaperonin**

*Enzymes and Protein Folding:* At least two types of enzymes, protein disulfide isomerase and peptidyl prolyl isomerase, catalyze protein folding by breaking and re-forming covalent bonds.

**protein disulfide isomerase, peptidyl prolyl isomerase**

*Protein Cleavage:* Proteolysis is an important step in the processing of many proteins. For example, secreted proteins and proteins incorporated into most eukaryotic organelles are targeted to their destinations by amino-terminal sequences that are removed by proteolytic cleavage as the polypeptide chain crosses the membrane.

**proteolysis, signal sequence, signal peptidase**

*Glycosylation:* Many eukaryotic proteins, particularly secreted and plasma membrane proteins, are modified by the addition of carbohydrates in the endoplasmic reticulum and Golgi apparatus.

**glycosylation, glycoprotein, dolichol phosphate**

*Attachment of Lipids:* Covalently attached lipids frequently target and anchor proteins to the plasma membrane.

**N-myristoylation, prenylation, palmitoylation, glycolipid, glycosylphosphatidylinositol (GPI) anchor**

## REGULATION OF PROTEIN FUNCTION

*Regulation by Small Molecules:* Many proteins are regulated by the binding of small molecules, such as amino acids and nucleotides, that induce changes in protein conformation and activity.

**allosteric regulation**

*Protein Phosphorylation:* Reversible phosphorylation, which controls the activities of a wide variety of cellular proteins, results from the action of protein kinases and phosphatases.

**protein kinase, protein-serine/threonine kinase, protein-tyrosine kinase, protein phosphatase**

*Protein-Protein Interactions:* Interactions between polypeptide chains are important in the regulation of allosteric enzymes and other cellular proteins.

## PROTEIN DEGRADATION

*The Ubiquitin-Proteasome Pathway:* The major pathway of selective protein degradation in eukaryotic cells uses ubiquitin as a marker that targets proteins for rapid proteolysis by the proteasome.

**ubiquitin, proteasome**

*Lysosomal Proteolysis:* Lysosomal proteases degrade extracellular proteins taken up by endocytosis and are responsible for the slow degradation of cytoplasmic organelles and long-lived cytosolic proteins. Some proteins are targeted for selective degradation in lysosomes as a response to cell starvation.

**lysosome, autophagy**

# QUESTIONS

**1.** You wish to express a cloned eukaryotic cDNA in bacteria. What type of sequence must you add in order for the mRNA to be translated on prokaryotic ribosomes?

**2.** What effect would an inhibitor of polyadenylation be expected to have on protein synthesis in fertilized eggs?

**3.** Why is a member of the Hsp70 family required for selective lysosomal degradation of proteins in starved cells, but not for the degradation of proteins taken up by autophagy?

**4.** You are interested in studying a protein expressed on the surface of liver cells. How could treatment of these cells with a phospholipase enable you to determine whether your protein is a transmembrane protein or is attached to the cell surface by a GPI anchor?

**5.** You are studying an enzyme that is activated by phosphorylation of a serine residue at position 59. How would mutation of this serine to threonine be expected to affect the enzyme's activity? What effect would you predict following mutation of the serine to alanine?

# REFERENCES AND FURTHER READING

## General References

Branden, C. and J. Tooze. 1991. *Introduction to Protein Structure.* New York: Garland Publishing.

Hill, W. E., A. Dahlberg, R. A. Garrett, P. B. Moore, D. Schlessinger and J. R.Warner, eds. 1990. *The Ribosome: Structure, Function, and Evolution.* Washington: American Society for Microbiology.

## Translation of mRNA

Chen, J.-J. and I. M. London. 1995. Regulation of protein synthesis by heme-regulated eIF-2α kinase. Trends Biochem. Sci. 20: 105–108. [R]

Crick, F. H. C. 1966. Codon-anticodon pairing: The wobble hypothesis. *J. Mol. Biol.* 19: 548–555. [P]

Curtis, D., R. Lehmann and P. D. Zamore. 1995. Translational regulation in development. *Cell* 81: 171–178. [R]

Gold, L. 1988. Posttranscriptional regulatory mechanisms in *Escherichia coli. Ann. Rev. Biochem.* 57: 199–233. [R]

Hershey, J. W. B. 1991. Translational control in mammalian cells. *Ann. Rev. Biochem.* 60: 717–755. [R]

Illangeskare, M., G. Sanchez, T. Nickles and M. Yarus. 1995. Aminoacyl-RNA synthesis catalyzed by an RNA. *Science* 267: 643–647. [P]

Klausner, R. D., T. A. Rouault and J. B. Harford. 1993. Regulating the fate of mRNA: The control of cellular iron metabolism. *Cell* 72: 19–28. [R]

Kozak, M. 1990. The scanning model for translation: An update. *J. Cell Biol.* 108: 229–241. [R]

Kozak, M. 1992. Regulation of translation in eukaryotic systems. *Ann. Rev. Cell Biol.* 8: 197–225. [R]

Merrick, W. C. 1992. Mechanism and regulation of eukaryotic protein synthesis. *Microbiol. Rev.* 56: 291–315. [R]

Noller, H. F. 1991. Ribosomal RNA and translation. *Ann. Rev. Biochem.* 60: 191–227. [R]

Noller, H. F., V. Hoffarth and L. Zimniak. 1992. Unusual resistance of peptidyl transferase to protein extraction procedures. *Science* 256: 1416–1419. [P]

Pause, A., G. J. Belsham, A.-C. Gingras, O. Donze, T.-A. Lin, , J. C. Lawrence Jr., and N. Sonenberg. 1994. Insulin-dependent stimulation of protein synthesis by phosphorylation of a regulator of 5′-cap function. *Nature* 371: 762–767. [P]

Piccirilli, J. A., T. S. McConnell, A. J. Zaug, H. F. Noller and T. R. Cech. 1992. Aminoacyl esterase activity of the *Tetrahymena* ribozyme. *Science* 256: 1420–1424. [P]

Rhoads, R. E. 1993. Regulation of eukaryotic protein synthesis by initiation factors. *J. Biol. Chem.* 268: 3017–3020. [R]

Saks, M. E., J. R. Sampson and J. N. Abelson. 1994. The transfer RNA identity problem: A search for rules. *Science* 263: 191–197. [R]

Samaha, R. R., R. Green and H. F. Noller. 1995. A base pair between tRNA and 23S rRNA in the peptidyl transferase centre of the ribosome. *Nature* 377: 309–314. [P]

Stern, S., T. Powers, L.-M. Changchien and H. F. Noller. 1989. RNA-protein interactions in 30S ribosomal subunits: folding and function of 16S rRNA. *Science* 244: 783–790. [P]

## Protein Folding and Processing

Bardwell, J. C. A. and J. Beckwith. 1993. The bonds that tie: Catalyzed disulfide bond formation. *Cell* 74: 769–771. [R]

Braig, K., Z. Otwinowski, R. Hegde, D. C. Bolsvert, A. Joachimiak, A. L. Horwich and P. B. Sigler. 1994. The crystal structure of the bacterial chaperonin GroEL at 2.8 Å. *Nature* 371: 578–586. [P]

Casey, P. J. 1995. Protein lipidation in cell signaling. *Science* 268: 221–225. [R]

Clarke, S. 1992. Protein isoprenylation and methylation at carboxy-terminal cysteine residues. *Ann. Rev. Biochem.* 61: 355–386. [R]

Dalbey, R. E. and G. von Heijne. 1992. Signal peptidases in prokaryotes and eukaryotes—a new protease family. *Trends Biochem. Sci.* 17: 474–478. [R]

Ellis, J. R. and S. M. van der Vies. 1991. Molecular chaperones. *Ann. Rev. Biochem.* 60: 321–347. [R]

Englund, P. T. 1993. The structure and biosynthesis of glycosyl phosphatidylinositol protein anchors. *Ann. Rev. Biochem.* 62: 121–138. [R]

Freedman, R. B., T. R. Hirst and M. F. Tuite. 1994. Protein disulphide isomerase: Building bridges in protein folding. *Trends Biochem. Sci.* 19: 331–336. [R]

Gething, M.-J. and J. Sambrook. 1992. Protein folding in the cell. *Nature* 355: 33–45. [R]

Gibbs, J. B., A. Oliff and N. E. Kohl. 1994. Farnesyltransferase inhibitors: Ras research yields a potential cancer therapeutic. *Cell* 77: 175–178. [R]

Gierasch, L. M. 1989. Signal sequences. *Biochemistry* 28: 923–930. [R]

Hart, G. W., R. S. Haltiwanger, G. D. Holt and W. G. Kelly. 1989. Glycosylation in the nucleus and cytoplasm. *Ann. Rev. Biochem.* 58: 841–874. [R]

Hartl, F. U. 1996. Molecular chaperones in cellular protein folding. *Nature* 381: 571–580. [R]

Hartl, F. U., R. Hlodan and T. Langer. 1994. Molecular chaperones in protein folding: the art of avoiding sticky situations. *Trends Biochem. Sci.* 19: 20–25. [R]

Hirschberg, C. B. and M. D. Snider. 1987. Topography of glycosylation in the rough endoplasmic reticulum and Golgi apparatus. *Ann. Rev. Biochem.* 56: 63–87. [R]

Johnson, D. R., R. S. Bhatnagar, L. J. Knoll and J. I. Gordon. 1994. Genetic and biochemical studies of protein N-myristoylation. *Ann. Rev. Biochem.* 63: 869–914. [R]

Kornfield, R. and S. Kornfield. 1985. Assembly of asparagine-linked oligosaccharides. *Ann. Rev. Biochem.* 45: 631–664. [R]

Rapaport, T. A. 1992. Transport of proteins across the endoplasmic reticulum membrane. *Science* 258: 931–936. [R]

Sanders, S. L. and R. Schekman. 1992. Polypeptide translocation across the endoplasmic reticulum membrane. *J. Biol. Chem.* 267: 13791–13794. [R]

Towler, D. A., J. I. Gordon, S. P. Adams and L. Glaser. 1988. The biology and enzymology of eukaryotic protein acylation. *Ann. Rev. Biochem.* 57: 69–99. [R]

Udenfriend, S. and K. Kodukula. 1995. How glycosylphosphatidylinositol-anchored membrane proteins are made. *Ann. Rev. Biochem.* 64: 563–591. [R]

## Regulation of Protein Function

Charbonneau, H. and N. K. Tonks. 1992. 1002 protein phosphatases? *Ann. Rev. Cell Biol.* 8: 463–493. [R]

Cohen, P. 1989. The structure and regulation of protein phosphatases. *Ann. Rev. Biochem.* 58: 453–508. [R]

Edelman, A. M., D. K. Blumenthal and E. G. Krebs. 1987. Protein serine/threonine kinases. *Ann. Rev. Biochem.* 56: 567–613. [R]

Fischer, E. H. and E. G. Krebs. 1989. Commentary on "The phosphorylase b to a converting enzyme of rabbit skeletal muscle." *Biochim. Biophys. Acta* 1000: 297–301. [R]

Hanks, S. K., A. M. Quinn and T. Hunter. 1988. The protein kinase family: Conserved features and deduced phylogeny of the catalytic domains. *Science* 241: 42–52. [R]

Hunter, T. 1995. Protein kinases and phosphatases: The yin and yang of protein phosphorylation and signaling. *Cell* 80: 225–236. [R]

Kantrowitz, E. R. and W. N. Lipscomb. 1990. *Escherichia coli* aspartate transcarbamoylase: The molecular basis for a concerted allosteric transition. *Trends. Biochem. Sci.* 15: 53–59. [R]

Milburn, M. V., L. Tong, A. M. DeVos, A. Brunger, Z. Yamaizumi, S. Nishimura and S.-H. Kim. 1990. Molecular switch for signal transduction: structural differences between active and inactive forms of protooncogenic Ras proteins. *Science* 247: 939–945. [P]

Monod, J., J.-P. Changeux and F. Jacob. 1963. Allosteric proteins and cellular control systems. *J. Mol. Biol.* 6: 306–329. [P]

Taylor, S. S., D. R. Knighton, J. Zheng, L. F. R. Eyck and J. M. Sowadski. 1992. Structural framework for the protein kinase family. *Ann. Rev. Cell Biol.* 8: 429–462. [R]

## Protein Degradation

Ciechanover, A. 1994. The ubiquitin-proteasome proteolytic pathway. *Cell* 79: 13—21. [R]

Deshaies, R. J. 1995. Make it or break it: The role of ubiquitin-dependent proteolysis in cellular regulation. *Trends Cell Biol.* 5: 428–434. [R]

Dice, J. F. 1990. Peptide sequences that target cytosolic proteins for lysosomal proteolysis. *Trends Biochem. Sci.* 15: 305–309. [R]

Glotzer, M., A. W. Murray and M. W. Kirschner. 1991. Cyclin is degraded by the ubiquitin pathway. *Nature* 349: 132–138. [P]

Goldberg, A. L. 1995. Functions of the proteasome: The lysis at the end of the tunnel. *Science* 268: 522–523. [R]

Jentsch, S. and S. Schlenker. 1995. Selective protein degradation: A journey's end within the proteasome. *Cell* 82: 881–884. [R]

Peters, J.-M. 1994. Proteasomes: Protein degradation machines of the cell. *Trends Biochem. Sci.* 19: 377–382. [R]

Bruce Heinemann, *Leaves, Seattle, Washington*, 1992

# *PART III* Cell Structure and Function

# The Nucleus

THE PRESENCE OF A NUCLEUS IS THE PRINCIPAL FEATURE that distinguishes eukaryotic from prokaryotic cells. By housing the cell's genome, the nucleus serves both as the repository of genetic information and as the cell's control center. DNA replication, transcription, and RNA processing all take place within the nucleus, with only the final stage of gene expression (translation) localized to the cytoplasm.

By separating the genome from the cytoplasm, the nuclear envelope allows gene expression to be regulated by mechanisms that are unique to eukaryotes. Whereas prokaryotic mRNAs are translated while their transcription is still in process, eukaryotic mRNAs undergo posttranscriptional processing (e.g., splicing) before being transported from the nucleus to the cytoplasm. The presence of a nucleus thus allows gene expression to be regulated by posttranscriptional mechanisms, such as alternative splicing. By limiting the access of proteins to the genetic material, the nuclear envelope also provides novel opportunities for the control of gene expression at the level of transcription. For example, the expression of some eukaryotic genes is controlled by the regulated transport of transcription factors from the cytoplasm into the nucleus—a form of transcriptional regulation unavailable to prokaryotes. The separation of the genome from the site of mRNA translation thus plays a central role in eukaryotic gene expression.

## THE NUCLEAR ENVELOPE AND TRAFFIC BETWEEN THE NUCLEUS AND CYTOPLASM

The nuclear envelope separates the contents of the nucleus from the cytoplasm and provides the structural framework of the nucleus. The nuclear membranes, acting as barriers that prevent the free passage of molecules between the nucleus and the cytoplasm, maintain the nucleus as a distinct biochemical compartment. The sole channels through the nuclear envelope are provided by the nuclear pore complexes, which allow the regulated exchange of molecules between the nucleus and cytoplasm. The selective traffic of proteins and RNAs through the nuclear pore complexes not only establishes the internal composition of the nucleus, but also plays a critical role in regulating eukaryotic gene expression.

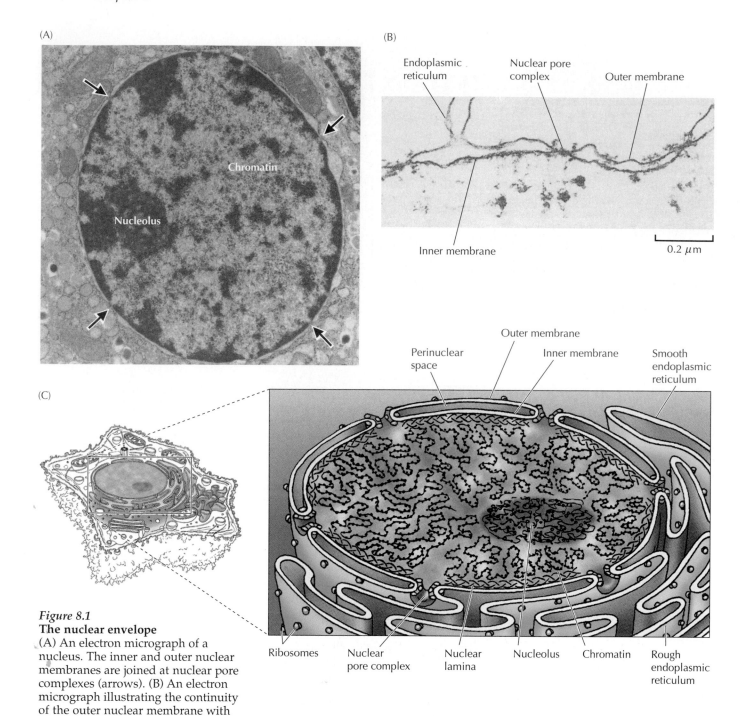

(A)

(B)

Endoplasmic reticulum

Nuclear pore complex

Outer membrane

Chromatin

Nucleolus

Inner membrane

0.2 μm

(C)

Outer membrane

Perinuclear space

Inner membrane

Smooth endoplasmic reticulum

Ribosomes   Nuclear pore complex   Nuclear lamina   Nucleolus   Chromatin   Rough endoplasmic reticulum

**Figure 8.1**
**The nuclear envelope**
(A) An electron micrograph of a nucleus. The inner and outer nuclear membranes are joined at nuclear pore complexes (arrows). (B) An electron micrograph illustrating the continuity of the outer nuclear membrane with the endoplasmic reticulum. (C) Schematic of the nuclear envelope. The inner nuclear membrane is lined by the nuclear lamina, which serves as an attachment site for chromatin. (A, David M. Phillips/Photo Researchers, Inc.; B courtesy of Larry Gerace, Scripps Research Institute.)

## Structure of the Nuclear Envelope

The **nuclear envelope** has a complex structure, consisting of two nuclear membranes, an underlying nuclear lamina, and nuclear pore complexes (Figure 8.1). The nucleus is surrounded by a system of two concentric membranes, called the inner and outer **nuclear membranes**. The outer nuclear membrane is continuous with the endoplasmic reticulum, so the space between the inner and outer nuclear membranes is directly connected with the lumen of the endoplasmic reticulum. The nuclear membranes are associated with enzymes characteristic of the endoplasmic reticulum and are functionally similar to the membranes of the endoplasmic reticulum (see Chapter 9). Moreover, like the rough endoplasmic reticulum, the outer nuclear membrane has ribosomes bound to its cytoplasmic surface.

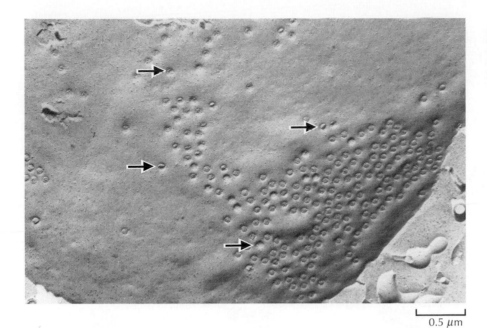

*Figure 8.2*
**Electron micrograph showing nuclear pores** Many nuclear pores (arrows) are visible in this freeze-fracture preparation of the nuclear envelope. (Photo Researchers, Inc.)

0.5 μm

Most important, however, is the fact that the two nuclear membranes act as a barrier that separates the contents of the nucleus from the cytoplasm. Like other cell membranes, the nuclear membranes are phospholipid bilayers, which are permeable only to small nonpolar molecules (see Figure 2.49). Other molecules are unable to diffuse through the phospholipid bilayer. The inner and outer nuclear membranes are joined at nuclear pore complexes, the sole channels through which small polar molecules and macromolecules are able to travel through the nuclear envelope (Figure 8.2). As discussed in the next section, the nuclear pore complex is a complicated structure that is responsible for the selective traffic of proteins and RNAs between the nucleus and the cytoplasm.

Underlying the inner nuclear membrane is the **nuclear lamina**, a fibrous meshwork that provides structural support to the nucleus (Figure 8.3). The

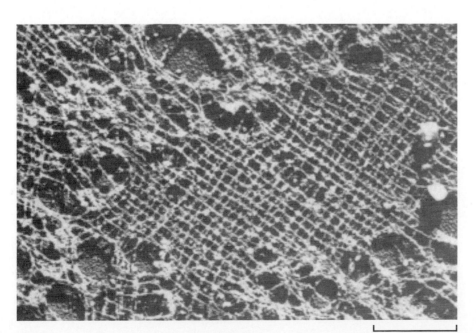

*Figure 8.3*
**Electron micrograph of the nuclear lamina** The lamina is a meshwork of filaments underlying the inner nuclear membrane. (Courtesy of Larry Gerace, Scripps Research Institute.)

0.5 μm

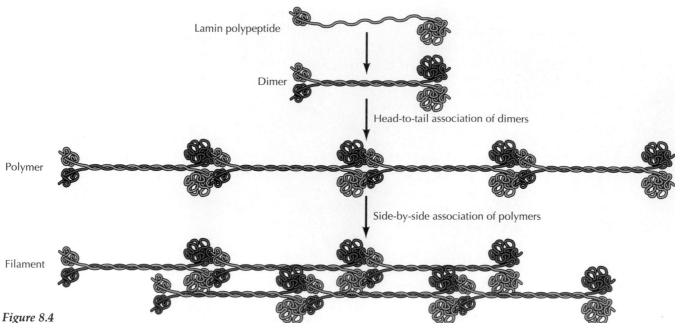

Lamin polypeptide

Dimer

Head-to-tail association of dimers

Polymer

Side-by-side association of polymers

Filament

**Figure 8.4**
**Model of lamin assembly**   The lamin polypeptides form dimers in which the central $\alpha$-helical regions of two polypeptide chains are wound around each other. Further assembly may involve the head-to-tail association of dimers to form linear polymers and the side-by-side association of polymers to form filaments.

nuclear lamina is composed of one or more related proteins called **lamins**. Most mammalian cells, for example, contain four different lamins, designated A, $B_1$, $B_2$, and C. All the lamins are 60- to 80-kilodalton (kd) fibrous proteins that are related to the intermediate filament proteins of the cytoskeleton (see Chapter 11). Like other intermediate filament proteins, the lamins associate with each other to form filaments (Figure 8.4). The first stage of this association is the interaction of two lamins to form a dimer in which the $\alpha$-helical regions of two polypeptide chains are wound around each other in a structure called a coiled coil. These lamin dimers then associate with each other to form the filaments that make up the nuclear lamina. In addition, the lamins bind to integral membrane proteins, which may help organize the lamin filaments into a meshwork and mediate their attachment to the inner nuclear membrane. The membrane association of B-type lamins is further facilitated by the posttranslational addition of lipid—in particular, prenylation of C-terminal cysteine residues (see Figure 7.30).

In addition to providing structural support to the nucleus, the nuclear lamina is thought to serve as a site of chromatin attachment. Chromatin within the nucleus is organized into large loops of DNA, some of which appear to be bound to the nuclear envelope. The lamins bind chromatin and may help mediate this interaction.

### The Nuclear Pore Complex

The **nuclear pore complexes** are the only channels through which small polar molecules, ions, and macromolecules (proteins and RNAs) are able to travel between the nucleus and the cytoplasm. The nuclear pore complex is an extremely large structure with a diameter of about 120 nm and an estimated molecular mass of approximately 125 million daltons—about 30 times the size of a ribosome. By controlling the traffic of molecules between the nucleus and cytoplasm, the nuclear pore complex plays a fundamental role in the physiology of all eukaryotic cells. RNAs that are synthesized in the nucleus must be efficiently exported to the cytoplasm, where they function in protein synthesis. Conversely, proteins required for nuclear functions (e.g., transcription factors) must be transported into the nucleus from their sites of synthesis in the cytoplasm. The regulated traffic of proteins and RNAs through the nuclear pore complex thus determines the composition of the nucleus and plays a key role in gene expression.

Depending on their size, molecules can travel through the nuclear pore complex by one of two different mechanisms (Figure 8.5). Small molecules (less than approximately 20 kd) pass rapidly across the nuclear envelope in either direction: cytoplasm to nucleus or nucleus to cytoplasm. These molecules diffuse passively through open aqueous channels in the nuclear pore complex. On the basis of the size of molecules that are capable of passing freely between the nucleus and cytoplasm, these open channels are estimated to have a diameter of approximately 10 nm. Most proteins and RNAs, however, are too large to pass through these open channels and are therefore unable to travel freely across the nuclear envelope. Instead, these macromolecules pass through the nuclear pore complex by an active process in which appropriate proteins and RNAs are recognized and selectively transported in only one direction (nucleus to cytoplasm or cytoplasm to nucleus). The traffic of these molecules occurs through regulated channels in the nuclear pore complex that, in response to appropriate signals, can open to a diameter of at least 25 nm—a size large enough to accommodate large ribonucleoprotein complexes, such as ribosomal subunits. It is through these regulated channels that nuclear proteins are selectively imported from the cytoplasm to the nucleus while RNAs are exported from the nucleus to the cytoplasm.

Visualization of nuclear pore complexes by electron microscopy reveals a structure with eightfold symmetry organized around a large central channel (Figure 8.6), which is the route through which proteins and RNAs cross the nuclear envelope. Detailed structural studies, including computer-based image analysis, have led to the development of three-dimensional models of the nuclear pore complex (Figure 8.7). These studies indicate that the nuclear pore complex consists of an assembly of eight spokes arranged around a central channel. The spokes are connected to rings at the nuclear and cytoplasmic surfaces, and the spoke-ring assembly is anchored within the nuclear envelope at sites of fusion between the inner and outer nuclear

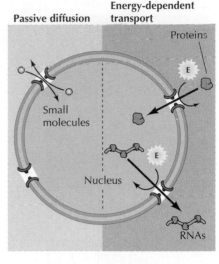

**Figure 8.5**
**Molecular traffic through nuclear pore complexes** Small molecules are able to pass rapidly through open channels in the nuclear pore complex by passive diffusion. In contrast, macromolecules are transported by a selective, energy-dependent mechanism that acts predominantly to import proteins to the nucleus and export RNAs to the cytoplasm.

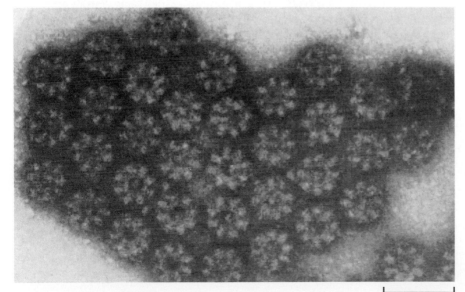

0.2 μm

**Figure 8.6**
**Electron micrograph of nuclear pore complexes** In this face-on view, isolated nuclear pore complexes appear to consist of eight structural subunits surrounding a central channel. (Courtesy of Larry Gerace, Scripps Research Institute.)

(A) Cytoplasmic surface view

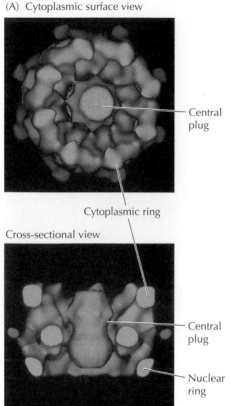

Central plug

Cytoplasmic ring

Cross-sectional view

Central plug

Nuclear ring

(B)

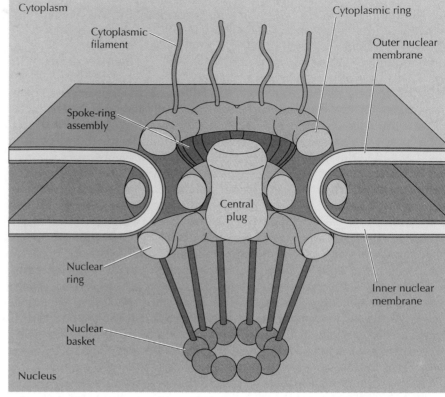

Cytoplasm

Cytoplasmic filament

Spoke-ring assembly

Central plug

Nuclear ring

Nuclear basket

Nucleus

Cytoplasmic ring

Outer nuclear membrane

Central plug

Inner nuclear membrane

*Figure 8.7*
**Model of the nuclear pore complex**
(A) Three-dimensional views of the cytoplasmic surface and a cross section of the nuclear pore complex obtained by cryoelectron microscopy. The spoke-ring assembly is blue and the central plug is pink. (B) Schematic of the nuclear pore complex. The complex consists of an assembly of eight spokes attached to rings on the cytoplasmic and nuclear sides of the nuclear envelope. The spoke-ring assembly surrounds a central channel containing the central plug. Cytoplasmic filaments extend from the cytoplasmic ring, and filaments forming the nuclear basket extend from the nuclear ring. (A from C. W. Akey and M. Radermacher, 1993, *J. Cell Biol.* 122: 1.)

membranes. Protein filaments extend from both the cytoplasmic and nuclear rings, forming a distinct basketlike structure on the nuclear side.

The central channel formed by the spoke-ring assembly is approximately 40 nm in diameter, which is wide enough to accommodate the largest particles able to cross the nuclear envelope. This central channel often appears to contain a structure called the central plug, the nature of which is unclear. Some investigators think the central plug is an intrinsic component of the nuclear pore complex that regulates the transport of molecules through the central channel. Alternatively, the central plug, rather than being part of the nuclear pore complex *per se*, may simply consist of macromolecules in the act of being transported through the central channel.

In addition to revealing the large central channel, detailed structural analysis of nuclear pore complexes has shown that assembly of the spoke-ring complex creates eight smaller channels between the spokes. These smaller channels have diameters of approximately 10 nm and are thought to represent the open channels through which small molecules diffuse. The free diffusion of small molecules and the selective transport of macromolecules thus appear to be routed through structurally distinct channels of the nuclear pore complex.

At the molecular level, the structure and function of the components of the nuclear pore complex are only beginning to be understood. On the basis of its size, the nuclear pore complex is estimated to be composed of a hundred or more distinct proteins, only a fraction of which have been identified. The mechanism by which the components of the nuclear pore complex selectively transport molecules across the nuclear envelope thus remains to be determined.

## Selective Import of Proteins to the Nucleus

The basis for selective traffic across the nuclear envelope is best understood for proteins that are imported from the cytoplasm to the nucleus. Such proteins are responsible for all aspects of genome structure and function; they include histones, DNA polymerases, RNA polymerases, transcription factors, splicing factors, and many others. These proteins are targeted to the nucleus by specific amino acid sequences, called **nuclear localization signals**, that direct their transport through the nuclear pore complex.

The first nuclear localization signal to be mapped in detail was characterized by Alan Smith and colleagues in 1984. These investigators studied simian virus 40 (SV40) T antigen, a virus-encoded protein that initiates viral DNA replication in infected cells (see Chapter 5). As expected for a replication protein, T antigen normally is localized to the nucleus. The signal responsible for its nuclear localization was first identified by the finding that mutation of a single lysine residue prevents nuclear import, resulting instead in the accumulation of T antigen in the cytoplasm. Subsequent studies defined the T antigen nuclear localization signal as the seven-amino-acid sequence Pro-Lys-Lys-Lys-Arg-Lys-Val. Not only was this sequence necessary for the nuclear transport of T antigen, but its addition to other, normally cytoplasmic, proteins was also sufficient to direct their accumulation in the nucleus.

Nuclear localization signals have since been identified in many other proteins. Most of these sequences, like that of T antigen, are short stretches rich in basic amino acid residues (lysine and arginine). In many cases, however, the amino acids that form the nuclear localization signal are close together but not immediately adjacent to each other. For example, the nuclear localization signal of nucleoplasmin (a protein involved in chromatin assembly) consists of two parts: a Lys-Arg pair followed by four lysines located ten amino acids farther upstream (Figure 8.8). Both the Lys-Arg and Lys-Lys-Lys-Lys sequences are required for nuclear targeting, but the ten amino acids between these sequences can be mutated without affecting nuclear localization. Because this nuclear localization sequence is composed of two separated elements, it is referred to as bipartite. Similar bipartite motifs appear to function as the localization signals of many nuclear proteins; thus they may be more common than the simpler nuclear localization signal of T antigen.

Protein import through the nuclear pore complex can be operationally divided into two steps, distinguished by whether they require energy

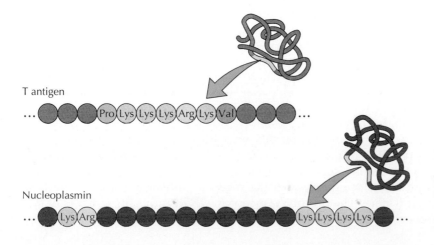

T antigen

Nucleoplasmin

**Figure 8.8**
**Nuclear localization signals**   The T antigen nuclear localization signal is a single stretch of amino acids. In contrast, the nuclear localization signal of nucleoplasmin is bipartite, consisting of a Lys-Arg sequence, followed by a Lys-Lys-Lys-Lys sequence located ten amino acids farther downstream.

# KEY EXPERIMENT

## Identification of Nuclear Localization Signals

**A Short Amino Acid Sequence Able to Specify Nuclear Location**

Daniel Kalderon, Bruce L. Roberts, William D. Richardson, and
Alan E. Smith
National Institute for Medical Research, Mill Hill, London
*Cell, Volume 39, 1984, 499–509*

### The Context

Maintaining the nucleus as a distinct biochemical compartment requires a mechanism by which proteins are segregated between the nucleus and the cytoplasm. Studies in the 1970s established that small molecules diffuse rapidly across the nuclear envelope, but that most proteins are unable to do so. It therefore appeared likely that nuclear proteins are specifically recognized and selectively imported to the nucleus from their sites of synthesis on cytoplasmic ribosomes.

Earlier experiments of Günter Blobel and his colleagues had established that proteins are targeted to the endoplasmic reticulum by signal sequences consisting of short stretches of amino acids (see Chapter 9). In this 1984 paper, Alan Smith and his colleagues extended this principle to the targeting of nuclear proteins by identifying a short amino acid sequence that serves as a nuclear localization signal.

### The Experiments

The viral protein SV40 T antigen was used as a model for studies of nuclear localization in animal cells. T antigen is a 94-kd protein that is required for

SV40 DNA replication and is normally localized to the nucleus of SV40-infected cells. Previous studies in both Alan Smith's laboratory and in the laboratory of Janet Butel (Lanford and Butel, 1984, *Cell* 37: 801–813) had shown that mutation of Lys-128 to either Thr or Asn prevented the normal nuclear accumulation of T antigen in both rodent and monkey cells. Rather than being transported to the nucleus, these mutant T antigens remained in the cytoplasm, suggesting that Lys-128 was part of a nuclear

localization signal. Kalderon and colleagues tested this hypothesis using two distinct experimental approaches.

First, they determined the effects of different deletions on the subcellular localization of T antigen. Mutant T antigens bearing deletions that eliminated amino acids either between residues 1 and 126 or between residue 136 and the C terminus were found to accumulate normally in the nucleus. In contrast, a mutant with a deletion of amino acids 127 to 132 remained in the cytoplasm. Thus, the amino acid sequence extending from residue 127 to 132 appeared to be responsible for nuclear localization of T antigen.

To determine whether this amino acid sequence was able to target other proteins to the nucleus, the investigators constructed chimeras in which the T antigen amino acid sequence

(A)

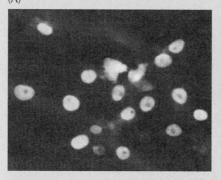

(B)

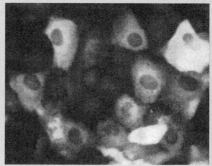

Cells were microinjected with plasmid DNAs encoding chimeric proteins in which SV40 amino acids were fused to pyruvate kinase. Cellular localization of the fusion proteins was then determined by immunofluorescence microscopy. (A) The fusion protein contains an intact SV40 nuclear localization signal (amino acids 126 to 132). (B) The nuclear localization signal has been inactivated by deletion of amino acids 131 and 132.

(Figure 8.9). In the absence of ATP, proteins that contain nuclear localization signals bind to the nuclear pore complex but do not pass through the pore. In this initial step, nuclear localization signals are recognized by a cytosolic receptor protein, and the receptor-substrate complex binds to the nuclear pore. The prototype receptor, called importin, consists of two subunits: One subunit (importin-$\alpha$) binds to the nuclear localization sig-

## Identification of Nuclear Localization Signals (continued)

was fused to proteins that were normally cytoplasmic. These experiments established that the addition of T antigen amino acids 126 to 132 to either β-galactosidase or pyruvate kinase is sufficient to specify the nuclear accumulation of these otherwise cytoplasmic proteins (see figure). This short amino acid sequence of SV40 T antigen thus functions as a nuclear localization signal, which is both necessary and sufficient to target proteins for nuclear import.

### The Impact

As Kalderon and colleagues suggested in their 1984 paper, the nuclear

localization signal of SV40 T antigen has proved to "represent a prototype of similar sequences in other nuclear proteins." By targeting proteins for nuclear import, these signals are key to establishing the biochemical identity of the nucleus and maintaining the fundamental division of eukaryotic cells into nuclear and cytoplasmic compartments. Nuclear localization signals are now known to be recognized by cytoplasmic receptors that transport their substrate proteins to the nuclear pore complex. Although the mechanism of transport through the nuclear pore complex remains to be elucidated, the identification of

Alan Smith

nuclear localization signals was a key advance in understanding nuclear protein import.

nal of substrate proteins, while a second subunit (importin-β) appears to mediate association with the nuclear pore complex.

The second step in nuclear import, translocation through the nuclear pore complex, is an energy-dependent process that requires ATP and GTP. Importin-α appears to be transported through the nuclear pore in association with its substrate protein; importin-β dissociates during translocation. One of the proteins required for translocation is a small GTP-binding protein called Ran, which is related to the Ras proteins (see Figure 7.35). It is

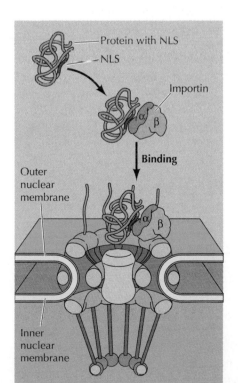

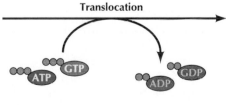

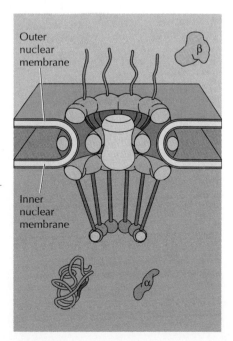

**Figure 8.9**
**Protein import through the nuclear pore complex** Proteins are transported through the nuclear pore complex in two steps. A protein with a nuclear localization sequence (NLS) is first recognized and transported to the nuclear pore complex by a cytoplasmic receptor (importin). The protein and the α subunit of importin are then translocated through the nuclear pore complex in a second, energy-requiring step.

thought that Ran may promote dissociation of the importin subunits, but further details of the mechanism of protein transport through the nuclear pore complex remain to be elucidated.

Some proteins remain within the nucleus following their import from the cytoplasm. Others, called shuttling proteins, constantly move back and forth between the nucleus and the cytoplasm. Some of these proteins act as carriers in the transport of other molecules, such as RNAs; others coordinate nuclear and cytoplasmic functions (e.g., by regulating the activities of transcription factors). Recent studies have identified specific amino acid sequences, nuclear export signals, that are responsible for the rapid export of some shuttling proteins from the nucleus to the cytoplasm. Like nuclear localization signals in nuclear import, nuclear export signals are thought to be recognized by receptors within the nucleus that direct protein transport through the nuclear pore complex.

### Regulation of Nuclear Protein Import

An intriguing aspect of the transport of proteins into the nucleus is that it is another level at which the activities of nuclear proteins can be controlled. Transcription factors, for example, are functional only when they are present in the nucleus, so regulation of their import to the nucleus is a novel means of controlling gene expression. As will be discussed in Chapter 13, the regulated nuclear import of both transcription factors and protein kinases plays an important role in controlling the behavior of cells in response to changes in the environment, because it provides a mechanism by which signals received at the cell surface can be transmitted to the nucleus.

In one mechanism of regulation, transcription factors (or other proteins) associate with cytoplasmic proteins that mask their nuclear localization signals; because their signals are no longer recognizable, these proteins remain in the cytoplasm. A good example is provided by the transcription factor NF-$\kappa$B, which activates transcription of immunoglobulin-$\kappa$ light chains in B lymphocytes (Figure 8.10). In unstimulated cells, NF-$\kappa$B is found as an inactive complex with an inhibitory protein (I$\kappa$B) in the cytoplasm. Binding to I$\kappa$B appears to mask the NF-$\kappa$B nuclear localization signal, thus preventing NF-$\kappa$B from being transported into the nucleus. In stimulated cells, I$\kappa$B is phosphorylated and degraded by ubiquitin-mediated proteolysis, allowing NF-$\kappa$B to enter the nucleus and activate transcription of its target genes.

*Figure 8.10*
**Regulation of nuclear import of transcription factors** The transcription factor NF-$\kappa$B is maintained as an inactive complex with I$\kappa$B, which masks its nuclear localization sequence (NLS), in the cytoplasm. In response to appropriate extracellular signals, I$\kappa$B is phosphorylated and degraded by proteolysis, allowing the import of NF-$\kappa$B to the nucleus. In contrast, the yeast transcription factor SWI5 is maintained in the cytoplasm by phosphorylation in the vicinity of its nuclear localization sequence. Regulated dephosphorylation exposes the NLS and allows SWI5 to be transported to the nucleus at the appropriate stage of the cell cycle.

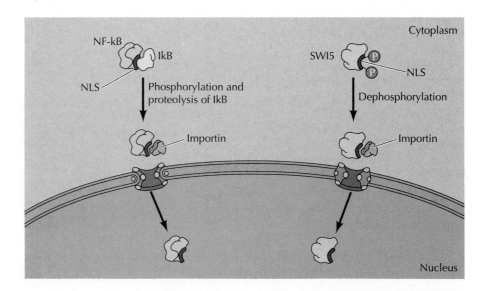

The nuclear import of other transcription factors is regulated directly by their phosphorylation, rather than by association with inhibitory proteins. For example, the yeast transcription factor SWI5 is imported into the nucleus only at a specific stage of the cell cycle (see Figure 8.10). Otherwise, SWI5 is retained in the cytoplasm as a result of phosphorylation at serine residues adjacent to its nuclear localization signal, preventing nuclear import. Regulated dephosphorylation of these sites activates SWI5 at the appropriate stage of the cell cycle by permitting its translocation to the nucleus.

### Transport of RNAs

Whereas many proteins are selectively transported from the cytoplasm into the nucleus, most RNAs are exported from the nucleus to the cytoplasm. Since proteins are synthesized in the cytoplasm, the export of mRNAs, rRNAs, and tRNAs is a critical step in gene expression in eukaryotic cells. Like protein import, the export of RNAs through nuclear pore complexes is an active, energy-dependent process that requires the Ran GTP-binding protein. Details of the mechanism of RNA export, however, are less well understood than those of protein import.

RNAs are transported across the nuclear envelope as RNA-protein complexes, which in some cases are large enough to visualize by electron microscopy (Figure 8.11). Since the substrates for transport are ribonucleoprotein complexes rather than naked RNAs, the signals that direct nuclear export may be present either on the RNAs themselves or on the proteins bound to them. Pre-mRNAs and mRNAs are associated with a set of at least 20 proteins (forming **heterogeneous nuclear ribonucleoproteins**, or **hnRNPs**) throughout their processing in the nucleus and eventual transport to the cytoplasm. At least one of these hnRNP proteins is a shuttling protein that contains a nuclear export signal and may function as a carrier of mRNAs during their export to the cytoplasm. As discussed in a later section of this chapter, ribosomal RNAs are assembled with ribosomal proteins in the nucleolus, and intact ribosomal subunits are then transported to the cytoplasm. Their export from the nucleus appears to be mediated by signals present on ribosomal proteins. For tRNAs, the specific proteins or signals that mediate nuclear export remain to be identified.

In contrast to mRNAs, tRNAs, and rRNAs, which function in the cytoplasm, the snRNAs function within the nucleus as components of the RNA processing machinery. Perhaps surprisingly, these RNAs are initially transported from the nucleus to the cytoplasm, where they associate with proteins to form functional snRNPs and then return to the nucleus (Figure 8.12). Proteins that bind to the 5′ caps of some snRNAs appear to be

*Figure 8.11*
**Transport of a ribonucleoprotein complex**   Insect salivary gland cells produce large ribonucleoprotein complexes (RNPs), which contain 35 to 40 kb of RNA and have a total mass of approximately 30 million daltons. This series of electron micrographs shows the attachment of such an RNP to a nuclear pore complex (A) and the unfolding of the RNA during its translocation to the cytoplasm (B-D). (From H. Mehlin et al., 1992, *Cell* 69: 605.)

(A)    (B)    (C)    (D)

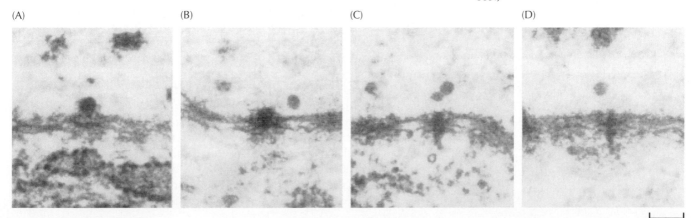

0.1 μm

*Figure 8.12*
**Transport of snRNAs between nucleus and cytoplasm**   Small nuclear RNAs are initially exported from the nucleus to the cytoplasm, where they associate with proteins to form snRNPs. The assembled snRNPs are then transported back into the nucleus.

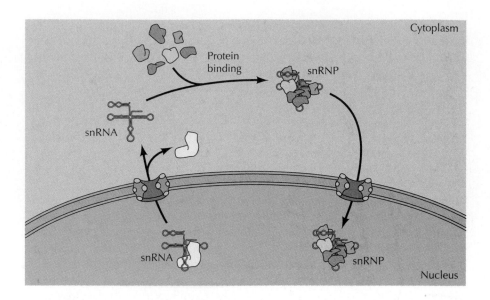

*Figure 8.12*
**Transport of snRNAs between nucleus and cytoplasm**   Small nuclear RNAs are initially exported from the nucleus to the cytoplasm, where they associate with proteins to form snRNPs. The assembled snRNPs are then transported back into the nucleus.

involved in the export of the snRNAs to the cytoplasm, whereas sequences present on the snRNP proteins are responsible for the transport of snRNPs from the cytoplasm to the nucleus.

## INTERNAL ORGANIZATION OF THE NUCLEUS

The nucleus is more than a container in which chromatin, RNAs, and nuclear proteins move freely in aqueous solution. Instead, the nucleus appears to have an internal structure that organizes the genetic material and localizes some nuclear functions to discrete sites. The most obvious aspect of the internal organization of the nucleus is the nucleolus, which, as discussed in the following section, is the site at which the rRNA genes are transcribed and ribosomal subunits are assembled. Additional elements of internal nuclear structure are suggested by the organization of chromosomes and by the potential localization of functions such as DNA replication and pre-mRNA processing to distinct nuclear domains.

### Chromosomes and Higher-Order Chromatin Structure

Chromatin becomes highly condensed during mitosis to form the compact metaphase chromosomes that are distributed to daughter nuclei (see Figure 4.12). During interphase, some of the chromatin (**heterochromatin**) remains highly condensed and is transcriptionally inactive; the remainder of the chromatin (**euchromatin**) is decondensed and distributed throughout the nucleus (Figure 8.13). Cells contain two types of heterochromatin. Constitutive heterochromatin contains DNA sequences that are never transcribed, such as the satellite sequences present at centromeres. Facultative heterochromatin contains sequences that are not transcribed in the cell being examined, but are transcribed in other cell types. Consequently, the amount of facultative heterochromatin varies depending on the transcriptional activity of the cell.

The phenomenon of **X chromosome inactivation** provides an example of the role of heterochromatin in gene expression. In many animals, including humans, females have two X chromosomes, and males have one X and one Y chromosome. The X chromosome contains thousands of genes that are not present on the much smaller Y chromosome (see Figure 4.30). Thus, females have twice as many X chromosome genes as males have. Despite

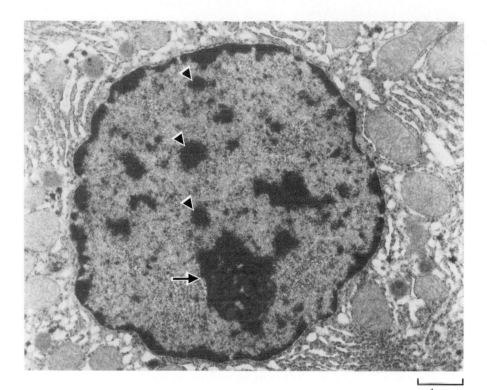

1 μm

*Figure 8.13*
**Heterochromatin in interphase nuclei**
The euchromatin is distributed
throughout the nucleus. The heterochromatin is indicated by arrowheads,
and the nucleolus by an arrow. (Courtesy of Ada L. Olins and Donald E.
Olins, Oak Ridge National Laboratory.)

this difference, female and male cells contain equal amounts of the proteins encoded by X chromosome genes. This results from a dosage compensation mechanism in which one of the two X chromosomes in female cells is inactivated by being converted to heterochromatin early in development. Consequently, only one copy of the X chromosome is available for transcription in either female or male cells.

Although interphase chromatin appears to be uniformly distributed, the chromosomes are actually arranged in an organized fashion and divided into discrete functional domains that play an important role in regulating gene expression. The nonrandom distribution of chromatin within the interphase nucleus was first suggested in 1885 by C. Rabl, who proposed that each chromosome occupies a distinct territory, with centromeres and telomeres attached to opposite sides of the nuclear envelope (Figure 8.14). This basic model of chromosome organization was confirmed nearly a hundred years later (in 1984) by detailed studies of polytene chromosomes in *Drosophila* salivary glands. Rather than randomly winding around one another, each chromosome was found to occupy a discrete region of the nucleus (Figure 8.15). The chromosomes are closely associated with the nuclear envelope at many sites, with their centromeres and telomeres clustered at opposite poles. Individual chromosomes occupy distinct territories within the nuclei of mammalian cells as well, although the positions of centromeres and telomeres in some cells vary from the Rabl pattern.

Like the DNA in metaphase chromosomes (see Figure 4.13), the chromatin in interphase nuclei appears to be organized into looped domains

(A)

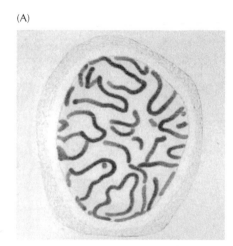

(B)

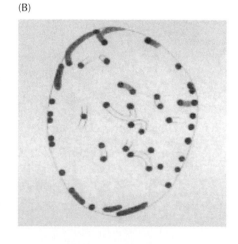

*Figure 8.14*
**Chromosome organization**   Reproduction of hand-drawn sketches of chromosomes in salamander cells. (A) Complete chromosomes. (B) Telomeres only (located at the nuclear membrane). (From C. Rabl, 1885, *Morphologisches Jahrbuch* 10: 214.)

(A)

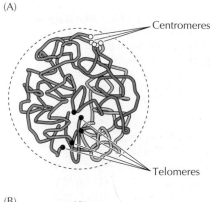

Centromeres

Telomeres

(B)

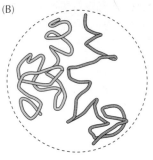

*Figure 8.15*
**Organization of *Drosophila* chromosomes**   (A) A model of the nucleus, showing the five chromosome arms in different colors. The positions of telomeres and centromeres are indicated. (B) The two arms of chromosome 3 are shown to illustrate the topological separation between chromosomes. (From D. Mathog et al., 1984, *Nature* 308: 414.)

containing approximately 50 to 100 kb of DNA. A good example of this looped-domain organization is provided by the highly transcribed chromosomes of amphibian oocytes, in which actively transcribed regions of DNA can be visualized as extended loops of decondensed chromatin (Figure 8.16). These chromatin domains appear to represent discrete functional units, which independently regulate gene expression.

The effects of chromosome organization on gene expression have been demonstrated by a variety of experiments showing that the position of a gene in chromosomal DNA affects the level at which the gene is expressed. For example, the transcriptional activity of genes introduced into transgenic mice depends on their sites of integration in the mouse genome. This effect of chromosomal position on gene expression can be alleviated by sequences known as **locus control regions**, which result in a high level of expression of the introduced genes irrespective of their site of integration. In contrast to transcriptional enhancers (see Chapter 6), locus control regions stimulate only transfected genes that are integrated into chromosomal DNA; they do not affect the expression of unintegrated plasmid DNAs in transient assays. In addition, rather than affecting individual promoters, locus control regions appear to activate large chromosome domains, presumably by inducing long-range alterations in chromatin structure.

The separation between chromosomal domains is maintained by **boundary elements**, which prevent the chromatin structure of one domain from spreading to its neighbors. In addition, sequences known as **insulators** act as barriers that prevent enhancers in one domain from acting on promoters located in an adjacent domain. Like locus control regions, insulators function only in the context of chromosomal DNA, suggesting that they regulate higher-order chromatin structure or control the association of chromatin domains with other nuclear components. Although the mechanisms of action of locus control regions and insulators remain to be elucidated, their functions clearly indicate the importance of higher-order chromatin organization in the control of eukaryotic gene expression.

### Functional Domains within the Nucleus

An internal organization of the nucleus is indicated also by the localization of some nuclear processes to distinct regions of the nucleus. Rather than taking place throughout the nucleus, DNA replication, pre-mRNA processing,

*Figure 8.16*
**Looped chromatin domains**   Light micrograph of a chromosome of amphibian oocytes, showing decondensed loops of actively transcribed chromatin extending from an axis of highly condensed nontranscribed chromatin. (Courtesy of Joseph Gall, Yale University.)

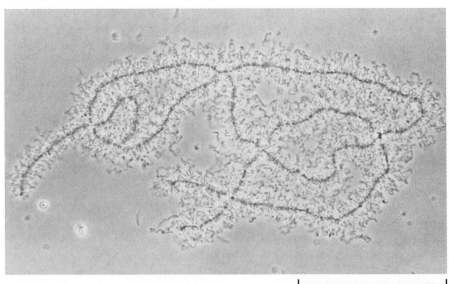

0.1 mm

*Figure 8.17*
**Clustered sites of DNA replication**   Newly replicated DNA was labeled by a brief exposure of cells to bromodeoxyuridine, which is incorporated into DNA in place of thymidine. This substitution allows detection of newly synthesized DNA by immunofluorescence following staining of the nuclei with an antibody against bromodeoxyuridine. Note that the newly replicated DNA is present in discrete clusters distributed throughout the nucleus. (Courtesy of Ronald Berezney, SUNY, Buffalo.)

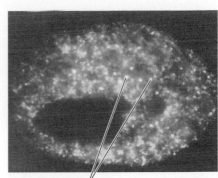

Fluorescent foci of DNA replication

and mRNA transport may be localized to discrete subnuclear structures or domains. The nature and function of these nuclear substructures are not yet clear, however, and understanding the organization of functional domains within the nucleus is an incompletely explored area of cell biology.

The nuclei of mammalian cells appear to contain clustered sites of DNA replication within which the replication of multiple DNA molecules takes place. These discrete sites of DNA replication have been defined by experiments in which newly synthesized DNA was visualized within cell nuclei (Figure 8.17). This was accomplished by labeling cells with bromodeoxyuridine—an analog of thymidine that can be incorporated into DNA and then detected by staining with fluorescent antibodies. In such experiments, the newly replicated DNA was detected in approximately 200 discrete clusters distributed throughout the nucleus. Since approximately 4000 origins of replication are active in a diploid mammalian cell at any given time, each of these clustered sites of DNA replication must contain approximately 40 replication forks. Thus, DNA replication appears to take place in large structures that contain multiple replication complexes organized into distinct functional domains, which have been called replication factories.

Actively transcribed genes appear to be distributed throughout the nucleus, but studies of the organization of components of the splicing machinery and of the localization of mRNAs have suggested that pre-mRNA processing and the transport of mRNA to the nuclear pore complex may also take place within organized structural domains. The localization of splicing components to discrete domains within the nucleus has been demonstrated by immunofluorescent staining with antibodies against snRNPs and splicing factors (Figure 8.18). Rather than being distributed uniformly throughout the nucleus, these components of the splicing apparatus are concentrated in 20 to 50 discrete structures termed nuclear speckles. On the basis of this concentration of splicing components, these speckles have been proposed to be distinct nuclear compartments in which pre-mRNA is processed. However, some scientists have suggested that the speckles could be sites of storage or accumulation of splicing components, rather than sites of active pre-mRNA splicing. Consistent with the latter possibility is evidence showing that at least some pre-mRNA splicing takes place at the sites of transcription, which are distinct from the speckles. The nature of the speckles and the organization of potential splicing domains within the nucleus thus remain unresolved.

The visualization by *in situ* hybridization of transcripts of individual genes has provided independent evidence supporting the view that tran-

*Figure 8.18*
**Localization of splicing components**   Staining with immunofluorescent antibodies indicates that splicing factors are concentrated in discrete domains within the nucleus. (Courtesy of David L. Spector, Cold Spring Harbor Laboratory.)

## MOLECULAR MEDICINE

# *Systemic Lupus Erythematosus*

### The Disease

Systemic lupus erythematosus is a chronic autoimmune disease in which patients produce antibodies against their own normal cell constituents, particularly nuclear antigens. The disease affects women about ten times more frequently than it affects men and has an incidence of about one in 700 among women between the ages of 20 and 60. The disease is characterized by the formation of large amounts of antibody-antigen complexes (immune complexes) that are deposited in tissues and blood vessels throughout the body. The tissue damage and inflammation resulting from deposition of these immune complexes produces a variety of clinical manifestations, including skin rashes, arthritis, and kidney disease. Damage to the kidney is particularly common and can be severe enough to cause kidney failure and death.

### Molecular and Cellular Basis

The fundamental abnormality in systemic lupus erythematosus is the production of autoantibodies that react with constituents of normal cells. These antibodies are frequently directed against components of the nucleus, and the presence of antinuclear antibodies is the principal diagnostic test for the disease. The autoantibodies detected in systemic lupus erythematosus patients are directed against many different nuclear targets, including DNA, histones, nonhistone chromosomal proteins, and snRNPs. Indeed, the snRNPs involved in pre-mRNA splicing were first characterized by the use of antibodies obtained from systemic lupus erythematosus patients.

Because so many different autoantibodies are produced, this disease is thought to result from a defect in immune regulation that leads to a general stimulation of B lymphocytes. Genetic factors contribute to the risk of disease, but the basic cause of the breakdown in immune regulation that leads to systemic lupus erythematosus remains unknown.

### Prevention and Treatment

In the absence of understanding the fundamental defects that cause the production of autoantibodies, there are no currently available approaches for either preventing or curing systemic lupus erythematosus. Instead, current treatments are restricted to arresting the pathological developments of the disease. Anti-inflammatory drugs help reduce the inflamma-

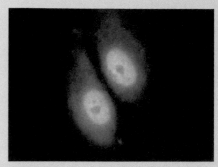

Immunofluorescent staining of nuclei with sera from a patient with systemic lupus erythematosus. (Courtesy of Joan Steitz, Yale University.)

tion and tissue damage resulting from the deposition of immune complexes. In life-threatening cases, patients are also treated with immunosuppressive drugs that inhibit the proliferation of lymphocytes and blunt the immune response. It is hoped that continuing research will lead to an understanding of the defects that result in autoimmunity, ultimately allowing the development of more effective therapies based on restoring normal immune function.

### Reference

Tan, E. M. 1989. Antinuclear antibodies: Diagnostic markers for autoimmune diseases and probes for cell biology. *Adv. Immunol.* 44: 93–151.

scription, pre-mRNA processing, and the transport of mRNAs to the nuclear envelope are associated with an organized nuclear substructure. In particular, hybridization with fluorescent probes for individual mRNAs reveals that specific transcripts accumulate in highly localized foci or tracks (Figure 8.19). Some scientists believe that these tracks represent newly transcribed pre-mRNAs that are associated with a nuclear substructure during their orderly processing and transport to the nuclear pore complex. Alternatively, it has been suggested that pre-mRNA tracks may simply represent nascent transcripts that are still associated with template DNA. Once released from the DNA, mRNAs could then travel rapidly to the nuclear periphery by simple diffusion. Thus, while some studies have suggested the involvement of organized nuclear structures in pre-mRNA processing and transport to the nuclear membrane, a clear understanding of the structural organization of these fundamental processes within the nucleus remains a challenge for future research.

### The Nuclear Matrix

Some scientists believe that a central component of the internal organization of the nucleus is the **nuclear matrix**, which is defined as the structural skeleton of the nucleus. The concept of the nuclear matrix was proposed in 1975 on the basis of experiments in which nuclei were treated with DNase to digest most of the DNA and extracted with high salt buffers to remove histones and other nuclear proteins. Such treatments left a residual framework that maintained the size and shape of the original nucleus (Figure 8.20). This network of insoluble material consisted of three components: the nuclear lamina, a residual nucleolar structure, and an internal network of granular fibers (the nuclear matrix).

The significance of the nuclear matrix detected in this manner remains a highly controversial issue. Advocates of its importance believe that it provides an internal structural framework for the nucleus, analogous to the role of the cytoskeleton as the structural framework of the cell. In addition, it has been proposed that the nuclear matrix serves to organize and anchor functional domains of the nucleus, including chromatin loops, DNA replication factories, splicing domains, and structures involved in mRNA transport.

Many scientists, however, are not convinced that the nuclear matrix exists in the living cell. They believe that the matrix observed after extraction of nuclei may be the result of artificial aggregation of proteins and nucleic acids during preparation. Although the nuclear matrix is visualized after extraction of nuclei by several different methods, its molecular composition has not been clearly defined. In the absence of definitive characterization of its structural components, the nuclear matrix remains an area of dispute among cell biologists.

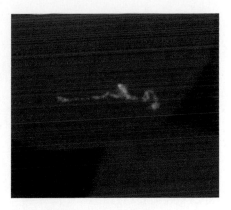

*Figure 8.19*
**RNA tracks**   Probes for a specific mRNA are hybridized to nuclei and detected by yellow-green fluorescence. The red fluorescence distributed throughout the nuclei represents staining of total DNA. The specific transcripts are highly localized in tracks. (From J. B. Lawrence et al., 1989, *Cell* 57: 493.)

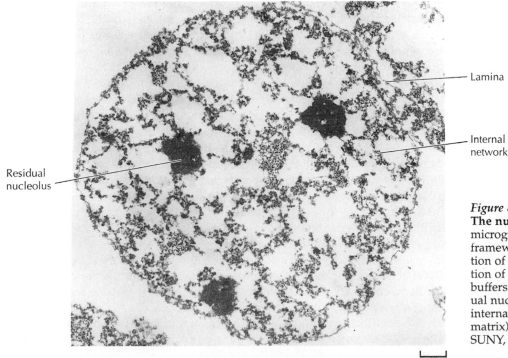

Lamina

Internal network

Residual nucleolus

1 μm

*Figure 8.20*
**The nuclear matrix**   An electron micrograph showing the residual framework of the nucleus after digestion of DNA with DNase and extraction of soluble proteins with high salt buffers. This treatment leaves a residual nuclear lamina, nucleolus, and internal fibrous network (the nuclear matrix). (Courtesy of Ronald Berezney, SUNY, Buffalo.)

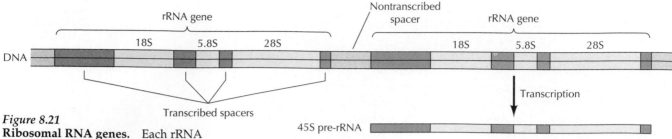

**Figure 8.21**
**Ribosomal RNA genes.** Each rRNA gene is a single transcription unit containing the 18S, 5.8S, and 28S rRNAs and transcribed spacer sequences. The rRNA genes are organized in tandem arrays, separated by nontranscribed spacer DNA.

## THE NUCLEOLUS

The most prominent substructure within the nucleus is the **nucleolus** (see Figure 8.1), which is the site of rRNA transcription and processing, and of ribosome assembly. As discussed in the preceding chapter, cells require large numbers of ribosomes to meet their needs for protein synthesis. Actively growing mammalian cells, for example, contain 5 million to 10 million ribosomes that must be synthesized each time the cell divides. The nucleolus is a ribosome production factory, designed to fulfill the need for large-scale production of rRNAs and assembly of the ribosomal subunits.

### Ribosomal RNA Genes and the Organization of the Nucleolus

The nucleolus, which is not surrounded by a membrane, is organized around the chromosomal regions that contain the genes for the 5.8S, 18S, and 28S rRNAs. Eukaryotic ribosomes contain four types of RNA, designated the 5S, 5.8S, 18S, and 28S rRNAs (see Figure 7.4). The 5.8S, 18S, and 28S rRNAs are transcribed as a single unit within the nucleolus by RNA polymerase I, yielding a 45S ribosomal precursor RNA (Figure 8.21). The 45S pre-rRNA is processed to the 18S rRNA of the 40S (small) ribosomal subunit and to the 5.8S and 28S rRNAs of the 60S (large) ribosomal subunit. Transcription of the 5S rRNA, which is also found in the 60S ribosomal subunit, takes place outside of the nucleolus and is catalyzed by RNA polymerase III.

To meet the need for transcription of large numbers of rRNA molecules, all cells contain multiple copies of the rRNA genes. The human genome, for example, contains about 200 copies of the gene that encodes the 5.8S, 18S, and 28S rRNAs and approximately 2000 copies of the gene that encodes 5S rRNA. The genes for 5.8S, 18S, and 28S rRNAs are clustered in tandem arrays on five different human chromosomes (chromosomes 13, 14, 15, 21, and 22); the 5S rRNA genes are present in a single tandem array on chromosome 1.

The importance of ribosome production is particularly evident in oocytes, in which the rRNA genes are amplified to support the synthesis of the large numbers of ribosomes required for early embryonic development. In *Xenopus* oocytes, the rRNA genes are amplified approximately 2000-fold, resulting in about 1 million copies per cell. These amplified rRNA genes are distributed to thousands of nucleoli (Figure 8.22), which support the accumulation of nearly $10^{12}$ ribosomes per oocyte.

Morphologically, nucleoli consist of three distinguishable regions: the fibrillar center, dense fibrillar component, and granular component (Figure 8.23). These different regions are thought to represent the sites of progressive stages of rRNA transcription, processing, and ribosome assembly. The rRNA genes are located and transcribed in the fibrillar centers and dense fibrillar component of the nucleolus, although which of these regions is the site of transcription remains controversial. Processing and assembly of the pre-rRNA with ribosomal proteins is initiated in the dense fibrillar component and continues in the granular component, which contains nearly com-

**Figure 8.22**
**Nucleoli in amphibian oocytes.** The amplified rRNA genes of *Xenopus* oocytes are clustered in multiple nucleoli (darkly stained spots). (From D. D. Brown and I. B. Dawid, 1968, *Science* 160: 272.)

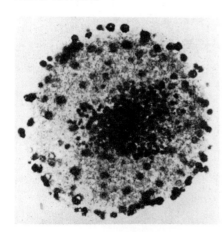

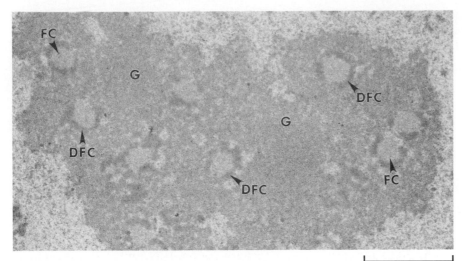

*Figure 8.23*
**Structure of the nucleolus**   An electron micrograph illustrating the fibrillar center (FC), dense fibrillar component (DFC), and granular component (G) of a nucleolus. (Courtesy of David L. Spector, Cold Spring Harbor Laboratory.)

pleted preribosomal subunits, ready for export to the cytoplasm.

Following each cell division, nucleoli form around the chromosomal regions that contain the 5.8S, 18S, and 28S rRNA genes, which are therefore called **nucleolar organizing regions**. The formation of nucleoli requires the transcription of 45S pre-rRNA, which appears to lead to the fusion of small prenucleolar bodies that contain processing factors and other components of the nucleolus. In most cells, the initially separate nucleoli then fuse to form a single nucleolus. The size of the nucleolus depends on the metabolic activity of the cell, with large nucleoli found in cells that are actively engaged in protein synthesis. This variation is due primarily to differences in the size of the granular component, reflecting the levels of ribosome synthesis.

### Transcription and Processing of rRNA

Each nucleolar organizing region contains a cluster of tandemly repeated rRNA genes that are separated from each other by nontranscribed spacer DNA. These genes are very actively transcribed by RNA polymerase I, allowing their transcription to be readily visualized by electron microscopy (Figure 8.24). In such electron micrographs, each of the tandemly arrayed rRNA genes is surrounded by densely packed growing RNA chains, forming a structure that looks like a Christmas tree. The high density of growing RNA chains reflects that of RNA polymerase molecules, which are present at a maximal density of approximately one polymerase per hundred base pairs of template DNA.

*Figure 8.24*
**Transcription of rRNA genes**   An electron micrograph of nucleolar chromatin, showing three rRNA genes separated by nontranscribed spacer DNA. Each rRNA gene is surrounded by an array of growing RNA chains, resulting in a Christmas tree appearance. (Courtesy of O. L. Miller, Jr.)

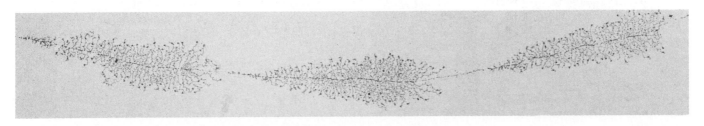

The primary transcript of the rRNA genes is the large 45S pre-rRNA, which contains the 18S, 5.8S, and 28S rRNAs as well as transcribed spacer regions (Figure 8.25). External transcribed spacers are present at both the 5' and 3' ends of the pre-rRNAs, and two internal transcribed spacers lie between the 18S, 5.8S, and 28S rRNA sequences. The initial processing step is a cleavage within the external transcribed spacer near the 5' end of the pre-rRNA, which takes place during the early stages of transcription. This cleavage requires the U3 snRNP (see below) that remains attached to the 5' end of the pre-rRNA, forming the characteristic knobs seen in Figure 8.24. Once transcription is complete, the external transcribed spacer at the 3' end of the molecule is removed. In human cells, this step is followed by a cleavage at the 5' end of the 5.8S region, yielding separate precursors to the 18S and 5.8S + 28S rRNAs. Additional cleavages then result in formation of the mature rRNAs. Processing follows a similar pattern in other species, although there are differences in the order of some of the cleavages.

In addition to cleavage, the processing of pre-rRNA involves a substantial amount of base modification resulting both from the addition of methyl groups to specific bases and ribose residues and from the conversion of uridine to pseudouridine (see Figure 6.38). In animal cells, pre-rRNA processing involves the methylation of approximately a hundred ribose residues and ten bases, in addition to the formation of about a hundred pseudouridines. Most of these modifications occur during or shortly after synthesis of the pre-rRNA, although a few take place at later stages of pre-rRNA processing.

The processing of pre-rRNA requires the action of both proteins and RNAs that are localized to the nucleolus. The involvement of small nuclear RNAs (snRNAs) in pre-mRNA splicing was discussed in Chapter 6. Five snRNAs (U1, U2, U4, U5, and U6) function in pre-mRNA processing as components of the spliceosome (see Figure 6.43). Many other species of small nuclear RNAs are found in nucleoli, where they are thought to be involved in the processing of pre-rRNA. The total number of nucleolar snRNAs is estimated to range from 50 to 100, although most of these nucleolar snRNAs have not yet been characterized and the functions of only a few are known (Table 8.1). The most abundant nucleolar snRNA is U3,

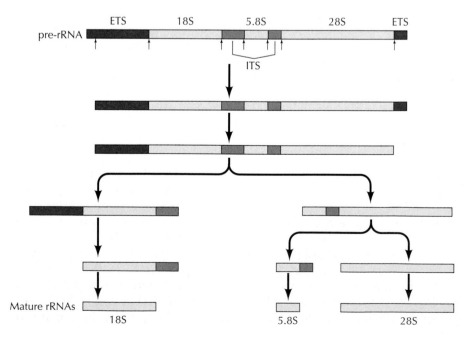

*Figure 8.25*
**Processing of pre-rRNA** The 45S pre-rRNA transcript contains external transcribed spacers (ETS) at both ends and internal transcribed spacers (ITS) between the sequences of 18S, 5.8S, and 28S rRNAs. The pre-rRNA is processed via a series of cleavages (illustrated for human pre-rRNA) to yield the mature rRNA species.

*Table 8.1* Examples of Nucleolar snRNAs

| snRNA | Size[a] | Function |
|---|---|---|
| U3 | 206–228 | Initial processing of 45S pre-rRNA |
| U8 | 136–140 | Processing of 5.8S and 28S rRNAs |
| U14 | 86–96 | Processing of 18S rRNA |
| U22 | 125 | Processing of 18S rRNA |
| MRP | 260–280 | Processing of 5.8S rRNA |

[a]The sizes are indicated in number of bases and are for the snRNAs of animal and plant cells.

which is present in about 200,000 copies per cell. As already noted, U3 is required for the initial cleavage of pre-rRNA within the 5′ external transcribed spacer sequences and may also be involved in subsequent steps of pre-rRNA processing. Another abundant nucleolar snRNA, U8, is required for cleavage both upstream of the 5.8S rRNA and at the 3′ end of the 28S rRNA. Processing of the 18S rRNA requires two other nucleolar snRNAs, U14 and U22, and cleavage upstream of the 5.8S rRNA requires an snRNA designated MRP, which is related to the ribozyme RNase P involved in tRNA processing (see Figure 6.38).

Although some nucleolar snRNAs (including U3 and U8) are transcribed normally, many nucleolar snRNAs are encoded within the introns of protein-coding genes. For example, U14 snRNA is encoded within the intron of the gene encoding the heat-shock protein Hsc70. These snRNAs are thought to be released from excised introns of the host pre-mRNAs by the action of endonucleases. Like the spliceosomal snRNAs, the nucleolar snRNAs are complexed with proteins, forming snRNPs. Individual nucleolar snRNPs appear to consist of single snRNAs associated with eight to ten proteins. The individual snRNPs then assemble on the pre-rRNA to form processing complexes in a manner analogous to the formation of spliceosomes on pre-mRNA.

Although cleavage of pre-rRNA is a major function of some nucleolar snRNPs, other snRNPs may play additional roles in pre-rRNA processing and ribosome assembly. Most of the nucleolar snRNAs contain short sequences complementary to 18S or 28S rRNA. By interacting with pre-rRNA, the nucleolar snRNAs could act as RNA chaperones throughout pre-rRNA maturation and ribosome assembly. For example, nucleolar snRNAs could play important roles in pre-rRNA folding, base modification, binding of ribosomal proteins, and export of assembled ribosomal subunits from the nucleolus. The nature and extent of nucleolar snRNA participation in these aspects of ribosome assembly remain to be elucidated.

## Ribosome Assembly

The formation of ribosomes involves the assembly of the ribosomal precursor RNA with both ribosomal proteins and 5S rRNA (Figure 8.26). The genes that encode ribosomal proteins are transcribed outside of the nucleolus by RNA polymerase II, yielding mRNAs that are translated on cytoplasmic ribosomes. The ribosomal proteins are then transported from the cytoplasm to the nucleolus, where they are assembled with rRNAs to form preribosomal particles. Although the genes for 5S rRNA are also transcribed outside of the nucleolus, in this case by RNA polymerase III, 5S rRNAs similarly are assembled into preribosomal particles within the nucleolus.

The association of ribosomal proteins with rRNA begins while the pre-rRNA is still being synthesized, and more than half of the ribosomal pro-

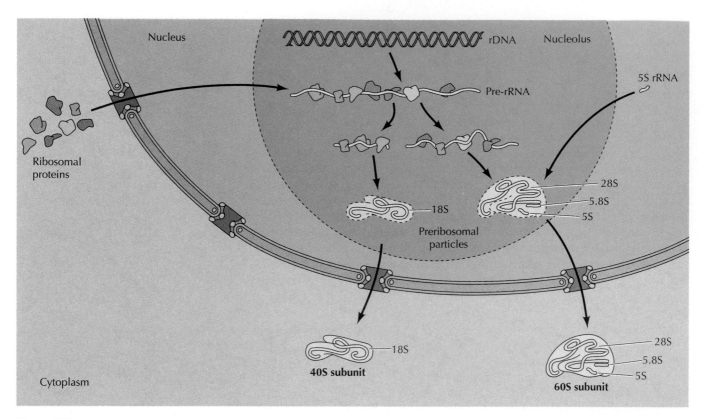

*Figure 8.26*
**Ribosome assembly**   Ribosomal proteins are imported to the nucleolus from the cytoplasm and begin to assemble on pre-rRNA prior to its cleavage. As the pre-rRNA is processed, additional ribosomal proteins and the 5S rRNA (which is synthesized elsewhere in the nucleus) assemble to form preribosomal particles. The final steps of maturation follow the export of preribosomal particles to the cytoplasm, yielding the 40S and 60S ribosomal subunits.

teins are complexed with the pre-rRNA prior to its cleavage. The remaining ribosomal proteins and the 5S rRNA are incorporated into preribosomal particles as cleavage of the pre-rRNA proceeds. The smaller ribosomal subunit, which contains only the 18S rRNA, matures more rapidly than the larger subunit, which contains 28S, 5.8S, and 5S rRNAs. Consequently, most of the preribosomal particles in the nucleolus represent precursors to the large subunit. The final stages of ribosome maturation follow the export of preribosomal particles to the cytoplasm, forming the active 40S and 60S subunits of eukaryotic ribosomes.

During the formation of ribosomes there is a considerable amount of molecular traffic between the nucleus and the cytoplasm. To assemble 10 million ribosomes in 20 hours (the doubling time of a typical rapidly proliferating mammalian cell), the nucleus must import more than 500 million ribosomal proteins, or nearly 10,000 per second, through its nuclear pore complexes. The assembled preribosomal subunits must then be transported from the nucleolus to the cytoplasm. Several major nucleolar proteins have been found to shuttle rapidly between the nucleolus and the cytoplasm. These shuttling proteins may act as carriers, either during the import of ribosomal proteins to the nucleus or during the export of ribosomal subunits to the cytoplasm.

## THE NUCLEUS DURING MITOSIS

A unique feature of the nucleus is that it disassembles and re-forms each time most cells divide. At the beginning of mitosis, the chromosomes condense, the nucleolus disappears, and the nuclear envelope breaks down, resulting in the release of most of the contents of the nucleus into the cytoplasm. At the end of mitosis, the process is reversed: The chromosomes decondense, and nuclear envelopes re-form around the separated sets of daughter chromosomes. Chapter 14 presents a comprehensive discussion of mitosis; in this section we will consider the mechanisms involved in the disassembly and re-formation of the nucleus. The process is controlled largely by reversible phosphorylation and dephosphorylation of nuclear proteins resulting from the action of the **Cdc2** protein kinase, which is a critical regulator of mitosis in all eukaryotic cells.

### Dissolution of the Nuclear Envelope

In most cells, the disassembly of the nuclear envelope marks the end of the prophase of mitosis (Figure 8.27). However, this disassembly of the nucleus is not a universal feature of mitosis and does not occur in all cells. Some unicellular eukaryotes (e.g., yeasts) undergo so-called closed mitosis, in which the nuclear envelope remains intact (Figure 8.28). In closed mitosis, the daughter chromosomes migrate to opposite poles of the nucleus, which then divides in two. The cells of higher eukaryotes, however, usually undergo open mitosis, which is characterized by breakdown of the nuclear envelope. The daughter chromosomes then migrate to opposite poles of the mitotic spindle, and new nuclei reassemble around them.

Disassembly of the nuclear envelope, which parallels a similar breakdown of the endoplasmic reticulum, involves changes in all three of its components: The nuclear membranes are fragmented into vesicles, the nuclear pore complexes dissociate, and the nuclear lamina depolymerizes. The best understood of these events is depolymerization of the nuclear

Prophase

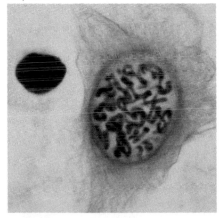

Prometaphase

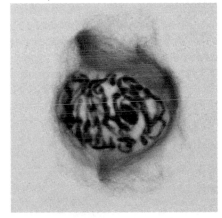

*Figure 8.27*
**The nucleus during mitosis** Micrographs illustrating the progressive stages of mitosis in a plant cell. During prophase, the chromosomes condense, the nucleolus disappears, and the nuclear envelope breaks down. At metaphase, the condensed chromosomes align on the center of the spindle. The daughter chromosomes then move to opposite poles of the spindle (anaphase), and during telophase the chromosomes decondense and the nuclei re-form. Chromosomes are stained blue and spindle microtubules are stained red. (Courtesy of Andrew S. Bajer, University of Oregon.)

Metaphase

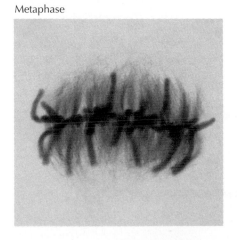

Anaphase

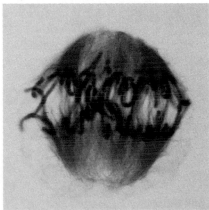

Telophase

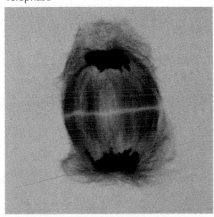

*Figure 8.28*
**Closed and open mitosis**   In closed mitosis, the nuclear envelope remains intact and chromosomes migrate to opposite poles of a spindle within the nucleus. In open mitosis, the nuclear envelope breaks down and then reforms around the two sets of separated chromosomes.

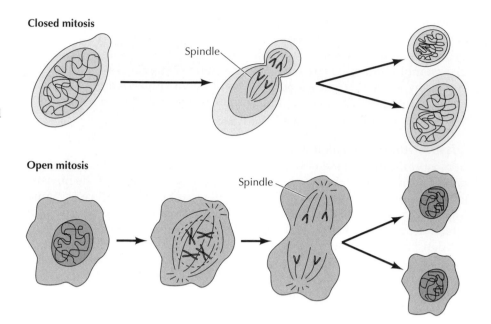

**Closed mitosis**

Spindle

**Open mitosis**

Spindle

lamina—the meshwork of filaments underlying the nuclear membrane. The nuclear lamina is composed of fibrous proteins, lamins, which associate with each other to form filaments. Disassembly of the nuclear lamina results from phosphorylation of the lamins, which causes the filaments to break down into individual lamin dimers (Figure 8.29). Phosphorylation of the lamins is catalyzed by the Cdc2 protein kinase, which was introduced in Chapter 7 (see Figure 7.40) and will be discussed in detail in Chapter 14 as a central regulator of mitosis. Cdc2 (as well as other protein kinases activated in mitotic cells) phosphorylates all the different types of lamins, and treatment of isolated nuclei with Cdc2 has been shown to be sufficient to induce depolymerization of the nuclear lamina. Moreover, the requirement for lamin phosphorylation in the breakdown of the nuclear lamina has been demonstrated directly by the construction of mutant lamins that can

*Figure 8.29*
**Dissolution of the nuclear lamina**
The nuclear lamina consists of a meshwork of lamin filaments. At mitosis, Cdc2 and other protein kinases phosphorylate the lamins, causing the filaments to dissociate into free lamin dimers.

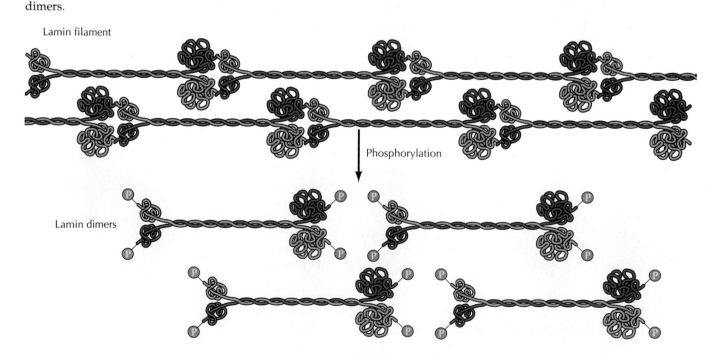

Lamin filament

Phosphorylation

Lamin dimers

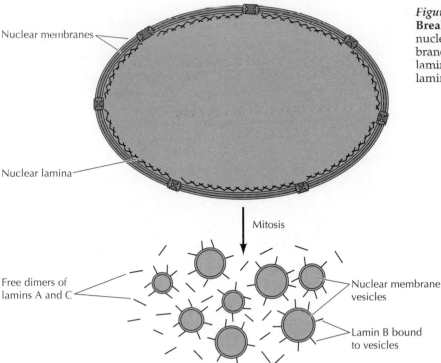

*Figure 8.30*
**Breakdown of the nuclear membrane**    As the nuclear lamina dissociates, the nuclear membrane fragments into vesicles. The B-type lamins remain bound to these vesicles, while lamins A and C are released as free dimers.

Nuclear membranes

Nuclear lamina

Mitosis

Free dimers of lamins A and C

Nuclear membrane vesicles

Lamin B bound to vesicles

no longer be phosphorylated. When genes encoding these mutant lamins were introduced into cells, their expression was found to block normal breakdown of the nuclear lamina as the cells entered mitosis.

In concert with dissolution of the nuclear lamina, the nuclear membrane fragments into vesicles (Figure 8.30). The B-type lamins remain associated with these vesicles, but lamins A and C dissociate from the nuclear membrane and are released as free dimers in the cytosol. Although dissolution of the nuclear lamina is thought to trigger nuclear membrane breakdown, other events may also be involved. The nuclear pore complexes, for example, appear to dissociate into subunits, and phosphorylation of one or more nuclear pore proteins may play a role in disassembly of the complex. Integral nuclear membrane proteins are also phosphorylated at mitosis, and phosphorylation of these proteins may be important in dissociation of the nuclear membrane from both chromosomes and the nuclear lamina. However, the role of these protein phosphorylations in nuclear membrane breakdown is not yet understood.

### Chromosome Condensation

The other major change in nuclear structure during mitosis is chromosome condensation. The interphase chromatin, which is already packaged into nucleosomes, condenses approximately a thousandfold further to form the compact chromosomes seen in mitotic cells (Figure 8.31). This condensation is needed to allow the chromosomes to move along the mitotic spindle without becoming tangled or broken during their distribution to daughter cells. DNA in this highly condensed state can no longer be transcribed, so all RNA synthesis stops during mitosis. As the chromosomes condense and transcription ceases, the nucleolus also disappears.

The condensed DNA in metaphase chromosomes appears to be organized into large loops, each encompassing about a hundred kilobases of DNA, which are attached to a protein scaffold (see Figure 4.13). Despite its

*Figure 8.31*
**Chromosome condensation**   Electron micrograph showing the condensation of individual chromosomes during the prophase of mitosis. (K. G. Murti/Visuals Unlimited.)

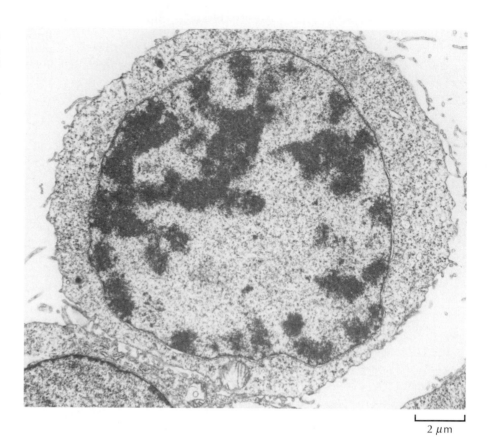

2 μm

fundamental importance, the mechanism of chromosome condensation during mitosis is not understood. The basic unit of chromatin structure is the nucleosome, which consists of 146 base pairs of DNA wrapped around a histone core containing two molecules each of histones H2A, H2B, H3, and H4 (see Figure 4.8). One molecule of histone H1 is bound to the DNA as it enters and exits each nucleosome core particle, and interactions between these H1 molecules are involved in the folding of chromatin into higher-order, more compact structures. Histone H1 is a substrate for the Cdc2 protein kinase and is phosphorylated during mitosis of most cells, consistent with its phosphorylation playing a role in mitotic chromosome condensation. However, the consequences of histone H1 phosphorylation on chromatin structure are not clearly defined and there are some cells in which the phosphorylation of histone H1 fails to correlate with chromosome condensation. In addition, histone H1 is not required for chromosome condensation in cell extracts. Therefore, while histone H1 phosphorylation may be involved, the phosphorylation of other proteins probably also plays a role in signaling mitotic chromosome condensation.

### Re-formation of the Interphase Nucleus

During the completion of mitosis (telophase), two new nuclei form around the separated sets of daughter chromosomes (see Figure 8.27). Chromosome decondensation and reassembly of the nuclear envelope appear to be signaled by inactivation of Cdc2, which was responsible for initiating mitosis by phosphorylating cellular target proteins, including the lamins and histone H1. The progression from metaphase to anaphase involves the activation of a ubiquitin-mediated proteolysis system that inactivates Cdc2 by degrading its regulatory subunit, cyclin B (see Figure 7.40). Inactivation of Cdc2 leads to the dephosphorylation of the proteins that were phosphorylated at the initiation of mitosis, resulting in exit from mitosis and the re-formation of interphase nuclei.

The initial step in re-formation of the nuclear envelope is the binding of the vesicles formed during nuclear membrane breakdown to the surface of chromosomes (Figure 8.32). These vesicles then fuse to form a double membrane around the chromosomes. This is followed by reassembly of the

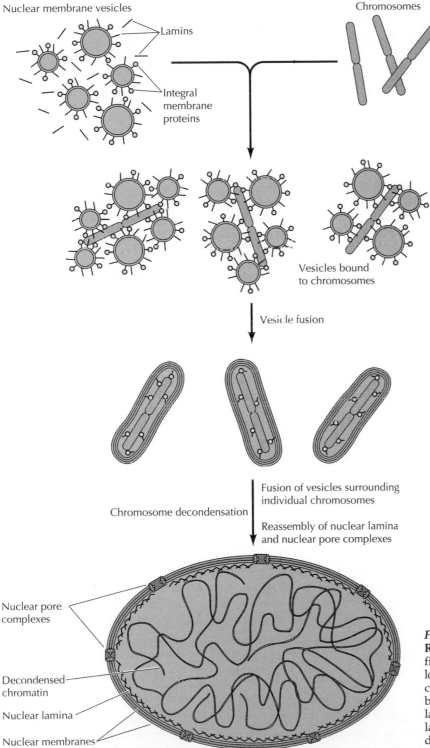

*Figure 8.32*
**Re-formation of the nuclear envelope**  The first step in reassembly of the nuclear envelope is the binding of membrane vesicles to chromosomes, which may be mediated by both integral membrane proteins and B-type lamins. The vesicles then fuse, the nuclear lamina reassembles, and the chromosomes decondense.

nuclear pore complexes, reformation of the nuclear lamina, and chromosome decondensation. The vesicles first fuse to form membranes around individual chromosomes, which then fuse with each other to form a complete single nucleus.

Nuclear reassembly can be studied most easily using *in vitro* systems in which cell-free extracts of frog eggs or mitotic mammalian cells are able to reassemble nuclear envelopes around chromatin or DNA. Analysis of these systems indicates that the initial binding of membrane vesicles to chromosomes is regulated by dephosphorylation of one or more vesicle proteins, which then bind to chromatin. However, the identity of the chromatin-binding proteins has not been determined, and it appears that both lamins and integral membrane proteins of the inner nuclear membrane may participate in the binding of membrane vesicles to chromosomes at the end of mitosis.

The initial re-formation of the nuclear envelope around condensed chromosomes excludes cytoplasmic molecules from the newly assembled nucleus. The new nucleus is then able to expand via the selective import of nuclear proteins from the cytoplasm. Because nuclear localization signals are not cleaved from proteins that are imported to the nucleus, the same nuclear proteins that were released into the cytoplasm following disassembly of the nuclear envelope at the beginning of mitosis can be reimported into the new nuclei formed after mitosis. The nucleolus, too, re-forms as the chromosomes decondense and transcription of the rRNA genes begins, completing the return from mitosis to an interphase nucleus.

## KEY TERMS

nuclear envelope, nuclear membrane, nuclear lamina, lamin

nuclear pore complex

nuclear localization signal

heterogeneous nuclear ribonucleoproteins (hnRNPs)

# Summary

## THE NUCLEAR ENVELOPE AND TRAFFIC BETWEEN THE NUCLEUS AND CYTOPLASM

*Structure of the Nuclear Envelope:* The nuclear envelope separates the contents of the nucleus from the cytoplasm, maintaining the nucleus as a distinct biochemical compartment that houses the genetic material and serves as the site of transcription and RNA processing in eukaryotic cells. The nuclear envelope consists of the inner and outer nuclear membranes, which are joined at nuclear pore complexes, and an underlying nuclear lamina.

*The Nuclear Pore Complex:* Nuclear pore complexes are large structures that provide the only routes through which molecules can travel between the nucleus and cytoplasm. Small molecules are able to diffuse freely through open channels in the nuclear pore complex. Macromolecules are selectively transported in an energy-dependent process.

*Selective Import of Proteins to the Nucleus:* Proteins destined for import to the nucleus contain nuclear localization signals that direct their transport through the nuclear pore complex. Many proteins constantly shuttle back and forth between the nucleus and the cytoplasm.

*Regulation of Nuclear Protein Import:* The activity of some proteins, such as transcription factors, is controlled by regulation of their import to the nucleus.

*Transport of RNAs:* RNAs are transported through the nuclear pore complex as ribonucleoprotein complexes. Messenger RNAs, ribosomal

RNAs, and transfer RNAs are exported from the nucleus to function in protein synthesis. Small nuclear RNAs are initially transported from the nucleus to the cytoplasm, where they associate with proteins to form snRNPs; then they return to the nucleus.

## INTERNAL ORGANIZATION OF THE NUCLEUS

*Chromosomes and Higher-Order Chromatin Structure:* The interphase nucleus contains transcriptionally inactive, highly condensed heterochromatin as well as decondensed euchromatin. Interphase chromosomes are organized within the nucleus and divided into large looped domains that function as independent units.

heterochromatin, euchromatin, X chromosome inactivation, locus control region, insulator, boundary element

*Functional Domains within the Nucleus:* Some nuclear processes, such as DNA replication and pre-mRNA metabolism, may be localized to discrete functional domains.

*The Nuclear Matrix:* Some scientists believe that an internal fibrous network (the nuclear matrix) serves as the structural framework of the nucleus. However, the components of the nuclear matrix have not yet been defined, and other scientists think that the matrix may be an aggregate of proteins and nucleic acids formed during extraction procedures rather than a structure that exists in the living cell.

nuclear matrix

## THE NUCLEOLUS

*Ribosomal RNA Genes and the Organization of the Nucleolus:* The nucleolus is organized around the genes for ribosomal RNAs. It is the site of rRNA transcription and processing, and of ribosome assembly.

nucleolus, nucleolar organizing region

*Transcription and Processing of rRNA:* The primary transcript of the rRNA genes is 45S pre-rRNA, which is processed to yield 18S, 5.8S, and 28S rRNAs. Processing of pre-rRNA is mediated by nucleolar snRNAs, which may also function as RNA chaperones during the assembly of ribosomal subunits.

*Ribosome Assembly:* Ribosomal subunits are assembled within the nucleolus from rRNAs and ribosomal proteins.

## THE NUCLEUS DURING MITOSIS

*Dissolution of the Nuclear Envelope:* Entry into mitosis is signaled by activation of the Cdc2 protein kinase. In most cells, the nuclear envelope breaks down at the end of prophase. Depolymerization of the nuclear lamina results from phosphorylation of the lamins by Cdc2 and other protein kinases.

Cdc2

*Chromosome Condensation:* Phosphorylation of histone H1 is correlated with the condensation of mitotic chromosomes, but the mechanism of chromosome condensation has not been established.

*Re-formation of the Interphase Nucleus:* Inactivation of Cdc2 at the end of mitosis leads to re-formation of the nuclear envelope and chromosome decondensation. Nuclear proteins are then selectively imported through nuclear pore complexes.

## QUESTIONS

**1.** You are studying a transcription factor that is regulated by phosphorylation of serine residues that inactivate its nuclear localization signal. How would mutating these serines to alanines affect subcellular localization of the transcription factor and expression of its target gene?

**2.** How would mutational inactivation of the nuclear export signal of a shuttling protein affect its subcellular distribution?

**3.** Consider the construct (diagrammed below) in which a single enhancer (E) acts on two separate promoters (P1 and P2). How would insertion of an insulator between E and P1 affect transcription from P1 and P2? What about insertion of an insulator between P1 and P2?

——E———P1———P2——

**4.** Treatment of cells with actinomycin D (a general inhibitor of transcription) prevents the formation of nucleoli following mitosis. Treatment with $\alpha$-amanitin (a specific inhibitor of RNA polymerase II) does not have this effect. Why?

**5.** What posttranslational modification of the B-type lamins facilitates their association with the nuclear membrane? How might this explain the difference in behavior of B-type lamins compared to that of A- and C-type lamins during mitosis?

## REFERENCES AND FURTHER READING

### General Reference

Newport, J. W. and D. J. Forbes. 1987. The nucleus: Structure, function, and dynamics. *Ann. Rev. Biochem.* 56: 535–565. [R]

### The Nuclear Envelope and Traffic between the Nucleus and Cytoplasm

Adam, S. A. 1995. The importance of importin. *Trends Cell Biol.* 5: 189–191. [R]

Akey, C. W. and M. Radermacher. 1993. Architecture of the *Xenopus* nuclear pore complex revealed by three-dimensional cryoelectron microscopy. *J. Cell Biol.* 122: 1–19. [P]

Davis, L. I. 1995. The nuclear pore complex. *Ann. Rev. Biochem.* 64: 865–896. [R]

Dingwall, C. and R. A. Laskey. 1991. Nuclear targeting sequences—a consensus? *Trends Biochem. Sci.* 16: 478–481. [R]

Dingwall, C. and R. Laskey. 1992. The nuclear membrane. *Science* 258: 942–947. [R]

Forbes, D. J. 1992. Structure and function of the nuclear pore. *Ann. Rev. Cell Biol.* 8: 495–527. [R]

Franke, W. W. 1987. Nuclear lamins and cytoplasmic intermediate filament proteins: A growing multigene family. *Cell* 48: 3–4. [R]

Gerace, L. 1995. Nuclear export signals and the fast track to the cytoplasm. *Cell* 82: 341–344. [R]

Gerace, L. and B. Burke. 1988. Functional organization of the nuclear envelope. *Ann. Rev. Cell Biol.* 4: 353–374. [R]

Gerace, L. and R. Foisner. 1994. Integral membrane proteins and dynamic organization of the nuclear envelope. *Trends Cell Biol.* 4: 127–131. [R]

Gorlich, D., F. Vogel, A. D. Mills, E. Hartmann and R. A. Laskey. 1995. Distinct functions for the two importin subunits in nuclear protein import. *Nature* 377: 246–248. [P]

Hinshaw, J. E., B. O. Carragher and R. A. Milligan. 1992. Architecture and design of the nuclear pore complex. *Cell* 69: 1133–1141. [P]

Izaurralde, E. and I. W. Mattaj. 1995. RNA export. *Cell* 81: 153–159. [R]

Kalderon, D., B. L. Roberts, W. D. Richardson and A. E. Smith. 1984. A short amino acid sequence able to specify nuclear location. *Cell* 39: 499–509. [P]

Lanford, R. E. and J. S. Butel. 1984. Construction and characterization of an SV40 mutant defective in nuclear transport of T antigen. *Cell* 37: 801–813. [P]

Michael, W. M., M. Choi and G. Dreyfuss. 1995. A nuclear export signal in hnRNP A1: A signal-mediated, temperature-dependent nuclear protein export pathway. *Cell* 83: 415–422. [P]

Moore, M. S. and G. Blobel. 1994. A G protein involved in nucleocytoplasmic transport: the role of Ran. *Trends Biochem. Sci.* 19: 211–216. [R]

Pante, N. and U. Aebi. 1993. The nuclear pore complex. *J. Cell Biol.* 122: 977–984. [R]

Pinol-Roma, S. and G. Dreyfuss. 1993. hnRNP proteins: Localization and transport between the nucleus and cytoplasm. *Trends Cell Biol.* 3: 151–155. [R]

Powers, M. A. and D. J. Forbes. 1994. Cytosolic factors in nuclear transport: What's importin? *Cell* 79: 931–934. [R]

Rout, M. P. and S. R. Wente. 1994. Pores for thought: Nuclear pore complex proteins. *Trends Cell Biol.* 4: 357–365. [R]

Vandromme, M., C. Gauthier-Rouviere, N. Lamb and A. Fernandez. 1996. Regulation of transcription factor localization: Fine-tuning of gene expression. *Trends Biochem. Sci.* 21: 59–64. [R]

### Internal Organization of the Nucleus

Berezney, R. and D. S. Coffey. 1975. Nuclear protein matrix: Association with newly synthesized DNA. *Science* 189: 291–293. [P]

Corces, V. G. 1995. Chromatin insulators: Keeping enhancers under control. *Nature* 376: 462–463. [R]

Dillon, N. and F. Grosveld. 1993. Transcriptional regulation of multigene loci: Multilevel control. *Trends Genet.* 9: 134–137. [R]

Gasser, S. M. and U. K. Laemmli. 1987. A glimpse at chromosomal order. *Trends Genet.* 3: 16–22. [R]

Gilson, E., T. Laroche and S. M. Gasser. 1993. Telomeres and the functional architecture of the nucleus. *Trends Cell Biol.* 3: 128–134. [R]

Hozak, P. and P. R. Cook. 1994. Replication factories. *Trends Cell Biol.* 4: 48–49. [R]

Kramer, J., Z. Zachar and P. M. Bingham. 1994. Nuclear pre-mRNA metabolism: Channels and tracks. *Trends Cell Biol.* 4: 35–37. [R]

Lyon, M. F. 1992. Some milestones in the history of X-chromosome inactivation. *Ann. Rev. Genet.* 26: 17–28. [R]

Manuelidis, L. 1990. A view of interphase chromosomes. *Science* 250: 1533–1540. [R]

Mathog, D., M. Hochstrasser, Y. Gruenbaum, H. Saumweber and J. Sedat. 1984. Characteristic folding pattern of polytene chromosomes in *Drosophila* salivary gland nuclei. *Nature* 308: 414–421. [P]

Rosbash, M. and R. H. Singer. 1993. RNA travel: Tracks from DNA to cytoplasm. *Cell* 75: 399–401. [R]

Spector, D. L. 1993. Macromolecular domains within the cell nucleus. *Ann. Rev. Cell Biol.* 9: 265–315. [R]

Wolffe, A. P. 1994. Insulating chromatin. *Curr. Biol.* 4: 85–87. [R]

Xing, Y. and J. B. Lawrence. 1993. Nuclear RNA tracks: Structural basis for transcription and splicing? *Trends Cell Biol.* 3: 346–353. [R]

Zhang, G., K. L. Taneja, R. H. Singer and M. R. Green. 1994. Localization of pre-mRNA splicing in mammalian nuclei. *Nature* 372: 809–812. [P]

## *The Nucleolus*

Borer, R. A., C. F. Lehner, H. M. Eppenberger and E. A. Nigg. 1989. Major nucleolar proteins shuttle between nucleus and cytoplasm. *Cell* 56: 379–390. [P]

Kass, S., K. Tyc, J. A. Steitz and B. Sollner-Webb. 1990. The U3 small nucleolar ribonucleoprotein functions in the first step of preribosomal RNA processing. *Cell* 60: 897–908. [P]

Maxwell, E. S. and M. J. Fournier. 1995. The small nucleolar RNAs. *Ann. Rev. Biochem.* 35: 897–934. [R]

Miller, O. L., Jr. and B. Beatty. 1969. Visualization of nucleolar genes. *Science* 164: 955–957. [P]

Scheer, U., M. Thiry and G. Goessens. 1993. Structure, function and assembly of the nucleolus. *Trends Cell Biol.* 3: 236–241. [R]

Sollner-Webb, B. 1993. Novel intron-encoded small nucleolar RNAs. *Cell* 75: 403–405. [R]

Sollner-Webb, B. and E. B. Mougey. 1991. News from the nucleolus: rRNA gene expression. *Trends Biochem. Sci.* 16: 58–62. [R]

Sommerville, J. 1986. Nucleolar structure and ribosome biogenesis. *Trends Biochem. Sci.* 11: 438–442. [R]

Steitz, J. A. and K. T. Tycowski. 1995. Small RNA chaperones for ribosome biogenesis. *Science* 270: 1626–1627. [R]

Warner, J. R. 1990. The nucleolus and ribosome formation. *Curr. Opin. Cell Biol.* 2: 521–527. [R]

## *The Nucleus during Mitosis*

Foisner, R. and L. Gerace. 1993. Integral membrane proteins of the nuclear envelope interact with lamins and chromosomes, and binding is modulated by mitotic phosphorylation. *Cell* 73: 1267–1279. [P]

Heald, R. and F. McKeon. 1990. Mutations of phosphorylation sites in lamin A that prevent nuclear lamina disassembly in mitosis. *Cell* 61: 579–589. [P]

Lourim, D. and G. Krohne. 1994. Lamin-dependent nuclear envelope reassembly following mitosis: An argument. *Trends Cell Biol.* 4: 314–318. [R]

Murray, A. and T. Hunt. 1993. *The Cell Cycle: An Introduction.* New York: W. H. Freeman.

Ohsumi, K., C. Katagiri and T. Kishimoto. 1993. Chromosome condensation in *Xenopus* mitotic extracts without histone H1. *Science* 262: 2033–2035. [P]

Peter, M., J. Nakagawa, M. Doreé, J. C. Labbé and E. A. Nigg. 1990. *In vitro* disassembly of the nuclear lamina and M phase-specific phosphorylation of lamins by cdc2 kinase. *Cell* 61: 591–602. [P]

Peterson, C. L. 1994. The SMC family: Novel motor proteins for chromosome condensation? *Cell* 79: 389–392. [R]

Pfaller, R., C. Smythe and J. W. Newport. 1991. Assembly/disassembly of the nuclear envelope membrane: Cell cycle-dependent binding of nuclear membrane vesicles to chromatin *in vitro. Cell* 65: 209–217. [P]

Ward, G. E. and M. W. Kirschner. 1990. Identification of cell cycle-regulated phosphorylation sites on nuclear lamin C. *Cell* 61: 561–577. [P]

Wolffe, A. 1995. *Chromatin: Structure and Function.* 2nd ed. New York: Academic Press.

# 9

# Protein Sorting and Transport

## The Endoplasmic Reticulum, Golgi Apparatus, and Lysosomes

IN ADDITION TO THE PRESENCE OF A NUCLEUS, eukaryotic cells are distinguished from prokaryotic cells by the presence of membrane-enclosed organelles within their cytoplasm. These organelles provide discrete compartments in which specific cellular activities take place, and the resulting subdivision of the cytoplasm allows eukaryotic cells to function efficiently in spite of their large size (about a thousand times the volume of bacteria).

Because of the complex internal organization of eukaryotic cells, the sorting and targeting of proteins to their appropriate destinations are considerable tasks. The first step of protein sorting takes place while translation is still in progress. Proteins destined for the endoplasmic reticulum, the Golgi apparatus, lysosomes, the plasma membrane, and secretion from the cell are synthesized on ribosomes that are bound to the membrane of the endoplasmic reticulum. As translation proceeds, the polypeptide chains are transported into the endoplasmic reticulum, where protein folding and processing take place. From the endoplasmic reticulum, proteins are transported in vesicles to the Golgi apparatus, where they are further processed and sorted for transport to lysosomes, the plasma membrane, or secretion from the cell. The endoplasmic reticulum, Golgi apparatus, and lysosomes are thus distinguished from other cytoplasmic organelles by their common involvement in protein processing and connection by vesicular transport.

## THE ENDOPLASMIC RETICULUM

The **endoplasmic reticulum** (ER) is a network of membrane-enclosed tubules and sacs (cisternae) that extends from the nuclear membrane throughout the cytoplasm (Figure 9.1). The entire endoplasmic reticulum is enclosed by a continuous membrane and is the largest organelle of most eukaryotic cells. Its membrane may account for about half of all cell membranes, and the space enclosed by the ER (the lumen, or cisternal space) may represent about 10% of the total cell volume. As discussed below, there are two distinct types of ER that perform different functions within the cell. The **rough ER**, which is covered by ribosomes on its outer surface, func-

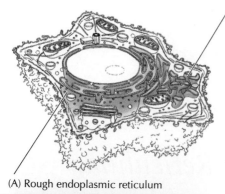

(A) Rough endoplasmic reticulum

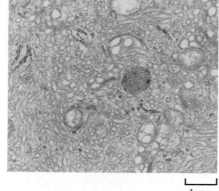

(B) Smooth endoplasmic reticulum

1 μm

*Figure 9.1*
**The endoplasmic reticulum (ER).**
(A) Electron micrograph of rough ER in rat liver cells. Ribosomes are attached to the cytosolic face of the ER membrane. (B) Electron micrograph of smooth ER in corpus lutem, which is active in steroid hormone synthesis. (A, Richard Rodewald/Biological Photo Service; B, Barry F. King/Biological Photo Service.)

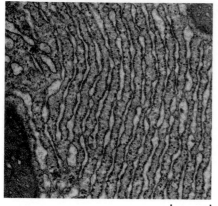

1 μm

tions in protein processing. The **smooth ER** is not associated with ribosomes and is involved in lipid, rather than protein, metabolism.

## The Endoplasmic Reticulum and Protein Secretion

The role of the endoplasmic reticulum in protein processing and sorting was first demonstrated by George Palade and his colleagues in the 1960s (Figure 9.2). These investigators studied the fate of newly synthesized proteins in specialized cells of the pancreas (pancreatic acinar cells) that secrete digestive enzymes into the small intestine. Because most proteins synthesized by these cells are secreted, Palade and coworkers were able to study the pathway taken by secreted proteins simply by labeling newly synthe-

*Figure 9.2*
**The secretory pathway**   Pancreatic acinar cells, which secrete most of their newly synthesized proteins into the digestive tract, were labeled with radioactive amino acids to study the intracellular pathway taken by secreted proteins. After a short incubation with radioactive amino acids (3-minute label), autoradiography revealed that newly synthesized proteins were localized to the rough ER. Following further incubation with nonradioactive amino acids (a chase), proteins were found to move from the ER to the Golgi apparatus and then, within secretory vesicles, to the plasma membrane and cell exterior.

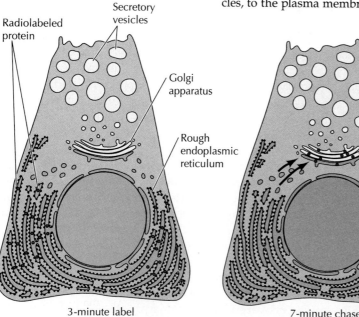
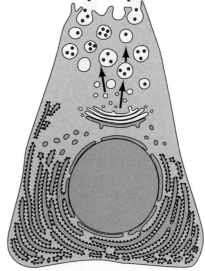

Radiolabeled protein

Secretory vesicles

Golgi apparatus

Rough endoplasmic reticulum

3-minute label          7-minute chase          120-minute chase

sized proteins with radioactive amino acids. The location of the radiolabeled proteins within the cell was then determined by autoradiography, revealing the cellular sites involved in the events leading to protein secretion. After a brief exposure of pancreatic acinar cells to radioactive amino acids, newly synthesized proteins were detected in the rough ER, which was therefore identified as the site of synthesis of proteins destined for secretion. If the cells were then incubated for a short time in media containing nonradioactive amino acids (a process known as a chase), the radiolabeled proteins were detected in the Golgi apparatus. Following longer chase periods, the radiolabeled proteins traveled from the Golgi apparatus to the cell surface in **secretory vesicles**, which then fused with the plasma membrane to release their contents outside of the cell.

These experiments defined a pathway taken by secreted proteins, the **secretory pathway**: rough ER →Golgi →secretory vesicles →cell exterior. Further studies extended these results and demonstrated that this pathway is not restricted to proteins destined for secretion from the cell. Plasma membrane and lysosomal proteins also travel from the rough ER to the Golgi and then to their final destinations. Still other proteins travel through the initial steps of the secretory pathway but are then retained and function within either the ER or the Golgi apparatus.

The entrance of proteins into the ER thus represents a major branch point for the traffic of proteins within eukaryotic cells, which corresponds to the synthesis of proteins on either free or membrane-bound ribosomes (Figure 9.3). Proteins destined to remain in the cytosol or to be incorporated into the nucleus, mitochondria, chloroplasts, or peroxisomes are synthesized on free ribosomes and released into the cytosol when their translation is complete. In contrast, most proteins destined for secretion or incorporation into the ER, Golgi apparatus, lysosomes, or plasma membrane are synthesized on membrane-bound ribosomes and transferred into the rough ER as their translation proceeds.

**Free ribosomes in cytosol**

**Membrane-bound ribosomes**

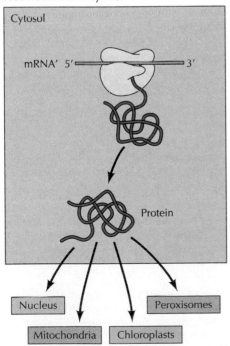

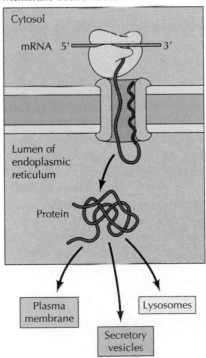

*Figure 9.3*
**Overview of protein sorting** Proteins synthesized on free ribosomes either remain in the cytosol or are transported to the nucleus, mitochondria, chloroplasts, or peroxisomes. In contrast, proteins synthesized on membrane-bound ribosomes are translocated into the ER while their translation is in progress. They may be either retained within the ER or transported to the Golgi apparatus and, from there, to lysosomes, the plasma membrane, or the cell exterior via secretory vesicles.

## Targeting Proteins to the Endoplasmic Reticulum

Ribosomes are targeted for binding to the ER membrane by the amino acid sequence of the polypeptide chain being synthesized, rather than by intrinsic properties of the ribosome itself. Free and membrane-bound ribosomes are functionally indistinguishable, and all protein synthesis initiates on ribosomes that are free in the cytosol. Ribosomes engaged in the synthesis of proteins that are destined for secretion are then targeted to the endoplasmic reticulum by a **signal sequence** at the amino terminus of the growing polypeptide chain. These signal sequences are short stretches of hydrophobic amino acids that are cleaved from the polypeptide chain during its transfer into the ER lumen.

The general role of signal sequences in targeting proteins to their appropriate locations within the cell was first elucidated by studies of the import of secretory proteins into the ER. These experiments used *in vitro* preparations of rough ER, which were isolated from cell extracts by density-gradient centrifugation (Figure 9.4). When cells are disrupted, the ER breaks up into small vesicles called **microsomes**. Because the vesicles derived from the rough ER are covered with ribosomes, they can be separated from similar vesicles derived from the smooth ER or from other membranes (e.g., the plasma membrane). In particular, the large amount of RNA within ribosomes increases the density of the membrane vesicles to which they are attached, allowing purification of vesicles derived from the rough ER (rough microsomes) by equilibrium centrifugation in density gradients.

David Sabatini and Günter Blobel first proposed in 1971 that the signal for ribosome attachment to the ER was an amino acid sequence near the amino terminus of the growing polypeptide chain. This hypothesis was supported by the results of *in vitro* translation of mRNAs encoding secreted proteins, such as immunoglobulins (Figure 9.5). If an mRNA encoding a secreted protein was translated on free ribosomes *in vitro*, it was found that the protein produced was slightly larger than the normal secreted protein. If microsomes were added to the system, however, the *in vitro*-translated protein was incorporated into the microsomes and cleaved to the correct size. These experiments led to a more detailed formulation of the signal hypothesis, which proposed that an amino-terminal leader sequence targets the polypeptide chain to the microsomes and is then cleaved by a microsomal protease. Many subsequent findings have substantiated this model, including recombinant DNA experiments demonstrating that addition of a signal sequence to a normally nonsecreted protein is sufficient to direct the incorporation of the recombinant protein into the rough ER.

The mechanism by which secretory proteins are targeted to the ER is now well understood. The signal sequences span about 20 amino acids, including a stretch of hydrophobic residues, usually at the amino terminus of the polypeptide chain (Figure 9.6). As they emerge from the ribosome, signal sequences are recognized and bound by a **signal recognition particle (SRP)** consisting of six polypeptides and a small cytoplasmic RNA (**7SL**

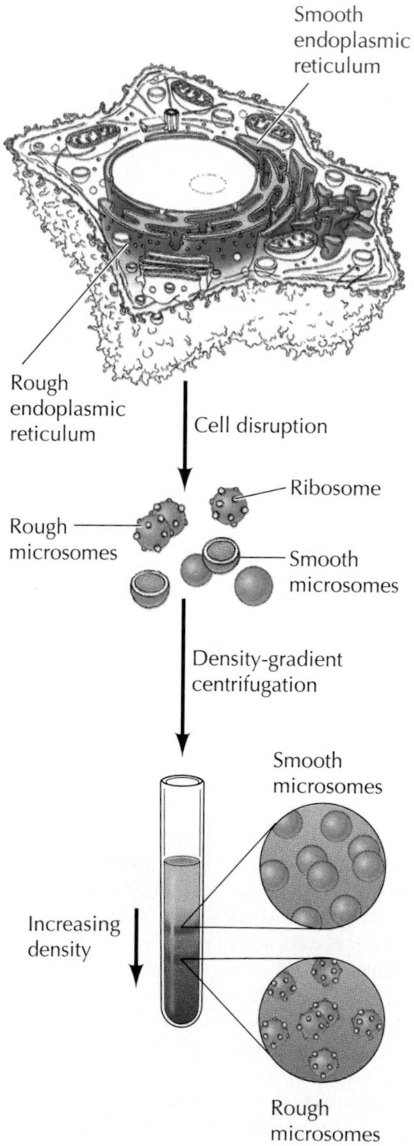

Smooth endoplasmic reticulum

Rough endoplasmic reticulum

Cell disruption

Rough microsomes

Ribosome

Smooth microsomes

Density-gradient centrifugation

Smooth microsomes

Increasing density

Rough microsomes

**Figure 9.4**
**Isolation of rough ER**   When cells are disrupted, the ER fragments into small vesicles called microsomes. The microsomes derived from the rough ER (rough microsomes) are lined with ribosomes on their outer surface. Because ribosomes contain large amounts of RNA, the rough microsomes are denser than smooth microsomes and can be isolated by equilibrium density-gradient centrifugation.

**Translation on free ribosomes**

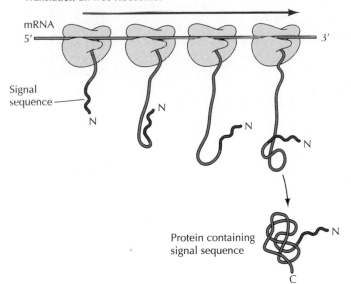

mRNA
5'                                                              3'

Signal
sequence

N

N

N

N

Protein containing
signal sequence

N

C

**Translation with microsomes present**

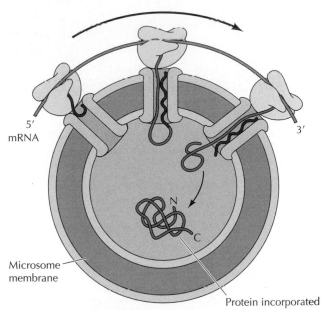

5'
mRNA

3'

N

C

Microsome
membrane

Protein incorporated
into microsomes

*Figure 9.5*
**Incorporation of secretory proteins into microsomes** Secretory proteins are targeted to the ER by a signal sequence at their amino (N) terminus, which is removed during incorporation of the growing polypeptide chain into the ER. This was demonstrated by experiments showing that translation of secretory protein mRNAs on free ribosomes yielded proteins that retained their signal sequences and were therefore slightly larger than the normal secreted proteins. However, when microsomes were added to the system, the growing polypeptide chains were incorporated into the microsomes and the signal sequences were removed by proteolytic cleavage.

RNA). Binding of the SRP inhibits further translation and targets the complex (the SRP, ribosome, and growing polypeptide chain) to the rough ER by binding to the SRP receptor on the ER membrane (Figure 9.7). Binding to the receptor releases the SRP from both the ribosome and the signal sequence of the nascent polypeptide chain. The ribosome then binds to a protein translocation complex in the ER membrane, and the signal sequence is inserted into a membrane channel. Translation is then able to resume, and the growing polypeptide chain is translocated across the membrane into the ER. As translocation proceeds, the signal sequence is cleaved by **signal peptidase** and the polypeptide is released into the lumen of the ER.

A variety of experiments, including analyses of ion conductance and studies with fluorescent probes, have demonstrated that polypeptide chains are translocated across the ER membrane through aqueous channels formed by transmembrane proteins. These channels appear to be opened by the signal sequence and are then maintained in the open configuration by ribosome binding, allowing the growing polypeptide chain to cross the membrane as translation proceeds. Both biochemical and genetic studies have identified the Sec61 complex, which consists of three membrane-spanning proteins, as a major component of the ER protein-conducting channel in both yeasts and mammalian cells. Sec61 binds the ribosome to the ER membrane, and the growing polypeptide chain probably is transferred directly from the ribosome into a Sec61 membrane channel. Particularly clear evidence for the key role of Sec61 comes from *in vitro* experiments by Tom Rapoport and colleagues, who were able to reconstitute a functional protein translocation apparatus by incorporating Sec61 and the SRP receptor into lipid vesicles. The yeast and mammalian Sec61 proteins are closely related to the plasma membrane proteins that translocate secreted polypeptides in *E. coli*, demonstrating a striking conservation of the protein secretion machinery in prokaryotic and eukaryotic cells.

*Figure 9.6*
**The signal sequence of growth hormone** Most signal sequences contain a stretch of hydrophobic amino acids, preceded by basic residues (e.g., arginine).

Cleavage site of
signal peptidase

Met Ala Thr Gly Ser Arg Thr Ser Leu Leu Leu Ala Phe Gly Leu Leu Cys Leu Pro Trp Leu Gln Glu Gly Ser Ala Phe Pro Thr

## KEY EXPERIMENT

# The Signal Hypothesis

**Transfer of Proteins across Membranes. I. Presence of Proteolytically Processed and Unprocessed Nascent Immunoglobulin Light Chains on Membrane-Bound Ribosomes of Murine Myeloma**

Günter Blobel and Bernhard Dobberstein
Rockefeller University, New York
*Journal of Cell Biology, 1975, Volume 67, pages 835-851*

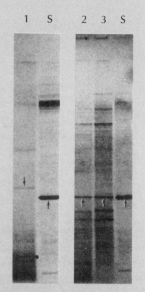

### The Context

How are specific polypeptide chains transferred across the appropriate membranes? Studies in the 1950s and 1960s indicated that secreted proteins are synthesized on membrane-bound ribosomes and transferred across the membrane during their synthesis. However, this did not explain why ribosomes engaged in the synthesis of secreted proteins attach to membranes while ribosomes synthesizing cytosolic proteins do not. A hypothesis to explain this difference was first suggested by Günter Blobel and David Sabatini in 1971. At that time, they proposed that (1) mRNAs to be translated on membrane-bound ribosomes contain a unique set of codons just 3' of the initiation site, (2) translation of these codons yields a unique sequence at the amino terminus of the growing polypeptide chain (the signal sequence), and (3) the signal sequence triggers attachment of the ribosome to the membrane. In 1975, Blobel and Dobberstein reported a series of experiments that provided critical support for this notion. In addition, they proposed "a somewhat more detailed version of this hypoth-

esis, henceforth referred to as the signal hypothesis."

### The Experiments

Myelomas are cancers of B lymphocytes that actively secrete immunoglobulins, so they provide a good model for studies of secreted proteins. Previous studies in Cesar Milstein's laboratory had shown that the proteins produced by *in vitro* translation of immunoglobulin light-chain mRNA contain about 20 amino acids at their amino terminus that are not present in the secreted light chains. This result led to the suggestion that these amino acids direct binding of the ribosome to the membrane. To test this idea, Blobel and Dobberstein investigated the synthesis of light chains by membrane-bound ribosomes from myeloma cells.

As expected from earlier work, *in vitro* translation of light-chain mRNA on free ribosomes yielded a protein that was larger than the secreted light chain (see figure). In contrast, *in vitro* translation of mRNA associated with membrane-bound ribosomes from myeloma cells yielded a protein that was the same size as the normally

*In vitro* translation of immunoglobulin light-chain mRNA on free ribosomes (lane 1) yields a product that migrates slower than secreted light chains (lane S) in gel electrophoresis. In contrast, light chains synthesized by *in vitro* translation on membrane-bound ribosomes (lane 2) are the same size as secreted light chains. In addition, the products of *in vitro* translation on membrane-bound ribosomes were unaffected by subsequent digestion with proteases (lane 3), indicating that they were protected from the proteases by insertion into microsomes.

secreted light chain. Moreover, the light chains synthesized by ribosomes that remained bound to microsomes were resistant to digestion by added proteases, indicating that the light chains had been transferred into the microsomes.

## Insertion of Proteins into the ER Membrane

Proteins destined for secretion or residence within the lumen of the ER, Golgi apparatus, or lysosomes are translocated across the ER membrane and released into the lumen of the ER as already described. However, proteins destined for incorporation into the plasma membrane or the membranes of the ER, Golgi, or lysosomes are initially inserted into the ER membrane instead of being released into the lumen. From the ER mem-

## The Signal Hypothesis *(continued)*

These results indicated that an amino-terminal signal sequence is removed by a microsomal protease as growing polypeptide chains are transferred across the membrane. The results were interpreted in terms of a more detailed version of the signal hypothesis. As stated by Blobel and Dobberstein, "the essential feature of the signal hypothesis is the occurrence of a unique sequence of codons, located immediately to the right of the initiation codon, which is present only in those mRNAs whose translation products are to be transferred across a membrane."

### The Impact

The selective transfer of proteins across membranes is critical to the maintenance of the membrane-enclosed organelles of eukaryotic cells. To maintain the identity of these organelles, proteins must be translocated specifically across the appropriate membranes. The signal hypothesis provided the conceptual basis for understanding this phenomenon. Not only has this basic model been firmly substantiated for the transfer of secreted proteins into the endoplasmic reticulum, but it also has provided the framework for understanding the targeting of proteins to all the other

Günter Blobel

membrane-enclosed compartments of the cell, thereby impacting virtually all areas of cell biology.

brane, they proceed to their final destination along the same pathway as that of secretory proteins: ER → Golgi → plasma membrane or lysosomes. These proteins are transported along this pathway as membrane components, however, rather than as soluble proteins.

Integral membrane proteins are embedded in the membrane by hydrophobic regions that span the phospholipid bilayer (see Figure 2.48). The

*Figure 9.7*
**Targeting secretory proteins to the ER**
Step 1: As the signal sequence emerges from the ribosome, it is recognized and bound by the signal recognition particle (SRP). Step 2: The SRP escorts the complex to the ER membrane, where it binds to the SRP receptor. Step 3: The SRP is released, the ribosome binds to a membrane translocation complex, and the signal sequence is inserted into a membrane channel. Step 4: Translation resumes, and the growing polypeptide chain is translocated across the membrane. Step 5: Cleavage of the signal sequence by signal peptidase releases the polypeptide into the lumen of the ER.

3' mRNA

5'

Signal sequence

3'

5'

**Step 1**

SRP

**Step 2**

5' 3'

**Step 3**

5' 3'

**Step 4**

5' 3'

SRP receptor

Endoplasmic reticulum lumen

Translocation complex

Signal peptidase

**Step 5**

membrane-spanning portions of these proteins are usually α helical regions consisting of 20 to 25 hydrophobic amino acids. The formation of an α helix maximizes hydrogen bonding between the peptide bonds, and the hydrophobic amino acid side chains interact with the fatty acid tails of the phospholipids. However, different integral membrane proteins differ in how they are inserted (Figure 9.8). For example, whereas some integral membrane proteins span the membrane only once, others have multiple membrane-spanning regions. In addition, some proteins are oriented in the membrane with their amino terminus on the cytosolic side; others have their carboxy terminus exposed to the cytosol. These orientations of proteins inserted into the ER, Golgi, lysosomal, and plasma membranes are established as the growing polypeptide chains are translocated into the ER. The lumen of the ER is topologically equivalent to the exterior of the cell, so the domains of plasma membrane proteins that are exposed on the cell surface correspond to the regions of polypeptide chains that are translocated into the ER (Figure 9.9).

The most straightforward mode of insertion into the ER membrane results in the synthesis of transmembrane proteins oriented with their carboxy termini exposed to the cytosol (Figure 9.10). These proteins have a normal amino-terminal signal sequence, which is cleaved by signal peptidase during translocation of the polypeptide chain across the ER membrane. They are then anchored in the membrane by a second membrane-spanning α helix in the middle of the protein. This sequence, called a stop-transfer sequence, blocks further translocation of the polypeptide chain across the ER membrane. The ribosome is released from the translocation apparatus, and the carboxy-terminal portion of the growing polypeptide chain continues to be synthesized in the cytosol. The insertion of these proteins in the membrane thus involves the sequential action of two distinct elements: a cleavable amino-terminal signal sequence that initiates translocation across the membrane and a stop-transfer sequence that anchors the protein in the membrane.

Proteins can also be anchored in the ER membrane by internal signal sequences that are not cleaved by signal peptidase (Figure 9.11). These internal signal sequences are recognized by the SRP and brought to the ER membrane as already discussed. Because they are not cleaved by signal peptidase, however, these signal sequences act as transmembrane α helices that anchor proteins in the ER membrane. Importantly, internal signal sequences can be oriented so as to direct the translocation of either the

*Figure 9.8* **Orientations of membrane proteins**  Integral membrane proteins span the membrane via α helical regions of 20 to 25 hydrophobic amino acids, which can be inserted in a variety of orientations. The two proteins at left and center each span the membrane once, but they differ in whether the amino (N) or carboxy (C) terminus is on the cytosolic side. On the right is an example of a protein that has multiple membrane-spanning regions.

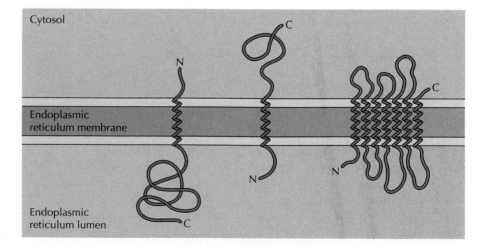

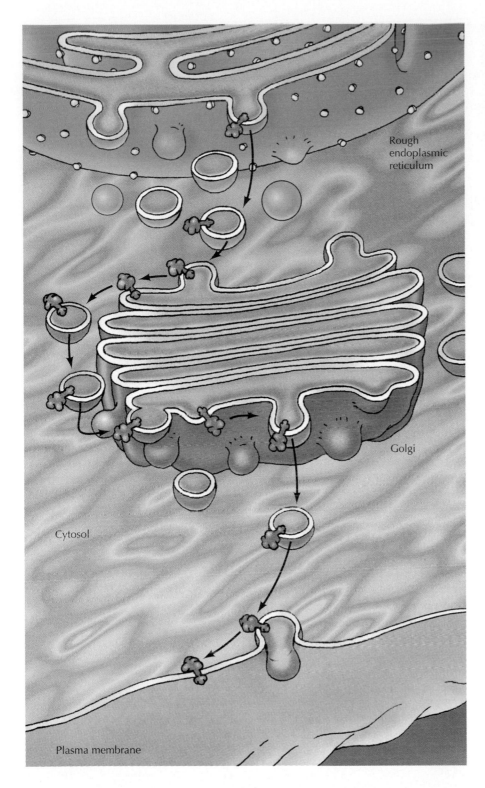

*Figure 9.9*
**Topology of the secretory pathway**
The lumens of the endoplasmic reticulum and Golgi apparatus are topologically equivalent to the exterior of the cell. Consequently, those portions of polypeptide chains that are translocated into the ER are exposed on the cell surface following transport to the plasma membrane.

Rough endoplasmic reticulum

Golgi

Cytosol

Plasma membrane

amino or carboxy terminus of the polypeptide chain across the membrane. Therefore, depending on the orientation of the signal sequence, proteins inserted into the membrane by this mechanism can have either their amino or carboxy terminus exposed to the cytosol.

Proteins that span the membrane multiple times are thought to be inserted as a result of an alternating series of internal signal sequences and

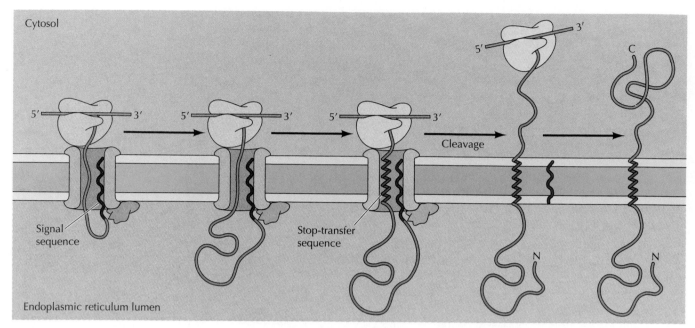

**Figure 9.10**
**Insertion of a membrane protein with a cleavable signal sequence and a single stop-transfer sequence** The signal sequence is cleaved as the polypeptide chain crosses the membrane, so the amino terminus of the polypeptide chain is exposed in the ER lumen. However, translocation of the poly-peptide chain across the membrane is halted by a stop-transfer sequence that anchors the protein in the membrane. The ribosome is released from the membrane and continued translation results in a membrane-spanning protein with its carboxy terminus on the cytosolic side.

**Figure 9.11**
**Insertion of membrane proteins with internal noncleavable signal sequences** Internal noncleavable signal sequences can lead to the insertion of polypeptide chains in either orientation in the ER membrane. (A) The signal sequence directs insertion of the polypeptide such that its amino terminus is exposed on the cytosolic side. The remainder of the polypeptide chain is translocated into the ER as translation proceeds. The signal sequence is not cleaved, so it acts as a membrane-spanning sequence that anchors the protein in the membrane with its carboxy terminus in the lumen of the ER. (B) Other internal signal sequences are oriented to direct the transfer of the amino-terminal portion of the polypeptide across the membrane. Continued translation results in a protein that spans the ER membrane with its amino terminus in the lumen and its carboxy terminus in the cytosol. Note that this orientation is the same as that resulting from insertion of a protein that contains a cleavable signal sequence followed by a stop-transfer sequence (see Figure 9.10).

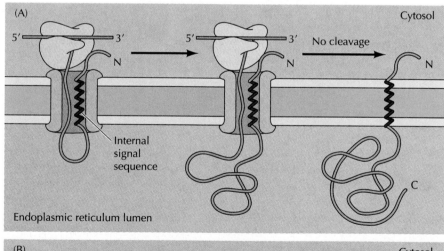

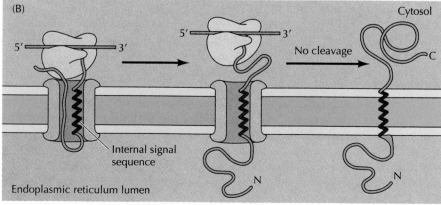

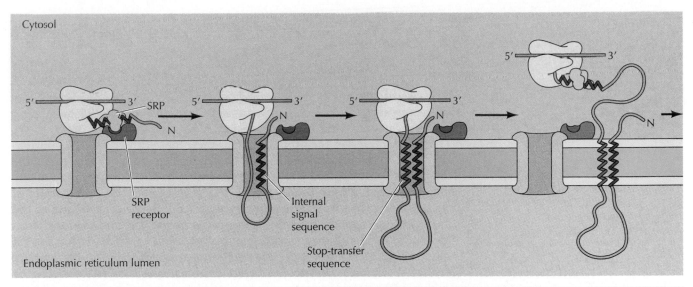

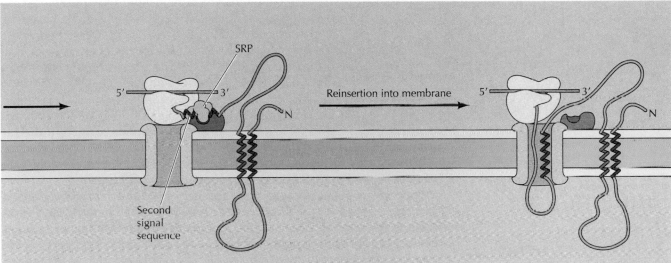

**Figure 9.12**
**Insertion of a protein that spans the membrane multiple times**   In this example, an internal signal sequence results in insertion of the polypeptide chain with its amino terminus on the cytosolic side of the membrane. A stop-transfer sequence then causes the polypeptide chain to form a loop within the lumen of the ER, and translation continues in the cytosol. A second internal signal sequence triggers reinsertion of the polypeptide chain into the ER membrane, forming a loop in the cytosol. The process can be repeated many times, resulting in the insertion of proteins with multiple membrane-spanning regions.

stop-transfer sequences. For example, an internal signal sequence can result in membrane insertion of a polypeptide chain with its amino terminus on the cytosolic side (Figure 9.12). If a stop-transfer sequence is then encountered, the polypeptide will form a loop in the ER lumen, and protein synthesis will continue on the cytosolic side of the membrane. If a second signal sequence is encountered, the growing polypeptide chain will again be inserted into the ER, forming another looped domain on the cytosolic side of the membrane. This can be followed by yet another stop-transfer sequence and so forth, so that an alternating series of signal and stop-transfer sequences can result in the insertion of proteins that span the membrane multiple times, with looped domains exposed on both the lumenal and cytosolic sides.

### Protein Folding and Processing in the ER

The folding of polypeptide chains into their correct three-dimensional conformations, the assembly of polypeptides into multisubunit proteins, and the covalent modifications involved in protein processing were discussed in Chapter 7. For proteins that enter the secretory pathway, many of these events occur either during translocation across the ER membrane or within the ER lumen. One such processing event is the proteolytic cleavage of the signal peptide as the polypeptide chain is translocated across the ER membrane. The ER is also the site of protein folding, assembly of multisubunit proteins, disulfide bond formation, the initial stages of glycosylation, and the addition of glycolipid anchors to some plasma membrane proteins. Indeed, the primary role of lumenal ER proteins is to catalyze the folding and assembly of newly translocated polypeptides.

As already discussed, proteins are translocated across the ER membrane as unfolded polypeptide chains while their translation is still in progress. These polypeptides, therefore, fold into their three-dimensional conformations within the ER, assisted by the molecular chaperones that facilitate the folding of polypeptide chains (see Chapter 7). For example, one of the major proteins within the ER lumen is a member of the Hsp70 family of chaperones called BiP (for *b*inding *p*rotein). BiP is thought to bind to the unfolded polypeptide chain as it crosses the membrane and then mediates protein folding and the assembly of multisubunit proteins within the ER (Figure 9.13). Correctly assembled proteins are released from BiP and are available for transport to the Golgi apparatus. Abnormally folded or improperly assembled proteins, however, remain bound to BiP and are consequently retained within the ER or degraded, rather than being transported farther along the secretory pathway.

The formation of disulfide bonds between the side chains of cysteine residues is an important aspect of protein folding and assembly within the ER. These bonds do not form in the cytosol, which is characterized by a reducing environment that maintains cysteine residues in their reduced (—SH) state. In the ER, however, an oxidizing environment promotes disulfide (S—S) bond formation, and disulfide bonds formed in the ER play important roles in the structure of secreted and cell surface proteins. Disulfide bond formation is facilitated by the enzyme **protein disulfide isomerase** (see Figure 7.21), which is located in the ER lumen.

Proteins are also glycosylated on specific asparagine residues (*N*-linked glycosylation) within the ER while their translation is still in process (Figure 9.14). As discussed in Chapter 7, oligosaccharide units consisting of 14 sugar residues are added to acceptor asparagine residues of growing polypeptide chains as they are translocated into the ER. The oligosaccharide is synthesized on a lipid (dolichol) carrier anchored in the ER membrane. It is then transferred as a unit to acceptor asparagine residues in the consensus sequence Asn-X-Ser/Thr by a membrane-bound enzyme called oligosaccharyl trans-

*Figure 9.13*
**Protein folding in the ER**   The molecular chaperone BiP binds to polypeptide chains as they cross the ER membrane and facilitates protein folding and assembly within the ER.

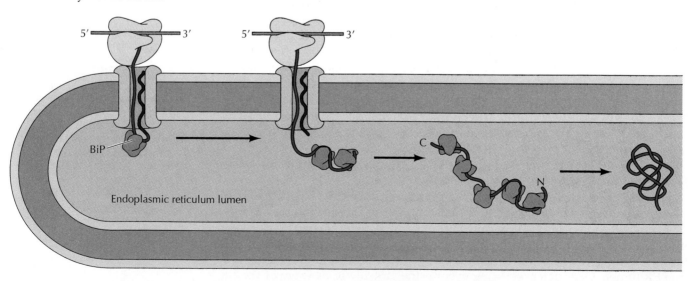

5'  3'      5'  3'

BiP

C

N

Endoplasmic reticulum lumen

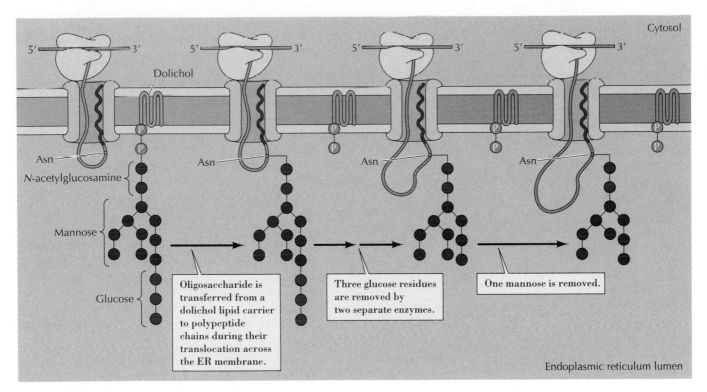

**Figure 9.14**
**Protein glycosylation in the ER**

ferase. Four sugar residues (three glucose and one mannose) are removed while the protein is still within the ER, and the protein is modified further after being transported to the Golgi apparatus (discussed later in this chapter).

Some proteins are anchored in the plasma membrane by glycolipids rather than by membrane-spanning regions of the polypeptide chain. Because these membrane-anchoring glycolipids contain phosphatidylinositol, they are called **glycosylphosphatidylinositol (GPI) anchors**, the structure of which was illustrated in Figure 7.32. The GPI anchors are assembled in the ER membrane. They are then added immediately after completion of protein synthesis to the carboxy terminus of some proteins anchored in the membrane by a C-terminal membrane-spanning region (Figure 9.15). The transmembrane region of the protein is exchanged for the GPI anchor, so these proteins remain attached to the membrane only by their associated glycolipid. Like transmembrane proteins, they are transported to the cell surface as membrane components via the secretory pathway. Their orientation within the ER dictates that GPI-anchored proteins are exposed on the outside of the cell, with the GPI anchor mediating their attachment to the plasma membrane.

## The Smooth ER and Lipid Synthesis

In addition to its activities in the processing of secreted and membrane proteins, the ER is the major site at which membrane lipids are synthesized in eukaryotic cells. Because they are extremely hydrophobic, lipids are synthesized in association with already existing cellular membranes rather than in the aqueous environment of the cytosol. Although some lipids are synthesized in association with other membranes, most are synthesized in the ER. They are then transported from the ER to their ultimate destinations either in vesicles or by carrier proteins, as discussed later in this chapter and in Chapter 10.

*Figure 9.15*
**Addition of GPI anchors** Glycosyl-phosphatidylinositol (GPI) anchors contain two fatty acid chains, an oligosaccharide portion consisting of inositol and other sugars, and ethanolamine (see Figure 7.32 for a more detailed structure). The GPI anchors are assembled in the ER and added to polypeptides anchored in the membrane by a carboxy-terminal membrane-spanning region. The membrane-spanning region is cleaved, and the new carboxy terminus is joined to the NH$_2$ group of ethanolamine immediately after translation is completed, leaving the protein attached to the membrane by the GPI anchor.

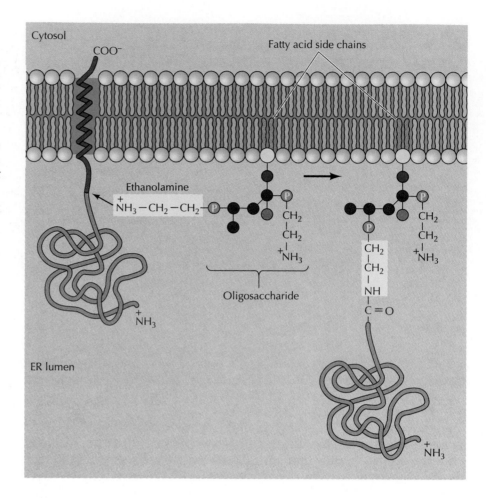

The membranes of eukaryotic cells are composed of three main types of lipids: phospholipids, glycolipids, and cholesterol. Most of the phospholipids, which are the basic structural components of the membrane, are derived from glycerol. They are synthesized on the cytosolic side of the ER membrane, from water-soluble cytosolic precursors (Figure 9.16). Fatty acids are first transferred from coenzyme A carriers to glycerol-3-phosphate by a membrane-bound enzyme, and the resulting phospholipid (phosphatidic acid) is inserted into the membrane. Enzymes on the cytosolic face of the ER membrane then catalyze the addition of different polar head groups, resulting in formation of phosphatidylcholine, phosphatidylserine, phosphatidylethanolamine, or phosphatidylinositol.

The synthesis of these phospholipids on the cytosolic side of the ER membrane allows the hydrophobic fatty acid chains to remain buried in the membrane while membrane-bound enzymes catalyze their reactions with

*Figure 9.16* ▶

**Synthesis of phospholipids** Glycerol phospholipids are synthesized in the ER membrane from cytosolic precursors. Two fatty acids linked to coenzyme A (CoA) carriers are first joined to glycerol-3-phosphate, yielding phosphatidic acid, which is simultaneously inserted into the membrane. A phosphatase then converts phosphatidic acid to diacylglycerol. The attachment of different polar head groups to diacylglycerol then results in formation of phosphatidylcholine, phosphatidylethanolamine, or phosphatidylserine. Phosphatidylinositol is formed from phosphatidic acid, rather than from diacylglycerol.

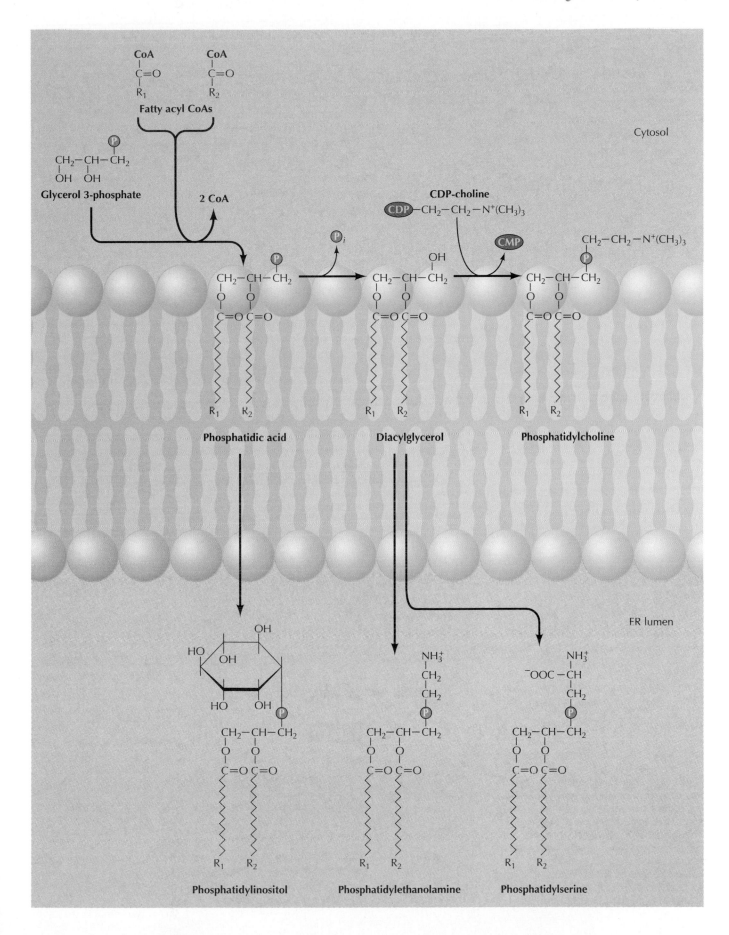

water-soluble precursors (e.g., CDP-choline) in the cytosol. Because of this topography, however, new phospholipids are added only to the cytosolic half of the ER membrane (Figure 9.17). To maintain a stable membrane, some of these newly synthesized phospholipids must therefore be transferred to the other (lumenal) half of the ER bilayer. This transfer does not occur spontaneously because it requires the passage of a polar head group through the membrane. Instead, membrane proteins called **flippases** catalyze the rapid translocation of phospholipids across the ER membrane, resulting in even growth of both halves of the bilayer.

In addition to its role in synthesis of the glycerol phospholipids, the ER also serves as the major site of synthesis of two other membrane lipids: cholesterol and ceramide (Figure 9.18). As discussed later, ceramide is converted to either glycolipids or sphingomyelin (the only membrane phospholipid not derived from glycerol) in the Golgi apparatus. The ER is thus responsible for synthesis of either the final products or the precursors of all the major lipids of eukaryotic membranes.

Smooth ER is abundant in cell types that are particularly active in lipid metabolism. For example, steroid hormones are synthesized (from cholesterol) in the ER, so large amounts of smooth ER are found in steroid-producing cells, such as those in the testis and ovary. In addition, smooth ER

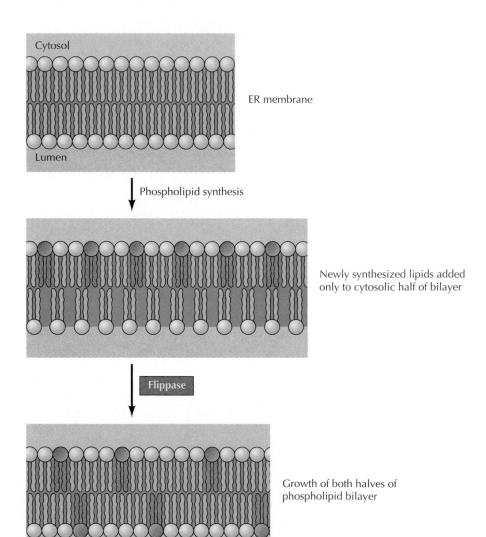

*Figure 9.17*
**Translocation of phospholipids across the ER membrane** Because phospholipids are synthesized on the cytosolic side of the ER membrane, they are added only to the cytosolic half of the bilayer. They are then translocated across the membrane by phospholipid flippases, resulting in even growth of both halves of the phospholipid bilayer.

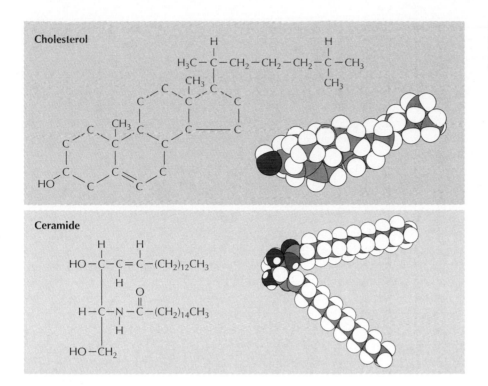

*Figure 9.18*
**Structure of cholesterol and ceramide**
The hydrogens attached to the ring
carbons of cholesterol are not shown.

is abundant in the liver, where it contains enzymes that metabolize various lipid-soluble compounds. These detoxifying enzymes inactivate a number of potentially harmful drugs (e.g., phenobarbital) by converting them to water-soluble compounds that can be eliminated from the body in the urine. The smooth ER is thus involved in multiple aspects of the metabolism of lipids and lipid-soluble compounds.

## Export of Proteins and Lipids from the ER

Both proteins and lipids travel along the secretory pathway in transport vesicles, which bud from the membrane of one organelle and then fuse with the membrane of another. Thus, molecules are exported from the ER in vesicles that bud from the ER and fuse with the Golgi membrane (Figure 9.19). Subsequent steps in the secretory pathway involve vesicular transport between different compartments of the Golgi and from the Golgi to lysosomes or the plasma membrane. In each case, proteins within the lumen of one organelle are packaged into the budding transport vesicle and then released into the lumen of the recipient organelle following vesicle fusion. Membrane proteins and lipids are transported similarly, and it is noteworthy that their topological orientation is maintained as they travel from one membrane-enclosed organelle to another. For example, the domains of a protein exposed on the cytosolic side of the ER membrane will also be exposed on the cytosolic side of the Golgi and plasma membranes, whereas protein domains exposed on the lumenal side of the ER membrane will be exposed on the lumenal side of the Golgi and on the exterior of the cell (see Figure 9.9).

While most proteins travel from the ER to the Golgi, some proteins must be retained within the ER rather than proceeding along the secretory pathway. In particular, proteins that function within the ER (including BiP, signal peptidase, protein disulfide isomerase, and other enzymes discussed earlier) must be retained within that organelle. Export to the Golgi versus retention

*Figure 9.19*
**Vesicular transport from the ER to the Golgi** Proteins and lipids are carried from the ER to the Golgi in transport vesicles that bud from the membrane of the ER and then fuse with the membrane of the Golgi. Lumenal ER proteins are taken up by the vesicles and released into the lumen of the Golgi. Membrane proteins maintain the same orientation in the Golgi as in the ER.

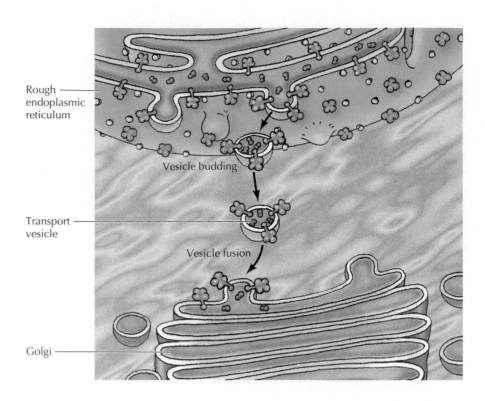

Rough endoplasmic reticulum

Vesicle budding

Transport vesicle

Vesicle fusion

Golgi

in the ER is thus the first branch point encountered by proteins being sorted to their correct destinations in the secretory pathway. Similar branch points arise at each subsequent stage of transport, such as retention in the Golgi versus export to lysosomes or the plasma membrane. In each case, specific localization signals target proteins to their correct intracellular destinations.

The distinction between proteins exported from and those retained in the ER can be governed by targeting sequences that specifically mark proteins as either (1) destined for transport to the Golgi or (2) destined for retention in the ER. Many proteins are retained in the ER lumen as a result of the presence of the targeting sequence Lys-Asp-Glu-Leu (KDEL, in the single-letter code) at their carboxy terminus. If this sequence is deleted from a protein that is normally retained in the ER (e.g., BiP), the mutated protein is instead transported to the Golgi and secreted from the cell. Conversely, addition of the KDEL sequence to the carboxy terminus of proteins that are normally secreted causes them to be retained in the ER. The retention of some transmembrane proteins in the ER is similarly dictated by short C-terminal sequences that contain two lysine residues (KKXX sequences).

Interestingly, the KDEL and KKXX signals do not prevent soluble ER proteins from being packaged into vesicles and carried to the Golgi. Instead, these signals cause resident ER proteins to be selectively retrieved from the Golgi and returned to the ER via a recycling pathway (Figure 9.20). Proteins bearing the KDEL and KKXX sequences appear to bind to specific recycling receptors in the Golgi membrane and are then selectively transported back to the ER.

The action of the KDEL and KKXX sequences as retention/retrieval signals indicates that there is a nonselective bulk flow of proteins through the secretory pathway leading from the ER to the cell surface. Retention/retrieval signals prevent proteins from being transported through this bulk flow pathway, thereby retaining marked proteins within the ER. It also appears, however, that proteins destined for export from the ER are selec-

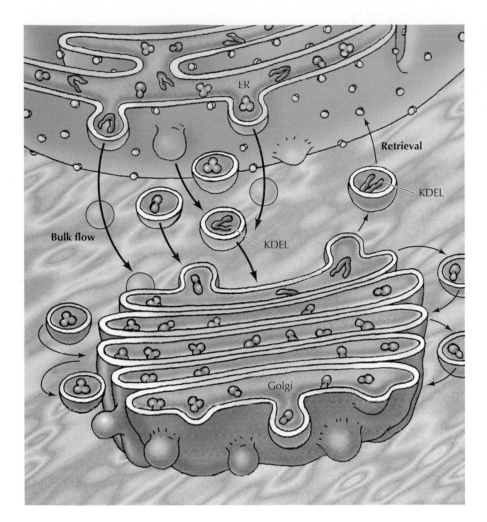

*Figure 9.20*
**Retrieval of resident ER proteins**
Proteins destined to remain in the lumen of the ER are marked by the sequence Lys-Asp-Glu-Leu (KDEL) at their carboxy terminus. These proteins are exported from the ER to the Golgi in the nonselective bulk flow of proteins through the secretory pathway, but they are recognized by a receptor in the Golgi and selectively returned to the ER.

tively packaged into transport vesicles targeted to the Golgi. Protein export from the ER may thus be controlled not only by retention/retrieval signals but also by targeting signals that mediate selective transport to the Golgi apparatus.

## THE GOLGI APPARATUS

The **Golgi apparatus**, or **Golgi complex**, functions as a factory in which proteins received from the ER are further processed and sorted for transport to their eventual destinations: lysosomes, the plasma membrane, or secretion. In addition, as noted earlier, glycolipids and sphingomyelin are synthesized within the Golgi. In plant cells, the Golgi apparatus further serves as the site at which the complex polysaccharides of the cell wall are synthesized. The Golgi apparatus is thus involved in processing the broad range of cellular constituents that travel along the secretory pathway.

### Organization of the Golgi

Morphologically the Golgi is composed of flattened membrane-enclosed sacs (cisternae) and associated vesicles (Figure 9.21). A striking feature of the Golgi apparatus is its distinct polarity in both structure and function. Proteins from the ER enter at its *cis* face (entry face), which is convex and usually oriented toward the nucleus. They are then transported through

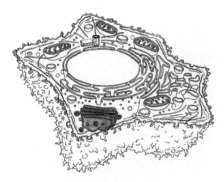

**Figure 9.21**
**Electron micrograph of a Golgi apparatus**    The Golgi apparatus consists of a stack of flattened cisternae and associated vesicles. Proteins and lipids from the ER enter the Golgi apparatus at its *cis* face and exit at its *trans* face. (Garry T. Cole/Biological Photo Service.)

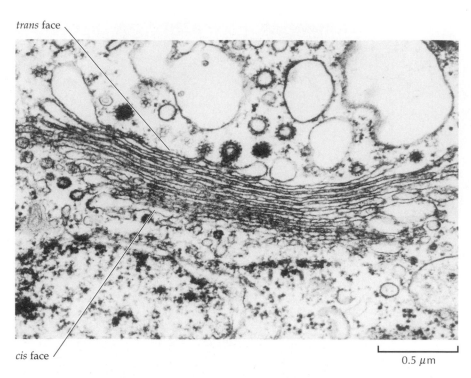

*trans* face

*cis* face

0.5 μm

the Golgi and exit from its concave *trans* face (exit face). As they pass through the Golgi, proteins are modified and sorted for transport to their eventual destinations within the cell.

Distinct processing and sorting events appear to take place in an ordered sequence within different regions of the Golgi complex, so the Golgi is usually considered to consist of multiple discrete compartments. Although the number of such compartments has not been established, the simplest model views the Golgi as consisting of three functionally distinct regions: the *cis* **Golgi network**, the **Golgi stack** (which may be divided into separate subcompartments), and the *trans* **Golgi network** (Figure 9.22). Transport vesicles carry proteins between these discrete compartments of the Golgi (from *cis* to *trans*) as processing and sorting proceed. In addition, membrane tubules may connect some of the Golgi cisternae.

Proteins enter the Golgi apparatus at the *cis* Golgi network, which serves primarily to receive transport vesicles from the ER and to sort their contents. Proteins marked for residence within the ER are recognized and returned to the ER by the recycling pathway shown in Figure 9.20. Other proteins are carried by transport vesicles to the Golgi stack, the stacked cisternae in the middle of the Golgi complex, which is the site of most metabolic activities of the Golgi apparatus. Transport vesicles then carry the modified proteins, lipids, and polysaccharides from the Golgi stack to the *trans* Golgi network, where the final stages of protein processing are completed. The *trans* Golgi network then acts as a sorting and distribution center, directing molecular traffic to lysosomes, the plasma membrane, or the cell exterior.

### Protein Glycosylation within the Golgi

Protein processing within the Golgi involves the modification and synthesis of the carbohydrate portions of glycoproteins. One of the major aspects of this processing is the modification of the *N*-linked oligosaccharides that were added to proteins in the ER. As discussed earlier in this chapter, proteins are modified within the ER by the addition of an oligosaccharide con-

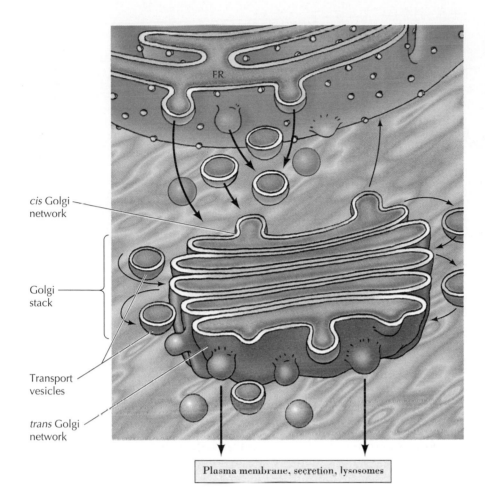

cis Golgi
network

Golgi
stack

Transport
vesicles

trans Golgi
network

ER

Plasma membrane, secretion, lysosomes

**Figure 9.22**
**Regions of the Golgi apparatus** The *cis* Golgi network is the site at which vesicles from the ER enter the Golgi and resident ER proteins are returned via the recycling pathway. The Golgi stack corresponds to the cisternae in the middle of the Golgi complex and is the site of most protein modifications. Proteins are then carried to the *trans* Golgi network, where they are sorted for transport to the plasma membrane, secretion, or lysosomes.

sisting of 14 sugar residues (see Figure 9.14). Three glucose residues and one mannose are then removed while the polypeptides are still in the ER. Following transport to the Golgi apparatus, the *N*-linked oligosaccharides of these glycoproteins are subject to extensive further modifications.

*N*-linked oligosaccharides are processed within the Golgi apparatus in an ordered sequence of reactions (Figure 9.23). The first modification of proteins destined for secretion or for the plasma membrane is the removal of three additional mannose residues. This is followed by the sequential addition of an *N*-acetylglucosamine, the removal of two more mannoses, and the addition of a fucose and two more *N*-acetylglucosamines. Finally, three galactose and three sialic acid residues are added, with the last of these reactions taking place in the *trans* Golgi network. As noted in Chapter 7, different glycoproteins are modified to different extents during their passage through the Golgi, depending on both the structure of the protein and on the amount of processing enzymes that are present within the Golgi complexes of different types of cells. Consequently, proteins can emerge from the Golgi with a variety of different *N*-linked oligosaccharides.

The processing of the *N*-linked oligosaccharide of lysosomal proteins differs from that of secreted and plasma membrane proteins. Rather than the initial removal of three mannose residues, proteins destined for incorporation into lysosomes are modified by mannose phosphorylation. In the first step of this reaction, *N*-acetylglucosamine phosphates are added to specific mannose residues, probably while the protein is still in the *cis*

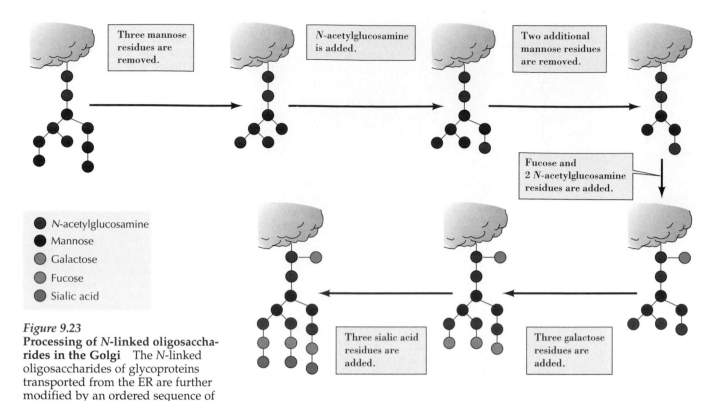

*Figure 9.23*
**Processing of *N*-linked oligosaccharides in the Golgi**  The *N*-linked oligosaccharides of glycoproteins transported from the ER are further modified by an ordered sequence of reactions in the Golgi.

Golgi network (Figure 9.24). This is followed by removal of the *N*-acetylglucosamine group, leaving **mannose-6-phosphate** residues on the *N*-linked oligosaccharide. Because of this modification, these residues are not removed during further processing. Instead, these phosphorylated mannose residues are specifically recognized by a mannose-6-phosphate receptor in the *trans* Golgi network, which directs the transport of these proteins to lysosomes.

The phosphorylation of mannose residues is thus a critical step in sorting lysosomal proteins to their correct intracellular destination. The specificity of this process resides in the enzyme that catalyzes the first step in the reaction sequence—the selective addition of *N*-acetylglucosamine phosphates to lysosomal proteins. This enzyme recognizes a structural determi-

*Figure 9.24*
**Targeting of lysosomal proteins by phosphorylation of mannose residues**
Proteins destined for incorporation into lysosomes are specifically recognized and modified by the addition of phosphate groups to the 6 position of mannose residues. In the first step of the reaction, *N*-acetylglucosamine phosphates are transferred to mannose residues from UDP-*N*-acetylglucosamine. The *N*-acetylglucosamine groups are then removed, leaving mannose-6-phosphates.

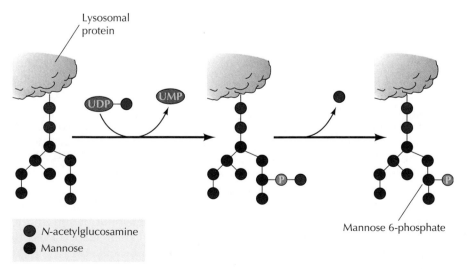

nant that is present on lysosomal proteins but not on proteins destined for the plasma membrane or secretion. This recognition determinant is not a simple sequence of amino acids; rather, it is formed in the folded protein by the juxtaposition of amino acid sequences from different regions of the polypeptide chain. In contrast to the signal sequences that direct protein translocation to the ER, the recognition determinant that leads to mannose phosphorylation, and thus ultimately targets proteins to lysosomes, depends on the three-dimensional conformation of the folded protein. Such determinants are called **signal patches**, in contrast to the linear targeting signals discussed earlier in this chapter.

Proteins can also be modified by the addition of carbohydrates to the side chains of acceptor serine and threonine residues within specific sequences of amino acids (*O*-linked glycosylation) (see Figure 7.28). These modifications take place in the Golgi apparatus by the sequential addition of single sugar residues. The serine or threonine is usually linked directly to *N*-acetylgalactosamine, to which other sugars can then be added. In some cases, these sugars are further modified by the addition of sulfate groups.

## Lipid and Polysaccharide Metabolism in the Golgi

In addition to its activities in processing and sorting glycoproteins, the Golgi apparatus functions in lipid metabolism—in particular, in the synthesis of glycolipids and sphingomyelin. As discussed earlier, the glycerol phospholipids, cholesterol, and ceramide are synthesized in the ER. Sphingomyelin and glycolipids are then synthesized from ceramide in the Golgi apparatus (Figure 9.25). Sphingomyelin (the only nonglycerol phospholipid

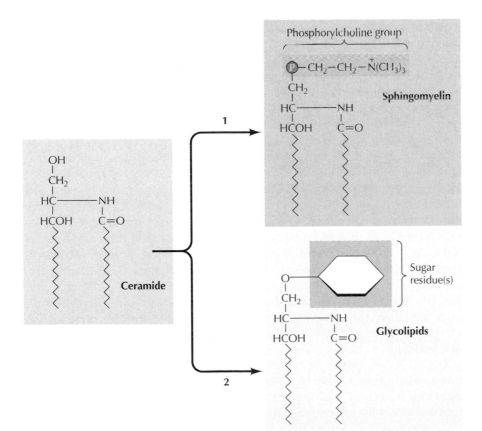

*Figure 9.25*
**Synthesis of sphingomyelin and glycolipids**   Ceramide, which is synthesized in the ER, is converted either to sphingomyelin (a phospholipid) or to glycolipids in the Golgi apparatus. In the first reaction, a phosphorylcholine group is transferred from phosphatidylcholine to ceramide. Alternatively, a variety of different glycolipids can be synthesized by the addition of one or more sugar residues (e.g., glucose).

in cell membranes) is synthesized by the transfer of a phosphorylcholine group from phosphatidylcholine to ceramide. Alternatively, the addition of carbohydrates to ceramide can yield a variety of different glycolipids.

Sphingomyelin is synthesized on the lumenal surface of the Golgi, but glucose is added to ceramide on the cytosolic side. Glucosylceramide then apparently flips, however, and additional carbohydrates are added on the lumenal side of the membrane. Neither sphingomyelin nor the glycolipids are then able to translocate across the Golgi membrane, so they are found only in the lumenal half of the Golgi bilayer. Following vesicular transport, they are correspondingly localized to the exterior half of the plasma membrane, with their polar head groups exposed on the cell surface. As will be discussed in Chapter 12, the oligosaccharide portions of glycolipids are important surface markers in cell-cell recognition.

In plant cells, the Golgi apparatus has the additional task of serving as the site where complex polysaccharides of the cell wall are synthesized. As discussed further in Chapter 12, the plant cell wall is composed of three major types of polysaccharides. Cellulose, the predominant constituent, is a simple linear polymer of glucose residues. It is synthesized at the cell surface by enzymes in the plasma membrane. The other cell wall polysaccharides (hemicelluloses and pectins), however, are complex, branched chain molecules that are synthesized in the Golgi apparatus and then transported in vesicles to the cell surface. The synthesis of these cell wall polysaccharides is a major cellular function, and as much as 80% of the metabolic activity of the Golgi apparatus in plant cells may be devoted to polysaccharide synthesis.

### Protein Sorting and Export from the Golgi Apparatus

Proteins, as well as lipids and polysaccharides, are transported from the Golgi apparatus to their final destinations through the secretory pathway. This involves the sorting of proteins into different kinds of transport vesicles, which bud from the *trans* Golgi network and deliver their contents to the appropriate cellular locations (Figure 9.26). In the absence of specific targeting signals, proteins are carried to the plasma membrane by bulk flow, which transports proteins nonselectively from the ER to the Golgi and then to the cell surface. This pathway accounts for the incorporation of new proteins and lipids into the plasma membrane, as well as for the continuous secretion of proteins from the cell. To be diverted from the bulk flow pathway, proteins must be specifically targeted to other destinations, such as lysosomes.

Proteins that function within the Golgi apparatus must be retained within that organelle, rather than being transported along the secretory pathway. One mechanism responsible for retention of membrane proteins within the Golgi appears to be based on their transmembrane domains. In particular, most Golgi membrane proteins have relatively short transmembrane $\alpha$ helices of only about 15 amino acids, and these short membrane-spanning regions may contribute to the retention of these proteins in the Golgi. In addition, like the KDEL and KKXX sequences of resident ER proteins, signals in the cytoplasmic tails of some Golgi proteins mediate the retrieval of these proteins from subsequent compartments along the secretory pathway.

The bulk flow pathway, which operates in all cells, leads to continual unregulated protein secretion. However, some cells also possess a distinct regulated secretory pathway in which specific proteins are secreted in response to environmental signals. Examples of regulated secretion include

*Figure 9.26*
**Transport from the Golgi apparatus**   Proteins are sorted in the *trans* Golgi network and transported in vesicles to their final destinations. In the absence of specific targeting signals, proteins are carried to the plasma membrane by bulk flow. Alternatively, proteins can be diverted from the bulk flow pathway and targeted to other destinations, such as lysosomes or regulated secretion from the cells.

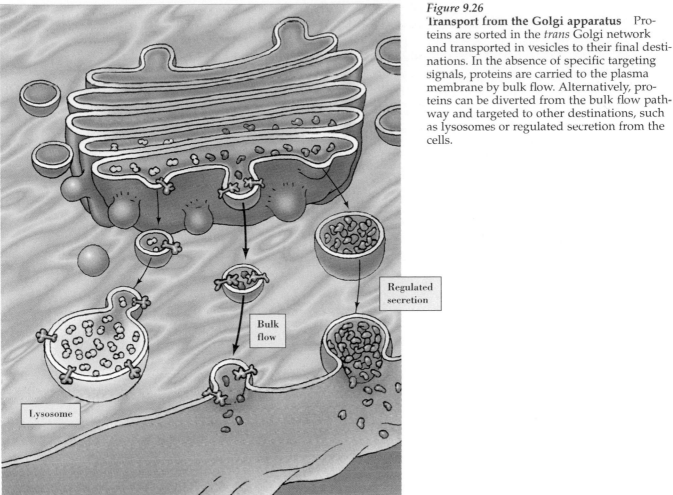

the release of hormones from endocrine cells, the release of neurotransmitters from neurons, and the release of digestive enzymes from the pancreatic acinar cells discussed at the beginning of this chapter (see Figure 9.2). Proteins are sorted into the regulated secretory pathway in the *trans* Golgi network, where they are packaged into specialized secretory vesicles. These secretory vesicles, which are larger than other transport vesicles, store their contents until specific signals direct their fusion with the plasma membrane. For example, the digestive enzymes produced by pancreatic acinar cells are stored in secretory vesicles until the presence of food in the stomach and small intestine triggers their secretion. The sorting of proteins into the regulated secretory pathway appears to involve the recognition of signal patches shared by multiple proteins that enter this pathway. These proteins selectively aggregate in the *trans* Golgi network and are then released by budding as secretory vesicles.

A further complication in the transport of proteins to the plasma membrane arises in many epithelial cells, which are polarized when they are organized into tissues. The plasma membrane of such cells is divided into two separate regions, the **apical domain** and the **basolateral domain**, that contain specific proteins related to their particular functions. For example, the apical membrane of intestinal epithelial cells faces the lumen of

*Figure 9.27*
**Transport to the plasma membrane of polarized cells** The plasma membranes of polarized epithelial cells are divided into apical and basolateral domains. In this example (intestinal epithelium), the apical surface of the cell faces the lumen of the intestine, the lateral surfaces are in contact with neighboring cells, and the basal surface rests on a sheet of extracellular matrix (the basal lamina). The apical membrane is characterized by the presence of microvilli, which facilitate the absorption of nutrients by increasing surface area. Specific proteins are targeted to either the apical or basolateral membranes in the *trans* Golgi network. Tight junctions between neighboring cells maintain the identity of the apical and basolateral membranes by preventing the diffusion of proteins between these domains.

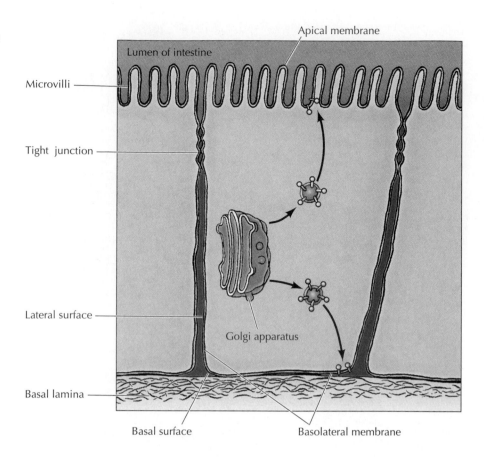

the intestine and is specialized for the efficient absorption of nutrients; the remainder of the cell is covered by the basolateral membrane (Figure 9.27). In some types of epithelia, membrane proteins are sorted in the *trans* Golgi network for selective transport to these distinct domains of the plasma membrane. The targeting signals involved are not yet fully characterized, although the presence of a GPI anchor is one signal that directs proteins to the apical domain.

The sorting of proteins for transport to lysosomes is the best-characterized diversion from the bulk flow pathway. As already discussed, lumenal lysosomal proteins are marked by mannose-6-phosphates that are formed by modification of their *N*-linked oligosaccharides shortly after entry into the Golgi apparatus. A specific receptor in the membrane of the *trans* Golgi network then recognizes these mannose-6-phosphate residues. The resulting complexes of receptor plus lysosomal enzyme are packaged into transport vesicles destined for lysosomes. Lysosomal membrane proteins are targeted by sequences in their cytoplasmic tails, rather than by mannose-6-phosphates.

In yeasts and plant cells, which lack lysosomes, proteins are transported from the Golgi apparatus to an additional destination: the **vacuole** (Figure 9.28). Vacuoles assume the functions of lysosomes in these cells as well as performing a variety of other tasks, such as the storage of nutrients and the maintenance of turgor pressure and osmotic balance. In contrast to lysosomal targeting, proteins are directed to vacuoles by short peptide sequences instead of carbohydrate markers. One signal responsible for targeting to the yeast vacuole has been identified as the sequence Gln-Arg-Pro-Leu, but other vacuolar proteins appear to possess different targeting signals.

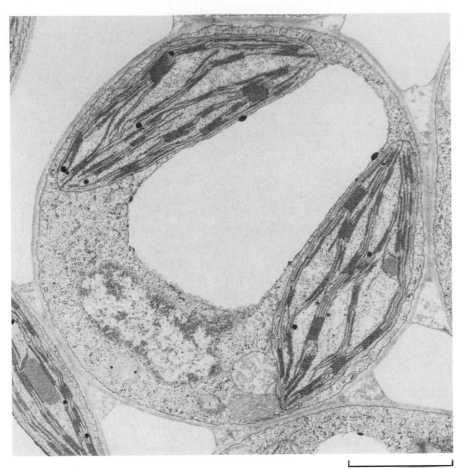

2 μm

*Figure 9.28*
**A plant cell vacuole** The large central vacuole functions as a lysosome in addition to storing nutrients and maintaining osmotic balance. (E. H. Newcombe/Biological Photo Service.)

## THE MECHANISM OF VESICULAR TRANSPORT

As is evident from the preceding sections of this chapter, transport vesicles play a central role in the traffic of molecules between different membrane-enclosed compartments of the secretory pathway. As discussed in Chapter 12, vesicles are similarly involved in the transport of materials taken up at the cell surface. Vesicular transport is thus a major cellular activity, responsible for molecular traffic between a variety of specific membrane-enclosed compartments. The selectivity of such transport is therefore key to maintaining the functional organization of the cell. For example, lysosomal enzymes must be transported specifically from the Golgi apparatus to lysosomes—not to the plasma membrane or to the ER. Some of the signals that target proteins to specific organelles, such as lysosomes, were discussed earlier in this chapter. These proteins are transported within vesicles, so the specificity of transport is based on the selective packaging of the intended cargo into vesicles that recognize and fuse only with the appropriate target membrane. Because of the central importance of vesicular transport to the organization of eukaryotic cells, understanding the molecular mechanisms that control vesicle packaging, budding, and fusion is a major area of current research.

### Experimental Approaches to Understanding Vesicular Transport

Progress toward elucidating the molecular mechanisms of vesicular transport has been made by three distinct experimental approaches: (1) isolation of yeast mutants that are defective in protein transport and sorting; (2)

reconstitution of vesicular transport in cell-free systems; and (3) biochemical analysis of synaptic vesicles, which are responsible for the regulated secretion of neurotransmitters by neurons. Each of these experimental systems has distinct advantages for understanding particular aspects of the transport process. Most important, however, is the fact that results from all three of these avenues of investigation have converged, indicating that similar molecular mechanisms regulate secretion in cells as different as yeasts and mammalian neurons.

As in other areas of cell biology, yeasts have proved to be advantageous in studying the secretory pathway because they are readily amenable to genetic analysis. In particular, Randy Schekman and his colleagues have pioneered the isolation of yeast mutants defective in vesicular transport. These include mutants that are defective at various stages of protein secretion (*sec* mutants), mutants that are unable to transport proteins to the vacuole, and mutants that are unable to retain resident ER proteins. The isolation of such mutants in yeasts led directly to the molecular cloning and analysis of the corresponding genes, thereby identifying a number of proteins involved in various steps of the secretory pathway. For example, the role of Sec61 as a major component of the protein translocation channel in the endoplasmic reticulum was discussed earlier in this chapter.

Biochemical studies of vesicular transport using reconstituted systems have complemented these genetic studies and have enabled the direct isolation of transport proteins from mammalian cells. The first cell-free transport system was developed by James Rothman and colleagues, who analyzed protein transport between compartments of the Golgi apparatus (Figure 9.29). The experimental design exploited a mutant mammalian cell line that lacked the enzyme required to transfer *N*-acetylglucosamine residues to the *N*-linked oligosaccharide at an early stage of its modification in the Golgi apparatus (see Figure 9.23). Consequently, the glycoproteins produced by this mutant cell line lacked added *N*-acetylglucosamine units. However, if Golgi stacks isolated from the mutant cell line were incubated with stacks isolated from normal cells, *N*-acetylglucosamine residues were added to glycoproteins synthesized by the mutant cells. A variety of experiments established that this resulted from vesicular transport of proteins from the Golgi stacks of the mutant cell line to the Golgi stacks of normal cells, so the addition of *N*-acetylglucosamine provided a readily detectable marker for vesicular transport in this reconstituted system. Similar reconstituted systems have been developed to analyze transport between other compartments, including transport from the ER to the Golgi and transport from the Golgi to secretory vesicles, vacuoles, and the plasma membrane. The development of these *in vitro* systems has enabled biochemical studies of the transport process and functional analysis of proteins identified by mutations in yeasts, as well as direct isolation of some of the proteins involved in vesicle budding and fusion.

Critical insights into the molecular mechanisms of vesicular transport have also come from studies of synaptic transmission in neurons, which represents a highly specialized form of regulated secretion. A synapse is the junction of a neuron with another cell, which may be either another neuron or an effector, such as a muscle cell. Information is transmitted across the synapse by chemical neurotransmitters, such as acetylcholine, which are stored in **synaptic vesicles**. Stimulation of the transmitting neuron triggers the fusion of synaptic vesicles with the plasma membrane, causing neurotransmitters to be released and stimulating the postsynaptic neuron or

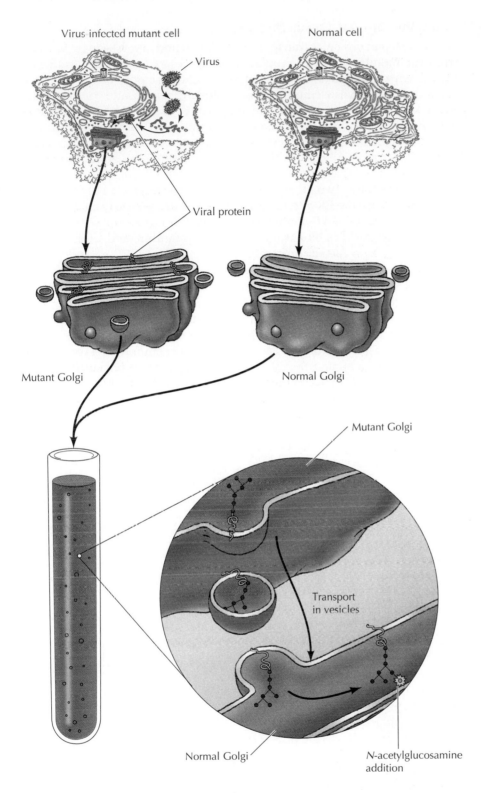

*Figure 9.29*
**Reconstituted vesicular transport**
Golgi stacks prepared from a virus-infected mutant cell line unable to catalyze the addition of *N*-acetylglucosamine to *N*-linked oligosaccharides are mixed with Golgi stacks from a normal cell line. Because the mutant cell line is infected with a virus, the proteins it synthesizes can be specifically detected. Transport of these proteins to normal Golgi stacks is signaled by the addition of *N*-acetylglucosamine.

effector cell. Synaptic vesicles are extremely abundant in the brain, allowing them to be purified in large amounts for biochemical analysis. As discussed later, relationships between some of the proteins isolated from synaptic vesicles and those identified by yeast genetics and biochemical reconstitution experiments have provided important insights into the molecular mechanisms of vesicle fusion.

### Coat Proteins and Vesicle Budding

The first step in vesicular transport is the formation of a vesicle by budding from the membrane. The cytoplasmic surfaces of transport vesicles are coated with proteins, and it appears to be the assembly of these protein coats that drives vesicle budding by distorting membrane conformation. Three kinds of coated vesicles, which appear to function in different types of vesicular transport, have been characterized. The first to be described were the **clathrin-coated vesicles**, which are responsible for the uptake of extracellular molecules from the plasma membrane by endocytosis (see Chapter 12) as well as the transport of molecules from the *trans* Golgi network to lysosomes. Two other types of coated vesicles have been identified as budding from the ER and Golgi complex. These vesicles are called non-clathrin-coated or **COP-coated vesicles** (COP stands for *coat protein*); COPI-coated vesicles bud from the Golgi apparatus, and COPII-coated vesicles bud from the endoplasmic reticulum. The COPII-coated vesicles transport their cargo from the endoplasmic reticulum to the Golgi, but the transport steps mediated by COPI-coated vesicles have not yet been fully defined: they may mediate transport between Golgi stacks, recycling from the Golgi to the ER, and possibly other transport processes.

The coats of clathrin-coated vesicles are composed of two types of proteins, which assemble on the cytosolic side of membranes (Figure 9.30). **Clathrin** plays a structural role by assembling into a basketlike lattice structure that distorts the membrane and drives vesicle budding. The binding of clathrin to membranes is mediated by a second class of proteins,

**Figure 9.30**
**Incorporation of lysosomal proteins into clathrin-coated vesicles**   Proteins targeted for lysosomes are marked by mannose-6-phosphates, which bind to mannose-6-phosphate receptors in the *trans* Golgi network. The mannose-6-phosphate receptors span the Golgi membrane and serve as binding sites for cytosolic adaptins, which in turn bind clathrin. Clathrins consist of three protein chains that associate with each other to form a basketlike lattice that distorts the membrane and drives vesicle budding.

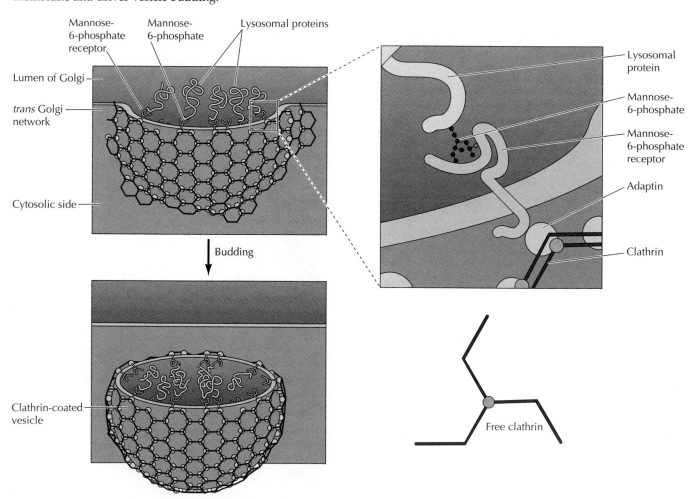

called **adaptins**. Different adaptins are responsible for the assembly of clathrin-coated vesicles at the plasma membrane and at the *trans* Golgi network, and it is the adaptins that are involved in selecting the specific molecules to be incorporated into the vesicles. For example, the adaptin involved in budding from the *trans* Golgi network binds to the cytosolic portion of the mannose-6-phosphate receptor, thereby directing proteins destined for lysosomes into clathrin-coated vesicles.

The coats of COPI- and COPII-coated vesicles are composed of distinct protein complexes. Interestingly, components of the COPI coat interact with the KKXX motif responsible for the retrieval of ER proteins from the Golgi apparatus, consistent with a role for COPI-coated vesicles in recycling from the Golgi to the ER. Although the proteins of COP-coated vesicles are not as well characterized as clathrin and the adaptins, the components of COPI and COPII coats probably perform analogous functions in vesicle budding.

The budding of both clathrin-coated and COPI-coated vesicles from the *trans* Golgi network also requires a GTP-binding protein called **ARF** (ADP-ribosylation factor). ARF was initially of interest because it was found to be related to the Ras proteins, which have been studied extensively because of their involvement as oncogenes in human cancers. As discussed in Chapter 7, the Ras proteins are examples of a large group of proteins (including several of the translation factors involved in protein synthesis) whose activities are regulated by the alternate binding of GTP or GDP. Further studies of ARF revealed that ARF is involved in secretion. In particular, ARF bound to GTP associates with Golgi membranes and is required for the binding of either COPI coat components or clathrin adaptins (Figure 9.31). ARF then

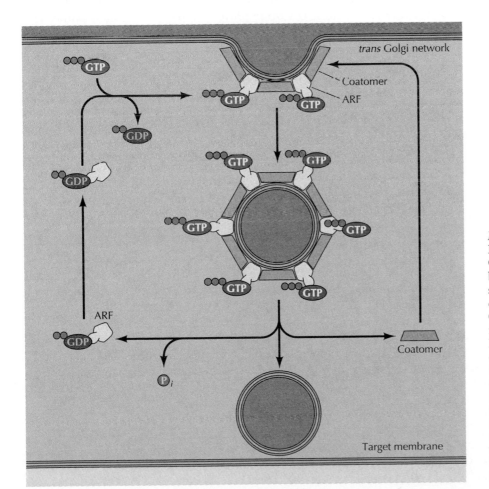

**Figure 9.31**
**Role of ARF in the formation of COP-coated vesicles** ARF alternates between GTP-bound and GDP-bound states. When bound to GTP, ARF associates with the membrane of the *trans* Golgi network and promotes the binding of COP coat protein (coatomer), leading to vesicle budding. Hydrolysis of the bound GTP then converts ARF to the GDP-bound state, leading to disassembly of the vesicle coat prior to fusion with the target membrane. The GDP-bound ARF is then reconverted to the GTP-bound state by the action of a Golgi membrane protein that promotes the exchange of free GTP for the bound GDP, leading to another cycle of coatomer assembly.

hydrolyzes its bound GTP following vesicle budding, leading to the conversion of ARF to the GDP-bound state and the dissociation of coat proteins from the vesicle membrane.

Several other Ras-related GTP-binding proteins have also been found to be involved in secretion. These include additional proteins identified by genetic analysis in yeasts and more than 30 Ras-related proteins (called Rab proteins) that have been implicated in vesicular transport in mammalian cells. One of the yeast proteins, SAR1, functions analogously to ARF in the budding of COPII-coated vesicles from the endoplasmic reticulum. Other GTP-binding proteins, including the Rab proteins, are instead thought to regulate vesicle fusion, as discussed next.

## Vesicle Fusion

The fusion of a transport vesicle with its target involves two types of events. First, the transport vesicle must specifically recognize the correct target membrane; for example, a vesicle carrying lysosomal enzymes has to deliver its cargo only to lysosomes. Second, the vesicle and target membranes must fuse, thereby delivering the contents of the vesicle to the target organelle. Specific recognition between a vesicle and its target is thought to be mediated by interactions between unique pairs of transmembrane proteins. Fusion between the vesicle and target membranes is then brought about by the action of general fusion proteins.

Proteins involved in vesicle fusion were initially identified in James Rothman's laboratory by biochemical analysis of reconstituted vesicular transport systems from mammalian cells (see Figure 9.29). The first such protein to be isolated was called NSF (for *N*-ethylmaleimide-sensitive *fusion*). NSF is a soluble cytosolic protein that binds to membranes as a complex with other proteins called SNAPs (for soluble *NSF attachment proteins*). Both NSF and SNAPs isolated from mammalian cells have counterparts in yeasts, and further studies indicated that these proteins are involved in a variety of different vesicle fusion events, including transport from the ER to the Golgi, transport between Golgi compartments, transport from the Golgi to the plasma membrane, and fusion of synaptic vesicles with nerve terminals. It thus appears that NSF and SNAPs are components of a general membrane fusion apparatus.

Additional experiments led Rothman and his colleagues to propose a general model that accounts for the specificity of vesicle fusion by suggesting that NSF and SNAPs bind to families of specific membrane receptors, called SNAP receptors, or SNAREs. According to the **SNARE hypothesis**, interactions between specific SNAREs on the vesicle (v-SNAREs) and target (t-SNAREs) membranes are responsible for the specificity of vesicle fusion (Figure 9.32). Following specific vesicle-target interaction, the SNARE complex recruits NSF and SNAPs, leading to fusion of the vesicle and target membranes.

This hypothesis was first proposed following the isolation of SNAREs from bovine brain on the basis of their ability to bind to a protein complex consisting of NSF and SNAPs. Importantly, the SNAREs isolated by this approach turned out to be proteins that had previously been found to be associated with synapses. Moreover, one of these SNAREs was a component of the membrane of synaptic vesicles (a v-SNARE), while another was localized to the plasma membranes of neurons (a t-SNARE). In addition, these SNAREs were related to a series of genes previously identified as secretion mutants in yeasts, which appeared to define pairs of proteins required for specific steps of the secretory pathway. For example, transport from the ER to the Golgi requires SNAREs that are located on both the vesicle and target membranes.

**Figure 9.32**
**The SNARE hypothesis** Binding between specific pairs of v-SNAREs and t-SNAREs on the vesicle and target membranes, respectively, is responsible for vesicle-target recognition. The SNARE complex then recruits the general fusion proteins, NSF and SNAPs, resulting in membrane fusion.

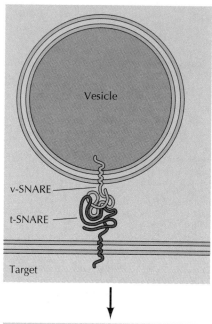

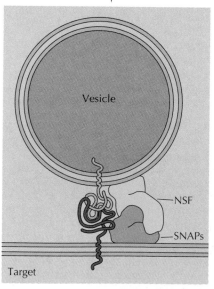

The involvement of such pairs of SNAREs at distinct steps in the secretory pathway led to the hypothesis that specific interactions between v-SNAREs and t-SNAREs are responsible for the specificity of vesicle-target binding. These interactions are additionally thought to be regulated by the **Rab** GTP-binding proteins, which are also essential for vesicle transport. It should be noted that the details of this process remain to be understood, and other mechanisms for vesicle-target recognition may also be discovered. However, the SNARE hypothesis provides an important central framework for understanding the molecular mechanisms of vesicle docking and fusion.

## LYSOSOMES

**Lysosomes** are membrane-enclosed organelles that contain an array of enzymes capable of breaking down all types of biological polymers—proteins, nucleic acids, carbohydrates, and lipids. Lysosomes function as the digestive system of the cell, serving both to degrade material taken up from outside the cell and to digest obsolete components of the cell itself. In their simplest form, lysosomes are visualized as dense spherical vacuoles, but they can display considerable variation in size and shape as a result of differences in the materials that have been taken up for digestion (Figure 9.33). Lysosomes thus represent morphologically diverse organelles defined by the common function of degrading intracellular material.

### Lysosomal Acid Hydrolases

Lysosomes contain about 50 different degradative enzymes that can hydrolyze proteins, DNA, RNA, polysaccharides, and lipids. Mutations in the genes that encode these enzymes are responsible for more than 30 different human genetic diseases, which are called **lysosomal storage diseases** because undegraded material accumulates within the lysosomes of affected individuals. Most of these diseases result from deficiencies in single lysosomal enzymes. For example, Gaucher's disease (the most common of these disorders) results from a mutation in the gene that encodes a lysosomal enzyme required for the breakdown of glycolipids. An intriguing exception is I-cell disease, which is caused by a deficiency in the enzyme

*Figure 9.33*
**Electron micrograph of lysosomes and mitochondria in a mammalian cell.** Lysosomes are indicated by arrows. (Visuals Unlimited/K.G. Murti.)

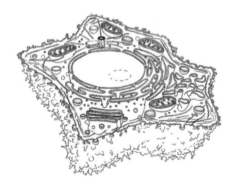

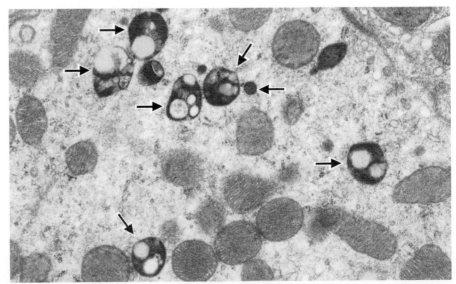

0.5 μm

# MOLECULAR MEDICINE

## Gaucher's Disease

### The Disease

Gaucher's disease is the most common of the lysosomal storage diseases, which are caused by a failure of lysosomes to degrade substances that they normally break down. The resulting accumulation of nondegraded compounds leads to an increase in the size and number of lysosomes within the cell, eventually resulting in cellular malfunction and pathological consequences to affected organs. Gaucher's disease is found primarily in the Jewish population, where it has a frequency of about 1 in 2500 individuals. There are three types of the disease, which differ in severity and nervous system involvement. In the most common form of the disease (type I), the nervous system is not involved; the disease is manifest as spleen and liver enlargement and development of bone lesions. Many patients with this form of the disease have no serious symptoms, and their life span is unaffected. The more severe forms of the disease (types II and III) are much rarer and found in both Jewish and non-Jewish populations. The most devastating is type II disease, in which extensive neurological involvement is evident in infancy and patients die early in life. Type III disease, intermediate in severity between types I and II, is characterized by the onset of neurological symptoms (including dementia and spasticity) by about age ten.

### Molecular and Cellular Basis

Gaucher's disease is caused by deficiency of the lysosomal enzyme glucocerebrosidase, which catalyzes the hydrolysis of glucocerebroside to glucose and ceramide (see figure). This enzyme deficiency was demonstrated in 1965, and the responsible gene was cloned in 1985. Since then, more than 30 different mutations responsible for

Gaucher's disease have been identified. Interestingly, the severity of the disease can be largely predicted from the nature of these mutations. For example, patients with a mutation leading to the relatively conservative amino acid substitution of serine for asparagine have type I disease, whereas patients with a mutation leading to substitution of proline for leucine have more severe enzyme deficiencies and develop type II or III disease.

Except for the very rare type II and III forms of the disease, the only cells affected in Gaucher's disease are macrophages. Because their function is to eliminate aged and damaged cells by phagocytosis, macrophages continually ingest large amounts of lipids, which are normally degraded in lysosomes. For example, phagocytosis by macrophages of the human spleen and liver is responsible for the daily elimination of approximately $10^{11}$ red blood cells. Deficiencies of glucocerebrosidase are therefore particularly evident in macrophages of these organs, consistent with spleen and liver abnormalities being the primary manifestation in most cases of Gaucher's disease.

### Prevention and Treatment

Gaucher's disease is a prime example of a disease that can be treated by enzyme replacement therapy, in which exogenous administration of an enzyme is used to correct an enzyme defect. This approach to treatment of lysosomal storage diseases was suggested by Christian deDuve in the 1960s, based on the idea that exogenously administered enzymes might be taken up by endocytosis and transported to lysosomes. In type I Gaucher's disease, this approach is particularly attractive because the single target cell is the macrophage. In the 1970s, it was discovered that macrophages express cell surface receptors that bind mannose residues on extracellular glycoproteins and then internalize these proteins by endocytosis. This finding suggested that exogenously administered glucocerebrosidase could be specifically targeted to macrophages by modifications that would expose mannose residues. Enzyme prepared from human placenta was appropriately modified, and clinical studies have clearly demonstrated its effectiveness in the treatment of Gaucher's disease.

Unfortunately, enzyme replacement therapy for Gaucher's disease is expensive: The cost of enzyme alone

The enzyme deficiency in Gaucher's disease prevents the hydrolysis of glucocerebroside to glucose and ceramide.

that catalyzes the first step in the tagging of lysosomal enzymes with mannose-6-phosphate in the Golgi apparatus (see Figure 9.24). The result is a general failure of lysosomal enzymes to be incorporated into lysosomes.

All of the lysosomal enzymes are acid hydrolases, which are active at the acidic pH (about 5) that is maintained within lysosomes but not at the neutral pH (about 7.2) characteristic of the rest of the cytoplasm (Figure 9.34). The requirement of these lysosomal hydrolases for acidic pH provides double protection against uncontrolled digestion of the contents of the cytosol; even if the lysosomal membrane were to break down, the released acid hydrolases would be inactive at the neutral pH of the cytosol. To maintain their acidic internal pH, lysosomes must actively concentrate $H^+$ ions (protons). This is accomplished by a proton pump in the lysosomal membrane, which actively transports protons into the lysosome from the cytosol. This pumping requires expenditure of energy in the form of ATP hydrolysis, since it maintains approximately a hundredfold higher $H^+$ concentration inside the lysosome.

### Endocytosis and Lysosome Formation

One of the major functions of lysosomes is the digestion of material taken up from outside the cell by **endocytosis**, which is discussed in detail in Chapter 12. However, the role of lysosomes in the digestion of material taken up by endocytosis relates not only to the function of lysosomes but also to their formation. In particular, lysosomes are formed by the fusion of transport vesicles budded from the *trans* Golgi network with endosomes, which contain molecules taken up by endocytosis at the plasma membrane.

The formation of lysosomes thus represents an intersection between the secretory pathway, through which lysosomal proteins are processed, and the endocytic pathway, through which extracellular molecules are taken up at the cell surface (Figure 9.35). Material from outside the cell is taken up in clathrin-coated endocytic vesicles, which bud from the plasma membrane and then fuse with early **endosomes**. Membrane components are then recycled to the plasma membrane (discussed in detail in Chapter 12) and the early endosomes gradually mature into late endosomes, which are the precursors to lysosomes. One of the important changes during endosome maturation is the lowering of the internal pH to about 5.5, which plays a key role in the delivery of lysosomal acid hydrolases from the *trans* Golgi network.

As discussed earlier, acid hydrolases are targeted to lysosomes by mannose-6-phosphate residues, which are recognized by mannose-6-phosphate receptors in the *trans* Golgi network and packaged into clathrin-coated vesicles. Following removal of the clathrin coat, these transport vesicles fuse with late endosomes, and the acidic internal pH causes the hydrolases to dissoci-

*Figure 9.34*
**Organization of the lysosome** Lysosomes contain a variety of acid hydrolases that are active at the acidic pH maintained within the lysosome, but not at the neutral pH of the cytosol. The acidic internal pH of lysosomes results from the action of a proton pump in the lysosomal membrane, which imports protons from the cytosol coupled to ATP hydrolysis.

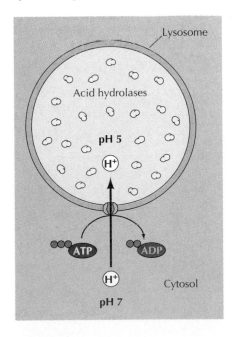

*Figure 9.35*
**Endocytosis and lysosome formation**
Molecules are taken up from outside the cell in endocytic vesicles, which fuse with early endosomes. Membrane components are recycled as the early endosomes mature into late endosomes. Transport vesicles carrying acid hydrolases from the Golgi apparatus then fuse with late endosomes, which mature into lysosomes as they acquire a full complement of lysosomal enzymes. The acid hydrolases dissociate from the mannose-6-phosphate receptor when the transport vesicles fuse with late endosomes, and the receptors are recycled to the Golgi apparatus.

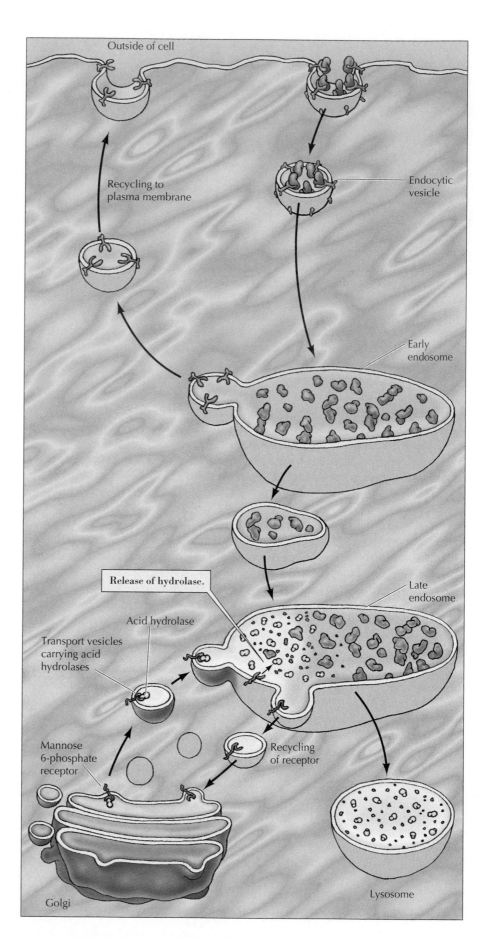

Outside of cell

Recycling to plasma membrane

Endocytic vesicle

Early endosome

Release of hydrolase.

Late endosome

Acid hydrolase

Transport vesicles carrying acid hydrolases

Mannose 6-phosphate receptor

Recycling of receptor

Golgi

Lysosome

ate from the mannose-6-phosphate receptor (see Figure 9.35). The hydrolases are thus released into the lumen of the endosome, while the receptors remain in the membrane and are eventually recycled to the Golgi. Late endosomes then mature into lysosomes as they acquire a full complement of acid hydrolases, which digest the molecules originally taken up by endocytosis.

## Phagocytosis and Autophagy

In addition to degrading molecules taken up by endocytosis, lysosomes digest material derived from two other routes: phagocytosis and autophagy (Figure 9.36). In **phagocytosis**, specialized cells, such as macrophages, take up and degrade large particles, including bacteria, cell debris, and aged cells that need to be eliminated from the body. Such large particles are taken up in phagocytic vacuoles (**phagosomes**) which then fuse with lysosomes, resulting in digestion of their contents. The lysosomes formed in this way (**phagolysosomes**) can be quite large and heterogeneous, since their size and shape is determined by the content of material that is being digested (Figure 9.37).

Lysosomes are also responsible for **autophagy**, the gradual turnover of the cell's own components. The first step of autophagy appears to be the enclosure of an organelle (e.g., a mitochondrion) in membrane derived

*Figure 9.36*
**Lysosomes in phagocytosis and autophagy**   In phagocytosis, large particles (such as bacteria) are taken up into phagocytic vacuoles or phagosomes. In autophagy, internal organelles (such as mitochondria) are enclosed by membrane fragments from the ER, forming autophagosomes. Both phagosomes and autophagosomes fuse with lysosomes to form large phagolysosomes, in which their contents are digested.

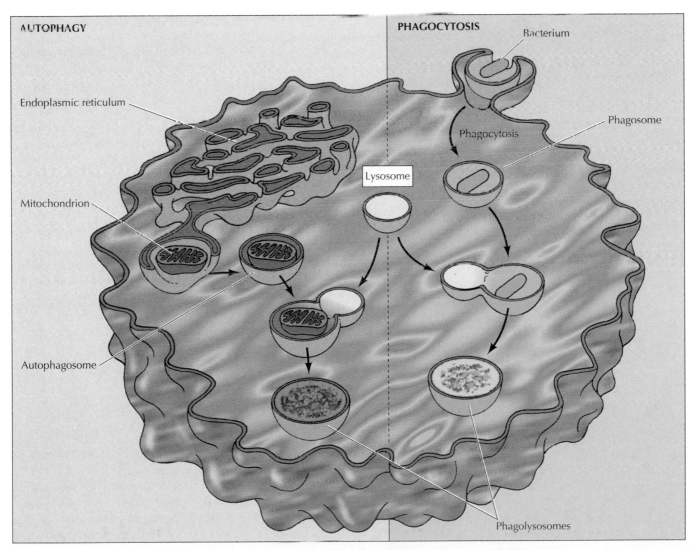

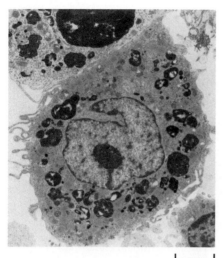

*Figure 9.37*
**Electron micrograph of phagolysosomes in a macrophage** (Fred E. Hossler/Visuals Unlimited.)

1 μm

from the ER. The resulting vesicle (an **autophagosome**) then fuses with a lysosome, and its contents are digested (see Figure 9.36). As discussed in Chapter 7, autophagy is responsible for the gradual turnover of cytoplasmic organelles. In addition, autophagy plays an important role in development by participating in the remodeling of tissues during differentiation.

## KEY TERMS

endoplasmic reticulum (ER), rough ER, smooth ER, secretory vesicle, secretory pathway

signal sequence, microsome, signal recognition particle (SRP), 7SL RNA, signal peptidase

protein disulfide isomerase, glycosylphosphatidylinositol (GPI) anchor

flippase

## Summary

### THE ENDOPLASMIC RETICULUM

*The Endoplasmic Reticulum and Protein Secretion:* The endoplasmic reticulum is the first branch point in protein sorting. Proteins destined for secretion, lysosomes, or the plasma membrane are translated on membrane-bound ribosomes and transferred into the rough ER as their translation proceeds.

*Targeting Proteins to the Endoplasmic Reticulum:* Ribosomes engaged in the synthesis of secreted proteins are targeted to the endoplasmic reticulum by signal sequences at the amino terminus of the polypeptide chain. Growing polypeptide chains are then translocated into the ER through protein channels and released into the ER lumen by cleavage of the signal sequence.

*Insertion of Proteins into the ER Membrane:* Integral membrane proteins of the plasma membrane or the membranes of the ER, Golgi apparatus, and lysosomes are initially inserted into the membrane of the ER. Rather than being translocated into the ER lumen, these proteins are anchored by membrane-spanning $\alpha$ helices that stop the transfer of the growing polypeptide chain across the membrane.

*Protein Folding and Processing in the ER:* Polypeptide chains are folded into their correct three-dimensional conformations within the ER. The ER is also the site of *N*-linked glycosylation and addition of GPI anchors.

*The Smooth ER and Lipid Synthesis:* The ER is the major site of lipid synthesis in eukaryotic cells, and smooth ER is abundant in cells that are active in lipid metabolism and detoxification of lipid-soluble drugs.

*Export of Proteins and Lipids from the ER:* Proteins and lipids are transported in vesicles from the ER to the Golgi apparatus. Resident ER proteins are marked by sequences that signal their return from the Golgi to the ER by a recycling pathway. Other targeting sequences may mediate the selective packaging of exported proteins into vesicles that transport them to the Golgi.

## THE GOLGI APPARATUS

*Organization of the Golgi:* The Golgi apparatus functions in protein processing and sorting as well as in the synthesis of lipids and polysaccharides. Proteins are transported from the endoplasmic reticulum to the *cis* Golgi network. From there they are transported to the Golgi stack, which represents the site of most metabolic activities of the Golgi apparatus. Modified proteins are transported from the Golgi stack to the *trans* Golgi network, where they are sorted and packaged in vesicles for transport to lysosomes, the plasma membrane, or the exterior of the cell.

Golgi apparatus, Golgi complex, *cis* Golgi network, Golgi stack, *trans* Golgi network

*Protein Glycosylation within the Golgi:* The N-linked oligosaccharides added to proteins in the ER are modified within the Golgi. Those proteins destined for lysosomes are specifically phosphorylated on mannose residues, and mannose-6-phosphate serves as a targeting signal that directs their transport to lysosomes from the *trans* Golgi network. O-linked glycosylation also takes place within the Golgi.

mannose-6-phosphate, signal patch

*Lipid and Polysaccharide Metabolism in the Golgi:* The Golgi apparatus is the site of synthesis of glycolipids, sphingomyelin, and the complex polysaccharides of plant cell walls.

*Protein Sorting and Export from the Golgi Apparatus:* Proteins are sorted in the *trans* Golgi network for packaging into transport vesicles targeted for secretion, the plasma membrane, lysosomes, or yeast and plant vacuoles. In some polarized epithelial cells, proteins are specifically targeted to the apical and basolateral domains of the plasma membrane.

apical domain, basolateral domain, vacuole

## THE MECHANISM OF VESICULAR TRANSPORT

*Experimental Approaches to Understanding Vesicular Transport:* The mechanism of vesicular transport has been elucidated through studies of yeast mutants, reconstituted cell-free systems, and synaptic vesicles.

synaptic vesicle

*Coat Proteins and Vesicle Budding:* The cytoplasmic surfaces of vesicles are coated with proteins that drive vesicle budding and select the specific molecules to be transported.

clathrin-coated vesicle, COP-coated vesicle, clathrin, adaptin, ARF

*Vesicle Fusion:* Vesicle binding to the correct target membrane is thought to be mediated by interactions between pairs of transmembrane proteins. Fusion between the vesicle and target membranes is then brought about by general fusion proteins that are recruited to the complex.

SNARE hypothesis, Rab

## LYSOSOMES

*Lysosomal Acid Hydrolases:* Lysosomes contain an array of acid hydrolases that degrade proteins, nucleic acids, polysaccharides, and lipids. These enzymes function specifically at the acidic pH maintained within lysosomes.

lysosome, lysosomal storage disease

*Endocytosis and Lysosome Formation:* Extracellular molecules taken up by endocytosis are transported to endosomes, which mature to lysosomes as lysosomal acid hydrolases are delivered from the Golgi.

endocytosis, endosome

*Phagocytosis and Autophagy:* Lysosomes are responsible for the degradation of large particles taken up by phagocytosis and for the gradual digestion of the cell's own components by autophagy. During development, autophagy plays an important role in tissue remodeling.

phagocytosis, phagosome, phagolysosome, autophagy, autophagosome

## QUESTIONS

1. You have generated a recombinant cDNA in which a signal sequence has been added to the amino terminus of a nuclear protein. Where would you predict the protein encoded by this cDNA to be localized?

2. Sec61 is a critical component of the protein channel through the ER membrane. In Sec61 mutant cells, what is the fate of proteins that are normally localized to the Golgi apparatus?

3. Why are the carbohydrate groups of glycoproteins always exposed on the surface of the cell?

4. What effect would the addition of a lysosome-targeting signal have on the subcellular localization of a protein that is normally cytosolic? How would it affect localization of a protein that is normally secreted?

5. What is the predicted fate of lysosomal acid hydrolases in I-cell disease, in which cells are deficient in the enzyme required for formation of mannose-6-phosphate residues?

## REFERENCES AND FURTHER READING

### The Endoplasmic Reticulum

Abeijon, C. and C. B. Hirschberg. 1992. Topography of glycosylation reactions in the endoplasmic reticulum. *Trends. Biochem. Sci.* 17: 32–36. [R]

Balch, W. E., J. M. McCaffery, H. Plutner and M. G. Farquhar. 1994. Vesicular stomatitis virus glycoprotein is sorted and concentrated during export from the endoplasmic reticulum. *Cell* 76: 841–852. [P]

Bishop, W. R. and R. M. Bell. 1988. Assembly of phospholipids into cellular membranes: Biosynthesis, transmembrane movement, and intracellular translocation. *Ann. Rev. Cell Biol.* 4: 580–610. [R]

Blobel, G. and B. Dobberstein. 1975. Transfer of proteins across the membrane. I. Presence of proteolytically processed and unprocessed nascent immunoglobulin light chains on membrane-bound ribosomes of murine myeloma. *J. Cell Biol.* 67: 835–851. [P]

Crowley, K. S., S. Liao, V. E. Worrell, G. D. Reinhart and A. E. Johnson. 1994. Secretory proteins move through the endoplasmic reticulum membrane via an aqueous, gated pore. *Cell* 78: 461–471. [P]

Englund, P. T. 1993. The structure and biosynthesis of glycosyl phosphatidylinositol anchors. *Ann. Rev. Biochem.* 62: 121–138. [R]

Fiedler, K. and K. Simons. 1995. The role of N-glycans in the secretory pathway. *Cell* 81: 309–312. [R]

Gilmore, R. 1993. Protein translocation across the endoplasmic reticulum: A tunnel with toll booths at entry and exit. *Cell* 75: 589–592. [R]

Görlich, D. and T. A. Rapoport. 1993. Protein translocation into proteoliposomes reconstituted from purified components of the endoplasmic reticulum membrane. *Cell* 75: 615–630. [P]

Hendrick, J. P. and F.-U. Hartl. 1993. Molecular chaperone functions of heat-shock proteins. *Ann. Rev. Biochem.* 62: 349–384. [R]

Hurtley, S. M. and A. Helenius. 1989. Protein oligomerization in the endoplasmic reticulum. *Ann. Rev. Cell Biol.* 5: 277–307. [R]

Kent, C. 1995. Eukaryotic phospholipid biosynthesis. *Ann. Rev. Biochem.* 64: 315–343. [R]

Menon, A. K. 1995. Flippases. *Trends Cell Biol.* 5: 355–360. [R]

Palade, G. 1975. Intracellular aspects of the process of protein synthesis. *Science* 189: 347–358. [R]

Pelham, H. R. B. 1989. Control of protein exit from the endoplasmic reticulum. *Ann. Rev. Cell Biol.* 5: 1–23. [R]

Rapoport, T. A. 1992. Transport of proteins across the endoplasmic reticulum membrane. *Science* 258: 931–936. [R]

Sanders, S. L. and R. Schekman. 1992. Polypeptide translocation across the endoplasmic reticulum membrane. *J. Biol. Chem.* 267: 13791–13794. [R]

Simon, S. M. and G. Blobel. 1991. A protein-conducting channel in the endoplasmic reticulum. *Cell* 65: 371–380. [P]

Singer, S. J. 1990. The structure and insertion of integral proteins in membranes. *Ann. Rev. Cell Biol.* 6: 247–296. [R]

Udenfriend, S. and K. Kodukula. 1995. How glycosylphosphatidylinositol anchored membrane proteins are made. *Ann. Rev. Biochem.* 64: 563–591. [R]

Walter, P. and A. E. Johnson. 1994. Signal sequence recognition and protein targeting to the endoplasmic reticulum membrane. *Ann. Rev. Cell Biol.* 10: 87–119. [R]

### The Golgi Apparatus

Baranski, T. J., P. L. Faust and S. Kornfeld. 1990. Generation of a lysosomal enzyme targeting signal in the secretory protein pepsinogen. *Cell* 63: 281–291. [P]

Burgess, T. L. and R. B. Kelly. 1987. Constitutive and regulated secretion of proteins. *Ann. Rev. Cell Biol.* 3: 243–293. [R]

Chrispeels, M. J. and N. V. Raikhel. 1992. Short peptide domains target proteins to plant vacuoles. *Cell* 68: 613–616. [R]

Conibear, E. and T. H. Stevens. 1995. Vacuolar biogenesis in yeast: Sorting out the sorting proteins. *Cell* 83: 513–516. [R]

Driouich, A., L. Faye and L. A. Staehelin. 1993. The plant Golgi apparatus: A factory for complex polysaccharides and glycoproteins. *Trends Biochem. Sci.* 18: 210–214. [R]

Hirschberg, C. B. and C. B. Snider. 1987. Topography of glycosylation in the rough endoplasmic reticulum and Golgi apparatus. *Ann. Rev. Biochem.* 56: 63–87. [R]

Hurtley, S. M. 1992. Golgi localization signals. *Trends Biochem. Sci.* 17: 2–3. [R]

Kornfeld, R. and S. Kornfeld. 1985. Assembly of asparagine-linked oligosaccharides. *Ann. Rev. Biochem.* 54: 631–664. [R]

Kornfeld, S. and I. Mellman. 1989. The biogenesis of lysosomes. *Ann. Rev. Cell Biol.* 5: 483–525. [R]

Machamer, C. E. 1993. Targeting and retention of Golgi membrane proteins. *Curr. Opin. Cell Biol.* 5: 606–612. [R]

Mellman, I. and K. Simons. 1992. The Golgi complex: *In vitro veritas? Cell* 68: 829–840. [R]

Nothwehr, S. F. and T. H. Stevens. 1994. Sorting of membrane proteins in the yeast secretory pathway. *J. Biol. Chem.* 269: 10185–10188. [R]

Pelham, H. R. B. and S. Munro. 1993. Sorting of membrane proteins in the secretory pathway. *Cell* 75: 603–605. [R]

Rodriguez-Boulan, E. and S. K. Powell. 1992. Polarity of epithelial and neuronal cells. *Ann. Rev. Cell Biol.* 8: 395–427. [R]

Rothman, J. E. 1994. Mechanisms of intracellular protein transport. *Nature* 355: 409–415. [R]

van Meer, G. 1989. Lipid traffic in animal cells. *Ann. Rev. Cell Biol.* 5: 247–275. [R]

### The Mechanism of Vesicular Transport

Balch, W. E., W. G. Dunphy, W. A. Braeli and J. E. Rothman. 1984. Reconstitution of the transport of protein between successive compartments of the Golgi measured by the coupled incorporation of N-acetylglucosamine. *Cell* 39: 405–416. [P]

Ferro-Novick, S. and R. Jahn. 1994. Vesicle fusion from yeast to man. *Nature* 370: 191–193 [R]

Fischer von Mallard, G., B. Stahl, C. Li, T. C. Südhof and R. Jahn. 1994. Rab proteins in regulated exocytosis. *Trends Biochem. Sci.* 19: 164–168. [R]

Mallabiabarrena, A. and V. Malhotra. 1995. Vesicle biogenesis: The coat connection. *Cell* 83: 667–669. [R]

Mellman, I. 1995. Enigma variations: Protein mediators of membrane fusion. *Cell* 82: 869–872. [R]

Novick, P. and P. Brennwald. 1993. Friends and family: The role of the Rab GTPases in vesicular traffic. *Cell* 75: 597–601. [R]

Novick, P., C. Field and R. Schekman. 1980. Identification of 23 complementation groups required for post-translational events in the yeast secretory pathway. *Cell* 21: 205–215. [P]

Nuoffer, C. and W. E. Balch. 1994. GTPases: Multifunctional molecular switches regulating vesicular traffic. *Ann. Rev. Biochem.* 63: 949–990. [R]

O'Connor, V., G. J. Augustine and H. Betz. 1994. Synaptic vesicle exocytosis: Molecules and models. *Cell* 76: 785–787. [R]

Pearse, B. M. F. and M. S. Robinson. 1990. Clathrin, adaptors, and sorting. *Ann. Rev. Cell Biol.* 6: 151–171. [R]

Pelham, H. R. B. 1994. About turn for the COPs? *Cell* 79. 1125–1127. [R]

Pryer, N. K., L. J. Wuestehube and R. Schekman. 1992. Vesicle-mediated protein sorting. *Ann. Rev. Biochem.* 61: 471–516. [R]

Rothman, J. E. 1994. Mechanisms of intracellular protein transport. *Nature* 372: 55–62. [R]

Rothman, J. E. and F. T. Wieland. 1996. Protein sorting by transport vesicles. *Science* 272: 227–234. [R]

Schekman, R. and L. Orci. 1996. Coat proteins and vesicle budding. *Science* 271: 1526–1533. [R]

Söllner, T., S. W. Whiteheart, M. Brunner, H. Erdjument-Bromage, S. Geromanos, P. Tempst and J. E. Rothman. 1993. SNAP receptors implicated in vesicle targeting and fusion. *Nature* 362: 318–324. [P]

Whiteheart, S. W. and E. W. Kubalek. 1995. SNAPs and NSF: General members of the fusion apparatus. *Trends Cell Biol.* 5: 64–68. [R]

## Lysosomes

Dunn, W. A., Jr. 1994. Autophagy and related mechanisms of lysosome-mediated protein degradation. *Trends Cell Biol.* 4: 139–143. [R]

Fukuda, M. 1991. Lysosomal membrane glycoproteins. *J. Biol. Chem.* 266: 21327–21330. [R]

Gruenberg, J. and K. E. Howell. 1989. Membrane traffic in endocytosis: Insights from cell-free assays. *Ann. Rev. Cell Biol.* 5: 453–481. [R]

Kornfeld, S. 1992. Structure and function of the mannose-6-phosphate/insulinlike growth factor II receptors. *Ann. Rev. Biochem.* 61: 307–330. [R]

Kornfeld, S. and I. Mellman. 1989. The biogenesis of lysosomes. *Ann. Rev. Cell Biol.* 5: 483–525. [R]

Neufeld, E. F. 1991. Lysosomal storage diseases. *Ann. Rev. Biochem.* 60: 257–280. [R]

# 10

## Bioenergetics and Metabolism

### Mitochondria, Chloroplasts, and Peroxisomes

I N ADDITION TO BEING INVOLVED IN PROTEIN SORTING AND TRANSPORT, cytoplasmic organelles provide specialized compartments in which a variety of metabolic activities take place. The generation of metabolic energy is a major activity of all cells, and two cytoplasmic organelles are specifically devoted to energy metabolism and the production of ATP. Mitochondria are responsible for generating most of the useful energy derived from the breakdown of lipids and carbohydrates, and chloroplasts use energy captured from sunlight to generate both ATP and the reducing power needed to synthesize carbohydrates from $CO_2$ and $H_2O$. The third organelle discussed in this chapter, the peroxisome, contains enzymes involved in a variety of different metabolic pathways, including the breakdown of fatty acids and the metabolism of a by-product of photosynthesis.

Mitochondria, chloroplasts, and peroxisomes differ from the organelles discussed in the preceding chapter not only in their functions, but also in their mechanism of assembly. Rather than being synthesized on membrane-bound ribosomes and translocated into the endoplasmic reticulum, proteins destined for peroxisomes, mitochondria, and chloroplasts are synthesized on free ribosomes in the cytosol and imported into their target organelles as completed polypeptide chains. Mitochondria and chloroplasts also contain their own genomes, which include some genes that are transcribed and translated within the organelle. Protein sorting to the cytoplasmic organelles discussed in this chapter is thus distinct from the pathways of vesicular transport that connect the endoplasmic reticulum, Golgi apparatus, lysosomes, and plasma membrane.

## MITOCHONDRIA

**Mitochondria** play a critical role in the generation of metabolic energy in eukaryotic cells. As reviewed in Chapter 2, they are responsible for most of the useful energy derived from the breakdown of carbohydrates and fatty acids, which is converted to ATP by the process of oxidative phosphorylation. Most mitochondrial proteins are translated on free cytosolic ribosomes and imported into the organelle by specific targeting signals. In addition, mitochondria are unique among the cytoplasmic organelles already discussed in that they contain their own DNA, which encodes tRNAs, rRNAs, and some mitochondrial proteins. The assembly of mitochondria thus involves proteins encoded by their own genomes and translated within the organelle, as well as proteins encoded by the nuclear genome and imported from the cytosol.

### Organization and Function of Mitochondria

Mitochondria are surrounded by a double-membrane system, consisting of inner and outer mitochondrial membranes separated by an intermembrane space (Figure 10.1). The inner membrane forms numerous folds (**cristae**), which extend into the interior (or **matrix**) of the organelle. Each of these components plays distinct functional roles, with the matrix and inner membrane representing the major working compartments of mitochondria.

The matrix contains the mitochondrial genetic system as well as the enzymes responsible for the central reactions of oxidative metabolism (Figure 10.2). As discussed in Chapter 2, the oxidative breakdown of glucose and fatty acids is the principal source of metabolic energy in animal cells. The initial stages of glucose metabolism (glycolysis) occur in the cytosol, where glucose is converted to pyruvate (see Figure 2.32). Pyruvate is then transported into mitochondria, where its complete oxidation to $CO_2$ yields the bulk of usable energy (ATP) obtained from glucose metabolism. This involves the initial oxidation of pyruvate to acetyl CoA, which is then broken down to $CO_2$ via the citric acid cycle (see Figures 2.33 and 2.34). The oxidation of fatty acids also yields acetyl CoA (see Figure 2.36), which is similarly metabolized by the citric acid cycle in mitochondria. The enzymes of the citric acid cycle (located in the matrix of mitochondria) thus are central players in the oxidative breakdown of both carbohydrates and fatty acids.

The oxidation of acetyl CoA to $CO_2$ is coupled to the reduction of $NAD^+$ and FAD to NADH and $FADH_2$, respectively. Most of the energy derived from oxidative metabolism is then produced by the process of oxidative phosphorylation (discussed in detail in the next section), which takes place

*Figure 10.1*
**Structure of a mitochondrion**   Mitochondria are bounded by a two membrane system, consisting of inner and outer membranes. Folds of the inner membrane (cristae) extend into the matrix. (Micrograph by K. R. Porter / Photo Researchers, Inc.)

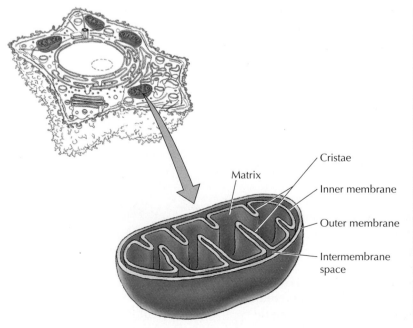

Cristae

Matrix

Inner membrane

Outer membrane

Intermembrane space

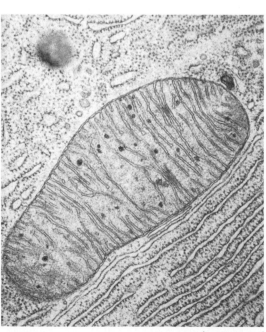

0.5 $\mu$m

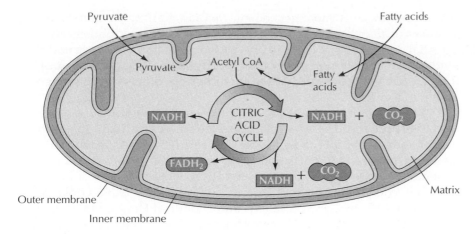

*Figure 10.2*
**Metabolism in the matrix of mitochondria** Pyruvate and fatty acids are imported from the cytosol and converted to acetyl CoA in the mitochondrial matrix. Acetyl CoA is then oxidized to $CO_2$ via the citric acid cycle, the central pathway of oxidative metabolism.

in the inner mitochondrial membrane. The high-energy electrons from NADH and $FADH_2$ are transferred through a series of carriers in the membrane to molecular oxygen. The energy derived from these electron transfer reactions is converted to potential energy stored in a proton gradient across the membrane, which is then used to drive ATP synthesis. The inner mitochondrial membrane thus represents the principal site of ATP generation, and this critical role is reflected in its structure. First, its surface area is substantially increased by its folding into cristae. In addition, the inner mitochondrial membrane contains an unusually high percentage (greater than 70%) of proteins, which are involved in oxidative phosphorylation as well as in the transport of metabolites (e.g., pyruvate and fatty acids) between the cytosol and mitochondria. Otherwise, the inner membrane is impermeable to most ions and small molecules—a property critical to maintaining the proton gradient that drives oxidative phosphorylation.

In contrast to the inner membrane, the outer mitochondrial membrane is freely permeable to small molecules. This is because it contains proteins called **porins**, which form channels that allow the free diffusion of molecules smaller than about 6000 daltons. The composition of the intermembrane space is therefore similar to the cytosol with respect to ions and small molecules. Consequently, the inner mitochondrial membrane is the functional barrier to the passage of small molecules between the cytosol and the matrix and maintains the proton gradient that drives oxidative phosphorylation.

## The Genetic System of Mitochondria

Mitochondria contain their own genetic system, which is separate and distinct from the nuclear genome of the cell. As reviewed in Chapter 1, mitochondria are thought to have evolved from bacteria that developed a symbiotic relationship in which they lived within larger cells (**endosymbiosis**). The genetic system of present-day mitochondria has thus descended from the genome of the original endosymbiotic bacteria that evolved into eukaryotic organelles.

Mitochondrial genomes are usually circular DNA molecules, like those of bacteria, which are present in multiple copies per organelle. They vary considerably in size between different species. The genomes of human and most other animal mitochondria are only about 16 kb, but substantially larger mitochondrial genomes are found in yeasts (approximately 80 kb) and plants (200 to 2000 kb). However, these larger mitochondrial genomes are composed predominantly of noncoding sequences and do not appear to

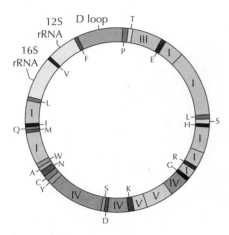

**Figure 10.3**
**The human mitochondrial genome**
The genome contains 13 protein-coding sequences, which are designated as components of respiratory complexes I, III, IV, or V. In addition, the genome contains genes for 12S and 16S rRNAs and for 22 tRNAs, which are designated by the one-letter code for the corresponding amino acid. The region of the genome designated "D loop" contains an origin of DNA replication and transcriptional promoter sequences.

contain significantly more genetic information. Rather, the mitochondrial genomes of all eukaryotes encode only a small number of proteins that are essential components of the oxidative phosphorylation system. In addition, mitochondrial genomes encode all of the ribosomal RNAs and most of the transfer RNAs needed for translation of these protein-coding sequences within mitochondria.

The human mitochondrial genome encodes 13 proteins involved in electron transport and oxidative phosphorylation (Figure 10.3). In addition, human mitochondrial DNA encodes 16S and 12S rRNAs and 22 tRNAs, which are required for translation of the proteins encoded by the organelle genome. The two rRNAs are the only RNA components of animal and yeast mitochondrial ribosomes, in contrast to the three rRNAs of bacterial ribosomes (23S, 16S, and 5S). Plant mitochondrial DNAs, however, also encode a third rRNA of 5S. The mitochondria of plants and protozoans also differ in importing and utilizing tRNAs encoded by the nuclear as well as the mitochondrial genome, whereas in animal mitochondria, all the tRNAs are encoded by the organelle.

The small number of tRNAs encoded by the mitochondrial genome highlights an important feature of the mitochondrial genetic system—the use of a slightly different genetic code, which is distinct from the "universal" genetic code used by both prokaryotic and eukaryotic cells (Table 10.1). As discussed in Chapter 3, there are 64 possible triplet codons, of which 61 encode the 20 different amino acids incorporated into proteins (see Table 3.1). Many tRNAs in both prokaryotic and eukaryotic cells are able to recognize more than a single codon in mRNA because of "wobble," which allows some mispairing between the tRNA anticodon and the third position of certain complementary codons (see Figure 7.3). However, at least 30 different tRNAs are required to translate the universal code according to the wobble rules. Yet human mitochondrial DNA encodes only 22 tRNA species, and these are the only tRNAs used for translation of mitochondrial mRNAs. This is accomplished by an extreme form of wobble in which U in the anticodon of the tRNA can pair with any of the four bases in the third codon position of mRNA, allowing four codons to be recognized by a single tRNA. In addition, some codons specify different amino acids in mitochondria than in the universal code.

Like the DNA of nuclear genomes, mitochondrial DNA can be altered by mutations, which are frequently deleterious to the organelle. Since almost all the mitochondria of fertilized eggs are contributed by the oocyte rather than by the sperm, germ-line mutations in mitochondrial DNA are transmitted to the next generation by the mother. Such mutations have been associated with a number of diseases. For example, Leber's hereditary optic neuropathy, a disease that leads to blindness, can be caused by mutations in mitochondrial genes that encode components of the electron transport chain. Other mutations in mitochondrial genes are thought to contribute to some cases of degenerative neurological diseases, such as Parkinson's disease and Alzheimer's disease.

### Protein Import and Mitochondrial Assembly

In contrast to the RNA components of the mitochondrial translation apparatus (rRNAs and tRNAs), the mitochondrial genome does not encode the proteins required for DNA replication, transcription, or translation. Instead, all the genes that encode proteins required for the replication and expression of mitochondrial DNA (including, with one exception in yeasts, the protein components of mitochondrial ribosomes) are contained in the nucleus. In addition, the nucleus contains the genes that encode most of

**Table 10.1** Differences between the Universal and Mitochondrial Genetic Codes

| Codon | Universal code | Human mitochondrial code |
|-------|----------------|--------------------------|
| UGA   | STOP           | Trp                      |
| AGA   | Arg            | STOP                     |
| AGG   | Arg            | STOP                     |
| AUA   | Ile            | Met                      |

Other codons vary from the universal code in yeast and plant mitochondria.

the mitochondrial proteins required for oxidative phosphorylation and all of the enzymes involved in mitochondrial metabolism (e.g., enzymes of the citric acid cycle). The proteins encoded by these genes are synthesized on free cytosolic ribosomes and imported into mitochondria as completed polypeptide chains. Because of the double-membrane structure of mitochondria, the import of proteins is considerably more complicated than the transfer of a polypeptide across a single phospholipid bilayer. Proteins targeted to the matrix have to cross both the inner and outer mitochondrial membranes, while other proteins need to be sorted to distinct compartments within the organelle (e.g., the intermembrane space).

The import of proteins to the matrix is the best-understood aspect of mitochondrial protein sorting (Figure 10.4). Most proteins are targeted to

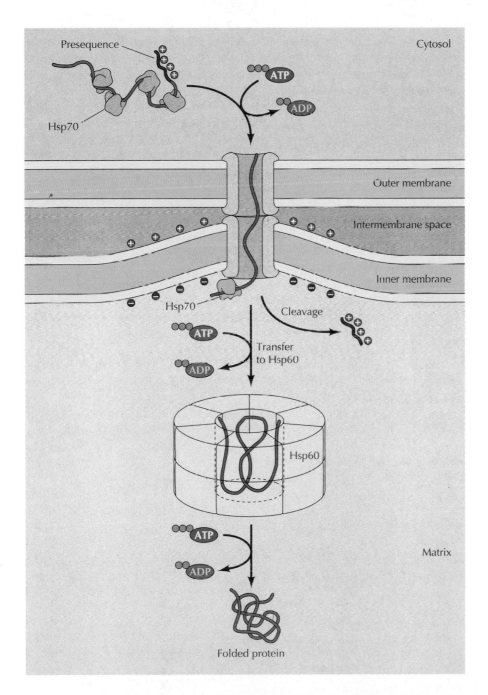

*Figure 10.4*
**Import of proteins into mitochondria**
Proteins are targeted for mitochondria by an amino-terminal presequence containing positively charged amino acids. Proteins are maintained in a partially unfolded state by association with a cytosolic Hsp70 and are recognized by a receptor on the surface of mitochondria. The unfolded polypeptide chains are then translocated through both the outer and inner membranes at regions of close contact between them. The voltage component of the electrochemical gradient is required for translocation across the inner membrane. The presequence is cleaved by a matrix protease, and a mitochondrial Hsp70 binds the polypeptide chain as it crosses the inner membrane, driving further protein translocation. A mitochondrial Hsp60 then facilitates folding of the imported polypeptide within the matrix.

## MOLECULAR MEDICINE

# Diseases of Mitochondria: Leber's Hereditary Optic Neuropathy

### The Disease

Leber's hereditary optic neuropathy (LHON) is a rare inherited disease that results in blindness because of degeneration of the optic nerve. Vision loss usually occurs between the ages of 15 and 35, and is generally the only manifestation of the disease. Not all individuals who inherit the genetic defects responsible for LHON develop the disease, and females are affected less frequently than males. This propensity to affect males might suggest that LHON is an X-linked disease. This is not the case, however, because males never transmit LHON to their offspring. Instead, the inheritance of LHON is entirely by maternal transmission. This characteristic is consistent with cytoplasmic rather than nuclear inheritance of LHON, since the cytoplasm of fertilized eggs is derived almost entirely from the oocyte.

### Molecular and Cellular Basis

In 1988 Douglas Wallace and his colleagues identified a mutation in the mitochondrial DNA of LHON patients. This mutation (at base pair 11778) affects one of the subunits of complex I of the electron transport chain (NADH dehydrogenase), resulting in the substitution of a histidine for an arginine. The 11778 mutation accounts for approximately half of all cases of LHON. Three other mutations of mitochondrial DNA have also been identified as primary causes of LHON. Two of these mutations affect other subunits of complex I, while the third affects cytochrome *b*, which is a component of complex III (see figure). Together, these four mutations account for more than 80% of LHON cases. A fifth mutation (at base pair 14459), affecting a complex I subunit, can cause either LHON or muscular disorders.

The mutations causing LHON reduce the capacity of mitochondria to carry out oxidative phosphorylation and generate ATP. This has the greatest effect on those tissues that are most dependent on oxidative phosphorylation, so defects in components of mitochondria can lead to clinical manifestations in specific organs, rather than to systemic disease. The central nervous system (including the brain and optic nerve) is most highly dependent on oxidative metabolism, consistent with blindness being the primary clinical manifestation resulting from the mitochondrial DNA mutations responsible for LHON.

As already noted, inheritance of LHON mutations does not always

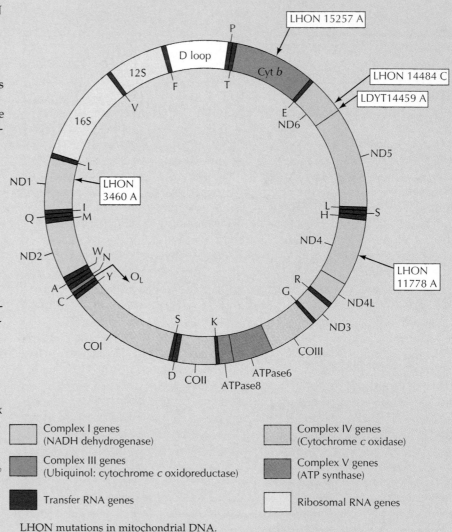

LHON mutations in mitochondrial DNA.

---

### Diseases of Mitochondria: Leber's Hereditary Optic Neuropathy *(continued)*

lead to development of the disease; only about 10% of females and 50% of males possessing a mutation suffer vision loss. One factor that may contribute to this low incidence of disease among carriers of LHON mutations is that each cell contains thousands of copies of mitochondrial DNA, which can be present in mixtures of mutant and normal mitochondria. These mitochondria are randomly distributed to daughter cells at cell division, so the population of mitochondria can change as cells divide, leading to the formation of cells containing either greater or lesser proportions of mutant organelles. Importantly however, many individuals who bear predominantly mutant mitochondrial DNAs still fail to develop the disease. Thus, additional genetic or environmental factors, which have yet to be identified, appear to play a significant role in the development of LHON.

**Prevention and Treatment**

The identification of the mitochondrial DNA mutations responsible for LHON allows molecular diagnosis of the disease, which can be important in establishing a definitive diagnosis of patients without a family history. However, the detection of mutations in mitochondrial DNA is of little value for screening members of affected families or for family planning. This contrasts to the utility of detecting inherited mutations of nuclear genes, where molecular analysis can determine whether a family member or embryo has inherited a mutant or wild-type allele. In LHON, however, mutant mitochondria are present in large numbers and are maternally transmitted to all offspring. As noted above, not all such offspring develop the disease, but this cannot be predicted by genetic analysis.

The finding that LHON is caused by mutations of mitochondrial DNA

suggests the potential of new therapies. One approach is metabolic therapy intended to enhance oxidative phosphorylation by administration of substrates or cofactors in the electron transport pathway, such as succinate or coenzyme Q. Another possibility that has been considered for treatment of LHON is gene therapy designed to relocate a normal gene allele to the nucleus. An appropriate targeting signal would be added to direct the gene product to mitochondria, where it could substitute for the defective mitochondrial-encoded protein.

**References**

Brown, M. D., D. S. Voljavec, M. T. Lott, I. MacDonald and D. C. Wallace. 1992. Leber's hereditary optic neuropathy: A model for mitochondrial neurodegenerative diseases. *FASEB J.* 6: 2791–2799.

Riordan-Eva, P. and A. E. Harding. 1995. Leber's hereditary optic neuropathy: The clinical relevance of different mitochondrial DNA mutations. *J. Med. Genet.* 32: 81–87.

---

mitochondria by amino-terminal sequences of 15 to 35 amino acids (called **presequences**) that are removed by proteolytic cleavage following their import into the organelle. The presequences of mitochondrial proteins, first characterized by Gottfried Schatz, contain multiple positively charged amino acid residues, usually in an amphipathic $\alpha$ helix. The first step in protein import is thought to be recognition of these presequences by receptors that target them to the surface of mitochondria. The presequences are then inserted into a protein complex that directs translocation across the outer membrane. The presequences are then transferred to a distinct protein complex in the inner membrane at areas in which the two membranes are in close contact. Continuing protein translocation requires the electrochemical potential established across the inner mitochondrial membrane during electron transport. As discussed in the next section of this chapter, the transfer of high-energy electrons from NADH and $FADH_2$ to molecular oxygen is coupled to the transfer of protons from the mitochondrial matrix to the intermembrane space. Since protons are charged particles, this transfer establishes an electric potential across the inner membrane, with the matrix being negative. During protein import, this electric potential drives membrane insertion and translocation of the positively charged presequence.

To be translocated across the mitochondrial membrane, proteins must be at least partially unfolded. Consequently, protein import into mitochondria requires molecular chaperones in addition to the membrane proteins involved in translocation (see Figure 10.4). On the cytosolic side, members of the Hsp70 family of chaperones maintain proteins in a partially unfolded state so that they can be inserted into the mitochondrial membrane. As they cross the inner membrane, the unfolded polypeptide chains bind to another member of the Hsp70 family within the mitochondrial matrix. Binding to the matrix Hsp70 promotes further protein translocation across the membrane, and the matrix Hsp70 appears to act as a motor that drives protein import. The polypeptide is then transferred to a chaperone of the Hsp60 family (a chaperonin), within which protein folding takes place. Since these interactions of polypeptide chains with molecular chaperones depend on ATP, protein import requires ATP both outside and inside the mitochondria, in addition to the electric potential across the inner membrane.

As noted above, some mitochondrial proteins are targeted to the outer membrane, inner membrane, or intermembrane space rather than to the matrix, so additional mechanisms are needed to direct these proteins to the correct submitochondrial compartment. The simplest case is the sorting of outer membrane proteins, which interact with receptors on the surface of mitochondria and are then inserted directly into the membrane. However, the sorting of proteins to the internal mitochondrial compartments (the inner membrane and intermembrane space) is more complex and may involve several different mechanisms.

One hypothesis is that some proteins destined for the inner membrane or intermembrane space are first imported into the mitochondrial matrix and then transported from there to their eventual destination (Figure 10.5). A typical positively charged presequence directs the import of these proteins to the matrix, but this sequence is followed by a second hydrophobic signal sequence that is similar to the signal sequences of proteins targeted for the endoplasmic reticulum (ER) or for secretion in bacteria. The positively charged presequence is cleaved following import to the matrix, and the hydrophobic signal sequence then targets the proteins back to the inner membrane. This can lead to either incorporation into the membrane or export to the intermembrane space, with removal of the hydrophobic signal sequence by a second proteolytic cleavage. This pathway of protein sorting seems to reflect the evolutionary origin of mitochondria, since it resembles

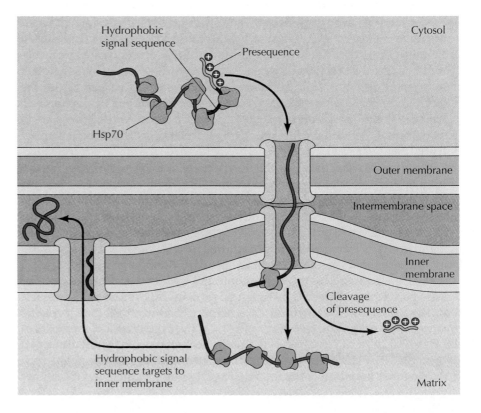

*Figure 10.5*
**Conservative model of protein sorting to the intermembrane space**  Proteins targeted for the intermembrane space contain a positively charged presequence followed by a hydrophobic signal sequence. In the conservative sorting pathway, the presequence directs import of the protein to the matrix, as depicted in Figure 10.4. Removal of the presequence within the matrix then exposes the hydrophobic signal sequence, which targets the protein back across the inner membrane. The hydrophobic signal sequence is cleaved as the protein is exported to the intermembrane space.

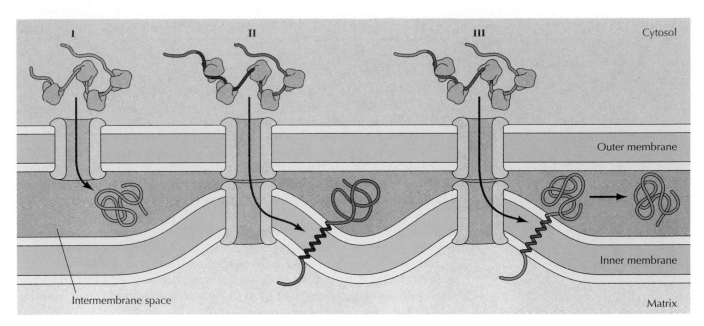

*Figure 10.6*
**Nonconservative sorting mechanisms** Proteins can be targeted to the inner membrane and intermembrane space by several nonconservative mechanisms that do not involve their prior import to the matrix. These include direct translocation across the outer membrane to the intermembrane space (I) and direct incorporation into the inner membrane (II). In addition, some proteins contain hydrophobic stop-transfer sequences that halt translocation across the inner membrane and are then cleaved to release the proteins into the intermembrane space (III).

protein secretion in bacteria; it is therefore referred to as conservative sorting.

However, many proteins of the mitochondrial inner membrane and intermembrane space do not follow the conservative sorting pathway, instead employing several possible nonconservative sorting mechanisms (Figure 10.6). Some proteins, including cytochrome *c*, are directly transferred across the outer membrane into the intermembrane space. Other proteins are transferred across the outer membrane and then directly inserted into the inner membrane, without first having been imported to the matrix. Yet other proteins are transferred across the outer membrane, inserted into the inner membrane, and then released into the intermembrane space by cleavage of hydrophobic stop-transfer sequences. It thus appears that proteins can be targeted to the inner membrane and intermembrane space by several different pathways, the relative importance of which is an area of active investigation.

Not only the proteins, but also the phospholipids of mitochondrial membranes are imported from the cytosol. In animal cells, phosphatidylcholine and phosphatidylethanolamine are synthesized in the ER and carried to mitochondria by **phospholipid transfer proteins**, which extract single phospholipid molecules from the membrane of the ER. The lipid can then be transported through the aqueous environment of the cytosol, buried in a hydrophobic binding site of the protein, and released when the complex reaches a new membrane, such as that of mitochondria. The mitochondria then synthesize phosphatidylserine from phosphatidylethanolamine, in addition to catalyzing the synthesis of the unusual phospholipid **cardiolipin**, which contains four fatty acid chains (Figure 10.7).

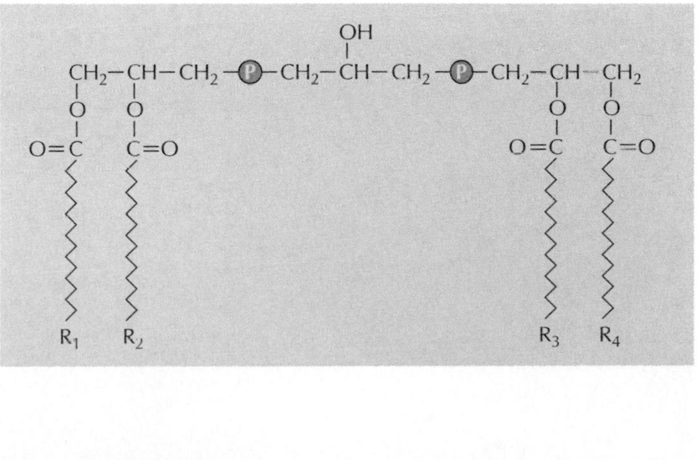

*Figure 10.7*
**Structure of cardiolipin** Cardiolipin is an unusual "double" phospholipid, containing four fatty acid chains, that is found primarily in the inner mitochondrial membrane.

## THE MECHANISM OF OXIDATIVE PHOSPHORYLATION

Most of the usable energy obtained from the breakdown of carbohydrates or fats is derived by oxidative phosphorylation, which takes place within mitochondria. For example, the breakdown of glucose by glycolysis and the citric acid cycle yields a total of four molecules of ATP, ten molecules of NADH, and two molecules of $FADH_2$ (see Chapter 2). Electrons from NADH and $FADH_2$ are then transferred to molecular oxygen, coupled to the formation of an additional 32 to 34 ATP molecules by oxidative phosphorylation. Electron transport and oxidative phosphorylation are critical activities of protein complexes in the inner mitochondrial membrane, which ultimately serve as the major source of cellular energy.

### The Electron Transport Chain

During **oxidative phosphorylation**, electrons derived from NADH and $FADH_2$ combine with $O_2$, and the energy released from these oxidation/reduction reactions is used to drive the synthesis of ATP from ADP. The transfer of electrons from NADH to $O_2$ is a very energy-yielding reaction,

*Figure 10.8*
**Transport of electrons from NADH**

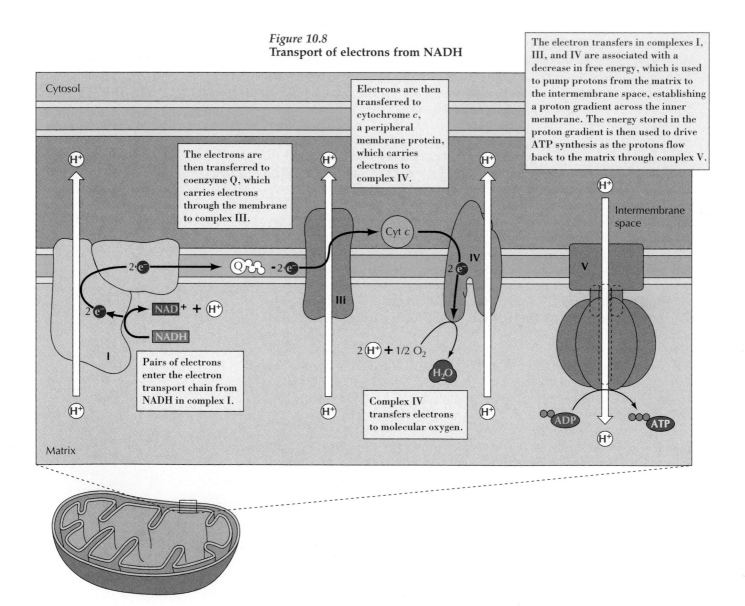

with $\Delta G^{\circ\prime} = -52.5$ kcal/mol for each pair of electrons transferred. To be harvested in usable form, this energy must be produced gradually, by the passage of electrons through a series of carriers, which constitute the **electron transport chain**. These carriers are organized into four complexes in the inner mitochondrial membrane. A fifth protein complex then serves to couple the energy-yielding reactions of electron transport to ATP synthesis.

Electrons from NADH enter the electron transport chain in complex I, which consists of nearly 40 polypeptide chains (Figure 10.8). These electrons are initially transferred from NADH to flavin mononucleotide and then, through an iron-sulfur carrier, to coenzyme Q—an energy-yielding process with $\Delta G^{\circ\prime} = -16.6$ kcal/mol. **Coenzyme Q** (also called **ubiquinone**) is a small, lipid-soluble molecule that carries electrons from complex I through the membrane to complex III, which consists of about ten polypeptides. In complex III, electrons are transferred from cytochrome *b* to cytochrome *c*—an energy-yielding reaction with $\Delta G^{\circ\prime} = -10.1$ kcal/mol. **Cytochrome *c***, a peripheral membrane protein bound to the outer face of the inner membrane, then carries electrons to complex IV (**cytochrome oxidase**), where they are finally transferred to $O_2$ ($\Delta G^{\circ\prime} = -25.8$ kcal/mol).

A distinct protein complex (complex II), which consists of four polypeptides, receives electrons from the citric acid cycle intermediate, succinate (Figure 10.9). These electrons are transferred to $FADH_2$, rather than to NADH, and then to coenzyme Q. From coenzyme Q, electrons are trans-

*Figure 10.9*
**Transport of electrons from $FADH_2$**
Electrons from succinate enter the electron transport chain via $FADH_2$ in complex II. They are then transferred to coenzyme Q and carried through the rest of the electron transport chain as described in Figure 10.8. The transfer of electrons from $FADH_2$ to coenzyme Q is not associated with a significant decrease in free energy, so protons are not pumped across the membrane at complex II.

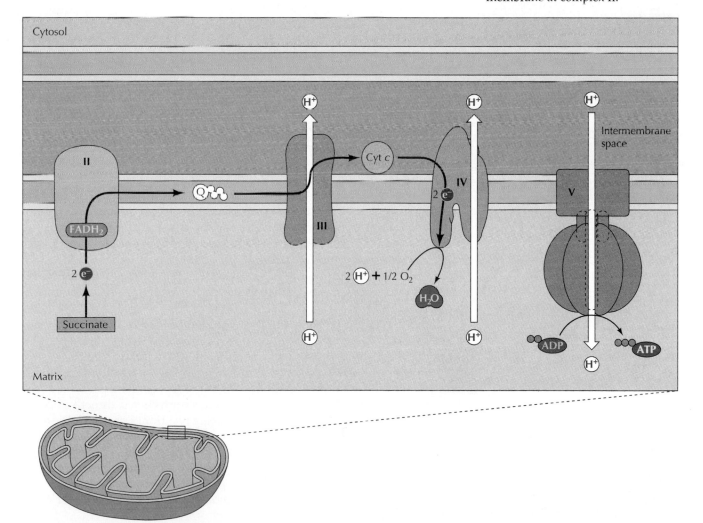

# KEY EXPERIMENT

## *The Chemiosmotic Theory*

**Coupling of Phosphorylation to Electron and Hydrogen Transfer by a Chemi-osmotic Type of Mechanism**

Peter Mitchell
University of Edinburgh, Edinburgh, Scotland
*Nature, 1961, Volume 191, pages 144–148*

### The Context

By the 1950s it had been clearly established that oxidative phosphorylation involved the stepwise transfer of electrons through a series of carriers to molecular oxygen. But how the energy derived from these electron transfer reactions was converted to ATP remained a mystery. The natural assumption was that ADP was converted to ATP by direct transfer of high-energy phosphate groups from some other intermediate, as was known to occur during glycolysis. Thus, it was postulated that high-energy intermediates were produced as a result of electron transfer reactions, and that these intermediates

then drove ATP synthesis by phosphate group transfer.

The search for these postulated high-energy intermediates became a central goal of research during the 1950s and 1960s. But despite many false leads, no such intermediates were to be found. Moreover, several features of oxidative phosphorylation were difficult to reconcile with the orthodox hypothesis that ATP synthesis was driven by simple phosphate group transfer. In particular, phosphorylation was closely associated with membranes and was inhibited by a variety of compounds that disrupted membrane structure. These considerations led Peter Mitchell to propose a fundamen-

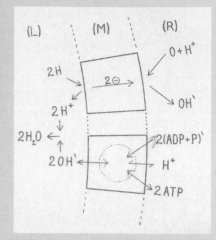

Mitchell's representation of chemiosmotic coupling between an electron transport system (top) and an ATP-generating system (bottom) in a membrane (M) enclosing aqueous phase L within aqueous phase R.

tally different mechanism of energy coupling, in which ATP synthesis was driven by an electrochemical gradient

---

ferred to complex III and then to complex IV as already described. In contrast to the transfer of electrons from NADH to coenzyme Q at complex I, the transfer of electrons from $FADH_2$ to coenzyme Q is not associated with a significant decrease in free energy and, therefore, is not coupled to ATP synthesis. Consequently, the passage of electrons derived from $FADH_2$ through the electron transport chain yields free energy only at complexes III and IV.

The free energy derived from the passage of electrons through complexes I, III, and IV is harvested by being coupled to the synthesis of ATP. Importantly, the mechanism by which the energy derived from these electron transport reactions is coupled to ATP synthesis is fundamentally different from the synthesis of ATP during glycolysis or the citric acid cycle. In the latter cases, a high-energy phosphate is transferred directly to ADP from the other substrate of an energy-yielding reaction. For example, in the final reaction of glycolysis, the high-energy phosphate of phosphoenolpyruvate is transferred to ADP, yielding pyruvate plus ATP (see Figure 2.32). Such direct transfer of high-energy phosphate groups does not occur during electron transport. Instead, the energy derived from electron transport is coupled to the generation of a proton gradient across the inner mito-

## The Chemiosmotic Theory *(continued)*

across a membrane rather than by the elusive high-energy intermediates sought by other investigators.

### The Experiments

The fundamental proposal of the chemiosmotic hypothesis was that the "intermediate" that coupled electron transport to ATP synthesis was a proton electrochemical gradient across the membrane. Mitchell postulated that such a gradient was produced by electron transport, and that the flow of protons back across the membrane in the energetically favorable direction was then coupled to ATP synthesis (see figure).

The hypothesis of chemiosmotic coupling clearly explained the lack of success in identifying a chemical high-energy intermediate, as well as the fact that intact membranes were required to synthesize ATP. Yet it was a radical concept that went against the biochemical dogma of the time. In a concluding paragraph of this 1961 paper, Mitchell took a philosophical view of his revolutionary proposal:

*In the exact sciences, cause and effect are no more than events linked in sequence. Biochemists now generally accept the idea that metabolism is the* cause of membrane transport. The underlying thesis of the hypothesis put forward here is that if the processes that we call metabolism and transport represent events in a sequence, not only can metabolism be the cause of transport, but also transport can be the cause of metabolism.

### The Impact

Mitchell's hypothesis was greeted with skepticism and remained the subject of acrimonious debate for more than a decade. However, the wealth of supporting evidence obtained by Mitchell and his colleagues, as well as by other investigators, eventually led to general acceptance of the chemiosmotic hypothesis—which became known instead as the chemiosmotic theory. It is now accepted not only as the basis for the generation of ATP during oxidative phosphorylation and photosynthesis in bacteria, mitochondria, and chloroplasts, but also for the energy-requiring transport of a variety of molecules across cell membranes.

Mitchell's work was recognized with a Nobel prize in 1978. The lecture he delivered on that occasion began as follows:

Peter Mitchell

*Although I had hoped that the chemiosmotic rationale of vectorial metabolism and biological energy transfer might one day come to be generally accepted, it would have been presumptuous of me to expect it to happen. Was it not Max Planck who remarked that a new scientific idea does not triumph by convincing its opponents, but rather because its opponents eventually die? The fact that what began as the chemiosmotic hypothesis has now been acclaimed as the chemiosmotic theory ... has therefore both astonished and delighted me, particularly because those who were formerly my most capable opponents are still in the prime of their scientific lives.*

---

chondrial membrane. The potential energy stored in this gradient is then harvested by a fifth protein complex, which couples the energetically favorable flow of protons back across the membrane to the synthesis of ATP.

### Chemiosmotic Coupling

The mechanism of coupling electron transport to ATP generation, **chemiosmotic coupling**, is a striking example of the relationship between structure and function in cell biology. The hypothesis of chemiosmotic coupling was first proposed in 1961 by Peter Mitchell, who suggested that ATP is generated by the use of energy stored in the form of proton gradients across biological membranes, rather than by direct chemical transfer of high-energy groups. Biochemists were initially highly skeptical of this proposal, and the chemiosmotic hypothesis took more than a decade to win general acceptance in the scientific community. Overwhelming evidence eventually accumulated in its favor, however, and chemiosmotic coupling is now recognized as a general mechanism of ATP generation, operating not only in mitochondria but also in chloroplasts and in bacteria, where ATP is generated via a proton gradient across the plasma membrane.

Electron transport through complexes I, III, and IV is coupled to the transport of protons out of the interior of the mitochondrion (see Figure 10.8). Thus, the energy-yielding reactions of electron transport are coupled to the pumping of protons from the matrix to the intermembrane space, which establishes a proton gradient across the inner membrane. Complexes I and III each pump four protons per pair of electrons transferred, while two protons per pair of electrons are transferred across the membrane by complex IV. In total, then, ten protons are transferred across the membrane for each pair of electrons derived from NADH, while six protons are transferred per pair of electrons derived from FADH$_2$ (see Figure 10.9). Although the molecular mechanisms have not been fully characterized, proton pumping plays the critical role of converting the energy derived from the oxidation/reduction reactions of electron transport to the potential energy stored in a proton gradient.

Because protons are electrically charged particles, the potential energy stored in the proton gradient is electric as well as chemical in nature. The electric component corresponds to the voltage difference across the inner mitochondrial membrane, with the matrix of the mitochondrion negative and the intermembrane space positive. The corresponding free energy is given by the equation

$$\Delta G = -F\Delta V$$

where $F$ is the Faraday constant and $\Delta V$ is the membrane potential. The additional free energy corresponding to the difference in proton concentration across the membrane is given by the equation

$$\Delta G = RT \ln \frac{[H^+]_i}{[H^+]_o}$$

where $[H^+]_i$ and $[H^+]_o$ refer, respectively, to the proton concentrations inside and outside the mitochondria.

In metabolically active cells, protons are typically pumped out of the matrix such that the proton gradient across the inner membrane corresponds to about one pH unit, or a tenfold lower concentration of protons

*Figure 10.10*
**The electrochemical nature of the proton gradient**   Since protons are positively charged, the proton gradient established across the inner mitochondrial membrane has both chemical and electric components. The chemical component is the proton concentration, or pH, gradient, which corresponds to about a tenfold higher concentration of protons on the cytosolic side of the inner mitochondrial membrane (a difference of one pH unit). In addition, there is an electric potential across the membrane, resulting from the net increase in positive charge on the cytosolic side.

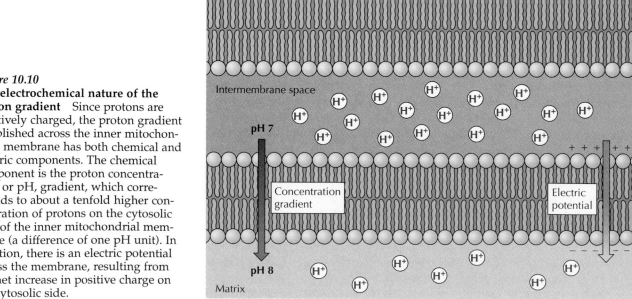

within mitochondria (Figure 10.10). The pH of the mitochondrial matrix is therefore about 8, compared to the neutral pH (approximately 7) of the cytosol and intermembrane space. This gradient also generates an electric potential of approximately 0.14 V across the membrane, with the matrix negative. Both the pH gradient and the electric potential drive protons back into the matrix from the cytosol, so they combine to form an **electrochemical gradient** across the inner mitochondrial membrane, corresponding to a $\Delta G$ of approximately –5 kcal/mol per proton.

Because the phospholipid bilayer is impermeable to ions, protons are able to cross the membrane only through a protein channel. This restriction allows the energy in the electrochemical gradient to be harnessed and converted to ATP as a result of the action of the fifth complex involved in oxidative phosphorylation, complex V, or **ATP synthase** (see Figure 10.8). ATP synthase is organized into two structurally distinct components, $F_0$ and $F_1$ (Figure 10.11). The $F_0$ portion spans the inner membrane and provides a channel through which protons are able to flow back from the intermembrane space to the matrix. The energetically favorable return of protons to the matrix is coupled to ATP synthesis by the $F_1$ subunit, which catalyzes the synthesis of ATP from ADP and phosphate ions ($P_i$). The free energy required for ATP synthesis corresponds to a $\Delta G$ of approximately +12 kcal/mol, so the flow of approximately three protons back across the membrane is needed to drive the synthesis of each molecule of ATP. Because ten protons are pumped across the membrane per pair of electrons derived from NADH, the oxidation of one molecule of NADH leads to the synthesis of approximately three molecules of ATP. However, the oxidation of FADH$_2$, which results in the transfer of six protons, drives the synthesis of only two ATP molecules. The free energy contributed to the gradient at each of the three complexes that transfer protons across the membrane (I, III, and IV) is thus sufficient to drive the synthesis of one ATP molecule.

## Transport of Metabolites across the Inner Membrane

In addition to driving the synthesis of ATP, the potential energy stored in the electrochemical gradient drives the transport of small molecules into and out of mitochondria. For example, the ATP synthesized within mitochondria has to be exported to the cytosol, while ADP and $P_i$ need to be imported from the cytosol for ATP synthesis to continue. The electrochemical gradient generated by proton pumping provides energy required for the transport of these molecules and other metabolites that need to be concentrated within mitochondria (Figure 10.12).

The transport of ATP and ADP across the inner membrane is mediated by an integral membrane protein, the adenine nucleotide translocator, which transports one molecule of ADP into the mitochondrion in exchange for one molecule of ATP transferred from the mitochondrion to the cytosol. Because ATP carries more negative charge than ADP (–4 compared to –3), this exchange is driven by the voltage component of the electrochemical gradient. Since the proton gradient establishes a positive charge on the cytosolic side of the membrane, the export of ATP in exchange for ADP is energetically favorable.

The synthesis of ATP within the mitochondrion requires phosphate ions ($P_i$) as well as ADP, so $P_i$ must also be imported from the cytosol. This is mediated by another membrane transport protein, which imports phosphate ($H_2PO_4^-$) and exports hydroxyl ions ($OH^-$). This exchange is electrically neutral because both phosphate and hydroxyl ions have a charge of –1. However, the exchange is driven by the proton concentration gradient; the higher pH within mitochondria corresponds to a higher concentration

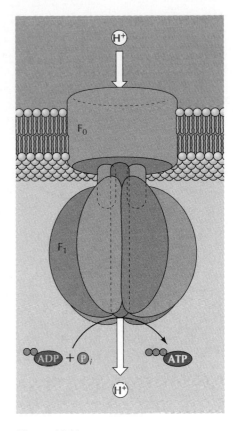

*Figure 10.11*
**Structure of ATP synthase**  The mitochondrial ATP synthase (complex V) consists of two multisubunit components, $F_0$ and $F_1$, which are linked by a slender stalk. $F_0$ spans the lipid bilayer, forming a channel through which protons can cross the membrane. $F_1$ harvests the free energy derived from proton movement down the electrochemical gradient by catalyzing the synthesis of ATP.

*Figure 10.12*
**Transport of metabolites across the mitochondrial inner membrane** The transport of small molecules across the inner membrane is mediated by membrane-spanning transport proteins and driven by the electrochemical gradient. For example, ATP is exported from mitochondria to the cytosol by a transporter that exchanges it for ADP. The voltage component of the electrochemical gradient drives this exchange: ATP carries a greater negative charge (−4) than ADP (−3), so ATP is exported from the mitochondrial matrix to the cytosol while ADP is imported to mitochondria. In contrast, the transport of phosphate ($P_i$) and pyruvate is coupled to an exchange for hydroxyl ions ($OH^-$); in this case, the pH component of the electrochemical gradient drives the export of hydroxyl ions, coupled to the transport of $P_i$ and pyruvate into mitochondria.

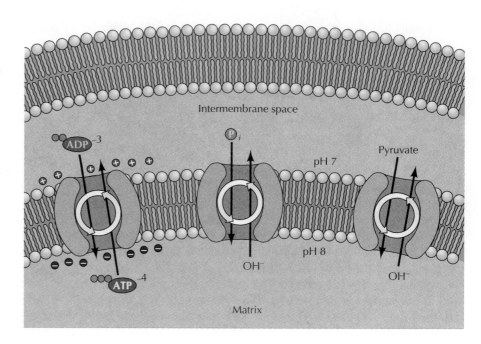

of hydroxyl ions, favoring their translocation to the cytosolic side of the membrane.

Energy from the electrochemical gradient is similarly used to drive the transport of other metabolites into mitochondria. For example, the import of pyruvate from the cytosol (where it is produced by glycolysis) is mediated by a transporter that exchanges pyruvate for hydroxyl ions. Other intermediates of the citric acid cycle are able to shuttle between mitochondria and the cytosol by similar exchange mechanisms.

## CHLOROPLASTS AND OTHER PLASTIDS

**Chloroplasts**, the organelles responsible for photosynthesis, are in many respects similar to mitochondria. Both chloroplasts and mitochondria function to generate metabolic energy, evolved by endosymbiosis, contain their own genetic systems, and replicate by division. However, chloroplasts are larger and more complex than mitochondria, and they perform several critical tasks in addition to the generation of ATP. Most importantly, chloroplasts are responsible for the photosynthetic conversion of $CO_2$ to carbohydrates. In addition, chloroplasts synthesize amino acids, fatty acids, and the lipid components of their own membranes. The reduction of nitrite ($NO_2^-$) to ammonia ($NH_3$), an essential step in the incorporation of nitrogen into organic compounds, also occurs in chloroplasts. Moreover, chloroplasts are only one of several types of related organelles (plastids) that play a variety of roles in plant cells.

### The Structure and Function of Chloroplasts

Plant chloroplasts are large organelles (5 to 10 $\mu$m long) that, like mitochondria, are bounded by a double membrane called the chloroplast envelope (Figure 10.13). In addition to the inner and outer membranes of the envelope, chloroplasts have a third internal membrane system, called the thylakoid membrane. The **thylakoid membrane** forms a network of flattened discs called thylakoids, which are frequently arranged in stacks called grana. Because of this three-membrane structure, the internal organization

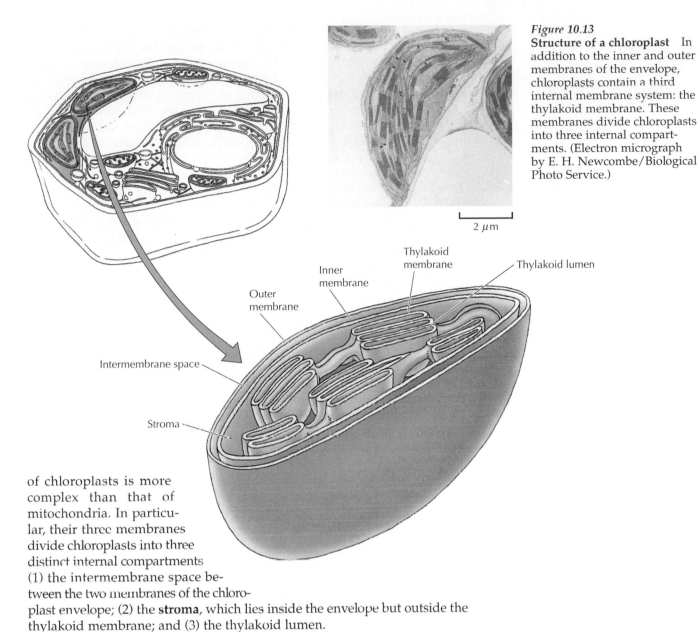

*Figure 10.13*
**Structure of a chloroplast** In addition to the inner and outer membranes of the envelope, chloroplasts contain a third internal membrane system: the thylakoid membrane. These membranes divide chloroplasts into three internal compartments. (Electron micrograph by E. H. Newcombe/Biological Photo Service.)

2 µm

Outer membrane

Inner membrane

Thylakoid membrane

Thylakoid lumen

Intermembrane space

Stroma

of chloroplasts is more complex than that of mitochondria. In particular, their three membranes divide chloroplasts into three distinct internal compartments (1) the intermembrane space between the two membranes of the chloroplast envelope; (2) the **stroma**, which lies inside the envelope but outside the thylakoid membrane; and (3) the thylakoid lumen.

Despite this greater complexity, the membranes of chloroplasts have clear functional similarities with those of mitochondria—as expected, given the role of both organelles in the chemiosmotic generation of ATP. The outer membrane of the chloroplast envelope, like that of mitochondria, contains porins and is therefore freely permeable to small molecules. In contrast, the inner membrane is impermeable to ions and metabolites, which are therefore able to enter chloroplasts only via specific membrane transporters. These properties of the inner and outer membranes of the chloroplast envelope are similar to the inner and outer membranes of mitochondria: In both cases the inner membrane restricts the passage of molecules between the cytosol and the interior of the organelle. The chloroplast stroma is also equivalent in function to the mitochondrial matrix: It contains the chloroplast genetic system and a variety of metabolic enzymes, including those responsible for the critical conversion of $CO_2$ to carbohydrates during photosynthesis.

The major difference between chloroplasts and mitochondria, in terms of both structure and function, is the thylakoid membrane. This membrane is of central importance in chloroplasts, where it fills the role of the inner mitochondrial membrane in electron transport and the chemiosmotic generation of ATP (Figure 10.14). The inner membrane of the chloroplast envelope (which is not folded into cristae) does not function in photosynthesis. Instead, the chloroplast electron transport system is located in the thylakoid membrane, and protons are pumped across this membrane from the stroma to the thylakoid lumen. The resulting electrochemical gradient then drives ATP synthesis as protons cross back into the stroma. In terms of its role in generation of metabolic energy, the thylakoid membrane of chloroplasts is thus equivalent to the inner membrane of mitochondria.

## The Chloroplast Genome

Like mitochondria, chloroplasts contain their own genetic system, reflecting their evolutionary origins from photosynthetic bacteria. The genomes of chloroplasts are similar to those of mitochondria in that they consist of circular DNA molecules present in multiple copies per organelle. However, chloroplast genomes are larger and more complex than those of mitochondria, ranging from 120 to 160 kb and containing approximately 120 genes.

The chloroplast genomes of several plants have been completely sequenced, leading to the identification of many of the genes contained in the organelle DNAs. These chloroplast genes encode both RNAs and proteins involved in gene expression, as well as a variety of proteins that function in photosynthesis (Table 10.2). Both the ribosomal and transfer RNAs used for translation of chloroplast mRNAs are encoded by the organelle genome. These include four rRNAs (23S, 16S, 5S, and 4.5S) and 30 tRNA species. In contrast to the smaller number of tRNAs encoded by the mitochondrial

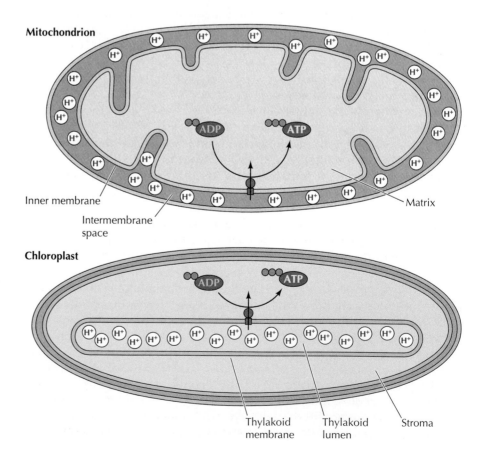

**Figure 10.14**
**Chemiosmotic generation of ATP in chloroplasts and mitochondria**  In mitochondria, electron transport generates a proton gradient across the inner membrane, which is then used to drive ATP synthesis in the matrix. In chloroplasts, the proton gradient is generated across the thylakoid membrane and used to drive ATP synthesis in the stroma.

*Table 10.2* Genes Encoded by Chloroplast DNA

| Function | Number of genes |
| --- | :---: |
| **Genes for the genetic apparatus** | |
| rRNAs (23S, 16S, 5S, 4.5S) | 4 |
| tRNAs | 30 |
| Ribosomal proteins | 21 |
| RNA polymerase subunits | 4 |
| **Genes for photosynthesis** | |
| Photosystem I | 5 |
| Photosystem II | 12 |
| Cytochrome *bf* complex | 4 |
| ATP synthase | 6 |
| Ribulose bisphosphate carboxylase | 1 |

Sequence analysis indicates that chloroplast genomes contain about 30 genes in addition to those listed here. Some of these encode proteins involved in respiration, but most remain to be identified.

genome, the chloroplast tRNAs are sufficient to translate all the mRNA codons according to the universal genetic code. In addition to these RNA components of the translation system, the chloroplast genome encodes about 20 ribosomal proteins, which represent approximately a third of the proteins of chloroplast ribosomes. Some subunits of RNA polymerase are also encoded by chloroplasts, although additional RNA polymerase subunits and other factors needed for chloroplast gene expression are encoded in the nucleus.

The chloroplast genome also encodes approximately 30 proteins that are involved in photosynthesis, including components of photosystems I and II, of the cytochrome *bf* complex, and of ATP synthase. In addition, one of the subunits of ribulose bisphosphate carboxylase (rubisco) is encoded by chloroplast DNA. Rubisco is the critical enzyme that catalyzes the addition of $CO_2$ to ribulose-1,5-bisphosphate during the Calvin cycle (see Figure 2.39). Not only is it the major protein component of the chloroplast stroma, but it is also thought to be the single most abundant protein on Earth, so it is noteworthy that one of its subunits is encoded by the chloroplast genome.

## Import and Sorting of Chloroplast Proteins

Although chloroplasts encode more of their own proteins than mitochondria, about 90% of chloroplast proteins are still encoded by nuclear genes. As with mitochondria, these proteins are synthesized on cytosolic ribosomes and then imported into chloroplasts as completed polypeptide chains. They must then be sorted to their appropriate location within chloroplasts—an even more complicated task than protein sorting in mitochondria, since chloroplasts contain three separate membranes that divide them into three distinct internal compartments.

Protein import into chloroplasts generally resembles mitochondrial protein import (Figure 10.15). Proteins are targeted for import into chloroplasts by N-terminal sequences of 30 to 100 amino acids, called **transit peptides**, which direct protein translocation across the two membranes of the chloroplast envelope and are then removed by proteolytic cleavage. As in mitochondria, proteins are transported across the inner and outer chloroplast membranes at regions of close contact between them. Translocation across the membranes presumably requires unfolding of the polypeptide, so molecular chaperones on both the cytosolic and stromal sides of the envelope are thought to be involved in protein import, which requires energy in the

*Figure 10.15*
**Protein import into the chloroplast
stroma**   Proteins are targeted for
import into chloroplasts by a transit
peptide at their amino terminus. The
transit peptide directs polypeptide
translocation through the chloroplast
envelope at regions of close contact
between the outer and inner membranes. This peptide is then removed
by proteolytic cleavage within the stroma. Both cytosolic and chloroplast
chaperones (Hsp60 and Hsp70) are
required for protein import.

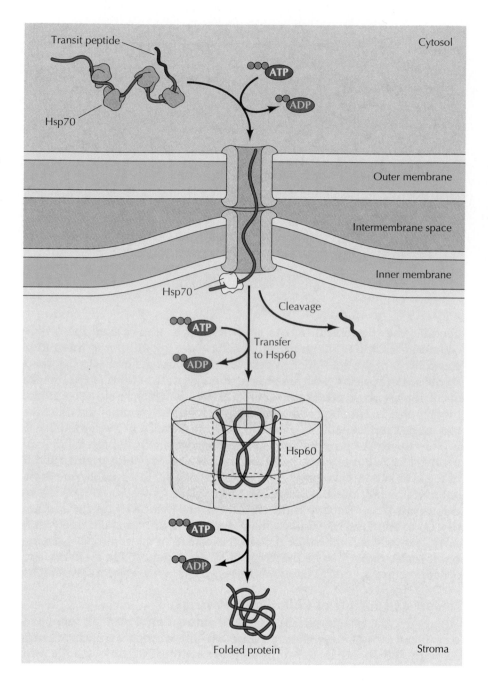

form of ATP. In contrast to the presequences of mitochondrial import, however, transit peptides are not positively charged and the translocation of
polypeptide chains into chloroplasts does not require an electric potential
across the membrane.

Proteins incorporated into the thylakoid lumen are transported to their
destination in two steps (Figure 10.16). They are first imported into the
stroma, as already described, and are then targeted for translocation across
the thylakoid membrane by a second hydrophobic signal sequence, which
is exposed following cleavage of the transit peptide. The hydrophobic signal sequence directs translocation of the polypeptide across the thylakoid
membrane and is finally removed by a second proteolytic cleavage within
the lumen. This two-step mechanism of targeting proteins to the thylakoid
lumen is very similar to the postulated conservative sorting of proteins to

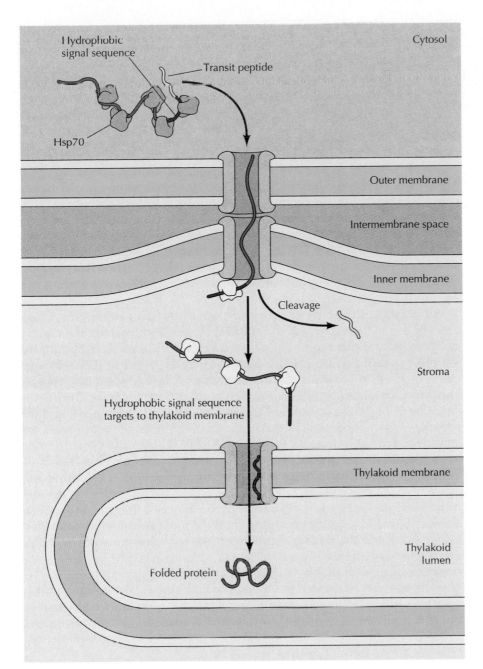

*Figure 10.16*
**Import of proteins into the thylakoid lumen** Proteins are imported into the thylakoid lumen in two steps. The first step is import into the chloroplast stroma, as illustrated in Figure 10.15. Cleavage of the transit peptide then exposes a second hydrophobic signal sequence, which directs protein translocation across the thylakoid membrane.

the intermembrane space of mitochondria (see Figure 10.5).

The pathways of protein sorting to the other four compartments of chloroplasts—the inner and outer membranes, thylakoid membrane, and intermembrane space—are less well established. As with mitochondria, proteins appear to be inserted directly into the outer membrane of the chloroplast envelope by interactions with surface receptors. In contrast, proteins destined for either the thylakoid membrane or the inner membrane of the chloroplast envelope are initially targeted for import into the stroma by N-terminal transit peptides. Following cleavage of the transit peptides, these proteins are then targeted for insertion into the appropriate membrane by other sequences, which are not yet well characterized. Finally, neither the sequences that target proteins to the intermembrane space nor the pathways by which they travel to that destination have been identified.

## Other Plastids

Chloroplasts are only one, albeit the most prominent, member of a larger family of plant organelles called **plastids**. All plastids contain the same genome as chloroplasts, but they differ in both structure and function. Chloroplasts are specialized for photosynthesis and are unique in that they contain the internal thylakoid membrane system. Other plastids, which are involved in different aspects of plant cell metabolism, are bounded by the two membranes of the plastid envelope but lack both the thylakoid membranes and other components of the photosynthetic apparatus.

The different types of plastids are frequently classified according to the kinds of pigments they contain. Chloroplasts are so named because they contain chlorophyll. **Chromoplasts** (Figure 10.17A) lack chlorophyll but contain carotenoids; they are responsible for the yellow, orange, and red colors of some flowers and fruits, although their precise function in cell metabolism is not clear. **Leucoplasts** are nonpigmented plastids, which store a variety of energy sources in nonphotosynthetic tissues. **Amyloplasts** (Figure 10.17B) and **elaioplasts** are examples of leucoplasts that store starch and lipids, respectively.

All plastids, including chloroplasts, develop from **proplastids**, small (0.5 to 1 $\mu$m in diameter) undifferentiated organelles present in the rapidly dividing cells of plant roots and shoots. Proplastids then develop into the various types of mature plastids according to the needs of differentiated cells. In addition, mature plastids are able to change from one type to another. Chromoplasts develop from chloroplasts, for example, during the ripening of fruit (e.g., tomatoes). During this process, chlorophyll and the thylakoid membranes break down, while new types of carotenoids are synthesized.

An interesting feature of plastids is that their development is controlled both by environmental signals and by intrinsic programs of cell differentiation. In the photosynthetic cells of leaves, for example, proplastids develop into chloroplasts (Figure 10.18). During this process, the thylakoid membrane is formed by vesicles budding from the inner membrane of the plastid envelope and the various components of the photosynthetic apparatus are synthesized and assembled. However, chloroplasts develop only in the presence of light. If plants are kept in the dark, the development of proplastids in leaves is arrested at an intermediate stage (called **etioplasts**), in which a semicrystalline array of tubular internal membranes has formed but chlorophyll has not been synthesized (Figure 10.19). If dark-grown

**Figure 10.17**
**Electron micrographs of chromoplasts and amyloplasts**   (A) Chromoplasts contain lipid droplets in which carotenoids are stored. (B) Amyloplasts contain large starch granules. (A, Biophoto Associates/Photo Researchers, Inc.; B, Dr. Jeremy Burgess/Photo Researchers, Inc.)

(A)

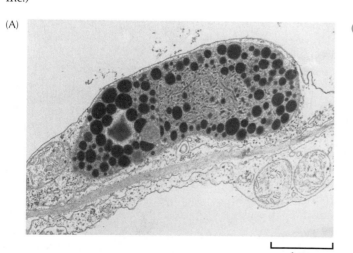

1 $\mu$m

(B)

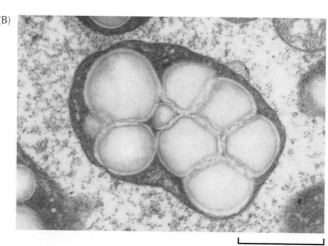

1 $\mu$m

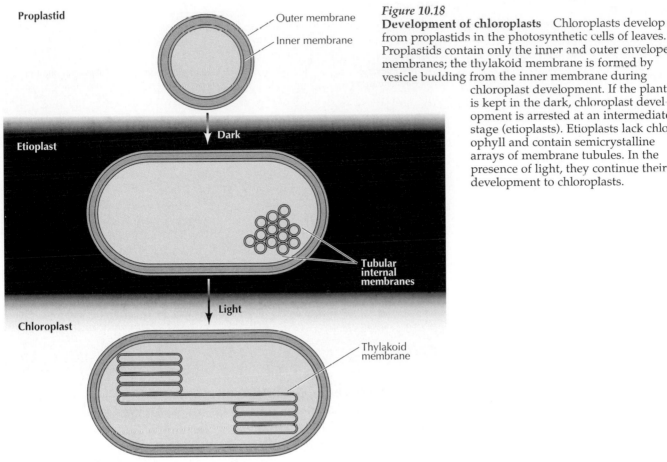

Proplastid

Outer membrane
Inner membrane

Dark

Etioplast

Tubular
internal
membranes

Light

Chloroplast

Thylakoid
membrane

*Figure 10.18*
**Development of chloroplasts**   Chloroplasts develop from proplastids in the photosynthetic cells of leaves. Proplastids contain only the inner and outer envelope membranes; the thylakoid membrane is formed by vesicle budding from the inner membrane during chloroplast development. If the plant is kept in the dark, chloroplast development is arrested at an intermediate stage (etioplasts). Etioplasts lack chlorophyll and contain semicrystalline arrays of membrane tubules. In the presence of light, they continue their development to chloroplasts.

plants are then exposed to light, the etioplasts continue their development to chloroplasts. It is noteworthy that this dual control of plastid development involves the coordinated expression of genes within both the plastid and nuclear genomes. The mechanisms responsible for such coordinated gene expression are largely unknown, and their elucidation represents a challenging problem in plant molecular biology.

## PHOTOSYNTHESIS

During photosynthesis, energy from sunlight is harvested and used to drive the synthesis of glucose from $CO_2$ and $H_2O$. By converting the energy of sunlight to a usable form of potential chemical energy, photosynthesis is the ultimate source of metabolic energy for all biological systems. Photosynthesis takes place in two distinct stages. In the light reactions, energy from sunlight drives the synthesis of ATP and NADPH, coupled to the formation of $O_2$ from $H_2O$. In the dark reactions, so named because they do not require sunlight, the ATP and NADPH produced by the light reactions drive glucose synthesis. In eukaryotic cells, both the light and dark reactions of photosynthesis occur within chloroplasts—the light reactions in the thylakoid membrane and the dark reactions within the stroma. This section discusses the light reactions of photosynthesis, which are related to oxidative phosphorylation in mitochondria. The dark reactions were discussed in detail in Chapter 2.

*Figure 10.19*
**Electron micrograph of an etioplast** (Micrograph by John N. A. Lott/Biological Photo Service.)

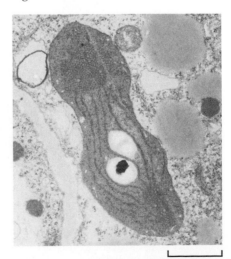

1 µm

*Figure 10.20*
**Organization of a photocenter**   Each photocenter consists of hundreds of antenna pigment molecules, which absorb photons and transfer energy to a reaction center chlorophyll. The reaction center chlorophyll then transfers its excited electron to an acceptor in the electron transport chain. The reaction center illustrated is that of photosystem II, in which electrons are transferred from the reaction center chlorophyll to pheophytin and then to quinones ($Q_A$, $Q_B$, and $QH_2$).

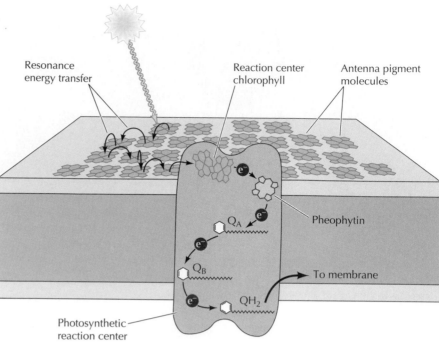

*Figure 10.21*
**Structure of a photosynthetic reaction center**   The reaction center of *R. viridis* consists of three transmembrane proteins (purple, blue, and beige) and a *c*-type cytochrome (green). Chlorophylls and other prosthetic groups are colored yellow. (Courtesy of Johann Deisenhofer, University of Texas Medical Center and The Nobel Foundation, 1989.)

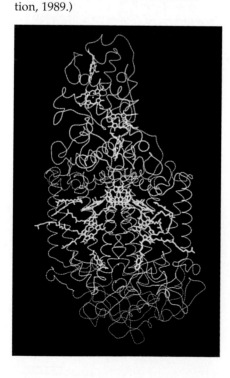

## Electron Flow through Photosystems I and II

Sunlight is absorbed by photosynthetic pigments, the most abundant of which in plants are the **chlorophylls**. Absorption of light excites an electron to a higher energy state, thus converting the energy of sunlight to potential chemical energy. The photosynthetic pigments are organized into **photocenters** in the thylakoid membrane, each of which contains hundreds of pigment molecules (Figure 10.20). The many pigment molecules in each photocenter act as antennae to absorb light and transfer the energy of their excited electrons to a chlorophyll molecule that serves as a reaction center. The reaction center chlorophyll then transfers its high-energy electron to an acceptor molecule in an electron transport chain. High-energy electrons are then transferred through a series of membrane carriers, coupled to the synthesis of ATP and NADPH.

The best characterized photosynthetic reaction center is that of the bacterium *Rhodopseudomonas viridis*, the structure of which was determined by Johann Deisenhofer, Hartmut Michel, Robert Huber, and their colleagues in 1985 (Figure 10.21). The reaction center consists of three transmembrane polypeptides, bound to a *c*-type cytochrome on the exterior side of the membrane. Energy from sunlight is captured by a pair of chlorophyll molecules known as the special pair. Electrons are then transferred from the special pair to another pair of chlorophylls and from there to other prosthetic groups (pheophytins and quinones). From there the electrons are transferred to a cytochrome *bc* complex in which electron transport is coupled to the generation of a proton gradient. The electrons are then transferred to the reaction center cytochrome and finally returned to the chlorophyll special pair. The reaction center thus converts the energy of sunlight to high-energy electrons, the potential energy of which is converted to a proton gradient by the cytochrome *bc* complex.

The proteins involved in the light reactions of photosynthesis in plants are organized into five complexes in the thylakoid membrane (Figure 10.22). Two of these complexes are photosystems (**photosystems I and II**),

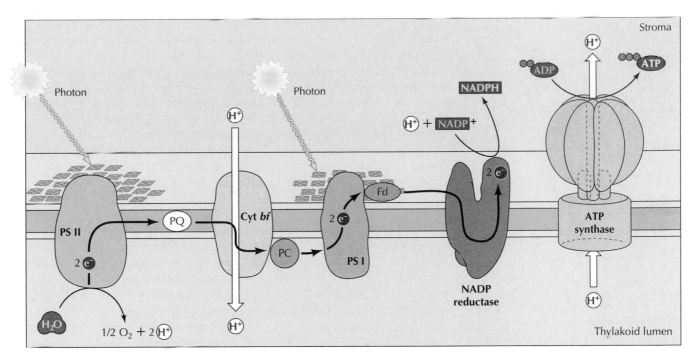

Stroma

Photon

Photon

NADPH

$H^+$ + NADP$^+$

$H^+$

ADP

ATP

Fd

2 $e^-$

PS II

Cyt *bf*

PQ

PC

2 $e^-$

PS I

2 $e^-$

ATP synthase

NADP reductase

$H_2O$

$1/2\ O_2 + 2\ H^+$

$H^+$

$H^+$

$H^+$

Thylakoid lumen

**Figure 10.22**
**Electron transport and ATP synthesis during photosynthesis** Five protein complexes in the thylakoid membrane function in electron transport and the synthesis of ATP and NADPH. Photons are absorbed by complexes of pigment molecules associated with photosystems I and II (PS I and PS II). At photosystem II, energy derived from photon absorption is used to split a water molecule within the thylakoid lumen. Electrons are then carried by plastoquinone (PQ) to the cytochrome *bf* complex, where they are transferred to a lower energy state and protons are pumped into the thylakoid lumen. Electrons are then transferred to photosystem I by plastocyanin (PC). At photosystem I, energy derived from light absorption again generates high-energy electrons, which are used to reduce NADP$^+$ to NADPH in the stroma. ATP synthase then uses the energy stored in the proton gradient to convert ADP to ATP.

in which light is absorbed and transferred to reaction center chlorophylls. High-energy electrons are then transferred through a series of carriers in both photosystems and in a third protein complex, the **cytochrome *bf* complex**. As in mitochondria, these electron transfers are coupled to the transfer of protons into the thylakoid lumen, thereby establishing a proton gradient across the thylakoid membrane. The energy stored in this proton gradient is then harvested by a fourth protein complex in the thylakoid membrane, ATP synthase, which (like the mitochondrial enzyme) couples proton flow back across the membrane to the synthesis of ATP.

One important difference between electron transport in chloroplasts and that in mitochondria is that the energy derived from sunlight during photosynthesis not only is converted to ATP but also is used to generate the NADPH required for subsequent conversion of $CO_2$ to carbohydrates. This is accomplished by the use of two different photosystems in the light reactions of photosynthesis, one to generate ATP and the other to generate NADPH. Electrons are transferred sequentially between the two photosystems, with photosystem I acting to generate NADPH and photosystem II acting to generate ATP.

The pathway of electron flow starts at photosystem II, which is homologous to the photosynthetic reaction center of *R. viridis* already described. However, at photosystem II the energy derived from absorption of photons is used to split water molecules to molecular oxygen and protons (see Figure 10.22). This reaction takes place within the thylakoid lumen, so the release of protons from $H_2O$ establishes a proton gradient across the thylakoid membrane. The high-energy electrons derived from this process are transferred through a series of carriers to plastoquinone, a lipid-soluble carrier similar to coenzyme Q (ubiquinone) of mitochondria. Plastoquinone carries electrons from photosystem II to the cytochrome *bf* complex, within which electrons are transferred to plastocyanin and additional protons are pumped into the thylakoid lumen. Electron transport through photosystem II is thus coupled to establishment of a proton gradient, which drives the chemiosmotic synthesis of ATP.

From photosystem II, electrons are carried by plastocyanin (a peripheral membrane protein) to photosystem I, where the absorption of additional photons again generates high-energy electrons. Photosystem I, however, does not act as a proton pump; instead, it uses these high-energy electrons to reduce $NADP^+$ to NADPH. The reaction center chlorophyll of photosystem I transfers its excited electrons through a series of carriers to ferrodoxin, a small protein on the stromal side of the thylakoid membrane. The enzyme **NADP reductase** then transfers electrons from ferrodoxin to $NADP^+$, generating NADPH. The passage of electrons through photosystems I and II thus generates both ATP and NADPH, which are used by the Calvin cycle enzymes in the chloroplast stroma to convert $CO_2$ to carbohydrates (see Figure 2.39).

### Cyclic Electron Flow

A second electron transport pathway, called **cyclic electron flow**, produces ATP without the synthesis of NADPH, thereby supplying additional ATP for other metabolic processes. In cyclic electron flow, light energy harvested at photosystem I is used for ATP synthesis rather than NADPH synthesis (Figure 10.23). Instead of being transferred to $NADP^+$, high-energy electrons from photosystem I are transferred to the cytochrome *bf* complex. Electron transfer through the cytochrome *bf* complex is then coupled, as in photosystem II, to the establishment of a proton gradient across the thylakoid membrane. Plastocyanin then returns these electrons to photosystem I in a lower energy state, completing a cycle of electron transport in which light energy harvested at photosystem I is used to pump protons at the cytochrome *bf* complex. Electron transfer from photosystem I can thus generate either ATP or NADPH, depending on the metabolic needs of the cell.

### ATP Synthesis

The ATP synthase of the thylakoid membrane is similar to the mitochondrial enzyme. However, the energy stored in the proton gradient across the thylakoid membrane, in contrast to the inner mitochondrial mem-

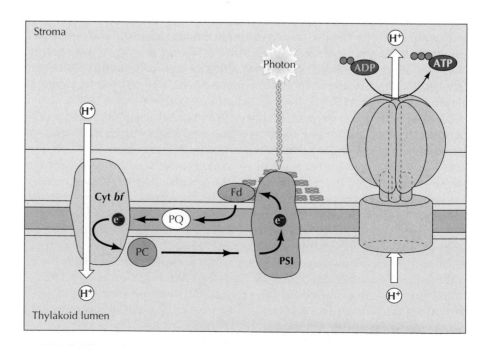

*Figure 10.23*
**The pathway of cyclic electron flow** Light energy absorbed at photosystem I (PSI) is used for ATP synthesis rather than NADPH synthesis. High-energy electrons generated by photon absorption are transferred to the cytochrome *bf* complex rather than to $NADP^+$. At the cytochrome *bf* complex, electrons are transferred to a lower energy state and protons are pumped into the thylakoid lumen. The electrons are then returned to photosystem I by plastocyanin (PC).

brane, is almost entirely chemical in nature. This is because the thylakoid membrane, although impermeable to protons, differs from the inner mitochondrial membrane in being permeable to other ions, particularly $Mg^{2+}$ and $Cl^-$. The free passage of these ions neutralizes the voltage component of the proton gradient, so the energy derived from photosynthesis is conserved mainly as the difference in proton concentration (pH) across the thylakoid membrane. However, because the thylakoid lumen is a closed compartment, this difference in proton concentration can be quite large, corresponding to a differential of more than three pH units between the stroma and the thylakoid lumen. Because of the magnitude of this pH differential, the total free energy stored across the thylakoid membrane is similar to that stored across the inner mitochondrial membrane. In both chloroplasts and mitochondria, therefore, the passage of approximately three protons back across the membrane is sufficient to drive the synthesis of one molecule of ATP.

For each pair of electrons transported, two protons are transferred across the thylakoid membrane at photosystem II and two to four protons at the cytochrome *bf* complex. Since approximately three protons are needed to drive the synthesis of one molecule of ATP, passage of each pair of electrons through photosystems I and II by noncyclic electron flow yields between 1.3 and 2 ATP molecules. Cyclic electron flow has a lower yield, corresponding to between 0.67 and 1.3 ATP molecules per pair of electrons.

## PEROXISOMES

**Peroxisomes** are small, membrane-enclosed organelles (Figure 10.24) that contain enzymes involved in a variety of metabolic reactions, including several aspects of energy metabolism. Although peroxisomes are morphologically similar to lysosomes, they are assembled, like mitochondria and chloroplasts, from proteins that are synthesized on free ribosomes and then imported into peroxisomes as completed polypeptide chains. Although peroxisomes do not contain their own genomes, they are similar to mitochondria and chloroplasts in that they replicate by division.

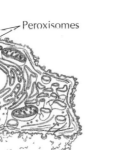
Peroxisomes

### Functions of Peroxisomes

Peroxisomes contain at least 50 different enzymes, which are involved in a variety of biochemical pathways in different types of cells.

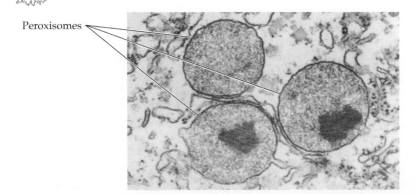
Peroxisomes

*Figure 10.24*
**Electron micrograph of peroxisomes**
Three peroxisomes from rat liver are shown. Two contain dense regions, which are paracrystalline arrays of the enzyme urate oxidase. (Don Fawcett/ Photo Researchers, Inc.)

0.5 μm

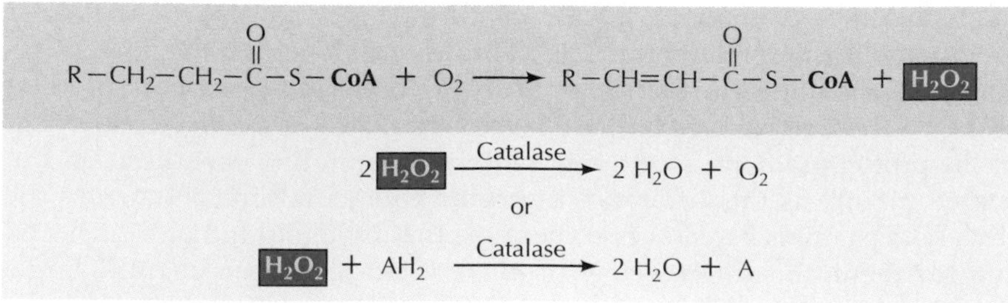

*Figure 10.25*
**Fatty acid oxidation in peroxisomes**
The oxidation of a fatty acid is accompanied by the production of hydrogen peroxide ($H_2O_2$) from oxygen. The hydrogen peroxide is decomposed by catalase, either by conversion to water or by oxidation of another organic compound (designated $AH_2$).

Peroxisomes originally were defined as organelles that carry out oxidation reactions leading to the production of hydrogen peroxide. Because hydrogen peroxide is harmful to the cell, peroxisomes also contain the enzyme **catalase**, which decomposes hydrogen peroxide either by converting it to water or by using it to oxidize another organic compound. A variety of substrates are broken down by such oxidative reactions in peroxisomes, including uric acid, amino acids, and fatty acids. The oxidation of fatty acids (Figure 10.25) is a particularly important example, since it provides a major source of metabolic energy. In animal cells, fatty acids are oxidized in both peroxisomes and mitochondria, but in yeasts and plants fatty acid oxidation is restricted to peroxisomes.

In addition to providing a compartment for oxidation reactions, peroxisomes are involved in lipid biosynthesis. In animal cells, cholesterol and dolichol are synthesized in peroxisomes as well as in the ER. In the liver, peroxisomes are also involved in the synthesis of bile acids, which are derived from cholesterol. In addition, peroxisomes contain enzymes required for the synthesis of **plasmalogens**—a family of phospholipids in which one of the hydrocarbon chains is joined to glycerol by an ether bond rather than an ester bond (Figure 10.26). Plasmalogens are important membrane components in some tissues, particularly heart and brain, although they are absent in others.

Peroxisomes play two particularly important roles in plants. First, peroxisomes in seeds are responsible for the conversion of stored fatty acids to carbohydrates, which is critical to providing energy and raw materials for growth of the germinating plant. This occurs via a series of reactions termed the **glyoxylate cycle**, which is a variant of the citric acid cycle (Figure 10.27). The peroxisomes in which this takes place are sometimes called **glyoxysomes**.

Second, peroxisomes in leaves are involved in **photorespiration**, which serves to metabolize a side product formed during photosynthesis (Figure 10.28). $CO_2$ is converted to carbohydrates during photosynthesis via a series of reactions called the Calvin cycle (see Figure 2.39). The first step is the addition of $CO_2$ to the five-carbon sugar ribulose-1,5-bisphosphate, yielding two molecules of 3-phosphoglycerate (three carbons each). However, the enzyme involved (ribulose bisphosphate carboxylase or rubisco) sometimes catalyzes the addition of $O_2$ instead of $CO_2$, producing one molecule of 3-phosphoglycerate and one molecule of phosphoglycolate (two carbons). This is a side reaction, and phosphoglycolate is not a useful metabolite. It is first converted to glycolate and then transferred to peroxisomes, where it is oxidized and converted to glycine. Glycine is then transferred to mitochondria, where two molecules of glycine are converted to one molecule of serine, with the loss of $CO_2$ and $NH_3$. The serine is then returned to peroxisomes, where it is converted to glycerate. Finally, the glycerate is transferred back to chloroplasts, where it reenters the Calvin cycle. Photorespiration does not appear to be beneficial for the plant, since it is essentially the reverse of photosynthesis—$O_2$ is consumed and $CO_2$ is

*Figure 10.26*
**Structure of a plasmalogen** The plasmalogen shown is analogous to phosphatidylcholine. However, one of the fatty acid chains is joined to glycerol by an ether, rather than an ester, bond.

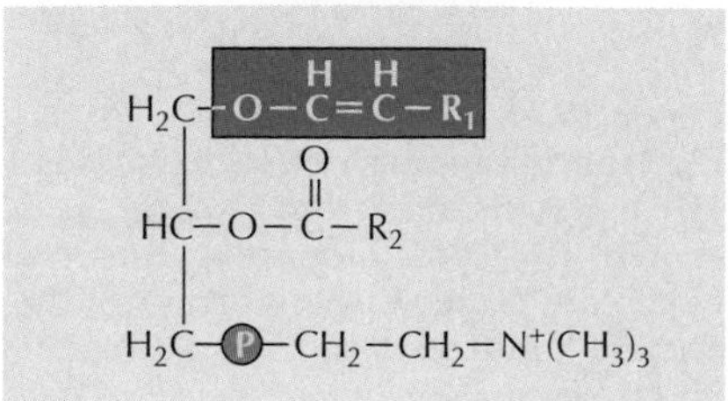

*Figure 10.27*
**The glyoxylate cycle** Plants are capable of synthesizing carbohydrates from fatty acids via the glyoxylate cycle, which is a variant of the citric acid cycle (see Figure 2.34). As in the citric acid cycle, acetyl CoA combines with oxaloacetate to form citrate, which is converted to isocitrate. However, instead of being degraded to $CO_2$ and $\alpha$-ketoglutarate, isocitrate is converted to succinate and glyoxylate. Glyoxylate then reacts with another molecule of acetyl CoA to yield malate, which is converted to oxaloacetate and used for glucose synthesis.

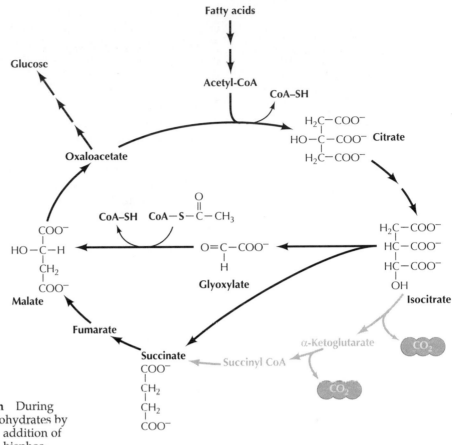

*Figure 10.28*
**Role of peroxisomes in photorespiration** During photosynthesis, $CO_2$ is converted to carbohydrates by the Calvin cycle, which initiates with the addition of $CO_2$ to the five-carbon sugar ribulose-1,5-bisphosphate. However, the enzyme involved sometimes catalyzes the addition of $O_2$ instead, resulting in production of the two-carbon compound phosphoglycolate. Phosphoglycolate is converted to glycolate, which is then transferred to peroxisomes, where it is oxidized and converted to glycine. Glycine is then transferred to mitochondria and converted to serine. The serine is returned to peroxisomes and converted to glycerate, which is transferred back to chloroplasts.

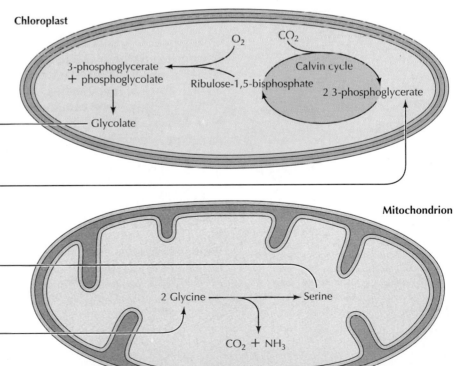

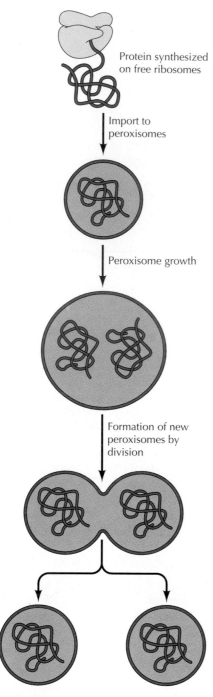

Protein synthesized on free ribosomes

Import to peroxisomes

Peroxisome growth

Formation of new peroxisomes by division

*Figure 10.29*
**Assembly of peroxisomes** Proteins destined for peroxisomes are synthesized on free ribosomes and imported into preexisting peroxisomes as completed polypeptide chains. Protein import results in peroxisome growth and the formation of new peroxisomes by division of old ones.

released without any gain of ATP. However, the occasional utilization of $O_2$ in place of $CO_2$ appears to be an inherent property of rubisco, so photorespiration is a general accompaniment of photosynthesis. Peroxisomes thus play an important role by allowing most of the carbon in glycolate to be recovered and utilized.

### Peroxisome Assembly

As already noted, the assembly of peroxisomes is fundamentally similar to that of mitochondria and chloroplasts, rather than to that of the endoplasmic reticulum, Golgi apparatus, and lysosomes. Proteins destined for peroxisomes are translated on free cytosolic ribosomes and then transported into peroxisomes as completed polypeptide chains (Figure 10.29). Phospholipids are also imported to peroxisomes, via phospholipid transfer proteins, from their major site of synthesis in the ER. The import of proteins and phospholipids results in peroxisome growth, and new peroxisomes are then formed by division of old ones.

Proteins are targeted to the interior of peroxisomes by at least two pathways, which are conserved from yeasts to humans. Most proteins are targeted to peroxisomes by the simple amino acid sequence Ser-Lys-Leu at their carboxy terminus (peroxisome targeting signal 1, or PTS1). Other proteins are targeted by a sequence of nine amino acids (PTS2) at their amino terminus, and some proteins may be targeted by alternative signals that have not yet been well defined. Distinct internal signals are also responsible for the targeting of integral membrane proteins to the peroxisome membrane. In contrast to the translocation of polypeptide chains across the membranes of the endoplasmic reticulum, mitochondria, and chloroplasts, targeting signals are usually not cleaved during the import of proteins into peroxisomes.

PTS1 and PTS2 are recognized by distinct receptors and then transferred to a translocation complex that mediates their transport across the peroxisome membrane. However, the mechanism of protein import into peroxisomes is less well characterized than the mechanisms of protein translocation across the membranes of the endoplasmic reticulum, mitochondria, or chloroplasts. Cytosolic Hsp70 has been implicated in protein import to peroxisomes, but the possible role of molecular chaperones within peroxisomes is unclear. Understanding peroxisomal protein import thus remains an active area of investigation.

Interestingly, some components of peroxisome import pathways have been identified not only as mutants of yeasts but also as mutations associated with serious human diseases involving disorders of peroxisomes. In some such diseases, only a single peroxisomal enzyme is deficient. However, in other diseases resulting from defects in peroxisome function, multiple peroxisomal enzymes fail to be imported to peroxisomes, instead being localized in the cytosol. The latter group of diseases results from deficiencies in the PTS1 or PTS2 pathways responsible for peroxisomal protein import. The prototypical example is Zellweger syndrome, which is lethal within the first ten years of life. Zellweger syndrome can result from mutations in at least ten different genes affecting peroxisomal protein import, one of which has been identified as the gene encoding the receptor for the peroxisome targeting signal PTS1.

## Summary

### MITOCHONDRIA

*Organization and Function of Mitochondria:* Mitochondria, which play a critical role in the generation of metabolic energy, are surrounded by a double-membrane system. The matrix contains the enzymes of the citric acid cycle; the inner membrane contains protein complexes involved in electron transport and oxidative phosphorylation. In contrast to the inner membrane, the outer membrane is freely permeable to small molecules.

mitochondria, crista, matrix, porin

*The Genetic System of Mitochondria:* Mitochondria contain their own genomes, which encode rRNAs, tRNAs, and some of the proteins that are involved in oxidative phosphorylation.

endosymbiosis

*Protein Import and Mitochondrial Assembly:* Most mitochondrial proteins are encoded by the nuclear genome. These proteins are translated on free ribosomes and imported into mitochondria as completed polypeptide chains. Positively charged presequences target proteins for import to the mitochondrial matrix. Phospholipids are carried to mitochondria from the endoplasmic reticulum by phospholipid transfer proteins.

presequence, phospholipid transfer protein, cardiolipin

### THE MECHANISM OF OXIDATIVE PHOSPHORYLATION

*The Electron Transport Chain:* Most of the energy derived from oxidative metabolism comes from the transfer of electrons from NADH and $FADH_2$ to $O_2$. In order to harvest this energy in usable form, electrons are transferred through a series of carriers organized into four protein complexes in the inner mitochondrial membrane.

oxidative phosphorylation, electron transport chain, coenzyme Q, ubiquinone, cytochrome *c*, cytochrome oxidase

*Chemiosmotic Coupling:* The energy-yielding reactions of electron transport are coupled to the generation of a proton gradient across the inner mitochondrial membrane. The potential energy stored in this gradient is harvested by a fifth protein complex, ATP synthase, which couples ATP synthesis to the energetically favorable return of protons to the mitochondrial matrix.

chemiosmotic coupling, electrochemical gradient, ATP synthase

*Transport of Metabolites across the Inner Membrane:* In addition to driving ATP synthesis, potential energy stored in the proton gradient drives the transport of ATP, ADP, and other metabolites into and out of mitochondria.

### CHLOROPLASTS AND OTHER PLASTIDS

*The Structure and Function of Chloroplasts:* Chloroplasts are large organelles that function in photosynthesis and a variety of other metabolic activities. Like mitochondria, chloroplasts are bounded by a double-membrane envelope. In addition, chloroplasts have an internal thylakoid membrane, which is the site of electron transport and the chemiosmotic generation of ATP.

chloroplast, thylakoid membrane, stroma

*The Chloroplast Genome:* Chloroplast genomes contain more than 100 genes, including genes encoding rRNAs, tRNAs, some ribosomal proteins, and some proteins involved in photosynthesis.

*Import and Sorting of Chloroplast Proteins:* Most chloroplast proteins are synthesized on free ribosomes in the cytosol and targeted for import

transit peptide

plastid, chromoplast, leucoplast, amyloplast, elaioplast, proplastid, etioplast

to chloroplasts by amino-terminal transit peptides. Proteins incorporated into the thylakoid lumen are first imported into the chloroplast stroma and then targeted for transport across the thylakoid membrane by a second hydrophobic signal sequence.

*Other Plastids:* The chloroplast is only one member of a family of related organelles, all of which contain the same genome. Other plastids serve to store energy sources, such as starch and lipids, and function in other aspects of plant metabolism.

## PHOTOSYNTHESIS

chlorophyll, photocenter, photosystem I, photosystem II, cytochrome *bf* complex, NADP reductase

*Electron Flow through Photosystems I and II:* During photosynthesis, energy from sunlight is harvested and converted to usable forms of potential chemical energy. Absorption of light by chlorophylls excites electrons to a higher energy state. These high-energy electrons are then transferred through a series of carriers organized into two photosystems and the cytochrome *bf* complex in the thylakoid membrane. The sequential flow of electrons through both photosystems is coupled to the synthesis of ATP at photosystem II and the reduction of $NADP^+$ to NADPH at photosystem I. Both ATP and NADPH are then used in the synthesis of carbohydrates from $CO_2$, which takes place in the chloroplast stroma.

cyclic electron flow

*Cyclic Electron Flow:* The alternative pathway of cyclic electron flow allows light energy harvested at photosystem I to be converted to ATP, rather than NADPH.

*ATP Synthesis:* The chemiosmotic synthesis of ATP is driven by a proton gradient across the thylakoid membrane.

## PEROXISOMES

peroxisome, catalase, plasmalogen, glyoxylate cycle, glyoxysome, photorespiration

*Functions of Peroxisomes:* Peroxisomes are small organelles, bounded by a single membrane, that contain enzymes involved in a variety of metabolic reactions, including fatty acid oxidation, the glyoxylate cycle, and photorespiration.

*Peroxisome Assembly:* Peroxisomal proteins are synthesized on free ribosomes in the cytosol and imported to peroxisomes as complete polypeptide chains. At least two types of signals target proteins to the interior of peroxisomes, but the mechanism of protein import is not well understood.

## QUESTIONS

**1.** How does the electrochemical gradient across the inner mitochondrial membrane contribute to protein import?

**2.** Assume that the electric potential across the inner mitochondrial membrane is dissipated, so the electrochemical gradient is composed solely of a proton concentration gradient corresponding to one pH unit. Calculate the free energy stored in this gradient. Under these conditions, approximately how many protons would be required

for the synthesis of one ATP molecule? For your calculation, use $R = 1.98 \times 10^{-3}$ kcal/mol/deg, $T = 298K$ (25°C), and note that $\ln(x) = 2.3 \log_{10}(x)$.

**3.** What is the topological relationship between the thylakoid lumen and the intermembrane space of chloroplasts?

**4.** Why are the transit peptides of chloroplast proteins, in contrast to the presequences of mitochondrial proteins, not positively charged?

**5.** How many high-energy electrons are required to drive the synthesis of one molecule of glucose during photosynthesis, coupled to the formation of six molecules of $O_2$? How many molecules of ATP and NADPH are generated by passage of these electrons through photosystems I and II?

**6.** What fraction of the carbon atoms converted to glycolate during photorespiration are salvaged by peroxisomes?

## REFERENCES AND FURTHER READING

### Mitochondria

Attardi, G. and G. Schatz. 1988. Biogenesis of mitochondria. *Ann. Rev. Cell Biol.* 4: 289–333. [R]

Clayton, D. A. 1991. Replication and transcription of vertebrate mitochondrial DNA. *Ann. Rev. Cell Biol.* 7: 453–478. [R]

Dietrich, A., J. H. Weil and L. Marechal-Drouard. 1992. Nuclear-encoded transfer RNAs in plant mitochondria. *Ann. Rev. Cell Biol.* 8: 115–131. [R]

Glick, B. S. 1995. Can Hsp70 proteins act as force-generating motors? *Cell* 80: 11–14. [R]

Glick, B. S., E. M. Beasley and G. Schatz. 1992. Protein sorting in mitochondria. *Trends Biochem. Sci.* 17: 453–459. [R]

Gray, M. W. 1989. Origin and evolution of mitochondrial DNA. *Ann. Rev. Cell Biol.* 5: 25–50. [R]

Hartl, F.-U. and W. Neupert. 1990. Protein sorting to mitochondria: Evolutionary conservations of folding and assembly. *Science* 247: 930–938. [R]

Lill, R. and W. Neupert. 1996. Mechanisms of protein import across the mitochondrial outer membrane. *Trends Cell Biol.* 6: 56–61 [R]

Luft, R. 1994. The development of mitochondrial medicine. *Proc. Natl. Acad. Sci. USA* 91: 8731–8738. [R]

Mihara, K. and T. Omura. 1996. Cytoplasmic chaperones in precursor targeting to mitochondria: The role of MSF and hsp70. *Trends Cell Biol.* 6: 104–108. [R]

Pfanner, N., E. A. Craig and M. Meijer. 1994. The protein import machinery of the mitochondrial inner membrane. *Trends Biochem. Sci.* 19: 368–372. [R]

Pfanner, N. and W. Neupert. 1990. The mitochondrial protein import apparatus. *Ann. Rev. Biochem.* 59: 331–353. [R]

Schatz, G. 1993. The protein import machinery of mitochondria. *Protein Sci.* 2: 141–146 [R]

Schatz, G. and B. Dobberstein. 1996. Common principles of protein translocation across membranes. *Science* 271: 1519–1526. [R]

Wallace, D. C. 1994. Mitochondrial DNA sequence variation in human evolution and disease. *Proc. Natl. Acad. Sci. USA* 91: 8739–8746. [R]

### The Mechanism of Oxidative Phosphorylation

Abrahams, J. P., A. G. W. Leslie, R. Lutter and J. E. Walker. 1994. Structure at 2.8 Å resolution of $F_1$-ATPase from bovine heart mitochondria. *Nature* 370: 621–628. [P]

Capaldi, R. A. 1990. Structure and function of cytochrome oxidase. *Ann. Rev. Biochem.* 59: 569–596. [R]

Hatefi, Y. 1985. The mitochondrial electron transport and oxidative phosphorylation system. *Ann. Rev. Biochem.* 54: 1015–1069. [R]

Iwata, S., C. Ostermeier, B. Ludwig, and H. Michel. 1995. Structure at 2.8 Å resolution of cytochrome *c* oxidase from *Paracoccus denitrificans*. *Nature* 376: 660–669. [P]

Mitchell, P. 1979. Keilin's respiratory chain concept and its chemiosmotic consequences. *Science* 206: 1148–1159. [R]

Nicholls, D. G. and S. J. Ferguson. 1992. *Bioenergetics* 2. London: Academic Press.

Pedersen, P. L. and L. M. Amzel. 1993. ATP synthases: Structure, reaction center, mechanism, and regulation of one of nature's most unique machines. *J. Biol. Chem.* 268: 9937–9940. [R]

Racker, E. 1980. From Pasteur to Mitchell: A hundred years of bioenergetics. *Fed. Proc.* 39: 210–215. [R]

Trumpower, B. L. and R. B. Gennis. 1994. Energy transduction by cytochrome complexes in mitochondrial and bacterial respiration: The enzymology of coupling electron transfer reactions to transmembrane proton translocation. *Ann. Rev. Biochem.* 63: 675–716. [R]

### Chloroplasts and Other Plastids

Douce, R. and J. Joyard. 1990. Biochemistry and function of the plastid envelope. *Ann. Rev. Cell Biol.* 6: 173–216. [R]

Ellis, R. J. 1990. Molecular chaperones: The plant connection. *Science* 250: 954–959. [R]

Gray, J. C. and P. E. Row. 1995. Protein translocation across chloroplast envelope membranes. *Trends Cell Biol.* 5: 243–247. [R]

Gruissem, W. 1989. Chloroplast gene expression: How plants turn their plastids on. *Cell* 56: 161–170. [R]

Keegstra, K. 1989. Transport and routing of proteins into chloroplasts. *Cell* 56: 247–253. [R]

Rochaix, J.-D. 1992. Post-transcriptional steps in the expression of chloroplast genes. *Ann. Rev. Cell Biol.* 8: 1–28. [R]

Schnell, D. J. 1995. Shedding light on the chloroplast protein import machinery. *Cell* 83: 521–524. [R]

Subramanian, A. R. 1993. Molecular genetics of chloroplast ribosomal proteins. *Trends Biochem. Sci.* 18: 177–180. [R]

Sugiura, M. 1992. The chloroplast genome. *Plant Mol. Biol.* 19: 149–168. [R]

Theg, S. M. and S. V. Scott. 1993. Protein import into chloroplasts. *Trends Cell Biol.* 3: 186–190. [R]

### Photosynthesis

Arnon, D. I. 1984. The discovery of photosynthetic phosphorylation. *Trends Biochem. Sci.* 9: 258–262. [R]

Barber, J. and B. Andersson. 1994. Revealing the blueprint of photosynthesis. *Nature* 370: 31–34. [R]

Bennett, J. 1979. The protein that harvests sunlight. *Trends Biochem. Sci.* 4: 268–271. [R]

Deisenhofer, J., O. Epp, K. Miki, R. Huber, and H. Michel. 1985. Structure of the protein subunits in the photosynthetic reaction centre of *Rhodopseudomonas viridis*. *Nature* 318: 618–624. [P]

Deisenhofer, J. and H. Michel. 1991. Structures of bacterial photosynthetic reaction centers. *Ann. Rev. Cell Biol.* 7: 1–23. [R]

Kuhlbrandt, W., D. N. Wang and Y. Fujiyoshi. 1994. Atomic model of plant light-harvesting complex by electron crystallography. *Nature* 367: 614–621. [P]

Nicholls, D. G. and S. J. Ferguson. 1992. *Bioenergetics* 2. San Diego, CA: Academic Press.

Rögner, M., E. J. Boekema and J. Barber. 1996. How does photosystem 2 split water? The structural basis of efficient energy conversion. *Trends Biochem. Sci.* 21: 44–49. [R]

Youvan, D. C. and B. L. Marrs. 1987. Molecular mechanisms of photosynthesis. *Sci. Am.* 256(6): 42–48. [R]

### Peroxisomes

McNew, J. A. and J. M. Goodman. 1996. The targeting and assembly of peroxisomal proteins: Some old rules do not apply. *Trends Biochem. Sci.* 21: 54–58. [R]

Purdue, P. E. and P. B. Lazarow. 1994. Peroxisomal biogenesis: Multiple pathways of protein import. *J. Biol. Chem.* 269: 30065–30068. [R]

Rachubinski, R. A. and S. Subramani. 1995. How proteins penetrate peroxisomes. *Cell* 83: 525–528. [R]

Subramani, S. 1993. Protein import into peroxisomes and biogenesis of the organelle. *Ann. Rev. Cell Biol.* 9: 445–478. [R]

Tolbert, N. E. 1981. Metabolic pathways in peroxisomes and glyoxysomes. *Ann. Rev. Biochem.* 50: 133–157. [R]

Van den Bosch, H., R. B. H. Schutgens, R. J. A. Wanders and J. M. Tager. 1992. Biochemistry of peroxisomes. *Ann. Rev. Biochem.* 61: 157–197. [R]

# 11

# The Cytoskeleton and Cell Movement

THE MEMBRANE-ENCLOSED ORGANELLES discussed in the preceding chapters constitute one level of the organizational substructure of eukaryotic cells. A further level of organization is provided by the cytoskeleton, which consists of a network of protein filaments extending throughout the cytoplasm of all eukaryotic cells. The cytoskeleton provides a structural framework for the cell, serving as a scaffold that determines cell shape and the general organization of the cytoplasm. In addition to playing this structural role, the cytoskeleton is responsible for cell movements. These include not only the movements of entire cells, but also the internal transport of organelles and other structures (such as mitotic chromosomes) through the cytoplasm. Importantly, the cytoskeleton is much less rigid and permanent than its name implies. Rather, it is a dynamic structure that is continually reorganized as cells move and change shape, for example, during cell division.

The cytoskeleton is composed of three principal types of protein filaments: actin filaments, intermediate filaments, and microtubules, which are held together and linked to subcellular organelles and the plasma membrane by a variety of accessory proteins. This chapter discusses the structure and organization of each of these three major components of the cytoskeleton, as well as their roles in cell motility, organelle transport, cell division, and other types of cell movements.

## STRUCTURE AND ORGANIZATION OF ACTIN FILAMENTS

The major cytoskeletal protein of most cells is **actin**, which polymerizes to form actin filaments—thin, flexible fibers approximately 7 nm in diameter and up to several micrometers in length (Figure 11.1). Within the cell, actin filaments (also called **microfilaments**) are organized into higher-order structures, forming bundles or three-dimensional networks with the properties of semisolid gels. The assembly and disassembly of actin filaments, their crosslinking into bundles and networks, and their association with other cell structures (such as the plasma membrane) are regulated by a variety of actin-binding proteins, which are critical components of the actin

*Figure 11.1*
**Actin filaments** Electron micrograph of actin filaments. (Courtesy of Roger Craig, University of Massachusetts Medical Center.)

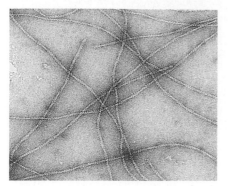

50 nm

cytoskeleton. Actin filaments are particularly abundant beneath the plasma membrane, where they form a network that provides mechanical support, determines cell shape, and allows movement of the cell surface, thereby enabling cells to migrate, engulf particles, and divide.

## Assembly and Disassembly of Actin Filaments

Actin was first isolated from muscle cells, in which it constitutes approximately 20% of total cell protein, in 1942. Although actin was initially thought to be uniquely involved in muscle contraction, it is now known to be an extremely abundant protein (typically 5 to 10% of total protein) in all types of eukaryotic cells. Yeasts have only a single actin gene, but higher eukaryotes have several distinct types of actin, which are encoded by different members of the actin gene family. Mammals, for example, have at least six distinct actin genes: Four are expressed in different types of muscle and two are expressed in nonmuscle cells. All of the actins, however, are very similar in amino acid sequence and have been highly conserved throughout the evolution of eukaryotes. Yeast actin, for example, is 90% identical in amino acid sequence to the actins of mammalian cells.

The three-dimensional structures of both individual actin molecules and actin filaments were determined in 1990 by Kenneth Holmes, Wolfgang Kabsch, and their colleagues. Individual actin molecules are globular proteins of 375 amino acids (43 kd). Each actin monomer (**globular [G] actin**) has tight binding sites that mediate head-to-tail interactions with two other actin monomers, so actin monomers polymerize to form filaments (**filamentous [F] actin**) (Figure 11.2). Each monomer is rotated by 166° in the filaments,

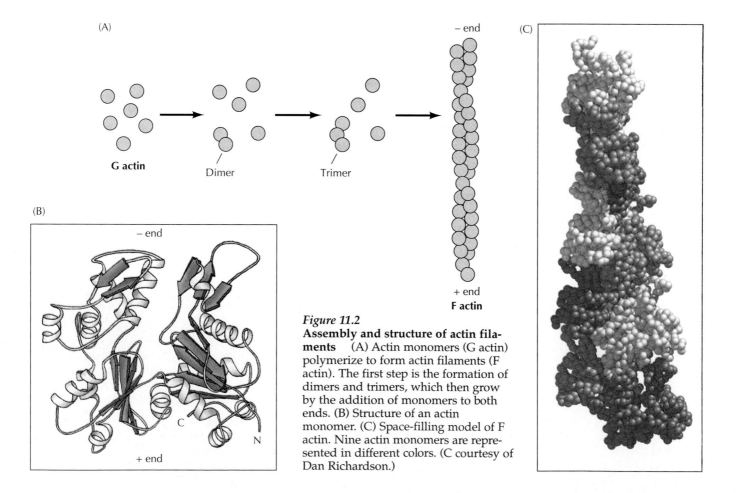

*Figure 11.2*
**Assembly and structure of actin filaments** (A) Actin monomers (G actin) polymerize to form actin filaments (F actin). The first step is the formation of dimers and trimers, which then grow by the addition of monomers to both ends. (B) Structure of an actin monomer. (C) Space-filling model of F actin. Nine actin monomers are represented in different colors. (C courtesy of Dan Richardson.)

which therefore have the appearance of a double-stranded helix. Because all the actin monomers are oriented in the same direction, actin filaments have a distinct polarity and their ends (called the plus and minus ends) are distinguishable from one another. This polarity of actin filaments is important both in their assembly and in establishing a unique direction of myosin movement relative to actin, as discussed later in the chapter.

The assembly of actin filaments can be studied *in vitro* by regulation of the ionic strength of actin solutions. In solutions of low ionic strength, actin filaments depolymerize to monomers. Actin then polymerizes spontaneously if the ionic strength is increased to physiological levels. The first step in actin polymerization (called nucleation) is the formation of a small aggregate consisting of three actin monomers. Actin filaments are then able to grow by the reversible addition of monomers to both ends, but one end (the plus end) elongates five to ten times faster than the minus end. The actin monomers also bind ATP, which is hydrolyzed to ADP following filament assembly. Although ATP is not required for polymerization, actin monomers to which ATP is bound polymerize more readily than those to which ADP is bound. As discussed below, ATP binding and hydrolysis play a key role in regulating the assembly and dynamic behavior of actin filaments.

Because actin polymerization is reversible, filaments can depolymerize by the dissociation of actin subunits, allowing actin filaments to be broken down when necessary (Figure 11.3). Thus, an apparent equilibrium exists between actin monomers and filaments, which is dependent on the concentration of free monomers. The rate at which actin monomers are incorporated into filaments is proportional to their concentration, so there is a critical concentration of actin monomers at which the rate of their polymerization into filaments equals the rate of dissociation. At this critical concentration, monomers and filaments are in apparent equilibrium.

As noted earlier, the two ends of an actin filament grow at different rates, with monomers being added to the fast-growing end (the plus end) five to ten times faster than to the slow-growing (minus) end. Because ATP-actin dissociates less readily than ADP-actin, this results in a difference in the critical concentration of monomers needed for polymerization at the two ends. This difference can result in the phenomenon known as **treadmilling**, which illustrates the dynamic behavior of actin filaments (Figure 11.4). For the sys-

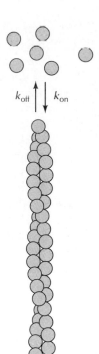

**Figure 11.3**
**Reversible polymerization of actin monomers** Actin polymerization is a reversible process, in which monomers both associate with and dissociate from the ends of actin filaments. The rate of subunit dissociation ($k_{off}$) is independent of monomer concentration, while the rate of subunit association is proportional to the concentration of free monomers and given by $C \times k_{on}$ ($C$ = concentration of free monomers). An apparent equilibrium is reached at the critical concentration of monomers ($C_c$), where $k_{off} = C_c \times k_{on}$.

**Figure 11.4**
**Treadmilling** The minus ends grow less rapidly than the plus ends of actin filaments. This difference in growth rate is reflected in a difference in the critical concentration for addition of monomers to the two ends of the filament. Actin bound to ATP associates with the rapidly growing plus ends, and the ATP bound to actin is then hydrolyzed to ADP. Because ADP-actin dissociates from filaments more readily than ATP-actin, the critical concentration of actin monomers is higher for addition to the minus end than to the plus end of actin filaments. Treadmilling takes place at monomer concentrations intermediate between the critical concentrations for the plus and minus ends. Under these conditions, there is a net dissociation of monomers (bound to ADP) from the minus end, balanced by the addition of monomers (bound to ATP) to the plus end.

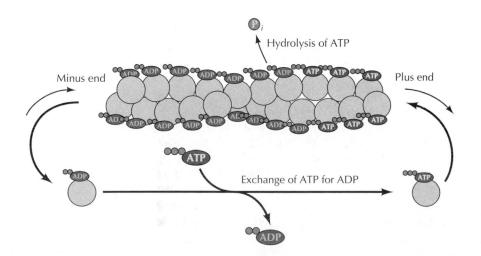

tem to be at an overall steady state, the concentration of free actin monomers must be intermediate between the critical concentrations required for polymerization at the plus and minus ends of the actin filaments. Under these conditions, there is a net loss of monomers from the minus end, which is balanced by a net addition to the plus end. Treadmilling requires ATP, with ATP-actin polymerizing at the plus end of filaments while ADP-actin dissociates from the minus end. Although the role of treadmilling in the cell is unclear, it may reflect the dynamic assembly and disassembly of actin filaments required for cells to move and change shape.

It is noteworthy that several drugs useful in cell biology act by binding to actin and affecting its polymerization. For example, the **cytochalasins** bind to the plus ends of actin filaments and block their elongation. This results in changes in cell shape as well as inhibition of some types of cell movements (e.g., cell division following mitosis), indicating that actin polymerization is required for these processes. Another drug, **phalloidin**, binds tightly to actin filaments and prevents their dissociation into individual actin molecules. Phalloidin labeled with a fluorescent dye is frequently used to visualize actin filaments by fluorescence microscopy.

Within the cell, both the assembly and disassembly of actin filaments are regulated by **actin-binding proteins** (Figure 11.5). These proteins can act either by binding and sequestering actin monomers, thereby preventing their incorporation into filaments, or by binding to the ends of filaments and preventing further monomer addition. Regulation of actin polymerization by these proteins allows the cell to maintain a pool of unpolymerized actin that is substantially higher (about 100 $\mu M$) than the critical concentration of actin monomers for polymerization *in vitro* (approximately 0.2 $\mu M$). An intracellular stockpile of actin monomers is thus available to be incorporated into filaments as needed.

*Figure 11.5*
**Effects of actin-binding proteins on filament assembly and disassembly** Thymosin and profilin are examples of proteins that bind actin monomers and regulate filament assembly. Both thymosin and profilin sequester ADP-bound monomers and prevent their assembly into filaments. In addition, profilin can enhance monomer polymerization by stimulating the exchange of bound ADP for ATP. Other actin-binding proteins can fragment filaments and/or cap their ends. Gelsolin, for example, severs filaments and then binds to the newly formed plus end, blocking further polymerization.

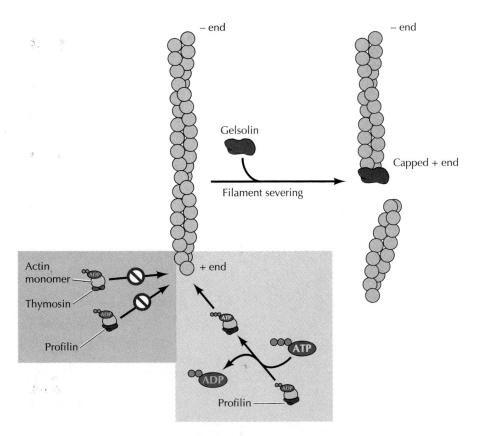

The most abundant of the actin-binding proteins is **thymosin**, a small (about 5 kd) protein that appears to be the major protein responsible for sequestering actin monomers and preventing their assembly into filaments. Another actin-binding protein, **profilin**, also binds actin monomers and can similarly prevent their incorporation into filaments. Conversely, however, profilin can promote monomer incorporation into filaments by stimulating the exchange of bound ADP for ATP. This exchange results in formation of ATP-actin monomers, which, as already discussed, are more rapidly assembled into filaments. Profilin also can associate with plasma membrane phospholipids (phosphatidylinositides) that are involved in cell signaling (discussed in Chapter 13), suggesting that profilin may serve as a regulatory molecule that controls actin polymerization, and hence changes in the cytoskeleton, in response to extracellular signals.

Assembly of actin filaments can also be regulated by **capping proteins** that bind to the ends of actin filaments and prevent the loss or addition of actin monomers. In addition, some of these proteins actively promote the disassembly of actin filaments. An example is provided by **gelsolin**, a widely distributed actin-binding protein that fragments filaments and then remains bound to the plus end, serving as a cap that blocks further filament growth. The filament-severing activity of gelsolin is activated by $Ca^{2+}$ and can therefore be stimulated by a variety of extracellular signals that transiently increase the concentration of $Ca^{2+}$ within the cytoplasm. Like profilin, gelsolin also binds to phosphatidylinositides, which inhibit its severing activity and dissociate it from actin filaments, leading to filament polymerization. Regulation of gelsolin by both $Ca^{2+}$ and phospholipids may thus provide a further link between extracellular signals and the actin cytoskeleton.

## Organization of Actin Filaments

Individual actin filaments are assembled into two general types of structures, called **actin bundles** and **actin networks**, which play different roles in the cell (Figure 11.6). In bundles, the actin filaments are crosslinked into closely packed parallel arrays. In networks, the actin filaments are loosely crosslinked in orthogonal arrays that form three-dimensional meshworks with the properties of semisolid gels. The formation of these structures is governed by a variety of actin-binding proteins that crosslink actin filaments in distinct patterns.

All of the actin-binding proteins involved in crosslinking contain at least two domains that bind actin, allowing them to bind and crosslink two different actin filaments. The nature of the association between these filaments is then determined by the size and shape of the crosslinking proteins (see Figure 11.6). The proteins that crosslink actin filaments into bundles (called **actin-bundling proteins**) usually are small rigid proteins that force the filaments to align closely with one another. In contrast, the proteins that organize actin filaments into networks tend to be large flexible proteins that can crosslink perpendicular filaments. These actin-crosslinking proteins appear to be modular proteins consisting of related structural

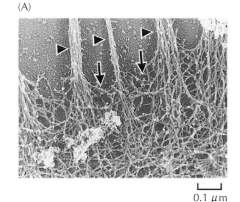

(A)

0.1 μm

(B)
**Bundle**

Actin filaments

Crosslinking protein

**Network**

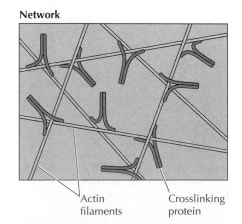

Actin filaments

Crosslinking protein

**Figure 11.6**
**Actin bundles and networks** (A) Electron micrograph of actin bundles (arrowheads) projecting from the actin network (arrows) underlying the plasma membrane of a macrophage. The bundles support cell surface projections called microspikes or filopodia (see Figure 11.17). (B) Schematic organization of bundles and networks. Actin filaments in bundles are crosslinked into parallel arrays by small proteins that align the filaments closely with one another. In contrast, networks are formed by large flexible proteins that crosslink orthogonal filaments. (A courtesy of John H. Hartwig, Brigham & Women's Hospital.)

**Parallel bundle**

Actin filaments

Ca²⁺-binding domain

ABD    ABD

Fimbrin

**Contractile bundle**

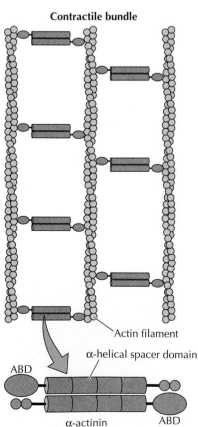

Actin filament

α-helical spacer domain

ABD

α-actinin    ABD

*Figure 11.7*
**Actin-bundling proteins**   Actin filaments are associated into two types of bundles by different actin-bundling proteins. Fimbrin has two adjacent actin-binding domains (ABD) and crosslinks actin filaments into closely packed parallel bundles in which the filaments are approximately 14 nm apart. In contrast, the two separated actin-binding domains of α-actinin dimers crosslink filaments into more loosely spaced contractile bundles in which the filaments are separated by 40 nm. Both fimbrin and α-actinin contain two related Ca²⁺-binding domains and α-actinin contains four repeated α-helical spacer domains.

units. In particular, the actin-binding domains of many of these proteins are similar in structure. They are separated by spacer sequences that vary in length and flexibility, and it is these differences in the spacer sequences that are responsible for the distinct crosslinking properties of different actin-binding proteins.

There are two structurally and functionally distinct types of actin bundles, involving different actin-bundling proteins (Figure 11.7). The first type of bundle, containing closely spaced actin filaments aligned in parallel, supports projections of the plasma membrane, such as microvilli (see Figures 11.15 and 11.16). In these bundles, all the filaments have the same polarity, with their plus ends adjacent to the plasma membrane. An example of a bundling protein involved in the formation of these structures is **fimbrin**, which was first isolated from intestinal microvilli and later found in surface projections of a wide variety of cell types. Fimbrin is a 68-kd protein, containing two adjacent actin-binding domains. It binds to actin filaments as a monomer, holding two parallel filaments close together.

The second type of actin bundle is composed of filaments that are more loosely spaced and are capable of contraction, such as the actin bundles of the contractile ring that divides cells in two following mitosis. The looser structure of these bundles (which are called **contractile bundles**) reflects the properties of the crosslinking protein **α-actinin**. In contrast to fimbrin, α-actinin binds to actin as a dimer, each subunit of which is a 102-kd protein containing a single actin-binding site. Filaments crosslinked by α-actinin are consequently separated by a greater distance than those crosslinked by fimbrin (40 nm apart instead of 14 nm). The increased spacing between filaments allows the motor protein myosin to interact with the actin filaments in these bundles, which (as discussed later) enables them to contract.

The actin filaments in networks are held together by large actin-binding proteins, such as **filamin** (Figure 11.8). Filamin (also called actin-binding protein or ABP-280) binds actin as a dimer of two 280-kd subunits. The actin-binding domains and dimerization domains are at opposite ends of each subunit, so the filamin dimer is a flexible V-shaped molecule with actin-binding domains at the ends of each arm. As a result, filamin forms cross-links between orthogonal actin filaments, creating a loose three-dimensional meshwork. As discussed in the next section, such networks of actin filaments underlie the plasma membrane and support the surface of the cell.

### Association of Actin Filaments with the Plasma Membrane

Actin filaments are highly concentrated at the periphery of the cell, where they form a three-dimensional network beneath the plasma membrane (see Figure 11.6). This network of actin filaments and associated actin-binding proteins (called the **cell cortex**) determines cell shape and is involved in a variety of cell surface activities, including movement. The association of the actin cytoskeleton with the plasma membrane is thus central to cell structure and function.

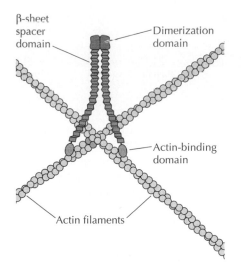

β-sheet spacer domain

Dimerization domain

Actin-binding domain

Actin filaments

*Figure 11.8*
**Actin networks and filamin**   Filamin is a dimer of two large (280-kd) subunits, forming a flexible V-shaped molecule that crosslinks actin filaments into orthogonal networks. The carboxy-terminal dimerization domain is separated from the amino-terminal actin-binding domain by repeated β-sheet spacer domains.

Red blood cells (erythrocytes) have proven particularly useful for studies of both the plasma membrane (discussed in the next chapter) and the cortical cytoskeleton. The principal advantage of red blood cells for these studies is that they contain no nucleus or internal organelles, so their plasma membrane and associated proteins can be easily isolated without contamination by the various internal membranes that are abundant in other cell types. In addition, human erythrocytes lack other cytoskeletal components (microtubules and intermediate filaments), so the cortical cytoskeleton is the principal determinant of their distinctive shape as biconcave discs (Figure 11.9).

The major protein that provides the structural basis for the cortical cytoskeleton in erythrocytes is the actin-binding protein **spectrin**, which is related to filamin (Figure 11.10). Erythrocyte spectrin is a tetramer consisting of two distinct polypeptide chains, called α and β, with molecular weights of 240 and 220 kd, respectively. The β chain has a single actin-binding domain at its amino terminus. The α and β chains associate later-

*Figure 11.9*
**Morphology of red blood cells**   Scanning electron micrograph of red blood cells illustrating their biconcave shape. (Omikron/Photo Researchers, Inc.)

5 μm

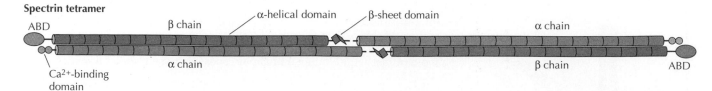

Spectrin tetramer

ABD    β chain    α-helical domain    β-sheet domain    α chain

Ca²⁺-binding domain    α chain    β chain    ABD

**Figure 11.10**
**Structure of spectrin**   Spectrin is a tetramer consisting of two $\alpha$ and two $\beta$ chains. Each $\beta$ chain has a single actin-binding domain (ABD) at its amino terminus. Both $\alpha$ and $\beta$ chains contain multiple repeats of $\alpha$-helical spacer domains, which separate the two actin-binding domains of the tetramer. The $\alpha$ chain has two $Ca^{2+}$ binding domains at its carboxy terminus.

**Figure 11.11**
**Association of the erythrocyte cortical cytoskeleton with the plasma membrane**   The plasma membrane is associated with a network of spectrin tetramers crosslinked by short actin filaments in association with protein 4.1. The spectrin-actin network is linked to the membrane by ankyrin, which binds to both spectrin and the abundant transmembrane protein band 3. An additional link may be provided by the binding of protein 4.1 to band 3 and glycophorin.

ally to form dimers, which then join head to head to form tetramers with two actin-binding domains separated by approximately 200 nm. The ends of the spectrin tetramers then associate with short actin filaments, resulting in the spectrin-actin network that forms the cortical cytoskeleton of red blood cells (Figure 11.11). The major link between the spectrin-actin network and the plasma membrane is provided by a protein called **ankyrin**, which binds both to spectrin and to the cytoplasmic domain of an abundant transmembrane protein called band 3. An additional link between the spectrin-actin network and the plasma membrane may be provided by protein 4.1, which binds to spectrin-actin junctions as well as recognizing the cytoplasmic domains of band 3 and glycophorin (another abundant transmembrane protein).

Other types of cells contain linkages between the cortical cytoskeleton and the plasma membrane that are similar to those observed in red blood cells. Proteins related to spectrin (nonerythroid spectrin is also called **fodrin**), ankyrin, and protein 4.1 are expressed in a wide range of cell types, where they may fulfill functions analogous to those described for erythrocytes. For example, the spectrin-related protein filamin (see Figure 11.8) constitutes a major link between actin filaments and the plasma membrane of blood platelets. Another member of this group of spectrin-related proteins is **dystrophin**, which is of particular interest because it is the product of the gene responsible for two types of muscular dystrophy (Duchenne's and Becker's). These X-linked inherited diseases result in progressive degeneration of skeletal muscle, and patients with the more severe form of the disease (Duchenne's muscular dystrophy) usually die in their teens or early twenties. Molecular cloning of the gene responsible for this disorder revealed that it encodes a large protein (427 kd) that is either absent or abnormal in patients with Duchenne's or Becker's muscular dys-

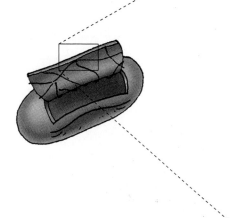

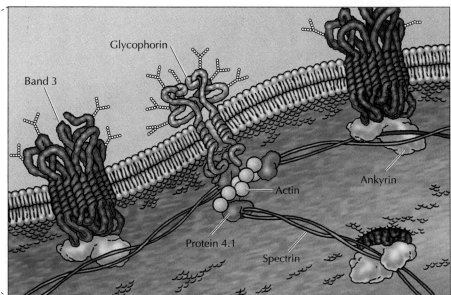

Glycophorin

Band 3

Ankyrin

Actin

Protein 4.1

Spectrin

## MOLECULAR MEDICINE

# Muscular Dystrophy and the Cytoskeleton

### The Disease

The muscular dystrophies are a group of hereditary diseases characterized by the progressive loss of muscle cells. Duchenne's muscular dystrophy (DMD) is the most common and severe form, affecting approximately one out of every 3500 male children. Children with DMD are usually unaffected until three to five years of age, when muscular weakness is first observed. Progressive loss of muscle then relentlessly continues, usually leaving affected children confined to wheelchairs by age 12 and frequently leading to death as a result of respiratory failure by the late teens or early twenties. Becker's muscular dystrophy (BMD) is a less common form of the disease, with an incidence of one in 30,000 male births. BMD is also less severe than DMD, and many BMD patients are able to live normal lives with minimal restrictions.

### Molecular and Cellular Basis

The much higher incidence of DMD and BMD in boys than in girls initially suggested that both diseases result from recessive sex-linked genes. This hypothesis was confirmed by genetic studies, which localized the DMD/BMD gene to a specific region of the X chromosome. On the basis of its chromosomal position, the gene responsible for DMD and BMD was cloned by the research groups of Lou Kunkel and Ron Worton in 1986. Sequence analysis established that it encodes a 427-kd protein, called dys-

trophin, which is related to spectrin. Dystrophin is linked to the plasma membrane of muscle cells by a complex of transmembrane proteins. These transmembrane proteins in turn bind to components of the extracellular matrix, so dystrophin plays a key role in anchoring the cytoskeleton of muscle cells to the extracellular matrix. This anchorage is thought to stabilize the plasma membrane and enable the cell to withstand the stress of muscle contraction. The mutations responsible for DMD or BMD result either in the absence of dystrophin or in the expression of an abnormal protein, respectively, consistent with the severity of disease in DMD and BMD patients.

### Prevention and Treatment

Identification of the DMD/BMD gene has allowed the development of sensitive diagnostic tests, such as the detection of mutations by the polymerase chain reaction. Women carrying mutant alleles of the DMD/BMD gene can thus be identified, and their pregnancies monitored to determine if the mutant gene has been transmitted to the fetus. Because of the sensitivity of PCR analysis, mutations can also be detected in early embryos derived by *in vitro* fertilization procedures prior to their return to the uterus. The combination of genetic counseling and prenatal diagnosis thus has the potential of preventing the transmission of inherited DMD/BMD cases.

Unfortunately, the effectiveness of this approach is limited by the fact

that approximately one-third of DMD cases are not inherited, but result instead from new mutations. These cases cannot be predicted by genetic analysis, so development of an effective therapy remains a compelling goal. Present efforts are focused on the use of gene therapy to restore dystrophin expression in muscle or to increase the expression of other dystrophin-related proteins that might compensate for its absence. Although treatment of muscular dystrophy remains a goal of future research, understanding its molecular basis has revolutionized diagnosis and permitted rational approaches to the development of new therapies.

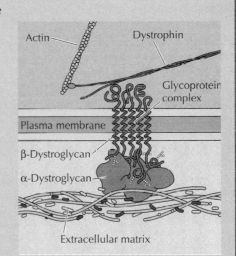

Dystrophin binds to actin and to a complex of transmembrane proteins, which link the cytoskeleton of muscle cells to the extracellular matrix.

trophy, respectively. The sequence of dystrophin further indicated that it is related to spectrin, with a single actin-binding domain at its amino terminus and a membrane-binding domain at its carboxy terminus. Like spectrin, dystrophin forms dimers that link actin filaments to transmembrane proteins of the muscle cell plasma membrane. These transmembrane proteins in turn link the cytoskeleton to the extracellular matrix, which plays an important role in maintaining cell stability during muscle contraction.

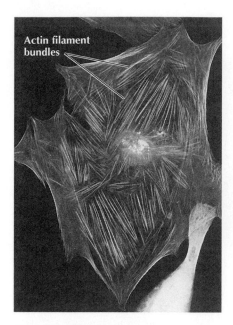

Actin filament bundles

**Figure 11.12**
**Stress fibers and focal adhesions**
Fluorescence microscopy of a human fibroblast in which actin filaments have been been stained with a fluorescent dye. Stress fibers are revealed as bundles of actin filaments anchored at sites of cell attachment to the culture dish surface (focal adhesions). (Don Fawcett/Photo Researchers, Inc.)

In contrast to the uniform surface of red blood cells, most cells have specialized regions of the plasma membrane that form contacts with adjacent cells, tissue components, or other substrates (such as the surface of a culture dish). These regions also serve as attachment sites for bundles of actin filaments that anchor the cytoskeleton to areas of cell contact. These attachments of actin filaments are particularly evident in fibroblasts maintained in tissue culture (Figure 11.12). Such cultured fibroblasts secrete extracellular matrix proteins (discussed in Chapter 12) that stick to the plastic surface of the culture dish. The fibroblasts then attach to the culture dish via the binding of transmembrane proteins (called **integrins**) to the extracellular matrix. The sites of attachment are discrete regions (called **focal adhesions**) that also serve as attachment sites for large bundles of actin filaments called **stress fibers**.

Stress fibers are contractile bundles of actin filaments, crosslinked by α-actinin, that anchor the cell and exert tension against the substratum. They are attached to the plasma membrane at focal adhesions via interactions with integrin. These associations, which are complex and not well understood, may be mediated by several other proteins, including **talin**, **vinculin**, and **tensin** (Figure 11.13). For example, both talin and α-actinin may bind to the cytoplasmic domains of integrins. Talin also binds to vinculin, which in turn may interact with actin, α-actinin, and tensin (which also binds to actin). Other proteins found at focal adhesions may also participate in the attachment of actin filaments, and a combination of these interactions may be responsible for the linkage of actin filaments to the plasma membrane.

The actin cytoskeleton is similarly anchored to regions of cell-cell contact called **adherens junctions** (Figure 11.14). In sheets of epithelial cells, these junctions form a continuous beltlike structure (called an **adhesion belt**) around each cell in which an underlying contractile bundle of actin filaments is linked to the plasma membrane. Contact between cells at adherens junctions is mediated by transmembrane proteins called **cadherins**, which are discussed further in Chapter 12. The cadherins form a complex with three cytoplasmic proteins called **α-catenin, β-catenin,** and **γ-catenin,** which in turn mediate the attachment of actin bundles. In particular, β- and γ-catenin are thought to bind to both the cytoplasmic domains of cadherins and to α-catenin. α-Catenin, which is related to vinculin, is then thought to serve as a cytoskeletal linking protein.

**Figure 11.13**
**Attachment of stress fibers to the plasma membrane at focal adhesions**
Focal adhesions are mediated by the binding of integrins to proteins of the extracellular matrix. Stress fibers (bundles of actin filaments crosslinked by α-actinin) are then bound to the cytoplasmic domain of integrins by complex associations involving a number of proteins. One possible association is illustrated, in which talin binds to both integrin and vinculin, which in turn binds to actin and α-actinin. A number of other proteins (not shown) are also present at focal adhesions and may be involved in anchoring stress fibers to the plasma membrane.

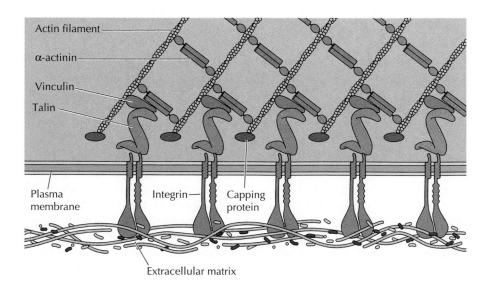

Actin filament

α-actinin

Vinculin

Talin

Plasma membrane

Integrin

Capping protein

Extracellular matrix

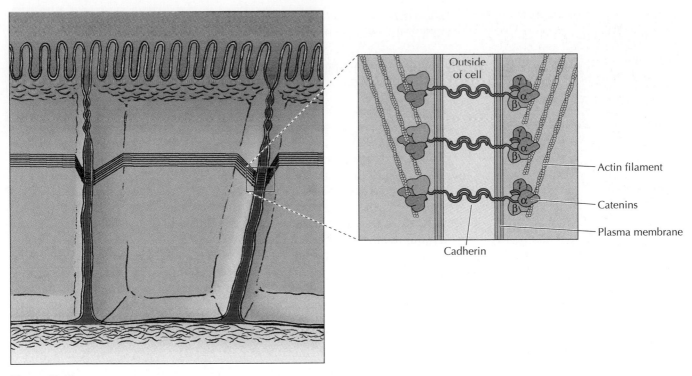

**Figure 11.14**
**Attachment of actin bundles to adherens junctions** Cell-cell contacts at adherens junctions are mediated by cadherins, which serve as sites of attachment of actin bundles. In sheets of epithelial cells, these junctions form a continuous belt of actin filaments around each cell. The cadherins are transmembrane proteins that bind catenins (designated $\alpha$, $\beta$, and $\gamma$) to their cytoplasmic domains. The $\beta$- and $\gamma$-catenins are thought to interact with both cadherin and $\alpha$-catenin, which serves as a link to actin filaments.

## Protrusions of the Cell Surface

The surfaces of most cells have a variety of protrusions or extensions that are involved in cell movement, phagocytosis, or specialized functions such as absorption of nutrients. Most of these cell surface extensions are based on actin filaments, which are organized into either relatively permanent or rapidly rearranging bundles or networks.

The best-characterized of these actin-based cell surface protrusions are **microvilli,** fingerlike extensions of the plasma membrane that are particularly abundant on the surfaces of cells involved in absorption, such as the epithelial cells lining the intestine (Figure 11.15). The microvilli of these cells form a layer on the apical surface (called a **brush border**) that consists of approximately a thousand microvilli per cell and increases the exposed surface area available for absorption by 10 to 20-fold. In addition to their role in absorption, specialized forms of microvilli, the **stereocilia** of auditory hair cells, are responsible for hearing by detecting sound vibrations.

Their abundance and ease of isolation have facilitated detailed structural analysis of intestinal microvilli, which contain closely packed parallel bundles of 20 to 30 actin filaments (Figure 11.16). The filaments in these bundles are crosslinked in part by fimbrin, an actin-bundling protein (discussed earlier) that is present in surface projections of a variety of cell types. However, the major actin-bundling protein in intestinal microvilli is **villin,** a 95-kd protein present in microvilli of only a few specialized types of cells, such as those lining the intestine and kidney

**Figure 11.15**
**Electron micrograph of microvilli** The microvilli (arrows) of intestinal epithelial cells are fingerlike projections of the plasma membrane. They are supported by actin bundles anchored in a dense region of the cortex called the terminal web. (Fred E. Hossler/Visuals Unlimited.)

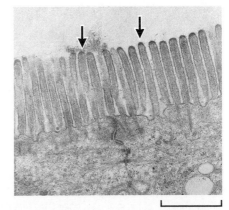

1 $\mu$m

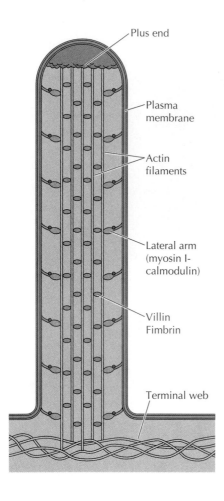

Plus end

Plasma membrane

Actin filaments

Lateral arm (myosin I-calmodulin)

Villin Fimbrin

Terminal web

*Figure 11.16*
**Organization of microvilli**   The core actin filaments of microvilli are crosslinked into closely packed bundles by fimbrin and villin. They are attached to the plasma membrane along their length by lateral arms, consisting of myosin I and calmodulin. The plus ends of the actin filaments are embedded in a cap of unidentified proteins at the tip of the microvillus.

tubules. Along their length, the actin bundles of microvilli are attached to the plasma membrane by lateral arms consisting of the calcium-binding protein calmodulin in association with myosin I, which may be involved in movement of the plasma membrane along the actin bundle of the microvillus. At their base, the actin bundles are anchored in a spectrin-rich region of the actin cortex called the terminal web, which crosslinks and stabilizes the microvilli.

In contrast to microvilli, many surface protrusions are transient structures that form in response to environmental stimuli. Several types of these structures extend from the leading edge of a moving cell and are involved in cell locomotion (Figure 11.17). **Pseudopodia** are extensions of moderate width, based on actin filaments crosslinked into a three-dimensional network, that are responsible for phagocytosis and for the movement of amoebas across a surface. **Lamellipodia** are broad, sheetlike extensions at the leading edge of fibroblasts, which similarly contain a network of actin filaments. Many cells also extend **microspikes** or **filopodia**, thin projections of the plasma membrane supported by actin bundles. The formation and retraction of these structures is based on the regulated assembly and disassembly of actin filaments, as discussed in the following section.

*Figure 11.17*
**Examples of cell surface projections involved in phagocytosis and movement**
(A) Scanning electron micrograph showing pseudopodia of a macrophage engulfing a tumor cell during phagocytosis. (B) An amoeba with several extended pseudopodia. (C) A tissue culture cell illustrating lamellipodia (L) and microspikes (arrow). (A, K. Wassermann/Visuals Unlimited; B, Stanley Flegler/Visuals Unlimited; C, Don Fawcett/Photo Researchers, Inc.)

(A)

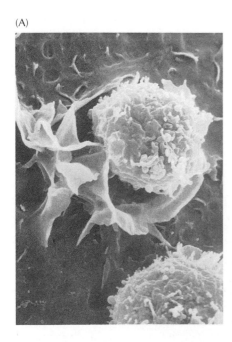

(B)

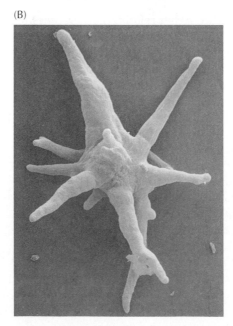

(C)

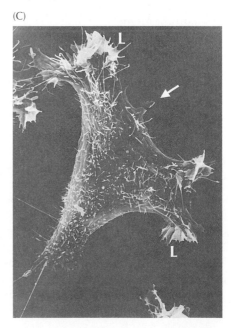

## ACTIN, MYOSIN, AND CELL MOVEMENT

Actin filaments, usually in association with **myosin**, are responsible for many types of cell movements. Myosin is the prototype of a **molecular motor**—a protein that converts chemical energy in the form of ATP to mechanical energy, thus generating force and movement. The most striking variety of such movement is muscle contraction, which has provided the model for understanding actin-myosin interactions and the motor activity of myosin molecules. However, interactions of actin and myosin are responsible not only for muscle contraction but also for a variety of movements of nonmuscle cells, including cell division, so these interactions play a central role in cell biology. Moreover, the actin cytoskeleton is responsible for the crawling movements of cells across a surface, which appear to be driven directly by actin polymerization as well as actin-myosin interactions.

### Muscle Contraction

Muscle cells are highly specialized for a single task, contraction, and it is this specialization in structure and function that has made muscle the prototype for studying movement at the cellular and molecular levels. There are three distinct types of muscle cells in vertebrates: skeletal muscle, which is responsible for all voluntary movements; cardiac muscle, which pumps blood from the heart; and smooth muscle, which is responsible for involuntary movements of organs such as the stomach, intestine, uterus, and blood vessels. In both skeletal and cardiac muscle, the contractile elements of the cytoskeleton are present in highly organized arrays that give rise to characteristic patterns of cross-striations. It is the characterization of these structures in skeletal muscle that has led to our current understanding of muscle contraction, and other actin-based cell movements, at the molecular level.

Skeletal muscles are bundles of **muscle fibers**, which are single large cells (approximately 50 $\mu$m in diameter and up to several centimeters in length) formed by the fusion of many individual cells during development (Figure 11.18). Most of

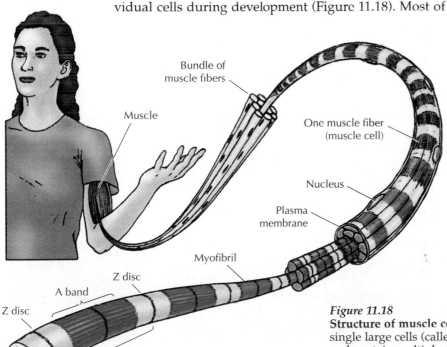

**Figure 11.18**
**Structure of muscle cells** Muscles are composed of bundles of single large cells (called muscle fibers) that form by cell fusion and contain multiple nuclei. Each muscle fiber contains many myofibrils, which are bundles of actin and myosin filaments organized into a chain of repeating units called sarcomeres.

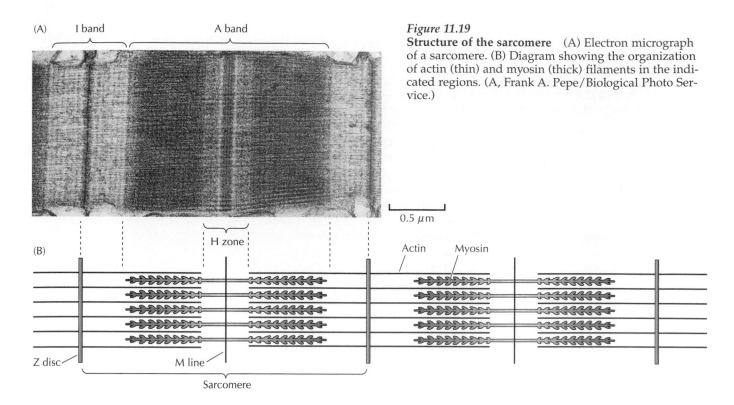

(A) I band       A band

0.5 μm

(B)

H zone

Actin    Myosin

Z disc       M line

Sarcomere

*Figure 11.19*
**Structure of the sarcomere** (A) Electron micrograph of a sarcomere. (B) Diagram showing the organization of actin (thin) and myosin (thick) filaments in the indicated regions. (A, Frank A. Pepe/Biological Photo Service.)

the cytoplasm consists of **myofibrils**, which are cylindrical bundles of two types of filaments: thick filaments of myosin (about 15 nm in diameter) and thin filaments of actin (about 7 nm in diameter). Each myofibril is organized as a chain of contractile units called **sarcomeres**, which are responsible for the striated appearance of skeletal and cardiac muscle.

The sarcomeres (which are approximately 2.3 μm long) consist of several distinct regions, discernible by electron microscopy, which provided critical insights into the mechanism of muscle contraction (Figure 11.19). The ends of each sarcomere are defined by the Z disc. Within each sarcomere, dark bands (called A bands because they are *a*nisotropic when viewed with polarized light) alternate with light bands (called I bands for *i*sotropic). These bands correspond to the presence or absence of myosin filaments. The I bands contain only thin (actin) filaments, whereas the A bands contain thick (myosin) filaments. The myosin and actin filaments overlap in peripheral regions of the A band, whereas a middle region (called the H zone) contains only myosin. The actin filaments are attached at their plus ends to the Z disc, which includes the crosslinking protein α-actinin. The myosin filaments are anchored at the M line in the middle of the sarcomere.

Two additional proteins (**titin** and **nebulin**) also contribute to sarcomere

*Figure 11.20*
**Titin and nebulin** Molecules of titin extend from the Z disc to the M line and act as springs to keep myosin filaments centered in the sarcomere. Molecules of nebulin extend from the Z disc and are thought to determine the length of associated actin filaments.

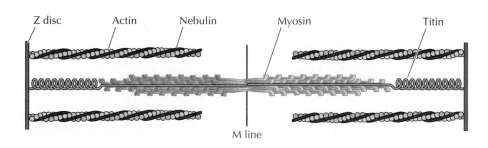

Z disc    Actin    Nebulin      Myosin      Titin

M line

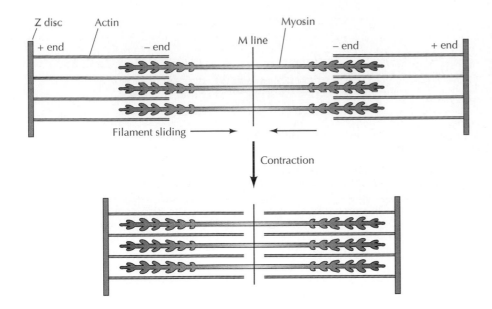

*Figure 11.21*
**Sliding-filament model of muscle contraction** The actin filaments slide past the myosin filaments toward the middle of the sarcomere. The result is shortening of the sarcomere without any change in filament length.

structure and stability (Figure 11.20). Titin is an extremely large protein (3000 kd), and single titin molecules extend from the M line to the Z disc. These long molecules of titin are thought to act like springs that keep the myosin filaments centered in the sarcomere and maintain the resting tension that allows a muscle to snap back if overextended. Nebulin filaments are associated with actin and are thought to regulate the assembly of actin filaments by acting as rulers that determine their length.

The basis for understanding muscle contraction is the **sliding filament model**, first proposed in 1954 both by Andrew Huxley and Ralph Niedergerke and by Hugh Huxley and Jean Hanson (Figure 11.21). During muscle contraction, each sarcomere shortens, bringing the Z discs closer together. There is no change in the width of the A band, but both the I bands and the H zone almost completely disappear. These changes are explained by the actin and myosin filaments sliding past one another, so that the actin filaments move into the A band and H zone. Muscle contraction thus results from an interaction between the actin and myosin filaments that generates their movement relative to one another. The molecular basis for this interaction is the binding of myosin to actin filaments, allowing myosin to function as a motor that drives filament sliding.

The type of myosin present in muscle (**myosin II**) is a very large protein (about 500 kd) consisting of two identical heavy chains (about 200 kd each) and four light chains (about 20 kd each) (Figure 11.22). Each heavy chain consists of a globular head region and a long $\alpha$-helical tail. The $\alpha$-helical tails of two heavy chains twist around each other in a coiled-coil structure to form a dimer, and a pair of light chains associates with each head region to form the complete myosin II molecule.

The thick filaments of muscle consist of several hundred myosin molecules, associated in a parallel staggered array by interactions between their tails (Figure 11.23). The globular heads of myosin bind actin, forming crossbridges between the thick and thin filaments. It is important to note that the orientation of myosin molecules in the thick filaments reverses at the M line of the sarcomere. The polarity of actin filaments (which are attached to Z discs at their plus ends) similarly reverses at the M line, so the relative orientation of myosin and actin filaments is the same on both halves of the

*Figure 11.22*
**Myosin II** The myosin II molecule consists of two heavy chains and four light chains. The heavy chains have globular head regions and long $\alpha$-helical tails, which coil around each other to form dimers.

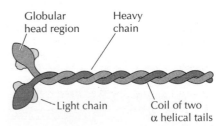

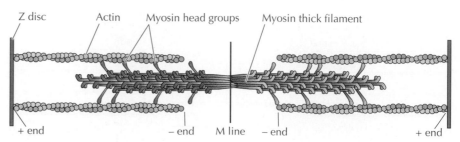

**Figure 11.23**
**Organization of myosin thick filaments**    Thick filaments are formed by the association of several hundred myosin II molecules in a staggered array. The globular heads of myosin bind actin, forming cross-bridges between the myosin and actin filaments. The orientation of both actin and myosin filaments reverses at the M line, so their relative polarity is the same on both sides of the sarcomere.

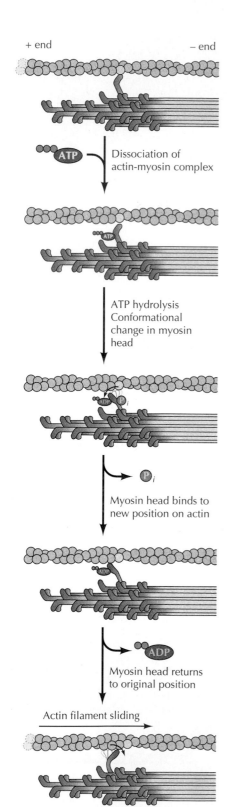

sarcomere. As discussed later, the motor activity of myosin moves its head groups along the actin filament in the direction of the plus end. This movement slides the actin filaments from both sides of the sarcomere toward the M line, shortening the sarcomere and resulting in muscle contraction.

In addition to binding actin, the myosin heads bind and hydrolyze ATP, which provides the energy to drive filament sliding. This translation of chemical energy to movement is mediated by changes in the shape of myosin resulting from ATP binding. The generally accepted model (the swinging-cross-bridge model) is that ATP hydrolysis drives repeated cycles of interaction between myosin heads and actin. During each cycle, conformational changes in myosin result in the movement of myosin heads along actin filaments.

Although the molecular mechanisms are still not fully understood, a plausible working model for myosin function has been derived both from *in vitro* studies of myosin movement along actin filaments (a system developed by James Spudich and Michael Sheetz) and from determination of the three-dimensional structure of myosin by Ivan Rayment and his colleagues (Figure 11.24). The cycle starts with myosin (in the absence of ATP) tightly bound to actin. ATP binding dissociates the myosin-actin complex and the hydrolysis of ATP induces a conformational change in myosin that appears to displace the myosin head region by about 10 nm. The products of hydrolysis (ADP and $P_i$) remain bound to the myosin head, which is said to be in the "cocked" position. The myosin head then rebinds at a new position on the actin filament, resulting in the release of $P_i$. This release triggers the "power stroke," in which ADP is released and the myosin head returns to its initial conformation, thereby sliding the actin filament toward the M line of the sarcomere.

The contraction of skeletal muscle is triggered by nerve impulses, which stimulate the release of $Ca^{2+}$ from the **sarcoplasmic reticulum**—a specialized network of internal membranes, similar to the endoplasmic reticulum, that stores high concentrations of $Ca^{2+}$ ions. The release of $Ca^{2+}$ from the sarcoplasmic reticulum increases the concentration of $Ca^{2+}$ in the cytosol from approximately $10^{-7}$ to $10^{-5}$ *M*. The increased $Ca^{2+}$ concentration sig-

**Figure 11.24**
**Model for myosin action**    The binding of ATP dissociates myosin from actin. ATP hydrolysis then induces a conformational change that displaces the myosin head group. This is followed by binding of the myosin head to a new position on the actin filament and release of $P_i$. ADP is then released and the return of the myosin head to its original conformation drives actin filament sliding.

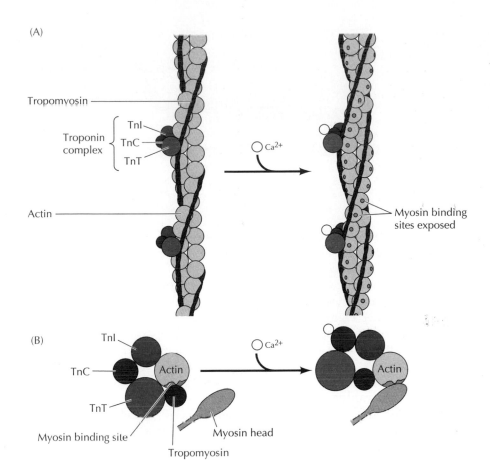

(A)

Tropomyosin

Troponin complex { TnI, TnC, TnT }

Actin

Ca²⁺

Myosin binding sites exposed

(B)

TnI

TnC

TnT

Actin

Myosin binding site

Myosin head

Tropomyosin

Ca²⁺

Actin

**Figure 11.25**
**Association of tropomyosin and troponins with actin filaments**
(A) Tropomyosin binds lengthwise along actin filaments and, in striated muscle, is associated with a complex of three troponins: troponin I (TnI), troponin C (TnC), and troponin T (TnT). In the absence of $Ca^{2+}$, the tropomyosin-troponin complex blocks the binding of myosin to actin. Binding of $Ca^{2+}$ to TnC shifts the complex, relieving this inhibition and allowing contraction to proceed. (B) Cross-sectional view.

nals muscle contraction via the action of two accessory proteins bound to the actin filaments: **tropomyosin** and **troponin** (Figure 11.25). Tropomyosin is a fibrous protein that binds lengthwise along the groove of actin filaments. In striated muscle, each tropomyosin molecule is bound to troponin, which is a complex of three polypeptides: troponin C ($Ca^{2+}$-binding), troponin I (inhibitory), and troponin T (tropomyosin-binding). When the concentration of $Ca^{2+}$ is low, the complex of the troponins with tropomyosin blocks the interaction of actin and myosin, so the muscle does not contract. At high concentrations, $Ca^{2+}$ binding to troponin C shifts the position of the complex, relieving this inhibition and allowing contraction to proceed.

### Contractile Assemblies of Actin and Myosin in Nonmuscle Cells

Contractile assemblies of actin and myosin, resembling small-scale versions of muscle fibers, are present also in nonmuscle cells. As in muscle, the actin filaments in these contractile assemblies are interdigitated with bipolar filaments of myosin II, consisting of 15 to 20 myosin II molecules, which produce contraction by sliding the actin filaments relative to one another (Figure 11.26). The actin filaments in contractile bundles in nonmuscle cells are also associated with tropomyosin, which facilitates their interaction with myosin II, probably by competing with filamin for binding sites on actin.

Two examples of contractile assemblies in nonmuscle cells, stress fibers and adhesion belts, were discussed earlier with respect to attachment of the actin cytoskeleton to regions of cell-substrate and cell-cell contacts (see Figures 11.13 and 11.14). The contraction of stress fibers produces tension across the cell, allowing the cell to pull on a substrate (e.g., the extracellular matrix) to which it is anchored. The contraction of adhesion belts alters

*Figure 11.26*
**Contractile assemblies in nonmuscle cells**   Bipolar filaments of myosin II produce contraction by sliding actin filaments in opposite directions.

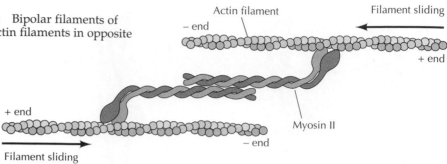

*Figure 11.27*
**Cytokinesis**   Following completion of mitosis (nuclear division), a contractile ring consisting of actin filaments and myosin II divides the cell in two.

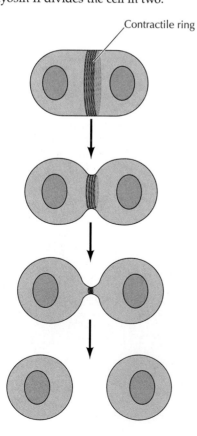

the shape of epithelial cell sheets: a process that is particularly important during embryonic development, when sheets of epithelial cells fold into structures such as tubes.

The most dramatic example of actin-myosin contraction in nonmuscle cells, however, is provided by **cytokinesis**—the division of a cell into two following mitosis (Figure 11.27). Toward the end of mitosis in animal cells, a **contractile ring** consisting of actin filaments and myosin II assembles just underneath the plasma membrane. Its contraction pulls the plasma membrane progressively inward, constricting the center of the cell and pinching it in two. Interestingly, the thickness of the contractile ring remains constant as it contracts, implying that actin filaments disassemble as contraction proceeds. The ring then disperses completely following cell division.

The regulation of actin-myosin contraction in striated muscle, discussed earlier, is mediated by the binding of $Ca^{2+}$ to troponin. In nonmuscle cells and in smooth muscle, however, contraction is regulated primarily by phosphorylation of one of the myosin light chains, called the regulatory light chain (Figure 11.28). Phosphorylation of the myosin light chain in these cells has at least two effects: It promotes the assembly of myosin into filaments and it increases myosin catalytic activity, enabling contraction to proceed. The enzyme that catalyzes this phosphorylation, called **myosin light-chain kinase**, is itself regulated by association with the $Ca^{2+}$-binding protein **calmodulin**. Increases in cytosolic $Ca^{2+}$ promote the binding of calmodulin to the kinase, resulting in phosphorylation of the myosin light chain. Increases in cytosolic $Ca^{2+}$ are thus responsible, albeit indirectly, for activating myosin in smooth muscle and nonmuscle cells, as well as in striated muscle.

## Unconventional Myosins

In addition to myosin II ("conventional" two-headed myosin), several other types of myosin are found in nonmuscle cells. In contrast to myosin II, these "unconventional" myosins do not form filaments and therefore are not involved in contraction. They may, however, be involved in a variety of other kinds of cell movements, such as the transport of membrane vesicles and organelles along actin filaments and the crawling movements of cells across a surface.

The best-studied of these unconventional myosins are members of the **myosin I** family (Figure 11.29). The myosin I proteins contain a globular head group that acts as a molecular motor, like that of myosin II. However, members of the myosin I family are much smaller molecules (about 110 kd in mammalian cells) that lack the long tail of myosin II and do not form dimers. Their tails can instead bind to other structures, such as membrane vesicles or organelles. The movement of myosin I along an actin filament

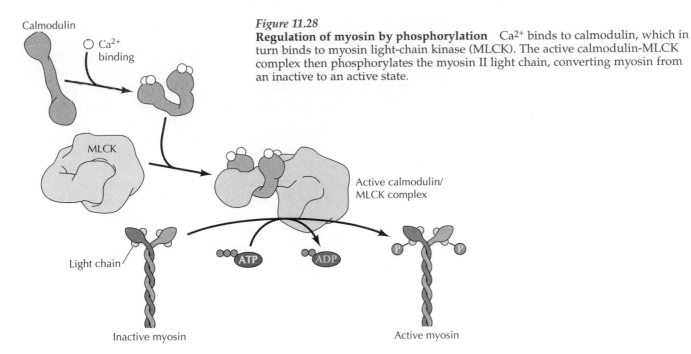

Calmodulin

Ca²⁺ binding

MLCK

Active calmodulin/ MLCK complex

Light chain

ATP

ADP

Inactive myosin

Active myosin

*Figure 11.28*
**Regulation of myosin by phosphorylation** Ca²⁺ binds to calmodulin, which in turn binds to myosin light-chain kinase (MLCK). The active calmodulin-MLCK complex then phosphorylates the myosin II light chain, converting myosin from an inactive to an active state.

can then transport its attached cargo. One function of myosin I, discussed earlier, is to form the lateral arms that link actin bundles to the plasma membrane of intestinal microvilli (see Figure 11.16). In these structures, the motor activity of myosin I may move the plasma membrane along the actin bundles, toward the tip of the microvillus. Additional functions of myosin I may be in the transport of vesicles and organelles along actin filaments and in movement of the plasma membrane during cell locomotion, as discussed next.

In addition to myosins I and II, at least nine other classes of unconventional myosins (III through XI) have been identified. Many of these are expressed in the same cells, and their functions remain to be determined. An interesting example is myosin V, which was identified both by biochemical purification from chicken brain and by the identification of mouse and yeast genes encoding similar proteins. The yeast myosin V mutants are defective in vesicle transport, suggesting a role for myosin V as a carrier of vesicles along actin filaments.

## Cell Crawling

The crawling movements of cells across a surface represent a basic form of cell locomotion, employed by a wide variety of different kinds of cells. Examples include the movements of amoebas, the migration of embryonic cells during development, the invasion of tissues by white blood cells to fight infection, the migration of cells involved in wound healing, and the spread of cancer cells during the metastasis of malignant tumors. Similar types of movement are also responsible for phagocytosis and for the extension of nerve cell processes during development of the nervous system. All of these movements are based on the dynamic properties of the actin cytoskeleton, although the detailed mechanisms involved remain to be fully understood.

Cell crawling involves a coordinated cycle of movements, which can be viewed in three stages. First, protrusions such as pseudopodia, lamellipodia, or microspikes (see Figure 11.17) must be extended from the leading

*Figure 11.29*
**Myosin I** Myosin I contains a head group similar to myosin II, but it has a comparatively short tail and does not form dimers or filaments. Although it cannot induce contraction, myosin I can move along actin filaments (toward the plus end), carrying a variety of cargoes (such as membrane vesicles) attached to its tail.

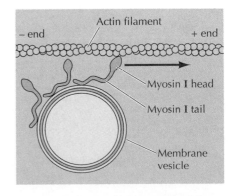

Actin filament

– end

+ end

Myosin I head

Myosin I tail

Membrane vesicle

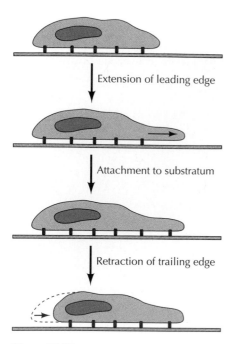

*Figure 11.30*
**Cell crawling**  The crawling movements of cells across a surface can be viewed as three stages of coordinated movements: (1) extension of the leading edge, (2) attachment of the leading edge to the substratum, and (3) retraction of the rear of the cell into the cell body.

edge of the cell (Figure 11.30). Second, these extensions must attach to the substratum across which the cell is migrating. Finally, the trailing edge of the cell must dissociate from the substratum and retract into the cell body.

A variety of experiments indicate that extension of the leading edge involves the polymerization and crosslinking of actin filaments. For example, inhibition of actin polymerization (e.g., by treatment with cytochalasin) blocks the formation of cell surface protrusions. However, the mechanisms responsible for generating the force required for extension of these structures have not been established. One possibility is that actin polymerization at the leading edge of the cell drives the formation of surface protrusions, either directly or as a result of osmotic swelling of the polymerizing actin gel. Alternatively, the formation of surface protrusions might be driven by myosin motors. If so, this process does not appear to involve the action of myosin II, since genetic manipulations of the cellular slime mold *Dictyostelium* have shown that inactivating the myosin II gene does not prevent pseudopodium formation. However, it remains possible that myosin I (or other unconventional myosins) might generate the force required to push the plasma membrane forward, in the direction of cell movement.

Following their extension, protrusions from the leading edge must attach to the substratum in order to function in cell locomotion. For slow-moving cells, such as fibroblasts, attachment involves the formation of focal adhesions (see Figure 11.13). Cells moving more rapidly, such as amoebas or white blood cells, form more diffuse contacts with the substratum, the molecular composition of which is not known.

The third stage of cell crawling, retraction of the trailing edge, is the least understood. The attachments of the trailing edge to the substratum are broken, and the rear of the cell recoils into the cell body. The process appears to require the development of tension between the front and rear of the cell, generating contractile force that eventually pulls the rear of the cell forward. This aspect of cell locomotion is impaired in myosin II-deficient *Dictyostelium* mutants, consistent with a role for myosin II in contracting the actin cortex and generating the force required for retraction of the trailing edge.

## INTERMEDIATE FILAMENTS

**Intermediate filaments** have a diameter of about 10 nm, which is intermediate between the diameters of the two other principal elements of the cytoskeleton, actin filaments (about 7 nm) and microtubules (about 25 nm). In contrast to actin filaments and microtubules, the intermediate filaments are not directly involved in cell movements. Instead, they appear to play basically a structural role by providing mechanical strength to cells and tissues.

### Intermediate Filament Proteins

Whereas actin filaments and microtubules are polymers of single types of proteins (actin and tubulin, respectively), intermediate filaments are composed of a variety of proteins that are expressed in different types of cells. More than 50 different intermediate filament proteins have been identified and classified into six groups based on similarities between their amino acid sequences (Table 11.1). Types I and II consist of two groups of **keratins**, each consisting of about 15 different proteins, which are expressed in epithelial cells. Each type of epithelial cell synthesizes at least one type I (acidic) and one type II (neutral/basic) keratin, which copolymerize to form filaments. Some type I and II keratins (called **hard keratins**) are used for production of structures such as hair, nails, and horns. The other type I and II keratins

*Table 11.1* Intermediate Filament Proteins

| Type | Protein | Size (kd) | Site of expression |
|------|---------|-----------|--------------------|
| I | Acidic keratins (~15 proteins) | 40–60 | Epithelial cells |
| II | Neutral or basic keratins (~15 proteins) | 50–70 | Epithelial cells |
| III | Vimentin | 54 | Fibroblasts, white blood cells, and other cell types |
| | Desmin | 53 | Muscle cells |
| | Glial fibrillary acidic protein | 51 | Glial cells |
| | Peripherin | 57 | Peripheral neurons |
| IV | Neurofilament proteins | | |
| | NF-L | 60–70 | Neurons |
| | NF-M | 105–110 | Neurons |
| | NF-H | 135–150 | Neurons |
| | $\alpha$-Internexin | 66 | Neurons |
| V | Nuclear lamins | 60–75 | Nuclear lamina of all cell types |
| VI | Nestin | 200 | Stem cells of central nervous system |

(**soft keratins**) are abundant in the cytoplasm of epithelial cells, with different keratins being expressed in various differentiated cell types.

The type III intermediate filament proteins include **vimentin**, which is found in a variety of different kinds of cells, including fibroblasts, smooth muscle cells, and white blood cells. Another type III protein, **desmin**, is specifically expressed in muscle cells, where it connects the Z discs of individual contractile elements. A third type III intermediate filament protein is specifically expressed in glial cells, and a fourth is expressed in neurons of the peripheral nervous system.

The type IV intermediate filament proteins include the three **neurofilament (NF) proteins** (designated NF-L, NF-M, and NF-H for *light, medium,* and *heavy,* respectively). These proteins form the major intermediate filaments of many types of mature neurons. They are particularly abundant in the axons of motor neurons and are thought to play a critical role in supporting these long, thin processes, which can extend more than a meter in length. Another type IV protein ($\alpha$-internexin) is expressed at an earlier stage of neuron development, prior to expression of the neurofilament proteins. The single type VI intermediate filament protein (nestin) is expressed even earlier during the development of neurons, in stem cells of the central nervous system.

The type V intermediate filament proteins are the nuclear lamins, which are found in most eukaryotic cells. Rather than being part of the cytoskeleton, the nuclear lamins are components of the nuclear envelope (see Figure 8.3). They also differ from the other intermediate filament proteins in that they assemble to form an orthogonal meshwork underlying the nuclear membrane.

Despite considerable diversity in size and amino acid sequence, the various intermediate filament proteins share a common structural organization (Figure 11.31). All of the intermediate filament proteins have a central $\alpha$-helical rod domain of approximately 310 amino acids (350 amino acids in the nuclear lamins). This central rod domain is flanked by amino- and carboxy-terminal domains, which vary among the different intermediate fila-

*Figure 11.31*
**Structure of intermediate filament proteins** Intermediate filament proteins contain a central $\alpha$-helical rod domain of approximately 310 amino acids (350 amino acids in the nuclear lamins). The N-terminal head and C-terminal tail domains vary in size and shape.

ment proteins in size, sequence, and secondary structure. As discussed next, the $\alpha$-helical rod domain plays a central role in filament assembly, while the variable head and tail domains presumably determine the specific functions of the different intermediate filament proteins.

## Assembly of Intermediate Filaments

The first stage of filament assembly is the formation of dimers in which the central rod domains of two polypeptide chains are wound around each other in a coiled-coil structure, similar to that formed by myosin II heavy chains (Figure 11.32). The dimers then associate in a staggered antiparallel fashion to form tetramers, which can assemble end to end to form protofilaments. The final intermediate filament contains approximately eight protofilaments wound around each other in a ropelike structure. Because they are assembled from antiparallel tetramers, both ends of intermediate filaments are equivalent. Consequently, in contrast to actin filaments and microtubules, intermediate filaments are apolar; they do not have distinct plus and minus ends.

Filament assembly requires interactions between specific types of intermediate filament proteins. For example, keratin filaments are always assembled from heterodimers containing one type I and one type II polypeptide. In contrast, the type III proteins can assemble into filaments containing only a single polypeptide (e.g., vimentin) or consisting of two different type III proteins (e.g., vimentin plus desmin). The type III proteins do not, however, form copolymers with the keratins. Among the type IV proteins, $\alpha$-internexin can assemble into filaments by itself, whereas the three neurofilament proteins copolymerize to form heteropolymers.

*Figure 11.32*
**Assembly of intermediate filaments** The central rod domains of two polypeptides wind around each other in a coiled-coil structure to form dimers. Dimers then associate in a staggered antiparallel fashion to form tetramers. Tetramers associate end to end to form protofilaments and laterally to form filaments. Each filament contains approximately eight protofilaments wound around each other in a ropelike structure.

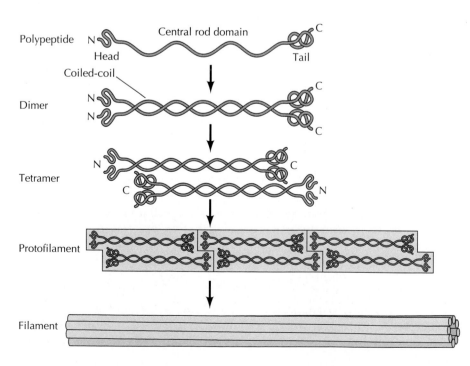

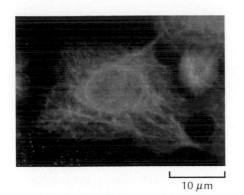

*Figure 11.33*
**Intracellular organization of keratin filaments** Micrograph of epithelial cells stained with fluorescent antibodies to keratin (green). Nuclei are stained blue. The keratin filaments extend from a ring surrounding the nucleus to the plasma membrane. (Nancy Kedersha/Immunogen/Photo Researchers, Inc.)

10 μm

Intermediate filaments are generally more stable than actin filaments or microtubules and do not exhibit the dynamic behavior associated with these other elements of the cytoskeleton (e.g., the treadmilling of actin filaments illustrated in Figure 11.4). However, intermediate filament proteins are frequently modified by phosphorylation, which can regulate their assembly and disassembly within the cell. The clearest example is phosphorylation of the nuclear lamins (see Figure 8.29), which results in disassembly of the nuclear lamina and breakdown of the nuclear envelope during mitosis. Cytoplasmic intermediate filaments, such as vimentin, are also phosphorylated at mitosis, which can lead to their disassembly and reorganization in dividing cells.

### Intracellular Organization of Intermediate Filaments

Intermediate filaments form an elaborate network in the cytoplasm of most cells, extending from a ring surrounding the nucleus to the plasma membrane (Figure 11.33). Both keratin and vimentin filaments attach to the nuclear envelope, apparently serving to position and anchor the nucleus within the cell. In addition, vimentin filaments can associate with the plasma membrane by binding to ankyrin, which was discussed earlier in this chapter as the membrane attachment site of spectrin. Thus, vimentin filaments might form a direct link between the nucleus and the plasma membrane. This intermediate filament network is probably also supported by association with microtubules, since drugs that induce microtubule disassembly cause the vimentin intermediate filaments to collapse around the nucleus.

The keratin filaments of epithelial cells are tightly anchored to the plasma membrane at two areas of specialized cell contacts, **desmosomes** and **hemidesmosomes**. Desmosomes are junctions between adjacent cells, at which cell-cell contacts are mediated by transmembrane proteins related to the cadherins. On their cytoplasmic side, desmosomes are associated with a characteristic dense plaque of intracellular proteins, to which keratin filaments are attached (Figure 11.34). Hemidesmosomes are morphologically similar junctions between epithelial cells and underlying connective tissue, at which keratin filaments bind to a type of integrin. Desmosomes

(A)

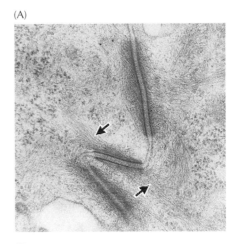

(B)

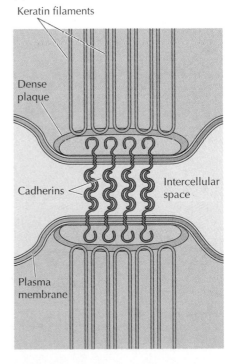

*Figure 11.34*
**Attachment of keratin filaments to desmosomes** (A) Electron micrograph illustrating keratin filaments (arrows) attached to the dense plaques of intracellular protein on both sides of a desmosome. (B) Schematic of a desmosome. The keratin filaments make hairpin turns at their sites of attachment. (A, Don Fawcett/Photo Researchers, Inc.)

and hemidesmosomes thus anchor intermediate filaments to regions of cell-cell and cell-substratum contact, respectively, similar to the attachment of the actin cytoskeleton to the plasma membrane at adherens junctions and focal adhesions. It is important to note that the keratin filaments anchored to both sides of desmosomes serve as a mechanical link between adjacent cells in an epithelial layer, thereby providing mechanical stability to the entire tissue.

Desmin and the neurofilaments play specialized roles in muscle and nerve cells, respectively. Desmin connects the individual actin-myosin assemblies of muscle cells both to one another and to the plasma membrane, thereby linking the actions of individual contractile elements. Neurofilaments are the major intermediate filaments in most mature neurons. They are particularly abundant in the long axons of motor neurons, where they are thought to play an important role in providing mechanical support for these long, thin extensions of the cell.

### Functions of Keratins and Neurofilaments: Diseases of the Skin and Nervous System

Although intermediate filaments have long been thought to provide structural support to the cell, direct evidence for their function has only recently been obtained. Some cells in culture make no intermediate filament proteins, indicating that these proteins are not required for the growth of cells *in vitro*. Similarly, injection of cultured cells with antibody against vimentin disrupts intermediate filament networks without affecting cell growth or movement. Therefore, it has been thought that intermediate filaments are most needed to strengthen the cytoskeleton of cells in the tissues of multicellular organisms, where they are subjected to a variety of mechanical stresses that do not affect cells in the isolated environment of a culture dish.

Experimental evidence for such an *in vivo* role of intermediate filaments was first provided in 1991 by studies in the laboratory of Elaine Fuchs. These investigators used transgenic mice to investigate the *in vivo* effects of expressing a keratin deletion mutant encoding a truncated polypeptide that disrupted the formation of normal keratin filaments (Figure 11.35). This mutant keratin gene was introduced into transgenic mice, where it was expressed in basal cells of the epidermis and disrupted formation of a normal keratin cytoskeleton. This resulted in the development of severe skin abnormalities, including blisters due to epidermal cell lysis following mild mechanical trauma, such as rubbing of the skin. The skin abnormalities of these transgenic mice thus provided direct support for the presumed role of keratins in providing mechanical strength to epithelial cells in tissues.

These experiments also pointed to the molecular basis of a human genetic disease, epidermolysis bullosa simplex (EBS). Like the transgenic mice expressing mutant keratin genes, patients with this disease develop skin blisters resulting from cell lysis after minor trauma. This similarity prompted studies of the keratin genes in EBS patients, leading to the demonstration that EBS is caused by keratin gene mutations that interfere with the normal assembly of keratin filaments. Thus, both experimental studies in transgenic mice and molecular analysis of a human genetic disease have demonstrated the role of keratins in allowing skin cells to withstand mechanical stress. Continuing studies have shown that mutations in other keratins are responsible for at least three other inherited skin diseases, which are similarly characterized by abnormal fragility of epidermal cells.

Other studies in transgenic mice have implicated abnormalities of neurofilaments in diseases of motor neurons, particularly amyotrophic lateral sclerosis (ALS). ALS, known as Lou Gehrig's disease and the disease afflict-

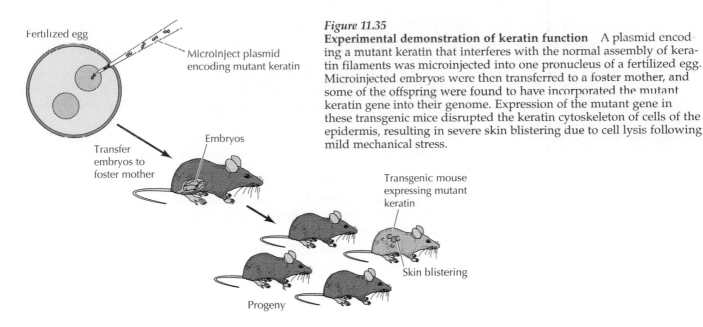

Fertilized egg

Microinject plasmid encoding mutant keratin

Transfer embryos to foster mother

Embryos

Transgenic mouse expressing mutant keratin

Skin blistering

Progeny

**Figure 11.35**
**Experimental demonstration of keratin function**   A plasmid encoding a mutant keratin that interferes with the normal assembly of keratin filaments was microinjected into one pronucleus of a fertilized egg. Microinjected embryos were then transferred to a foster mother, and some of the offspring were found to have incorporated the mutant keratin gene into their genome. Expression of the mutant gene in these transgenic mice disrupted the keratin cytoskeleton of cells of the epidermis, resulting in severe skin blistering due to cell lysis following mild mechanical stress.

ing the renowned physicist Stephen Hawking, results from progressive loss of motor neurons, which in turn leads to muscle atrophy, paralysis, and eventual death. ALS and other types of motor neuron disease are characterized by the accumulation and abnormal assembly of neurofilaments, suggesting that neurofilament abnormalities might contribute to these pathologies. Consistent with this possibility, overexpression of NF-L or NF-H in transgenic mice has been found to result in the development of a condition similar to ALS. Although the mechanism involved remains to be understood, these experiments clearly suggest the involvement of neurofilaments in the pathogenesis of motor neuron disease.

## MICROTUBULES

**Microtubules**, the third principal component of the cytoskeleton, are rigid hollow rods approximately 25 nm in diameter. Like actin filaments, microtubules are dynamic structures that undergo continual assembly and disassembly within the cell. They function both to determine cell shape and in a variety of cell movements, including some forms of cell locomotion, the intracellular transport of organelles, and the separation of chromosomes during mitosis.

### Structure, Assembly, and Dynamic Instability of Microtubules

In contrast to intermediate filaments, which are composed of a variety of different fibrous proteins, microtubules are composed of a single type of globular protein, called **tubulin**. Tubulin is a dimer consisting of two closely related 55-kd polypeptides, $\alpha$-tubulin and $\beta$-tubulin. Like actin, both $\alpha$- and $\beta$-tubulin are encoded by small families of related genes. In addition, a third type of tubulin ($\gamma$-tubulin) is specifically localized to the centrosome, where it plays a critical role in initiating microtubule assembly (discussed shortly).

Tubulin dimers polymerize to form microtubules, which generally consist of 13 linear protofilaments assembled around a hollow core (Figure 11.36). The protofilaments, which are composed of head-to-tail arrays of tubulin dimers, are arranged in parallel. Consequently, microtubules (like

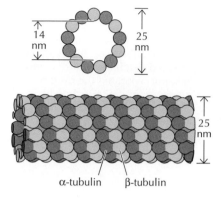

*Figure 11.36*
**Structure of microtubules** Dimers of $\alpha$- and $\beta$-tubulin polymerize to form microtubules, which are composed of 13 protofilaments assembled around a hollow core.

actin filaments) are polar structures with two distinct ends: a fast-growing plus end and a slow-growing minus end. This polarity is an important consideration in determining the direction of movement along microtubules, just as the polarity of actin filaments defines the direction of myosin movement.

Tubulin dimers can depolymerize as well as polymerize, and microtubules can undergo rapid cycles of assembly and disassembly. Both $\alpha$- and $\beta$-tubulin bind GTP, which functions analogously to the ATP bound to actin to regulate polymerization. In particular, the GTP bound to $\beta$-tubulin (though not that bound to $\alpha$-tubulin) is hydrolyzed to GDP fol-

*Figure 11.37*
**Dynamic instability of microtubules** Dynamic instability results from the hydrolysis of GTP bound to $\beta$-tubulin following polymerization, which reduces its binding affinity for adjacent molecules. Growth of microtubules continues as long as there is a high concentration of tubulin bound to GTP. New GTP-bound tubulin molecules are then added more rapidly than GTP is hydrolyzed, so a GTP cap is retained at the growing end. However, if GTP is hydrolyzed more rapidly than new subunits are then added, the presence of GDP-bound tubulin at the end of the microtubule leads to disassembly and shrinkage. Only the plus ends of microtubules are illustrated.

**High concentration of tubulin bound to GTP**

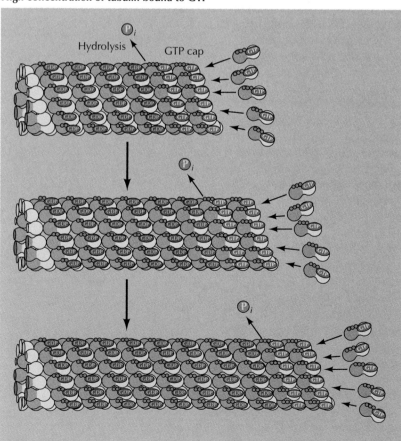

**Low concentration of tubulin bound to GTP**

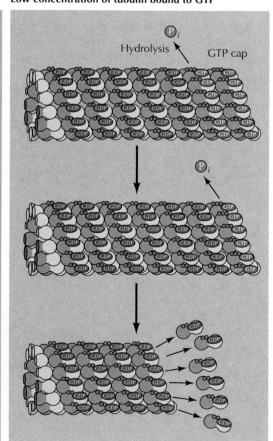

lowing polymerization. This GTP hydrolysis weakens the binding affinity of tubulin for adjacent molecules, thereby favoring depolymerization and resulting in the dynamic behavior of microtubules. Like actin filaments (see Figure 11.4), microtubules undergo treadmilling, a dynamic behavior in which tubulin molecules bound to GDP are continually lost from the minus end and replaced by the addition of tubulin molecules bound to GTP to the plus end of the same microtubule. In microtubules, GTP hydrolysis also results in the behavior known as **dynamic instability**, in which individual microtubules alternate between cycles of growth and shrinkage (Figure 11.37). Whether a microtubule grows or shrinks is determined by the rate of tubulin addition relative to the rate of GTP hydrolysis. As long as new GTP-bound tubulin molecules are added more rapidly than GTP is hydrolyzed, the microtubule retains a GTP cap at its plus end and microtubule growth continues. However, if the rate of polymerization slows, the GTP bound to tubulin at the plus end of the microtubule will be hydrolyzed to GDP. If this occurs, the GDP-bound tubulin will dissociate, resulting in rapid depolymerization and shrinkage of the microtubule.

Dynamic instability, described by Tim Mitchison and Marc Kirschner in 1984, results in the continual and rapid turnover of most microtubules, which have half-lives of only several minutes within the cell. As discussed later, this rapid turnover of microtubules is particularly critical for the remodeling of the cytoskeleton that occurs during mitosis. Because of the central role of microtubules in mitosis, drugs that affect microtubule assembly are useful not only as experimental tools in cell biology but also in the treatment of cancer. **Colchicine** and **colcemid** are examples of commonly used experimental drugs that bind tubulin and inhibit microtubule polymerization, which in turn blocks mitosis. Two related drugs (**vincristine** and **vinblastine**) are used in cancer chemotherapy because they selectively inhibit rapidly dividing cells. Another useful drug, **taxol**, stabilizes microtubules rather than inhibiting their assembly. Such stabilization also blocks cell division, and taxol is used as an anticancer agent as well as an experimental tool.

### The Centrosome and Microtubule Organization

The microtubules in most cells extend outward from a **microtubule-organizing center**, in which the minus ends of microtubules are anchored. In animal cells, the major microtubule-organizing center is the **centrosome**, which is located adjacent to the nucleus near the center of interphase (nondividing) cells (Figure 11.38). During mitosis, microtubules similarly extend outward from duplicated centrosomes to form the mitotic spindle, which is responsible for the separation and distribution of chromosomes to daughter cells. The centrosome thus plays a key role in determining the intracellular organization of microtubules, although most details of its function remain a mystery.

The centrosome serves as the initiation site for the assembly of microtubules, which grow outward from the centrosome toward the periphery of the cell. This can be clearly visualized in cells that have been treated with colcemid to disassemble their microtubules (Figure 11.39). When the drug is removed, the cells recover and new microtubules can be seen growing outward from the centrosome. Importantly, the initiation of microtubule growth at the centrosome estab-

*Figure 11.38*
**Intracellular organization of microtubules** The minus ends of microtubules are anchored in the centrosome. In interphase cells, the centrosome is located near the nucleus and microtubules extend outward to the cell periphery. During mitosis, duplicated centrosomes separate and microtubules reorganize to form the mitotic spindle.

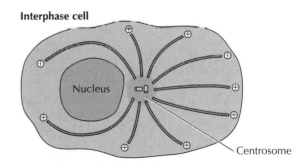

**Interphase cell**

Nucleus

Centrosome

**Mitotic cell**

Chromosomes

Centrosome

*Figure 11.39*
**Growth of microtubules from the centrosome** Microtubules in mouse fibroblasts are visualized by immunofluorescence microscopy using an antibody against tubulin. (A) The distribution of microtubules in a normal interphase cell. (B) This cell was treated with colcemid for one hour to disassemble microtubules. The drug was then removed and the cell allowed to recover for 30 minutes, allowing the visualization of new microtubules growing out of the centrosome. (From M. Osborn and K. Weber, 1976. *Proc. Natl. Acad. Sci. USA* 73: 867.)

(A)

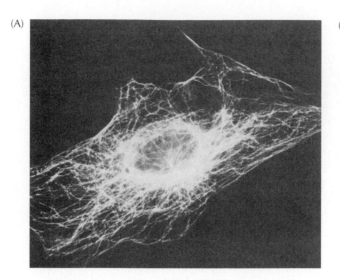

(B)

10 μm

lishes the polarity of microtubules within the cell. In particular, microtubules grow by the addition of tubulin to their plus ends, which extend outward from the centrosome toward the cell periphery.

The centrosomes of most animal cells contain a pair of **centrioles**, oriented perpendicular to each other, surrounded by amorphous **pericentriolar material** (Figure 11.40). The centrioles are cylindrical structures consisting of nine triplets of microtubules, similar to the basal bodies of cilia and flagella (discussed later in the chapter). Although centrioles are probably the precursors of basal bodies, they appear to be irrelevant for the function of the centrosome. Centrioles do not appear to be required for the assembly or organization of microtubules, and they are not found in plant cells, many unicellular eukaryotes, and some animal cells (such as mouse eggs). The microtubules that emanate from the centrosome terminate in the pericentriolar material, not the centrioles, and it is the pericentriolar material that initiates microtubule assembly.

A number of proteins associated with centrosomes have been identified, but in most cases their potential roles in microtubule assembly have not been determined. A notable exception, however, is γ-tubulin, a minor species of tubulin first identified in fungi. Subsequent studies have shown that γ-tubulin is specifically localized to centrosomes of a variety

*Figure 11.40*
**Structure of centrosomes** (A) Electron micrograph of a centrosome showing microtubules radiating from the pericentriolar material that surrounds a pair of centrioles. (B) Transverse section of a centriole illustrating its nine triplets of microtubules. (A, David M. Phillips/Visuals Unlimited; B, Don Fawcett, Photo Researchers, Inc.)

(A)

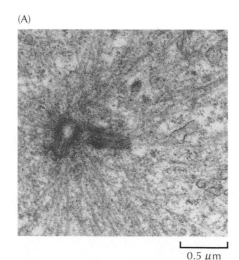

0.5 μm

(B)

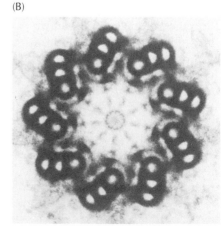

0.1 μm

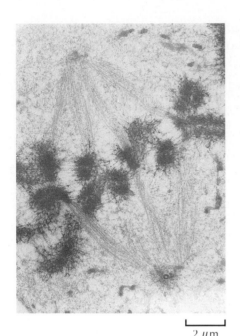

*Figure 11.41*
**Electron micrograph of the mitotic spindle** The spindle microtubules are attached to condensed chromosomes at metaphase. (From C. L. Rieder and S. S. Bowser, 1985, *J. Histochem. Cytochem.* 33: 165/Biological Photo Service.)

2 μm

of cells, where it appears to play a key role in microtubule nucleation. For example, microinjection of cells with antibody against γ-tubulin prevents nucleation of new microtubules, and γ-tubulin is required for assembly of functional centrosomes in cell-free extracts. In addition, γ-tubulin complexes purified from *Xenopus* eggs have been shown to nucleate microtubule growth *in vitro*. These γ-tubulin complexes are ring structures that have a diameter of 25 to 28 nm (similar to that of a microtubule) and are estimated to contain 10 to 13 γ-tubulin molecules. They are thought to bind α- and β-tubulins and serve as nucleation sites for further microtubule assembly.

## Reorganization of Microtubules during Mitosis

As noted earlier, microtubules completely reorganize during mitosis, providing a dramatic example of the importance of their dynamic instability. The microtubule array present in interphase cells disassembles and the free tubulin subunits are reassembled to form the **mitotic spindle**, which is responsible for the separation of daughter chromosomes (Figure 11.41). This restructuring of the microtubule cytoskeleton is directed by duplication of the centrosome to form two separate microtubule-organizing centers at opposite poles of the mitotic spindle.

The centrioles and other components of the centrosome are duplicated in interphase cells, but they remain together on one side of the nucleus until the beginning of mitosis (Figure 11.42). The two centrosomes then separate and move to opposite sides of the nucleus, forming the two poles

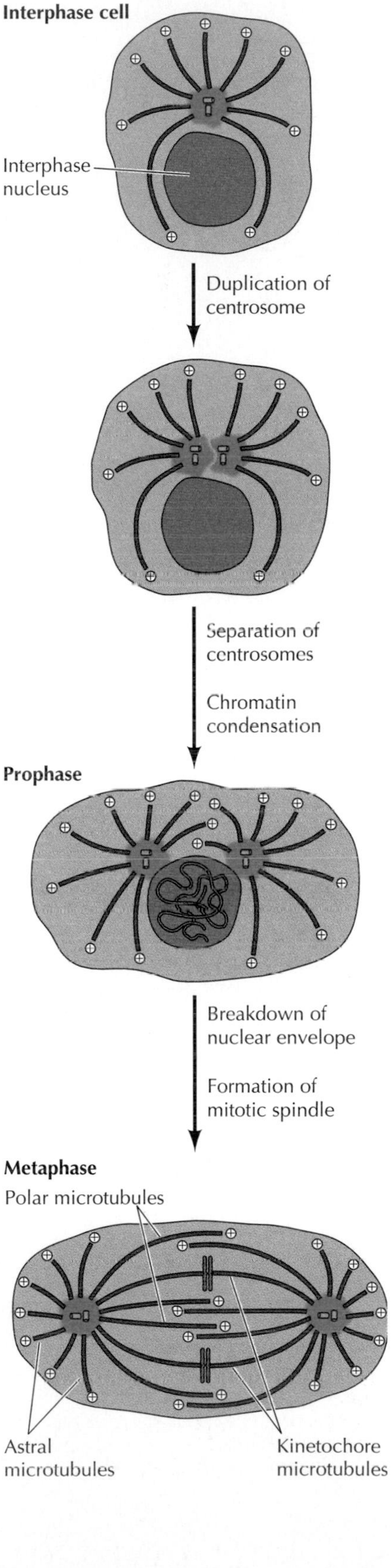

**Interphase cell**

Interphase nucleus

Duplication of centrosome

Separation of centrosomes

Chromatin condensation

**Prophase**

Breakdown of nuclear envelope

Formation of mitotic spindle

**Metaphase**

Polar microtubules

Astral microtubules

Kinetochore microtubules

*Figure 11.42*
**Formation of the mitotic spindle** The centrioles and centrosomes duplicate during interphase. During prophase of mitosis, the duplicated centrosomes separate and move to opposite sides of the nucleus. The nuclear envelope then disassembles, and microtubules reorganize to form the mitotic spindle. Kinetochore microtubules are attached to the condensed chromosomes, polar microtubules overlap with each other in the center of the cell, and astral microtubules extend outward to the cell periphery. At metaphase, the condensed chromosomes are aligned at the center of the spindle.

of the mitotic spindle. As the cell enters mitosis, the dynamics of microtubule assembly and disassembly also change dramatically. First, the rate of microtubule disassembly increases about tenfold, resulting in overall depolymerization and shrinkage of microtubules. At the same time, the number of microtubules emanating from the centrosome increases by five- to tenfold. In combination, these changes result in disassembly of the interphase microtubules and the outgrowth of large numbers of short microtubules from the centrosomes.

As first proposed by Marc Kirschner and Tim Mitchison in 1986, formation of the mitotic spindle involves the selective stabilization of some of the microtubules radiating from the centrosomes. These microtubules are of three types, two of which make up the mitotic spindle. **Kinetochore microtubules** attach to the condensed chromosomes of mitotic cells at their centromeres, which are associated with specific proteins to form the kinetochore (see Figure 4.16). Attachment to the kinetochore stabilizes these microtubules, which, as discussed below, play a critical role in separation of the mitotic chromosomes. The second type of microtubules found in the mitotic spindle (**polar microtubules**) are not attached to chromosomes. Instead, the polar microtubules emanating from the two centrosomes are stabilized by overlapping with each other in the center of the cell. **Astral microtubules** extend outward from the centrosomes to the cell periphery and have freely exposed plus ends. As discussed later, both the polar and astral microtubules also contribute to chromosome movement by pushing the spindle poles apart.

As mitosis proceeds, the condensed chromosomes first align on the metaphase plate and then separate, with the two chromatids of each chromosome being pulled to opposite poles of the spindle. Chromosome movement is mediated by motor proteins associated with the spindle microtubules, as will be discussed shortly. In the final stage of mitosis, nuclear envelopes re-form, the chromosomes decondense, and cytokinesis takes place. Each daughter cell then contains one centrosome, which nucleates the formation of a new network of interphase microtubules.

### Stabilization of Microtubules and Cell Polarity

Because of their inherent dynamic instability, most microtubules are frequently disassembled within the cell. This dynamic behavior can, however, be modified by the interactions of microtubules with other proteins. Such interactions allow the cell to stabilize microtubules in particular locations and provide an important mechanism for determining cell shape and polarity.

The dynamic behavior of microtubules is regulated by a variety of **microtubule-associated proteins (MAPs)**. A large number of MAPs have been identified, and they vary depending on the type of cell. The best-characterized MAPs are those isolated from mammalian brain, including MAP-1, MAP-2, and tau proteins. These proteins bind to microtubules and inhibit the dissociation of tubulin subunits. In addition to stabilizing microtubules, they can mediate their association with other elements of the cytoskeleton, such as intermediate filaments.

A good example of the role of stable microtubules in determining cell polarity is provided by nerve cells, which consist of two distinct types of processes (axons and dendrites) extending from a cell body (Figure 11.43). Both axons and dendrites are supported by stable microtubules, together with the neurofilaments discussed in the preceding section of this chapter. However, the microtubules in axons and dendrites are organized differently and associated with distinct MAPs. In axons, the microtubules are all oriented with their plus ends away from the cell body, similar to the gen-

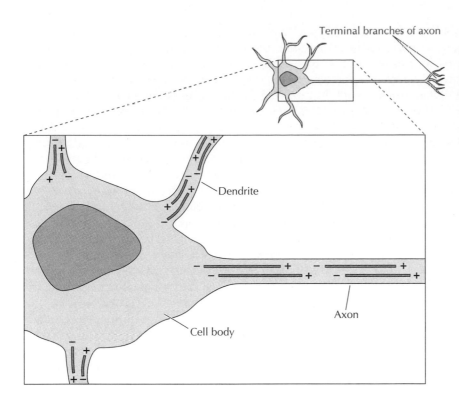

**Figure 11.43**
**Organization of microtubules in nerve cells** Two distinct types of processes extend from the cell body of nerve cells (neurons). Dendrites are short processes that receive stimuli from other nerve cells. The single long axon then carries impulses from the cell body to other cells, which may be either other neurons or an effector cell, such as a muscle. Stable microtubules in both axons and dendrites terminate in the cytoplasm rather than being anchored in the centrosome. In dendrites, microtubules are oriented in both directions, with their plus ends pointing both toward and away from the cell body. In contrast, all of the axon microtubules are oriented with their plus ends pointing toward the tip of the axon.

eral orientation of microtubules in other cell types. The minus ends of most of the microtubules in axons, however, are not anchored in the centrosome; instead, both the plus and minus ends of these microtubules terminate in the cytoplasm of the axon. In dendrites, the microtubules are oriented in both directions; some plus ends point toward the cell body and some point toward the cell periphery. These distinct microtubule arrangements are paralleled by differences in MAPs: Axons contain tau proteins, but no MAP-2, whereas dendrites contain MAP-2, but no tau proteins.

The distinctive differences in MAP-2 and tau distribution suggest that these microtubule-associated proteins may be responsible for the organization of stable microtubules in axons and dendrites. Consistent with this hypothesis, interference with tau expression in cultured neuronal cells has been shown to specifically inhibit the development of axons. Conversely, abnormal expression of tau in fibroblasts results in the reorganization of microtubules into axonlike projections. It thus appears that tau plays a direct role in the stabilization and organization of microtubules in nerve cell axons, providing a clear example of the function of a microtubule-associated protein in establishing cell shape and polarity.

## MICROTUBULE MOTORS AND MOVEMENTS

Microtubules are responsible for a variety of cell movements, including the intracellular transport and positioning of membrane vesicles and organelles, the separation of chromosomes at mitosis, and the beating of cilia and flagella. As discussed for actin filaments earlier in this chapter, movement along microtubules is based on the action of motor proteins that utilize energy derived from ATP hydrolysis to produce force and movement. Members of two large families of motor proteins—the **kinesins** and the **dyneins**—are responsible for powering the variety of movements in which microtubules participate.

## Identification of Microtubule Motor Proteins

Kinesin and dynein are distinct motor proteins that move along microtubules in opposite directions—kinesin toward the plus end and dynein toward the minus end (Figure 11.44). The first of these microtubule motor proteins to be identified was dynein, which was isolated by Ian Gibbons in 1965. The purification of this form of dynein (called **axonemal dynein**) was facilitated because it is a highly abundant protein in cilia, just as the abundance of myosin facilitated its isolation from muscle cells. The identification of other microtubule-based motors, however, was more problematic because the proteins responsible for processes such as chromosome movement and organelle transport are present at comparatively low concentrations in the cytoplasm. Isolation of these proteins therefore depended on the development of new experimental methods to detect the activity of molecular motors in cell-free systems.

The development of *in vitro* assays for cytoplasmic motor proteins was based on the use of **video-enhanced microscopy**, developed by Robert Allen and Shinya Inoué in the early 1980s, to study the movement of membrane vesicles and organelles along microtubules in squid axons. In this method, a video camera is used to increase the contrast of images obtained with the light microscope, substantially improving the detection of small objects and allowing the movement of organelles to be followed in living cells. Using this approach, Allen, Scott Brady, and Ray Lasek demonstrated that organelle movements also took place in a cell-free system in which the plasma membrane had been removed and a cytoplasmic extract had been spread on a glass slide. These observations led to the development of an *in vitro* reconstructed system, which provided an assay capable of detecting cellular proteins responsible for organelle movement. In 1985 Brady, as well as Ronald Vale, Thomas Reese, and Michael Sheetz, capitalized on these developments to identify kinesin as a novel microtubule motor protein, present in both squid axons and bovine brain.

Further studies demonstrated that kinesin translocates along microtubules in only a single direction—toward the plus end. Because the plus ends of microtubules in axons are all oriented away from the cell body (see Figure 11.43), the movement of kinesin in this direction transports vesicles and organelles away from the cell body, toward the tip of the axon. Within intact axons, however, vesicles and organelles also had been observed to

*Figure 11.44*
**Microtubule motor proteins** Kinesin and dynein move in opposite directions along microtubules, toward the plus and minus ends, respectively. Kinesin consists of two heavy chains, wound around each other in a coiled-coil structure, and two light chains. The globular head domains of the heavy chains bind microtubules and are the motor domains of the molecule. Dynein consists of two or three heavy chains (two are shown here) in association with multiple light and intermediate chains. The globular head domains of the heavy chains are the motor domains.

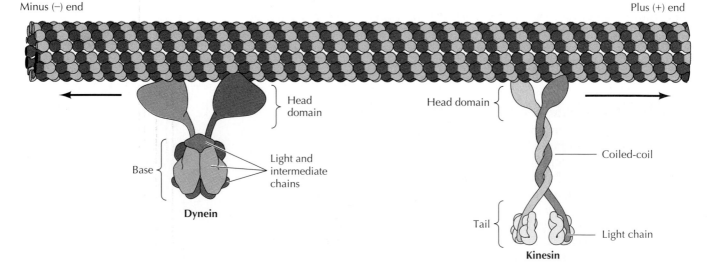

Minus (−) end          Plus (+) end

Head domain

Base

Light and intermediate chains

**Dynein**

Head domain

Coiled-coil

Tail

Light chain

**Kinesin**

move back toward the cell body, implying that a different motor protein might be responsible for movement along microtubules in the opposite direction—toward the minus end. Consistent with this prediction, further experiments showed that a protein previously identified as the microtubule-associated protein MAP-1C was in fact a motor protein that moved along microtubules in the minus end direction. Subsequent analysis demonstrated that MAP-1C is related to the dynein isolated from cilia (axonemal dynein), so MAP-1C is now referred to as **cytoplasmic dynein**.

Kinesin is a molecule of approximately 360 kd, consisting of two heavy chains (each about 110 kd) and two light chains (60 to 70 kd each) (see Figure 11.44). The heavy chains have long $\alpha$-helical regions that wind around each other in a coiled-coil structure. The amino-terminal globular head domains of the heavy chains are the motor domains of the molecule: They bind to both microtubules and ATP, the hydrolysis of which provides the energy required for movement. Although the motor domain of kinesin (approximately 340 amino acids) is much smaller than that of myosin (about 850 amino acids), X-ray crystallography indicates that the kinesin and myosin motor domains are structurally similar, suggesting that kinesin and myosin utilize similar molecular mechanisms to couple movement with the hydrolysis of ATP. The tail portion of the kinesin molecule consists of the light chains in association with the carboxy-terminal domains of the heavy chains. This portion of kinesin is responsible for binding to other cell components (such as membrane vesicles and organelles) that are transported along microtubules by the action of kinesin motors.

Dynein is an extremely large molecule (up to 2000 kd), which consists of two or three heavy chains (each about 500 kd) complexed with a variable number of light and intermediate polypeptides, which range from 14 to 120 kd (see Figure 11.44). As in kinesin, the heavy chains form globular ATP-binding motor domains that are responsible for movement along microtubules. The basal portion of the molecule, including the light and intermediate chains, is thought to bind to other subcellular structures, such as organelles and vesicles.

Like the myosins, both kinesin and dynein define families of related motor proteins. Following the initial isolation of kinesin in 1985, a variety of kinesin-related proteins have been identified. These proteins are similar to kinesin in their motor domains but differ in the sequences of their carboxy-terminal tails. It therefore seems likely that these distinct kinesin-related proteins are responsible for the movements of different types of "cargo" along microtubules. Interestingly, not all of the kinesin-related proteins move along microtubules in the plus end direction. For example, the *Drosophila* kinesin-related protein Ncd moves instead toward microtubule minus ends. There are also several types of axonemal dynein, although only single genes encoding the heavy chain of cytoplasmic dynein have been found in several organisms, including yeasts and mice.

## Organelle Transport and Intracellular Organization

One of the major roles of microtubules is to transport membrane vesicles and organelles through the cytoplasm of eukaryotic cells. As already discussed, such cytoplasmic organelle transport is particularly evident in nerve cell axons, which may extend more than a meter in length. Ribosomes are present only in the cell body and dendrites, so proteins, membrane vesicles, and organelles (e.g., mitochondria) must be transported from the cell body to the axon. Via video-enhanced microscopy, the transport of membrane vesicles and organelles in both directions can be visualized along axon microtubules, where kinesin and dynein carry their cargoes to and from the

# KEY EXPERIMENT

## *The Isolation of Kinesin*

### Identification of a Novel Force-Generating Protein, Kinesin, Involved in Microtubule-Based Motility

Ronald D. Vale, Thomas S. Reese, and Michael P. Sheetz

National Institute of Neurological and Communicative Disorders and Stroke, Marine Biological Laboratory, Woods Hole, MA; University of Connecticut Health Center, Farmington, CT; Stanford University School of Medicine, Stanford, CA

*Cell, Volume 42, 1985, pages 39-50*

### The Context

The transport and positioning of cytoplasmic organelles is key to the organization of eukaryotic cells, so understanding the mechanisms responsible for vesicle and organelle transport is a fundamental question in cell biology. In 1982, Robert Allen, Scott Brady, Ray Lasek, and their colleagues used video-enhanced microscopy to visualize the movement of organelles along cytoplasmic filaments in squid giant axons, both *in vivo* and in a cell-free system. These filaments were then identified as microtubules by electron microscopy, but the motor proteins responsible for organelle movement were unknown: The only microtubule motor identified at the time was axonemal dynein, which was present

only in cilia and flagella. In 1985, Ronald Vale, Thomas Reese, and Michael Sheetz described the isolation of a novel motor protein, kinesin, which was responsible for the movement of organelles along microtub-

Binding of a motor protein to microtubules in the presence of AMP-PNP. Protein samples were analyzed by electrophoresis through a polyacrylamide gel, stained, and photographed. Molecular weights of marker proteins are indicated in kilodaltons at the left. Lane a represents purified microtubules, in which only tubulin (55 kd) is detected. Lane b is a soluble extract of squid axon cytoplasm, containing many different polypeptides. This soluble extract was incubated with microtubules either without (lane c) or with (lane d) AMP-PNP. Microtubules were then recovered and incubated with ATP to release bound proteins, which were subjected to electrophoresis. Note that a 110-kd polypeptide was specifically bound to microtubules in the presence of AMP-PNP and then released by ATP. The proteins bound in the presence of AMP-PNP and released by ATP were also found to induce microtubule movement (indicated by "+" above the gel).

ules. Similar experiments were reported at the same time by Scott Brady (*Nature* 317: 73–75, 1985).

### The Experiments

Two experimental strategies were key to the isolation of kinesin. The first, based on the work of Allen and his colleagues, was the use of an *in vitro* system in which motor protein activity could be detected. Vale, Reese, and Sheetz used a

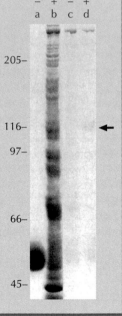

tips of the axons, respectively. For example, secretory vesicles containing neurotransmitters are carried from the Golgi apparatus to the terminal branches of the axon by kinesin. In the reverse direction, cytoplasmic dynein transports endocytic vesicles from the axon back to the cell body.

Microtubules similarly transport membrane vesicles and organelles in other types of cells. Because microtubules are usually oriented with their minus end anchored in the centrosome and their plus end extending toward the cell periphery, kinesin and cytoplasmic dynein are thought to transport vesicles and organelles in opposite directions through the cytoplasm (Figure 11.45). In addition to transporting membrane vesicles in the endocytic and secretory pathways, microtubules and associated motor proteins position membrane-enclosed organelles (such as the endoplasmic reticulum, Golgi apparatus, and lysosomes) within the cell. For example,

## The Isolation of Kinesin *(continued)*

system in which proteins in the cytoplasm of axons were found to power the movement of microtubules across the surface of glass coverslips. This cell-free system provided a sensitive and rapid functional assay for the activity of kinesin as a molecular motor.

The second important approach used in the isolation of kinesin was to take advantage of its binding to microtubules. Lasek and Brady (*Nature* 316: 645–647, 1985) had made the key observation that *in vitro* movements of microtubules require ATP and are inhibited by an ATP analog (adenylyl imidodiphosphate, AMP-PNP) that cannot be hydrolyzed and therefore does not provide a usable source of energy. In the presence of AMP-PNP, organelles remained attached to microtubules, suggesting that the motor protein responsible for organelle movement might also remain bound to microtubules under these conditions.

On the basis of these findings, Vale and colleagues incubated microtubules with cytoplasmic proteins from squid axon in the presence of AMP-PNP. The microtubules were then recovered and incubated with ATP to release proteins that were specifically bound in the presence of the nonhydrolyzable ATP analog. This experiment identified a 110-kd polypeptide that was bound to microtubules in the presence of AMP-PNP

and released by subsequent incubation with ATP (see figure). Furthermore, the proteins bound to microtubules and then released by ATP were also shown to support the movement of microtubules *in vitro*. Binding to microtubules in the presence of AMP-PNP thus provided an efficient approach to isolation of the motor protein, which was shown by further biochemical studies to contain the 110-kd polypeptide complexed to polypeptides of 60 to 70 kd. By a similar approach, a related protein was also purified from bovine brain. The authors concluded that these proteins "represent a novel class of motility proteins that are structurally as well as enzymatically distinct from dynein, and we propose to call these translocators kinesin (from the Greek *kinein*, to move)."

### The Impact

The movement of vesicles and organelles along microtubules is fundamental to the organization of eukaryotic cells, so the motor proteins responsible for these movements play critical roles in cell biology. Subsequent experiments using *in vitro* assays similar to those described here established that kinesin moves along microtubules in the plus end direction, whereas cytoplasmic dynein is responsible for the transport of vesicles and organelles toward microtubule minus ends. Moreover, a large

R. D. Vale

T. S. Reese

M. P. Sheetz

family of kinesin-related proteins has since been identified. In addition to their roles in vesicle and organelle transport, members of these families of motor proteins are responsible for the separation and distribution of chromosomes during mitosis. The identification of kinesin thus opened the door to understanding a variety of microtubule-based movements that are critical to the structure and function of eukaryotic cells.

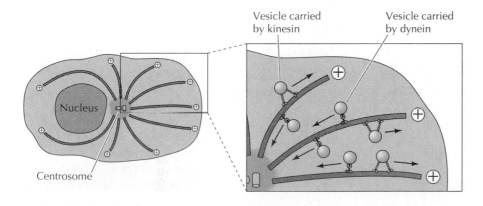

Vesicle carried by kinesin

Vesicle carried by dynein

Nucleus

Centrosome

*Figure 11.45*
**Transport of vesicles along microtubules** Kinesin transports vesicles and organelles in the direction of microtubule plus ends, which extend toward the cell periphery. In contrast, dynein carries its cargo in the direction of microtubule minus ends, which are anchored in the center of the cell.

the endoplasmic reticulum extends to the periphery of the cell in association with microtubules (Figure 11.46). Drugs that depolymerize microtubules cause the endoplasmic reticulum to retract toward the cell center, indicating that association with microtubules is required to maintain the endoplasmic reticulum in its extended state. This positioning of the endoplasmic reticulum is thought to involve the action of kinesin, which pulls the endoplasmic reticulum along microtubules in the plus end direction, toward the cell periphery.

Conversely, the Golgi apparatus is located in the center of the cell, near the centrosome. If microtubules are disrupted, either by a drug or when the cell enters mitosis, the Golgi apparatus breaks up into small vesicles that disperse throughout the cytoplasm. When the microtubules re-form, the Golgi apparatus also reassembles, with the Golgi vesicles apparently being transported to the center of the cell (toward the minus end of microtubules) by cytoplasmic dynein. Movement along microtubules is thus responsible not only for vesicle transport, but also for establishing the positions of membrane-enclosed organelles within the cytoplasm of eukaryotic cells.

### Separation of Mitotic Chromosomes

As discussed earlier in this chapter, microtubules reorganize at the beginning of mitosis to form the mitotic spindle, which plays a central role in cell division by distributing the duplicated chromosomes to daughter nuclei. This critical distribution of the genetic material takes place during anaphase of mitosis, when sister chromatids separate and move to opposite poles of the spindle. Chromosome movement proceeds by two distinct mechanisms, referred to as **anaphase A** and **anaphase B,** which involve different types of spindle microtubules and associated motor proteins.

(A)

(B)

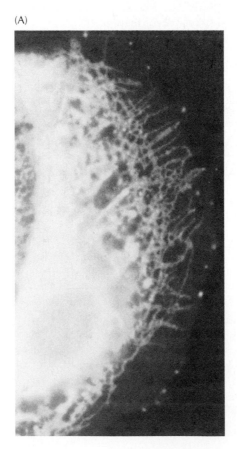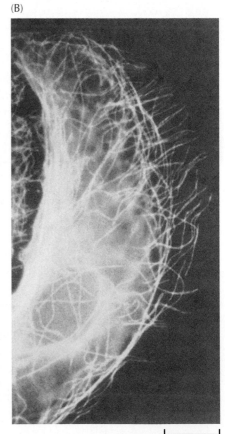

*Figure 11.46*
**Association of the endoplasmic reticulum with microtubules**   Fluorescence microscopy of the endoplasmic reticulum (A) and microtubules (B) in an epithelial cell. The endoplasmic reticulum is stained with a fluorescent dye and microtubules with an antibody against tubulin. Note the close correlation between the endoplasmic reticulum and microtubules at the periphery of the cell. (From M. Terasaki, L. B. Chen, and K. Fujiwara, 1986. *J. Cell Biol.* 103: 1557.)

10 µm

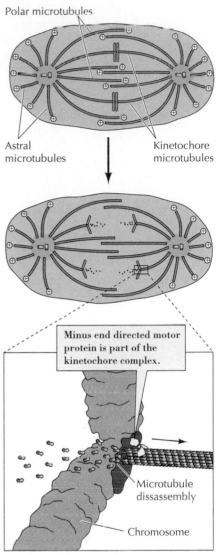

*Figure 11.47*
**Anaphase A chromosome movement** Chromosomes move toward the spindle poles along the kinetochore microtubules. Chromosome movement is thought to be driven (at least in part) by minus end-directed motor proteins associated with the kinetochore. The action of these motor proteins is coupled to disassembly and shortening of the kinetochore microtubules.

Anaphase A consists of the movement of chromosomes toward the spindle poles along the kinetochore microtubules, which shorten as chromosome movement proceeds (Figure 11.47). This type of chromosome movement is thought to be driven in part by a kinetochore-associated motor protein that translocates chromosomes along the spindle microtubules in the minus end direction, toward the centrosomes. The action of these motor proteins is coupled to disassembly and shortening of the kinetochore microtubules, and it is thought that motor activity and microtubule disassembly both contribute to chromosome movement. In addition to the minus end-directed kinetochore motors, anaphase A chromosome movement appears to be driven by plus end-directed motors and microtubule disassembly at the spindle poles.

Anaphase B refers to the separation of the spindle poles themselves (Figure 11.48). Spindle-pole separation is accompanied by elongation of the polar microtubules and is similar to the initial separation of duplicated centrosomes to form the spindle poles at the beginning of mitosis (see Figure 11.42). Spindle-pole separation may result from two types of mecha-

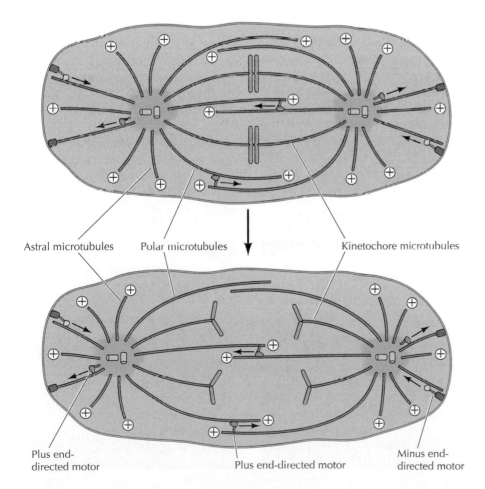

*Figure 11.48*
**Spindle pole separation in anaphase B** The separation of spindle poles results from two types of movement. First, overlapping polar microtubules slide past each other to push the spindle poles apart, probably as a result of the action of plus end-directed motor proteins. Second, the spindle poles are pulled apart by the astral microtubules. The driving force could be either a minus end-directed motor anchored to a cytoplasmic structure, such as the cell cortex, or a plus end-directed motor associated with the spindle pole.

nisms, involving both polar and astral microtubules. First, the overlapping polar microtubules can slide against one another, pushing the spindle poles apart. This movement may result from the action of plus end-directed motor proteins, which move polar microtubules toward the plus end of their overlapping microtubule and therefore away from the opposite spindle pole. Second, the spindle poles can be pulled apart by the astral microtubules. The mechanism responsible for this type of movement has not been established, but it could result from the action of cytoplasmic dynein anchored to the cell cortex or another structure in the cytoplasm. The translocation of such an anchored dynein motor along astral microtubules in the minus end direction would have the effect of pulling the spindle poles apart, toward the periphery of the cell. Alternatively, a motor protein associated with the spindle poles could move along astral microtubules in the plus end direction, which would also pull the spindle poles toward the cell periphery.

Chromosome separation, as well as formation of the mitotic spindle, thus appears to be mediated by several types of movements in which different classes of microtubules and associated motor proteins participate. Identifying these motor proteins and understanding the mechanisms that coordinate their activities remain challenges for future research.

## Cilia and Flagella

**Cilia** and **flagella** are microtubule-based projections of the plasma membrane that are responsible for movement of a variety of eukaryotic cells. Many bacteria also have flagella, but these prokaryotic flagella are quite different from those of eukaryotes. Bacterial flagella (which are not discussed further here) are protein filaments projecting from the cell surface, rather

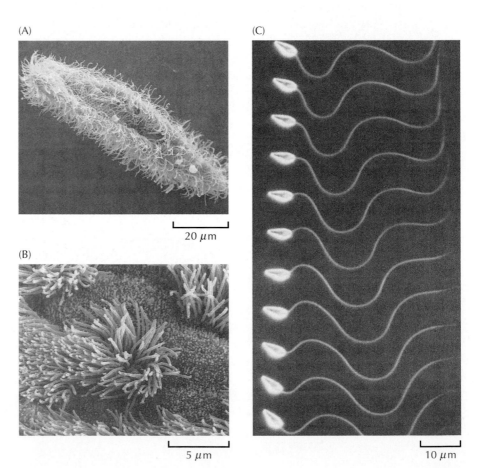

(A)

(B)

(C)

20 μm

5 μm

10 μm

*Figure 11.49*
**Examples of cilia and flagella**   (A) Scanning electron micrograph showing numerous cilia covering the surface of *Paramecium*. (B) Scanning electron micrograph of ciliated epithelial cells lining the surface of a trachea. (C) Multiple-flash photograph (500 flashes per second) showing the wavelike movement of a sea urchin sperm flagellum. (A, Stanley Flagler/Visuals Unlimited; B, Fred E. Hossler/Visuals Unlimited; C, C. J. Brokaw, California Institute of Technology.)

than projections of the plasma membrane supported by microtubules.

Eukaryotic cilia and flagella are very similar structures, each with a diameter of approximately 0.25 μm (Figure 11.49). Many cells are covered by numerous cilia, which are about 10 μm in length. Cilia beat in a coordinated back-and-forth motion, which either moves the cell through fluid or moves fluid over the surface of the cell. For example, the cilia of some protozoans (such as *Paramecium*) are responsible both for cell motility and for sweeping food organisms over the cell surface and into the oral cavity. In animals, an important function of cilia is to move fluid or mucus over the surface of epithelial cell sheets. A good example is provided by the ciliated cells lining the respiratory tract, which clear mucus and dust from the respiratory passages. Flagella differ from cilia in their length (they can be as long as 200 μm) and in their wavelike pattern of beating. Cells usually have only one or two flagella, which are responsible for the locomotion of a variety of protozoans and of sperm.

The fundamental structure of both cilia and flagella is the **axoneme**, which is composed of microtubules and their associated proteins (Figure 11.50). The microtubules are arranged in a characteristic "9 + 2" pattern in which a central pair of microtubules is surrounded by nine outer microtubule doublets. The two fused microtubules of each outer doublet are distinct: One (called the A tubule) is a complete microtubule consisting of 13 protofilaments; the other (the B tubule) is incomplete, containing only 10 or 11 protofilaments fused to the A tubule. The outer microtubule doublets are connected to the central pair by radial spokes and to each other by links of a protein called **nexin**. In addition, two arms of dynein are attached to each A tubule, and it is the motor activity of these axonemal dyneins that drives the beating of cilia and flagella.

The minus ends of the microtubules of cilia and flagella are anchored in a **basal body**, which is similar in structure to a centriole and contains nine triplets of microtubules (Figure 11.51). Centrioles were discussed earlier as components of the centrosome, in which their function is uncertain. Basal bodies, however, play a clear role in organization of the axoneme microtubules. Namely, each of the outer microtubule doublets of the axoneme is formed by extension of two of the microtubules present in the triplets of the basal body. Basal bodies thus serve to initiate the growth of axonemal microtubules, as well as anchoring cilia and flagella to the surface of the cell.

The movements of cilia and flagella result from the sliding of outer

*Figure 11.50*
**Structure of the axoneme of cilia and flagella**   (A) Computer-enhanced electron micrograph of a cross section of the axoneme of a rat sperm flagellum. (B) Schematic cross section of an axoneme. The nine outer doublets consist of one complete (A) and one incomplete (B) microtubule, containing only 10 or 11 protofilaments. The outer doublets are joined to each other by nexin links and to the central pair of microtubules by radial spokes. Each outer microtubule doublet is associated with inner and outer dynein arms. (A, K. G. Murti/Visuals Unlimited.)

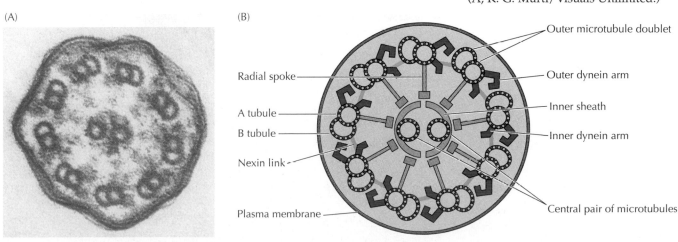

(A)

0.1 μm

(B)

Radial spoke

A tubule

B tubule

Nexin link

Plasma membrane

Outer microtubule doublet

Outer dynein arm

Inner sheath

Inner dynein arm

Central pair of microtubules

(A)

(B)

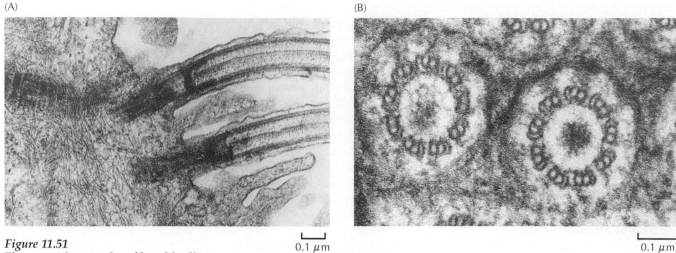

0.1 μm

0.1 μm

*Figure 11.51*
**Electron micrographs of basal bodies**
(A) A longitudinal view of cilia anchored in basal bodies. (B) A cross section of basal bodies. Each basal body consists of nine triplets of microtubules. (A, Conly L. Reider/Biological Photo Service; B, W. L. Dentler, Biological Photo Service.)

microtubule doublets relative to one another, powered by the motor activity of axonemal dynein (Figure 11.52). The dynein bases bind to the A tubules while the dynein head groups bind to the B tubules of adjacent doublets. Movement of the dynein head group in the minus end direction then causes the A tubule of one doublet to slide toward the basal end of the adjacent B tubule. Because the microtubule doublets in an axoneme are connected by nexin links, the sliding of one doublet along another causes them to bend, forming the basis of the beating movements of cilia and flagella. It is apparent, however, that the activities of dynein molecules in different regions of the axoneme must be carefully regulated to produce the coordinated beating of cilia and the wavelike oscillations of flagella—a process about which little is currently understood.

*Figure 11.52*
**Movement of microtubules in cilia and flagella**   The bases of dynein arms are attached to A tubules, and the motor head groups interact with the B tubules of adjacent doublets. Movement of the dynein head groups in the minus end direction (toward the base of the cilium) then causes the A tubule of one doublet to slide toward the base of the adjacent B tubule. Because both microtubule doublets are connected by nexin links, this sliding movement forces them to bend.

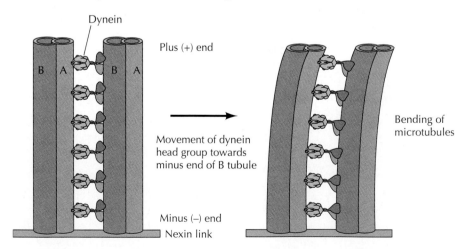

## Summary

### STRUCTURE AND ORGANIZATION OF ACTIN FILAMENTS

*Assembly and Disassembly of Actin Filaments:* Actin filaments are formed by the head-to-tail polymerization of actin monomers into a helix. A variety of actin-binding proteins regulate the assembly and disassembly of actin filaments within the cell.

*Organization of Actin Filaments:* Actin filaments are crosslinked by actin-binding proteins to form bundles or three-dimensional networks.

*Association of Actin Filaments with the Plasma Membrane:* A network of actin filaments and other cytoskeletal proteins underlies the plasma membrane and determines cell shape. Actin bundles also attach to the plasma membrane and anchor the cell at regions of cell-cell and cell-substratum contact.

*Protrusions of the Cell Surface:* Actin filaments support permanent protrusions of the cell surface, such as microvilli, as well as transient extensions that are responsible for phagocytosis and cell locomotion.

### ACTIN, MYOSIN, AND CELL MOVEMENT

*Muscle Contraction:* Studies of muscle established the role of myosin as a motor protein that uses the energy derived from ATP hydrolysis to generate force and movement. Muscle contraction results from the sliding of actin and myosin filaments past each other. ATP hydrolysis drives repeated cycles of interaction between myosin and actin, during which conformational changes in the myosin head group result in its movement along actin filaments.

*Contractile Assemblies of Actin and Myosin in Nonmuscle Cells:* Assemblies of actin and myosin II are responsible for a variety of movements of nonmuscle cells, including cytokinesis.

*Unconventional Myosins:* Other types of myosin that do not function in contraction serve to transport membrane vesicles and organelles along actin filaments.

*Cell Crawling:* Cell crawling is a complex process in which extensions of the plasma membrane are formed by polymerization of actin filaments at the leading edge of the cell. These extensions then attach to the substratum, and the trailing edge retracts into the cell body. Myosin motors appear to play a role in retraction of the trailing edge and may also function in extension of the leading edge.

### INTERMEDIATE FILAMENTS

*Intermediate Filament Proteins:* Intermediate filaments are polymers of more than 50 different proteins that are expressed in various types of cells. They are not involved in cell movement, but provide mechanical support to cells and tissues.

**actin, microfilament, globular (G) actin, filamentous (F) actin, treadmilling, cytochalasin, phalloidin, actin-binding protein, thymosin, profilin, capping protein, gelsolin**

**actin bundle, actin network, actin-bundling protein, fimbrin, contractile bundle, $\alpha$-actinin, filamin**

**cell cortex, spectrin, ankyrin, fodrin, dystrophin, integrin, focal adhesion, stress fiber, talin, vinculin, tensin, adherens junction, adhesion belt, cadherin, $\alpha$-catenin, $\beta$-catenin, $\gamma$-catenin**

**microvillus, brush border, stereocilium, villin, pseudopodium, lamellipodium, microspike, filopodium**

**myosin, molecular motor, muscle fiber, myofibril, sarcomere, titin, nebulin, sliding filament model, myosin II, sarcoplasmic reticulum, tropomyosin, troponin**

**cytokinesis, contractile ring, myosin light-chain kinase, calmodulin**

**myosin I**

**intermediate filament, keratin, hard keratin, soft keratin, vimentin, desmin, neurofilament (NF) protein**

**desmosome, hemidesmosome**

*Assembly of Intermediate Filaments:* Intermediate filaments are formed from dimers of two polypeptide chains wound around each other in a coiled-coil structure. The dimers then associate to form tetramers, which assemble into protofilaments. Intermediate filaments are formed from protofilaments wound around one another in a ropelike structure.

*Intracellular Organization of Intermediate Filaments:* Intermediate filaments form a network extending from a ring surrounding the nucleus to the plasma membrane of most cell types. In epithelial cells, intermediate filaments are anchored to the plasma membrane at regions of specialized cell contacts (desmosomes and hemidesmosomes). Intermediate filaments also play specialized roles in muscle and nerve cells.

*Functions of Keratins and Neurofilaments: Diseases of the Skin and Nervous System:* The importance of intermediate filaments in providing mechanical strength to cells in tissues has been demonstrated by the introduction of mutant keratin genes into transgenic mice. Similar keratin gene mutations are responsible for some human skin diseases, and abnormalities of neurofilaments have been implicated in the development of motor neuron disease.

## MICROTUBULES

**microtubule, tubulin, dynamic instability, colchicine, colcemid, vincristine, vinblastine, taxol**

*Structure, Assembly, and Dynamic Instability of Microtubules:* Microtubules are formed by the reversible polymerization of tubulin. They display dynamic instability and undergo continual cycles of assembly and disassembly as a result of GTP hydrolysis following tubulin polymerization.

**microtubule-organizing center, centrosome, centriole, pericentriolar material**

*The Centrosome and Microtubule Organization:* The microtubules in most cells extend outward from a microtubule-organizing center, or centrosome, located near the center of the cell. In animal cells, the centrosome usually contains a pair of centrioles surrounded by pericentriolar material. The growth of microtubules is initiated in the pericentriolar material, which then serves to anchor their minus ends.

**mitotic spindle, kinetochore microtubule, polar microtubule, astral microtubule**

*Reorganization of Microtubules during Mitosis:* Microtubules reorganize at the beginning of mitosis to form the mitotic spindle, which is responsible for chromosome separation.

**microtubule-associated protein (MAP)**

*Stabilization of Microtubules and Cell Polarity:* Selective stabilization of microtubules by microtubule-associated proteins can determine cell shape and polarity, such as the extension of nerve cell axons and dendrites.

## MICROTUBULE MOTORS AND MOVEMENTS

**kinesin, dynein, axonemal dynein, video-enhanced microscopy, cytoplasmic dynein**

*Identification of Microtubule Motor Proteins:* Two families of motor proteins, the kinesins and the dyneins, are responsible for movement along microtubules. Kinesin and most kinesin-related proteins move in the plus end direction, whereas the dyneins move toward microtubule minus ends.

*Organelle Transport and Intracellular Organization:* Movement along microtubules transports membrane vesicles and organelles through the cytoplasm, as well as positioning cytoplasmic organelles within the cell.

*Separation of Mitotic Chromosomes:* During anaphase of mitosis, daughter chromosomes separate and move to opposite poles of the mitotic spindle. Chromosome separation results from several types of movements in which different classes of spindle microtubules and motor proteins participate.

anaphase A, anaphase B

*Cilia and Flagella:* Cilia and flagella are microtubule-based extensions of the plasma membrane. Their movements result from the sliding of microtubules, driven by the action of dynein motors.

cilium, flagellum, axoneme, nexin, basal body

## QUESTIONS

**1.** Would cytochalasin affect the movement of cells that extend pseudopodia or that utilize flagella for locomotion?

**2.** Why is the polarity of actin filaments important to muscle contraction?

**3.** Would you expect mutations of keratin genes to affect fibroblasts?

**4.** Which aspect of cell division would be affected by colchicine: chromosome segregation or cytokinesis?

**5.** How would elimination of nexin affect the beating of cilia?

## REFERENCES AND FURTHER READING

### General References

Amos, L. A. and W. B. Amos. 1991. *Molecules of the Cytoskeleton.* New York: Guilford Press.

Bray, D. 1992. *Cell Movements.* New York: Garland Publishing.

### Structure and Organization of Actin Filaments

Bennett, V. and D. M. Gilligan. 1993. The spectrin-based membrane skeleton and micron-scale organization of the plasma membrane. *Ann. Rev. Cell Biol.* 9: 27–66. [R]

Bretscher, A. 1991. Microfilament structure and function in the cortical cytoskeleton. *Ann. Rev. Cell Biol.* 7: 337–374. [R]

Burridge, K., K. Fath, T. Kelly, G. Nuckolls, and C. Turner. 1988. Focal adhesions: Transmembrane junctions between the extracellular matrix and the cytoskeleton. *Ann. Rev. Cell Biol.* 4: 487–525. [R]

Campbell, K. P. 1995. Three muscular dystrophies: Loss of cytoskeleton-extracellular matrix linkage. *Cell* 80: 675–679. [R]

Cowin, P. 1994. Unraveling the cytoplasmic interactions of the cadherin superfamily. *Proc. Natl. Acad. Sci. USA* 91: 10759–10761. [R]

Holmes, K. C., D. Popp, W. Gebhard and W. Kabsch. 1990. Atomic model of the actin filament. *Nature* 347: 44–49. [P]

Jokusch, B. M., P. Bubeck, K. Giehl, M. Kroemker, J. Moschner, M. Rothkegel, M. Rüdiger, K. Schlüter, G. Stanke and J. Winkler. 1995. The molecular architecture of focal adhesions. *Ann. Rev. Cell Dev. Biol.* 11: 379–416. [R]

Kabsch, W., H. G. Mannherz, D. Suck, E. F. Pai and K. C. Holmes. 1990. Atomic structure of the actin: DNase I complex. *Nature* 347: 37–44. [P]

Luna, E. J. and A. L. Hitt. 1992. Cytoskeleton-plasma membrane interactions. *Science* 258: 955–964. [R]

Machesky, L. M. and T. D. Pollard. 1993. Profilin as a potential mediator of membrane-cytoskeleton communication. *Trends Cell Biol.* 3: 381–385. [R]

Matsudaira, P. 1991. Modular organization of actin crosslinking proteins. *Trends Biochem. Sci.* 16: 87–92. [R]

Pollard, T. D., S. Almo, S. Quirk, V. Vinson and E. E. Lattman. 1994. Structure of actin binding proteins: Insights about function at atomic resolution. *Ann. Rev. Cell Biol.* 10: 207–249. [R]

Pumplin, D. W. and R. J. Bloch. 1993. The membrane skeleton. *Trends Cell Biol.* 3: 113–117. [R]

Schafer, D. A. and J. A. Cooper. 1995. Control of actin assembly at filament ends. *Ann. Rev. Cell Dev. Biol.* 11: 497–518. [R]

Stossel, T. P. 1989. From signal to pseudopod: How cells control cytoplasmic actin assembly. *J. Biol. Chem.* 264: 18261–18264. [R]

Theriot, J. A. and T. J. Mitchison. 1993. The three faces of profilin. *Cell* 75: 835–838. [R]

### Actin, Myosin, and Cell Movement

Condeelis, J. 1993. Life at the leading edge: The formation of cell protrusions. *Ann. Rev. Cell Biol.* 9: 411–444. [R]

Endow, S. A. and M. A. Titus. 1992. Genetic approaches to molecular motors. *Ann. Rev. Cell Biol.* 8: 29–66. [R]

Finer, J. T., R. M. Simmons and J. A. Spudich. 1994. Single myosin molecule mechanics: Piconewton forces and nanometre steps. *Nature* 368: 113–119. [P]

Huxley, A. F. and R. Niedergerke. 1954. Interference microscopy of living muscle fibres. *Nature* 173: 971–973. [P]

Huxley, H. E. 1969. The mechanism of muscle contraction. *Science* 164: 1356–1366. [R]

Huxley, H.E. and J. Hanson. 1954. Changes in the cross-striations of muscle contraction and their structural interpretation. *Nature* 173: 973–976. [P]

Lauffenburger, D. A. and A. F. Horwitz. 1996. Cell migration: A physically integrated molecular process. *Cell* 84: 359–369. [R]

Lee, J., A. Ishihara and K. Jacobson. 1993. How do cells move along surfaces? *Trends Cell Biol.* 3: 366–370. [R]

Mitchison, T. J. and L. P. Cramer. 1996. Actin-based cell motility and cell locomotion. *Cell* 84: 371–379. [R]

Mooseker, M. S. and R. E. Cheney. 1995. Unconventional myosins. *Ann. Rev. Cell Dev. Biol.* 11: 633–675. [R]

Rayment, I. and H. M. Holden. 1994. The three-dimensional structure of a molecular motor. *Trends Biochem. Sci.* 19: 129–134. [R]

Rayment, I., H. M. Holden, M. Whittaker, C. B. Yohn, M. Lorenz, K. C. Kolmes and R. A. Milligan. 1993. Structure of the actin-myosin complex and its implications for muscle contraction. *Science* 261: 58–65. [P]

Rayment, I., W. R. Rypniewski, K. Schmidt-Base, R. Smith, D. R. Tomchick, M. M. Benning, D. A. Winkelmann, G. Wesenberg and H. M. Holden. 1993. Three-dimensional structure of myosin subfragment-1: A molecular motor. *Science* 261: 50–58. [P]

Spudich, J. A. 1994. How molecular motors work. *Nature* 372: 515– 518. [R]

Stossel, T. P. 1993. On the crawling of animal cells. *Science* 260: 1086–1094. [R]

Tan, J. L., S. Ravid and J. A. Spudich. 1992. Control of nonmuscle myosins by phosphorylation. *Ann. Rev. Biochem.* 61: 721–759. [R]

Trinick, J. 1994. Titin and nebulin: Protein rulers in muscle? *Trends Biochem. Sci.* 19: 405–409. [R]

Vale, R. D. 1994. Getting a grip on myosin. *Cell* 78: 733–737. [R]

Zot, A. S. and J. D. Potter. 1987. Structural aspects of troponin-tropomyosin regulation of skeletal muscle contraction. *Ann. Rev. Biophys. Biophys. Chem.* 16: 535–559. [R]

## Intermediate Filaments

Bonifas, J. M., A. L. Rothman and E. H. Epstein Jr. 1991. Epidermolysis bullosa simplex: Evidence in two families for keratin gene abnormalities. *Science* 254: 1202–1205. [P]

Brown, R. H., Jr. 1995. Amyotrophic lateral sclerosis: Recent insights from genetics and transgenic mice. *Cell* 80: 687–692. [R]

Coulombe, P. A., M. E. Hutton, A. Letai, A. Hebert, A. S. Paller and E. Fuchs. 1991. Point mutations in human keratin 14 genes of epidermolysis bullosa simplex patients: Genetic and functional analyses. *Cell* 66: 1301–1311. [P]

Fuchs, E. 1995. Keratins and the skin. *Ann. Rev. Cell Dev. Biol.* 11: 123–153. [R]

Fuchs, E., Y. Chan, A. S. Paller and Q.-C. Yu. 1994. Cracks in the foundation: Keratin filaments and genetic disease. *Trends Cell Biol.* 4: 321–326. [R]

Schwarz, M. A., K. Owaribe, J. Kartenbeck and W. W. Franke. 1990. Desmosomes and hemidesmosomes: Constitutive molecular components. *Ann. Rev. Cell Biol.* 6: 461–491. [R]

Skalli, O., Y.-H. Chou and R. D. Goldman. 1992. Intermediate filaments: Not so tough after all. *Trends Cell Biol.* 2: 308–312. [R]

Steinert, P. M. and D. R. Roop. 1988. Molecular and cellular biology of intermediate filaments. *Ann. Rev. Biochem.* 57: 593–625. [R]

Vassar, R., P. A. Coulombe, L. Degenstein, K. Albers and E. Fuchs. 1991. Mutant keratin expression in transgenic mice causes marked abnormalities resembling a human genetic skin disease. *Cell* 64: 365–380. [P]

## Microtubules

Brinkley, R. R. 1985. Microtubule organizing centers. *Ann. Rev. Cell Biol.* 1: 145–172. [R]

Caceres, A. and K. S. Kosik. 1990. Inhibition of neurite polarity by tau antisense oligonucleotides in cerebellar neurons. *Nature* 343: 461–463 [P]

Gelfand, V. I. and A. D. Bershadsky. 1991. Microtubule dynamics: Mechanism, regulation, and function. *Ann. Rev. Cell Biol.* 7: 93–116. [R]

Hyman, A. A. and E. Karsenti. 1996. Morphogenetic properties of microtubules and mitotic spindle assembly. *Cell* 84: 401–410 [R].

Kalt, A. and M. Schliwa. 1993. Molecular components of the centrosome. *Trends Cell Biol.* 3: 118–128. [R]

Kellogg, D. R., M. Moritz and B. M. Alberts. 1994. The centrosome and cellular organization. *Ann. Rev. Biochem.* 63: 639–674. [R]

Kirschner, M. and T. Mitchison. 1986. Beyond self-assembly: From microtubules to morphogenesis. *Cell* 45: 329–342. [R]

Lee, G. 1993. Non-motor microtubule-associated proteins. *Curr. Opin. Cell Biol.* 5: 88–94. [R]

Mitchison, T. J. 1992. Compare and contrast actin filaments and microtubules. *Mol. Biol. Cell* 3: 1309–1315. [R]

Mitchison, T. and M. Kirschner. 1984. Dynamic instability of microtubule growth. *Nature* 312: 237–242. [P]

Mitchison, T. and M. Kirschner. 1984. Microtubule assembly nucleated by isolated centrosomes. *Nature* 312: 232–237. [P]

Oakley, B. R., C. E. Oakley, Y. Yoon and M. K. Jung. 1990. γ-Tubulin is a component of the spindle pole body that is essential for microtubule function in *Aspergillus nidulans. Cell* 61: 1289–1301. [P]

Osborn, M. and K. Weber. 1976. Cytoplasmic microtubules in tissue culture cells appear to grow from an organizing structure towards the plasma membrane. *Proc. Natl. Acad. Sci. USA* 73: 867–871. [P]

Zheng, Y., M. L. Wong, B. Alberts and T. Mitchison. 1995. Nucleation of microtubule assembly by a γ-tubulin-containing ring complex. *Nature* 378: 578–583. [P]

## Microtubule Motors and Movements

Asai, D. J. and C. J. Brokaw. 1993. Dynein heavy chain isoforms and axonemal motility. *Trends Cell Biol.* 3: 398–403. [R]

Block, S. M. 1995. Nanometres and piconewtons: The macromolecular mechanics of kinesin. *Trends Cell Biol.* 5: 169–175. [R]

Brady, S. T. 1985. A novel brain ATPase with properties expected for the fast axonal motor. *Nature* 317: 73–75. [P]

Brady, S. T. 1995. A kinesin medley: Biochemical and functional heterogeneity. *Trends Cell Biol.* 5: 159–164. [R]

Brady, S. T., R. J. Lasek and R. D. Allen. 1982. Fast axonal transport in extruded axoplasm from squid giant axon. *Science* 218: 1129–1131. [P]

Desai, A. and T. J. Mitchison. 1995. A new role for motor proteins as couplers to depolymerizing microtubules. *J. Cell Biol.* 128: 1–4. [R]

Endow, S. A. and M. A. Titus. 1992. Genetic approaches to molecular motors. *Ann. Rev. Cell Biol.* 8: 29–66. [R]

Fuller, M. T. and P. G. Wilson. 1992. Force and counterforce in the mitotic spindle. *Cell* 71: 547–550. [R]

Gibbons, I. R. 1981. Cilia and flagella of eukaryotes. *J. Cell Biol.* 91: 107s–124s. [R]

Gibbons, I. R. and A. Rowe. 1965. Dynein: A protein with adenosine triphosphatase activity from cilia. *Science* 149: 424–426. [P]

Haimo, L. T. 1995. Regulation of kinesin-directed movements. *Trends Cell Biol.* 5: 165–168. [R]

Holzbaur, E. L. F. and R. B. Vallee. 1994. Dyneins: Molecular structure and cellular function. *Ann. Rev. Cell Biol.* 10: 339–372. [R]

Kull, F. J., E. P. Sablin, R. Lau, R. J. Fletterick and R. D. Vale. 1996. Crystal structure of the kinesin motor domain reveals a structural similarity to myosin. *Nature* 380: 550–555. [P]

Lasek, R. J. and S. T. Brady. 1985. Attachment of transported vesicles to microtubules in axoplasm is facilitated by AMP-PNP. *Nature* 316: 645–647. [P]

Porter, M. E. and K. A. Johnson. 1991. Dynein structure and function. *Ann. Rev. Cell Biol.* 5: 119–151. [R]

Salmon, E. D. 1995. VE-DIC light microscopy and the discovery of kinesin. *Trends Cell Biol.* 5: 154–157. [R]

Saunders, W. S. 1993. Mitotic spindle pole separation. *Trends Cell Biol.* 3: 432–437. [R]

Vale, R. D. 1987. Intracellular transport using microtubule-based motors. *Ann. Rev. Cell Biol.* 3: 347–378. [R]

Vale, R. D. 1992. Microtubule motors: Many new models off the assembly line. *Trends Biochem. Sci.* 17: 300–304. [R]

Vale, R. D., T. S. Reese and M. P. Sheetz. 1985. Identification of a novel force-generating protein, kinesin, involved in microtubule-based motility. *Cell* 42: 39–50. [P]

Vallee, R. B. and M. P. Sheetz. 1996. Targeting of motor proteins. *Science* 271: 1539–1544. [R]

Vallee, R. B. and H. S. Shpetner. 1990. Motor proteins of cytoplasmic microtubules. *Ann. Rev. Biochem.* 59: 909–932. [R]

Walker, R. A. and M. P. Sheetz. 1993. Cytoplasmic microtubule-associated motors. *Ann. Rev. Biochem.* 62: 429–451. [R]

# 12

# The Cell Surface

ALL CELLS—BOTH PROKARYOTIC AND EUKARYOTIC—are surrounded by a plasma membrane, which defines the boundary of the cell and separates its internal contents from the environment. By serving as a selective barrier to the passage of molecules, the plasma membrane determines the composition of the cytoplasm. This ultimately defines the very identity of the cell, so the plasma membrane is one of the most fundamental structures of cellular evolution. Indeed, as discussed in Chapter 1, the first cell is thought to have arisen by the enclosure of self-replicating RNA in a membrane of phospholipids.

The plasma membranes of present-day cells are composed of both lipids and proteins. The basic structure of the plasma membrane is the phospholipid bilayer, which is impermeable to most water-soluble molecules. The passage of ions and most biological molecules across the plasma membrane is therefore mediated by proteins, which are responsible for the selective traffic of molecules into and out of the cell. Other proteins of the plasma membrane control the interactions between cells of multicellular organisms and serve as sensors through which the cell receives signals from its environment. The plasma membrane thus plays a dual role: It both isolates the cytoplasm and mediates interactions between the cell and its environment.

## STRUCTURE OF THE PLASMA MEMBRANE

Like all other cellular membranes, the plasma membrane consists of both lipids and proteins. The fundamental structure of the membrane is the phospholipid bilayer, which forms a stable barrier between two aqueous compartments. In the case of the plasma membrane, these compartments are the inside and the outside of the cell. Proteins embedded within the phospholipid bilayer carry out the specific functions of the plasma membrane, including selective transport of molecules and cell-cell recognition.

### The Phospholipid Bilayer

The plasma membrane is the most thoroughly studied of all cell membranes, and it is largely through investigations of the plasma membrane

467

that our current concepts of membrane structure have evolved. The plasma membranes of mammalian red blood cells (erythrocytes) have been particularly useful as a model for studies of membrane structure. Mammalian red blood cells do not contain nuclei or internal membranes, so they represent a source from which pure plasma membranes can be easily isolated for biochemical analysis. Indeed, studies of the red blood cell plasma membrane provided the first evidence that biological membranes consist of lipid bilayers. In 1925, two Dutch scientists (E. Gorter and R. Grendel) extracted the membrane lipids from a known number of red blood cells, corresponding to a known surface area of plasma membrane. They then determined the surface area occupied by a monolayer of the extracted lipid spread out at an air-water interface. The surface area of the lipid monolayer turned out to be twice that occupied by the erythrocyte plasma membranes, leading to the conclusion that the membranes consisted of lipid bilayers rather than monolayers.

The bilayer structure of the erythrocyte plasma membrane is clearly evident in high-magnification electron micrographs (Figure 12.1). The plasma membrane appears as two dense lines separated by an intervening space—a morphology frequently referred to as a "railroad track" appearance. This image results from the binding of the electron-dense heavy metals used as stains in transmission electron microscopy (see Chapter 1) to the polar head groups of the phospholipids, which therefore appear as dark lines. These dense lines are separated by the lightly stained interior portion of the membrane, which contains the hydrophobic fatty acid chains.

As discussed in Chapter 2, the plasma membranes of animal cells contain four major phospholipids (**phosphatidylcholine**, **phosphatidylethanolamine**, **phosphatidylserine**, and **sphingomyelin**), which together account for more than half of the lipid in most membranes. These phospholipids are asymmetrically distributed between the two halves of the membrane bilayer (Figure 12.2). The outer leaflet of the plasma membrane consists mainly of phosphatidylcholine and sphingomyelin, whereas phosphatidylethanolamine and phosphatidylserine are the predominant phospholipids of the inner leaflet. A fifth phospholipid, **phosphatidylinositol**, is also localized to the inner half of the plasma membrane. Although phosphatidylinositol is a quantitatively minor membrane component, it plays an important role in cell signaling, as discussed in the next chapter. The head groups of both phosphatidylserine and phosphatidylinositol are negatively charged, so their predominance in the inner leaflet results in a net

Membrane

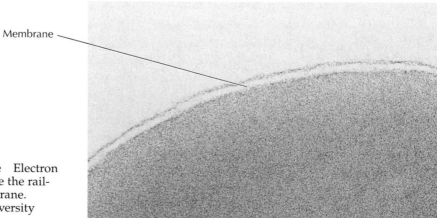

**Figure 12.1**
**Bilayer structure of the plasma membrane**   Electron micrograph of a human red blood cell. Note the railroad track appearance of the plasma membrane. (Courtesy of J. David Robertson, Duke University Medical Center.)

20 nm

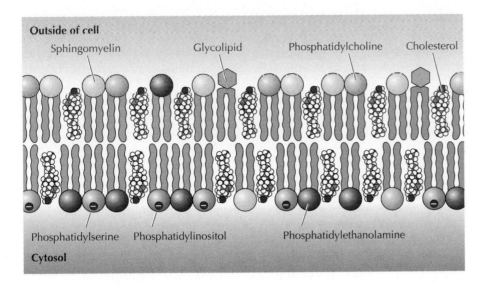

**Figure 12.2**
**Lipid components of the plasma membrane** The outer leaflet consists predominantly of phosphatidylcholine, sphingomyelin, and glycolipids, whereas the inner leaflet contains phosphatidylethanolamine, phosphatidylserine, and phosphatidylinositol. Cholesterol is distributed in both leaflets. The net negative charge of the head groups of phosphatidylserine and phosphatidylinositol is indicated. The structures of phospholipids, glycolipids, and cholesterol are shown in Figures 2.7, 2.8, and 2.9, respectively.

negative charge on the cytosolic face of the plasma membrane.

In addition to the phospholipids, the plasma membranes of animal cells contain **glycolipids** and **cholesterol**. The glycolipids are found exclusively in the outer leaflet of the plasma membrane, with their carbohydrate portions exposed on the cell surface. They are relatively minor membrane components, constituting only about 2% of the lipids of most plasma membranes. Cholesterol, on the other hand, is a major membrane constituent of animal cells, being present in about the same molar amounts as the phospholipids.

Two general features of phospholipid bilayers are critical to membrane function. First, the structure of phospholipids is responsible for the basic function of membranes as barriers between two aqueous compartments. Because the interior of the phospholipid bilayer is occupied by hydrophobic fatty acid chains, the membrane is impermeable to water-soluble molecules, including ions and most biological molecules. Second, bilayers of the naturally occurring phospholipids are viscous fluids, not solids. The fatty acids of most natural phospholipids have one or more double bonds, which introduce kinks into the hydrocarbon chains and make them difficult to pack together. The long hydrocarbon chains of the fatty acids therefore move freely in the interior of the membrane, so the membrane itself is soft and flexible. In addition, both phospholipids and proteins are free to diffuse laterally within the membrane—a property that is critical for many membrane functions.

Because of its rigid ring structure, cholesterol plays a distinct role in membrane structure. Cholesterol will not form a membrane by itself, but inserts into a bilayer of phospholipids with its polar hydroxyl group close to the phospholipid head groups (see Figure 12.2). Depending on the temperature, cholesterol has distinct effects on membrane fluidity. At high temperatures, cholesterol interferes with the movement of the phospholipid fatty acid chains, making the outer part of the membrane less fluid and reducing its permeability to small molecules. At low temperatures, however, cholesterol has the opposite effect: By interfering with interactions between fatty acid chains, cholesterol prevents membranes from freezing and maintains membrane fluidity. Although cholesterol is not present in bacteria, it is an essential component of animal cell plasma membranes. Plant cells also lack cholesterol, but they contain related compounds (sterols) that fulfill a similar function.

While cholesterol is required for animal cell growth, it can also play a highly deleterious role in human health. High levels of serum cholesterol can contribute to the development of plaques on the walls of arteries, resulting in arterial blockage with the potentially lethal consequences of heart attacks and strokes. Cholesterol thus plays a dual role in human physiology: On the one hand, it is required for membrane function; on the other hand, it contributes to the development of cardiovascular diseases that represent major causes of death in the United States.

### Membrane Proteins

While lipids are the fundamental structural elements of membranes, proteins are responsible for carrying out specific membrane functions. Most plasma membranes consist of approximately 50% lipid and 50% protein by weight, with the carbohydrate portions of glycolipids and glycoproteins constituting 5 to 10% of the membrane mass. Since proteins are much larger than lipids, this percentage corresponds to about one protein molecule per every 50 to 100 molecules of lipid. In 1972, Jonathan Singer and Garth Nicolson proposed the **fluid mosaic model** of membrane structure, which is now generally accepted as the basic paradigm for the organization of all biological membranes. In this model, membranes are viewed as two-dimensional fluids in which proteins are inserted into lipid bilayers (Figure 12.3).

Singer and Nicolson distinguished two classes of membrane-associated proteins, which they called **peripheral** and **integral membrane proteins**. Peripheral membrane proteins were operationally defined as proteins that dissociate from the membrane following treatments with polar reagents, such as solutions of extreme pH or high salt concentration, that do not disrupt the phospholipid bilayer. Once dissociated from the membrane, peripheral membrane proteins are soluble in aqueous buffers. These proteins are not inserted into the hydrophobic interior of the lipid bilayer. Instead, they are indirectly associated with membranes through protein-

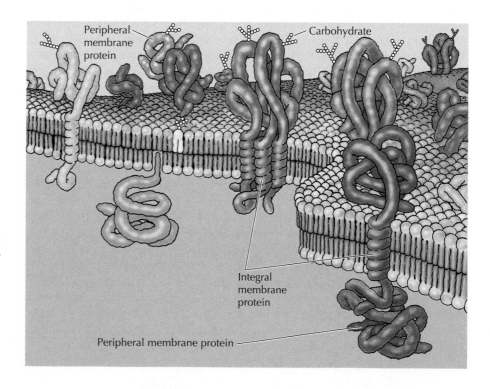

*Figure 12.3*
**Fluid mosaic model of the plasma membrane**   Integral membrane proteins are inserted into the lipid bilayer, whereas peripheral proteins are bound to the membrane indirectly by protein-protein interactions. Most integral membrane proteins are transmembrane proteins, with portions exposed on both sides of the lipid bilayer. The extracellular portions of these proteins are usually glycosylated, as are the peripheral membrane proteins bound to the external face of the membrane.

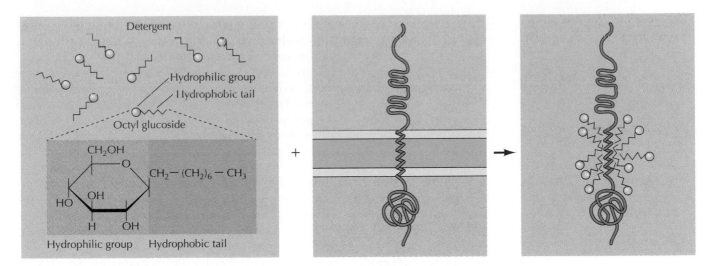

*Figure 12.4*
**Solubilization of integral membrane proteins by detergents**   Detergents (e.g., octyl glucoside) are amphipathic molecules containing hydrophilic head groups and hydrophobic tails. The hydrophobic tails bind to the hydrophobic regions of integral membrane proteins, forming detergent-protein complexes that are soluble in aqueous solution.

protein interactions. These interactions frequently involve ionic bonds, which are disrupted by extreme pH or high salt.

In contrast to the peripheral membrane proteins, integral membrane proteins can be released only by treatments that disrupt the phospholipid bilayer. Portions of these integral membrane proteins are inserted into the lipid bilayer, so they can be dissociated only by reagents that disrupt hydrophobic interactions. The most commonly used reagents for solubilization of integral membrane proteins are detergents, which are small amphipathic molecules containing both hydrophobic and hydrophilic groups (Figure 12.4). The hydrophobic portions of detergents displace the membrane lipids and bind to the hydrophobic portions of integral membrane proteins. Because the other end of the detergent molecule is hydrophilic, the detergent-protein complexes are soluble in aqueous solutions.

Many integral proteins are **transmembrane proteins**, which span the lipid bilayer with portions exposed on both sides of the membrane. These proteins can be visualized in electron micrographs of plasma membranes prepared by the freeze-fracture technique (see Figure 1.33). In these specimens, the membrane is split and separates into its two leaflets. Transmembrane proteins are then apparent as particles on the internal faces of the membrane (Figure 12.5).

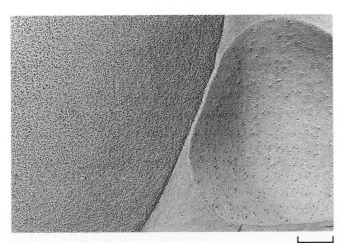

0.2 μm

*Figure 12.5*
**Freeze-fracture electron micrograph of human red blood cell membranes**   The particles in the membrane are transmembrane proteins. (Harold H. Edwards/Visuals Unlimited.)

The membrane-spanning portions of transmembrane proteins are usually $\alpha$ helices of 20 to 25 hydrophobic amino acids that are inserted into the membrane of the endoplasmic reticulum during synthesis of the polypeptide chain (see Figures 9.10, 9.11, and 9.12). These proteins are then transported in membrane vesicles from the endoplasmic reticulum to the Golgi apparatus, and from there to the plasma membrane. Carbohydrate groups are added to the polypeptide chains in both the endoplasmic reticulum and Golgi apparatus, so most transmembrane proteins of the plasma membrane are glycoproteins with their oligosaccharides exposed on the surface of the cell.

Studies of red blood cells have provided good examples of both peripheral and integral proteins associated with the plasma membrane. The membranes of human erythrocytes contain about a dozen major proteins, which were originally identified by gel electrophoresis of membrane preparations. Most of these are peripheral membrane proteins that have been identified as components of the cortical cytoskeleton, which underlies the plasma membrane and determines cell shape (see Chapter 11). For example, the most abundant peripheral membrane protein of red blood cells is spectrin, which is the major cytoskeletal protein of erythrocytes. Other peripheral membrane proteins of red blood cells include actin, ankyrin, and band 4.1. Ankyrin serves as the principal link between the plasma membrane and the cytoskeleton by binding to both spectrin and the integral membrane protein band 3 (see Figure 11.11). An additional link between the membrane and the cytoskeleton may be provided by band 4.1, which binds to the junctions of spectrin and actin, as well as to both band 3 and glycophorin (the other major integral membrane protein of erythrocytes).

The two major integral membrane proteins of red blood cells, glycophorin and band 3, provide well-studied examples of transmembrane protein structure (Figure 12.6). Glycophorin is a small glycoprotein of 131 amino acids, with a molecular weight of about 30,000, half of which is protein and half carbohydrate. Glycophorin crosses the membrane with a single membrane-spanning $\alpha$ helix of 23 amino acids, with its glycosylated amino-terminal portion exposed on the cell surface. Although glycophorin was one of the first transmembrane proteins to be characterized, its precise function remains unknown. In contrast, the function of the other major transmembrane protein of red blood cells is well understood. This protein, originally known as band 3, is the anion transporter responsible for the passage of bicarbonate ($HCO_3^-$) and chloride ($Cl^-$) ions across the red blood cell membrane. The band 3 polypeptide chain is 929 amino acids and is

**Figure 12.6**
**Integral membrane proteins of red blood cells**  Glycophorin (131 amino acids) contains a single transmembrane $\alpha$ helix. It is heavily glyocosylated, with oligosaccharides attached to 16 sites on the extracellular portion of the polypeptide chain. Band 3 (929 amino acids) has multiple transmembrane $\alpha$ helices and is thought to cross the membrane 14 times.

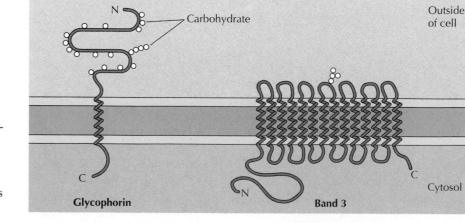

thought to have 14 membrane-spanning α-helical regions. Within the membrane, dimers of band 3 form globular structures containing internal channels through which ions are able to travel across the lipid bilayer.

Because of their amphipathic character, transmembrane proteins have proved difficult to crystallize, as required for three-dimensional structural analysis by X-ray diffraction. Consequently, the three-dimensional structures of only a few transmembrane proteins are known. The first transmembrane protein to be analyzed by X-ray crystallography was the photosynthetic reaction center of the bacterium *Rhodopseudomonas viridis*, whose structure was reported in 1985 (Figure 12.7). The reaction center contains three transmembrane proteins, designated L, M, and H (light, medium, and heavy) according to their apparent sizes indicated by gel electrophoresis. The L and M subunits each have five membrane-spanning α helices. The H subunit has only a single transmembrane α helix, with the bulk of the polypeptide chain on the cytosolic side of the membrane. The fourth subunit of the reaction center is a cytochrome, which is a peripheral membrane protein bound to the complex by protein-protein interactions.

Although most transmembrane proteins span the membrane by α-helical regions, this is not always the case. A well-characterized exception is provided by the **porins**—a class of proteins that form channels in the outer membranes of some bacteria. Many bacteria, including *E. coli*, have a dual membrane system in which the plasma membrane (or inner membrane) is surrounded by the cell wall and a distinct outer membrane (Figure 12.8). In contrast to the plasma membrane, the outer membrane is highly permeable to ions and small polar molecules (in the case of *E. coli*, with molecular weights up to 600). This permeability results from the porins, which form open aqueous channels through the lipid bilayer. As discussed in Chapter 10, proteins related to the bacterial porins are also found in the outer membranes of mitochondria and chloroplasts.

Structural analysis has indicated that the porins do not contain hydrophobic α-helical regions. Instead, they cross the membrane as β barrels, in which 16 β sheets fold up into a barrel-like structure enclosing an aqueous pore (Figure 12.9). The side chains of polar amino acids line the pore, whereas side chains of hydrophobic amino acids interact with the interior of the membrane. The porin monomers associate to form stable trimers, each of which contains three open channels through which polar molecules

*Figure 12.7*
**A bacterial photosynthetic reaction center** The reaction center consists of three transmembrane proteins, designated L (red), M (yellow), and H (green). The L and M subunits each have five transmembrane α helices, whereas the H subunit has only one. The fourth subunit of the reaction center is a cytochrome (white), which is a peripheral membrane protein.

*Figure 12.9*
**Structure of a porin monomer** Each monomer is a β barrel consisting of 16 antiparallel β strands (arrows). The top end of the molecule faces the external medium. (From H. Nikaido, 1994, *J. Biol. Chem.* 269: 3905.)

*Figure 12.8*
**Bacterial outer membranes** The plasma membrane of some bacteria is surrounded by a cell wall and a distinct outer membrane. The outer membrane contains porins, which form open aqueous channels allowing the free passage of ions and small molecules.

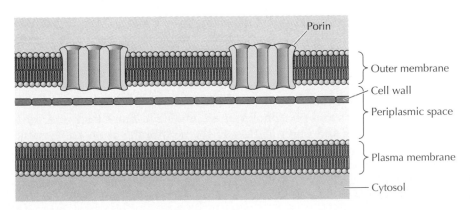

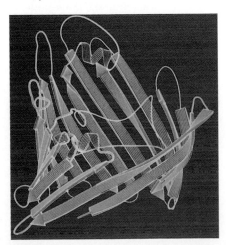

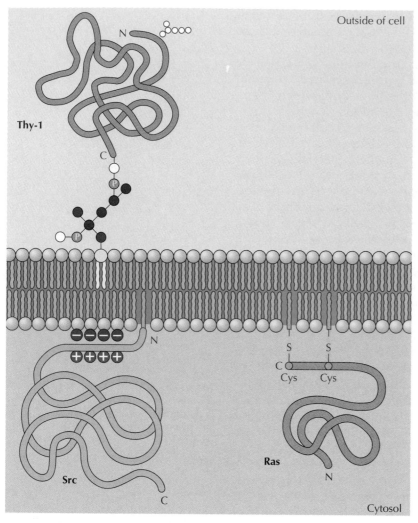

Outside of cell

Cytosol

**Figure 12.10**
**Examples of proteins anchored in the plasma membrane by lipids and glycolipids**  Some proteins (e.g., the lymphocyte protein Thy-1) are anchored in the outer leaflet of the plasma membrane by GPI anchors added to their C terminus in the endoplasmic reticulum. These proteins are glycosylated and exposed on the cell surface. Other proteins are anchored in the inner leaflet of the plasma membrane following their translation on free cytosolic ribosomes. The Ras protein illustrated is anchored by a prenyl group attached to the side chain of a C-terminal cysteine and by a palmitoyl group attached to a cysteine located five amino acids upstream. The Src protein is anchored by a myristoyl group attached to its N terminus. A positively charged region of Src also plays a role in membrane association, perhaps by interacting with the negatively charged head groups of phosphatidylserine. The structures of these lipid and glycolipid groups are illustrated in Figures 7.29 through 7.32.

can diffuse across the membrane.

In contrast to transmembrane proteins, a variety of proteins (many of which behave as integral membrane proteins) are anchored in the plasma membrane by covalently attached lipids or glycolipids (Figure 12.10). Members of one class of these proteins are inserted into the outer leaflet of the plasma membrane by **glycosylphosphatidylinositol (GPI) anchors**. GPI anchors are added to certain proteins that have been transferred into the endoplasmic reticulum and are anchored in the membrane by a C-terminal transmembrane region (see Figure 9.15). The transmembrane region is cleaved as the GPI anchor is added, so these proteins remain attached to the membrane only by the glycolipid. Since the polypeptide chains of GPI-anchored proteins are transferred into the endoplasmic reticulum, they are glycosylated and exposed on the surface of the cell following transport to the plasma membrane.

Other proteins are anchored in the inner leaflet of the plasma membrane by covalently attached lipids. Rather than being processed through the secretory pathway, these proteins are synthesized on free cytosolic ribosomes and then modified by the addition of lipids. These modifications include the addition of myristic acid (a 14-carbon fatty acid) to the amino terminus of the polypeptide chain, the addition of palmitic acid (16 carbons) to the side chains of cysteine residues, and the addition of prenyl groups (15 or 20 carbons) to the side chains of carboxy-terminal cysteine residues (see Figures 7.29, 7.30, and 7.31). In some cases, these proteins (many of which behave as peripheral membrane proteins) are targeted to the plasma membrane by positively charged regions of the polypeptide chain as well as by the attached lipids. These positively charged protein domains may interact with the negatively charged head groups of phosphatidylserine on the cytosolic face of the plasma membrane. It is noteworthy that many of the proteins anchored in the inner leaflet of the plasma membrane (including the Src and Ras proteins illustrated in Figure 12.10) play important roles in the transmission of signals from cell surface receptors to intracellular targets, as discussed in the next chapter.

## Mobility of Membrane Proteins

Membrane proteins and phospholipids are unable to move back and forth between the inner and outer leaflets of the membrane at an appreciable rate. However, because they are inserted into a fluid lipid bilayer, both proteins and lipids are able to diffuse laterally through the membrane. This lateral movement was first shown directly in an experiment reported by Larry Frye and Michael Edidin in 1970, which provided support for the fluid mosaic model. Frye and Edidin fused human and mouse cells in culture to produce human-mouse cell hybrids (Figure 12.11). They then analyzed the distribution of proteins in the membranes of these hybrid cells using antibodies that specifically recognize proteins of human and mouse origin. These antibodies were labeled with different fluorescent dyes, so the human and mouse proteins could be distinguished by fluorescence microscopy. Immediately after fusion, human and mouse proteins were localized to different halves of the hybrid cells. However, after a brief period of incubation at 37°C, the human and mouse proteins were completely intermixed over the cell surface, indicating that they moved freely through the plasma membrane.

However, not all proteins are able to diffuse freely through the membrane. In some cases, the mobility of membrane proteins is restricted by their association with the cytoskeleton. For example, a fraction of band 3 in the red blood cell membrane is immobilized as a result of its association with ankyrin and spectrin. In other cases, the mobility of membrane proteins may be restricted by their associations with other membrane proteins, with proteins on the surface of adjacent cells, or with the extracellular matrix.

In contrast to blood cells, epithelial cells are polarized when they are organized into tissues, with different parts of the cell responsible for performing distinct functions. Consequently, the plasma membranes of many epithelial cells are divided into distinct **apical** and **basolateral domains** that differ in function and protein composition (Figure 12.12). For example, epithelial cells of the small intestine function to absorb nutrients from the digestive tract. The apical surface of these cells, which faces the intestinal lumen, is therefore covered by microvilli and specialized for nutrient absorption. The basolateral surface, which faces underlying connective tissue and the blood supply, is specialized to mediate the transfer of absorbed nutrients into the circulation. In order to maintain these distinct functions,

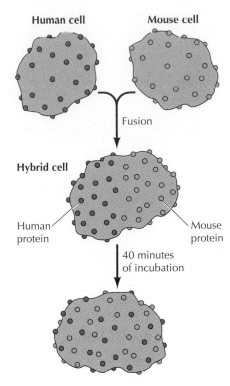

*Figure 12.11*
**Mobility of membrane proteins**
Human and mouse cells were fused to produce hybrid cells. The distribution of cell surface proteins was then analyzed using anti-human and anti-mouse antibodies labeled with different fluorescent dyes (red and green, respectively). The human and mouse proteins were detected in different halves of the hybrid cells immediately after fusion but had intermingled over the cell surface following 40 minutes of incubation.

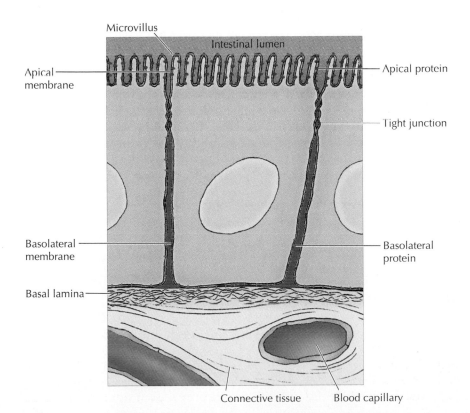

*Figure 12.12*
**A polarized intestinal epithelial cell**
The apical surface of the cell contains microvilli and is specialized for absorption of nutrients from the intestinal lumen. The basolateral surface is specialized for the transfer of absorbed nutrients to the underlying connective tissue, which contains blood capillaries. Tight junctions separate the apical and basolateral domains of the plasma membrane. Membrane proteins are free to diffuse within each domain but are not able to cross from one domain to the other.

the mobility of plasma membrane proteins must be restricted to the appropriate domains of the cell surface. At least part of the mechanism by which this occurs involves the formation of tight junctions (which are discussed later in this chapter) between adjacent cells of the epithelium. These junctions not only seal the space between cells but also serve as barriers to the movement of membrane lipids and proteins. As a result, proteins are able to diffuse within either the apical or basolateral domains of the plasma membrane but are not able to cross from one domain to the other.

### The Glycocalyx

As already discussed, the extracellular portions of plasma membrane proteins are generally glycosylated. Likewise, the carbohydrate portions of glycolipids are exposed on the outer face of the plasma membrane. Consequently, the surface of the cell is covered by a carbohydrate coat, known as the **glycocalyx**, formed by the oligosaccharides of glycolipids and transmembrane glycoproteins (Figure 12.13).

Part of the role of the glycocalyx is to protect the cell surface. In addition, the oligosaccharides of the glycocalyx serve as markers for a variety of cell-cell interactions. A well-studied example of these interactions is the adhesion of white blood cells (leukocytes) to the endothelial cells that line blood vessels—a process that allows the leukocytes to leave the circulatory system and mediate the inflammatory response in injured tissues. The initial step in adhesion between leukocytes and endothelial cells is mediated by a family of transmembrane proteins called **selectins**, which recognize specific carbohydrates on the cell surface (Figure 12.14). Two members of the selectin family (E-selectin and P-selectin) are expressed by endothelial cells and bind to specific oligosaccharides expressed on the surface of leukocytes. A different selectin (L-selectin) is expressed by leukocytes and recognizes an oligosaccharide on the surface of endothelial cells. The oligosaccharides exposed on the cell surface thus provide a set of markers that help identify the distinct cell types of multicellular organisms.

## TRANSPORT OF SMALL MOLECULES

The internal composition of the cell is maintained because the plasma membrane is selectively permeable to small molecules. Most biological molecules are unable to diffuse through the phospholipid bilayer, so the plasma membrane forms a barrier that blocks the free exchange of molecules between the cytoplasm and the external environment of the cell. Specific transport proteins (carrier proteins and channel proteins) then mediate

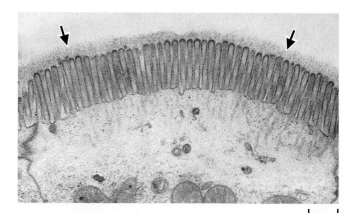

*Figure 12.13*
**The glycocalyx** An electron micrograph of intestinal epithelium illustrating the glycocalyx (arrows). (Don Fawcett/ Visuals Unlimited.)

1 μm

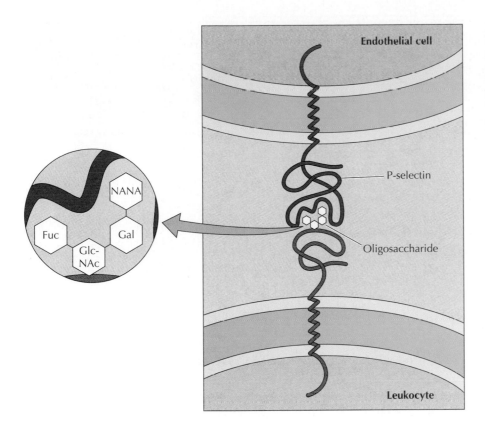

*Figure 12.14*
**Binding of selectins to oligosaccharides**   P-selectin is a transmembrane protein expressed by endothelial cells that binds to an oligosaccharide expressed on the surface of leukocytes. The oligosaccharide recognized by P-selectin contains *N*-acetylglucosamine (GlcNAc), fucose (Fuc), galactose (Gal), and sialic acid (*N*-acetylneuraminic acid, NANA).

the selective passage of small molecules across the membrane, allowing the cell to control the composition of its cytoplasm.

## Passive Diffusion

The simplest mechanism by which molecules can cross the plasma membrane is **passive diffusion**. During passive diffusion, a molecule simply dissolves in the phospholipid bilayer, diffuses across it, and then dissolves in the aqueous solution at the other side of the membrane. No membrane proteins are involved and the direction of transport is determined simply by the relative concentrations of the molecule inside and outside of the cell. The net flow of molecules is always down their concentration gradient—from a compartment with a high concentration to one with a lower concentration of the molecule.

Passive diffusion is thus a nonselective process by which any molecule able to dissolve in the phospholipid bilayer is able to cross the plasma membrane and equilibrate between the inside and outside of the cell. Importantly, only small, relatively hydrophobic molecules are able to diffuse across a phospholipid bilayer at significant rates (Figure 12.15). Thus, gases (such as $O_2$ and $CO_2$), hydrophobic molecules (such as benzene), and small polar but uncharged molecules (such as $H_2O$ and ethanol) are able to diffuse across the plasma membrane. Other biological molecules, however, are unable to dissolve in the hydrophobic interior of the phospholipid bilayer. Consequently, larger uncharged polar molecules such as glucose are unable to cross the plasma membrane by passive diffusion, as are charged molecules of any size (including small ions such as $H^+$, $Na^+$, $K^+$, and $Cl^-$). The passage of these molecules across the membrane instead requires the activity of specific transport and channel proteins, which therefore control the traffic of most biological molecules into and out of the cell.

*Figure 12.15*
**Permeability of phospholipid bilayers** Gases, hydrophobic molecules, and small polar uncharged molecules can diffuse through phospholipid bilayers. Larger polar molecules and charged molecules cannot.

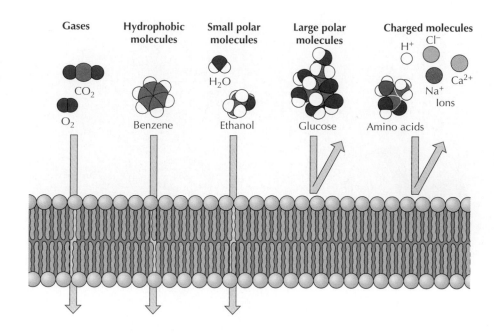

## Facilitated Diffusion and Carrier Proteins

**Facilitated diffusion**, like passive diffusion, involves the movement of molecules in the direction determined by their relative concentrations inside and outside of the cell. No external source of energy is provided, so molecules travel across the membrane in the direction determined by their concentration gradients and, in the case of charged molecules, by the electric potential across the membrane. However, facilitated diffusion differs from passive diffusion in that the transported molecules do not dissolve in the phospholipid bilayer. Instead, their passage is mediated by proteins that enable the transported molecules to cross the membrane without directly interacting with its hydrophobic interior. Facilitated diffusion therefore allows polar and charged molecules, such as carbohydrates, amino acids, nucleosides, and ions, to cross the plasma membrane.

Two classes of proteins that mediate facilitated diffusion are generally distinguished: carrier proteins and channel proteins. **Carrier proteins** bind specific molecules to be transported on one side of the membrane. They then undergo conformational changes that allow the molecule to pass through the membrane and be released on the other side. In contrast, **channel proteins** (see the next section) form open pores through the membrane, allowing the free diffusion of any molecule of the appropriate size and charge.

Carrier proteins are responsible for the facilitated diffusion of sugars, amino acids, and nucleosides across the plasma membranes of most cells. The uptake of glucose, which serves as a primary source of metabolic energy, is one of the most important transport functions of the plasma membrane, and the glucose transporter provides a well-studied example of a carrier protein. The glucose transporter was initially identified as a 55-kd protein in human red blood cells, in which it represents approximately 5% of total membrane protein. Subsequent isolation and sequence analysis of a cDNA clone revealed that the glucose transporter has 12 $\alpha$-helical transmembrane segments—a structure typical of many carrier proteins (Figure 12.16). These transmembrane $\alpha$ helices contain predominantly hydrophobic amino acids, but several also contain polar amino acid residues that are thought to form the glucose-binding site in the interior of the protein.

As with most other membrane proteins, the three-dimensional structure

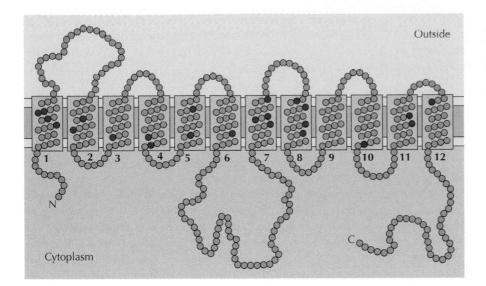

*Figure 12.16*
**Structure of the glucose transporter**
The glucose transporter has 12 trans-membrane $\alpha$ helices. Polar amino acid residues located within the phospholipid bilayer are indicated as dark purple circles. (Adapted from G. I. Bell, C. F. Burant, J. Takeda, and G. W. Gould, 1993, *J. Biol. Chem.* 268: 19161.)

of the glucose transporter is not known, so the molecular mechanism of glucose transport remains an open question. However, kinetic studies indicate that the glucose transporter functions by alternating between two conformational states (Figure 12.17). In the first conformation, a glucose-binding site faces the outside of the cell. The binding of glucose to this exterior site induces a conformational change in the transporter, such that the glucose-binding site now faces the interior of the cell. Glucose can then be released into the cytosol, followed by the return of the transporter to its original conformation.

Most cells, including erythrocytes, are exposed to extracellular glucose concentrations that are higher than those inside the cell, so facilitated diffusion results in the net inward transport of glucose. Once glucose is taken up by these cells it is rapidly metabolized, so intracellular glucose concentrations remain low and glucose continues to be transported into the cell from the extracellular fluids. Because the conformational changes of the

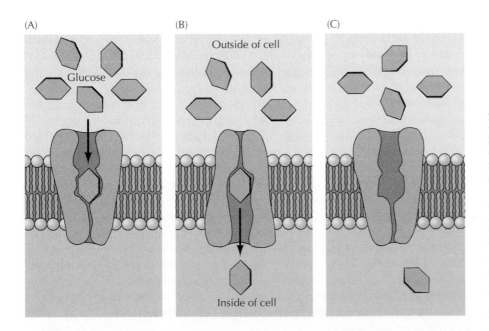

*Figure 12.17*
**Model for the facilitated diffusion of glucose** The glucose transporter alternates between two conformations in which a glucose-binding site is alternately exposed on the outside and the inside of the cell. In the first conformation shown (A), glucose binds to a site exposed on the outside of the plasma membrane. The transporter then undergoes a conformational change such that the glucose-binding site faces the inside of the cell and glucose is released into the cytosol (B). The transporter then returns to its original conformation (C).

glucose transporter are reversible, however, glucose can be transported in the opposite direction simply by reversing the steps in Figure 12.17. Such reverse flow occurs, for example, in liver cells, in which glucose is synthesized and released into the circulation.

### Ion Channels

In contrast to carrier proteins, channel proteins simply form open pores in the membrane, allowing small molecules of the appropriate size and charge to pass freely through the lipid bilayer. One group of channel proteins, discussed earlier, is the porins, which permit the free passage of ions and small polar molecules through the outer membranes of bacteria (see Figure 12.8). Channel proteins also permit the passage of molecules between cells connected at gap junctions, which are discussed later in the chapter. The plasma membranes of many cells also contain water channel proteins (aquaporins), through which water molecules are able to cross the membrane much more rapidly than they can diffuse through the phospholipid bilayer. The best-characterized channel proteins, however, are the **ion channels**, which mediate the passage of ions across plasma membranes. Although ion channels are present in the membranes of all cells, they have been especially well studied in nerve and muscle, where their regulated opening and closing is responsible for the transmission of electric signals.

Three properties of ion channels are central to their function (Figure 12.18). First, transport through channels is extremely rapid. More than a million ions per second flow through open channels—a flow rate approximately a thousand times greater than the rate of transport by carrier proteins. Second, ion channels are highly selective because narrow pores in the channel restrict passage to ions of the appropriate size and charge. Thus, specific channel proteins allow the passage of $Na^+$, $K^+$, $Ca^{2+}$, and $Cl^-$ across the membrane. Third, most ion channels are not permanently open. Instead, the opening of ion channels is regulated by "gates" that transiently open in response to specific stimuli. Some channels (called **ligand-gated channels**) open in response to the binding of neurotransmitters or other signaling molecules; others (**voltage-gated channels**) open in response to changes in electric potential across the plasma membrane.

The fundamental role of ion channels in the transmission of electric impulses was elucidated through a series of elegant experiments reported by Alan Hodgkin and Andrew Huxley in 1952. These investigators used the giant nerve cells of the squid as a model. The axons of these giant neurons have a diameter of about 1 mm, making it possible to insert electrodes and measure the changes in membrane potential that take place during the transmission of nerve impulses. Using this approach, Hodgkin and Huxley demonstrated that these changes in membrane potential result from the regulated opening and closing of $Na^+$ and $K^+$ channels in the plasma membrane. It subsequently became possible to study the activity of individual ion channels, using the **patch clamp technique** developed by Erwin Neher and Bert Sakmann in 1976 (Figure 12.19). In this method, a micropipette with a tip diameter of about 1 $\mu$m is used to isolate a small patch of membrane, allowing the flow of ions through a single channel to be analyzed and greatly increasing the precision with which the activities of ion channels can be studied.

The flow of ions through membrane channels is dependent on the establishment of ion gradients across the plasma membrane. All cells, including nerve and muscle, contain ion pumps (discussed in the next section) that use energy derived from ATP hydrolysis to actively transport ions across the plasma membrane. As a result, the ionic composition of the

*Figure 12.18*
**Model of an ion channel**   In the closed conformation, the flow of ions is blocked by a gate. Opening of the gate allows ions to flow rapidly through the channel. The channel contains a narrow pore that restricts passage to ions of the appropriate size and charge.

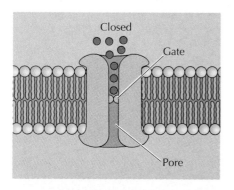

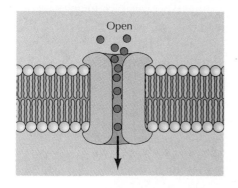

cytoplasm is substantially different from that of extracellular fluids (Table 12.1). For example, Na$^+$ is actively pumped out of cells while K$^+$ is pumped in. In the squid axon, therefore, the concentration of Na$^+$ is about 10 times higher in extracellular fluids than inside the cell, whereas the concentration of K$^+$ is approximately 20 times higher in the cytosol than in the surrounding medium.

Because ions are electrically charged, their transport results in the establishment of an electric gradient across the plasma membrane. With resting squid axons there is an electric potential of about 60 mV across the plasma membrane, with the inside of the cell negative with respect to the outside (Figure 12.20). This electric potential arises both from ion pumps and from the flow of ions through channels that are open in the resting cell plasma membrane. The plasma membrane of resting squid axons contains open K$^+$ channels, so it is more permeable to K$^+$ than to Na$^+$ or other ions. Consequently, the flow of K$^+$ makes the largest contribution to the resting membrane potential.

As discussed in Chapter 10, the flow of ions across a membrane is driven by both the concentration and voltage components of an electrochemical gradient. For example, the 20-fold higher concentration of K$^+$ inside the squid axon as compared to the extracellular fluid drives the flow of K$^+$ out of the cell. However, because K$^+$ is positively charged, this efflux of K$^+$ from the cell generates an electric potential across the membrane, with the inside of the cell becoming negatively charged. This membrane potential opposes the continuing flow of K$^+$ out of the cell, and the system approaches the equilibrium state, in which the membrane potential balances the K$^+$ concentration gradient.

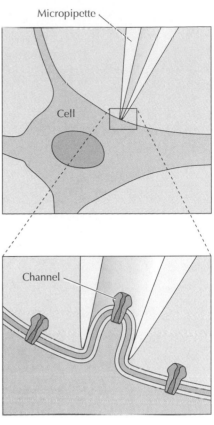

*Figure 12.19*
**The patch clamp technique** A small patch of membrane is isolated in the tip of a micropipette. Stimuli can then be applied from within the pipette, allowing the behavior of the trapped channel to be measured. (Adapted from E. Neher and B. Sakmann, 1992, *Sci. Am.* 266(3):44.)

*Figure 12.20*
**Ion gradients and resting membrane potential of the giant squid axon** Only the concentrations of Na$^+$ and K$^+$ are shown, because these are the ions that function in the transmission of nerve impulses. Na$^+$ is pumped out of the cell while K$^+$ is pumped in, so the concentration of Na$^+$ is higher outside than inside of the axon, whereas the concentration of K$^+$ is higher inside than out. The resting membrane is more permeable to K$^+$ than to Na$^+$ or other ions because it contains open K$^+$ channels. The flow of K$^+$ through these channels makes the major contribution to the resting membrane potential of –60 mV, which is therefore close to the K$^+$ equilibrium potential.

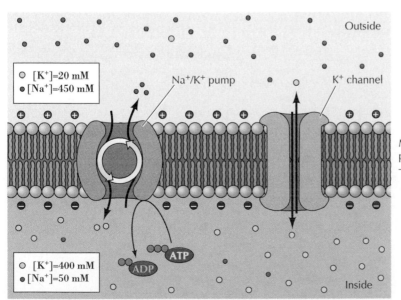

**Table 12.1** Extracellular and Intracellular Ion Concentrations

| | Concentration (m*M*) | |
| --- | --- | --- |
| **Ion** | **Intracellular** | **Extracellular** |
| **Squid axon** | | |
| K$^+$ | 400 | 20 |
| Na$^+$ | 50 | 440 |
| Cl$^-$ | 40–150 | 560 |
| Ca$^{2+}$ | 0.0001 | 10 |
| **Mammalian cell** | | |
| K$^+$ | 140 | 5 |
| Na$^+$ | 5–15 | 145 |
| Cl$^-$ | 4 | 110 |
| Ca$^{2+}$ | 0.0001 | 2.5–5 |

Quantitatively, the relationship between ion concentration and membrane potential is given by the **Nernst equation**:

$$V = \frac{RT}{zF} \ln \frac{C_o}{C_i}$$

where $V$ is the equilibrium potential in volts, $R$ is the gas constant, $T$ is the absolute temperature, $z$ is the charge of the ion, $F$ is Faraday's constant, and $C_o$ and $C_i$ are the concentrations of the ion outside and inside of the cell, respectively. An equilibrium potential exists separately for each ion, and the membrane potential is determined by the flow of all the ions that cross the plasma membrane. However, because resting squid axons are more permeable to $K^+$ than to $Na^+$ or other ions (including $Cl^-$), the resting membrane potential (–60 mV) is close to the equilibrium potential determined by the intracellular and extracellular $K^+$ concentrations (–75 mV).

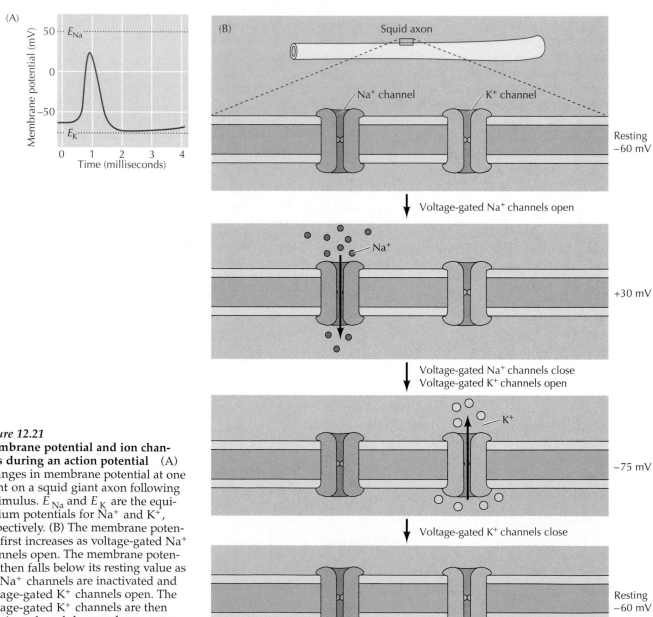

*Figure 12.21*
**Membrane potential and ion channels during an action potential** (A) Changes in membrane potential at one point on a squid giant axon following a stimulus. $E_{Na}$ and $E_K$ are the equilibrium potentials for $Na^+$ and $K^+$, respectively. (B) The membrane potential first increases as voltage-gated $Na^+$ channels open. The membrane potential then falls below its resting value as the $Na^+$ channels are inactivated and voltage-gated $K^+$ channels open. The voltage-gated $K^+$ channels are then inactivated, and the membrane potential returns to its resting value.

As nerve impulses (**action potentials**) travel along axons, the membrane depolarizes (Figure 12.21). The membrane potential changes from –60 mV to approximately +30 mV in less than a millisecond, after which it becomes negative again and returns to its resting value. These changes result from the rapid sequential opening and closing of voltage-gated Na⁺ and K⁺ channels. Relatively small initial changes in membrane potential (from –60 to about –40 mV) lead to the rapid opening of Na⁺ channels. This allows Na⁺ to flow into the cell, driven by both its concentration gradient and the membrane potential. The sudden entry of Na⁺ leads to a large change in membrane potential, which increases to nearly +30 mV, approaching the Na⁺ equilibrium potential of approximately +50 mV. At this time, the Na⁺ channels are inactivated and voltage-gated K⁺ channels open, substantially increasing the permeability of the membrane to K⁺. K⁺ then flows rapidly out of the cell, driven by both the membrane potential and the K⁺ concentration gradient, leading to a rapid decrease in membrane potential to about –75 mV (the K⁺ equilibrium potential). The voltage-gated K⁺ channels are then inactivated and the membrane potential returns to its resting level of –60 mV, determined by the flow of K⁺ and other ions through the channels that remain open in unstimulated cells.

Depolarization of adjacent regions of the plasma membrane allows action potentials to travel down the length of nerve cell axons as electric signals, resulting in the rapid transmission of nerve impulses over long distances. For example, the axons of human motor neurons can be more than a meter long. The arrival of action potentials at the terminus of most neurons then signals the release of neurotransmitters, such as acetylcholine, which carry signals between cells at a synapse (Figure 12.22). Neurotransmitters released from presynaptic cells bind to receptors on the membranes of postsynaptic cells, where they act to open ligand-gated ion channels. One of the best-characterized of these channels is the acetylcholine receptor of muscle cells. Binding of acetylcholine opens a channel that is permeable to both Na⁺ and K⁺. This permits the rapid influx of Na⁺, which depolarizes the muscle cell membrane and triggers an action potential. The action potential then results in the opening of voltage-gated Ca²⁺ channels, leading to the increase in intracellular Ca²⁺ that signals contraction (see Figure 11.25).

The acetylcholine receptor, initially isolated from the electric organ of

*Figure 12.22*
**Signaling by neurotransmitter release at a synapse** The arrival of a nerve impulse at the terminus of the neuron signals the fusion of synaptic vesicles with the plasma membrane, resulting in the release of neurotransmitter from the presynaptic cell into the synaptic cleft. The neurotransmitter binds to receptors and opens ligand-gated ion channels in the target cell plasma membrane.

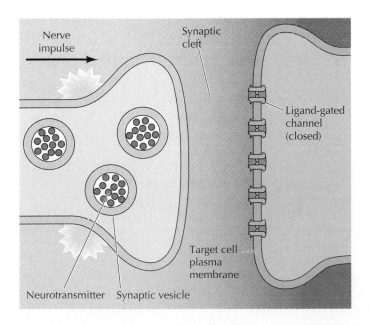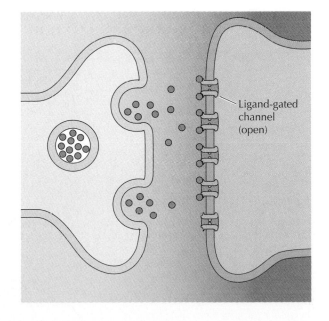

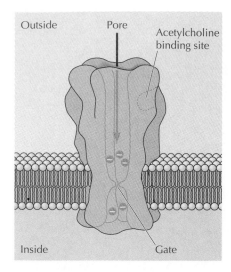

*Figure 12.23*
**Model of the acetylcholine receptor**
The receptor consists of five subunits arranged around a central pore. The binding of acetylcholine to a site in the extracellular region of the receptor induces allosteric changes that open the channel gate. The channel is lined by negatively charged amino acids that prevent the flow of negatively charged ions. (Adapted from N. Unwin, 1993, *Cell* 72/*Neuron* 10 (Suppl.):31.)

*Torpedo* rays in the 1970s, is the prototype of ligand-gated channels. The receptor consists of five subunits arranged as a cylinder in the membrane (Figure 12.23). In its closed state, the channel pore is thought to be blocked by the side chains of hydrophobic amino acids. The binding of acetylcholine induces a conformational change in the receptor such that these hydrophobic side chains shift out of the channel, opening a pore that allows the passage of positively charged ions, including $Na^+$ and $K^+$. However, the channel remains impermeable to negatively charged ions, such as $Cl^-$, because it is lined by negatively charged amino acids.

A greater degree of ion selectivity is displayed by the voltage-gated $Na^+$ and $K^+$ channels. $Na^+$ channels are more than ten times more permeable to $Na^+$ than to $K^+$, whereas $K^+$ channels are a hundred times more permeable to $K^+$ than to $Na^+$. The selectivity of the $Na^+$ channel can be explained, at least in part, on the basis of a narrow pore that acts as a size filter. The ionic radius of $Na^+$ is smaller than that of $K^+$, and it is thought that the $Na^+$ channel pore is narrow enough to interfere with the passage of $K^+$ or larger ions (Figure 12.24).

$K^+$ channels also have narrow pores, which prevent the passage of larger ions. However, since $Na^+$ has a smaller ionic radius, this does not account for the selective permeability of these channels to $K^+$. Instead, this selectivity is thought to result from interactions between $K^+$ and polar amino acid side chains lining the pore. These interactions displace the water molecules to which $K^+$ is bound, allowing dehydrated $K^+$ to pass through the pore (Figure 12.25). In contrast, a dehydrated $Na^+$ is too small to interact with these polar side chains. Consequently, $Na^+$ remains bound to water molecules in a hydrated complex that is too large to pass through the channel.

Voltage-gated $Na^+$, $K^+$, and $Ca^{2+}$ channels all belong to a large family of related proteins (Figure 12.26). $K^+$ channels consist of four identical subunits, each containing six transmembrane $\alpha$ helices. $Na^+$ and $Ca^{2+}$ channels consist of a single polypeptide chain, but each polypeptide contains four repeated domains that correspond to the $K^+$ channel subunits. Voltage gating is mediated by one of the transmembrane $\alpha$ helices, which contains multiple positively charged amino acids. Membrane depolarization induces the movement of these positive charges toward the outside of the cell, shifting the position of this transmembrane segment and opening the channel. Rapid inactivation of $Na^+$ and $K^+$ channels during the propagation of action potentials is then mediated by cytoplasmic portions of the polypeptide chain, which bind to the cytoplasmic mouth of the channel pore

*Figure 12.24*
**Ion selectivity of $Na^+$ channels** A narrow pore permits the passage of $Na^+$ bound to a single water molecule but interferes with the passage of $K^+$ or larger ions.

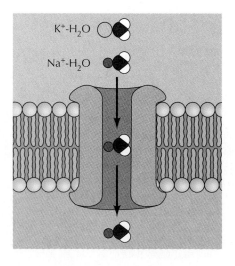
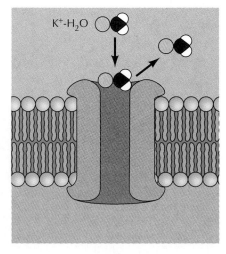

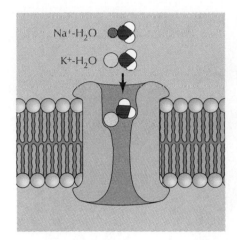

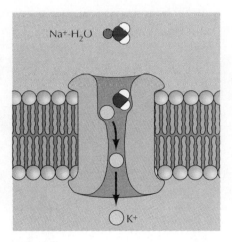

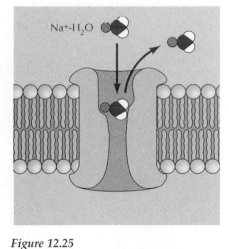

and prevent further ion flow (Figure 12.27).

A wide variety of ion channels (including $Ca^{2+}$ and $Cl^-$ channels) respond to different neurotransmitters or open and close with different kinetics following membrane depolarization. The concerted actions of these multiple channels are responsible for the complexities of signaling in the nervous system. Moreover, as discussed in the next chapter, the roles of ion channels are not restricted to the electrically excitable cells of nerve and muscle; they also play critical roles in signaling in other cell types. The regulated opening and closing of ion channels thus provides cells with a sensitive and versatile mechanism for responding to a variety of environmental stimuli.

*Figure 12.25*
**Selectivity of $K^+$ channels** The $K^+$ channel pore is just wide enough to allow the passage of $K^+$ from which all associated water molecules have been displaced as a result of interactions between $K^+$ and the side chains of polar amino acids lining the pore. $Na^+$ is too small to interact with these amino acid side chains, so it remains bound to water in a complex that is too large to pass through the channel pore.

*Figure 12.26*
**Structures of voltage-gated cation channels** The $K^+$, $Na^+$, and $Ca^{2+}$ channels belong to a family of related proteins. The $K^+$ channel is formed from the association of four identical subunits, one of which is shown. The $Na^+$ channel consists of a single polypeptide chain containing four repeated domains, each of which is similar to one $K^+$ channel subunit. The $Ca^{2+}$ channel is similar to the $Na^+$ channel. Each subunit or domain contains six membrane-spanning $\alpha$ helices. The $\alpha$ helix designated 4 contains multiple positively charged amino acids and acts as the voltage sensor that mediates channel opening in response to changes in membrane potential.

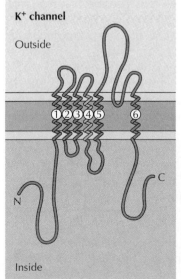

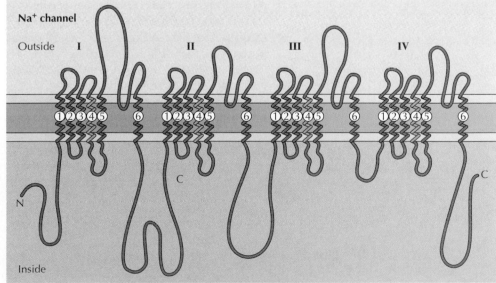

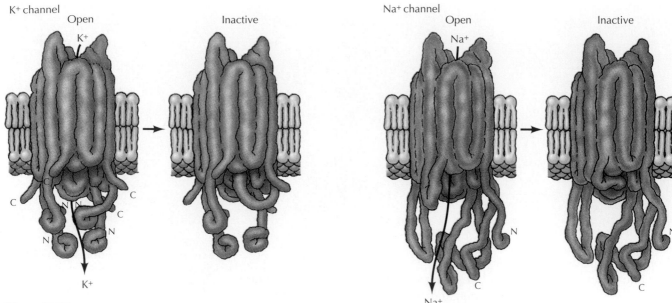

**Figure 12.27**
**Inactivation of K⁺ and Na⁺ channels**
Following voltage-gated opening, the K⁺ and Na⁺ channels are rapidly inactivated by the binding of cytoplasmic portions of the polypeptide chains to the pore. For the K⁺ channel, inactivation is mediated by a ball-and-chain mechanism with the ball corresponding to the amino terminus of the polypeptide chain. For the Na⁺ channel, inactivation is mediated by the intracellular loop connecting domains III and IV.

## *Active Transport Driven by ATP Hydrolysis*

The net flow of molecules by facilitated diffusion, through either carrier proteins or channel proteins, is always energetically downhill in the direction determined by electrochemical gradients across the membrane. In many cases, however, the cell must transport molecules against their concentration gradients. In **active transport**, energy provided by another coupled reaction (such as the hydrolysis of ATP) is used to drive the uphill transport of molecules in the energetically unfavorable direction.

The **ion pumps** responsible for maintaining gradients of ions across the plasma membrane provide important examples of active transport driven directly by ATP hydrolysis. As discussed earlier (see Table 12.1), the concentration of Na⁺ is approximately ten times higher outside than inside of cells, whereas the concentration of K⁺ is higher inside than out. These ion gradients are maintained by the **Na⁺-K⁺ pump** (also called the **Na⁺-K⁺ ATPase**), which uses energy derived from ATP hydrolysis to transport Na⁺ and K⁺ against their electrochemical gradients. This process is a result of ATP-driven conformational changes in the pump (Figure 12.28). First, Na⁺ ions bind to high-affinity sites inside the cell. This binding stimulates the hydrolysis of ATP and phosphorylation of the pump, inducing a conformational change that exposes the Na⁺-binding sites to the outside of the cell and reduces their affinity for Na⁺. Consequently, the bound Na⁺ is released into the extracellular fluids. At the same time, high-affinity K⁺-binding sites are exposed on the cell surface. The binding of extracellular K⁺ to these sites then stimulates hydrolysis of the phosphate group bound to the pump, which induces a second conformational change, exposing the K⁺-binding sites to the cytosol and lowering their binding affinity so that K⁺ is released inside the cell. The pump has three binding sites for Na⁺ and two for K⁺, so each cycle transports three Na⁺ and two K⁺ across the plasma membrane at the expense of one molecule of ATP.

The importance of the Na⁺-K⁺ pump is indicated by the fact that it is estimated to consume nearly 25% of the ATP utilized by many animal cells. One critical role of the Na⁺ and K⁺ gradients established by the pump is the propagation of electric signals in nerve and muscle. As will be dis-

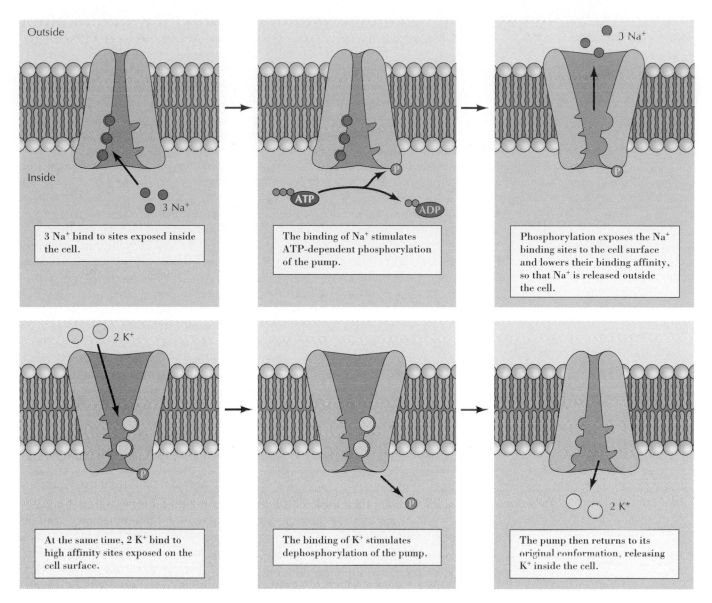

Outside

Inside

3 Na⁺ bind to sites exposed inside the cell.

The binding of Na⁺ stimulates ATP-dependent phosphorylation of the pump.

Phosphorylation exposes the Na⁺ binding sites to the cell surface and lowers their binding affinity, so that Na⁺ is released outside the cell.

At the same time, 2 K⁺ bind to high affinity sites exposed on the cell surface.

The binding of K⁺ stimulates dephosphorylation of the pump.

The pump then returns to its original conformation, releasing K⁺ inside the cell.

*Figure 12.28*
**Model for operation of the Na⁺-K⁺ pump**

cussed shortly, the $Na^+$ gradient established by the pump is also utilized to drive the active transport of a variety of other molecules. Yet another important role of the $Na^+$-$K^+$ pump in most animal cells is to maintain osmotic balance and cell volume. The cytoplasm contains a high concentration of organic molecules, including macromolecules, amino acids, sugars, and nucleotides. In the absence of a counterbalance, this would drive the inward flow of water by osmosis, which if unchecked would result in swelling and eventual bursting of the cell. The required counterbalance is provided by the ion gradients established by the $Na^+$-$K^+$ pump (Figure 12.29). In particular, the pump establishes a higher concentration of $Na^+$ outside than inside the cell. As already discussed, the flow of $K^+$ through open channels further establishes an electric potential across the plasma membrane. This membrane potential in turn drives $Cl^-$ out of the cell, so the concentration of $Cl^-$ (like that of $Na^+$) is about ten times higher in extracellular fluids than in the cytoplasm. These differences in ion concentrations balance the high concentrations of organic molecules inside cells, equalizing the osmotic pressure and preventing the net influx of water.

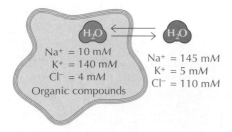

**Figure 12.29**
**Ion gradients across the plasma membrane of a typical mammalian cell** The concentrations of Na$^+$ and Cl$^-$ are higher outside than inside the cell, whereas the concentration of K$^+$ is higher inside than out. The low concentrations of Na$^+$ and Cl$^-$ balance the high intracellular concentration of organic compounds, equalizing the osmotic pressure and preventing the net influx of water.

The active transport of Ca$^{2+}$ across the plasma membrane is driven by a Ca$^{2+}$ pump that is structurally related to the Na$^+$-K$^+$ pump and is similarly powered by ATP hydrolysis. The Ca$^{2+}$ pump transports Ca$^{2+}$ out of the cell, so intracellular Ca$^{2+}$ concentrations are extremely low: approximately 0.1 $\mu M$, in comparison to extracellular concentrations of about 1 m$M$. This low intracellular concentration of Ca$^{2+}$ makes the cell sensitive to small increases in intracellular Ca$^{2+}$ levels. Such transient increases in intracellular Ca$^{2+}$ play important roles in cell signaling, as noted already with respect to muscle contraction (see Figure 11.25) and discussed further in the next chapter.

Similar ion pumps in the plasma membranes of bacteria, yeasts, and plant cells are responsible for the active transport of H$^+$ out of the cell. In addition, H$^+$ is actively pumped out of cells lining the stomach, resulting in the acidity of gastric fluids. Structurally distinct pumps are responsible for the active transport of H$^+$ into lysosomes and endosomes (see Figure 9.34). Yet a third type of H$^+$ pump is exemplified by the ATP synthases of mitochondria and chloroplasts: In these cases the pumps can be viewed as operating in reverse, with the movement of ions down the electrochemical gradient being used to drive ATP synthesis.

The largest family of membrane transporters consists of the **ABC transporters**, so called because they are characterized by highly conserved ATP-binding domains or *ATP-binding cassettes* (Figure 12.30). More than a hundred family members have been identified in both prokaryotic and eukaryotic cells. In bacteria, ABC transporters utilize energy derived from ATP hydrolysis to transport a wide range of molecules, including ions, sugars, and amino acids. In eukaryotic cells, the first ABC transporter was discovered as the product of a gene (called the multidrug resistance, or *mdr*, gene) that makes cancer cells resistant to a variety of drugs used in chemotherapy. The MDR transporter is normally expressed in cells of the liver, intestine, and kidneys, where it is thought to remove potentially toxic foreign compounds. It is also expressed in capillary endothelial cells of the brain, where it functions as a pump to protect the brain from toxic compounds. Unfortunately, the *mdr* gene is frequently expressed at high levels in cancer cells. This leads to the synthesis of elevated levels of the MDR transporter, which recognizes a variety of drugs and pumps them out of cells, thereby making the cancer cells resistant to a broad spectrum of chemothe-

**Figure 12.30**
**Structure of an ABC transporter** The basic structural unit of ABC transporters consists of six transmembrane domains followed by an ATP-binding cassette. In plasma membrane transporters (such as the MDR transporter illustrated), two of these units are fused so that the transporter contains 12 transmembrane domains and two ATP-binding sites.

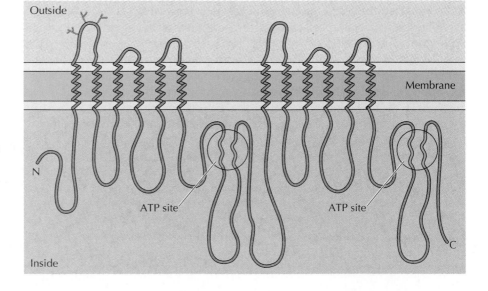

rapeutic agents and posing a major obstacle to successful cancer treatment.

Another medically important member of the ABC transporter family is the gene responsible for cystic fibrosis. Although it is a member of the ABC family, the product of this gene (called the cystic fibrosis transmembrane conductance regulator, or CFTR) functions as a $Cl^-$ channel in epithelial cells, and defective $Cl^-$ transport is characteristic of the disease. The CFTR $Cl^-$ channel is also unusual in that it appears to require both ATP hydrolysis and cAMP-dependent phosphorylation in order to open. The structural basis for the function of CFTR as a regulated ion channel remains to be elucidated by future research.

## Active Transport Driven by Ion Gradients

The ion pumps and ABC transporters discussed in the previous section utilize energy derived directly from ATP hydrolysis to transport molecules against their electrochemical gradients. Other molecules are transported against their concentration gradients using energy derived not from ATP hydrolysis but from the coupled transport of a second molecule in the energetically favorable direction. The $Na^+$ gradient established by the $Na^+$-$K^+$ pump provides a source of energy that is frequently used to power the active transport of sugars, amino acids, and ions in mammalian cells. The $H^+$ gradients established by the $H^+$ pumps of bacteria, yeast, and plant cells play similar roles.

The epithelial cells lining the intestine provide a good example of active transport driven by the $Na^+$ gradient. These cells use active-transport systems in the apical domains of their plasma membranes to take up dietary sugars and amino acids from the lumen of the intestine. The uptake of glucose, for example, is carried out by a transporter that coordinately transports two $Na^+$ and one glucose into the cell (Figure 12.31). The flow of $Na^+$ down its electrochemical gradient provides the energy required to take up dietary glucose and to accumulate high intracellular glucose concentrations. Glucose is then released into the underlying connective tissue (which contains blood capillaries) at the basolateral surface of the intestinal epithelium, where it is transported down its concentration gradient by facilitated diffu-

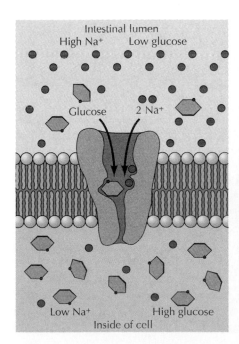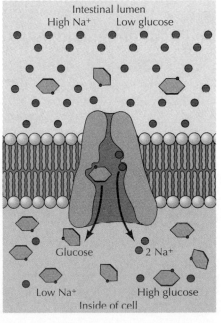

*Figure 12.31*
**Active transport of glucose**　Active transport driven by the $Na^+$ gradient is responsible for the uptake of glucose from the intestinal lumen. The transporter coordinately binds and transports one glucose and two $Na^+$ into the cell. The transport of $Na^+$ in the energetically favorable direction drives the uptake of glucose against its concentration gradient.

# MOLECULAR MEDICINE

## Cystic Fibrosis

### The Disease

Cystic fibrosis is a recessive genetic disease affecting children and young adults. It is the most common lethal inherited disease of Caucasians, with approximately one in 2500 newborns affected, although it is rare in other races. The characteristic dysfunction in cystic fibrosis is the production of abnormally thick sticky mucus by several types of epithelial cells, including the cells lining the respiratory and gastrointestinal tracts. The primary clinical manifestation is respiratory disease resulting from obstruction of the pulmonary airways by thick plugs of mucus, followed by the development of recurrent bacterial infections. In most patients, the pancreas is also involved because the pancreatic ducts are blocked by mucus. Sweat glands also function abnormally, and the presence of excessive salt in sweat is diagnostic of cystic fibrosis. Current management of the disease includes physical therapy to promote bronchial drainage, antibiotic administration, and pancreatic enzyme replacement. Although such treatment has extended the survival of affected individuals to about 30 years of age, cystic fibrosis is ultimately fatal, with lung disease being responsible for 95% of mortality.

### Molecular and Cellular Basis

The hallmark of cystic fibrosis is defective Cl⁻ transport in affected epithelia, including sweat ducts and the cells lining the respiratory tract. In 1984, it was demonstrated that Cl⁻ channels fail to function normally in epithelial cells from cystic fibrosis patients. The molecular basis of the disease was then elucidated in 1989, with the isolation of the cystic fibrosis gene as a molecular clone. The sequence of the gene revealed that it encodes a protein (called CFTR for

*cystic fibrosis transmembrane conductance regulator*) belonging to the ABC transporter family. A variety of subsequent studies then demonstrated that CFTR functions as a Cl⁻ channel and that the inherited mutations responsible for cystic fibrosis result directly in defective Cl⁻ transport.

### Prevention and Treatment

As with other inherited diseases, isolation of the cystic fibrosis gene opens the possibility of genetic screening to identify individuals carrying mutant alleles. In some populations, the frequency of heterozygote carriers of mutant genes is as high as one in 25 individuals, suggesting the possibility of general population screening to identify couples at risk and provide genetic counseling. In addition, understanding the function of CFTR as a Cl⁻ channel has suggested new approaches to treatment. One possibility is the use of drugs that stimulate the opening of other Cl⁻ channels in affected epithelia. Alternatively, gene therapy provides the potential of replacing normal CFTR genes in the respiratory epithelium of cystic fibrosis patients.

The potential of gene therapy has been supported by experiments demonstrating that introduction of a normal CFTR gene into cultured cells of cystic fibrosis patients is sufficient to restore Cl⁻ channel function. The possible application of gene therapy to cystic fibrosis is also enhanced by the accessibility of the epithelial cells lining the airway to aerosol delivery systems. Studies with experimental animals have demonstrated that viral vectors can transmit CFTR cDNA to the respiratory epithelium, and the first human trial of gene therapy for cystic fibrosis was initiated in 1993. Although

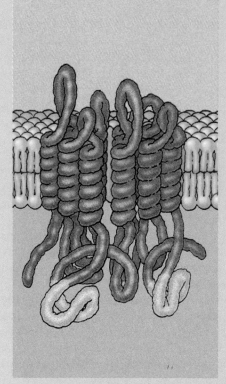

Model of the cystic fibrosis transmembrane conductance regulator (CFTR).

these studies are still in progress, current results indicate that normal CFTR cDNA can be transferred to the airway epithelial cells of cystic fibrosis patients, supporting the feasibility of ultimately developing an effective gene therapy protocol.

### REFERENCES

Collins, F. S. 1992. Cystic fibrosis: Molecular biology and therapeutic implications. *Science* 256:774–779.

Korst, R. J., N. G. McElvaney, C.-S. Chu, M. A. Rosenfeld, A. Mastrangeli, J. Hay, S. L. Brody, N. T. Eissa, C. Danel, H. A. Jaffe and R. G. Crystal. 1995. Gene therapy for the respiratory manifestations of cystic fibrosis. *Am. J. Respir. Crit. Care Med.* 151:S75–S87.

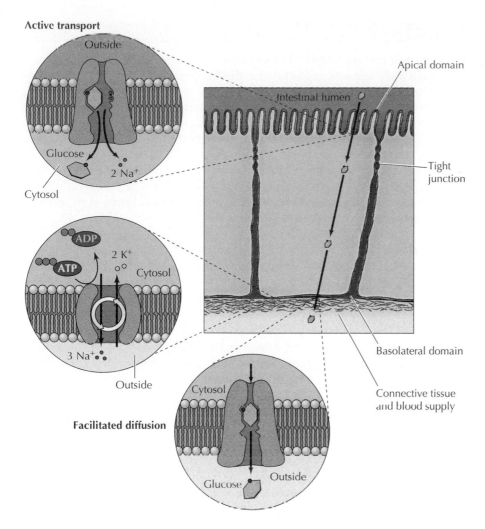

**Active transport**

Outside

Glucose

2 Na⁺

Cytosol

ADP

ATP

2 K⁺

Cytosol

3 Na⁺

Outside

**Facilitated diffusion**

Intestinal lumen

Apical domain

Tight junction

Basolateral domain

Connective tissue and blood supply

Cytosol

Glucose    Outside

*Figure 12.32*
**Glucose transport by intestinal epithelial cells**    A transporter in the apical domain of the plasma membrane is responsible for the active uptake of glucose (by cotransport with Na⁺) from the intestinal lumen. As a result, dietary glucose is absorbed and concentrated within intestinal epithelial cells. Glucose is then transferred from these cells to the underlying connective tissue and blood supply by facilitated diffusion, mediated by a transporter in the basolateral domain of the plasma membrane. The system is driven by the Na⁺-K⁺ pump, which is also found in the basolateral domain. Note that the uptake of glucose from the digestive tract and its transfer to the circulation is dependent on the restricted localization of glucose transporters mediating active transport and facilitated diffusion to the apical and basolateral domains of the plasma membrane, respectively.

sion (Figure 12.32). The uptake of glucose from the intestinal lumen and its release into the circulation thus provides a good example of the polarized function of epithelial cells, which results from the specific localization of active transport and facilitated diffusion carriers to the apical and basolateral domains of the plasma membrane, respectively.

The coordinate uptake of glucose and Na⁺ is an example of **symport**, the transport of two molecules in the same direction. In contrast, the facilitated diffusion of glucose is an example of **uniport**, the transport of only a single molecule. Active transport can also take place by **antiport**, in which two molecules are transported in opposite directions (Figure 12.33). For example, Ca²⁺ is exported from cells not only by the Ca²⁺ pump but also by an Na⁺-Ca²⁺ antiporter that transports Na⁺ into the cell and Ca²⁺ out. Another example is provided by the Na⁺-H⁺ exchange protein, which functions in the regulation of intracellular pH. The Na⁺-H⁺ antiporter couples the transport of Na⁺ into the cell with the export of H⁺, thereby removing excess H⁺ produced by metabolic reactions and preventing acidification of the cytoplasm.

*Figure 12.33*
**Examples of antiport**    Ca²⁺ and H⁺ are exported from cells by antiporters, which couple their export to the energetically favorable import of Na⁺.

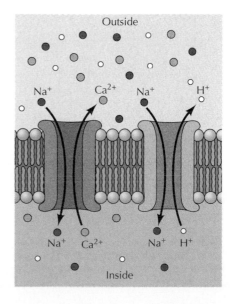

Outside

Na⁺    Ca²⁺    Na⁺    H⁺

Na⁺    Ca²⁺    Na⁺    H⁺

Inside

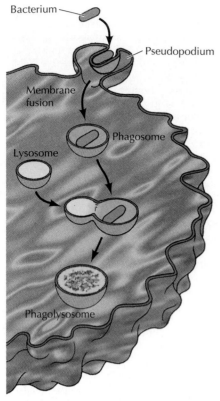

*Figure 12.34*
**Phagocytosis**   Binding of a bacterium to the cell surface stimulates the extension of a pseudopodium, which eventually engulfs the bacterium. Fusion of the pseudopodium membranes then results in formation of a large intracellular vesicle (a phagosome). The phagosome fuses with lysosomes to form a phagolysosome, within which the ingested bacterium is digested.

# ENDOCYTOSIS

The carrier and channel proteins discussed in the preceding section transport small molecules through the phospholipid bilayer. Eukaryotic cells are also able to take up macromolecules and particles from the surrounding medium by a distinct process called **endocytosis**. In endocytosis, the material to be internalized is surrounded by an area of plasma membrane, which then buds off inside the cell to form a vesicle containing the ingested material. The term "endocytosis" was coined by Christian deDuve in 1963 to include both the ingestion of large particles (such as bacteria) and the uptake of fluids or macromolecules in small vesicles. The former of these activities is known as **phagocytosis** (cell eating) and the latter as **pinocytosis** (cell drinking).

## Phagocytosis

During phagocytosis, cells engulf large particles such as bacteria, cell debris, or even intact cells (Figure 12.34). Binding of the particle to receptors on the surface of the phagocytic cell triggers the extension of pseudopodia—an actin-based movement of the cell surface, discussed in Chapter 11. The pseudopodia eventually surround the particle and their membranes fuse to form a large intracellular vesicle (>0.25 $\mu$m in diameter) called a **phagosome**. The phagosomes then fuse with lysosomes, producing **phagolysosomes** in which the ingested material is digested by the action of lysosomal acid hydrolases (see Chapter 9). During maturation of the phagolysosome, some of the internalized membrane proteins are recycled to the plasma membrane, as discussed in the next section for receptor-mediated endocytosis.

The ingestion of large particles by phagocytosis plays distinct roles in different kinds of cells (Figure 12.35). Many amoebas use phagocytosis to capture food particles, such as bacteria or other protozoans. In multicellular animals, the major roles of phagocytosis are to provide a defense against invading microorganisms and to eliminate aged or damaged cells from the body. In mammals, phagocytosis is the function of primarily two types of white blood cells, macrophages and neutrophils, which are fre-

(A)

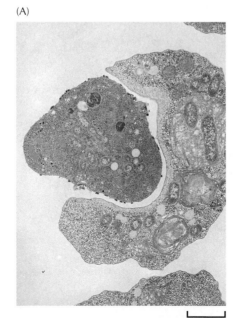

(B)

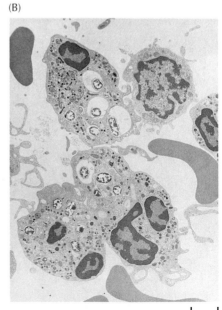

*Figure 12.35*
**Examples of phagocytic cells**   (A) An amoeba engulfing another protist. (B) White blood cells ingesting bacteria. (A, R. N. Band and H. S. Pankratz/ Biological Photo Service; B, David M. Phillips/Visuals Unlimited.)

2 $\mu$m

5 $\mu$m

(A)

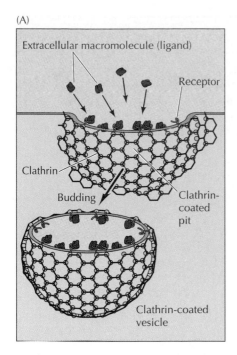

Extracellular macromolecule (ligand)

Receptor

Clathrin

Budding

Clathrin-coated pit

Clathrin-coated vesicle

(B)

Stage 1

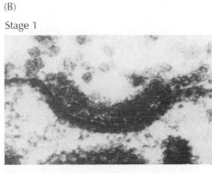

Stage 2

Stage 3

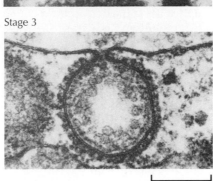

Stage 4

0.2 μm

*Figure 12.36*
**Clathrin-coated vesicle formation**
(A) Extracellular macromolecules (ligands) bind to cell surface receptors that are concentrated in clathrin-coated pits. These pits bud from the plasma membrane to form intracellular clathrin-coated vesicles. (B) Electron micrographs showing four stages in the formation of a clathrin-coated vesicle from a clathrin-coated pit. (B, M. M. Perry, 1979, *J. Cell Science* 34: 266.)

quently referred to as "professional phagocytes." Both macrophages and neutrophils play critical roles in the body's defense systems by eliminating microorganisms from infected tissues. In addition, macrophages eliminate aged or dead cells from tissues throughout the body. A striking example of the scope of this activity is provided by the macrophages of the human spleen and liver, which are responsible for the disposal of more than $10^{11}$ aged blood cells on a daily basis.

### Receptor-Mediated Endocytosis

In contrast to phagocytosis, which plays only specialized roles, pinocytosis is common among eukaryotic cells. The best-characterized form of this process is **receptor-mediated endocytosis**, which provides a mechanism for the selective uptake of specific macromolecules (Figure 12.36). The macromolecules to be internalized first bind to specific cell surface receptors. These receptors are concentrated in specialized regions of the plasma membrane, called **clathrin-coated pits**. These pits bud from the membrane to form small **clathrin-coated vesicles** containing the receptors and their bound macromolecules (**ligands**). The clathrin-coated vesicles then fuse with early endosomes, in which their contents are sorted for transport to lysosomes or recycling to the plasma membrane.

The uptake of cholesterol by mammalian cells has provided a key model for understanding receptor-mediated endocytosis at the molecular level. Cholesterol is transported through the bloodstream in the form of lipoprotein particles, the most common of which is called **low-density lipoprotein**, or **LDL** (Figure 12.37). Studies in the laboratories of Michael Brown and Joseph

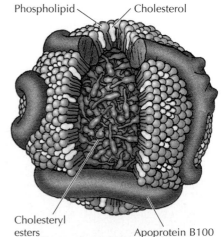

Phospholipid

Cholesterol

Cholesteryl esters

Apoprotein B100

*Figure 12.37*
**Structure of LDL** Each particle of LDL contains approximately 1500 molecules of cholesteryl esters in an oily core. The core is surrounded by a coat containing 500 molecules of cholesterol, 800 molecules of phospholipid, and one molecule of apoprotein B100.

## KEY EXPERIMENT

## *The LDL Receptor*

**Familial Hypercholesterolemia: Defective Binding of Lipoproteins to Cultured Fibroblasts Associated with Impaired Regulation of 3-Hydroxy-3-Methylglutaryl Coenzyme A Reductase Activity**

Michael S. Brown and Joseph L. Goldstein
University of Texas Southwestern Medical School, Dallas
*Proceedings of the National Academy of Science USA, 1974, Volume 71, pages 788–792*

### The Context

Familial hypercholesterolemia (FH) is a genetic disease in which patients have greatly elevated levels of serum cholesterol and suffer from heart attacks early in life. Michael Brown and Joseph Goldstein began their efforts to understand this disease in 1972 with the idea that cholesterol overproduction results from a defect in the control mechanisms that normally regulate cholesterol biosynthesis. Consistent with this hypothesis, they found that addition of LDL to the culture medium of normal human fibroblasts inhibits the activity of 3-hydroxy-3-methylglutaryl coenzyme A reductase (HMG-CoA reductase), the rate-limiting enzyme in the cholesterol biosynthetic pathway. In contrast, HMG-CoA reductase activity is unaffected by addition of LDL to the cells of FH patients, resulting in overproduction of cholesterol by FH cells.

Perhaps surprisingly, subsequent experiments indicated that this abnormality in HMG-CoA reductase regulation is not the result of a mutation in the HMG-CoA reductase gene. Instead, the abnormal regulation of HMG-CoA reductase appeared to be due to the inability of FH cells to extract cholesterol from LDL. In 1974, Brown and Goldstein demonstrated that the lesion in FH cells is a defect in LDL binding to a receptor on the cell surface. This identification of the LDL receptor led to a series of groundbreaking experiments in

which Brown, Goldstein, and their colleagues delineated the pathway of receptor-mediated endocytosis.

### The Experiments

In their 1974 paper, Brown and Goldstein reported the results of experiments in which they investigated the binding of radiolabeled LDL to fibroblasts from either normal individuals or FH patients. Small amounts of radioactive LDL were added to the culture media, and the amount of radioactivity bound to the cells was determined after varying times of incubation (see figure).

Increasing amounts of radioactive LDL bound to normal cells as a function of incubation time. Importantly, addition of excess unlabeled LDL reduced the binding of radioactive LDL, indicating that binding was due to a specific interaction of LDL with a limited number of sites on the cell surface. The specificity of the interaction was further demonstrated by the failure of excess amounts of other lipoproteins to interfere with LDL binding.

In contrast to these results with normal fibroblasts, the cells of FH patients failed to bind specifically radioactive LDL. It therefore appeared that normal fibroblasts possessed a specific LDL receptor that was absent or defective in FH cells. Brown and Goldstein concluded that the defect in LDL binding observed in FH cells "may represent the primary genetic lesion in this disorder," accounting for the inability of LDL to inhibit HMG-CoA reductase and the

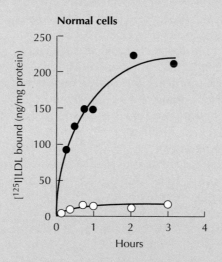

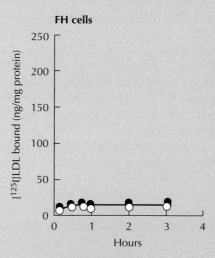

Time course of the binding of radioactive LDL to normal and FH cells. Cells were incubated with radioactive [$^{125}$I]LDL in the presence (open symbols) or absence (closed symbols) of excess unlabeled LDL. Cells were then harvested, and the amount of radioactive LDL bound was determined. Data are presented as nanograms of LDL bound per milligram of cell protein.

***The LDL Receptor*** (continued)

resultant overproduction of cholesterol. Additional experiments showed that the LDL bound to normal fibroblasts is associated with the membrane fraction of the cell, suggesting that the LDL receptor is a cell surface protein.

**The Impact**

Following this identification of the LDL receptor, Brown and Goldstein demonstrated that LDL bound to the cell surface is rapidly internalized and degraded to free cholesterol in lysosomes. In collaboration with Richard Anderson, they further established that the LDL receptor is internalized by endocytosis from coated pits. In addition, their early studies demonstrated that the LDL receptor is recycled to the plasma membrane after dissociation of its ligand inside the cell. Experiments that had been initiated with the goal of understanding the regulation of cholesterol biosynthesis thus led to the elucidation of a major pathway by which eukaryotic cells internalize specific macromolecules—a striking example of the way in which science and scientists can proceed in unanticipated but exciting new directions.

Michael Brown

Joseph Goldstein

Goldstein demonstrated that the uptake of LDL by mammalian cells requires the binding of LDL to a specific cell surface receptor that is concentrated in clathrin-coated pits and internalized by endocytosis. As discussed in the next section, the receptor is then recycled to the plasma membrane while LDL is transported to lysosomes, where cholesterol is released for use by the cell.

The key insights into this process came from studies of patients with the inherited disease known as familial hypercholesterolemia. Patients with this disease have very high levels of serum cholesterol and suffer heart attacks early in life. Brown and Goldstein found that cells of these patients are unable to internalize LDL from extracellular fluids, resulting in the accumulation of high levels of cholesterol in the circulation. Further experiments demonstrated that cells of normal individuals possess a receptor for LDL, which is concentrated in coated pits, and that familial hypercholesterolemia results from inherited mutations in the LDL receptor. These mutations are of two types. Cells from most patients with familial hypercholesterolemia simply failed to bind LDL, demonstrating that a specific cell surface receptor was required for LDL uptake. In addition, a few patients were identified whose cells bound LDL but were unable to internalize it. The LDL receptors of these patients failed to concentrate in coated pits, providing direct evidence for the central role of coated pits in receptor-mediated endocytosis.

The mutations that prevent the LDL receptor from concentrating in coated pits lie within the cytoplasmic tail of the receptor and can be as subtle as the change of a single tyrosine to cysteine (Figure 12.38). Further studies have defined the internalization signal of the LDL receptor as a sequence of six amino acids, including the essential tyrosine. Similar internalization signals, frequently including tyrosine residues, are found in the cytoplasmic tails of other receptors taken up via clathrin-coated pits. These internalization signals bind to adaptins, which in turn bind clathrin on the cytosolic side of the membrane (Figure 12.39), similar to the way in which clathrin-coated vesicles form during the transport of lysosomal hydrolases from the *trans* Golgi network (see Figure 9.30). Clathrin assembles into a basketlike structure that distorts the membrane, forming invaginated pits. Another protein, called dynamin, assembles into rings around the necks of these invaginated pits,

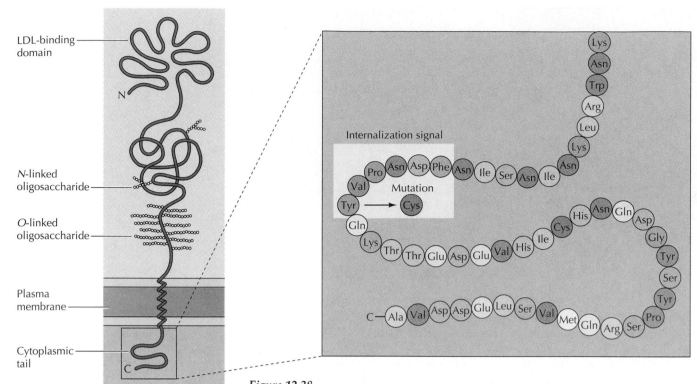

**LDL-binding domain**

N

**N-linked oligosaccharide**

**O-linked oligosaccharide**

**Plasma membrane**

**Cytoplasmic tail**

C

*Figure 12.38*
**The LDL receptor** The LDL receptor includes 700 extracellular amino acids, a transmembrane α helix of 22 amino acids, and a cytoplasmic tail of 50 amino acids. The N-terminal 292 amino acids constitute the LDL-binding domain. Six amino acids within the cytoplasmic tail define the internalization signal, first recognized because the mutation of Tyr to Cys in a case of familial hypercholesterolemia prevented concentration of the receptor in coated pits.

*Figure 12.39*
**Formation of clathrin-coated pits** (A) Adaptor proteins (adaptin) bind both to clathrin and to the internalization signals present in the cytoplasmic tails of receptors. (B) Electron micrograph of a clathrin-coated pit showing the basketlike structure of the clathrin network. (B, courtesy of John E. Heuser, Washington University School of Medicine.)

eventually constricting and pinching off coated vesicles inside the cell.

Receptor-mediated endocytosis is a major activity of the plasma membranes of eukaryotic cells. More than 20 different receptors have been shown to be selectively internalized by this pathway. Extracellular fluids are also incorporated into coated vesicles as they bud from the plasma membrane, so receptor-mediated endocytosis results in the nonselective

(A)

(B)

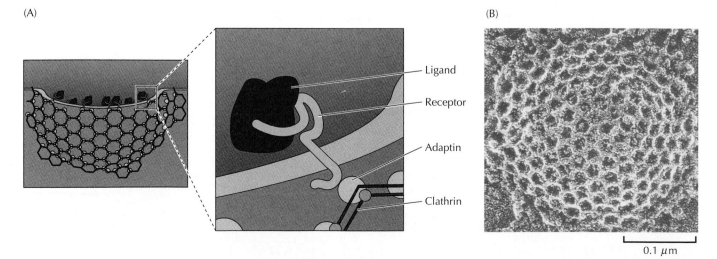

Ligand

Receptor

Adaptin

Clathrin

0.1 μm

uptake of extracellular fluids and their contents (**fluid phase endocytosis**), in addition to the internalization of specific macromolecules. Coated pits typically occupy 1 to 2% of the surface area of the plasma membrane and are estimated to have a lifetime of 1 to 2 minutes. From these figures, one can calculate that receptor mediated endocytosis results in the internalization of an area of cell surface equivalent to the entire plasma membrane approximately every 2 hours.

A variety of studies indicate that cells also possess clathrin-independent endocytosis pathways. For example, fluids and some membrane-bound molecules continue to be endocytosed under experimental conditions that inhibit endocytosis from clathrin-coated pits. However, the mechanisms of such clathrin-independent endocytosis are not yet understood. One possibility is that clathrin-independent endocytosis is mediated by small invaginations of the plasma membrane (50 to 80 nm in diameter) called **caveolae** (Figure 12.40). Caveolae possess a distinct coat (formed by a protein called caveolin) and have been implicated in a variety of transport processes, including endocytosis. However, the budding of caveolae to form endocytic vesicles has not been demonstrated. On the other hand, non-clathrin-coated endocytic vesicles of larger size (about 95 nm in diameter) have been reported, although the origin of these vesicles is unclear. In addition, large vesicles (0.15 to 5.0 $\mu$m in diameter) can mediate the uptake of fluids in a process known as **macropinocytosis**. Thus, while clathrin-dependent endocytosis clearly provides a major pathway for the uptake of both fluids and specific macromolecules, several possible clathrin-independent mechanisms remain to be characterized.

## Protein Trafficking in Endocytosis

Following their internalization, clathrin-coated vesicles rapidly shed their coats and fuse with early **endosomes,** which are vesicles with tubular extensions located at the periphery of the cell. The specificity of fusion of endocytic vesicles with endosomes is determined by interactions between complementary pairs of transmembrane proteins of the vesicle and target membranes (v-SNAREs and t-SNAREs) and by Rab GTP-binding proteins, as discussed in Chapter 9. The early endosomes serve as a sorting compartment, from which molecules taken up by endocytosis are either recycled to the plasma membrane or transported to lysosomes for degradation. In addition, the early endosomes of polarized cells can transfer endocytosed proteins between different domains of the plasma membrane—for example, between the apical and basolateral domains of epithelial cells.

An important feature of early endosomes is that they maintain an acidic internal pH (about 6.0 to 6.2) as the result of the action of a membrane H$^+$ pump. This acidic pH leads to the dissociation of many ligands from their

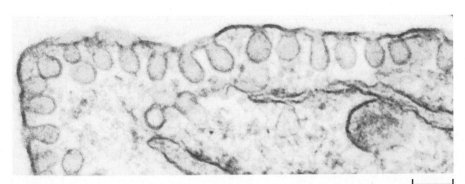

*Figure 12.40*
**Caveolae**  Electron micrograph of a human fibroblast showing caveolae in the plasma membrane. (Courtesy of R. G. W. Anderson/University of Texas Southwestern Medical School-Dallas.)

0.2 $\mu$m

*Figure 12.41*
**Sorting in early endosomes** LDL
bound to its receptor is internalized in
clathrin-coated vesicles, which shed
their coats and fuse with early endosomes. At the acidic pH of early endosomes, LDL dissociates from its receptor and the endocytosed materials are
sorted for degradation in lysosomes or
recycling to the plasma membrane.
LDL is transported from early to late
endosomes in large carrier vesicles
that move along microtubules. Transport vesicles carrying lysosomal
hydrolases from the Golgi apparatus
then fuse with late endosomes, which
mature to lysosomes where LDL is
degraded and cholesterol is released.
In contrast, the LDL receptor is recycled from early endosomes to the plasma membrane.

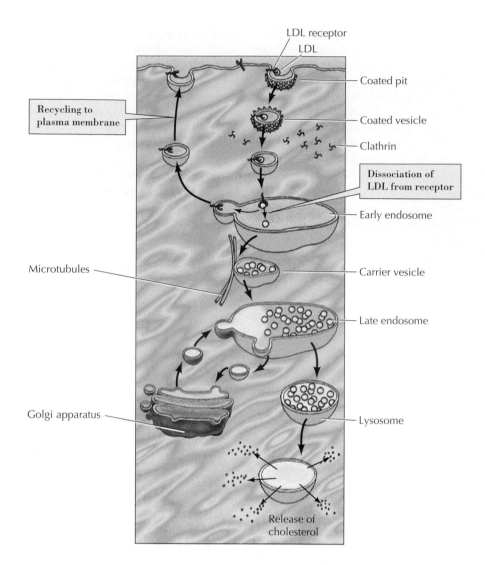

receptors within the early endosome compartment. Following this uncou-
pling, the receptors and their ligands can be transported to different intracellular destinations. A classic example is provided by LDL, which dissociates from its receptor within early endosomes (Figure 12.41). The receptor
is then returned to the plasma membrane via transport vesicles that bud
from the tubular extensions of endosomes. In contrast, LDL is transported
(along with other soluble contents of the endosome) to lysosomes, where
its degradation releases cholesterol.

Recycling to the plasma membrane is the major fate of membrane proteins taken up by receptor-mediated endocytosis, with many receptors (like
the LDL receptor) being returned to the plasma membrane following dissociation of their bound ligands in early endosomes. The recycling of these
receptors results in the continuous internalization of their ligands. Each
LDL receptor, for example, makes a round-trip from the plasma membrane
to endosomes and back approximately every 10 minutes. The importance
of the recycling pathway is further emphasized by the magnitude of membrane traffic resulting from endocytosis. As already noted, approximately
50% of the plasma membrane is internalized by receptor-mediated endocytosis every hour and must therefore be replaced at an equivalent rate.
Most of this replacement is the result of receptor recycling; only about 5%

of the cell surface is newly synthesized per hour.

Ligands and membrane proteins destined for degradation in lysosomes are transported from early endosomes to late endosomes, which are located near the nucleus (see Figure 12.41). Transport from early to late endosomes is mediated by the movement of large endocytic carrier vesicles along microtubules. The late endosomes are more acidic than early endosomes (pH about 5.5 to 6.0) and, as discussed in Chapter 9, are able to fuse with transport vesicles carrying lysosomal hydrolases from the Golgi apparatus. Late endosomes then mature into lysosomes as they acquire a full complement of lysosomal enzymes and become still more acidic (pH about 5). Within lysosomes, the endocytosed materials are degraded by the action of acid hydrolases.

Although many receptors (like the LDL receptor) are recycled to the plasma membrane, others follow different fates. Some are transported to lysosomes and degraded along with their ligands. For example, the cell surface receptors for several growth factors (discussed in the next chapter) are internalized following growth factor binding and eventually degraded in lysosomes. The effect of this process is to remove the receptor-ligand complexes from the plasma membrane, thereby terminating the response of the cell to growth factor stimulation—a phenomenon known as **receptor down-regulation**.

A specialized kind of recycling from endosomes plays an important role in the transmission of nerve impulses across synapses (Figure 12.42). As discussed earlier in this chapter, the arrival of an action potential at the terminus of most neurons signals the fusion of synaptic vesicles with the plasma membrane, releasing the neurotransmitters that carry the signal to postsynaptic cells. The empty synaptic vesicles are then recovered from the plasma membrane in clathrin-coated vesicles, which fuse with early endosomes. The synaptic vesicles are then regenerated directly by budding from endosomes. They accumulate a new supply of neurotransmitter and recycle to the plasma membrane, ready for the next cycle of synaptic transmission.

In polarized cells (e.g., epithelial cells), internalized receptors can also be transferred across the cell to the opposite domain of the plasma membrane—a process called **transcytosis**. For example, a receptor endocytosed

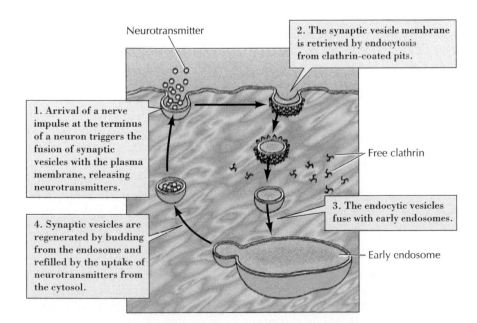

Neurotransmitter

2. The synaptic vesicle membrane is retrieved by endocytosis from clathrin-coated pits.

1. Arrival of a nerve impulse at the terminus of a neuron triggers the fusion of synaptic vesicles with the plasma membrane, releasing neurotransmitters.

Free clathrin

3. The endocytic vesicles fuse with early endosomes.

4. Synaptic vesicles are regenerated by budding from the endosome and refilled by the uptake of neurotransmitters from the cytosol.

Early endosome

*Figure 12.42*
**Recycling of synaptic vesicles**

*Figure 12.43*
**Protein sorting by transcytosis** A protein destined for the apical domain of the plasma membrane is first transported from the Golgi apparatus to the basolateral domain. It is then endocytosed and selectively transported to the apical domain from early endosomes.

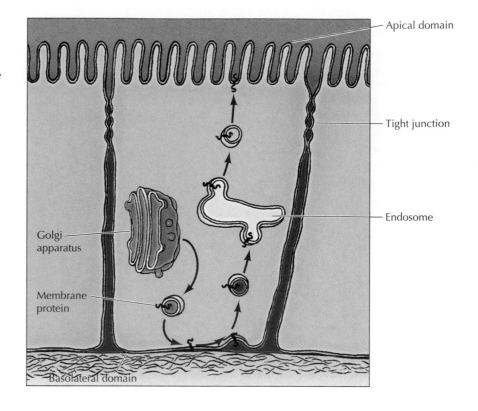

*Figure 12.44*
**Bacterial cell walls** The plasma membrane of Gram-negative bacteria is surrounded by a thin cell wall beneath the outer membrane. Gram-positive bacteria lack outer membranes and have thick cell walls.

**Gram-negative**

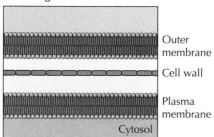

**Gram-positive**

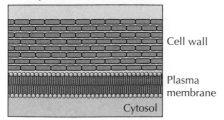

from the basolateral domain of the plasma membrane can be sorted in early endosomes for transport to the apical membrane. In some cells, this is an important mechanism for sorting membrane proteins (Figure 12.43). Rather than being sorted for delivery to the apical or basolateral domains in the *trans* Golgi network (see Figure 9.27), proteins are initially delivered to the basolateral membrane. Proteins targeted for the apical membrane are then transferred to that site by transcytosis. In addition, transcytosis provides a mechanism for the transfer of extracellular macromolecules across epithelial cell sheets. For example, many kinds of epithelial cells transport antibodies from the blood to a variety of secreted fluids, such as milk. The antibodies bind to receptors on the basolateral surface and are then transcytosed along with their receptors to the apical surface. The receptors are then cleaved, releasing the antibodies into extracellular secretions.

## CELL WALLS AND THE EXTRACELLULAR MATRIX

Although cell boundaries are defined by the plasma membrane, many cells are surrounded by an insoluble array of secreted macromolecules. Cells of bacteria, fungi, algae, and higher plants are surrounded by rigid cell walls, which are an integral part of the cell. Although not encased in cell walls, animal cells in tissues are closely associated with an extracellular matrix composed of proteins and polysaccharides. The extracellular matrix not only provides structural support to cells and tissues, but also plays important roles in regulating the behavior of cells in multicellular organisms.

### Bacterial Cell Walls

The rigid cell walls of bacteria determine cell shape and prevent the cell from bursting as a result of osmotic pressure. The structure of their cell walls divides bacteria into two broad classes that can be distinguished by a staining procedure known as the Gram stain, developed by Christian Gram in 1884 (Figure 12.44). As described earlier in this chapter, Gram-negative bacteria (such as *E. coli*) have a dual membrane system, in which the plasma membrane is surrounded by a permeable outer membrane. These bacteria have thin cell walls located between their inner and outer membranes. In contrast, Gram-positive bacteria (such as the common human pathogen *Staphylococcus aureus*) have only a single plasma membrane, which is surrounded by a much thicker cell wall.

Despite these structural differences, the principal component of the cell walls of both Gram-positive and Gram-negative bacteria is a **peptidoglycan** (Figure 12.45) consisting of linear polysaccharide chains crosslinked by

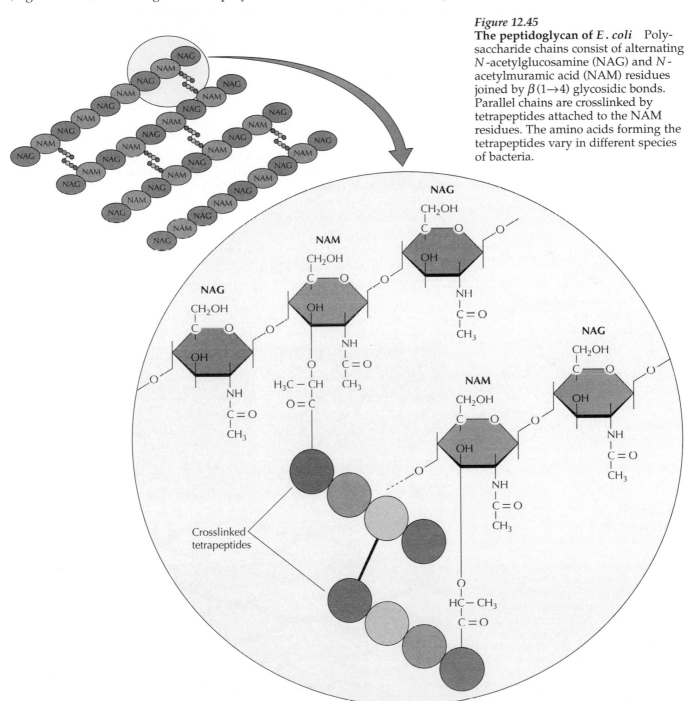

**Figure 12.45**
**The peptidoglycan of *E. coli*** Polysaccharide chains consist of alternating *N*-acetylglucosamine (NAG) and *N*-acetylmuramic acid (NAM) residues joined by $\beta\,(1{\rightarrow}4)$ glycosidic bonds. Parallel chains are crosslinked by tetrapeptides attached to the NAM residues. The amino acids forming the tetrapeptides vary in different species of bacteria.

(A) **Chitin**

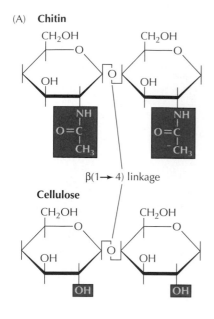

β(1→ 4) linkage

**Cellulose**

*Figure 12.46*
**Polysaccharides of fungal and plant cell walls** (A) Chitin (the principal polysaccharide of fungal cell walls) is a linear polymer of *N*-acetylglucosamine residues, whereas cellulose is a linear polymer of glucose. The carbohydrate monomers are joined by β (1→4) linkages, allowing the polysaccharides to form long, straight chains. (B) Parallel chains of cellulose associate to form microfibrils.

short peptides. Because of this crosslinked structure, the peptidoglycan forms a strong covalent shell around the entire bacterial cell. Interestingly, the unique structure of their cell walls also makes bacteria vulnerable to some antibiotics. Penicillin, for example, inhibits the enzyme responsible for forming cross-links between different strands of the peptidoglycan, thereby interfering with cell wall synthesis and blocking bacterial growth.

## Plant Cell Walls

In contrast to bacteria, the cell walls of eukaryotes (including fungi, algae, and higher plants) are composed principally of polysaccharides (Figure 12.46). The basic structural polysaccharide of fungal cell walls is **chitin** (a polymer of *N*-acetylglucosamine residues), which also forms the exoskeleton of arthropods (e.g., the shells of crabs). The cell walls of most algae and higher plants are composed principally of **cellulose**, which is the single most abundant polymer on Earth. Cellulose is a linear polymer of glucose residues, often containing more than 10,000 glucose monomers. The glucose residues are joined by β (1→4) linkages, which allow the polysaccharide to form long straight chains. Several dozen such chains then associate in parallel with one another to form **cellulose microfibrils**, which can extend for many micrometers in length.

(B)

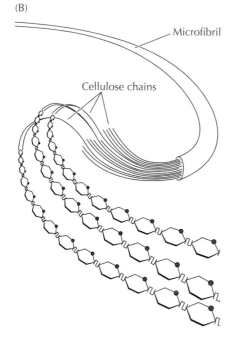

Microfibril

Cellulose chains

(A)

Hemicellulose

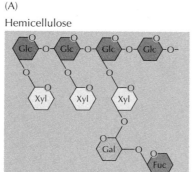

Pectin

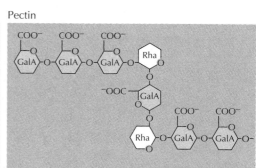

*Figure 12.47*
**Model of a plant cell wall** (A) Structures of a representative hemicellulose (xyloglucan) and pectin (rhamnogalacturonan). Xyloglucan consists of a backbone of glucose (Glc) residues with side chains of xylose (Xyl), galactose (Gal), and fucose (Fuc). The backbone of rhamnogalacturonan contains galacturonic acid (GalA) and rhamnose (Rha) residues, to which numerous side chains are also attached. (B) Hemicelluloses bind to the surface of cellulose microfibrils, forming a fibrous network that is embedded in a gel-like matrix of pectins.

(B)

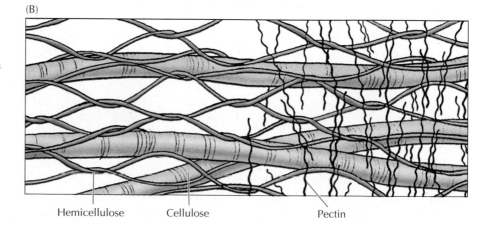

Hemicellulose      Cellulose      Pectin

**Figure 12.48**
**Primary and secondary cell walls** Secondary cell walls are laid down between the primary cell wall and the plasma membrane. Secondary walls frequently consist of three layers, which differ in the orientation of their cellulose microfibrils. Electron micrographs show cellulose microfibrils in primary and secondary cell walls. (Primary wall courtesy of F. C. Steward; secondary wall Biophoto Associates/Photo Researchers Inc.)

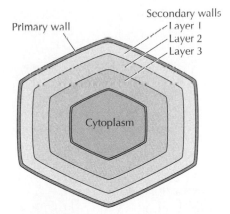

Primary wall

1 μm

Secondary wall

0.2 μm

Within the cell wall, cellulose microfibrils are embedded in a matrix consisting of proteins and two other types of polysaccharides: **hemicelluloses** and **pectins** (Figure 12.47). Hemicelluloses are highly branched polysaccharides that are hydrogen-bonded to the surface of cellulose microfibrils. This crosslinks the cellulose microfibrils into a network of tough, fibrous molecules, which is responsible for the mechanical strength of plant cell walls. Pectins are branched polysaccharides containing a large number of negatively charged galacturonic acid residues. Because of these multiple negative charges, pectins bind positively charged ions (such as $Ca^{2+}$) and trap water molecules to form gels. An illustration of their gel-forming properties is provided by the fact that jams and jellies are produced by the addition of pectins to fruit juice. In the cell wall, the pectins form a gel-like network that is interlocked with the crosslinked cellulose microfibrils. In addition, cell walls contain a variety of glycoproteins that are incorporated into the matrix and are thought to provide further structural support.

Both the structure and function of cell walls change as plant cells develop. The walls of growing plant cells (called **primary cell walls**) are relatively thin and flexible, allowing the cell to expand in size. Once cells have ceased growth, they frequently lay down **secondary cell walls** between the plasma membrane and the primary cell wall (Figure 12.48). Such secondary cell walls, which are both thicker and more rigid than primary walls, are particularly important in cell types responsible for conducting water and providing mechanical strength to the plant.

Primary and secondary cell walls differ in composition as well as in thickness. Primary cell walls contain approximately equal amounts of cellulose, hemicelluloses, and pectins. In contrast, the more rigid secondary walls generally lack pectin and contain 50 to 80% cellulose. Many secondary walls are further strengthened by **lignin**, a complex polymer of phenolic residues that is responsible for much of the strength and density of wood. The orientation of cellulose microfibrils also differs in primary and secondary cell walls. The cellulose fibers of primary walls appear to be randomly arranged, whereas those of secondary walls are highly ordered (see Figure 12.48). Secondary walls are frequently laid down in layers in which the cellulose fibers differ in orientation, forming a laminated structure that greatly increases cell wall strength.

One of the critical functions of plant cell walls is to prevent cell swelling as a result of osmotic pressure. In contrast to animal cells, plant cells do not maintain an osmotic balance between their cytosol and extracellular fluids. Consequently, osmotic pressure continually drives the flow of water into the cell. This water influx is tolerated by plant cells because their rigid cell walls prevent swelling and bursting. Instead, an internal hydrostatic pressure (called **turgor pressure**) builds up within the cell, eventually equalizing the osmotic pressure and preventing the further influx of water.

Turgor pressure is responsible for much of the rigidity of plant tissues, as is readily apparent from examination of a dehydrated, wilted plant. In addition, turgor pressure provides the basis for a form of cell growth that is unique to plants. In particular, plant cells frequently expand by taking up

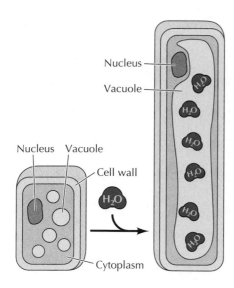

Nucleus

Vacuole

Nucleus   Vacuole

Cell wall

$H_2O$

Cytoplasm

*Figure 12.49*
**Expansion of plant cells**    Turgor pressure drives the expansion of plant cells by the uptake of water, which is accumulated in a large central vacuole.

water without synthesizing new cytoplasmic components (Figure 12.49). Cell expansion by this mechanism is signaled by plant hormones (**auxins**) that weaken a region of the cell wall, allowing turgor pressure to drive the expansion of the cell in that direction. As this occurs, the water that flows into the cell accumulates within a large central vacuole, so the cell expands without increasing the volume of its cytosol. Such expansion can result in a 10- to 100-fold increase in the size of plant cells during development.

As cells expand, new components of the cell wall are deposited outside the plasma membrane. Matrix components, including hemicelluloses and pectins, are synthesized in the Golgi apparatus and secreted. Cellulose, however, is synthesized by a plasma membrane enzyme complex (**cellulose synthase**). In expanding cells, the newly synthesized cellulose microfibrils are deposited at right angles to the direction of cell elongation—an orientation that is thought to play an important role in determining the direction of further cell expansion (Figure 12.50). Interestingly, the cellulose microfibrils in elongating cell walls are laid down in parallel to cortical microtubules underlying the plasma membrane. These microtubules appear to define the orientation of newly synthesized cellulose microfibrils, possibly by determining the direction of movement of the cellulose synthase complexes in the membrane. The cortical microtubules thus define the direction of cell wall growth, which in turn determines the direction of cell expansion and ultimately the shape of the entire plant.

### The Extracellular Matrix

Although animal cells are not surrounded by cell walls, many of the cells in tissues of multicellular organisms are embedded in an **extracellular matrix** consisting of secreted proteins and polysaccharides. The extracellular matrix fills the spaces between cells and binds cells and tissues together. One type of extracellular matrix is exemplified by the thin, sheetlike **basal**

*Figure 12.50*
**Cellulose synthesis during cell elongation**    New cellulose microfibrils, synthesized by a plasma membrane enzyme complex (cellulose synthase), are laid down at right angles to the direction of cell elongation. The direction of cellulose synthesis is parallel to microtubules beneath the plasma membrane.

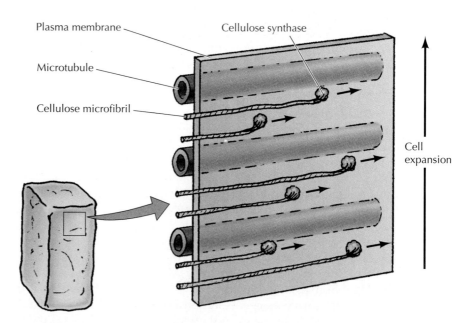

Plasma membrane

Cellulose synthase

Microtubule

Cellulose microfibril

Cell expansion

**laminae, or basement membranes,** upon which layers of epithelial cells rest (Figure 12.51). In addition to supporting sheets of epithelial cells, basal laminae surround muscle cells, adipose cells, and peripheral nerves. Extracellular matrix, however, is most abundant in connective tissues. For example, the loose connective tissue beneath epithelial cell layers consists predominantly of an extracellular matrix in which fibroblasts are distributed. Other types of connective tissue, such as bone, tendon, and cartilage, similarly consist largely of extracellular matrix, which is principally responsible for their structure and function.

Extracellular matrices are composed of tough fibrous proteins embedded in a gel-like polysaccharide ground substance—a design basically similar to that of plant cell walls. In addition to fibrous structural proteins and polysaccharides, the extracellular matrix contains adhesion proteins that link components of the matrix both to one another and to attached cells. The differences between the various types of extracellular matrix result from variations on this general theme. For example, tendons contain a high proportion of fibrous proteins, whereas cartilage contains a high concentration of polysaccharides that form a firm compression-resistant gel. In bone, the extracellular matrix is hardened by deposition of calcium phosphate crystals. The sheetlike structure of basal laminae also results from the utilization of matrix components that differ from those found in connective tissues.

The major structural protein of the extracellular matrix is **collagen**, which is the single most abundant protein in animal tissues. The collagens are a large family of proteins, containing at least 19 different members. They are characterized by the formation of triple helices in which three polypeptide chains are wound tightly around one another in a ropelike structure (Figure 12.52). The triple helix domains of the collagens consist of repeats of the amino acid sequence Gly-X-Y. A glycine (the smallest amino acid, with a side chain consisting only of a hydrogen) is required in every third position in order for the polypeptide chains to pack together close enough to form the collagen triple helix. Proline is frequently found in the X position and hydroxyproline in the Y position; because of their ring structure, these amino acids stabilize the helical conformations of the polypeptide chains.

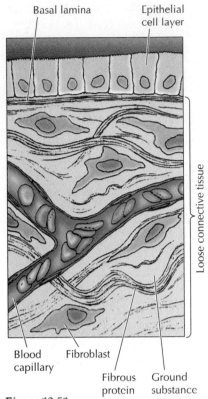

*Figure 12.51*
**Examples of extracellular matrix**
Sheets of epithelial cells rest on a thin layer of extracellular matrix called a basal lamina. Beneath the basal lamina is loose connective tissue, which consists largely of extracellular matrix secreted by fibroblasts. The extracellular matrix contains fibrous structural proteins embedded in a gel-like polysaccharide ground substance.

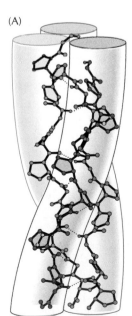

(A)

*Figure 12.52*
**Structure of collagen** (A) Three polypeptide chains coil around one another in a characteristic triple helix structure. (B) The amino acid sequence of a collagen triple helix domain consists of Gly-X-Y repeats, in which X is frequently proline and Y is frequently hydroxyproline (Hyp).

(B) **Amino acid sequence**

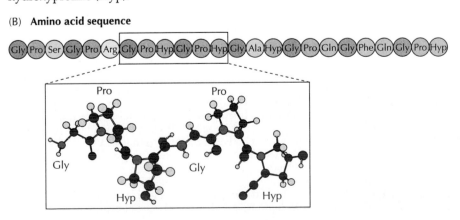

Proline

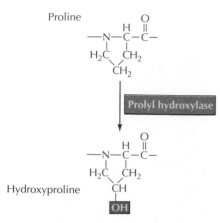

**Prolyl hydroxylase**

Hydroxyproline

**Figure 12.53**
**Formation of hydroxyproline** Prolyl hydroxylase converts proline residues in collagen to hydroxyproline.

*Table 12.2* Representative Members of the Collagen Family

| Collagen class | Types | Tissue distribution |
|---|---|---|
| Fibril-forming | I | Most connective tissues |
| | II | Cartilage and vitreous humor |
| | III | Extensible connective tissues (e.g., skin and lung) |
| | V | Tissues containing collagen I |
| | XI | Tissues containing collagen II |
| Fibril-associated | IX | Tissues containing collagen II |
| | XII | Tissues containing collagen I |
| | XIV | Tissues containing collagen I |
| | XVI | Many tissues |
| Network-forming | IV | Basal laminae |
| Anchoring filaments | VII | Attachments of basal laminae to underlying connective tissue |

The unusual amino acid hydroxyproline is formed within the endoplasmic reticulum by modification of proline residues that have already been incorporated into collagen polypeptide chains (Figure 12.53). Lysine residues in collagen are also frequently converted to hydroxylysines. The hydroxyl groups of these modified amino acids are thought to stabilize the collagen triple helix by forming hydrogen bonds between polypeptide chains. These amino acids are rarely found in other proteins, although hydroxyproline is also common in some of the glycoproteins of plant cell walls.

The most abundant type of collagen (type I collagen) is one of the fibril-forming collagens that are the basic structural components of connective tissues (Table 12.2). The polypeptide chains of these collagens consist of approximately a thousand amino acids or 330 Gly-X-Y repeats. After being secreted from the cell, these collagens assemble into **collagen fibrils** in which the triple helical molecules are associated in regular staggered arrays (Figure 12.54). These fibrils do not form within the cell, because the fibril-forming collagens are synthesized as soluble precursors (**procollagens**) that contain nonhelical segments at both ends of the polypeptide chain. Procollagen is cleaved to collagen after its secretion, so the assembly of collagen into fibrils takes place only outside the cell. The association of collagen molecules in fibrils is further strengthened by the formation of covalent cross-links between the side chains of lysine and hydroxylysine residues. Frequently, the fibrils further associate with one another to form collagen fibers, which can be several micrometers in diameter.

Several other types of collagen do not form fibrils but play distinct roles in various kinds of extracellular matrices. In addition to the fibril-forming collagens, connective tissues contain fibril-associated collagens, which bind to the surface of collagen fibrils and link them both to one another and to other matrix components. Basal laminae form from a different type of col-

(A)

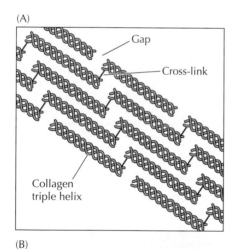

Gap

Cross-link

Collagen triple helix

(B)

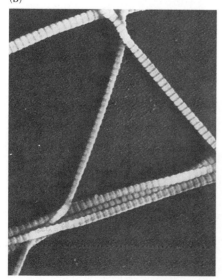

1 μm

**Figure 12.54**
**Collagen fibrils** (A) Collagen molecules assemble in a regular staggered array to form fibrils. The molecules overlap by one-fourth of their length, and there is a short gap between the N terminus of one molecule and the C terminus of the next. The assembly is strengthened by covalent cross-links between side chains of lysine or hydroxylysine residues, primarily at the ends of the molecules. (B) Electron micrograph of collagen fibrils. The staggered arrangement of collagen molecules and the gaps between them are responsible for characteristic cross-striations in the fibrils. (B, J. Gross, F. O. Sahmitt, and D. Fawcett/Visuals Unlimited.)

*Figure 12.55*
**Type IV collagen**   (A) The Gly-X-Y repeat structure of type IV collagen (yellow)
is interrupted by multiple nonhelical sequences (bars). (B) Electron micrograph of
a type IV collagen network. (B, P. D. Yurchenco and J. C. Schittny, 1990, *FASEB J.* 4:
1577.)

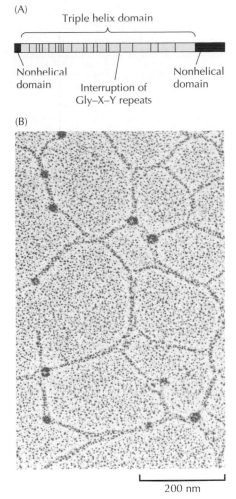

(A)

Triple helix domain

Nonhelical
domain

Interruption of
Gly–X–Y repeats

Nonhelical
domain

(B)

200 nm

lagen (type IV collagen), which is a network-forming collagen (Figure
12.55). The Gly-X-Y repeats of these collagens are frequently interrupted by
short nonhelical sequences. Because of these interruptions, the network-
forming collagens are more flexible than the fibril-forming collagens. Consequently, they assemble into two-dimensional crosslinked networks
instead of fibrils. Yet another type of collagen forms anchoring fibrils,
which link some basal laminae to underlying connective tissues.

Connective tissues also contain **elastic fibers**, which are particularly
abundant in organs that regularly stretch and then return to their original
shape. The lungs, for example, stretch each time a breath is inhaled and
return to their original shape with each exhalation. Elastic fibers are composed principally of a protein called **elastin**, which is crosslinked into a
network by covalent bonds formed between the side chains of lysine
residues (similar to those found in collagen). This network of crosslinked
elastin chains behaves like a rubber band, stretching under tension and
then snapping back when the tension is released.

The fibrous structural proteins of the extracellular matrix are embedded
in gels formed from polysaccharides called **glycosaminoglycans**, or **GAGs**,
which consist of repeating units of disaccharides (Figure 12.56). One sugar
of the disaccharide is either *N*-acetylglucosamine or *N*-acetylgalactosamine
and the second is usually acidic (either glucuronic acid or iduronic acid).
With the exception of hyaluronan, these sugars are modified by the addition of sulfate groups. Consequently, GAGs are highly negatively charged.
Like the pectins of plant cell walls, they bind positively charged ions and
trap water molecules to form hydrated gels, thereby providing mechanical
support to the extracellular matrix.

Hyaluronan is the only GAG that occurs as a single long polysaccharide
chain. All of the other GAGs are linked to proteins to form **proteoglycans**,

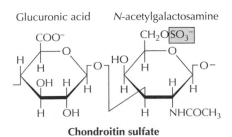

Glucuronic acid   *N*-acetylglucosamine

COO⁻

CH₂OH

**Hyaluronan**

Glucuronic acid   *N*-acetylgalactosamine

COO⁻

CH₂O$SO_3^-$

**Chondroitin sulfate**

Iduronic acid   *N*-acetylgalactosamine

$^-O_3S$   CH₂OH

COO⁻

**Dermatan sulfate**

Galactose   *N*-acetylglucosamine

CH₂OH

CH₂O$SO_3^-$

**Keratan sulfate**

*Figure 12.56*
**Major types of glycosaminoglycans**
Glycosaminoglycans consist of repeating disaccharide units. With the exception of hyaluronan, the sugars frequently contain sulfate. Heparan
sulfate is similar to heparin except that
it contains fewer sulfate groups.

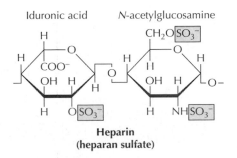

Iduronic acid   *N*-acetylglucosamine

CH₂O$SO_3^-$

COO⁻

O$SO_3^-$   NH$SO_3^-$

**Heparin**
**(heparan sulfate)**

Chondroitin    Keratan    Core protein
sulfate    sulfate

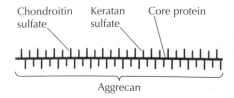

Aggrecan

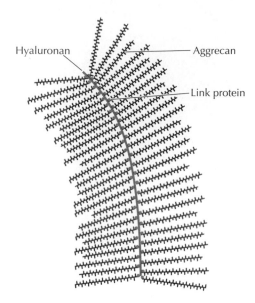

Hyaluronan    Aggrecan

Link protein

*Figure 12.57*
**Complexes of aggrecan and hyaluronan**    Aggrecan is a large proteoglycan consisting of keratan sulfate and chondroitin sulfate chains joined to a core protein. Multiple aggrecan molecules bind to long chains of hyaluronan, forming large complexes in the extracellular matrix of cartilage. This association is stabilized by link proteins.

which can consist of up to 95% polysaccharide by weight. Proteoglycans can contain as few as one or as many as more than a hundred GAG chains attached to serine residues of a core protein. A variety of core proteins (ranging from 10 to >500 kd) have been identified, so the proteoglycans are a diverse group of macromolecules. In addition to being components of the extracellular matrix, some proteoglycans are cell surface proteins that function in cell adhesion.

A number of proteoglycans interact with hyaluronan to form large complexes in the extracellular matrix. A well-characterized example is aggrecan, the major proteoglycan of cartilage (Figure 12.57). More than a hundred chains of chondroitin sulfate and approximately 30 chains of keratan sulfate are attached to a core protein of about 250 kd, forming a proteoglycan of about 3000 kd. Multiple aggrecan molecules then associate with chains of hyaluronan, forming large aggregates (>100,000 kd) that become trapped in the collagen network. Proteoglycans also interact with both collagen and other matrix proteins to form gel-like networks in which the fibrous structural proteins of the extracellular matrix are embedded. For example, perlecan (the major heparan sulfate proteoglycan of basal laminae) binds to both type IV collagen and to the adhesion protein laminin, which is discussed shortly.

Adhesion proteins, the third class of extracellular matrix constituents, are responsible for linking the components of the matrix both to one another and to the surfaces of cells. The prototype of these molecules is **fibronectin**, which is the principal adhesion protein of connective tissues. Fibronectin is a dimeric glycoprotein consisting of two polypeptide chains, each containing nearly 2500 amino acids (Figure 12.58). In the extracellular matrix, fibronectin is further crosslinked into fibrils by disulfide bonds. Fibronectin has binding sites for both collagen and GAGs, so it crosslinks these matrix components. A distinct site on the fibronectin molecule is recognized by cell surface receptors and is thus responsible for the attachment of cells to the extracellular matrix.

Basal laminae contain a distinct adhesion protein called **laminin** (Figure 12.59). Like type IV collagen, laminins can self-assemble into meshlike polymers. Such laminin networks are the major structural

*Figure 12.58*
**Structure of fibronectin**
Fibronectin is a dimer of similar polypeptide chains joined by disulfide bonds near the C terminus. Sites for binding to proteoglycans, cells, and collagen are indicated. The molecule also contains additional binding sites that are not shown.

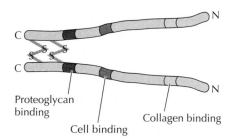

Proteoglycan binding    Cell binding    Collagen binding

**Figure 12.59**
**Structure of laminin**  Laminin consists of three polypeptide chains designated A, B1, and B2. Some of the binding sites for entactin, type IV collagen, proteoglycans, and cell surface receptors are indicated.

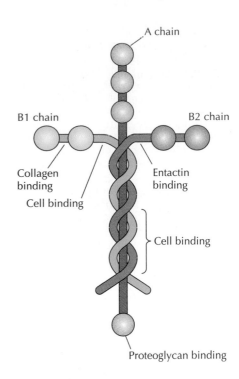

components of the basal laminae synthesized in very early embryos, which do not contain collagen. The laminins also have binding sites for cell surface receptors, type IV collagen, and perlecan. In addition, laminins are tightly associated with another adhesion protein, called **entactin** or **nidogen**, which also binds to type IV collagen. As a result of these multiple interactions, laminin, entactin, type IV collagen, and perlecan form crosslinked networks in the basal lamina.

The major cell surface receptors responsible for the attachment of cells to the extracellular matrix are the **integrins**. The integrins are a family of transmembrane proteins consisting of two subunits, designated $\alpha$ and $\beta$ (Figure 12.60). About 20 different integrins, formed from combinations of 14 known $\alpha$ subunits and 8 known $\beta$ subunits, have been identified. The integrins bind to short amino acid sequences present in multiple components of the extracellular matrix, including collagen, fibronectin, and laminin. The first such integrin-binding site to be characterized was the sequence Arg-Gly-Asp, which is recognized by several members of the integrin family. Other integrins, however, bind to distinct peptide sequences, such as the sequence Asp-Gly-Glu-Ala in collagens and laminin.

In addition to attaching cells to the extracellular matrix, the integrins serve as anchors for the cytoskeleton (Figure 12.61). The resulting linkage of the cytoskeleton to the extracellular matrix is responsible for the stability of cell-matrix junctions. Distinct interactions between integrins and the cytoskeleton are found at two types of cell-matrix junctions, **focal adhesions** and **hemidesmosomes**, which were discussed in Chapter 11. Focal adhesions attach a variety of cells, including fibroblasts, to the extracellular matrix. The cytoplasmic domains of the $\beta$ subunits of integrins at these cell-matrix junctions anchor the actin cytoskeleton by associating with bundles of actin filaments. Hemidesmosomes are specialized sites of epithelial cell attachment at which a specific integrin (designated $\alpha_6\beta_4$) interacts with intermediate filaments instead of with actin. The $\alpha_6\beta_4$ integrin binds to laminin, so hemidesmosomes anchor epithelial cells to the basal lamina.

**Figure 12.60**
**Structure of integrins**  The integrins are heterodimers of two transmembrane subunits, designated $\alpha$ and $\beta$. The $\alpha$ subunit binds divalent cations ($M^{2+}$). The matrix-binding region is composed of portions of both subunits.

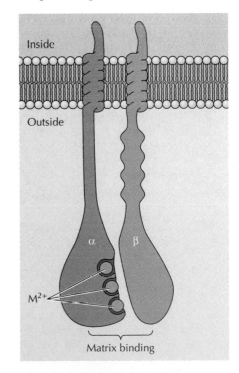

## CELL-CELL INTERACTIONS

Direct interactions between cells, as well as between cells and the extracellular matrix, are critical to the development and function of multicellular organisms. Some cell-cell interactions are transient, such as the interactions between cells of the immune system and the interactions that direct white blood cells to sites of tissue inflammation. In other cases, stable cell-cell junctions play a key role in the organization of cells in tissues. For example, several different types of stable cell-cell junctions are critical to the maintenance and function of epithelial cell sheets. Plant cells also associate with their neighbors not only by interactions between their cell walls, but also by specialized junctions between their plasma membranes.

### Cell Adhesion Proteins

Cell-cell adhesion is a selective process, such that cells adhere only to other cells of specific types. This selectivity was first demonstrated in classical studies of embryo development, which showed that cells from one tissue

*Figure 12.61*
**Junctions between cells and the extracellular matrix** Integrins mediate two types of stable junctions in which the cytoskeleton is linked to the extracellular matrix. In focal adhesions, bundles of actin filaments are an-chored to the $\beta$ subunits of most integrins via associations with a number of other proteins including $\alpha$-actinin, talin, and vinculin (see Figure 11.13). In hemidesmosomes, $\alpha_6\beta_4$ integrin links the basal lamina to intermediate filaments, which are attached to a dense plaque of intracellular pro-

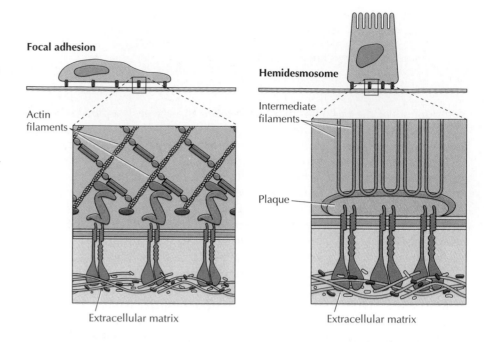

(e.g., liver) specifically adhere to cells of the same tissue rather than to cells of a different tissue (e.g., brain). Such selective cell-cell adhesion is mediated by transmembrane proteins called **cell adhesion molecules**, which can be divided into four major groups: the **selectins**, the **integrins**, the **immunoglobulin (Ig) superfamily** (so named because they contain structural domains similar to immunoglobulins), and the **cadherins** (Table 12.3). Cell adhesion mediated by the selectins, integrins, and cadherins requires $Ca^{2+}$ or $Mg^{2+}$, so many adhesive interactions between cells are $Ca^{2+}$- or $Mg^{2+}$-dependent.

The selectins, integrins, and members of the Ig superfamily mediate transient cell-cell adhesions, as illustrated by the interactions between leukocytes and endothelial cells during the migration of leukocytes from the circulation to sites of tissue inflammation (Figure 12.62). The selectins, which were discussed earlier in this chapter, recognize cell surface carbohydrates and mediate the initial adhesion of leukocytes to endothelial cells. This is followed by the formation of more stable adhesions, in which integrins on the surface of leukocytes bind to intercellular adhesion molecules (ICAMs), which are members of the Ig superfamily expressed on the surface of endothelial cells. The firmly attached leukocytes are then able to penetrate the walls of capillaries and enter the underlying tissue by migrating between endothelial cells.

The binding of ICAMs to integrins is an example of a **heterophilic**

*Table 12.3* Cell Adhesion Molecules

| Family | Ligands recognized | Stable cell junctions |
|---|---|---|
| Selectins | Carbohydrates | No |
| Integrins | Extracellular matrix | Focal adhesions and hemidesmosomes |
| | Members of Ig superfamily | No |
| Ig superfamily | Integrins | No |
| | Homophilic interactions | No |
| Cadherins | Homophilic interactions | Adherens junctions and desmosomes |

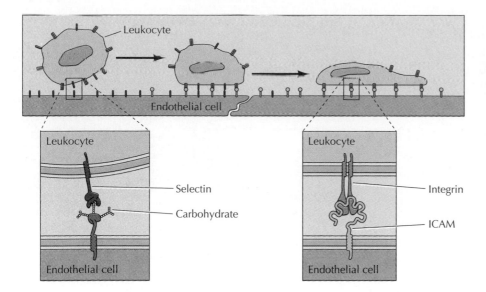

**Figure 12.62**
**Adhesion between leukocytes and endothelial cells**  Leukocytes leave the circulation at sites of tissue inflammation by interacting with the endothelial cells of capillary walls. The first step in this interaction is the binding of leukocyte selectins to carbohydrate ligands on the endothelial cell surface. This step is followed by more stable interactions between leukocyte integrins and members of the Ig superfamily (ICAMs) on endothelial cells.

**interaction**, in which an adhesion molecule on the surface of one cell (e.g., an ICAM) recognizes a different molecule on the surface of another cell (e.g., an integrin). Other members of the Ig superfamily mediate **homophilic interactions**, in which an adhesion molecule on the surface of one cell binds to the same molecule on the surface of another cell. Such homophilic binding leads to selective adhesion between cells of the same type. For example, nerve cell adhesion molecules (N-CAMs) are members of the Ig superfamily expressed on nerve cells, and homophilic binding between N-CAMs contributes to the formation of selective associations between nerve cells during development.

The fourth group of cell adhesion molecules, the cadherins, also display homophilic binding specificities. They are not only involved in selective adhesion between embryonic cells but are also primarily responsible for the formation of stable junctions between cells in tissues. For example, E-cadherin is expressed on epithelial cells, so homophilic interactions between E-cadherins lead to the selective adhesion of epithelial cells to one another. Different members of the cadherin family, such as N-cadherin (neural cadherin) and P-cadherin (placental cadherin), mediate selective adhesion of other cell types.

In contrast to the stable cell-matrix junctions discussed in the preceding section, the cell-cell interactions mediated by the selectins, integrins, and members of the Ig superfamily are transient adhesions in which the cytoskeletons of adjacent cells are not linked to one another. Stable adhesion junctions involving the cytoskeletons of adjacent cells are instead mediated by the cadherins. As discussed in Chapter 11, these cell-cell junctions are of two types: **adherens junctions** and **desmosomes**, in which cadherins or related proteins (desmogleins and desmocollins) are linked to actin bundles and intermediate filaments, respectively (Figure 12.63). The role of the cadherins in linking the cytoskeletons of adjacent cells is thus analogous to that of the integrins in forming stable junctions between cells and the extracellular matrix.

## Tight Junctions

In addition to the adhesion junctions mediated by the cadherins, two other types of specialized cell-cell junctions play key roles in animal tissues. **Tight junctions**, which are usually associated with adherens junctions and

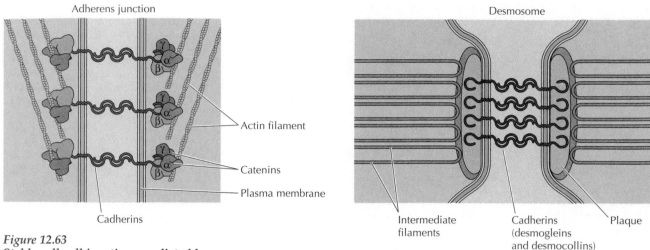

Adherens junction

Actin filament

Catenins

Plasma membrane

Cadherins

Desmosome

Intermediate
filaments

Cadherins
(desmogleins
and desmocollins)

Plaque

*Figure 12.63*
**Stable cell-cell junctions mediated by the cadherins**   Homophilic interactions between cadherins mediate two types of stable cell-cell adhesions. In adherens junctions, the cadherins are linked to bundles of actin filaments via the catenins (see Figure 11.14). In desmosomes, dense protein plaques link members of the cadherin superfamily (desmogleins and desmocollins) to intermediate filaments (see Figure 11.34).

desmosomes in a **junctional complex** (Figure 12.64), are critically important to the function of epithelial cell sheets as barriers between fluid compartments. For example, the intestinal epithelium separates the lumen of the intestine from the underlying connective tissue, which contains blood capillaries. Tight junctions play two roles in allowing epithelia to fulfill such barrier functions. First, tight junctions form seals that prevent the free passage of molecules (including ions) between the cells of epithelial sheets. Second, tight junctions separate the apical and basolateral domains of the plasma membrane by preventing the free diffusion of lipids and membrane proteins between them. Consequently, specialized transport systems in the apical and basolateral domains are able to control the traffic of molecules between distinct extracellular compartments, such as the transport of glucose between the intestinal lumen and the blood supply (see Figure 12.32).

(A)

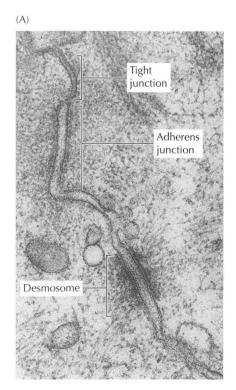

Tight junction

Adherens junction

Desmosome

*Figure 12.64*
**Tight junctions**   (A) Electron micrograph of epithelial cells joined by a junctional complex, including a tight junction, an adherens junction, and a desmosome. (B) Tight junctions are formed by interactions between strands of transmembrane proteins (occludin) on adjacent cells. (A, Don Fawcett/Photo Researchers Inc.)

(B)

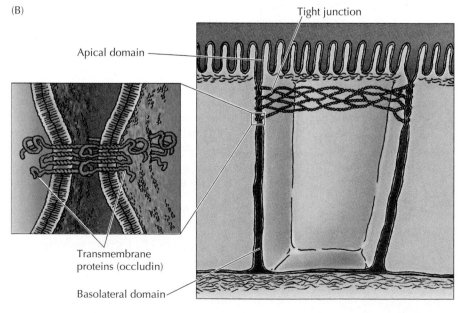

Tight junction

Apical domain

Transmembrane
proteins (occludin)

Basolateral domain

Tight junctions are the closest known contacts between adjacent cells. They were originally described as sites of apparent fusion between the outer leaflets of the plasma membranes, although it is now clear that the membranes do not fuse. Instead, tight junctions appear to be formed by a network of protein strands that continues around the entire circumference of the cell (see Figure 12.64). Each strand in these networks is thought to be composed of a transmembrane protein (called occludin) that binds to similar proteins on adjacent cells, thereby sealing the space between their plasma membranes.

## Gap Junctions

**Gap junctions**, which are found in most animal tissues, serve as direct connections between the cytoplasms of adjacent cells. They provide open channels through the plasma membrane, allowing ions and small molecules (less than approximately a thousand daltons) to diffuse freely between neighboring cells, but preventing the passage of proteins and nucleic acids. Consequently, gap junctions couple both the metabolic activities and the electric responses of the cells they connect. Most cells in animal tissues—including epithelial cells, endothelial cells, and the cells of cardiac and smooth muscle—communicate by gap junctions. In electrically excitable cells, such as heart muscle cells, the direct passage of ions through gap junctions couples and synchronizes the contractions of neighboring cells. Gap junctions also allow the passage of some intracellular signaling molecules, such as cAMP and $Ca^{2+}$, between adjacent cells, potentially coordinating the responses of cells in tissues.

Gap junctions are constructed of transmembrane proteins called **connexins** (Figure 12.65). Six connexins assemble to form a cylinder with an open aqueous pore in its center. Such an assembly of connexins in the plasma membrane of one cell then aligns with the connexins of an adjacent cell, forming an open channel between the two cytoplasms. The plasma membranes of the two cells are separated by a gap corresponding to the space occupied by the connexin extracellular domains—hence the term "gap junction," which was coined by electron microscopists.

## Plant Cell Adhesion and Plasmodesmata

Adhesion between plant cells is mediated by their cell walls rather than by transmembrane proteins. In particular, a specialized pectin-rich region of the cell wall called the **middle lamella** acts as a glue to hold adjacent cells together. Because of the rigidity of plant cell walls, stable associations between plant cells do not require the formation of cytoskeletal links, such as those provided by the desmosomes and adherens junctions of animal cells. However, adjacent plant cells communicate with each other through cytoplasmic connections called **plasmodesmata** (singular, plasmodesma), which function analogously to animal cell gap junctions.

(A)

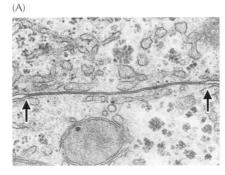

**Figure 12.65**
**Gap junctions** (A) Electron micrograph of a gap junction (arrows) between two liver cells. (B) Gap junctions consist of assemblies of six connexins, which form open channels through the plasma membranes of adjacent cells. (A, Don Fawcett and R. Wood/Photo Researchers Inc.)

(B)

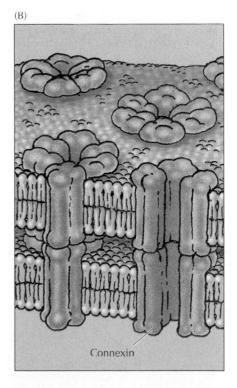

Connexin

(A)

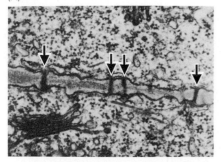

(B)

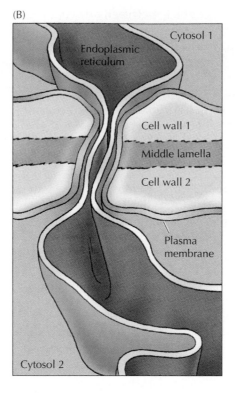

Cytosol 1

Endoplasmic
reticulum

Cell wall 1

Middle lamella

Cell wall 2

Plasma
membrane

Cytosol 2

*Figure 12.66*
**Plasmodesmata**    (A) Electron micrograph of plasmodesmata (arrows).
(B) At plasmodesmata, the plasma membranes of neighboring cells are continuous, forming cytoplasmic channels through the adjacent cell walls. An extension of the endoplasmic reticulum usually passes through the channel. (A, BioPhoto Associates/Photo Researchers Inc.)

0.2 µm

Despite their similarities in function, plasmodesmata are structurally unrelated to gap junctions. At each plasmodesma the plasma membrane of one cell is continuous with that of its neighbor, creating an open channel between the two cytosols (Figure 12.66). An extension of the smooth endoplasmic reticulum passes through the pore, leaving a ring of surrounding cytoplasm through which ions and small molecules are able to pass freely between the cells. In addition, plasmodesmata can expand in response to appropriate stimuli, permitting the regulated passage of macromolecules between adjacent cells.

## KEY TERMS

phosphatidylcholine, phosphatidylethanolamine, phosphatidylserine, sphingomyelin, phosphatidylinositol, glycolipids, cholesterol

fluid mosaic model, peripheral membrane proteins, integral membrane proteins, transmembrane proteins, porins, glycosylphosphatidylinositol (GPI) anchor

apical domain, basolateral domain

glycocalyx, selectin

passive diffusion

# Summary

## STRUCTURE OF THE PLASMA MEMBRANE

*The Phospholipid Bilayer:* The fundamental structure of the plasma membrane is a phospholipid bilayer, which also contains glycolipids and cholesterol.

*Membrane Proteins:* Associated proteins are responsible for carrying out specific membrane functions. Membranes are viewed as fluid mosaics in which proteins are inserted into phospholipid bilayers.

*Mobility of Membrane Proteins:* Proteins are free to diffuse laterally through the phospholipid bilayer. However, the mobility of some proteins is restricted by their associations with other molecules. In addition, tight junctions prevent proteins from moving between distinct plasma membrane domains of epithelial cells.

*The Glycocalyx:* The cell surface is covered by a carbohydrate coat called the glycocalyx. Cell surface carbohydrates serve as markers for cell-cell recognition.

## TRANSPORT OF SMALL MOLECULES

*Passive Diffusion:* Small hydrophobic molecules are able to cross the plasma membrane by diffusing through the phospholipid bilayer.

*Facilitated Diffusion and Carrier Proteins:* The passage of most biological molecules is mediated by carrier or channel proteins that allow polar and charged molecules to cross the plasma membrane without interacting with its hydrophobic interior.

facilitated diffusion, carrier protein, channel protein

*Ion Channels:* Ion channels mediate the rapid passage of selected ions across the plasma membrane. They are particularly well characterized in nerve and muscle cells, where they are responsible for the transmission of electric signals.

ion channel, ligand-gated channel, voltage-gated channel, patch clamp technique, Nernst equation, action potential

*Active Transport Driven by ATP Hydrolysis:* Energy derived from ATP hydrolysis can drive the transport of molecules against their electrochemical gradients.

active transport, ion pump, $Na^+$-$K^+$ pump, $Na^+$-$K^+$ ATPase, ABC transporter

*Active Transport Driven by Ion Gradients:* Ion gradients are frequently used as a source of energy to drive the active transport of other molecules.

symport, uniport, antiport

## ENDOCYTOSIS

*Phagocytosis:* Cells ingest large particles, such as bacteria and cell debris, by phagocytosis.

endocytosis, phagocytosis, pinocytosis, phagosome, phagolysosome

*Receptor-Mediated Endocytosis:* The best-characterized form of endocytosis is receptor-mediated endocytosis, which provides a mechanism for the selective uptake of specific macromolecules.

receptor-mediated endocytosis, clathrin-coated pit, clathrin-coated vesicle, ligand, low-density lipoprotein (LDL), fluid phase endocytosis, caveola, macropinocytosis

*Protein Trafficking in Endocytosis:* Molecules taken up by endocytosis are transported to endosomes, where they are sorted for recycling to the plasma membrane or degradation in lysosomes.

endosome, receptor down-regulation, transcytosis

## CELL WALLS AND THE EXTRACELLULAR MATRIX

*Bacterial Cell Walls:* The principal component of bacterial cell walls is a peptidoglycan consisting of polysaccharide chains crosslinked by short peptides.

peptidoglycan

*Plant Cell Walls:* The cell walls of fungi, algae, and higher plants are composed of fibrous polysaccharides (e.g., cellulose) embedded in a gel-like matrix of polysaccharides and proteins. Their rigid cell walls allow plant cells to expand rapidly by the uptake of water.

chitin, cellulose, cellulose microfibril, hemicellulose, pectin, primary cell wall, secondary cell wall, lignin, turgor pressure, auxin, cellulose synthase

*The Extracellular Matrix:* Animal cells in tissues are surrounded by an extracellular matrix consisting of secreted proteins and polysaccharides. Cell surface receptors bind to the extracellular matrix and anchor the cytoskeleton at cell-matrix junctions.

extracellular matrix, basal lamina, basement membrane, collagen, collagen fibril, procollagen, elastic fiber, elastin, glycosaminoglycan (GAG), proteoglycan, fibronectin, laminin, entactin, nidogen, integrin, focal adhesion, hemidesmosome

## CELL-CELL INTERACTIONS

*Cell Adhesion Proteins:* Selective cell-cell interactions are mediated by four major groups of cell adhesion proteins: selectins, integrins, immunoglobulin superfamily members, and cadherins. The cadherins link the cytoskeletons of adjacent cells at stable cell-cell junctions.

cell adhesion molecule, selectin, integrin, immunoglobulin (Ig) superfamily, cadherin, heterophilic interaction, homophilic interaction, adherens junction, desmosome

*Tight Junctions:* Tight junctions prevent the free passage of molecules between epithelial cells and separate the apical and basolateral domains of the plasma membrane.

tight junction, junctional complex

| gap junction, connexin | *Gap Junctions:* Gap junctions are open channels connecting the cytoplasms of adjacent cells. |
|---|---|
| middle lamella, plasmodesma | *Plant Cell Adhesion and Plasmodesmata:* Adjacent plant cells are linked by cytoplasmic connections called plasmodesmata. |

## QUESTIONS

**1.** What would be the consequences of incorporating porins into the plasma membrane instead of into the outer membrane of bacteria?

**2.** The concentration of K$^+$ is about 20 times higher inside squid axons than in extracellular fluids, generating an equilibrium membrane potential of –75 mV. What would be the expected equilibrium membrane potential if the K$^+$ concentration were only 10 times higher inside than outside the cell?

Why does the actual resting membrane potential (–60 mV) differ from the K$^+$ equilibrium potential of –75 mV?

**3.** An important function of the Na$^+$-K$^+$ pump in animal cells is the maintenance of osmotic equilibrium. Why is this unnecessary for plant cells?

**4.** An anticipated problem in gene therapy for cystic fibrosis is that normal CFTR genes would probably be

successfully transferred to only a fraction of the cells in the patient's respiratory epithelium. How is this potential pitfall affected by the presence of gap junctions?

**5.** What is the importance of selectively targeting different glucose transporters to the apical and basolateral domains of the plasma membrane of intestinal epithelial cells? What is the role of tight junctions in this process?

## REFERENCES AND FURTHER READING

### General References

Petty, H. R. 1993. *Molecular Biology of Membranes: Structure and Function.* New York: Plenum.

Yeagle, P. L. 1993. *The Membranes of Cells.* 2nd ed. San Diego, CA: Academic Press.

### Structure of the Plasma Membrane

Bennett, V. and D. M. Gilligan. 1993. The spectrin-based membrane skeleton and micron-scale organization of the plasma membrane. *Ann. Rev. Cell Biol.* 9: 27–66. [R]

Branton, D., C. M. Cohen and J. Tyler. 1981. Interaction of cytoskeletal proteins on the human erythrocyte membrane. *Cell* 24: 24–32. [P]

Bretscher, M. 1985. The molecules of the cell membrane. *Sci. Am.* 253(4): 100–108. [R]

Bretscher, M. S. and S. Munro. 1993. Cholesterol and the Golgi apparatus. *Science* 261: 1280–1281. [R]

Casey, P. J. 1995. Protein lipidation in cell signaling. *Science* 268: 221–225. [R]

Clarke, S. 1992. Protein isoprenylation and methylation at carboxy-terminal cysteine residues. *Ann. Rev. Biochem.* 61: 355–386. [R]

Diesenhofer, J., O. Epp, K. Miki, R. Huber and H. Michel. 1985. The structure of the protein subunits in the photosynthetic reaction centre of *Rhodopseudomonas viridis* at 3-Å resolution. *Nature* 318: 618–624. [P]

Englund, P. T. 1993. The structure and biosynthesis of glycosyl phosphatidylinositol anchors. *Ann. Rev. Biochem.* 62: 121–138. [R]

Frye, L. D. and M. Edidin. 1970. The rapid intermixing of cell surface antigens after formation of mouse-human heterokaryons. *J. Cell Sci.* 7: 319–335. [P]

Jacobson, K., E. D. Sheets and R. Simson. 1995. Revisiting the fluid mosaic model of membranes. *Science* 268: 1441–1442. [R]

Lasky, L. A. 1995. Selectin-carbohydrate interactions and the initiation of the inflammatory response. *Ann. Rev. Biochem.* 64: 113–139. [R]

Nikaido, H. 1994. Porins and specific diffusion channels in bacterial outer membranes. *J. Biol. Chem.* 269: 3905–3908. [R]

Rees, D. C., H. Komiya, T. O. Yeates, J. P. Allen and G. Feher. 1989. The bacterial photosynthetic reaction center as a model for membrane proteins. *Ann. Rev. Biochem.* 58: 607–633. [R]

Sharon, N. and H. Lis. 1993. Carbohydrates in cell recognition. *Sci. Am.* 268(1): 82–89. [R]

Singer, S. J. 1990. The structure and insertion of integral proteins in membranes. *Ann. Rev. Cell Biol.* 6: 247–296. [R]

Singer, S. J. and G. L. Nicolson. 1972. The fluid mosaic model of the structure of cell membranes. *Science* 175: 720–731. [P]

### Transport of Small Molecules

Bell, G. I., C. F. Burant, J. Takeda and G. W. Gould. 1993. Structure and function of mammalian facilitative sugar transporters. *J. Biol. Chem.* 268: 19161–19164. [R]

Carafoli, E. 1992. The Ca$^{2+}$ pump of the plasma membrane. *J. Biol. Chem.* 267: 2115–2118. [R]

Catterall, W. A. 1995. Structure and function of voltage-gated ion channels. *Ann. Rev. Biochem.* 64: 493–531. [R]

Chrispeels, M. J. and P. Agre. 1994. Aquaporins: Water channel proteins of plant and animal cells. *Trends Biochem. Sci.* 19: 421–425. [R]

Gottesman, M. M. and I. M. Pastan. 1993. Biochemistry of multidrug resistance mediated by the multidrug transporter. *Ann. Rev. Biochem.* 62: 385–427. [R]

Higgins, C. F. 1992. ABC transporters: From microorganisms to man. *Ann. Rev. Cell Biol.* 8: 67–113. [R]

Higgins, C. F. 1995. The ABC of channel regulation. *Cell* 82: 693–696. [R]

Hille, B. 1992. *Ionic Channels of Excitable Membranes.* 2nd ed. Sunderland, MA: Sinauer Associates.

Hodgkin, A. L. and A. F. Huxley. 1952. A quantitative description of membrane current and its application to conduction and excitation in nerve. *J. Physiol.* 117: 500–544. [P]

Jan, L. Y. and Y. N. Jan. 1992. Tracing the roots of ion channels. *Cell* 69: 715–718. [R]

Jan, L. Y. and Y. N. Jan. 1994. Potassium channels and their evolving gates. *Nature* 371: 119–122. [R]

Lanyi, J. K. 1995. Bacteriorhodopsin as a model for proton pumps. *Nature* 375: 461–463. [R]

Levitan, I. B. and L. K. Kaczmarek. 1991. *The Neuron: Cell and Molecular Biology.* New York: Oxford Univ. Press.

Lingrel, J. B. and T. Kuntzweiler. 1994. Na+,K+-ATPase. *J. Biol. Chem.* 269: 19659–19662. [R]

Miller, R. J. 1992. Voltage-sensitive $Ca^{2+}$ channels. *J. Biol. Chem.* 267: 1403–1406. [R]

Neher, E. and B. Sakmann. 1992. The patch clamp technique. *Sci. Am.* 266(3): 44–51. [R]

Sakmann, B. 1992. Elementary steps in synaptic transmission revealed by currents through single ion channels. *Science* 256: 503–512. [R]

Silverman, M. 1991. Structure and function of hexose transporters. *Ann. Rev. Biochem.* 60: 757–794. [R]

Unwin, N. 1993. Neurotransmitter action: Opening of ligand-gated ion channels. *Cell* 72/*Neuron* 10 (Suppl.): 31–41. [R]

Welsh, M. J. and A. E. Smith. 1993. Molecular mechanisms of CFTR chloride channel dysfunction in cystic fibrosis. *Cell* 73: 1251–1254. [R]

## Endocytosis

Beron, W., C. Alvarez-Dominguez, L. Mayorga and P. D. Stahl. 1995. Membrane trafficking along the phagocytic pathway. *Trends Cell Biol.* 5: 100–104. [R]

Brown, M. S. and J. L. Goldstein. 1986. A receptor-mediated pathway for cholesterol homeostasis. *Science* 232: 34–47. [R]

Gruenberg, J. and K. E. Howell. 1989. Membrane traffic in endocytosis: Insights from cell-free assays. *Ann. Rev. Cell Biol.* 5: 453–481. [R]

Mostov, K., G. Apodaca, B. Aroeti and C. Okamoto. 1992. Plasma membrane protein sorting in polarized epithelial cells. *J. Cell Biol.* 116: 577–583. [R]

Mostov, K. and N. E. Simister. 1985. Transcytosis. *Cell* 43: 389–390. [R]

Nelson, W. J. 1992. Regulation of cell surface polarity from bacteria to mammals. *Science* 258: 948–955. [R]

Parton, R. G. and K. Simons. 1995. Digging into caveolae. *Science* 269: 1398–1401. [R]

Pearse, B. M. F. and M. S. Robinson. 1990. Clathrin, adaptors, and sorting. *Ann. Rev. Cell Biol.* 6: 151–171. [R]

Rabinovitch, M. 1995. Professional and non-professional phagocytes: An introduction. *Trends Cell Biol.* 5: 85–87. [R]

Sudhof, T. C. 1995. The synaptic vesicle cycle: A cascade of protein-protein interactions. *Nature* 375: 645–653. [R]

Swanson, J. A. and C. Watts. 1995. Macropinocytosis. *Trends Cell Biol.* 5:424–428. [R]

Trowbridge, I. S., J. F. Collawn and C. R. Hopkins. 1993. Signal-dependent membrane protein trafficking in the endocytic pathway. *Ann. Rev. Cell Biol.* 9: 129–161. [R]

Van Deurs, B., P. K. Holm, K. Sandvig and S. H. Hansen. 1993. Are caveolae involved in clathrin-independent endocytosis? *Trends Cell Biol.* 3: 249–251. [R]

## Cell Walls and the Extracellular Matrix

Carpita, N. C. and D. M. Gibeaut. 1993. Structural models of primary cell walls in flowering plants: Consistency of molecular structure with the physical properties of the walls during growth. *Plant J.* 3: 1–30. [R]

Cyr, R. J. 1994. Microtubules in plant morphogenesis: Role of the cortical array. *Ann. Rev. Cell Biol.* 10: 153–180. [R]

Engel, J. 1992. Laminins and other strange proteins. *Biochemistry* 31: 10643–10651. [R]

Fosket, D. E. 1994. *Plant Growth and Development: A Molecular Approach.* San Diego: Academic Press.

Hynes, R. O. 1990. *Fibronectins.* New York: Springer-Verlag.

Hynes, R. O. 1992. Integrins: Versatility, modulation, and signaling in cell adhesion. *Cell* 69: 11–25. [R]

Kjellén, L. and U. Lindahl. 1991. Proteoglycans: Structures and interactions. *Ann. Rev. Biochem.* 60: 443–475. [R]

Loftus, J. C., J. W. Smith and M. H. Ginsberg. 1994. Integrin-mediated cell adhesion: The extracellular face. *J. Biol. Chem.* 269: 25235–25238. [R]

McDonald, J. A. 1988. Extracellular matrix assembly. *Ann. Rev. Cell Biol.* 4: 183–207. [R]

McNeil, M., A. G. Darvill, S. C. Fry and P. Albersheim. 1984. Structure and function of the primary cell walls of plants. *Ann. Rev. Biochem.* 53: 625–663. [R]

Mecham, R. P. 1991. Laminin receptors. *Ann. Rev. Cell Biol.* 7: 71–91. [R]

Neidhardt, F. C., J. L. Ingraham and M. Schaechter. 1990. *Physiology of the Bacterial Cell: A Molecular Approach.* Sunderland, MA.: Sinauer Associates.

Prockop, D. J. and K. I. Kivirikko. 1995. Collagens: Molecular biology, diseases, and potentials for therapy. *Ann. Rev. Biochem.* 64: 403–434. [R]

Ruoslahti, E. 1988. Fibronectin and its receptors. *Ann. Rev. Biochem.* 57: 375–413. [R]

Ruoslahti, E. 1988. Structure and biology of proteoglycans. *Ann. Rev. Cell Biol.* 4: 229–255. [R]

Talz, L. 1994. Expansins: Proteins that promote cell wall loosening in plants. *Proc. Natl. Acad. Sci. USA* 91: 7387–7389. [R]

Varner, J. E. and L.-S. Lin. 1989. Plant cell wall architecture. *Cell* 56: 231–239. [R]

Vertel, B. M. 1995. The ins and outs of aggrecan. *Trends Cell Biol.* 5: 458–464. [R]

Vuorio, E. and B. de Crombrugghe. 1990. The family of collagen genes. *Ann. Rev. Biochem.* 59: 837–872. [R]

Yurchenco, P. D. and J. C. Schnitty. 1990. Molecular architecture of basement membranes. *FASEB J.* 4: 1577–1590. [R]

## Cell-Cell Interactions

Beyer, E. C. 1993. Gap junctions. *Int. Rev. Cytol.* 137C: 1–37. [R]

Citi, S. 1993. The molecular organization of tight junctions. *J. Cell Biol.* 121: 485–489. [R]

Epel, B. L. 1994. Plasmodesmata: Composition, structure, and trafficking. *Plant Mol. Biol.* 26: 1343–1356. [R]

Geiger, B. and O. Ayalon. 1992. Cadherins. *Ann. Rev. Cell Biol.* 8: 307–332. [R]

Gumbiner, B. M. 1993. Breaking through the tight junction barrier. *J. Cell Biol.* 123: 1631–1633. [R]

Gumbiner, B. M. 1996. Cell adhesion: The molecular basis of tissue architecture and morphogenesis. *Cell* 84: 345–357. [R]

Kumar, N. M. and N. B. Gilula. 1996. The gap junction communication channel. *Cell* 84: 381–388. [R]

Schwarz, M. A., K. Owaribe, J. Kartenbeck and W. W. Franke. 1990. Desmosomes and hemidesmosomes: Constitutive molecular components. *Ann. Rev. Cell Biol.* 6: 461–491. [R]

Springer, T. A. 1990. Adhesion receptors of the immune system. *Nature* 346: 425–434. [R]

Springer, T. A. 1994. Traffic signals for lymphocyte recirculation and leukocyte emigration: The multistep paradigm. *Cell* 76: 301–314. [R]

Takeichi, M. 1991. Cadherin cell adhesion receptors as a morphogenetic regulator. *Science* 251: 1451–1455. [R]

Zambryski, P. 1995. Plasmodemata: Plant channels for molecules on the move. *Science* 270: 1943–1944. [R]

# PART IV Cell Regulation

Georgia O'Keefe, *Corn, Dark I*, 1924

# 13

# Cell Signaling

ALL CELLS RECEIVE AND RESPOND TO SIGNALS from their surroundings. Even the simplest bacteria sense and swim toward high concentrations of nutrients, such as glucose or amino acids. Many unicellular eukaryotes also respond to signaling molecules secreted by other cells, allowing cell-cell communication. Mating between yeast cells, for example, is signaled by peptides that are secreted by one cell and bind to receptors on the surface of another. It is in multicellular organisms, however, that cell-cell communication reaches its highest level of sophistication. Whereas the cells of prokaryotes and unicellular eukaryotes are largely autonomous, the behavior of each individual cell in multicellular plants and animals must be carefully regulated to meet the needs of the organism as a whole. This is accomplished by a variety of signaling molecules that are secreted or expressed on the surface of one cell and bind to receptors expressed by other cells, thereby integrating and coordinating the functions of the many individual cells that make up organisms as complex as human beings.

The binding of most signaling molecules to their receptors initiates a series of intracellular reactions that regulate virtually all aspects of cell behavior, including metabolism, movement, proliferation, and differentiation. Understanding the molecular mechanisms responsible for these pathways of cell signaling has thus become a major area of active research. Interest in this area is further heightened by the fact that most cancers arise as a result of a breakdown in the signaling pathways that control normal cell growth and differentiation. Conversely, many of our current insights into cell signaling mechanisms have come from the study of cancer cells—a striking example of the fruitful interplay between medicine and basic research in cell and molecular biology.

## SIGNALING MOLECULES AND THEIR RECEPTORS

Many different kinds of molecules transmit information between the cells of multicellular organisms. Although all these molecules act as ligands that bind to receptors expressed by their target cells, there is considerable variation in the structure and function of the different types of molecules that serve as signal transmitters. Structurally, the signaling molecules used by plants and animals range in complexity from simple gases to proteins. Some of these molecules carry signals over long distances, whereas others act locally to convey information between neighboring cells. In addition, signaling molecules differ in their mode of action on their target cells. Some signaling molecules are able to cross the plasma membrane and bind to intracellular receptors in the cytoplasm or nucleus, whereas most bind to receptors expressed on the target cell surface. The sections that follow discuss the major types of signaling molecules and the receptors with which they interact. Subsequent discussion in this chapter focuses on the mechanisms by which cell surface receptors then function to regulate cell behavior.

## Modes of Cell-Cell Signaling

The multiple varieties of cell-cell signaling are frequently divided into three general categories on the basis of the distance over which signals are transmitted (Figure 13.1). In **endocrine signaling**, the signaling molecules (**hormones**) are secreted by specialized endocrine cells and carried through the circulation to act on target cells at distant body sites. A classic example is provided by the steroid hormone estrogen, which is produced by the ovary and stimulates development and maintenance of the female reproductive system and secondary sex characteristics. In animals, more than 50 different hormones are produced by endocrine glands, including the pituitary, thyroid, parathyroid, pancreas, adrenal glands, and gonads.

In contrast to hormones, some signaling molecules act locally to affect the behavior of nearby cells. In **paracrine signaling**, a molecule released by one cell acts on neighboring target cells. An example is provided by the action of neurotransmitters in carrying signals between nerve cells at a synapse. In addition, some cells respond to signaling molecules that are expressed on the surface of an adjacent cell rather than being released into extracellular fluids. These kinds of signaling play critical roles in regulating cell-cell interactions, including the many interactions between different types of cells that take place during embryonic development.

Finally, some cells respond to signaling molecules that they themselves produce. One important example of such **autocrine signaling** is the response of cells of the vertebrate immune system to foreign antigens. Certain types of T lymphocytes respond to antigenic stimulation by synthesizing a growth factor that drives their own proliferation, thereby increasing the number of responsive T lymphocytes and amplifying the immune response. It is also noteworthy that abnormal autocrine signaling frequently contributes to the uncontrolled growth of cancer cells (see Chapter 15). In this situation, a cancer cell produces a growth factor to which it also responds, thereby continuously driving its own unregulated proliferation.

## Steroid Hormones and the Steroid Receptor Superfamily

As already noted, all signaling molecules act by binding to receptors expressed by their target cells. In many cases, these receptors are expressed

(A) Endocrine signaling

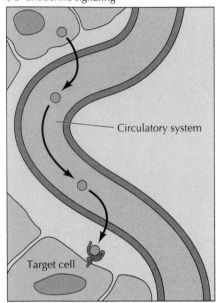

(B) Paracrine signaling

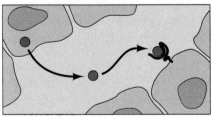

(C) Direct cell-to-cell signaling

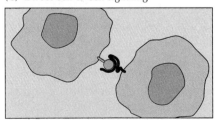

(D) Autocrine signaling

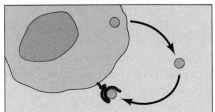

**Figure 13.1**
**Modes of cell-cell signaling** (A) In endocrine signaling, hormones are carried through the circulatory system to act on distant target cells. (B) In paracrine signaling, a molecule released from one cell acts locally to affect nearby target cells. (C) Rather than being released into extracellular fluids, some signaling molecules remain bound to the cell surface and act as ligands in direct cell-cell signaling. (D) In autocrine signaling, a cell produces a signaling molecule to which it also responds.

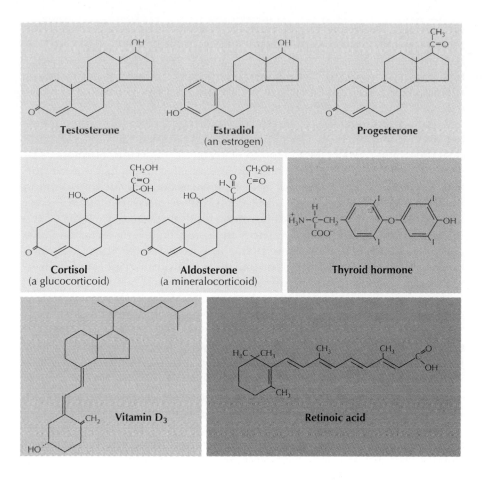

*Figure 13.2*

**Structure of steroid hormones, thyroid hormone, vitamin D₃, and retinoic acid**   The steroids include the sex hormones (testosterone, estrogen, and progesterone), glucocorticoids, and mineralocorticoids.

Testosterone

Estradiol
(an estrogen)

Progesterone

Cortisol
(a glucocorticoid)

Aldosterone
(a mineralocorticoid)

Thyroid hormone

Vitamin D₃

Retinoic acid

*Figure 13.3*

**Action of steroid hormones**   The steroid hormones diffuse across the plasma membrane and bind to intracellular receptors, which directly stimulate transcription of their target genes. The steroid hormone receptors bind DNA as dimers.

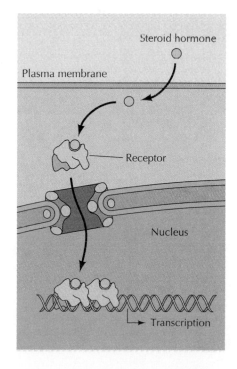

Steroid hormone

Plasma membrane

Receptor

Nucleus

Transcription

on the target cell surface, but some receptors are intracellular proteins located in the cytosol or the nucleus. These intracellular receptors respond to small hydrophobic signaling molecules that are able to diffuse across the plasma membrane. The **steroid hormones** are the classic examples of this group of signaling molecules, which also includes thyroid hormone, vitamin D₃, and retinoic acid (Figure 13.2).

The steroid hormones (including testosterone, estrogen, progesterone, the corticosteroids, and ecdysone) are all synthesized from cholesterol. **Testosterone**, **estrogen**, and **progesterone** are the sex steroids, which are produced by the gonads. The **corticosteroids** are produced by the adrenal gland. They include the **glucocorticoids**, which act on a variety of cells to stimulate production of glucose, and the **mineralocorticoids**, which act on the kidney to regulate salt and water balance. **Ecdysone** is an insect hormone that plays a key role in development by triggering the metamorphosis of larvae to adults.

Although thyroid hormone, vitamin D₃, and retinoic acid are both structurally and functionally distinct from the steroids, they share a common mechanism of action in their target cells. **Thyroid hormone** is synthesized from tyrosine in the thyroid gland; it plays important roles in development and regulation of metabolism. **Vitamin D₃** regulates $Ca^{2+}$ metabolism and bone growth. **Retinoic acid** and related compounds (**retinoids**) synthesized from vitamin A play important roles in vertebrate development.

Because of their hydrophobic character, the steroid hormones, thyroid hormone, vitamin D₃, and retinoic acid are able to enter cells by diffusing across the plasma membrane (Figure 13.3). Once inside the cell, they bind to intracellular receptors that are expressed by the hormonally responsive

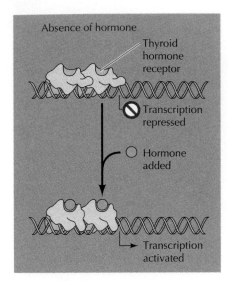

**Figure 13.4**
**Gene regulation by the thyroid hormone receptor**   Thyroid hormone receptor binds DNA in either the presence or absence of hormone. However, hormone binding changes the function of the receptor from a repressor to an activator of target gene transcription.

target cells. These receptors, which are members of a family of proteins known as the **steroid receptor superfamily**, are transcription factors that contain related domains for ligand binding, DNA binding, and transcriptional activation. Ligand binding regulates their function as activators or repressors of their target genes, so the steroid hormones and related molecules directly regulate gene expression.

Ligand binding has distinct effects on different receptors. Some members of the steroid receptor superfamily, such as the estrogen and glucocorticoid receptors, are unable to bind to DNA in the absence of hormone. The binding of hormone induces a conformational change in the receptor, allowing it to bind to regulatory DNA sequences and activate transcription of target genes. In other cases, the receptor binds DNA in either the presence or absence of hormone, but hormone binding alters the activity of the receptor as a transcriptional regulatory molecule. For example, thyroid hormone receptor acts as a repressor in the absence of hormone, but hormone binding converts it to an activator that stimulates transcription of thyroid hormone-inducible genes (Figure 13.4).

### Nitric Oxide

The simple gas **nitric oxide** (NO) is a major paracrine signaling molecule in the nervous, immune, and circulatory systems. Like the steroid hormones, NO is able to diffuse directly across the plasma membrane of its target cells. The molecular basis of NO action, however, is distinct from that of steroid action; rather than binding to a receptor that regulates transcription, NO alters the activity of intracellular target enzymes.

Nitric oxide is synthesized from the amino acid arginine by the enzyme nitric oxide synthase (Figure 13.5). Once synthesized, NO diffuses out of the cell and can act locally to affect nearby cells. Its action is restricted to such local effects because NO is extremely unstable, with a half-life of only a few seconds. One well-characterized example of NO action is signaling the dilation of blood vessels. The first step in this process is the release of neurotransmitters, such as acetylcholine, from the terminus of nerve cells in the blood vessel wall. These neurotransmitters act on endothelial cells to stimulate NO synthesis. NO then diffuses to neighboring smooth muscle cells where it reacts with iron bound to the active site of the enzyme guanylyl cyclase. This increases enzymatic activity, resulting in synthesis of the second messenger cyclic GMP (discussed later in this chapter), which induces muscle cell relaxation and blood vessel dilation. For example, NO is responsible for signaling the dilation of blood vessels that leads to penile erection. It is also interesting to note that the medical use of nitroglycerin in treatment of heart disease is based on its conversion to NO, which dilates coronary blood vessels and increases blood flow to the heart.

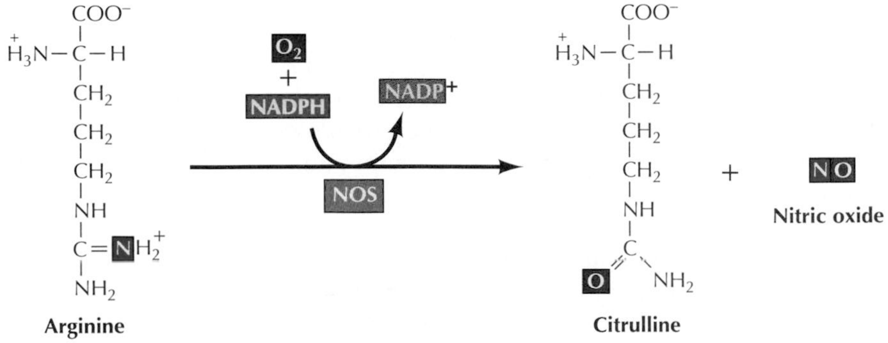

**Figure 13.5**
**Synthesis of nitric oxide**   The enzyme nitric oxide synthase (NOS) catalyzes the formation of nitric oxide from arginine.

## Neurotransmitters

The **neurotransmitters** carry signals between neurons or from neurons to other types of target cells (such as muscle cells). They are a diverse group of small hydrophilic molecules including acetylcholine, dopamine, epinephrine (adrenaline), serotonin, histamine, glutamate, glycine, and γ-aminobutyric acid (GABA) (Figure 13.6). The release of neurotransmitters is signaled by the arrival of an action potential at the terminus of a neuron (see Figure 12.22). The neurotransmitters then diffuse across the synaptic cleft and bind to receptors on the target cell surface. Note that some neurotransmitters can also act as hormones. For example, epinephrine functions both as a neurotransmitter and as a hormone produced by the adrenal gland to signal glycogen breakdown in muscle cells.

Because the neurotransmitters are hydrophilic molecules, they are unable to cross the plasma membrane of their target cells. Therefore, in contrast to steroid hormones and NO, the neurotransmitters act by binding to cell surface receptors. Many neurotransmitter receptors are ligand-gated ion channels, such as the acetylcholine receptor discussed in the preceding chapter (see Figure 12.23). Neurotransmitter binding to these receptors induces a conformational change that opens ion channels, directly resulting in changes in ion flux in the target cell. Other neurotransmitter receptors are coupled to G proteins—a major group of signaling molecules (discussed later in this chapter) that link cell surface receptors to a variety of intracellular responses. In the case of neurotransmitter receptors, the associated G proteins frequently act to indirectly regulate ion channel activity.

## Peptide Hormones and Growth Factors

The widest variety of signaling molecules in animals are peptides, ranging in size from only a few to more than a hundred amino acids. This group of signaling molecules includes peptide hormones, neuropeptides, and a diverse array of polypeptide growth factors (Table 13.1). Well-known examples of **peptide hormones** include insulin, glucagon, and the hormones produced by the pituitary gland (growth hormone, follicle-stimulating hormone, prolactin, and others).

**Neuropeptides** are secreted by some neurons instead of the small-molecule neurotransmitters discussed in the previous section. Some of these peptides, such as the **enkephalins** and **endorphins**, function not only as neurotransmitters at synapses but also as **neurohormones** that act on distant cells. The enkephalins and endorphins have been widely studied because of their activity as natural analgesics that decrease pain responses in the central nervous system. Discovered during studies of drug addiction, they are naturally occurring compounds that bind to the same receptors on the surface of brain cells as morphine does.

The polypeptide **growth factors** include a wide variety of signaling molecules that control animal cell growth and differentiation. The first of these factors (**nerve growth factor,** or **NGF**) was discovered by Rita Levi-Montalcini in the 1950s. NGF is a member of a family of polypeptides (called **neurotrophins**) that regulate the development and survival of neurons. During the course of experiments on NGF, Stanley Cohen serendipitously discovered an unrelated factor (called **epidermal growth factor,** or **EGF**) that stimulates cell proliferation. EGF, a 53-amino-acid polypeptide (Figure

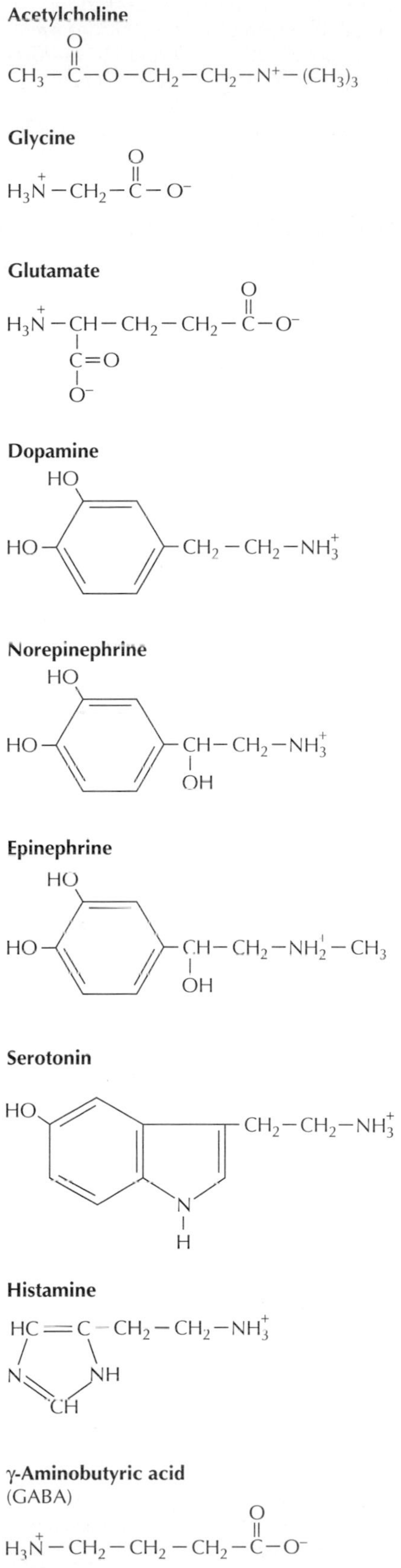

**Figure 13.6**
**Structure of representative neurotransmitters** The neurotransmitters are hydrophilic molecules that bind to cell surface receptors.

*Figure 13.7*
**Structure of epidermal growth factor (EGF)** EGF is a single polypeptide chain of 53 amino acids. Disulfide bonds between cysteine residues are indicated. (After G. Carpenter and S. Cohen, 1979, *Ann. Rev. Biochem.* 48: 193.)

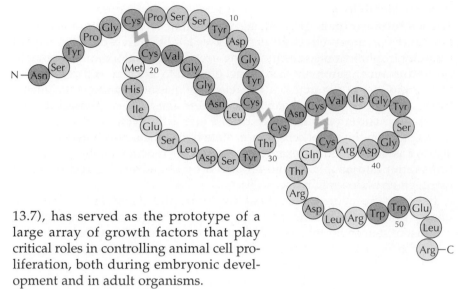

13.7), has served as the prototype of a large array of growth factors that play critical roles in controlling animal cell proliferation, both during embryonic development and in adult organisms.

A good example of growth factor action is provided by the activity of **platelet-derived growth factor (PDGF)** in wound healing. PDGF is stored in blood platelets and released during blood clotting at the site of a wound. It then stimulates the proliferation of fibroblasts in the vicinity of the clot,

*Table 13.1* Representative Peptide Hormones, Neuropeptides, and Growth Factors

| Signaling molecule | Size[a] | Activities[b] |
|---|---|---|
| **Peptide hormones** | | |
| Insulin | A = 21, B = 30 | Regulation of glucose uptake; stimulation of cell proliferation |
| Glucagon | 29 | Stimulation of glucose synthesis |
| Growth hormone | 191 | General stimulation of growth |
| Follicle-stimulating hormone (FSH) | $\alpha$ = 92, $\beta$ = 118 | Stimulation of the growth of oocytes and ovarian follicles |
| Prolactin | 198 | Stimulation of milk production |
| **Neuropeptides and neurohormones** | | |
| Substance P | 11 | Sensory synaptic transmission |
| Oxytocin | 9 | Stimulation of smooth muscle contraction |
| Vasopressin | 9 | Stimulation of water reabsorption in the kidney |
| Enkephalins | 5 | Analgesics |
| $\beta$-Endorphin | 31 | Analgesic |
| **Growth factors** | | |
| Nerve growth factor (NGF) | 118 | Differentiation and survival of neurons |
| Epidermal growth factor (EGF) | 53 | Proliferation of many types of cells |
| Platelet-derived growth factor (PDGF) | A = 125, B = 109 | Proliferation of fibroblasts and other cell types |
| Interleukin-2 | 133 | Proliferation of T lymphocytes |
| Erythropoietin | 166 | Development of red blood cells |

[a] Size is indicated in number of amino acids. Some hormones and growth factors consist of two different polypeptide chains, which are designated either A and B or $\alpha$ and $\beta$.

[b] Most of these hormones and growth factors possess other activities in addition to those indicated.

thereby contributing to regrowth of the damaged tissue. Members of another large group of polypeptide growth factors (called **cytokines**) regulate the development and differentiation of blood cells and control the activities of lymphocytes during the immune response. Other polypeptide growth factors (**membrane-anchored growth factors**) remain associated with the plasma membrane rather than being secreted into extracellular fluids, therefore functioning specifically as signaling molecules during direct cell-cell interactions.

Peptide hormones, neuropeptides, and growth factors are unable to cross the plasma membrane of their target cells, so they act by binding to cell surface receptors, as discussed later in this chapter. As might be expected from the critical roles of polypeptide growth factors in controlling cell proliferation, abnormalities in growth factor signaling are the basis for a variety of diseases, including many kinds of cancer. For example, abnormal expression of a close relative of the EGF receptor is an important factor in the development of many human breast and ovarian cancers.

## Eicosanoids

Several types of lipids serve as signaling molecules that, in contrast to the steroid hormones, act by binding to cell surface receptors. The most important of these molecules are members of a class of lipids called the **eicosanoids**, which includes **prostaglandins**, **prostacyclin**, **thromboxanes**, and **leukotrienes** (Figure 13.8). The eicosanoids are rapidly broken down

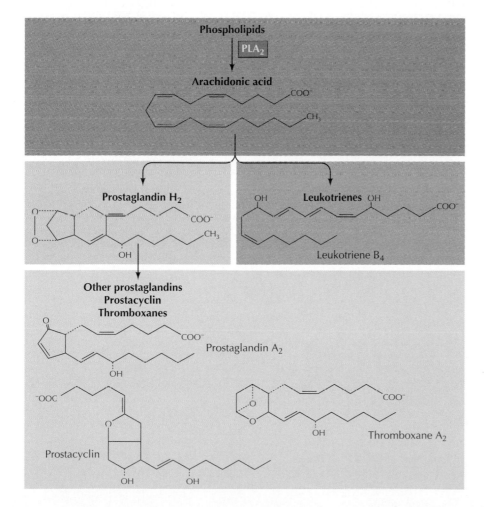

*Figure 13.8*
**Synthesis and structure of eicosanoids**   The eicosanoids include the prostaglandins, prostacyclin, thromboxanes, and leukotrienes. They are synthesized from arachidonic acid, which is formed by the hydrolysis of phospholipids catalyzed by phospholipase $A_2$ ($PLA_2$). Arachidonic acid can then be metabolized via two alternative pathways; one pathway leads to synthesis of prostaglandins, prostacyclin, and thromboxanes, while the other pathway leads to synthesis of leukotrienes.

and therefore act locally in autocrine or paracrine signaling pathways. They stimulate a variety of responses in their target cells, including blood platelet aggregation, inflammation, and smooth-muscle contraction.

All eicosanoids are synthesized from arachidonic acid, which is formed from phospholipids. The first step in the pathway leading to synthesis of either prostaglandins or thromboxanes is the conversion of arachidonic acid to prostaglandin $H_2$. Interestingly, the enzyme that catalyzes this reaction (prostaglandin synthase) is the target of aspirin and other nonsteroidal anti-inflammatory drugs. By inhibiting synthesis of the prostaglandins, aspirin reduces inflammation and pain. By inhibiting synthesis of thromboxane, aspirin also reduces platelet aggregation and blood clotting. Because of this activity, small daily doses of aspirin are frequently prescribed for prevention of strokes.

### Plant Hormones

Plant growth and development are regulated by a group of small molecules called **plant hormones**. The levels of these molecules within the plant are typically modified by environmental factors, such as light or infection, so they coordinate the responses of tissues in different parts of the plant to environmental signals.

The plant hormones are generally divided into five classes: **auxins**, **gibberellins**, **cytokinins**, **abscisic acid**, and **ethylene** (Figure 13.9). The first plant hormone to be identified was auxin, with the early experiments leading to its discovery having been performed by Charles Darwin in the 1880s. One of the effects of auxins is to induce plant cell elongation by weakening the cell wall (see Figure 12.49). In addition, auxins regulate many other aspects of plant development, including cell division and differentiation. The other plant hormones likewise have multiple effects in their target tissues, including stem elongation (gibberellins), fruit ripening (ethylene), cell division (cytokinins), and the onset of dormancy (abscisic acid).

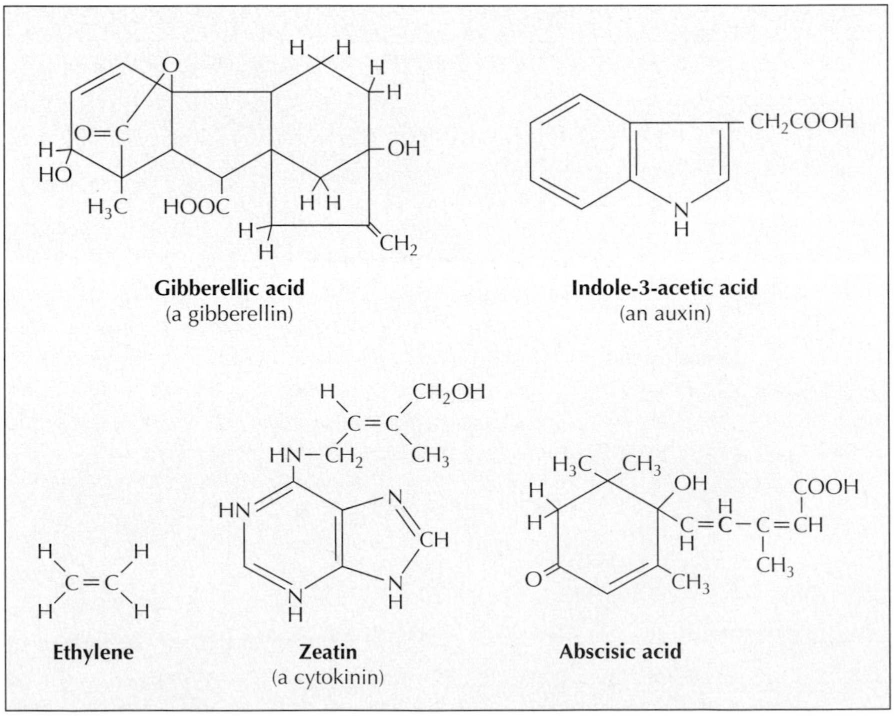

*Figure 13.9*
**Structure of plant hormones**

Our understanding of the molecular mechanisms of plant hormone action is less advanced than comparable studies of animal cells, and the receptors for plant hormones are just beginning to be identified and characterized. One area of noteworthy progress has been in understanding the mechanism by which plant cells respond to ethylene. Using the small weed *Arabidopsis* as a model, several of the genes required for ethylene responsiveness have been identified. These include at least one gene encoding a potential ethylene receptor and another encoding a protein related to the Raf protein kinase, which plays a key role in animal cell signaling pathways (discussed later in this chapter).

## FUNCTIONS OF CELL SURFACE RECEPTORS

As already reviewed, most ligands responsible for cell-cell signaling (including neurotransmitters, peptide hormones, and growth factors) bind to receptors on the surface of their target cells. Consequently, a major challenge in understanding cell-cell signaling is unraveling the mechanisms by which cell surface receptors transmit the signals initiated by ligand binding. As discussed in Chapter 12, some neurotransmitter receptors are ligand-gated ion channels that directly control ion flux across the plasma membrane. Other cell surface receptors, including the receptors for peptide hormones and growth factors, act instead by regulating the activity of intracellular proteins. These proteins then transmit signals from the receptor to a series of additional intracellular targets, frequently including transcription factors. Ligand binding to a receptor on the surface of the cell thus initiates a chain of intracellular reactions, ultimately reaching the target cell nucleus and resulting in programmed changes in gene expression. The functions of the major classes of cell surface receptors are discussed here, with the pathways of intracellular signaling downstream of these receptors being considered in the next section of this chapter.

### G Protein-Coupled Receptors

The largest family of cell surface receptors transmit signals to intracellular targets via the intermediary action of guanine nucleotide-binding proteins called **G proteins**. More than a thousand such **G protein-coupled receptors** have been identified, including the receptors for many neurotransmitters, neuropeptides, and peptide hormones. In addition, the G protein-coupled receptor family includes a large number of receptors that are responsible for smell, sight, and taste.

The G protein-coupled receptors are structurally and functionally related proteins characterized by seven membrane-spanning $\alpha$ helices (Figure 13.10). The binding of ligands to the extracellular domain of these receptors induces a conformational change that allows the cytosolic domain of the receptor to bind to a G protein associated with the inner face of the plasma membrane. This interaction activates the G protein, which then dissociates from the receptor and carries the signal to an intracellular target, which may be either an enzyme or an ion channel.

The discovery of G proteins came from studies of hormones (such as epinephrine) that regulate the synthesis of cyclic AMP (cAMP) in their target cells. As discussed later in the chapter, cAMP is an important second messenger that mediates cellular responses to a variety of hormones. In the 1970s, Martin Rodbell and his colleagues made the key observation that GTP is required for hormonal stimulation of adenylyl cyclase (the enzyme responsible for cAMP formation). This finding led to the discovery that a guanine nucleotide-binding protein (called a G protein) is an intermediary in

*Figure 13.10*
**Structure of a G protein-coupled receptor** The G protein-coupled receptors are characterized by seven transmembrane $\alpha$ helices.

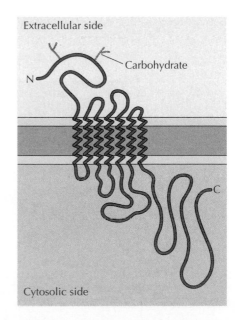

Extracellular side

Carbohydrate

N

C

Cytosolic side

*Figure 13.11*
**Hormonal activation of adenylyl cyclase** Binding of hormone promotes the interaction of the receptor with a G protein. The activated G protein $\alpha$ subunit then dissociates from the receptor and stimulates adenylyl cyclase, which catalyzes the conversion of ATP to cAMP.

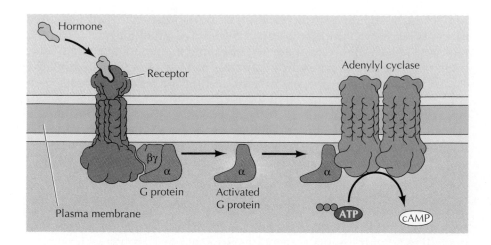

adenylyl cyclase activation (Figure 13.11). Since then, an array of G proteins have been found to act as physiological switches that regulate the activities of a variety of intracellular targets in response to extracellular signals.

G proteins consist of three subunits, designated $\alpha$, $\beta$, and $\gamma$ (Figure 13.12). They are frequently called **heterotrimeric G proteins** to distinguish them from other guanine nucleotide-binding proteins, such as the Ras proteins discussed later in the chapter. The $\alpha$ subunit binds guanine nucleotides, which regulate G protein activity. In the resting state, $\alpha$ is bound to GDP in a complex with $\beta$ and $\gamma$. Hormone binding induces a conforma-

*Figure 13.12*
**Regulation of G proteins**

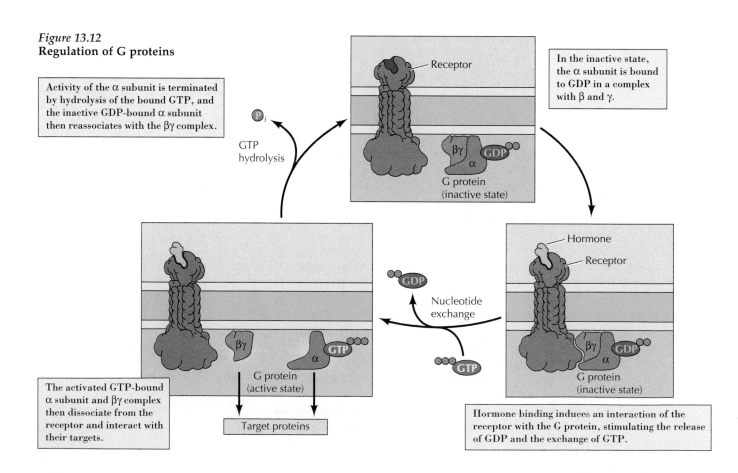

Activity of the $\alpha$ subunit is terminated by hydrolysis of the bound GTP, and the inactive GDP-bound $\alpha$ subunit then reassociates with the $\beta\gamma$ complex.

In the inactive state, the $\alpha$ subunit is bound to GDP in a complex with $\beta$ and $\gamma$.

The activated GTP-bound $\alpha$ subunit and $\beta\gamma$ complex then dissociate from the receptor and interact with their targets.

Hormone binding induces an interaction of the receptor with the G protein, stimulating the release of GDP and the exchange of GTP.

tional change in the receptor, such that the cytosolic domain of the receptor interacts with the G protein and stimulates the release of bound GDP and its exchange for GTP. The activated GTP-bound $\alpha$ subunit then dissociates from $\beta$ and $\gamma$, which remain together and function as a $\beta\gamma$ complex. Both the active GTP-bound $\alpha$ subunit and the $\beta\gamma$ complex then interact with their targets to elicit an intracellular response. The activity of the $\alpha$ subunit is terminated by hydrolysis of the bound GTP, and the inactive $\alpha$ subunit (now with GDP bound) then reassociates with the $\beta\gamma$ complex, ready for the cycle to start anew.

Mammalian genomes encode at least 16 different $\alpha$ subunits, 5 $\beta$ subunits, and 11 $\gamma$ subunits. Different G proteins associate with different receptors, so this array of G proteins couples receptors to distinct intracellular targets. For example, the G protein associated with the epinephrine receptor is called $G_s$ because its $\alpha$ subunit stimulates adenylyl cyclase (see Figure 13.11). Other G protein $\alpha$ and $\beta\gamma$ subunits act instead to inhibit adenylyl cyclase or to regulate the activities of other target enzymes.

In addition to regulating target enzymes, both the $\alpha$ and $\beta\gamma$ subunits of some G proteins directly regulate ion channels. A good example is provided by the action of the neurotransmitter acetylcholine on heart muscle, which is distinct from its effects on nerve and skeletal muscle. The acetylcholine receptor on nerve and skeletal muscle cells is a ligand-gated ion channel (see Figure 12.23). Heart muscle cells have a different acetylcholine receptor, which is G protein-coupled. This G protein is designated $G_i$ because its $\alpha$ subunit *i*nhibits adenylyl cyclase. In addition, the $G_i$ $\beta\gamma$ subunits act directly to open $K^+$ channels in the plasma membrane, which has the effect of slowing heart muscle contraction.

## Receptor Protein-Tyrosine Kinases

In contrast to the G protein-coupled receptors, other cell surface receptors are directly linked to intracellular enzymes. The largest family of such enzyme-linked receptors are the **receptor protein-tyrosine kinases**, which phosphorylate their substrate proteins on tyrosine residues. This family includes the receptors for most polypeptide growth factors, so protein-tyrosine phosphorylation has been particularly well studied as a signaling mechanism involved in the control of animal cell growth and differentiation. Indeed, the first protein-tyrosine kinase was discovered in 1980 during studies of the oncogenic proteins of animal tumor viruses, in particular Rous sarcoma virus, by Tony Hunter and Bartholomew Sefton. The EGF receptor was then found to function as a protein-tyrosine kinase by Stanley Cohen and his colleagues, clearly establishing protein-tyrosine phosphorylation as a key signaling mechanism in the response of cells to growth factor stimulation.

By now more than 50 receptor protein-tyrosine kinases have been identified, including the receptors for EGF, NGF, PDGF, insulin, and many other growth factors. All these receptors share a common structural organization: an N-terminal extracellular ligand-binding domain, a single transmembrane $\alpha$ helix, and a cytosolic C-terminal domain with protein-tyrosine kinase activity (Figure 13.13). Most of the receptor protein-tyrosine kinases consist of single polypeptides, although the insulin receptor and some related receptors are dimers consisting of two pairs of polypeptide chains. The binding of ligands (e.g., growth factors) to the extracellular domains of these receptors activates their cytosolic kinase domains, resulting in phosphorylation of both the receptors themselves and intracellular target proteins that propagate the signal initiated by growth factor binding.

# KEY EXPERIMENT

## The Src Protein-Tyrosine Kinase

### Transforming Gene Product of Rous Sarcoma Virus Phosphorylates Tyrosine

Tony Hunter and Bartholomew M. Sefton

The Salk Institute, San Diego, CA

*Proceedings of the National Academy of Science USA, 1980, Volume 77, pages 1311–1315*

### The Context

Following its isolation in 1911, Rous sarcoma virus (RSV) became the first virus that was generally accepted to cause tumors in animals (see the Molecular Medicine box in Chapter 1). Several features of RSV then made it an attractive model for studying the development of cancer. In particular, the small size of the RSV genome offered the hope of identifying specific viral genes responsible for inducing the abnormal proliferation that is characteristic of cancer cells. This goal was reached in the 1970s, when it was established that a single RSV gene (called *src* for *sarcoma*) is required for tumor induction. Importantly, a closely related *src* gene was also found to be part of the normal genetic complement of a variety of vertebrates, including humans. Since the viral Src protein is responsible for driving the uncontrolled proliferation of cancer cells, it appeared that understanding Src function would yield crucial insights into the molecular bases of both cancer induction and the regulation of normal cell proliferation.

In 1977, Ray Erikson and his colleagues identified the Src protein by immunoprecipitation (see Figure 3.32) with antisera from animals bearing RSV-induced tumors. Shortly thereafter, it was found that incubation of Src immunoprecipitates with radioactive ATP resulted in phosphorylation of the immunoglobulin molecules. Src therefore appeared to be a protein kinase, clearly implicating

protein phosphorylation in the control of cell proliferation.

All previously studied protein kinases phosphorylated serine or threonine residues, which were also the only phosphoamino acids to have been detected in animal cells. However, Walter Eckhardt and Tony Hunter had observed in 1979 that the oncogenic protein of another animal tumor virus (polyomavirus) was phosphorylated on a tyrosine residue. Hunter and Sefton therefore tested the possibility that Src might phosphorylate tyrosine, rather than serine/threonine, residues in its substrate proteins. Their experiments demonstrated that Src does indeed function as a protein-tyrosine kinase—an activity now recognized as playing a central role in cell signaling pathways.

### The Experiments

Hunter and Sefton identified the amino acid phosphorylated by Src by incubating Src immunoprecipitates with $^{32}$P-labeled ATP. The amino acid that was phosphorylated by Src in the substrate protein (in this case, immunoglobulin) therefore became radioactively labeled. The immunoglobulin was then isolated and hydrolyzed to yield individual amino acids, which were analyzed by electrophoresis and chromatography methods that separated phosphotyrosine, phosphoserine, and phosphothreonine (see figure). The radioactive amino acid detected in these experiments was phosphotyrosine, indicat-

ing that Src specifically phosphorylates tyrosine residues.

Further experiments showed that the normal cell Src protein, as well as viral Src, functions as a protein-tyrosine kinase in immunoprecipitation assays. In addition, Hunter and Sefton extended these *in vitro* experiments by demonstrating the presence of phosphotyrosine in proteins extracted from whole cells. In normal cells, phosphotyrosine accounted for only about 0.03% of total phosphoamino acids (the rest being phosphoserine and phosphothreonine), explaining why it had previously escaped detection. However, phosphotyrosine was about ten times more abundant in cells that were infected with RSV, suggesting that increased protein-tyrosine kinase

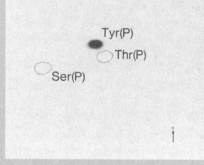

Identification of phosphotyrosine in immunoglobulin phosphorylated by Src. An immunoprecipitate containing RSV Src was incubated with [$^{32}$P]-ATP. The immunoglobulin was then isolated and hydrolyzed. Amino acids in the hydrolysate were separated by electrophoresis and chromatography on a cellulose thin-layer plate. The positions of $^{32}$P-labeled amino acids were determined by exposing the plate to X-ray film. Broken lines indicate the positions of unlabeled phosphoamino acids that were included as markers. Note that the principal $^{32}$P-labeled amino acid is phosphotyrosine.

## The Src Protein-Tyrosine Kinase (continued)

activity of the viral Src protein was responsible for its ability to induce abnormal cell proliferation.

### The Impact

The discovery that Src was a protein-tyrosine kinase both identified a new protein kinase activity and established this activity as being related to the control of cell proliferation. The results of Hunter and Sefton were followed by demonstrations that many other tumor virus proteins also function as protein-tyrosine kinases, generalizing the link between protein-tyrosine phosphorylation and the abnormal proliferation of cancer cells. Stanley Cohen and his colleagues further found that the EGF receptor is a protein-tyrosine kinase, directly implicating protein-tyrosine phosphorylation in the control of normal cell proliferation. Continuing studies have identified numerous additional receptor and nonreceptor protein-tyrosine kinases that function in a variety of cell signaling pathways. Studies of the mechanism by which a virus causes cancer in chickens thus revealed a previously unknown enzymatic activity that plays a central role in the signaling pathways that regulate animal cell growth and differentiation.

Tony Hunter

Bartholomew Sefton

The first step in signaling from most receptor protein-tyrosine kinases is ligand-induced receptor dimerization (Figure 13.14). Some growth factors, such as PDGF and NGF, are themselves dimers consisting of two identical polypeptide chains; these growth factors directly induce dimerization by simultaneously binding to two different receptor molecules. Other growth factors (such as EGF) are monomers but may have two distinct receptor binding sites that serve to crosslink receptors, as well as inducing confor-

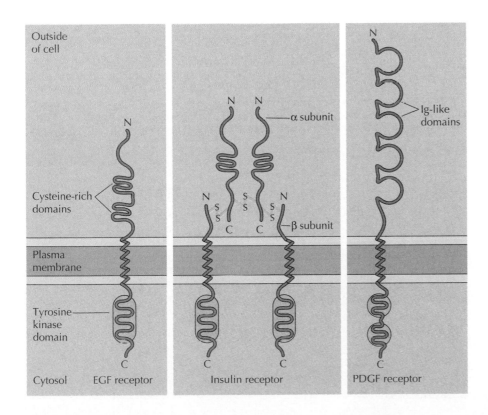

**Figure 13.13**
**Organization of receptor protein-tyrosine kinases** Each receptor consists of an N-terminal extracellular ligand-binding domain, a single transmembrane $\alpha$ helix, and a cytosolic C-terminal domain with protein-tyrosine kinase activity. The structures of three distinct subfamilies of receptor protein-tyrosine kinases are shown. The EGF receptor and insulin receptor both have cysteine-rich extracellular domains, whereas the PDGF receptor has immunoglobulin (Ig)-like domains. The PDGF receptor is also noteworthy in that its kinase domain is interrupted by an insert of approximately a hundred amino acids unrelated to those found in most other protein-tyrosine kinase catalytic domains. The insulin receptor is unusual in being a dimer of two pairs of polypeptide chains (designated $\alpha$ and $\beta$).

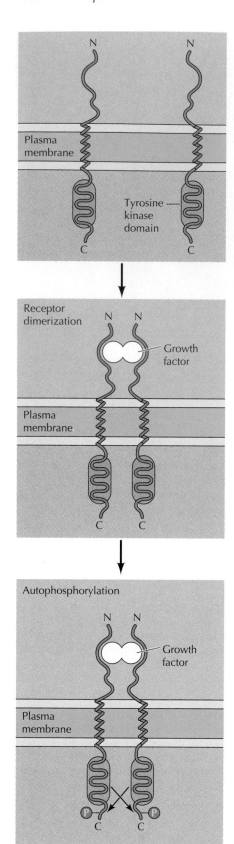

*Figure 13.14*
**Dimerization and autophosphorylation of receptor protein-tyrosine kinases**
Growth factor binding induces receptor dimerization, which results in receptor autophosphorylation as the two polypeptide chains phosphorylate one another.

mational changes that favor receptor dimerization.

Ligand-induced dimerization then leads to **autophosphorylation** of the receptor as the dimerized polypeptide chains cross-phosphorylate one another (see Figure 13.14). Such autophosphorylation plays two key roles in signaling from these receptors. First, phosphorylation of tyrosine residues within the catalytic domain may play a regulatory role by increasing receptor protein kinase activity. Second, phosphorylation of tyrosine residues outside of the catalytic domain creates specific binding sites for additional proteins that transmit intracellular signals downstream of the activated receptors.

The association of these downstream signaling molecules with receptor protein-tyrosine kinases is mediated by protein domains that bind to specific phosphotyrosine-containing peptides (Figure 13.15). The best-characterized of these domains are called **SH2 domains** (for *Src homology 2*) because they were first recognized in protein-tyrosine kinases related to Src, the oncogenic protein of Rous sarcoma virus. SH2 domains consist of approximately a hundred amino acids and bind to specific short peptide sequences containing phosphotyrosine residues (Figure 13.16). The resulting association of SH2-containing proteins with activated receptor protein-tyrosine kinases can have several effects: It localizes the SH2-containing proteins to the plasma membrane, leads to their association with other proteins, promotes their phosphorylation, and stimulates their enzymatic activities. The association of these proteins with autophosphorylated receptors thus represents the first step in the intracellular transmission of signals initiated by the binding of growth factors to the cell surface.

*Figure 13.15*
**Association of downstream signaling molecules with receptor protein-tyrosine kinases**   SH2 domains bind to specific phosphotyrosine-containing peptides of the activated receptors.

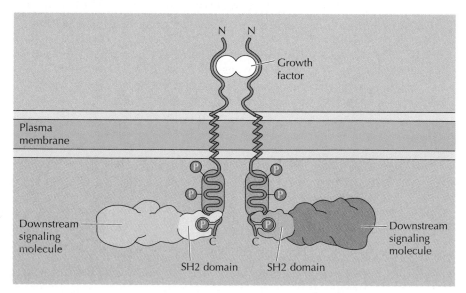

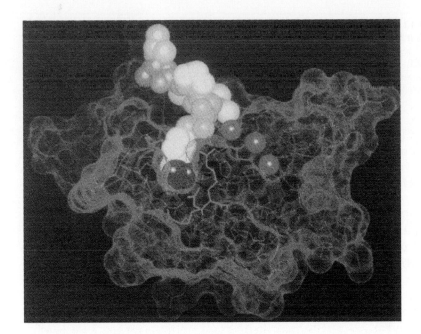

*Figure 13.16*
**Complex between an SH2 domain and a phosphotyrosine peptide** The polypeptide chain of the Src SH2 domain is shown in red with its surface indicated by green dots. Purple spheres indicate a groove on the surface. The three amino acid residues that interact with the phosphotyrosine are shown in blue. The phosphotyrosine-containing peptide is shown as a space-filling model. Yellow and white spheres indicate the backbone and side-chain atoms, respectively, and the phosphate group is shown in red. (From G. Waksman and 13 others, 1992, *Nature* 358: 646.)

## Cytokine Receptors and Nonreceptor Protein-Tyrosine Kinases

Rather than possessing intrinsic enzymatic activity, many receptors act by stimulating intracellular protein-tyrosine kinases with which they are noncovalently associated. This family of receptors (called the **cytokine receptor superfamily**) includes the receptors for most cytokines (e.g., interleukin-2 and erythropoietin) and for some polypeptide hormones (e.g., growth hormone). Like receptor protein-tyrosine kinases, the cytokine receptors contain N-terminal extracellular ligand-binding domains, single transmembrane α helices, and C-terminal cytosolic domains. However, the cytosolic domains of the cytokine receptors are devoid of any known catalytic activity. Instead, the cytokine receptors function in association with **nonreceptor protein-tyrosine kinases**, which are activated as a result of ligand binding.

The first step in signaling from cytokine receptors is thought to be ligand-induced receptor dimerization and cross-phosphorylation of the associated nonreceptor protein-tyrosine kinases (Figure 13.17). These activated kinases then phosphorylate the receptor, providing phosphotyrosine-binding sites for the recruitment of downstream signaling molecules that contain SH2 domains. Combinations of cytokine receptors plus associated nonreceptor protein-tyrosine kinases thus function analogously to the receptor protein-tyrosine kinases discussed in the previous section.

The nonreceptor protein-tyrosine kinases associated with the cytokine receptors fall into two major families. Many of these kinases are members of the **Src** family, which consists of Src and eight closely related proteins. As already noted, Src was initially identified as the oncogenic protein of Rous sarcoma virus and was the first protein shown to possess protein-tyrosine kinase activity, so it has played a pivotal role in experiments leading to our current understanding of cell signaling. In addition to Src family members, the cytokine receptors are associated with nonreceptor protein-tyrosine kinases belonging to the **Janus kinase**, or **JAK**, family. Members of the JAK family appear to be universally required for signaling from cytokine receptors, indicating that JAK family kinases play a critical role in coupling these receptors to the tyrosine phosphorylation of intracellular

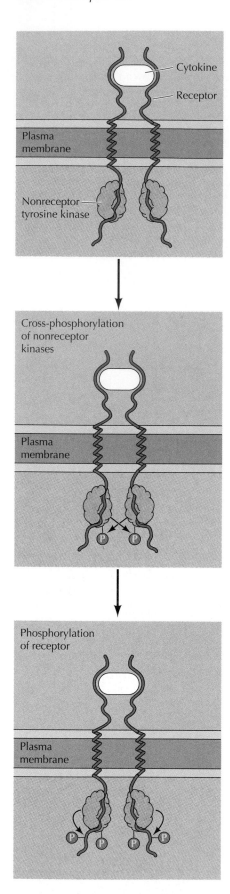

Cytokine

Receptor

Plasma membrane

Nonreceptor tyrosine kinase

Cross-phosphorylation of nonreceptor kinases

Plasma membrane

Phosphorylation of receptor

Plasma membrane

*Figure 13.17*
**Signaling from cytokine receptors** Ligand binding induces receptor dimerization and leads to the activation of associated nonreceptor protein-tyrosine kinases as a result of cross-phosphorylation. The activated kinases then phosphorylate tyrosine residues of the receptor, creating phosphotyrosine-binding sites for downstream signaling molecules.

targets. In contrast, members of the Src family play key roles in signaling from antigen receptors on B and T lymphocytes but do not appear to be required for signaling from most cytokine receptors.

## Receptors Linked to Other Enzymatic Activities

Although the vast majority of enzyme-linked receptors stimulate protein-tyrosine phosphorylation, some receptors are associated with other enzymatic activities. These receptors include protein-tyrosine phosphatases, protein-serine/threonine kinases, and guanylyl cyclases. The functions of most of these receptors are less well understood than those of either the G protein-coupled receptors or the receptors associated with protein-tyrosine kinase activity.

**Protein-tyrosine phosphatases** remove phosphate groups from phosphotyrosine residues, thus acting to counterbalance the effects of protein-tyrosine kinases. In many cases, protein-tyrosine phosphatases play negative regulatory roles in cell signaling pathways by terminating the signals initiated by protein-tyrosine phosphorylation. However, some protein-tyrosine phosphatases are cell surface receptors whose enzymatic activities play a positive role in cell signaling. A good example is provided by a receptor called CD45, which is expressed on the surface of T and B lymphocytes. Following antigen stimulation, CD45 is thought to dephosphorylate a specific phosphotyrosine that inhibits the enzymatic activity of Src family members. Thus, the CD45 protein-tyrosine phosphatase acts (somewhat paradoxically) to stimulate nonreceptor protein-tyrosine kinases.

The receptors for **transforming growth factor β (TGF-β)** and related polypeptides are protein kinases that phosphorylate serine or threonine, rather than tyrosine, residues on their substrate proteins. TGF-β is the prototype of a family of polypeptide growth factors that control proliferation and differentiation of a variety of cell types, generally inhibiting proliferation of their target cells. The cloning of the first receptor for a member of the TGF-β family in 1991 revealed that it is the prototype of a unique receptor family with a cytosolic **protein-serine/threonine kinase** domain. Since then, receptors for additional TGF-β family members have similarly been found to be protein-serine/threonine kinases. The binding of ligand to these receptors results in the association of two distinct polypeptide chains, which are encoded by different members of the TGF-β receptor family, to form heterodimers in which the receptor kinases cross-phosphorylate one another. At present, however, little is known about the downstream signaling molecules that are stimulated by members of the TGF-β receptor family.

Some peptide ligands bind to receptors whose cytosolic domains are guanylyl cyclases, which catalyze formation of cyclic GMP. As discussed earlier, nitric oxide also acts by stimulating guanylyl cyclase, but the target of nitric oxide is an intracellular enzyme rather than a transmembrane receptor. The receptor **guanylyl cyclases** have an extracellular ligand-binding domain, a single transmembrane α helix, and a cytosolic domain with catalytic activity. Ligand binding stimulates cyclase activity, leading to the formation of cyclic GMP—a second messenger whose intracellular effects are discussed in the next section of this chapter.

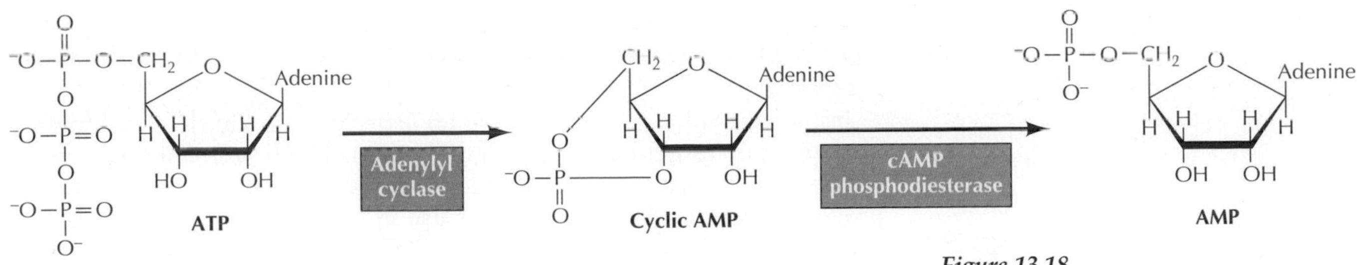

*Figure 13.18*
**Synthesis and degradation of cAMP**
Cyclic AMP is synthesized from ATP by adenylyl cyclase and degraded to AMP by cAMP phosphodiesterase.

Finally, some receptors have been identified that bind to cytoplasmic proteins with unknown biochemical activities. The signaling pathways stimulated by these receptors thus remain to be elucidated by further research. Important examples of such receptors include members of the Notch family, which have been found to play key roles in cell-cell signaling during development by genetic studies in *Drosophila* and *Caenorhabditis elegans*.

## PATHWAYS OF INTRACELLULAR SIGNAL TRANSDUCTION

Most cell surface receptors stimulate intracellular target enzymes, which may be either directly linked or indirectly coupled to receptors by G proteins. These intracellular enzymes serve as downstream signaling elements that propagate and amplify the signal initiated by ligand binding. In most cases, a chain of reactions transmits signals from the cell surface to a variety of intracellular targets—a process called **intracellular signal transduction**. The targets of such signaling pathways frequently include transcription factors that function to regulate gene expression. Intracellular signaling pathways thus connect the cell surface to the nucleus, leading to changes in gene expression in response to extracellular stimuli.

### The cAMP Pathway: Second Messengers and Protein Phosphorylation

Intracellular signaling was first elucidated by studies of the action of hormones such as epinephrine, which signals the breakdown of glycogen to glucose in anticipation of muscular activity. In 1958, Earl Sutherland discovered that the action of epinephrine was mediated by an increase in the intracellular concentration of **cyclic AMP (cAMP)**, leading to the concept that cAMP is a **second messenger** in hormonal signaling (the first messenger being the hormone itself). Cyclic AMP is formed from ATP by the action of **adenylyl cyclase** and degraded to AMP by **cAMP phosphodiesterase** (Figure 13.18). As discussed earlier, the epinephrine receptor is coupled to adenylyl cyclase via a G protein that stimulates enzymatic activity, thereby increasing the intracellular concentration of cAMP (see Figure 13.11).

How does cAMP then signal the breakdown of glycogen? This and most other effects of cAMP in animal cells are mediated by the action of **cAMP-dependent protein kinase**, or **protein kinase A**, an enzyme discovered by Donal Walsh and Ed Krebs in 1968. The inactive form of protein kinase A is a tetramer consisting of two catalytic and two regulatory subunits (Figure 13.19). Cyclic AMP binds to the regulatory subunits, leading to their disso-

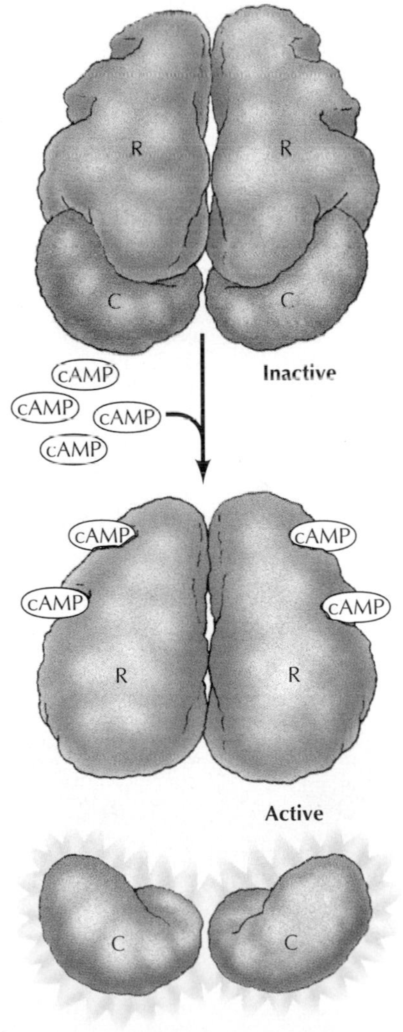

*Figure 13.19*
**Regulation of protein kinase A** The inactive form of protein kinase A consists of two regulatory (R) and two catalytic (C) subunits. Binding of cAMP to the regulatory subunits induces a conformational change that leads to dissociation of the catalytic subunits, which are then enzymatically active.

ciation from the catalytic subunits. The free catalytic subunits are then enzymatically active and able to phosphorylate serine residues on their target proteins.

In the regulation of glycogen metabolism, protein kinase A phosphorylates two key target enzymes (Figure 13.20). The first is another protein kinase, phosphorylase kinase, which is phosphorylated and activated by protein kinase A. Phosphorylase kinase in turn phosphorylates and activates glycogen phosphorylase, which catalyzes the breakdown of glycogen to glucose-1-phosphate. In addition, protein kinase A phosphorylates the enzyme glycogen synthase, which catalyzes glycogen synthesis. In this case, however, phosphorylation inhibits enzymatic activity. Elevation of cAMP and activation of protein kinase A thus blocks further glycogen synthesis at the same time as it stimulates glycogen breakdown.

The chain of reactions leading from the epinephrine receptor to glycogen phosphorylase provides a good illustration of signal amplification during intracellular signal transduction. Each molecule of epinephrine activates only a single receptor. However, each receptor may activate up to a hundred molecules of $G_s$. Each molecule of $G_s$ then stimulates the enzymatic activity of adenylyl cyclase, which can catalyze the synthesis of many molecules of cAMP. Signal amplification continues as each molecule of protein kinase A phosphorylates many molecules of phosphorylase kinase, which in turn phosphorylate many molecules of glycogen phosphorylase. Hormone binding to a small number of receptors thus leads to activation of a much larger number of intracellular target enzymes.

In many animal cells, increases in cAMP activate the transcription of specific target genes that contain a regulatory sequence called the **cAMP response element**, or **CRE** (Figure 13.21). In this case, the signal is carried

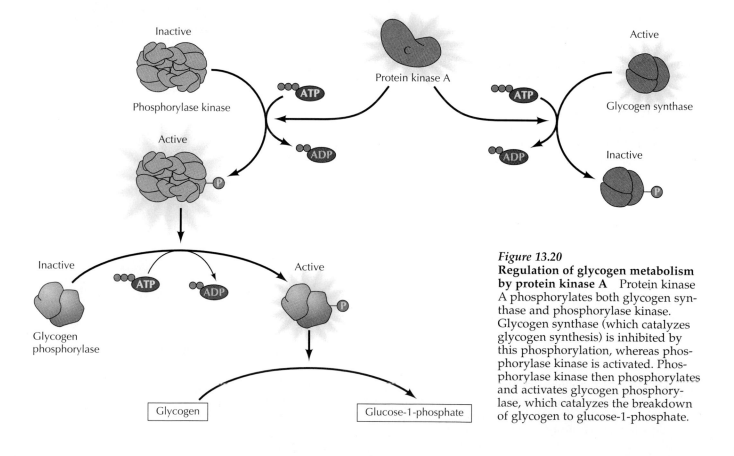

*Figure 13.20*
**Regulation of glycogen metabolism by protein kinase A**    Protein kinase A phosphorylates both glycogen synthase and phosphorylase kinase. Glycogen synthase (which catalyzes glycogen synthesis) is inhibited by this phosphorylation, whereas phosphorylase kinase is activated. Phosphorylase kinase then phosphorylates and activates glycogen phosphorylase, which catalyzes the breakdown of glycogen to glucose-1-phosphate.

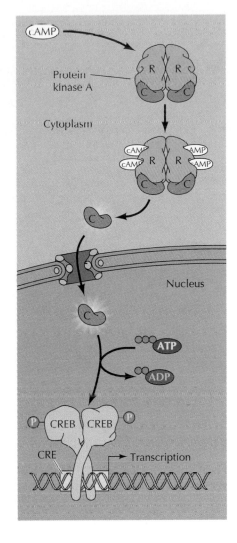

*Figure 13.21*
**Cyclic AMP-inducible gene expression**   The free catalytic subunit of protein kinase A translocates to the nucleus and phosphorylates the transcription factor CREB (CRE-binding protein), leading to expression of cAMP-inducible genes.

from the cytoplasm to the nucleus by the catalytic subunit of protein kinase A, which is able to enter the nucleus following its release from the regulatory subunit. Within the nucleus, protein kinase A phosphorylates a transcription factor called **CREB** (for CRE-*b*inding protein), leading to the activation of cAMP-inducible genes.

Regulation of gene expression by cAMP plays important roles in controlling the proliferation and differentiation of several types of animal cells. In addition, CREB has recently been found to play a critical role in learning and memory. A striking example is provided by the behavior of mice in which the gene that encodes CREB has been inactivated by homologous recombination. In contrast to their normal counterparts, these mutant mice are severely deficient in the ability to learn to avoid an electric shock in classical Pavlovian conditioning experiments. These studies, together with related experiments in *Drosophila* and the mollusk *Aplysia*, indicate that CREB plays a conserved role in the mechanisms of learning and memory—a finding that may open new approaches to a molecular understanding of these basic functions of the nervous system.

It is important to recognize that protein kinases, such as protein kinase A, do not function in isolation within the cell. To the contrary, protein phosphorylation is rapidly reversed by the action of protein phosphatases. Some protein phosphatases are transmembrane receptors, as discussed in the preceding section. A number of others are cytosolic enzymes that remove phosphate groups from either phosphorylated tyrosine or serine/threonine residues in their substrate proteins. These protein phosphatases serve to terminate the responses initiated by receptor activation of protein kinases. For example, the serine residues of proteins that are phosphorylated by protein kinase A are usually dephosphorylated by the action of a phosphatase called protein phosphatase 1 (Figure 13.22). The levels of phosphorylation of protein kinase A substrates (such as phosphorylase kinase and CREB) are thus determined by a balance between the intracellular activities of protein kinase A and protein phosphatases. Interestingly, one of the proteins phosphorylated by protein kinase A is an inhibitor of protein phosphatase 1, which is active only when phosphorylated, so protein kinase A not only catalyzes the phosphorylation of its substrate proteins but also indirectly inhibits their dephosphorylation.

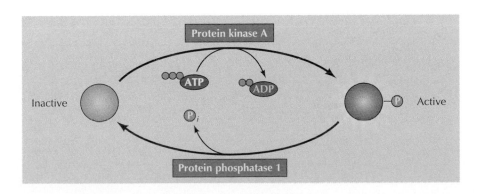

*Figure 13.22*
**Regulation of protein phosphorylation by protein kinase A and protein phosphatase 1**   The phosphorylation of target proteins by protein kinase A is reversed by the action of protein phosphatase 1.

Although most effects of cAMP are mediated by protein kinase A, cAMP can also directly regulate ion channels, independent of protein phosphorylation. Cyclic AMP functions in this way as a second messenger involved in sensing smells. Many of the odorant receptors in sensory neurons in the nose are G protein-coupled receptors that stimulate adenylyl cyclase, leading to an increase in intracellular cAMP. Rather than stimulating protein kinase A, cAMP in this system directly opens Na$^+$ channels in the plasma membrane, leading to membrane depolarization and initiation of a nerve impulse.

## Cyclic GMP

**Cyclic GMP (cGMP)** is also an important second messenger in animal cells, although its roles are not as clearly understood as those of cAMP. Cyclic GMP is formed from GTP by guanylyl cyclases and degraded to GMP by a phosphodiesterase. As discussed earlier in this chapter, different types of guanylyl cyclases are activated by both nitric oxide and peptide ligands. Stimulation of these guanylyl cyclases leads to elevated levels of cGMP, which then mediate biological responses, such as blood vessel dilation. The mechanism of cGMP action in these systems, however, remains unclear, although it appears to involve activation of a cGMP-dependent protein kinase.

The best-characterized role of cGMP is in the vertebrate eye, where it serves as the second messenger responsible for converting the visual signals received as light to nerve impulses. The photoreceptor in rod cells of the retina is a G protein-coupled receptor called **rhodopsin** (Figure 13.23). Rhodopsin is activated as a result of the absorption of light by the associated small molecule 11-*cis*-retinal, which then isomerizes to all-*trans*-retinal, inducing a conformational change in the rhodopsin protein. Rhodopsin then activates the G protein **transducin**, and the $\alpha$ subunit of transducin stimulates the activity of **cGMP phosphodiesterase**, leading to a decrease in the intracellular level of cGMP. This change in cGMP level in retinal rod cells is translated to a nerve impulse by a direct effect of cGMP on ion channels in the plasma membrane, similar to the action of cAMP in sensing smells.

## Phospholipids and Ca$^{2+}$

One of the most widespread pathways of intracellular signaling is based on the use of second messengers derived from the membrane phospholipid **phosphatidylinositol 4,5-bisphosphate (PIP$_2$)**. PIP$_2$ is a minor component of the plasma membrane, localized to the inner leaflet of the phospholipid bilayer (see Figure 12.2). A variety of hormones and growth

*Figure 13.23*
**Role of cGMP in photoreception**
Absorption of light by retinal activates the G protein-coupled receptor rhodopsin. The $\alpha$ subunit of transducin then stimulates cGMP phosphodiesterase, leading to a decrease in intracellular levels of cGMP.

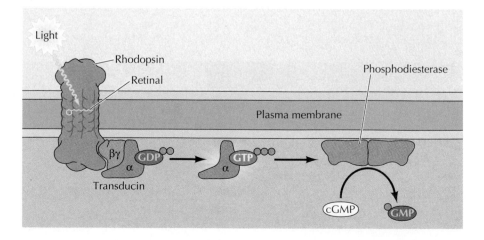

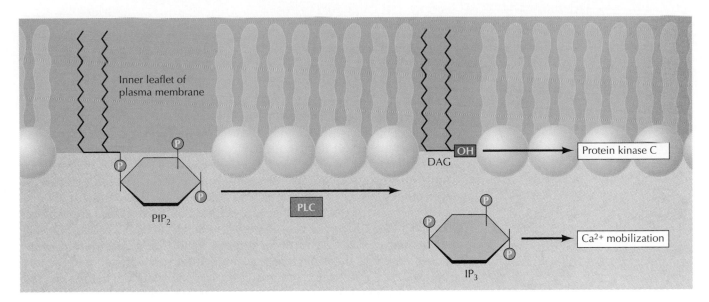

*Figure 13.24*
**Hydrolysis of PIP$_2$** Phospholipase C (PLC) catalyzes the hydrolysis of phosphatidylinositol 4,5-bisphosphate (PIP$_2$) to yield diacylglycerol (DAG) and inositol trisphosphate (IP$_3$). Diacylglycerol activates members of the protein kinase C family, and IP$_3$ signals the release of Ca$^{2+}$ from intracellular stores.

factors stimulate the hydrolysis of PIP$_2$ by **phospholipase C**—a reaction that produces two distinct second messengers, **diacylglycerol** and **inositol 1,4,5-trisphosphate (IP$_3$)** (Figure 13.24). Diacylglycerol and IP$_3$ stimulate distinct downstream signaling pathways (protein kinase C and Ca$^{2+}$ mobilization, respectively), so PIP$_2$ hydrolysis triggers a two-armed cascade of intracellular signaling.

It is noteworthy that the hydrolysis of PIP$_2$ is activated downstream of both G protein-coupled receptors and protein-tyrosine kinases. This occurs because one form of phospholipase C (PLC-$\beta$) is stimulated by G proteins, whereas a second (PLC-$\gamma$) contains SH2 domains that mediate its association with activated receptor protein-tyrosine kinases (Figure 13.25). This interaction localizes PLC-$\gamma$ to the plasma membrane as well as leading to

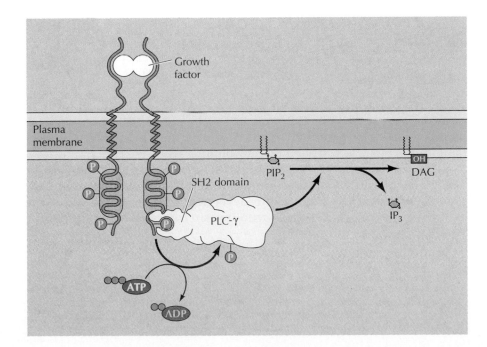

*Figure 13.25*
**Activation of phospholipase C by protein-tyrosine kinases** Phospholipase C-$\gamma$ (PLC-$\gamma$) binds to activated receptor protein-tyrosine kinases via its SH2 domains. Tyrosine phosphorylation increases PLC-$\gamma$ activity, stimulating the hydrolysis of PIP$_2$.

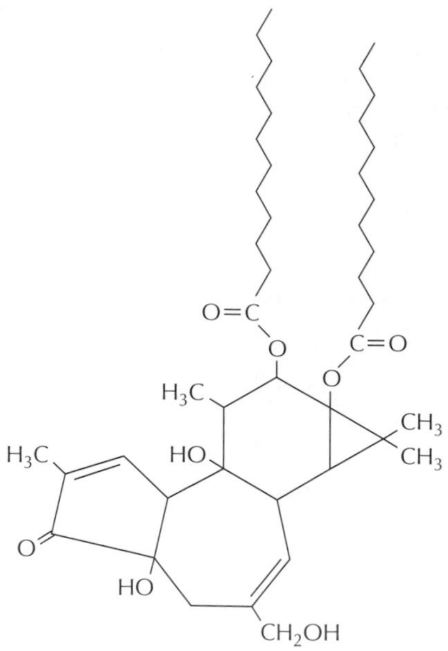

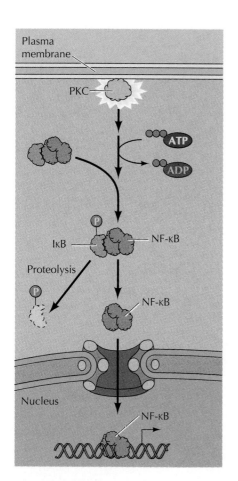

*Figure 13.26*
**Structure of a phorbol ester**    Phorbol esters stimulate protein kinase C by acting as analogs of diacylglycerol.

its tyrosine phosphorylation, which increases its catalytic activity.

The diacylglycerol produced by hydrolysis of PIP$_2$ activates protein-ser-ine/threonine kinases belonging to the **protein kinase C** family, many of which play important roles in the control of cell growth and differentiation. A good illustration of this role of protein kinase C is provided by the action of **phorbol esters** (Figure 13.26), which have been studied extensively because they promote the growth of tumors in animals. This tumor-pro-moting activity of the phorbol esters is based on their ability to stimulate protein kinase C by acting as analogs of diacylglycerol. Protein kinase C then activates other intracellular targets, including a cascade of protein kinases known as the MAP kinase pathway (discussed in detail in the next section), leading to transcription factor phosphorylation, changes in gene expression, and stimulation of cell proliferation.

Another notable target of protein kinase C signaling is the transcription factor **NF-κB**, which is involved in a variety of aspects of the immune response. The activity of NF-κB is regulated by translocation from the cytoplasm to the nucleus (Figure 13.27). In its inactive state, NF-κB is retained in the cytoplasm as a complex with an inhibitory subunit called **IκB**. Activation of protein kinase C leads indirectly to phosphorylation of IκB, which targets IκB for proteolytic degradation. Destruction of IκB allows NF-κB to translocate to the nucleus, where it regulates transcription of its target genes.

Whereas diacylglycerol remains associated with the plasma membrane, the other second messenger produced by PIP$_2$ cleavage, IP$_3$, is a small polar molecule that is released into the cytosol, where it acts to signal the release of Ca$^{2+}$ from intracellular stores (Figure 13.28). As noted in Chapter 12, the cytosolic concentration of Ca$^{2+}$ is maintained at an extremely low level (about 0.1 $\mu M$) as a result of Ca$^{2+}$ pumps that actively export Ca$^{2+}$ from the cell. Ca$^{2+}$ is pumped not only across the plasma membrane, but also into the endoplasmic reticulum, which therefore serves as an intracellular Ca$^{2+}$ store. IP$_3$ acts to release Ca$^{2+}$ from the endoplasmic reticulum by binding to receptors that are ligand-gated Ca$^{2+}$ channels. As a result, cytosolic Ca$^{2+}$ levels increase to about 1 $\mu M$, which affects the activities of a variety of tar-get proteins, including protein kinases and phosphatases. For example, some members of the protein kinase C family require Ca$^{2+}$ as well as diacylglycerol for their activation, so these protein kinases are regulated jointly by both arms of the PIP$_2$ signaling pathway.

Many of the effects of Ca$^{2+}$ are mediated by the Ca$^{2+}$-binding protein **calmodulin**, which is activated by Ca$^{2+}$ binding when the concentration of cytosolic Ca$^{2+}$ increases to about 0.5 $\mu M$ (Figure 13.29). Ca$^{2+}$/calmodulin then binds to a variety of target proteins, including protein kinases. One example of such a Ca$^{2+}$/calmodulin-dependent protein kinase is myosin light-chain kinase, which signals actin-myosin contraction by phosphory-

*Figure 13.27*
**Activation of NF-κB**    In its inactive state, NF-κB is retained in the cytoplasm as a complex with IκB. Stimulation of protein kinase C leads indirectly to IκB phos-phorylation, which signals its proteolytic degradation. NF-κB is then able to translocate to the nucleus and regulate target gene expression.

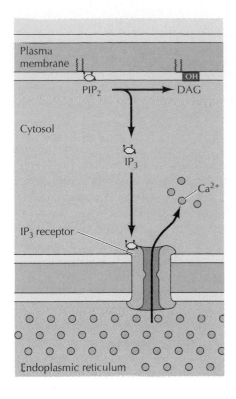

*Figure 13.28*
**Ca$^{2+}$ mobilization by IP$_3$** Ca$^{2+}$ is pumped from the cytosol into the endoplasmic reticulum, which therefore serves as an intracellular Ca$^{2+}$ store. IP$_3$ binds to receptors that are ligand-gated Ca$^{2+}$ channels in the endoplasmic reticulum membrane, thereby allowing the efflux of Ca$^{2+}$ to the cytosol.

lating one of the myosin light chains (see Figure 11.28). Other protein kinases that are activated by Ca$^{2+}$/calmodulin include members of the **CaM kinase II** family, which phosphorylate a number of different proteins, including metabolic enzymes, ion channels, and transcription factors. One form of CaM kinase II is particularly abundant in the nervous system, where it regulates the synthesis and release of neurotransmitters. In addition, CaM kinase II can regulate gene expression by phosphorylating transcription factors. Interestingly, one of the transcription factors phosphorylated by CaM kinase II is CREB, which (as discussed earlier) is phosphorylated at the same site by protein kinase A. This phosphorylation of CREB illustrates one of many intersections between the Ca$^{2+}$ and cAMP signaling pathways. Other examples include the regulation of adenylyl cyclases and phosphodiesterases by Ca$^{2+}$/calmodulin, the regulation of Ca$^{2+}$ channels by cAMP, and the phosphorylation of a number of target proteins by both protein kinase A and Ca$^{2+}$/calmodulin-dependent protein kinases. The cAMP and Ca$^{2+}$ signaling pathways thus function coordinately to regulate many cellular responses.

Ca$^{2+}$ is an extremely common second messenger, and it is important to note that IP$_3$-mediated release of Ca$^{2+}$ from the endoplasmic reticulum is not the only mechanism by which the intracellular concentration of Ca$^{2+}$ can be increased. One alternative pathway involves the entry of extracellular Ca$^{2+}$ through Ca$^{2+}$ channels in the plasma membrane. In many cells, the transient increase in intracellular Ca$^{2+}$ resulting from production of IP$_3$ is followed by a more sustained increase resulting from extracellular Ca$^{2+}$ entry. The entry of extracellular Ca$^{2+}$ is particularly important in the electrically excitable cells of nerve and muscle, in which voltage-gated Ca$^{2+}$ channels in the plasma membrane are opened by membrane depolarization (Figure 13.30). The resulting increases in intracellular Ca$^{2+}$ then trigger the further release of Ca$^{2+}$ from intracellular stores by activating distinct Ca$^{2+}$ channels known as **ryanodine receptors**. One effect of increases in intracellular Ca$^{2+}$ in neurons is to trigger the release of neurotransmitters, so Ca$^{2+}$ plays a critical role in converting electric to chemical signals in the nervous

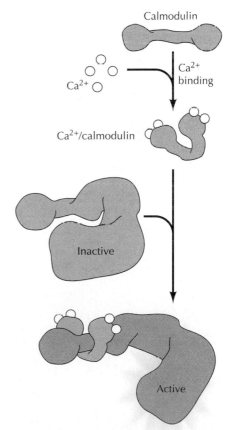

*Figure 13.29*
**Function of calmodulin** Calmodulin is a dumbbell-shaped protein with four Ca$^{2+}$-binding sites. The active Ca$^{2+}$/calmodulin complex binds to a variety of target proteins, including Ca$^{2+}$/calmodulin-dependent protein kinases.

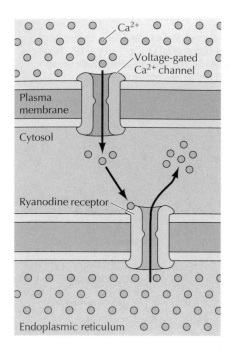

*Figure 13.30*
**Regulation of intracellular Ca²⁺ in electrically excitable cells**   Membrane depolarization leads to the opening of voltage-gated $Ca^{2+}$ channels in the plasma membrane, causing the influx of $Ca^{2+}$ from extracellular fluids. The resulting increase in intracellular $Ca^{2+}$ then signals the further release of $Ca^{2+}$ from intracellular stores by opening distinct $Ca^{2+}$ channels (ryanodine receptors) in the endoplasmic reticulum membrane. In muscle cells, ryanodine receptors in the sarcoplasmic reticulum may also be opened directly in response to membrane depolarization.

system. In muscle cells, $Ca^{2+}$ is stored in the sarcoplasmic reticulum, from which it is released by the opening of ryanodine receptors in response to changes in membrane potential. This release of stored $Ca^{2+}$ leads to large increases in cytosolic $Ca^{2+}$, which trigger muscle contraction (see Chapter 11). Cells thus utilize a variety of mechanisms to regulate intracellular $Ca^{2+}$ levels, making $Ca^{2+}$ a versatile second messenger that controls a wide range of cellular processes.

Although $PIP_2$ hydrolysis is the best-characterized phospholipid signaling pathway, other phospholipid-derived second messengers also play important roles. In addition to being cleaved by phospholipase C, $PIP_2$ can be phosphorylated on the 3 position of inositol by the enzyme **phosphatidylinositide (PI) 3-kinase** (Figure 13.31). Like phospholipase C, one form of PI 3-kinase is activated by G proteins, while a second has SH2 domains and is activated by association with receptor protein-tyrosine kinases. Phosphorylation of $PIP_2$ yields **phosphatidylinositol 3,4,5-trisphosphate ($PIP_3$)**, which is thought to function as a distinct second messenger. Although its targets have not yet been fully characterized, $PIP_3$ appears to stimulate protein kinases, possibly including some members of the protein kinase C family and a protein-serine/threonine kinase called Akt.

Second messengers can also be derived from other phospholipids. The hydrolysis of phosphatidylcholine is stimulated by a variety of growth factors, providing a second source of diacylglycerol, in addition to that derived from $PIP_2$. While $PIP_2$ hydrolysis is a transient response to growth factor stimulation, the hydrolysis of phosphatidylcholine typically persists for several hours, providing a sustained source of diacylglycerol that may be important in signaling long-term responses, such as cell proliferation. Sphingomyelin is also cleaved in response to a variety of extracellular stimuli, resulting in the formation of ceramide. Although its targets remain to be elucidated, a number of experiments implicate ceramide as an additional intracellular regulator of growth and differentiation.

## Ras, Raf, and the MAP Kinase Pathway

The MAP kinase pathway refers to a cascade of protein kinases that are highly conserved in evolution and play central roles in signal transduction in all eukaryotic cells, ranging from yeasts to humans. The central elements in the pathway are a family of protein-serine/threonine kinases called the **MAP kinases** (for *m*itogen-*a*ctivated *p*rotein kinases) that are activated in response to a variety of growth factors and other signaling molecules. In yeasts, MAP kinase pathways control a variety of cellular responses,

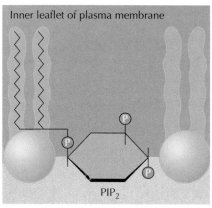

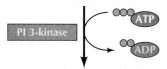

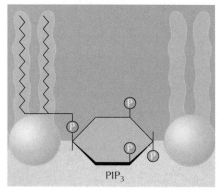

*Figure 13.31*
**Activity of PI 3-kinase**   PI 3-kinase phosphorylates the 3 position of inositol, converting $PIP_2$ to $PIP_3$.

including mating, cell shape, and sporulation. In higher eukaryotes (including *C. elegans, Drosophila,* frogs, and mammals), MAP kinases are ubiquitous regulators of cell growth and differentiation.

The best-characterized forms of MAP kinase in mammalian cells belong to the **ERK** (*extracellular signal-regulated kinase*) family. ERK activation plays a central role in signaling cell proliferation induced by growth factors that act through either protein-tyrosine kinase or G protein-coupled receptors. As already noted, protein kinase C can also activate the ERK pathway, which appears to be responsible for the stimulation of cell proliferation induced by phorbol ester tumor promoters. In addition, both the $Ca^{2+}$ and cAMP pathways intersect with ERK signaling, either stimulating or inhibiting the ERK pathway in different types of cells.

Activation of ERK is mediated by two upstream protein kinases, which are coupled to growth factor receptors by a GTP-binding protein called **Ras** (Figure 13.32). Activation of Ras leads to activation of the **Raf** protein-serine/threonine kinase, which phosphorylates and activates a second protein kinase called **MEK** (for *MAP kinase/ERK kinase*). MEK is a dual-specificity protein kinase that activates members of the ERK family by phosphorylation of both threonine and tyrosine residues separated by one amino acid (e.g., threonine-183 and tyrosine-185 of ERK2). Once activated, ERK phosphorylates a variety of targets, including other protein kinases and transcription factors.

The central role of the ERK pathway in mammalian cells emerged from studies of the Ras proteins, which were first identified as the oncogenic proteins of tumor viruses that cause sarcomas in rats (hence the name Ras, from *ra*t *s*arcoma virus). Interest in Ras intensified considerably in 1982, when mutations in *ras* genes were first implicated in the development of human cancers (discussed in Chapter 15). The importance of Ras in intracellular signaling was then indicated by experiments showing that microinjection of

*Figure 13.32*
**Activation of the ERK MAP kinases**
Stimulation of growth factor receptors leads to activation of the small GTP-binding protein Ras, which interacts with the Raf protein kinase. Raf phosphorylates and activates MEK, a dual-specificity protein kinase that activates ERK by phosphorylation on both threonine and tyrosine residues (Thr-183 and Tyr-185). ERK then phosphorylates a variety of nuclear and cytoplasmic target proteins.

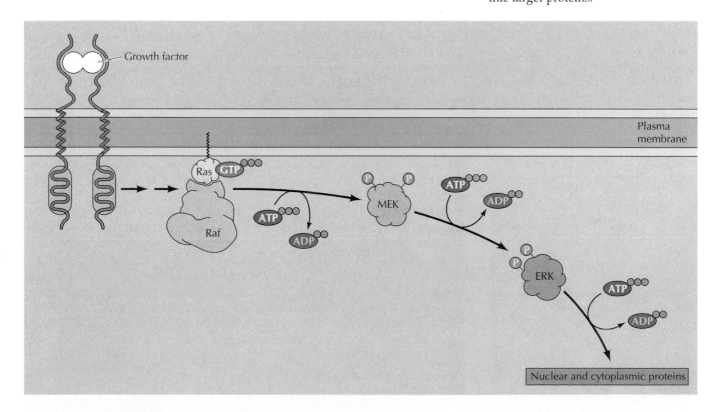

active Ras protein directly induces proliferation of normal mammalian cells. Conversely, interference with Ras function by either microinjection of anti-Ras antibody or expression of a dominant negative Ras mutant blocks growth factor-induced cell proliferation. Thus, Ras is not only capable of inducing the abnormal growth characteristic of cancer cells, but also appears to be required for the response of normal cells to growth factor stimulation.

The Ras proteins are guanine nucleotide-binding proteins that function analogously to the $\alpha$ subunits of G proteins, alternating between inactive GDP-bound and active GTP-bound forms (Figure 13.33). In contrast to the G protein $\alpha$ subunits, however, Ras functions as a monomer rather than in association with $\beta\gamma$ subunits. Ras activation is mediated by **guanine nucleotide exchange factors** that stimulate the release of bound GDP and its exchange for GTP. Activity of the Ras-GTP complex is then terminated by GTP hydrolysis, which is stimulated by the interaction of Ras-GTP with **GTPase-activating proteins**. It is interesting to note that the mutations of *ras* genes in human cancers have the effect of inhibiting GTP hydrolysis by the Ras proteins. These mutated Ras proteins therefore remain continuously in the active GTP-bound form, driving the unregulated proliferation of cancer cells even in the absence of growth factor stimulation.

The Ras proteins are prototypes of a large family of approximately 50 related proteins, frequently called **small GTP-binding proteins** because Ras and its relatives are about half the size of G protein $\alpha$ subunits. While the Ras proteins regulate cell growth and differentiation, other subfamilies of small GTP-binding proteins control distinct cellular activities. For example, the largest subfamily of small GTP-binding proteins (the ARF and Rab proteins) function to regulate vesicle trafficking, as discussed in Chapter 9.

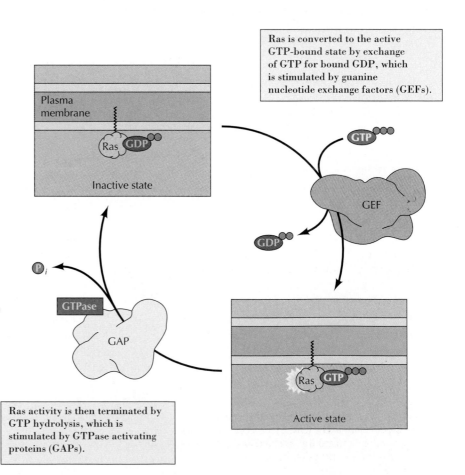

*Figure 13.33*
**Regulation of Ras proteins**  Ras proteins alternate between inactive GDP-bound and active GTP-bound states.

Ras is converted to the active GTP-bound state by exchange of GTP for bound GDP, which is stimulated by guanine nucleotide exchange factors (GEFs).

Ras activity is then terminated by GTP hydrolysis, which is stimulated by GTPase activating proteins (GAPs).

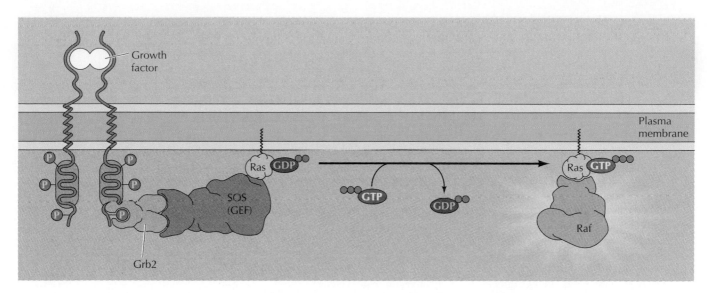

*Figure 13.34*
**Ras activation downstream of receptor protein-tyrosine kinases** A complex of Grb2 and the guanine nucleotide exchange factor Sos binds to a phosphotyrosine-containing sequence in the activated receptor via the Grb2 SH2 domain. This interaction recruits Sos to the plasma membrane, where it can stimulate Ras GDP/GTP exchange. The activated Ras-GTP complex then binds to the Raf protein kinase.

Other small GTP-binding proteins are involved in nuclear protein import (the Ran protein, discussed in Chapter 8) and organization of the cytoskeleton (the Rho subfamily, discussed later in this chapter).

The best understood mode of Ras activation is that mediated by receptor protein-tyrosine kinases (Figure 13.34). Autophosphorylation of these receptors results in their association with Ras guanine nucleotide exchange factors as a result of SH2-mediated protein interactions. One well-characterized example is provided by the guanine nucleotide exchange factor Sos, which is bound to the SH2-containing protein Grb2 in the cytosol of unstimulated cells. Tyrosine phosphorylation of receptors (or of other receptor-associated proteins) creates a binding site for the Grb2 SH2 domains. Association of Grb2 with activated receptors localizes Sos to the plasma membrane, where it is able to interact with Ras proteins, which are anchored to the inner leaflet of the plasma membrane by lipids attached to the Ras C terminus (see Figure 12.10). Sos then stimulates guanine nucleotide exchange, resulting in formation of the active Ras-GTP complex. In its active GTP-bound form, Ras interacts with effector proteins, including the Raf protein-serine/threonine kinase. Interestingly, this interaction with Ras does not directly stimulate Raf catalytic activity; rather it recruits Raf from the cytosol to the plasma membrane where Raf can be activated by other signals, probably including phosphorylation by other plasma membrane-associated protein kinases.

As already noted, activation of Raf initiates a protein kinase cascade leading to ERK activation. ERK then phosphorylates a variety of target proteins, including phospholipase $A_2$ (which catalyzes the formation of arachidonic acid; see Figure 13.8) and other protein kinases. Importantly, a fraction of activated ERK translocates to the nucleus, where it regulates transcription factors by phosphorylation (Figure 13.35). In this regard, it is notable that a primary response to growth factor stimulation is the rapid

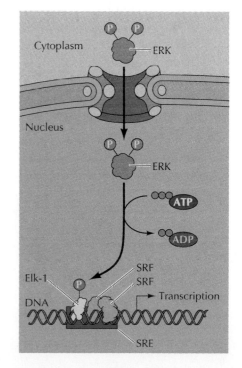

*Figure 13.35*
**Induction of immediate-early genes by ERK** Activated ERK translocates to the nucleus, where it phosphorylates the transcription factor Elk-1. Elk-1 binds to the serum response element (SRE) in a complex with serum response factor (SRF). Phosphorylation stimulates the activity of Elk-1 as a transcriptional activator, leading to immediate-early gene induction.

## MOLECULAR MEDICINE

# Cancer, Signal Transduction, and the ras Oncogenes

### The Disease

Cancer claims the lives of approximately one out of every four Americans, accounting for 500,000 deaths each year in the United States. There are more than a hundred different kinds of cancer, but some are more common than others. In this country, the most common lethal cancers are those of the lung and colon, which together account for about 40% of all cancer deaths. Other major contributors to cancer mortality include cancers of the breast, prostate, and pancreas, which are responsible for approximately 8%, 7%, and 5% of U.S. cancer deaths, respectively.

The common feature of all cancers is the unrestrained proliferation of cancer cells, which eventually spread throughout the body, invading normal tissues and organs and leading to death of the patient. Surgery and radiotherapy are effective treatments for localized cancers but are unable to reach cancer cells that have spread to distant body sites. Treatment of these cancers therefore requires the use of chemotherapeutic drugs. Unfortunately, the currently available chemotherapeutic agents are not specific for cancer cells. Most act by either damaging DNA or interfering with DNA synthesis, so they also kill rapidly dividing normal cells, such as the epithelial cells that line the digestive tract and the blood-forming cells of the bone marrow. The resulting toxicity of these drugs limits their effectiveness, and many cancers are not eliminated by doses of chemotherapy

that can be tolerated by the patient. Consequently, although major progress has been made in cancer treatment, nearly half of all patients diagnosed with cancer ultimately die of their disease.

### Molecular and Cellular Basis

The identification of viral genes that can convert normal cells to cancer cells, such as the *src* gene of RSV, provided the first demonstration that cancers can result from the action of specific genes (oncogenes). The subsequent discovery that viral oncogenes are related to genes of normal cells then engendered the hypothesis that non-virus-induced cancers (including most human cancers) might arise as a result of mutations in normal cell genes, giving rise to oncogenes of cellular rather than viral origin. Such cellular oncogenes were first identified in human cancers in 1981. Shortly thereafter, human oncogenes of bladder, lung, and colon cancers were found to be related to the *ras* genes previously identified in rat sarcoma viruses.

Although many different genes are now known to play critical roles in cancer development, mutations of the *ras* genes remain one of the most common genetic abnormalities in human tumors. Mutated *ras* oncogenes are found in about 15% of all human cancers, including approximately 25% of lung cancers, 50% of colon cancers, and more than 90% of pancreatic cancers. Moreover, the action of *ras* oncogenes has clearly linked the develop-

ment of human cancer to abnormalities in the signaling pathways that regulate cell proliferation. The mutations that convert normal *ras* genes to oncogenes substantially decrease GTP hydrolysis by the Ras proteins. Consequently, the mutated oncogenic Ras proteins remain locked in the active GTP-bound form, rather than alternating normally between inactive and active states in response to extracellular signals. The oncogenic Ras proteins thus continuously stimulate the MAP kinase pathway and drive cell proliferation, even in the absence of the growth factors that would be required to activate Ras and signal proliferation of normal cells.

### Prevention and Treatment

The discovery of mutated oncogenes in human cancers raises the possibility

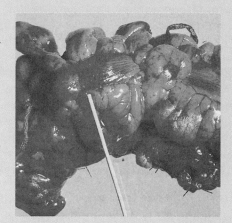

A human colon cancer. The *ras* oncogenes contribute to the development of about half of these cancers. (John Cabisco/Visuals Unlimited).

transcriptional induction of a family of 50 to 100 genes called **immediate-early genes**. The induction of a number of immediate-early genes is mediated by a regulatory sequence called the **serum response element (SRE)**, which is recognized by a complex of transcription factors including the **serum response factor (SRF)** and **Elk-1**. ERK phosphorylates and activates Elk-1, providing a direct link between the ERK family of MAP kinases and

*Figure 13.36*
**Pathways of MAP kinase activation in mammalian cells** In addition to ERK, mammalian cells contain JNK and p38 MAP kinases. Activation of JNK and p38 is mediated by protein kinase cascades parallel to that responsible for ERK activation. The protein kinase cascades leading to JNK and p38 activation may be mediated by small GTP-binding proteins other than Ras (e.g., Rac and Cdc42) and appear to be preferentially activated by cytokines or stress.

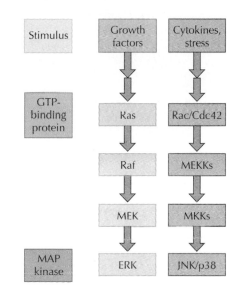

immediate-early gene induction. Many immediate-early genes themselves encode transcription factors, so their induction in response to growth factor stimulation leads to altered expression of a battery of other downstream genes, thereby establishing new programs of gene expression.

In addition to ERK, two other MAP kinases (called the JNK and p38 MAP kinases) have been identified in mammalian cells (Figure 13.36). Although the JNK and p38 MAP kinases are also stimulated by growth factors, they are preferentially activated in response to cytokines and environmental stress (e.g., ultraviolet irradiation). The JNK and p38 MAP kinases are activated not by MEK but by distinct dual-specificity kinases called MKKs (for *MAP kinase kinases*). The MKK that activates JNK is in turn activated by a protein kinase designated MEKK, rather than by Raf. Thus, distinct protein kinase cascades lead to activation of the ERK, JNK, and p38 MAP kinases. Although the pathways leading to JNK and p38 MAP kinase activation remain to be fully defined, it appears that activation of MEKK can be mediated indirectly by some members of the Rho family of small GTP-binding proteins (e.g., Rac and Cdc42), which may thus function analogously to Ras as activators of distinct MAP kinase pathways. Like ERK, the JNK MAP kinase can translocate to the nucleus and phosphorylate transcription factors that regulate immediate-early genes. Multiple MAP kinase pathways may thus function to coordinately regulate gene expression in response to distinct extracellular signals.

## The JAK/STAT Pathway

The MAP kinase pathway provides an indirect connection between the cell surface and the nucleus, in which a cascade of protein kinases ultimately leads to transcription factor phosphorylation. An alternative pathway, known as the **JAK/STAT pathway**, provides a much more immediate con-

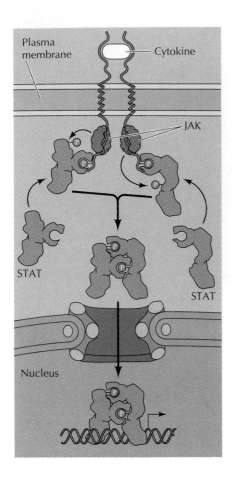

Plasma
membrane

Cytokine

JAK

STAT

STAT

Nucleus

**Figure 13.37**
**The JAK/STAT pathway**   The STAT proteins are transcription factors that contain SH2 domains that mediate their binding to phosphotyrosine-containing sequences. In unstimulated cells, STAT proteins are inactive in the cytosol. Stimulation of cytokine receptors leads to the binding of STAT proteins, where they are phosphorylated by the receptor-associated JAK protein-tyrosine kinases. The phosphorylated STAT proteins then dimerize and translocate to the nucleus, where they activate the transcription of target genes.

nection between protein-tyrosine kinases and transcription factors. In this pathway protein-tyrosine phosphorylation directly affects transcription factor localization and function (Figure 13.37).

The key elements in this pathway are the **STAT proteins** (signal *t*ransducers and *a*ctivators of *t*ranscription), which were originally identified in studies of cytokine receptor signaling. The STAT proteins are a family of transcription factors that contain SH2 domains. They are inactive in unstimulated cells, where they are localized to the cytoplasm. Stimulation of cytokine receptors leads to recruitment of STAT proteins, which bind via their SH2 domains to phosphotyrosine-containing sequences in the cytoplasmic domains of receptor polypeptides. Following their association with activated receptors, the STAT proteins are phosphorylated by members of the JAK family of nonreceptor protein-tyrosine kinases, which are associated with cytokine receptors. Tyrosine phosphorylation promotes the dimerization of STAT proteins, which then translocate to the nucleus, where they stimulate transcription of their target genes.

Further studies have shown that STAT proteins are also activated downstream of receptor protein-tyrosine kinases, where their phosphorylation may be catalyzed either by the receptors themselves or by associated nonreceptor kinases. The STAT transcription factors thus serve as direct links between both cytokine and growth factor receptors on the cell surface and regulation of gene expression in the nucleus.

## SIGNAL TRANSDUCTION AND THE CYTOSKELETON

The preceding sections focused on signaling pathways that regulate changes in metabolism or gene expression in response to hormones and growth factors. However, the functions of most cells are also directly affected by cell adhesion and the organization of the cytoskeleton. The receptors responsible for cell adhesion thus act to initiate intracellular signaling pathways that regulate other aspects of cell behavior, including gene expression. Conversely, growth factors frequently act to induce cytoskeletal alterations resulting in cell movement or changes in cell shape. Components of the cytoskeleton thus act as both receptors and targets in cell signaling pathways, integrating cell shape and movement with other cellular responses.

### Integrins and Signal Transduction

As discussed in Chapters 11 and 12, the integrins are the major receptors responsible for the attachment of cells to the extracellular matrix. At two types of cell-matrix junctions (focal adhesions and hemidesmosomes), the integrins also interact with components of the cytoskeleton to provide a stable linkage between the extracellular matrix and adherent cells (see Figure 12.61). In addition to this structural role, the integrins serve as receptors that activate intracellular signaling pathways, thereby controlling gene expression and other aspects of cell behavior in response to adhesive interactions.

Like members of the cytokine receptor superfamily, the integrins have

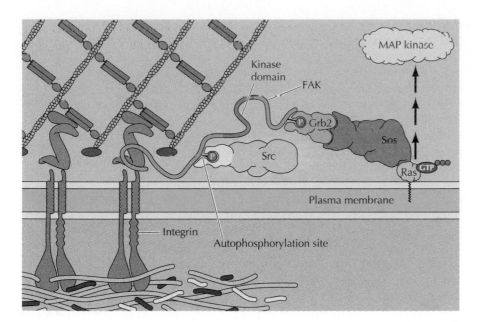

*Figure 13.38*
**Model for signaling from the FAK protein-tyrosine kinase** Binding of integrins to the extracellular matrix stimulates FAK activity, leading to its autophosphorylation. Src then binds to the FAK autophosphorylation site and phosphorylates FAK on additional tyrosine residues. These phosphotyrosines serve as binding sites for the Grb2-Sos complex, leading to activation of Ras and the MAP kinase cascade, as well as for additional downstream signaling molecules, including PI 3-kinase.

short cytoplasmic tails that lack any intrinsic enzymatic activity. However, protein-tyrosine phosphorylation is an early response to the interaction of integrins with extracellular matrix components, suggesting that the integrins are linked to nonreceptor protein-tyrosine kinases. In particular, a nonreceptor protein-tyrosine kinase called **FAK** (*f*ocal *a*dhesion *k*inase) plays a key role in integrin signaling (Figure 13.38). As its name implies, FAK is localized to focal adhesions and rapidly becomes tyrosine-phosphorylated following the binding of integrin to extracellular matrix components, such as fibronectin. Like other protein-tyrosine kinases, the activation of FAK is thought to involve autophosphorylation induced by the clustering of integrins bound to the extracellular matrix, although details of the mechanism of FAK activation remain to be established.

In addition to FAK, members of the Src family of nonreceptor protein-tyrosine kinases also associate with focal adhesions and are involved in integrin signaling. Interestingly, Src and FAK appear to function in association with each other as a result of the binding of the Src SH2 domain to an autophosphorylation site of FAK (see Figure 13.38). Src then phosphorylates additional sites on FAK, so these two nonreceptor protein-tyrosine kinases act jointly in signaling from integrin receptors.

As discussed earlier for growth factor receptors, tyrosine phosphorylation of FAK creates binding sites for the SH2 domains of other downstream signaling molecules, including PI 3-kinase and the Grb2-Sos complex. Recruitment of the Sos guanine nucleotide exchange factor leads to activation of Ras, which in turn couples integrins to activation of the ERK MAP kinase signaling pathway. Integrin activation of the FAK and Src nonreceptor protein-tyrosine kinases thus links cell adhesion to changes in gene expression and cell behavior that are analogous to those induced by the binding of growth factors to their cell surface receptors.

## Regulation of the Actin Cytoskeleton

Cellular responses to extracellular signals, including growth factors, frequently include changes in cell movement and cell shape. For example, growth factor-induced alterations in cell motility (as well as in cell proliferation) play critical roles in processes such as wound healing and embryonic development. As discussed in Chapter 11, these aspects of cell behavior are governed principally by the actin cytoskeleton. In particular, many types of cell movement are based on the dynamic assembly and disassembly of actin filaments underlying the plasma membrane. Remodeling of the actin cytoskeleton therefore represents a key element of the response of many cells to growth factors and other extracellular stimuli.

Studies of the response of fibroblasts to growth factor stimulation have demonstrated a critical role for members of the Rho subfamily of small GTP-binding proteins (including **Rho**, **Rac**, and **Cdc42**) in regulating different aspects of actin reorganization (Figure 13.39). The cytoskeletal alterations resulting from growth factor stimulation include the production of cell surface protrusions (filopodia, lamellipodia, and membrane ruffles) as well as the formation of focal adhesions and stress fibers. Microinjection of cells with specific mutants of different Rho family members has shown that Cdc42 induces the formation of microspikes or filopodia, Rac mediates the formation of lamellipodia and membrane ruffles, and Rho is responsible for the formation of focal adhesions and stress fibers. As discussed earlier in this chapter, members of the Rho family can also activate MAP kinase signaling pathways, so these small GTP-binding proteins function as dual regulators of both gene expression and cytoskeletal remodeling.

The activities of these Rho family members are linked to one another in a hierarchical fashion: Cdc42 activates Rac, and Rac then activates Rho (see Figure 13.39). This hierarchy suggests the intriguing possibility that these proteins coordinately control the cytoskeletal alterations responsible for cell movement. As discussed in Chapter 11, the crawling movements of cells across a surface can be viewed as a series of three types of events (see Figure 11.30). First, microspikes and lamellipodia are extended from the leading edge of the cell—the expected consequences of activation of Cdc42 and Rac, respectively. These extensions of the cell surface then form new attachments to the substratum, corresponding to the formation of focal adhesions mediated by Rho. Finally, the trailing edge of the cell retracts into the cell body—a process that may involve the formation of stress fibers, also mediated by Rho. The coordinated sequential action of these different Rho family members may thus help integrate the different aspects of actin remodeling responsible for the complex crawling motions of cells.

*Figure 13.39*
**Regulation of actin remodeling by Rho family proteins**   Different members of the Rho family regulate the polymerization of actin to produce filopodia (Cdc42), lamellipodia (Rac), and focal adhesions and stress fibers (Rho). The activities of these Rho family members are regulated in a hierarchical manner, such that Cdc42 activates Rac and Rac activates Rho. Fluorescence micrographs illustrate the distribution of actin following microinjection of fibroblasts with Cdc42, Rac, and Rho. (From C. D. Nobes and A. Hall, 1995, *Cell* 81: 53.)

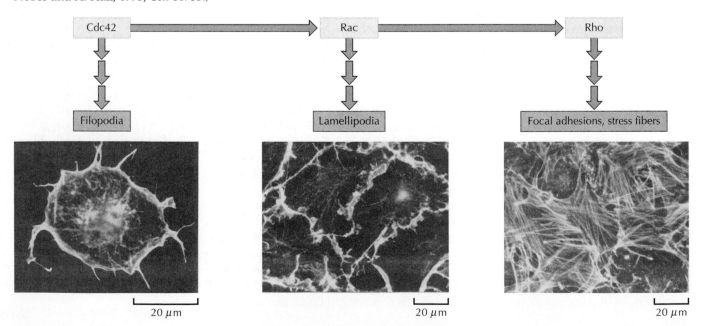

Although these members of the Rho family clearly play central roles in regulation of the cytoskeleton, many questions remain unanswered. First, the signaling pathways that regulate the activities of Rho family members in response to extracellular stimuli remain to be elucidated. Also unclear are the molecular mechanisms by which these small GTP-binding proteins ultimately regulate actin polymerization. Interestingly, however, Rac and Rho have been reported to stimulate the synthesis of $PIP_2$ from phosphatidylinositol 4-phosphate (PIP) by activating the enzyme PI 5-kinase. As mentioned in Chapter 11, $PIP_2$ can regulate the activities of several actin-binding proteins (e.g., gelsolin and profilin) that control the assembly of actin filaments. It is thus possible that changes in $PIP_2$ levels are responsible for the effects of Rac and Rho on actin polymerization, although this remains to be directly demonstrated.

## SIGNALING IN DEVELOPMENT AND DIFFERENTIATION

Understanding the molecular mechanisms that govern animal development is one of the major challenges of contemporary cell and molecular biology. Starting from only a single cell, the fertilized egg, all the diverse cell types of the body are produced and organized into tissues and organs. Both cell differentiation and the development of body structures must be regulated by intricate pathways of cell-cell signaling that coordinate the activities of individual cells and ultimately give rise to organisms as complex as human beings. Although a comprehensive discussion of developmental biology is beyond the scope of this book, it is noteworthy that considerable progress has been made in elucidating the signaling pathways responsible for some aspects of development and differentiation. Three examples of such signaling pathways are discussed here.

### Mesoderm Induction in Xenopus

Animal development initiates with the fusion of the sperm and the egg at fertilization. The fertilized egg then undergoes a series of rapid mitotic cleavages that divide its large volume of cytoplasm into smaller cells, forming a morula and then a blastula (Figure 13.40). The cells of the blastula then rearrange to form three germ layers, which give rise to different tissues. The outer germ layer (**ectoderm**) gives rise to skin and the nervous system; the inner germ layer (**endoderm**) gives rise to the cells lining the digestive tract and to internal organs such as the liver and pancreas; and the middle germ layer (**mesoderm**) gives rise to connective tissues, muscle, and blood cells. The cells of these three germ layers then interact with one another and undergo a series of rearrangements to produce the organs and tissues of the body.

Studies of the frog *Xenopus laevis*, which is widely used to study vertebrate development, have provided important insights into the signaling mechanisms responsible for germ layer formation. *Xenopus* eggs are extremely large cells (approximately 1 mm in diameter) that consist of distinct animal and vegetal poles, with the vegetal pole characterized by a high content of yolk protein. Cells isolated and cultured from these different regions of the embryo produce tissues derived from different germ layers. In particular, cells from the animal pole produce ectodermal tissues, whereas cells from the vegetal pole produce endodermal tissues. Mesoderm forms only if cells from both the animal and vegetal poles are cultured together. Under these conditions, mesodermal tissues are formed from the animal pole cells, indicating that signals from vegetal pole cells cause animal pole cells to alter their program of differentiation and give

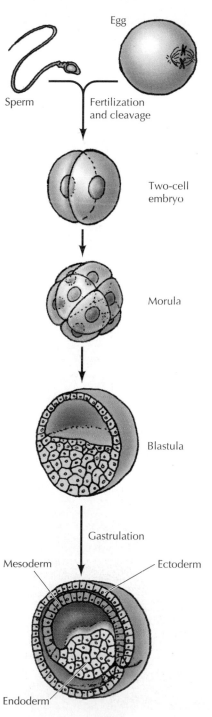

**Figure 13.40**
**Early animal development** The fertilized egg (e.g., of *Xenopus laevis*) undergoes a series of rapid cleavages to form a morula and blastula. The cells then undergo extensive rearrangements (gastrulation) to form the three germ layers (ectoderm, mesoderm, and endoderm).

Egg

Sperm    Fertilization and cleavage

Two-cell embryo

Morula

Blastula

Gastrulation

Mesoderm    Ectoderm

Endoderm

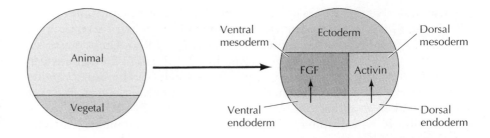

*Figure 13.41*
**Mesoderm induction** The *Xenopus* egg is divided into animal and vegetal hemispheres. Cells from the animal pole differentiate into ectoderm, and cells from the vegetal pole form endoderm. In addition, cells from the animal pole can be induced to form mesoderm in response to factors secreted by vegetal pole cells. Fibroblast growth factor (FGF) and activin play key roles in the induction of ventral and dorsal mesoderm, respectively.

rise to mesoderm instead of ectoderm.

Further studies of this system have indicated that mesoderm induction is mediated by growth factors that are secreted by the vegetal pole cells (Figure 13.41). At least two such signaling systems are involved, giving rise to different types of mesoderm in different regions of the embryo. In the ventral and lateral parts of the embryo, the secretion of fibroblast growth factor (FGF) by vegetal pole cells appears to play a key role in mesoderm formation. FGF stimulates a receptor protein-tyrosine kinase, and it has been shown that blocking this signaling pathway with dominant negative mutants of either the receptor, Ras, or Raf is sufficient to prevent mesoderm formation. Stimulation of the FGF receptor protein-tyrosine kinase and activation of the Ras/Raf/MAP kinase signaling pathway thus appears to be critical for ventral mesoderm induction.

Distinct factors secreted by vegetal cells are responsible for induction of dorsal mesoderm. A member of the TGF-β family, called activin, is thought to play a key role in this process, although other factors are probably also involved. Like other members of the TGF-β family, activin stimulates a receptor protein-serine/threonine kinase, the activation of which presumably leads (via currently unknown pathways) to changes in gene expression and cell differentiation. Although details of these signaling pathways remain to be understood, mesoderm induction in *Xenopus* provides a clear model for the action of at least two families of growth factors in early stages of embryonic development.

## Eye Development in Drosophila

The development of the compound eye of *Drosophila* provides a good example of the role of direct cell-cell signaling in differentiation, which has been elucidated largely by genetic analysis. The *Drosophila* compound eye consists of about 800 individual units, each of which contains eight photoreceptor neurons (R1 through R8) and 12 lens cells (Figure 13.42). The photoreceptor neurons develop in a fixed order, beginning with the differentiation of R8. R8 then induces two neighboring cells to become the R2 and R5 photoreceptors. Next, R2 induces neighboring cells to become R1 and R3, and R5 induces neighboring cells to become R4 and R6. The final step is differentiation of R7, which is induced by interaction with R8. Lens cells then develop from those cells that do not differentiate into photoreceptors.

The signaling pathway leading to development of the R7 cell has been characterized in detail, based on the isolation of mutant flies in which R7 fails to develop (Figure 13.43). One of these mutants (*sevenless*) results from defects in a gene encoding a receptor protein-tyrosine kinase that is expressed by precursors of R7 cells. Another mutant (called *boss*, which is short for *bride-of-sevenless*) results from defects in a gene encoding a cell surface protein expressed by R8 cells. Boss is the ligand for Sevenless, so direct cell-cell interaction between R8 and a precursor cell activates the Sevenless protein-tyrosine kinase. Further studies have shown that cell differentiation induced by sig-

*Figure 13.42*
**The *Drosophila* compound eye** (A) Scanning electron micrograph showing the compound eye, composed of about 800 individual units. (B) Each unit contains eight photoreceptor neurons (designated R1 through R8), which develop in a fixed order and pattern. (A Courtesy of T. Venkatesh, University of Michigan.)

(A)

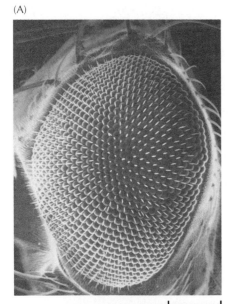

100 μm

(B)

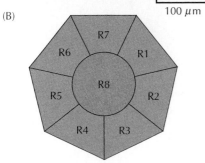

naling from Sevenless also requires Ras and Raf, leading to activation of the ERK MAP kinase pathway and resulting phosphorylation of transcription factors that mediate R7 differentiation.

### Vulval Induction in C. elegans

Development of the vulva in the nematode *C. elegans* is another example in which a pathway of cell-cell signaling has been elucidated by genetic analysis. In this system, a single cell first differentiates to become a gonadal anchor cell, which attaches the vulva to the uterus. The anchor cell then induces the differentiation of three precursor cells, which proliferate to form the 22 cells of the vulva.

Isolation of mutants in which the vulva fails to develop has allowed the characterization of several genes that are necessary for vulval induction, thereby delineating the pathway by which the anchor cell signals differentiation of the vulval precursor cells (Figure 13.44). One of these genes, *lin-3*, encodes a protein related to the mammalian growth factor EGF. The Lin-3 protein is secreted by anchor cells and binds to a receptor (Let-23) expressed on the surface of vulval precursor cells. Let-23 is a receptor protein-tyrosine kinase related to the mammalian EGF receptor. Other genes required for vulval induction include *let-60* and *lin-45*, which encode *C. elegans* Ras and Raf proteins, respectively. Vulval development in *C. elegans* thus involves growth factor stimulation of a receptor protein-tyrosine kinase and subsequent activation of the Ras/Raf signaling pathway, similar to the induction of ventral mesoderm in *Xenopus* and of the R7 photoreceptor in *Drosophila*.

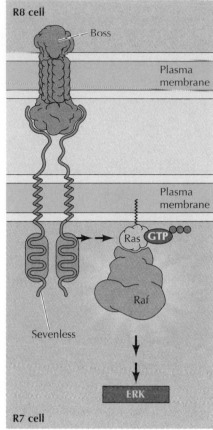

**Figure 13.43**
**Induction of R7 differentiation** Differentiation of the R7 photoreceptor neuron is induced by contact of a precursor cell with R8. The Boss protein on the R8 cell surface is the ligand for the Sevenless receptor protein-tyrosine kinase, which is expressed by R7 precursor cells. Stimulation of Sevenless activates the Ras/Raf/ERK MAP kinase pathway.

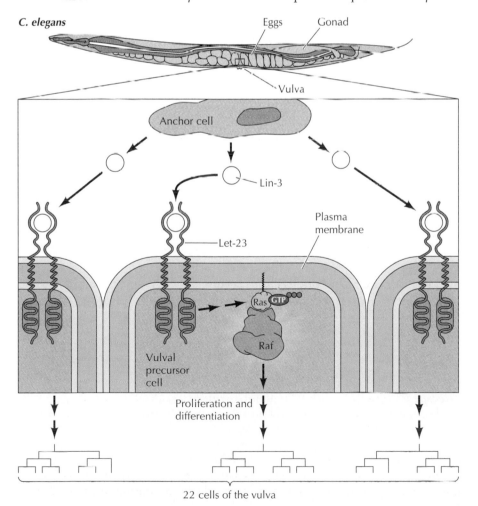

**Figure 13.44**
**Induction of the vulva in *C. elegans*** The gonadal anchor cell secretes Lin-3, which is related to EGF. Lin-3 stimulates Let-23, a receptor protein-tyrosine kinase expressed by vulval precursor cells. Activated Let-23 stimulates the Ras/Raf pathway, which induces three vulval precursor cells to proliferate and differentiate, forming the 22 cells of the vulva.

endocrine signaling, hormone, paracrine signaling, autocrine signaling

steroid hormone, testosterone, estrogen, progesterone, corticosteroid, glucocorticoid, mineralocorticoid, ecdysone, thyroid hormone, vitamin D$_3$, retinoic acid, retinoid, steroid receptor superfamily

nitric oxide

neurotransmitter

peptide hormone, neuropeptide, enkephalin, endorphin, neurohormone, growth factor, nerve growth factor (NGF), neurotrophin, epidermal growth factor (EGF), platelet-derived growth factor (PDGF), cytokine, membrane-anchored growth factor

eicosanoid, prostaglandin, prostacyclin, thromboxane, leukotriene

plant hormone, auxin, gibberellin, cytokinin, abscisic acid, ethylene

G protein, G protein-coupled receptor, heterotrimeric G protein

receptor protein-tyrosine kinase, autophosphorylation, SH2 domain

cytokine receptor superfamily, nonreceptor protein-tyrosine kinase, Src, Janus kinase (JAK)

protein-tyrosine phosphatase, transforming growth factor $\beta$ (TGF-$\beta$), protein-serine/threonine kinase, guanylyl cyclase

# *Summary*

## SIGNALING MOLECULES AND THEIR RECEPTORS

*Modes of Cell-Cell Signaling:* Most signaling molecules are secreted by one cell and bind to receptors expressed by a target cell. Cell-cell signaling is divided into three general categories (endocrine, paracrine, and autocrine signaling) based on the distance over which signals are transmitted.

*Steroid Hormones and the Steroid Receptor Superfamily:* The steroid hormones, thyroid hormone, vitamin D$_3$, and retinoic acid are small hydrophobic molecules that diffuse across the plasma membrane of their target cells and bind to intracellular receptors. Members of the steroid receptor superfamily function as transcription factors to directly regulate gene expression in response to ligand binding.

*Nitric Oxide:* Nitric oxide is an important paracrine signaling molecule in a variety of cell types.

*Neurotransmitters:* Neurotransmitters are small hydrophilic molecules that carry signals between neurons or between neurons and other target cells at a synapse. Many neurotransmitters bind to ligand-gated ion channels.

*Peptide Hormones and Growth Factors:* The widest variety of signaling molecules in animals are peptides, ranging from only a few to more than a hundred amino acids. This group of molecules includes peptide hormones, neuropeptides, and growth factors.

*Eicosanoids:* The eicosanoids are a class of lipids that function in paracrine and autocrine signaling.

*Plant Hormones:* Small molecules known as plant hormones regulate plant growth and development.

## FUNCTIONS OF CELL SURFACE RECEPTORS

*G Protein-Coupled Receptors:* The largest family of cell surface receptors, including the receptors for many hormones and neurotransmitters, transmit signals to intracellular targets via the intermediary action of G proteins.

*Receptor Protein-Tyrosine Kinases:* The receptors for most growth factors are protein-tyrosine kinases.

*Cytokine Receptors and Nonreceptor Protein-Tyrosine Kinases:* The receptors for many cytokines act in association with nonreceptor protein-tyrosine kinases.

*Receptors Linked to Other Enzymatic Activities:* Other kinds of cell surface receptors include protein-tyrosine phosphatases, protein-serine/threonine kinases, and guanylyl cyclases.

## PATHWAYS OF INTRACELLULAR SIGNAL TRANSDUCTION

*The cAMP Pathway: Second Messengers and Protein Phosphorylation:*
Cyclic AMP is an important second messenger in the response of animal cells to a variety of hormones and odorants. Most actions of cAMP are mediated by protein kinase A, which phosphorylates both metabolic enzymes and the transcription factor CREB.

*Cyclic GMP:* Cyclic GMP is also an important second messenger in animal cells. Its best-characterized role is in visual reception in the vertebrate eye.

*Phospholipids and Ca$^{2+}$:* Phospholipids and Ca$^{2+}$ are common second messengers activated downstream of both G protein-coupled receptors and protein-tyrosine kinases. Hydrolysis of phosphatidylinositol 4,5-bisphosphate (PIP$_2$) yields diacylglycerol and inositol 1,4,5-trisphosphate (IP$_3$), which activate protein kinase C and mobilize Ca$^{2+}$ from intracellular stores, respectively. Increased levels of cytosolic Ca$^{2+}$ then activate a variety of target proteins, including Ca$^{2+}$/calmodulin-dependent protein kinases. In electrically excitable cells of nerve and muscle, levels of cytosolic Ca$^{2+}$ are increased by the opening of voltage-gated Ca$^{2+}$ channels in the plasma membrane and ryanodine receptors in the endoplasmic and sarcoplasmic reticulum. In addition to being cleaved into diacylglycerol and IP$_3$, PIP$_2$ can be phosphorylated to the distinct second messenger PIP$_3$.

*Ras, Raf, and the MAP Kinase Pathway:* The MAP kinase pathway is a conserved chain of protein kinases activated downstream of a variety of extracellular signals. In animal cells, the best-characterized forms of MAP kinase are coupled to growth factor receptors by the small GTP-binding protein Ras, which initiates a protein kinase cascade leading to MAP kinase (ERK) activation. ERK then phosphorylates a variety of cytosolic and nuclear proteins, including transcription factors that mediate immediate-early gene induction.

*The JAK/STAT Pathway:* STAT proteins are transcription factors that contain SH2 domains and are activated directly by protein-tyrosine kinases associated with cytokine and growth factor receptors.

## SIGNAL TRANSDUCTION AND THE CYTOSKELETON

*Integrins and Signal Transduction:* Binding of integrins to the extracellular matrix stimulates the FAK nonreceptor protein-tyrosine kinase, leading to activation of MAP kinase and other downstream signaling pathways.

*Regulation of the Actin Cytoskeleton:* Growth factors induce alterations in cell movement and cell shape by remodeling the actin cytoskeleton. These cytoskeletal alterations are mediated by members of the Rho subfamily of small GTP-binding proteins.

---

intracellular signal transduction, cyclic AMP (cAMP), second messenger, adenylyl cyclase, cAMP phosphodiesterase, cAMP-dependent protein kinase, protein kinase A, cAMP response element (CRE), CREB

cyclic GMP (cGMP), rhodopsin, transducin, cGMP phosphodiesterase

phosphatidylinositol 4,5-bisphosphate (PIP$_2$), phospholipase C, diacylglycerol, inositol 1,4,5-trisphosphate (IP$_3$), protein kinase C, phorbol ester, NF-κB, IκB, calmodulin, CaM kinase II, ryanodine receptor, phosphatidylinositide (PI) 3-kinase, phosphatidylinositol 3,4,5-trisphosphate (PIP$_3$)

MAP kinase, ERK, Ras, Raf, MEK, guanine nucleotide exchange factor, GTPase-activating protein, small GTP-binding protein, immediate-early gene, serum response element (SRE), serum response factor (SRF), Elk-1

JAK/STAT pathway, STAT protein

FAK

Rho, Rac, Cdc42

| | |
|---|---|
| ectoderm, endoderm, mesoderm | **SIGNALING IN DEVELOPMENT AND DIFFERENTIATION**<br><br>*Mesoderm Induction in Xenopus:* The formation of mesoderm is induced by the action of secreted growth factors, including fibroblast growth factor and a member of the TGF-β family.<br><br>*Eye Development in Drosophila:* Differentiation of the R7 photoreceptor neuron is mediated by a cell-cell interaction that stimulates a receptor protein-tyrosine kinase, leading to activation of the Ras/Raf/MAP kinase signaling pathway.<br><br>*Vulval Induction in C. elegans:* A growth factor secreted by a gonadal anchor cell induces the differentiation of adjacent vulval precursor cells by stimulating a receptor protein-tyrosine kinase and activating the Ras/Raf pathway. |

## QUESTIONS

**1.** The proliferation of thyroid cells is stimulated by hormones that activate a receptor coupled to $G_s$. How would inhibitors of cAMP phosphodiesterase affect the proliferation of these cells?

**2.** The epinephrine receptor is coupled to $G_s$, whereas the acetylcholine receptor (on heart muscle cells) is coupled to $G_i$. Suppose you were to construct a recombinant molecule containing the extracellular sequences of the epinephrine receptor joined to the cytosolic sequences of the acetylcholine receptor. What effect would epinephrine have on cAMP levels in cells expressing such a recombinant receptor? What would be the effect of acetylcholine?

**3.** Platelet-derived growth factor (PDGF) is a dimer of two polypeptide chains. What would be the predicted effect of PDGF monomers on signaling from the PDGF receptor?

**4.** How would overexpression of protein phosphatase 1 affect the induction of cAMP-inducible genes in response to hormone stimulation of appropriate target cells? Would protein phosphatase 1 affect the function of cAMP-gated ion channels involved in odorant reception?

**5.** Protein kinase C-α (PKC-α) and protein kinase C-ε (PKC-ε) are two different members of the protein kinase C fami-

ly, which differ in their regulation. PKC-α requires both $Ca^{2+}$ and diacylglycerol for activation, whereas PKC-ε requires only diacylglycerol. How would hydrolysis of the phospholipids $PIP_2$ and phosphatidylcholine by phospholipase C affect the activities of these different PKC family members?

**6.** Dominant negative mutants of both Ras and Raf block growth factor-stimulated cell proliferation. The inhibitory effects of dominant negative Ras are overcome by expression of activated Raf. Would you expect activated Ras similarly to overcome the inhibitory effects of dominant negative Raf? How about activated MEK?

## REFERENCES AND FURTHER READING

### *Signaling Molecules and Their Receptors*

Arai, K., F. Lee, A. Miyajima, S. Miyatake, N. Arai and T. Yokota. 1990. Cytokines: Coordinators of immune and inflammatory responses. *Ann. Rev. Biochem.* 59: 783–836. [R]

Beato, M. 1989. Gene regulation by steroid hormones. *Cell* 56: 335–344. [R]

Bredt, D. S. and S. H. Snyder. 1994. Nitric oxide: A physiologic messenger molecule. *Ann. Rev. Biochem.* 63: 175–195. [R]

Burgess, W. H. and T. Maciag. 1989. The heparin-binding (fibroblast) growth factor family of proteins. *Ann. Rev. Biochem.* 58: 575–606. [R]

Carpenter, G. and S. Cohen. 1990. Epidermal growth factor. *J. Biol. Chem.* 265: 7709–7712. [R]

Ecker, J. R. 1995. The ethylene signal transduction pathway in plants. *Science* 268: 667–675. [R]

Evans, R. M. 1988. The steroid and thyroid hormone receptor superfamily. *Science* 240: 889–895. [R]

Fosket, D.E. 1994. *Plant Growth and Development: A Molecular Approach.* San Diego, CA: Academic Press.

Jessell, T. M. and E. R. Kandel. 1993. Synaptic transmission: A bidirectional and self-modifiable form of cell-cell communication. *Cell* 72/*Neuron* 10 (Suppl.): 1–30. [R]

Kastner, P., M. Mark and P. Chambon. 1995. Nonsteroid nuclear receptors: What are genetic studies telling us about their role in real life? *Cell* 83: 859–869. [R]

Levi-Montalcini, R. 1987. The nerve growth factor 35 years later. *Science* 237: 1154–1162. [R]

Mangelsdorf, D. J., C. Thummel, M. Beato, P. Herrlich, G. Schütz, K. Umesono, B. Blumberg, P. Kastner, M. Mark, P. Chambon and R. M. Evans. 1995. The nuclear receptor superfamily: The second decade. *Cell* 83: 835–839. [R]

Massagué, J. and A. Pandiella. 1993. Membrane-anchored growth factors. *Ann. Rev. Biochem.* 62: 515–541. [R]

Means, A. L. and L. J. Gudas. 1995. The roles of retinoids in vertebrate development. *Ann. Rev. Biochem.* 64: 201–233. [R]

Ross, R., E. W. Raines and D. F. Bowen-Pope. 1986. The biology of platelet-derived growth factor. *Cell* 46: 155–169. [R]

Schmidt, H. H. H. W. and U. Walter. 1994. NO at work. *Cell* 78: 919–925. [R]

Tsai, M.-J. and B. W. O'Malley. 1994. Molecular mechanisms of action of steroid/thyroid receptor superfamily members. *Ann. Rev. Biochem.* 63: 451–486. [R]

Vane, J. and R. Botting. 1987. Inflammation and the mechanism of action of anti-inflammatory drugs. *FASEB J.* 1: 89–96. [R]

## Functions of Cell Surface Receptors

Artavanis-Tsakonas, S., K. Matsuno and M. E. Fortini. 1995. Notch signaling. *Science* 268: 225–232. [R]

Derynck, R. 1994. TGF-β receptor-mediated signaling. *Trends Biochem. Sci.* 19: 548–553. [R]

Garbers, D. L. and D. G. Lowe. 1994. Guanylyl cyclase receptors. *J. Biol. Chem.* 269: 30741–30744. [R]

Heldin, C.-H. 1995. Dimerization of cell surface receptors in signal transduction. *Cell* 80: 213–223. [R]

Hepler, J. R. and A. G. Gilman. 1992. G proteins. *Trends Biochem. Sci.* 17: 383–387. [R]

Hunter, T. and B. M. Sefton. 1980. Transforming gene product of Rous sarcoma virus phosphorylates tyrosine. *Proc. Natl. Acad. Sci. USA* 77: 1311–1315. [P]

Ihle, J. N. 1995. Cytokine receptor signalling. *Nature* 377: 591–594. [R]

Lemmon, M. A. and J. Schlessinger. 1994. Regulation of signal transduction and signal diversity by receptor oligomerization. *Trends Biochem. Sci.* 19: 459–463. [R]

Massagué, J., L. Attisano and J. L. Wrana. 1994. The TGF-β family and its composite receptors. *Trends Cell Biol.* 4: 172–178. [R]

Neer, E. J. 1995. Heterotrimeric G proteins: Organizers of transmembrane signals. *Cell* 80: 249–257. [R]

Pawson, T. 1995. Protein modules and signalling networks. *Nature* 373: 573–580. [R]

Rodbell, M. 1980. The role of hormone receptors and GTP-regulatory proteins in membrane transduction. *Nature* 284: 17–22. [R]

Strader, C. D., T. M. Fong, M. R. Tota, D. Underwood and R. A. F. Dixon. 1994. Structure and function of G protein-coupled receptors. *Ann. Rev. Biochem.* 63: 101–132. [R]

Taniguchi, T. 1995. Cytokine signaling through nonreceptor protein tyrosine kinases. *Science* 268: 251–255. [R]

Van der Geer, P., T. Hunter and R. A. Lindberg. 1994. Receptor protein-tyrosine kinases and their signal transduction pathways. *Ann. Rev. Cell Biol.* 10: 251–337. [R]

Weiss, A. and D. R. Littman. 1994. Signal transduction by lymphocyte antigen receptors. *Cell* 76: 263–274. [R]

## Pathways of Intracellular Signal Transduction

Blenis, J. 1993. Signal transduction via the MAP kinases: Proceed at your own RSK. *Proc. Natl. Acad. Sci. USA* 90: 5889–5992. [R]

Clapham, D. E. 1995. Calcium signaling. *Cell* 80: 259–268. [R]

Cobb, M. H. and E. J. Goldsmith. 1995. How MAP kinases are regulated. *J. Biol. Chem.* 270: 14843–14846. [R]

Cooper, D. M. F., N. Mons and J. W. Karpen. 1995. Adenylyl cyclases and the interaction between calcium and cAMP signaling. *Nature* 374: 421–424. [R]

Daum, G., I. Eisenmann-Tappe, H.-W. Fries, J. Troppmair and U. R. Rapp. 1994. The ins and outs of Raf kinases. *Trends Biochem. Sci.* 19: 474–480. [R]

Divecha, N. and R. F. Irvine. 1995. Phospholipid signaling. *Cell* 80: 269–278. [R]

Exton, J. H. 1994. Phosphatidylcholine breakdown and signal transduction. *Biochim. Biophys. Acta* 1212: 26–42. [R]

Frank, D. A., and M. E. Greenberg. 1994. CREB: A mediator of long-term memory from mollusks to mammals. *Cell* 79: 5–8. [R]

Ghosh, A. and M. E. Greenberg. 1995. Calcium signaling in neurons: Molecular mechanisms and cellular consequences. *Science* 268: 239–247. [R]

Hannun, Y. A. 1994. The sphingomyelin cycle and the second messenger function of ceramide. *J. Biol. Chem.* 269: 3125–3128. [R]

Hanson, P. I. and H. Schulman. 1992. Neuronal Ca²⁺/calmodulin dependent protein kinases. *Ann. Rev. Biochem.* 61: 559–601. [R]

Herskowitz, I. 1995. MAP kinase pathways in yeast: For mating and more. *Cell* 80: 187–197. [R]

Hill, C. S. and R. Treisman. 1995. Transcriptional regulation by extracellular signals: Mechanisms and specificity. *Cell* 80: 199–211. [R]

Hunter, T. 1995. Protein kinases and phosphatases: The yin and yang of protein phosphorylation and signaling. *Cell* 80: 225–236. [R]

Hunter, T. and M. Karin. 1992. The regulation of transcription by phosphorylation. *Cell* 70: 375–387. [R]

Ihle, J. N. 1996. Cytokine receptor signalling. *Nature* 377: 591–594. [R]

Lincoln, T. M. and T. L. Cornwell. 1993. Intracellular cyclic GMP receptor proteins. *FASEB J.* 7: 328–338. [R]

Lowy, D. R. and B. M. Willumsen. 1993. Function and regulation of Ras. *Ann. Rev. Biochem.* 62: 851–891. [R]

Marshall, C. J. 1995. Specificity of receptor tyrosine kinase signaling: Transient versus sustained extracellular signal-regulated kinase activation. *Cell* 80: 179–185. [R]

Nishizuka, Y. 1992. Intracellular signaling by hydrolysis of phospholipids and activation of protein kinase C. *Science* 258: 607–614. [R]

Schindler, C. and J. E. Darnell Jr. 1995. Transcriptional responses to polypeptide ligands: The JAK-STAT pathway. *Ann. Rev. Biochem.* 64: 621–651. [R]

Schlessinger, J. 1993. How receptor tyrosine kinases activate Ras. *Trends Biochem. Sci.* 18: 273–275. [R]

Shenolikar, S. 1994. Protein serine/threonine phosphatases: New avenues for cell regulation. *Ann. Rev. Cell Biol.* 10: 55–86. [R]

Stryer, L. 1991. Visual excitation and recovery. *J. Biol. Chem.* 266: 10711–10714. [R]

Taussig, R. and A. G. Gilman. 1995. Mammalian membrane-bound adenylyl cyclases. *J. Biol. Chem.* 270: 1–4. [R]

Taylor, S. S., D. R. Knighton, J. Zheng, L. F. Ten Eyck and J. M. Sowadski. 1992. Structural framework for the protein kinase family. *Ann. Rev. Cell Biol.* 8: 429–462. [R]

Thanos, D. and T. Maniatis. 1995. NF-κB: A lesson in family values. *Cell* 80: 529–532. [R]

Vojtek, A. B. and J. A. Cooper. 1995. Rho family members: Activators of MAP kinase cascades. *Cell* 82: 527–529. [R]

Yarfitz, S. and J. B. Hurley. 1994. Transduction mechanisms of vertebrate and invertebrate photoreceptors. *J. Biol. Chem.* 269: 14329–14332. [R]

Zufall, F., S. Firestein and G. M. Shepherd. 1994. Cyclic nucleotide-gated ion channels and sensory transduction in olfactory receptor neurons. *Ann. Rev. Biophys. Biomol. Struct.* 23: 577–607. [R]

## Signal Transduction and the Cytoskeleton

Chong, L. D., A. Traynor-Kaplan, G. M. Bokoch, and M. A. Schwartz. 1994. The small GTP-binding protein Rho regulates a phosphatidylinositol 4-phosphate 5-kinase in mammalian cells. *Cell* 79: 507–513. [P]

Clark, E. A. and J. S. Brugge. 1995. Integrins and signal transduction pathways: The road taken. *Science* 268: 233–239. [R]

Hall, A. 1994. Small GTP-binding proteins and the regulation of the actin cytoskeleton. *Ann. Rev. Cell Biol.* 10: 31–54. [R]

Nobles, C. D. and A. Hall. 1995. Rho, Rac, and Cdc42 GTPases regulate the assembly of multimolecular focal complexes associated with actin stress fibers, lamellipodia, and filopodia. *Cell* 81: 53–62. [P]

Schaller, M. D. and J. T. Parsons. 1993. Focal adhesion kinase: An integrin-linked protein tyrosine kinase. *Trends Cell Biol.* 3: 258–262. [R]

### *Signaling in Development and Differentiation*

Ayama, E., T. J. Musci and M. Kirschner. 1991. Expression of a dominant negative mutant of the FGF receptor disrupts mesoderm formation in *Xenopus* embryos. *Cell* 66: 257–270. [P]

Dickson, B. 1995. Nuclear factors in sevenless signalling. *Trends Genet.* 11: 106–111. [R]

Gilbert, S. F. 1994. *Developmental Biology.* 4th ed. Sunderland, MA: Sinauer Associates.

Kessler, D. S. and D. A. Melton. 1994. Vertebrate embryonic induction: Mesodermal and neural patterning. *Science* 266: 596–604. [R]

Slack, J. M. W. 1991. *From Egg to Embryo: Regional Specification in Early Development.* 2d ed. Cambridge, England: Cambridge University Press.

Smith, J. C. 1993. Mesoderm-inducing factors in early vertebrate development. *EMBO J.* 12: 4463–4470. [R]

Sternberg, P. W. 1993. Intercellular signaling and signal transduction in *C. elegans. Ann. Rev. Genet.* 27: 497–521. [R]

Zipursky, S. L. and G. M. Rubin. 1994. Determination of neuronal cell fate: Lessons from the R7 neuron of *Drosophila. Ann. Rev. Neurosci.* 17: 373–397. [R]

# 14

# *The Cell Cycle*

S ELF-REPRODUCTION IS PERHAPS THE MOST FUNDAMENTAL characteristic of cells—as may be said for all living organisms. All cells reproduce by dividing in two, with each parental cell giving rise to two daughter cells on completion of each cycle of cell division. These newly formed daughter cells can themselves grow and divide, giving rise to a new cell population formed by the growth and division of a single parental cell and its progeny. In the simplest case, such cycles of growth and division allow a single bacterium to form a colony consisting of millions of progeny cells during overnight incubation on a plate of nutrient agar medium. In a more complex case, repeated cycles of cell growth and division result in the development of a single fertilized egg into the more than $10^{13}$ cells that make up the human body.

The division of all cells must be carefully regulated and coordinated with both cell growth and DNA replication in order to ensure the formation of progeny cells containing intact genomes. In eukaryotic cells, progression through the cell cycle is controlled by a series of protein kinases that have been conserved from yeasts to mammals. In higher eukaryotes, this cell cycle machinery is itself regulated by the growth factors that control cell proliferation, allowing the division of individual cells to be coordinated with the needs of the organism as a whole. Not surprisingly, defects in cell cycle regulation are a common cause of the abnormal proliferation of cancer cells, so studies of the cell cycle and cancer have become closely interconnected, similar to the relationship between studies of cancer and the cell signaling pathways discussed in Chapter 13.

## THE EUKARYOTIC CELL CYCLE

The division cycle of most cells consists of four coordinated processes: cell growth, DNA replication, distribution of the duplicated chromosomes to daughter cells, and cell division. In bacteria, cell growth and DNA replication take place throughout most of the cell cycle, and duplicated chromosomes are distributed to daughter cells in association with the plasma membrane. In eukaryotes, however, the cell cycle is more complex and consists of four discrete phases. Although cell growth is usually a continuous process, DNA is synthesized during only one phase of the cell cycle and the replicated chromosomes are then distributed to daughter nuclei by a complex series of events preceding cell division. Progression between these stages of the cell cycle is controlled by a conserved regulatory apparatus, which not only coordinates the different events of the cell cycle but also links the cell cycle with extracellular signals that control cell proliferation.

## Phases of the Cell Cycle

A typical eukaryotic cell cycle is illustrated by human cells in culture, which divide approximately every 24 hours. As viewed in the microscope, the cell cycle is divided into two basic parts: **mitosis** and **interphase**. Mitosis (nuclear division) is the most dramatic stage of the cell cycle, corresponding to the separation of daughter chromosomes and usually ending with cell division (**cytokinesis**). However, mitosis and cytokinesis last only about an hour, so approximately 95% of the cell cycle is spent in interphase—the period between mitoses. During interphase, the chromosomes are decondensed and distributed throughout the nucleus, so the nucleus appears morphologically uniform. At the molecular level, however, interphase is the time during which both cell growth and DNA replication occur in an orderly manner in preparation for cell division.

The cell grows at a steady rate throughout interphase, with most dividing cells doubling in size between one mitosis and the next. In contrast, DNA is synthesized during only a portion of interphase. The timing of DNA synthesis thus divides the cycle of eukaryotic cells into four discrete phases (Figure 14.1). The **M phase** of the cycle corresponds to mitosis, which is usually followed by cytokinesis. This phase is followed by the **G₁ phase** (gap 1), which corresponds to the interval (gap) between mitosis and initiation of DNA replication. During $G_1$, the cell is metabolically active and continuously

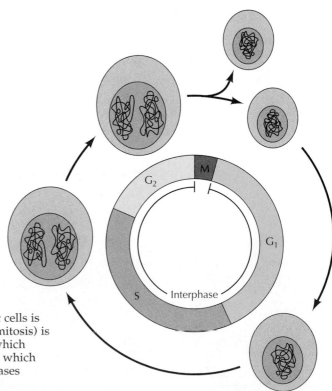

*Figure 14.1*
**Phases of the cell cycle**   The division cycle of most eukaryotic cells is divided into four discrete phases: M, $G_1$, S, and $G_2$. M phase (mitosis) is usually followed by cytokinesis. S phase is the period during which DNA replication occurs. The cell grows throughout interphase, which includes $G_1$, S, and $G_2$. The relative lengths of the cell cycle phases shown here are typical of rapidly replicating mammalian cells.

grows but does not replicate its DNA. $G_1$ is followed by **S phase** (synthesis), during which DNA replication takes place. The completion of DNA synthesis is followed by the $G_2$ **phase** (gap 2), during which cell growth continues and proteins are synthesized in preparation for mitosis.

The duration of these cell cycle phases varies considerably in different kinds of cells. For a typical rapidly proliferating human cell with a total cycle time of 24 hours, the $G_1$ phase might last about 11 hours, S phase about 8 hours, $G_2$ about 4 hours, and M about 1 hour. Other types of cells, however, can divide much more rapidly. Budding yeasts, for example, can progress through all four stages of the cell cycle in only about 90 minutes. Even shorter cell cycles (30 minutes or less) occur in early embryo cells shortly after fertilization of the egg (Figure 14.2). In this case, however, cell growth does not take place. Instead, these early embryonic cell cycles rapidly divide the egg cytoplasm into smaller cells. There is no $G_1$ or $G_2$ phase, and DNA replication occurs very rapidly in these early embryonic cell cycles, which therefore consist of very short S phases alternating with M phases.

In contrast to the rapid proliferation of embryonic cells, some cells in adult animals cease division altogether (e.g., nerve cells) and many other cells divide only occasionally, as needed to replace cells that have been lost because of injury or cell death. Cells of the latter type include skin fibroblasts, as well as the cells of many internal organs, such as the liver, kidney, and lung. As discussed further in the next section, these cells exit $G_1$ to enter a quiescent stage of the cycle called $G_0$, where they remain metabolically active but no longer proliferate unless called on to do so by appropriate extracellular signals.

Analysis of the cell cycle requires identification of cells at the different stages discussed above. Although mitotic cells can be distinguished microscopically, cells in other phases of the cycle ($G_1$, S, and $G_2$) must be identified by biochemical criteria. Cells in S phase can be readily identified because they incorporate radioactive thymidine, which is used exclusively for DNA synthesis (Figure 14.3). For example, if a population of rapidly proliferating human cells in culture is exposed to radioactive thymidine for a short period of time (e.g., 15 minutes) and then analyzed by autoradiography, about a third of the cells will be found to be radioactively labeled, corresponding to the fraction of cells in S phase.

Variations of such cell labeling experiments can also be used to determine the length of different stages of the cell cycle. For example, consider an experiment in which cells are exposed to radioactive thymidine for 15 minutes, after which the radioactive thymidine is removed and the cells are cultured for varying lengths of time prior to autoradiography. Radioactively labeled interphase cells that were in S phase during the time of exposure to radioactive thymidine will be observed for several hours as they

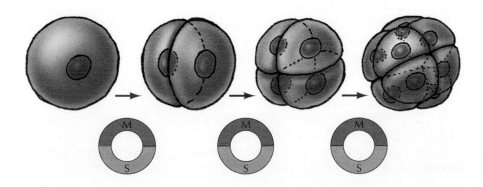

*Figure 14.2*
**Embryonic cell cycles**   Early embryonic cell cycles rapidly divide the cytoplasm of the egg into smaller cells. The cells do not grow during these cycles, which lack $G_1$ and $G_2$ and consist simply of short S phases alternating with M phases.

*Figure 14.3*
**Identification of S phase cells by incorporation of radioactive thymidine**    The cells were exposed to radioactive thymidine and analyzed by autoradiography. Labeled cells are indicated by arrows. (From D. W. Stacey et al, 1991, *Mol. Cell Biol.* 11: 4053.)

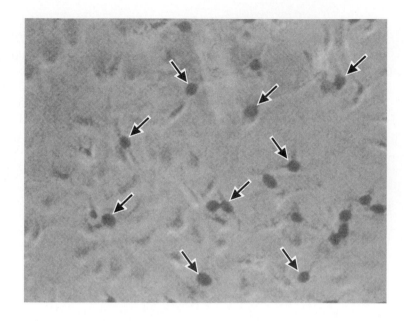

*Figure 14.4*
**Determination of cellular DNA content**    A population of cells is labeled with a fluorescent dye that binds DNA. The cells are then passed through a flow cytometer, which measures the fluorescence intensity of individual cells. The data are plotted as cell number versus fluorescence intensity, which is proportional to DNA content. The distribution shows two peaks, corresponding to cells with DNA contents of $2n$ and $4n$; these cells are in the $G_1$ and $G_2/M$ phases of the cycle, respectively. Cells in S phase have DNA contents between $2n$ and $4n$ and are distributed between these two peaks.

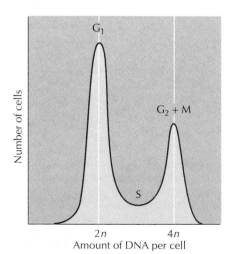

progress through the remainder of S and $G_2$. In contrast, radioactively labeled mitotic cells will not be observed until 4 hours after labeling. This 4-hour lag time corresponds to the length of $G_2$—the minimum time required for a cell that incorporated radioactive thymidine at the end of S phase to enter mitosis.

Cells at different stages of the cell cycle can also be distinguished by their DNA content (Figure 14.4). For example, animal cells in $G_1$ are diploid (containing two copies of each chromosome), so their DNA content is referred to as $2n$ ($n$ designates the haploid DNA content of the genome). During S phase, replication increases the DNA content of the cell from $2n$ to $4n$, so cells in S have DNA contents ranging from $2n$ to $4n$. DNA content then remains at $4n$ for cells in $G_2$ and M, decreasing to $2n$ after cytokinesis. Experimentally, cellular DNA content can be determined by incubation of cells with a fluorescent dye that binds to DNA, followed by analysis of the fluorescence intensity of individual cells in a **flow cytometer** or **fluorescence-activated cell sorter**, thereby distinguishing cells in the $G_1$, S, and $G_2/M$ phases of the cell cycle.

### Regulation of the Cell Cycle by Cell Growth and Extracellular Signals

The progression of cells through the division cycle is regulated by extracellular signals from the environment, as well as by internal signals that monitor and coordinate the various processes that take place during different cell cycle phases. An example of cell cycle regulation by extracellular signals is provided by the effect of growth factors on animal cell proliferation. In addition, different cellular processes, such as cell growth, DNA replication, and mitosis, all must be coordinated during cell cycle progression. This is accomplished by a series of control points that regulate progression through various phases of the cell cycle.

A major cell cycle regulatory point in many types of cells occurs late in $G_1$ and controls progression from $G_1$ to S. This regulatory point was first defined by studies of budding yeast (*Saccharomyces cerevisiae*), where it is known as **START** (Figure 14.5). Once cells have passed START, they are committed to entering S phase and undergoing one cell division cycle. However, passage through START is a highly regulated event in the yeast cell cycle, where it is

(A)

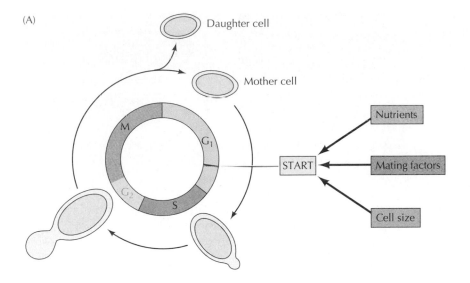

(B)

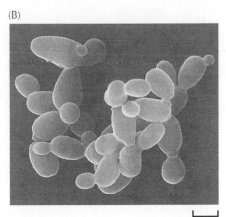

**Figure 14.5**  1 μm
**Regulation of the cell cycle of budding yeast** (A) The cell cycle of *Saccharomyces cerevisiae* is regulated primarily at a point in late G₁ called START. Passage through START is controlled by the availability of nutrients, mating factors, and cell size. Note that these yeasts divide by budding. Buds form just after START and continue growing until they separate from the mother cell after mitosis. The daughter cell formed from the bud is smaller than the mother cell and therefore requires more time to grow during the G₁ phase of the next cell cycle. Although G₁ and S phases occur normally, the mitotic spindle begins to form during S phase, so the cell cycle of budding yeast lacks a distinct G₂ phase. (B) Scanning electron micrograph of *S. cerevisiae*. The size of the bud reflects the position of the cell in the cycle. (David M. Phillips/Visuals Unlimited.)

controlled by external signals, such as the availability of nutrients, as well as by cell size. For example, if yeasts are faced with a shortage of nutrients, they arrest their cell cycle at START and enter a resting state rather than proceeding to S phase. Thus, START represents a decision point at which the cell determines whether sufficient nutrients are available to support progression through the rest of the division cycle. Polypeptide factors that signal yeast mating also arrest the cell cycle at START, allowing haploid yeast cells to fuse with one another instead of progressing to S phase.

In addition to serving as a decision point for monitoring extracellular signals, START is the point at which cell growth is coordinated with DNA replication and cell division. The importance of this regulation is particularly evident in budding yeasts, in which cell division produces progeny cells of very different sizes: a large mother cell and a small daughter cell. In order for yeast cells to maintain a constant size, the small daughter cell must grow more than the large mother cell does before they divide again. Thus, cell size must be monitored in order to coordinate cell growth with other cell cycle events. This regulation is accomplished by a control mechanism that requires each cell to reach a minimum size before it can pass START. Consequently, the small daughter cell spends a longer time in G₁ and grows more than the mother cell.

The proliferation of most animal cells is similarly regulated in the G₁ phase of the cell cycle. In particular, a decision point in late G₁, called the **restriction point** in animal cells, functions analogously to START in yeasts (Figure 14.6). In contrast to yeasts, however, the passage of animal cells through the cell cycle is regulated primarily by the extracellular growth factors that signal cell proliferation, rather than by the availability of nutrients. In the presence of the appropriate growth factors, cells pass the restriction point and enter S phase. Once it has passed through the restriction point, the cell is committed to proceed through S phase and the rest of

**Figure 14.6**
**Regulation of animal cell cycles by growth factors** The availability of growth factors controls the animal cell cycle at a point in late G₁ called the restriction point. If growth factors are not available during G₁, the cells enter a quiescent stage of the cycle called G₀.

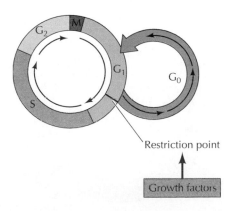

(A)

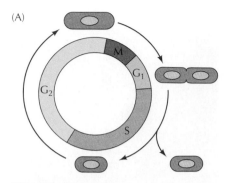

(B)

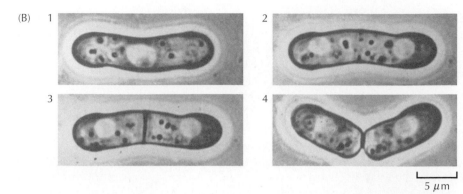

5 μm

**Figure 14.7**
**Cell cycle of fission yeast** (A) Fission yeasts grow by elongating at both ends and divide by forming a wall through the middle of the cell. In contrast to the cycle of budding yeasts, the cell cycle of fission yeasts has normal $G_1$, S, $G_2$, and M phases. Note that cytokinesis occurs in $G_1$. The length of the cell indicates its position in the cycle. (B) Light micrographs showing successive stages of mitosis and cytokinesis of *Schizosaccharomyces pombe*. (B, courtesy of C. F. Robinow, University of Western Ontario.)

the cell cycle, even in the absence of further growth factor stimulation. On the other hand, if appropriate growth factors are not available in $G_1$, progression through the cell cycle stops at the restriction point. Such arrested cells then enter a quiescent stage of the cell cycle called $G_0$, in which they can remain for long periods of time without proliferating. $G_0$ cells are metabolically active, although they cease growth and have reduced rates of protein synthesis. As already noted, many cells in animals remain in $G_0$ unless called on to proliferate by appropriate growth factors or other extracellular signals. For example, skin fibroblasts are arrested in $G_0$ until they are stimulated to divide as required to repair damage resulting from a wound. The proliferation of these cells is triggered by platelet-derived growth factor, which is released from blood platelets during clotting and signals the proliferation of fibroblasts in the vicinity of the injured tissue.

Although the proliferation of most cells is regulated primarily in $G_1$, some cell cycles are instead controlled principally in $G_2$. One example is the cell cycle of the fission yeast *Schizosaccharomyces pombe* (Figure 14.7). In contrast to *Saccharomyces cerevisiae*, the cell cycle of *S. pombe* is regulated primarily by control of the transition from $G_2$ to M, which is the principal point at which cell size and nutrient availability are monitored. In animals, the primary example of cell cycle control in $G_2$ is provided by oocytes. Vertebrate oocytes can remain arrested in $G_2$ for long periods of time (several decades in humans) until their progression to M phase is triggered by hormonal stimulation. Extracellular signals can thus control cell proliferation by regulating progression from the $G_2$ to M as well as the $G_1$ to S phases of the cell cycle.

**Figure 14.8**
**Cell cycle checkpoints** Several checkpoints function to ensure that complete genomes are transmitted to daughter cells. One major checkpoint arrests cells in $G_2$ in response to damaged or unreplicated DNA. The presence of damaged DNA also leads to cell cycle arrest at a checkpoint in $G_1$. Another checkpoint, in M phase, arrests mitosis if the daughter chromosomes are not properly aligned on the mitotic spindle.

## Cell Cycle Checkpoints

The controls discussed in the previous section regulate cell cycle progression in response to cell size and extracellular signals, such as nutrients and growth factors. In addition, the events that take place during different stages of the cell cycle must be coordinated with one another so that they occur in the appropriate order. For example, it is critically important that the cell not begin mitosis until replication of the genome has been completed. The alternative would be a catastrophic cell division, in which the daughter cells failed to inherit complete copies of the genetic material. In most cells, this coordination between different phases of the cell cycle is dependent on a system of checkpoints and feedback controls that prevent entry into the next phase of the cell cycle until the events of the preceding phase have been completed.

Several cell cycle checkpoints function to ensure that incomplete or damaged chromosomes are not replicated and passed on to daughter cells (Figure 14.8). One of the most clearly defined of these checkpoints occurs in $G_2$ and prevents the initiation of mitosis until DNA replication is completed. This $G_2$ checkpoint senses unreplicated DNA, which generates a signal that

Unreplicated or damaged DNA

Chromosome misalignment

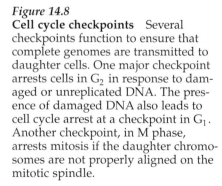
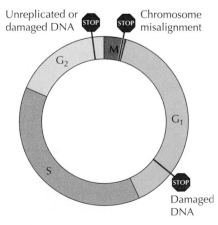

Damaged DNA

leads to cell cycle arrest. Operation of the $G_2$ checkpoint therefore prevents the initiation of M phase before completion of S phase, so cells remain in $G_2$ until the genome has been completely replicated. Only then is the inhibition of $G_2$ progression relieved, allowing the cell to initiate mitosis and distribute the completely replicated chromosomes to daughter cells.

Progression through the cell cycle is also arrested at the $G_2$ checkpoint in response to DNA damage, such as that resulting from irradiation. This arrest allows time for the damage to be repaired, rather than being passed on to daughter cells. Some yeast mutants are unable to arrest the cell cycle in $G_2$ in response to either unreplicated or damaged DNA, indicating that the same cell cycle checkpoint is responsible for $G_2$ arrest induced by either of these events. However, the molecular mechanisms by which unreplicated or damaged DNA signals arrest at the checkpoint are not yet understood.

DNA damage arrests the cell cycle not only in $G_2$, but also at a checkpoint in $G_1$. This $G_1$ arrest may allow repair of the damage to take place before the cell enters S phase, where the damaged DNA would be replicated. In mammalian cells, arrest at the $G_1$ checkpoint is mediated by the action of a protein known as **p53**, which is rapidly induced in response to damaged DNA (Figure 14.9). Interestingly, the gene encoding p53 is frequently mutated in human cancers. Loss of p53 function as a result of these mutations prevents $G_1$ arrest in response to DNA damage, so the damaged DNA is replicated and passed on to daughter cells instead of being repaired. This inheritance of damaged DNA results in an increased frequency of mutations and general instability of the cellular genome, which contributes to cancer development. Mutations in the *p53* gene are the most common genetic alterations in human cancers (see Chapter 15), illustrating the critical importance of cell cycle regulation in the life of multicellular organisms.

Another important cell cycle checkpoint that maintains the integrity of the genome occurs toward the end of mitosis (see Figure 14.8). This checkpoint monitors the alignment of chromosomes on the mitotic spindle, thus ensuring that a complete set of chromosomes is distributed accurately to the daughter cells. For example, the failure of one or more chromosomes to align properly on the spindle causes mitosis to arrest at metaphase, prior to the segregation of the newly replicated chromosomes to daughter nuclei. As a result of this checkpoint, the chromosomes do not separate until a complete complement of chromosomes has been organized for distribution to each daughter cell.

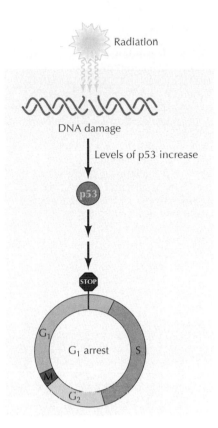

*Figure 14.9*
**Role of p53 in $G_1$ arrest induced by DNA damage**    DNA damage, such as that resulting from irradiation, leads to rapid increases in p53 levels. The protein p53 then signals cell cycle arrest at the $G_1$ checkpoint.

## Coupling of S Phase to M Phase

The $G_2$ checkpoint prevents the initiation of mitosis prior to the completion of S phase, thereby ensuring that incompletely replicated DNA is not distributed to daughter cells. It is equally important to ensure that the genome is replicated only once per cell cycle. Thus, once DNA has been replicated, control mechanisms must exist to prevent initiation of a new S phase prior to mitosis. These controls prevent cells in $G_2$ from reentering S phase and block the initiation of another round of DNA replication until after mitosis, at which point the cell has entered the $G_1$ phase of the next cell cycle.

Initial insights into this dependence of S phase on M phase came from cell fusion experiments of Potu Rao and Robert Johnson in 1970 (Figure 14.10). These investigators isolated cells in different phases of the cycle and then fused these cells to each other to form cell hybrids. When $G_1$ cells were fused with S phase cells, the $G_1$ nucleus immediately began to synthesize DNA. Thus, the cytoplasm of S phase cells contained factors that initiated DNA synthesis in the $G_1$ nucleus. Fusing $G_2$ cells with S phase cells, however, yielded a quite different result: The $G_2$ nucleus was unable to initiate DNA synthesis even in the presence of an S phase cytoplasm. It thus

Cells in S phase were fused either to cells in $G_1$ or to cells in $G_2$. When $G_1$ cells were fused with S phase cells, the $G_1$ nucleus immediately began to replicate DNA. In contrast, when $G_2$ cells were fused with S phase cells, only the S phase nucleus continued DNA replication. It therefore appeared that the $G_2$ nucleus had to pass through M and enter $G_1$ before another round of DNA replication could be initiated.

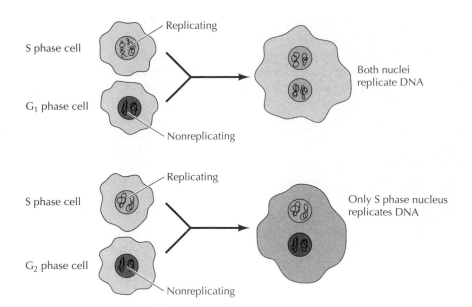

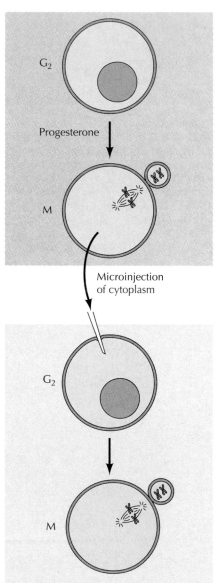

appeared that DNA synthesis in the $G_2$ nucleus was prevented by a mechanism that blocked rereplication of the genome until after mitosis had taken place. One model to account for this block is that a "licensing factor" required for DNA replication binds to the chromosomes during mitosis. The licensing factor is inactivated during replication and is unable to enter the nucleus again until the nuclear envelope breaks down during M phase, thereby preventing DNA replication until the $G_1$ phase of the next cell cycle. However, the molecular basis of this control remains to be established.

## REGULATORS OF CELL CYCLE PROGRESSION

One of the most exciting developments of the last decade has been the elucidation of the molecular mechanisms that control the progression of eukaryotic cells through the division cycle. Our current understanding of cell cycle regulation has emerged from a convergence of results obtained through experiments on organisms as diverse as yeasts, sea urchins, frogs, and mammals. These studies have revealed that the cell cycle of all eukaryotes is controlled by a conserved set of protein kinases, which are responsible for triggering the major cell cycle transitions.

### MPF: A Dimer of Cdc2 and Cyclin

Three initially distinct experimental approaches contributed to identification of the key molecules responsible for cell cycle regulation. The first of these avenues of investigation originated with studies of frog oocytes (Figure 14.11). These oocytes are arrested in the $G_2$ phase of the cell cycle until hormonal stimulation triggers their entry into the M phase of meiosis (dis-

*Figure 14.11*
**Identification of MPF** Frog oocytes are arrested in the $G_2$ phase of the cell cycle, and entry into M phase of meiosis is triggered by the hormone progesterone. In the experiment diagrammed here, $G_2$-arrested oocytes were microinjected with cytoplasm extracted from oocytes that had undergone the transition from $G_2$ to M. Such cytoplasmic transfers induced the $G_2$ to M transition in the absence of hormonal stimulation, demonstrating that a cytoplasmic factor (MPF) is sufficient to induce entry into the M phase of meiosis.

cussed later in this chapter). In 1971, two independent teams of researchers (Yoshio Masui and Clement Markert, as well as Dennis Smith and Robert Ecker) found that oocytes arrested in $G_2$ could be induced to enter M phase by microinjection of cytoplasm from oocytes that had been hormonally stimulated. It thus appeared that a cytoplasmic factor present in hormone-treated oocytes was sufficient to trigger the transition from $G_2$ to M in oocytes that had not been exposed to hormone. Because the entry of oocytes into meiosis is frequently referred to as oocyte maturation, this cytoplasmic factor was called **maturation promoting factor (MPF)**. Further studies showed, however, that the activity of MPF is not restricted to the entry of oocytes into meiosis. To the contrary, MPF is also present in somatic cells, where it induces entry into M phase of the mitotic cycle. Rather than being specific to oocytes, MPF thus appeared to act as a general regulator of the transition from $G_2$ to M.

The second approach to understanding cell cycle regulation was the genetic analysis of yeasts, pioneered by Lee Hartwell and his colleagues in the early 1970s. Studying the budding yeast *Saccharomyces cerevisiae*, these investigators identified temperature-sensitive mutants that were defective in cell cycle progression. The key characteristic of these mutants (called *cdc* for *cell division cycle* mutants) was that they underwent growth arrest at specific points in the cell cycle. For example, a particularly important mutant designated *cdc28* caused the cell cycle to arrest at START, indicating that the Cdc28 protein is required for passage through this critical regulatory point in $G_1$ (Figure 14.12). A similar collection of cell cycle mutants was isolated in the fission yeast *Schizosaccharomyces pombe* by Paul Nurse and his collaborators. These mutants included *cdc2*, which arrests the *S. pombe* cell cycle both in $G_1$ and at the $G_2$ to M transition (the major regulatory point in fission yeast). Further studies showed that *S. cerevisiae cdc28* and *S. pombe cdc2* are functionally homologous genes, which are required for passage through START as well as for entry into mitosis in both species of yeasts. To avoid confusion resulting from the difference in genetic nomenclature between *S. cerevisiae* and *S. pombe*, the protein encoded by both genes will be called

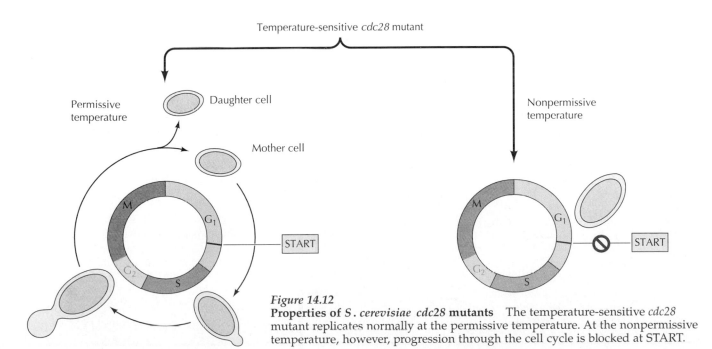

**Figure 14.12**
**Properties of *S. cerevisiae cdc28* mutants**   The temperature-sensitive *cdc28* mutant replicates normally at the permissive temperature. At the nonpermissive temperature, however, progression through the cell cycle is blocked at START.

# KEY EXPERIMENT

## The Discovery of MPF

**Cytoplasmic Control of Nuclear Behavior during Meiotic Maturation of Frog Oocytes**

Yoshio Masui and Clement L. Markert
Yale University, New Haven, CT
*Journal of Experimental Zoology, 1971, Volume 177, Pages 129–146*

### The Context

Nuclear transplantation and cell fusion experiments performed in the 1960s indicated that nuclei transferred into cells at different stages of the mitotic cell cycle generally adopt the behavior of the host cell. Thus, it appeared that mitotic activity of the nucleus was regulated by the cytoplasm. However, the existence of postulated cytoplasmic factors that control the mitotic activity of nuclei remained to be demonstrated by a direct experimental approach. This demonstration was provided by the studies of Masui and Markert, who investigated the role of cytoplasmic factors in regulating nuclear behavior during meiosis of frog oocytes.

Several features of frog oocyte meiosis suggested that it is controlled by cytoplasmic factors. In particular, meiosis of frog oocytes is arrested at the end of the prophase of meiosis I. Treatment with the hormone progesterone triggers the resumption of meiosis, which is equivalent to the $G_2$ to M transition of somatic cells. The oocytes then undergo a second arrest at the metaphase of meiosis II, where they remain until fertilization. Masui and Markert hypothesized that the effects of both hormone treatment and fertilization on meiosis were due to changes in the cytoplasm that secondarily controlled behavior of the nucleus. They tested this hypothesis directly by transferring cytoplasm from hormone-stimulated oocytes into unstimulated oocytes. These experiments demonstrated that a

cytoplasmic factor, which Masui and Markert named maturation promoting factor (MPF), is responsible for the induction of meiosis following hormone treatment.

### The Experiments

Because of both their large size and their ability to survive injection with glass micropipettes, frog oocytes appeared to provide a particularly suitable experimental system for testing the activity of cytoplasmic factors. The basic design of Masui and Markert's experiments was to remove cytoplasm from donor oocytes that had been treated with progesterone to induce the resumption of meiosis. Varying amounts of this cytoplasm were then injected into untreated recipient oocytes. The key result was that cytoplasm removed six or more hours after hormone treatment of donor oocytes induced the resumption of meiosis in injected recipients (see figure). In contrast, injection of cytoplasm from control oocytes that had not been exposed to progesterone had no effect on the recipients. It thus appeared that hormone-treated oocytes contained a cytoplasmic factor that could induce the resumption of meiosis in recipients that had never been exposed to progesterone.

Control experiments ruled out the possibility that progesterone itself is the meiosis-inducing factor in donor cytoplasm. In particular, it was demonstrated that the injection of progesterone into recipient oocytes fails to induce meiosis. Only external

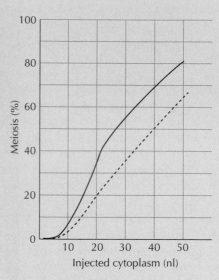

Recipient oocytes were injected with the indicated amounts of cytoplasm from oocytes that had been treated with progesterone. Donor cytoplasm was either withdrawn from the central region of oocytes with a micropipette (solid line) or prepared by homogenization of whole oocytes (dashed line). Results are presented as the percentage of injected oocytes that were induced to resume meiosis.

application of the hormone is effective, indicating that progesterone acts on a cell surface receptor to activate a distinct cytoplasmic factor. Similar experiments performed independently by Dennis Smith and Robert Ecker (The interaction of steroids with *Rana pipiens* oocytes in the induction of maturation. *Dev. Biol.* 25: 232–247, 1971) led to the same conclusion. Interestingly, the action of progesterone in this system is distinct from its action in most cells, where it diffuses across the plasma membrane and binds to an intracellular receptor (see Chapter 13). In oocytes, however, progesterone clearly acts on the cell surface to activate a distinct factor in the oocyte cytoplasm. Since the

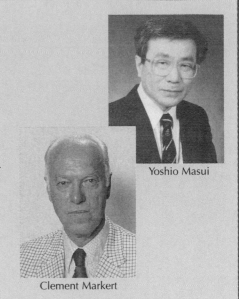

## The Discovery of MPF *(continued)*

resumption of oocyte meiosis is commonly referred to as oocyte maturation, Masui and Markert coined the term "maturation promoting factor" for their newly discovered regulator of meiosis.

### The Impact

Following its discovery in frog oocytes, MPF was also found to be present in somatic cells, where it induces the $G_2$ to M transition of mitosis. Thus, MPF appeared to be a general regulator of entry into the M phase of both mitotic and meiotic cell cycles. The eventual purification of MPF from frog oocytes in 1988 then converged with yeast genetics and studies of sea urchin embryos to reveal the identity of this critical cell cycle regulator. Namely, MPF was found to be a dimer of cyclin B and the Cdc2 protein kinase. Further studies have established that both cyclin B and Cdc2 are members of large families of proteins, with different cyclins and Cdc2-related protein kinases functioning analogously to MPF in the regulation of other cell cycle transitions. The discovery of MPF in frog oocytes thus paved the way to understanding a cell cycle regulatory apparatus that is conserved throughout all eukaryotes.

Yoshio Masui

Clement Markert

Cdc2 in this text. Further studies of *cdc2* yielded two important insights. First, molecular cloning and nucleotide sequencing revealed that *cdc2* encodes a protein kinase—the first indication of the prominent role of protein phosphorylation in regulating the cell cycle. Second, a human gene related to *cdc2* was identified and shown to function in yeasts, providing a dramatic demonstration of the conserved activity of this cell cycle regulator.

The third line of investigation that eventually converged with the identification of MPF and yeast genetics emanated from studies of protein synthesis in early sea urchin embryos. Following fertilization, these embryos go through a series of rapid cell divisions. Intriguingly, studies with protein synthesis inhibitors had revealed that entry into M phase of these embryonic cell cycles requires new protein synthesis. In 1983, Tim Hunt and his colleagues identified two proteins that display a periodic pattern of accumulation and degradation in sea urchin and clam embryos. These proteins accumulate throughout interphase and are then rapidly degraded toward the end of each mitosis (Figure 14.13). Hunt called these proteins **cyclins** (the two proteins were designated cyclin A and cyclin B) and suggested that they might function to induce mitosis, with their periodic accumula-

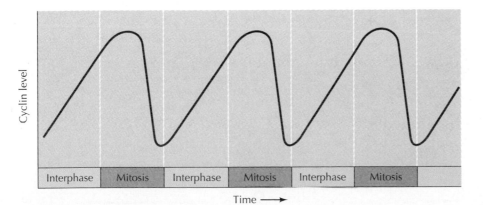

*Figure 14.13*
**Accumulation and degradation of cyclins in sea urchin embryos** The cyclins were identified as proteins that accumulate throughout interphase and are rapidly degraded toward the end of mitosis.

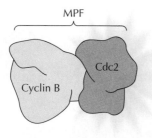

***Figure 14.14***
**Structure of MPF**   MPF is a dimer consisting of cyclin B and the Cdc2 protein kinase.

***Figure 14.15***
**MPF regulation**   Cdc2 forms complexes with cyclin B during S and G$_2$. Cdc2 is then phosphorylated on threonine-161, which is required for Cdc2 activity, as well as on threonine-14 and tyrosine-15, which inhibits Cdc2 activity. Dephosphorylation of Thr14 and Tyr15 activates MPF at the G$_2$ to M transition. MPF activity is then terminated toward the end of mitosis by proteolytic degradation of cyclin B.

tion and destruction controlling entry and exit from M phase. Direct support for such a role of cyclins was provided in 1986, when Joan Ruderman and her colleagues showed that microinjection of cyclin A into frog oocytes is sufficient to trigger the G$_2$ to M transition.

These initially independent approaches converged dramatically in 1988, when MPF was purified from frog eggs in the laboratory of James Maller. Molecular characterization of MPF in several laboratories then showed that this conserved regulator of the cell cycle is composed of two key subunits: Cdc2 and cyclin B (Figure 14.14). Cyclin B is a regulatory subunit required for catalytic activity of the Cdc2 protein kinase, consistent with the notion that MPF activity is controlled by the periodic accumulation and degradation of cyclin B during cell cycle progression.

A variety of further studies have confirmed this role of cyclin B, as well as demonstrating the regulation of MPF by phosphorylation and dephosphorylation of Cdc2 (Figure 14.15). In mammalian cells, cyclin B synthesis begins in S phase. Cyclin B then accumulates and forms complexes with Cdc2 throughout S and G$_2$. As these complexes form, Cdc2 is phosphorylated at two critical regulatory positions. One of these phosphorylations occurs on threonine-161 and is required for Cdc2 kinase activity. The second is a dual phosphorylation of threonine-14 and tyrosine-15, which inhibits Cdc2 activity and leads to the accumulation of inactive Cdc2/cyclin B complexes throughout S and G$_2$. The transition from G$_2$ to M is then brought about by activation of the Cdc2/cyclin B complex as a result of dephosphorylation of threonine-14 and tyrosine-15 by a protein phosphatase called Cdc25.

Once activated, the Cdc2 protein kinase phosphorylates a variety of target proteins that initiate the events of M phase, which are discussed later in this chapter. In addition, Cdc2 activity triggers the degradation of cyclin B, which occurs as a result of ubiquitin-mediated proteolysis. This proteolytic destruction of cyclin B then inactivates Cdc2, leading the cell to exit mitosis, undergo cytokinesis, and return to interphase.

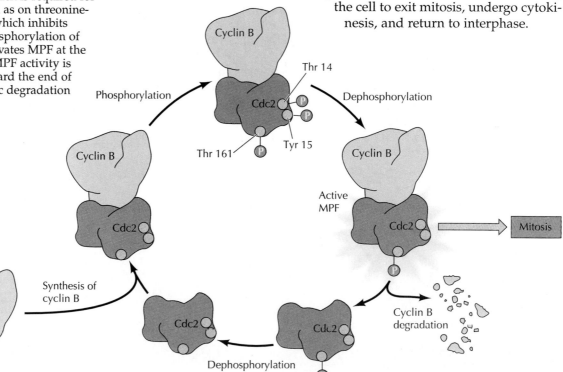

## Families of Cyclins and Cyclin-Dependent Kinases

The structure and function of MPF (Cdc2/cyclin B) provide not only a molecular basis for understanding entry and exit from M phase, but also the foundation for elucidating the regulation of other cell cycle transitions. The insights provided by characterization of the Cdc2/cyclin B complex have thus had a sweeping impact on understanding cell cycle regulation. In particular, further research has established that both Cdc2 and cyclin B are members of large families of related proteins, with different members of these families controlling progression through distinct phases of the cell cycle.

As discussed earlier, Cdc2 controls passage through START as well as entry into mitosis in yeasts. It does so, however, in association with distinct cyclins (Figure 14.16). In particular, the $G_2$ to M transition is driven by Cdc2 in association with the mitotic B-type cyclins (Clb1, Clb2, Clb3, and Clb4). Passage through START, however, is controlled by Cdc2 in association with a distinct class of cyclins called **$G_1$ cyclins** or **Cln's**. Cdc2 then associates with different B-type cyclins (Clb5 and Clb6), which are required for progression through S phase. These associations of Cdc2 with distinct B-type and $G_1$ cyclins presumably direct Cdc2 to phosphorylate different substrate proteins, as required for progression through specific phases of the cell cycle.

The cell cycles of higher eukaryotes are controlled not only by multiple cyclins, but also by multiple Cdc2-related protein kinases. These Cdc2-related kinases are known as **Cdk's** (for *cyclin-dependent kinases*). As the original member of this family, Cdc2 is also known as Cdk1, with other currently identified family members being designated Cdk2 through Cdk7.

These multiple members of the Cdk family associate with specific cyclins to drive progression through the different stages of the cell cycle (see Figure 14.16). For example, progression from $G_1$ to S is regulated principally by Cdk2 and Cdk4 (and in some cells Cdk6) in association with cyclins D and E. Complexes of Cdk2, Cdk4, and Cdk6 with the D-type cyclins (cyclin D1, D2, and D3) play a critical role in progression through the restriction point in $G_1$. Cyclin E is expressed later in $G_1$, and Cdk2/cyclin E complexes may play a role in the $G_1$ to S transition and initiation of DNA synthesis. Complexes of Cdk2 with cyclin A are also thought to function in both the initiation of DNA synthesis and the progression of cells through S phase. As already discussed, the transition from $G_2$ to M is then driven by complexes of Cdc2 with cyclin B.

The activity of Cdk's during cell cycle progression is regulated by at least four molecular mechanisms (Figure 14.17). As already discussed for Cdc2, the first level of regulation involves the association of Cdk's with their cyclin partners. Thus, the formation of specific Cdk/cyclin complexes is controlled by cyclin synthesis and degradation. Second, activation of Cdk/cyclin complexes requires phospho-

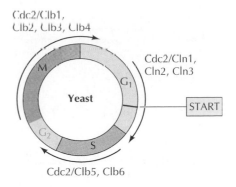

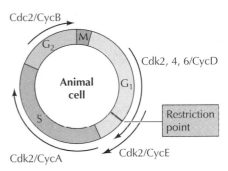

*Figure 14.16*
**Complexes of cyclins and cyclin-dependent kinases**   In yeast, passage through START is controlled by Cdc2 in association with $G_1$ cyclins (Cln1, Cln2, and Cln3). Complexes of Cdc2 with distinct B-type cyclins (Clb's) then regulate progression through S phase and entry into mitosis. In animal cells, progression through the $G_1$ restriction point is controlled by complexes of Cdk2, Cdk4, and Cdk6 with D-type cyclins. Cdk2/cyclin E complexes function later in $G_1$ and are required for the $G_1$ to S transition. Cdk2/cyclin A complexes are then required for progression through S phase, and Cdc2/cyclin B complexes drive the $G_2$ to M transition.

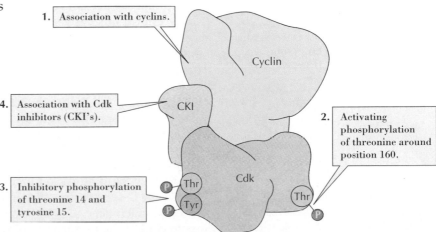

*Figure 14.17*
**Mechanisms of Cdk regulation**
The activities of Cdk's are regulated by four molecular mechanisms.

*Table 14.1* Cdk Inhibitors

| Inhibitor | Cdk/cyclin complex |
|---|---|
| **Animal cells** | |
| p15, p16[a] | Cdk4/cyclin D |
| | Cdk6/cyclin D |
| p21, p27[a] | Cdk4/cyclin D |
| | Cdk6/cyclin D |
| | Cdk2/cyclin A |
| | Cdk2/cyclin E |
| **Yeast** | |
| Far1 | Cdc2/Cln |
| p40 | Cdc2/Clb |

[a]Both p15/p16 and p21/p27 are pairs of closely related proteins.

rylation of a conserved Cdk threonine residue around position 160. This activating phosphorylation of the Cdk's is catalyzed by an enzyme called CAK (for *C*dk-*a*ctivating *k*inase), which may itself be composed of a Cdk (Cdk7) complexed with cyclin H. Complexes of Cdk7 and cyclin H are also associated with the transcription factor TFIIH, which is involved in DNA repair as well as in transcription by RNA polymerase II (see Chapters 5 and 6). It thus appears that this member of the Cdk family may participate in three distinct cellular activities: transcription, DNA repair, and cell cycle regulation.

In contrast to the activating phosphorylation by CAK, the third mechanism of Cdk regulation involves inhibitory phosphorylation of threonine and tyrosine residues near the Cdk amino terminus. In particular, both Cdc2 and Cdk2 are inhibited by phosphorylation of threonine-14 and tyrosine-15. These Cdk's are then activated by dephosphorylation of these residues by members of the Cdc25 family of protein phosphatases.

In addition to regulation of the Cdk's by phosphorylation, their activities can also be controlled by the binding of inhibitory proteins to Cdk/cyclin complexes. A family of such **Cdk inhibitors (CKIs)** has been identified, and different family members inhibit different Cdk's (Table 14.1). For example, one Cdk inhibitor (called p21) binds to a variety of Cdk/cyclin complexes, whereas other Cdk inhibitors control progression through discrete regulatory points of the cell cycle by binding selectively to specific Cdk/cyclin complexes. An example is the Cdk inhibitor called p16, which specifically arrests cells in $G_1$ by binding to complexes of Cdk4 and Cdk6 with cyclin D. Control of Cdk inhibitors thus provides an additional mechanism for regulating Cdk activity. The combined effects of these multiple modes of Cdk regulation are responsible for controlling cell cycle progression in response both to checkpoint controls and to the variety of extracellular stimuli that regulate cell proliferation.

### Growth Factors and the D-Type Cyclins

As discussed earlier, the proliferation of animal cells is regulated largely by a variety of extracellular growth factors that control the progression of cells through the restriction point in late $G_1$. In the absence of growth factors, cells are unable to pass the restriction point and become quiescent, frequently entering the resting state known as $G_0$, from which they can reenter the cell cycle in response to growth factor stimulation. This control of cell cycle progression by extracellular growth factors implies that the intracellular signaling pathways stimulated downstream of growth factor receptors (discussed in the preceding chapter) ultimately act to regulate components of the cell cycle machinery.

One critical link between growth factor signaling and cell cycle progression is provided by the D-type cyclins (Figure 14.18). Cyclin D synthesis is induced in response to growth factor stimulation, and the D-type cyclins continue to be synthesized as long as growth factors are present. However, the D-type cyclins are also rapidly degraded, so their intracellular concentrations rapidly fall if growth factors are removed. Thus, as long as growth factors are present through $G_1$, complexes of Cdk4/cyclin D drive cells through the restriction point. On the other hand, if growth factors are removed prior to this key regulatory point in the cell cycle, the levels of cyclin D rapidly fall and cells are unable to progress through $G_1$ to S, instead becoming quiescent and entering $G_0$. The inducibility and rapid turnover of D-type cyclins thus integrates growth factor signaling with the cell cycle machinery, allowing the availability of extracellular growth factors to control the progression of cells through $G_1$.

**Figure 14.18**
**Induction of D-type cyclins** Growth factors regulate cell cycle progression through the $G_1$ restriction point by inducing synthesis of D-type cyclins.

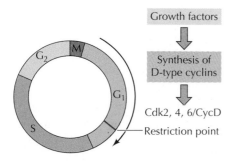

Growth factors

Synthesis of D-type cyclins

Cdk2, 4, 6/CycD

Restriction point

Since cyclin D is a critical target of growth factor signaling, it might be expected that defects in cyclin D regulation could contribute to the loss of growth regulation characteristic of cancer cells. Consistent with this expectation, many human cancers have been found to arise as a result of defects in cell cycle regulation, just as many others result from abnormalities in the intracellular signaling pathways activated by growth factor receptors (see Chapter 13). For example, mutations resulting in continual unregulated expression of cyclin D1 contribute to the development of a variety of human cancers, including lymphomas and breast cancers. Similarly, mutations that inactivate the Cdk inhibitors (e.g., p16) that bind to Cdk4/cyclin D complexes are commonly found in human cancer cells.

The connection between cyclin D, growth control, and cancer is further fortified by the fact that a key substrate protein of Cdk4/cyclin D complexes is itself frequently mutated in a wide array of human tumors. This protein, designated **Rb**, was first identified as the product of a gene responsible for retinoblastoma, a rare inherited childhood eye tumor (see Chapter 15). Further studies then showed that mutations resulting in the absence of functional Rb protein are not restricted to retinoblastoma but also contribute to a variety of common human cancers. Rb is the prototype of a **tumor suppressor gene**—a gene whose inactivation leads to tumor development. Whereas oncogene proteins such as Ras (see Chapter 13) and cyclin D drive cell proliferation, the proteins encoded by tumor suppressor genes act as brakes that slow down cell cycle progression. Additional examples of cell cycle regulators encoded by tumor suppressor genes include the Cdk inhibitors that bind Cdk4/cyclin D complexes and the important growth regulator p53, which was discussed earlier in this chapter.

Further studies of Rb have revealed that it plays a key role in coupling the cell cycle machinery to the expression of genes required for cell cycle progression and DNA synthesis (Figure 14.19). The activity of Rb is regulated by changes in its phosphorylation as cells progress through the cycle. In particular, Rb becomes phosphorylated by Cdk4/cyclin D complexes as cells pass through the restriction point in $G_1$. In its underphosphorylated form (present in $G_0$ or early $G_1$), Rb binds to members of the **E2F** family of transcription factors, which regulate expression of several genes involved in cell cycle progression and DNA replication, including genes encoding enzymes required for the synthesis of deoxyribonucleoside triphosphates. E2F binds to its target sequences in either the presence or absence of Rb.

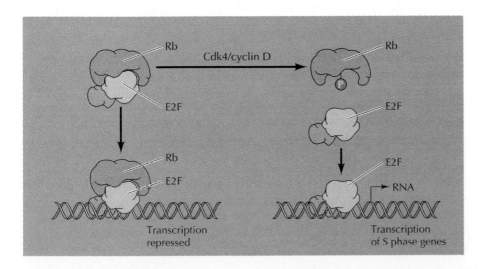

*Figure 14.19*
**Cell cycle regulation of Rb and E2F** In its underphosphorylated form, Rb binds to members of the E2F family, repressing transcription of E2F-regulated genes. Phosphorylation of Rb by Cdk4/cyclin D complexes results in its dissociation from E2F in late $G_1$. E2F then stimulates expression of its target genes, which encode proteins required for cell cycle progression and DNA synthesis.

However, Rb acts as a repressor, so the Rb/E2F complex suppresses transcription of E2F-regulated genes. Phosphorylation of Rb by Cdk4/cyclin D complexes results in its dissociation from E2F, which then activates transcription of its target genes. Rb thus acts as a molecular switch that converts E2F from a repressor to an activator of genes required for DNA synthesis and cell cycle progression. The control of Rb via Cdk4/cyclin D phosphorylation in turn couples this critical regulation of gene expression to the availability of growth factors in $G_1$.

Perhaps not surprisingly, the regulation of cell cycle progression by Rb appears to be considerably more complex than outlined here. As already noted, E2F refers to a family of at least five distinct transcription factors rather than to a single protein. In addition, other cellular proteins related to Rb have been identified, and their roles in cell cycle regulation remain to be elucidated. It is noteworthy that only three members of the E2F family interact with Rb; the other two E2Fs may instead be regulated by other Rb-related proteins. In addition, Rb and its relatives may act on other targets in addition to E2F. Despite this emerging complexity, however, regulation of E2F by Rb clearly represents at least one key mechanism responsible for controlling cell cycle progression and provides a paradigm for the integration of gene expression with the cell cycle machinery.

### Inhibitors of Cell Cycle Progression

Cell proliferation is regulated not only by growth factors but also by a variety of signals that act to inhibit cell cycle progression. For example, agents that damage DNA result in cell cycle arrest, presumably to allow time for the cell to repair the damage. In addition, cell contacts and a variety of extracellular factors act to inhibit rather than stimulate proliferation of their target cells. The effects of such inhibitory signals are also mediated by regulators of the cell cycle machinery, frequently via the induction of Cdk inhibitors.

A good example of the action of Cdk inhibitors is provided by cell cycle arrest in response to DNA damage, which is mediated by the protein p53 (discussed earlier in this chapter). The p53 protein is a transcriptional regulator that functions, at least in part, to stimulate expression of the Cdk inhibitor p21 (Figure 14.20). The p21 protein inhibits several Cdk/cyclin complexes, and its induction by p53 appears to represent at least one mechanism of p53-dependent cell cycle arrest following DNA damage. In addition to inhibiting cell cycle progression via its interaction with Cdk's, p21 may directly inhibit DNA replication. In particular, p21 binds to proliferating cell nuclear antigen (PCNA), which, as discussed in Chapter 5, is a subunit of DNA polymerase δ. Thus, p21 may play a dual role in cell cycle arrest induced by DNA damage, not only blocking cell cycle progression by inhibiting Cdk's but also directly inhibiting DNA replication in S phase cells.

The best-characterized extracellular inhibitor of animal cell proliferation is **TGF-β**—a polypeptide factor that inhibits the proliferation of a variety of types of epithelial cells by arresting cell cycle progression in $G_1$. This action of TGF-β appears to be mediated by induction of the Cdk inhibitors p15 and p27, which bind to Cdk4/cyclin D complexes. In the resulting absence of Cdk4 activity, Rb phosphorylation is blocked and the cell cycle is arrested in $G_1$.

Other antiproliferative signals function via different interactions with the cell cycle machinery, frequently involving induction of other Cdk inhibitors. For example, in some mammalian cells, the second messenger cAMP inhibits cell proliferation in $G_1$ by inducing the Cdk inhibitor p27. As discussed earlier in this chapter, yeast mating factors arrest budding yeast

**Figure 14.20**
**Induction of p21 by DNA damage**
DNA damage results in the elevation of intracellular levels of p53, which activates transcription of the gene encoding the Cdk inhibitor p21. In addition to inhibiting cell cycle progression by binding to Cdk/cyclin complexes, p21 may directly inhibit DNA synthesis by interacting with PCNA (a subunit of DNA polymerase δ).

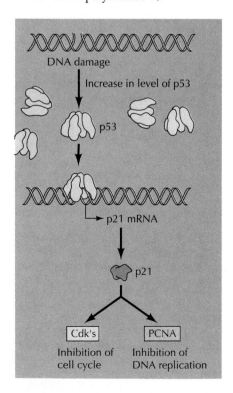

proliferation at START; they function by inducing a Cdk inhibitor called Far1, which inhibits Cdc2 complexed with $G_1$ cyclins. A variety of factors that inhibit cell proliferation thus act by inducing Cdk inhibitors—in effect, the converse of the action of growth factors in inducing synthesis of the D-type cyclins.

## THE EVENTS OF M PHASE

M phase is the most dramatic period of the cell cycle, involving a major reorganization of virtually all cell components. During mitosis (nuclear division), the chromosomes condense, the nuclear envelope of most cells breaks down, the cytoskeleton reorganizes to form the mitotic spindle, and the chromosomes move to opposite poles. Chromosome segregation is then usually followed by cell division (cytokinesis). Although many of these events have been discussed in previous chapters with respect to the structure and function of the nucleus and cytoskeleton, they are reviewed here in the context of a coordinated view of M phase and the action of MPF.

### Stages of Mitosis

Although many of the details of mitosis vary among different organisms, the fundamental processes that ensure the faithful segregation of sister chromatids are conserved in all eukaryotes. These basic events of mitosis include chromosome condensation, formation of the mitotic spindle, and attachment of chromosomes to the spindle microtubules. Sister chromatids then separate from each other and move to opposite poles of the spindle, followed by the formation of daughter nuclei.

Mitosis is conventionally divided into four stages—**prophase, metaphase, anaphase,** and **telophase**—which are illustrated for an animal cell in Figures 14.21 and 14.22. The beginning of prophase is marked by the appearance of condensed chromosomes, each of which consists of two sister chromatids (the daughter DNA molecules produced in S phase). These newly replicated DNA molecules remain intertwined throughout S and $G_2$, becoming untangled during the process of chromatin condensation. The condensed sister chromatids are then held together at the **centromere**, which (as discussed in Chapter 4) is a DNA sequence to which proteins bind to form the **kinetochore**—the site of eventual attachment of the spindle microtubules. In addition to chromosome condensation, cytoplasmic changes leading to the development of the mitotic spindle initiate during prophase. The **centrosomes** (which had duplicated during interphase) separate and move to opposite sides of the nucleus. There they serve as the two poles of the **mitotic spindle**, which begins to form during late prophase.

In higher eukaryotes the end of prophase corresponds to the breakdown of the nuclear envelope. As discussed in Chapter 8, however, nuclear envelope breakdown is not a universal feature of mitosis. In particular, yeasts and many other unicellular eukaryotes undergo "closed mitosis," in which the nuclear envelope remains intact (see Figure 8.28). In these cells the spindle pole bodies are embedded within the nuclear envelope, and the nucleus divides in two following migration of daughter chromosomes to opposite poles of the spindle.

Following completion of prophase, the cell enters **prometaphase**—a transition period between prophase and metaphase. During prometaphase the microtubules of the mitotic spindle attach to the kinetochores of condensed chromosomes. The kinetochores of sister chromatids are oriented on opposite sides of the chromosome, so they attach to microtubules emanating from

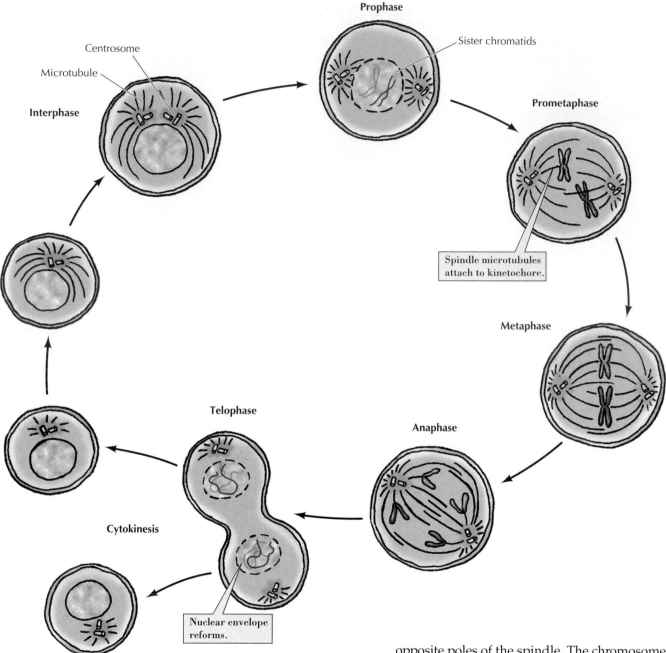

*Figure 14.21*
**Stages of mitosis in an animal cell**   During prophase, the chromosomes condense and centrosomes move to opposite sides of the nucleus, initiating formation of the mitotic spindle. Breakdown of the nuclear envelope then allows spindle microtubules to attach to the kinetochores of chromosomes. During prometaphase the chromosomes shuffle back and forth between the centrosomes and the center of the cell, eventually aligning in the center of the spindle (metaphase). At anaphase, the sister chromatids separate and move to opposite poles of the spindle. Mitosis then ends with re-formation of nuclear envelopes and chromosome decondensation during telophase, and cytokinesis yields two interphase daughter cells. Note that each daughter cell receives one centrosome, which duplicates prior to the next mitosis.

opposite poles of the spindle. The chromosomes shuffle back and forth until they eventually align on the metaphase plate in the center of the spindle. At this stage, the cell has reached metaphase.

Most cells remain only briefly at metaphase before proceeding to anaphase. The transition from metaphase to anaphase is triggered by breakage of the link between sister chromatids, which then separate and move to opposite poles of the spindle. Mitosis ends with telophase, during which nuclei re-form and the chromosomes decondense. Cytokinesis usually begins during late anaphase and is almost complete by the end of telophase, resulting in the formation of two interphase daughter cells.

**Mitosis**

Interphase

Early prophase

Late prophase

Prometaphase

Metaphase

Early anaphase

Late anaphase

Telophase

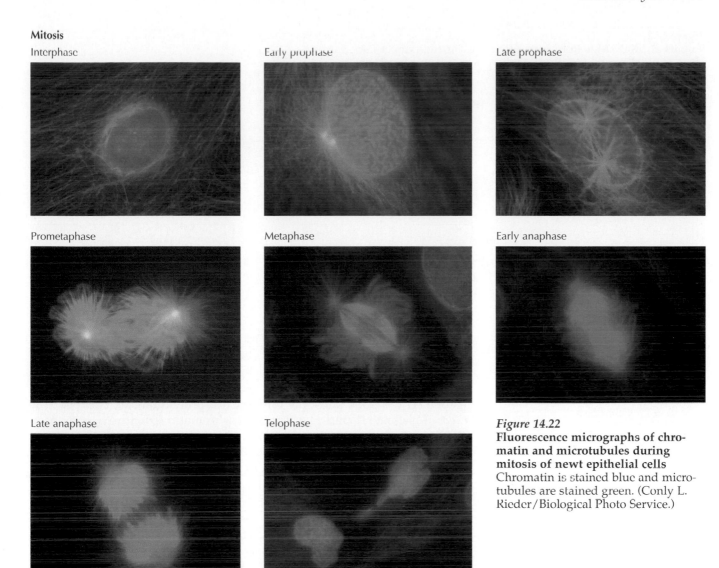

*Figure 14.22*
**Fluorescence micrographs of chromatin and microtubules during mitosis of newt epithelial cells** Chromatin is stained blue and microtubules are stained green. (Conly L. Rieder/Biological Photo Service.)

## MPF and Progression to Metaphase

Mitosis involves dramatic changes in multiple cellular components, leading to a major reorganization of the entire structure of the cell. As discussed earlier in this chapter, these events are initiated by activation of the MPF protein kinase (Cdc2/cyclin B). It appears that MPF not only acts as a master regulator of the M phase transition, phosphorylating and activating other downstream protein kinases, but also acts directly by phosphorylating some of the structural proteins involved in this cellular reorganization (Figure 14.23).

The condensation of interphase chromatin to form the compact chromosomes of mitotic cells is a key event in mitosis, critical in enabling the chromosomes to move along the mitotic spindle without becoming broken or tangled with one another. As discussed in Chapter 4, the chromatin in interphase nuclei condenses nearly a thousand fold during the formation of metaphase chromosomes. Such highly condensed chromatin cannot be transcribed, so transcription ceases as chromatin condensation takes place. Despite the fundamental importance of this event, we do not fully understand either the structure of metaphase chromosomes or the molecular mechanism of chromatin condensation. The enzyme topoisomerase II (see

*Figure 14.23*
**Targets of MPF** MPF induces multiple nuclear and cytoplasmic changes at the onset of M phase, both by activating other protein kinases and by phosphorylating structural proteins such as histone H1 and the nuclear lamins.

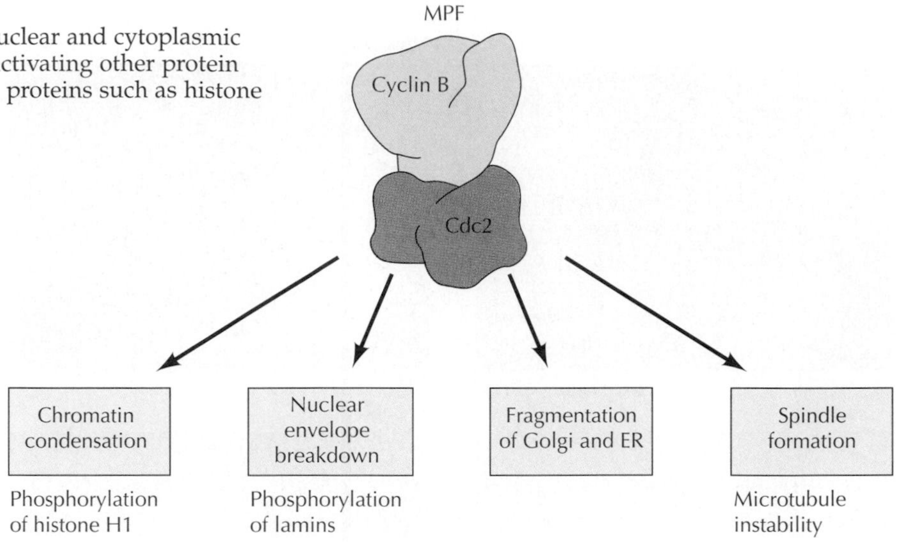

MPF

Cyclin B

Cdc2

| Chromatin condensation | Nuclear envelope breakdown | Fragmentation of Golgi and ER | Spindle formation |

Phosphorylation of histone H1

Phosphorylation of lamins

Microtubule instability

*Figure 14.24*
**Electron micrograph of microtubules attached to the kinetochore of a chromosome.** (Conly L. Rieder/ Biological Photo Service.)

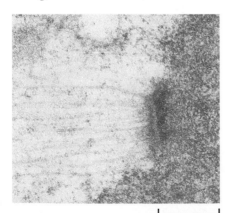

0.5 μm

Chapter 5) plays an important role by untangling the DNA of sister chromatids during condensation, but its role in chromosome structure is unclear. One molecular alteration that generally accompanies chromosome condensation is phosphorylation of **histone H1**, so it is noteworthy that histone H1 is a good substrate for the Cdc2 protein kinase. However, the effect of histone H1 phosphorylation on chromatin structure has not been established, so whether histone H1 is a functionally significant target of MPF remains an open question. New insights into the mechanism of chromatin condensation may emerge from studies of a recently identified family of motor proteins that appear to be required for chromosome condensation at mitosis. In addition to understanding the mechanism of action of these proteins, it will be of interest to determine how MPF regulates their activities.

Breakdown of the nuclear envelope, which is one of the most dramatic events of mitosis, represents the most clearly defined target for MPF action. As discussed in Chapter 8, Cdc2 phosphorylates the **lamins**, leading directly to depolymerization of the nuclear lamina (see Figure 8.29). This is followed by fragmentation of the nuclear membrane into small vesicles, which eventually fuse to form new daughter nuclei at telophase. The endoplasmic reticulum and Golgi apparatus similarly fragment into small vesicles, which can then be distributed to daughter cells at cytokinesis. Although the breakdown of these membranes is also induced by MPF, the molecular mechanism by which MPF leads to membrane fragmentation remains to be determined.

The reorganization of the cytoskeleton that culminates in formation of the mitotic spindle results from the dynamic instability of microtubules (see Chapter 11). At the beginning of prophase, the centrosomes move to opposite sides of the nucleus. The rise in MPF activity then induces a dramatic change in the dynamic behavior of microtubules. First, the rate of microtubule disassembly increases, resulting in depolymerization and shrinkage of the interphase microtubules. This disassembly is thought to result from phosphorylation of microtubule-associated proteins, either by MPF itself or by other MPF-activated protein kinases. In addition, the number of microtubules emanating from the centrosomes increases, so the interphase microtubules are replaced by large numbers of short microtubules radiating from the centrosomes.

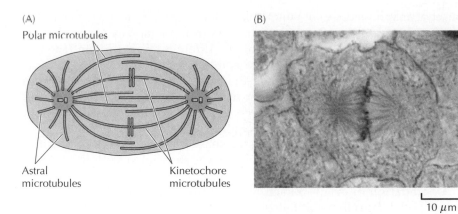

(A)
Polar microtubules

Astral microtubules

Kinetochore microtubules

(B)

10 μm

***Figure 14.25***
**The metaphase spindle** (A) The spindle consists of three kinds of microtubules. Kinetochore microtubules are attached to chromosomes, polar microtubules overlap in the center of the cell, and astral microtubules radiate from the centrosome to the cell periphery. (B) A whitefish cell at metaphase. (B, Michael Abbey/Photo Researchers, Inc.)

The breakdown of the nuclear envelope then allows some of the spindle microtubules to attach to chromosomes at their kinetochores (Figure 14.24), initiating the process of chromosome movement that characterizes prometaphase. The proteins assembled at the kinetochore include microtubule motors that direct the movement of chromosomes toward the minus ends of the spindle microtubules, which are anchored in the centrosome. The movement of chromosomes toward the poles of the spindle is also driven by motor proteins located at the centrosome itself. The action of these proteins, which draw chromosomes toward the centrosome, is opposed by a force (called the **polar wind**) that repels chromosomes from the spindle poles. Consequently, the chromosomes in prometaphase shuffle back and forth between the centrosomes and the center of the spindle. The nature of the polar wind is not clear; it may result either from the outgrowth of microtubules from the centrosome or from the action of motor proteins associated with the chromosome arms, which move the chromosomes toward the plus ends rather than the minus ends of spindle microtubules.

Microtubules from opposite poles of the spindle eventually attach to the two kinetochores of sister chromatids (which are located on opposite sides of the chromosome), and the balance of forces acting on the chromosomes leads to their alignment on the metaphase plate in the center of the spindle (Figure 14.25). As discussed in Chapter 11, the spindle consists of both **kinetochore microtubules**, which are attached to the chromosomes, and **polar microtubules**, which overlap with one another in the center of the cell. In addition, short **astral microtubules** radiate outward from the centrosomes toward the cell periphery.

### Proteolysis and the Inactivation of MPF: Anaphase and Telophase

As discussed earlier in this chapter, an important cell cycle checkpoint monitors the alignment of chromosomes on the metaphase spindle. Once this has been accomplished, the cell proceeds to initiate anaphase and complete mitosis. The progression from metaphase to anaphase results from activation of a ubiquitin-mediated proteolysis system (see Figure 7.40) that degrades cyclin B and thereby inactivates MPF. This ubiquitin degradation pathway is induced by MPF at the beginning of mitosis, so MPF ultimately triggers its own destruction. Proteolytic activity is restrained, however, until the cell passes the metaphase checkpoint, after which activation of the ubiquitin degradation system brings about the transition from metaphase to anaphase and progression through the rest of mitosis.

Interestingly, cyclin B is not the only protein that needs to be degraded in order for the cell to complete mitosis (Figure 14.26). In particular, the

***Figure 14.26***
**Targets of the cyclin B proteolysis system** The ubiquitin proteolysis system that degrades cyclin B is activated by MPF. Following passage through the metaphase checkpoint, this system brings about the transition from metaphase to anaphase by degrading an unknown target protein (X). It also degrades cyclin B, leading to inactivation of MPF, exit from mitosis, and cytokinesis.

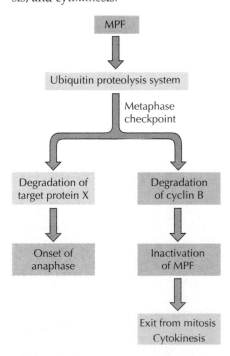

MPF

↓

Ubiquitin proteolysis system

Metaphase checkpoint

Degradation of target protein X → Onset of anaphase

Degradation of cyclin B → Inactivation of MPF → Exit from mitosis Cytokinesis

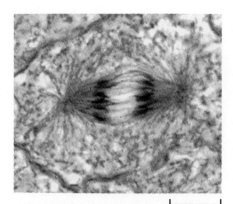

*Figure 14.27*
**A whitefish cell at anaphase**
(Michael Abbey/Photo Researchers,
Inc.)

onset of anaphase does not require either the degradation of cyclin B or the inactivation of MPF. Instead, the transition from metaphase to anaphase is initiated by proteolytic degradation of a different target protein. The key protein whose degradation is responsible for the onset of anaphase is unknown, but one interesting possibility is that it plays a direct role in holding sister chromatids together at the centromere. Its degradation would then break the linkage between sister chromatids, allowing them to segregate by moving to opposite poles of the spindle (Figure 14.27). The separation of chromosomes during anaphase then proceeds as a result of the action of several types of motor proteins associated with the spindle microtubules (see Figures 11.47 and 11.48).

Degradation of cyclin B itself leads to inactivation of MPF, which is required for the cell to exit mitosis and return to interphase. Many of the cellular changes involved in these transitions are simply the reversal of the events induced by MPF during entry into mitosis. For example, reassembly of the nuclear envelope, chromatin decondensation, and the return of microtubules to an interphase state probably result directly from loss of MPF activity and dephosphorylation of proteins that had been phosphorylated by MPF at the beginning of mitosis. As discussed next, inactivation of MPF also triggers cytokinesis.

## Cytokinesis

The completion of mitosis is usually accompanied by cytokinesis, giving rise to two daughter cells. Cytokinesis usually initiates in late anaphase and is triggered by the inactivation of MPF, thereby coordinating nuclear and cytoplasmic division of the cell. As discussed in Chapter 11, cytokinesis of animal cells is mediated by a **contractile ring** of actin and myosin II filaments that forms beneath the plasma membrane (Figure 14.28). The location of this ring is determined by the position of the mitotic spindle, so the cell is eventually cleaved in a plane that passes through the metaphase plate perpendicular to the spindle. Cleavage proceeds as contraction of the actin-myosin filaments pulls the plasma membrane inward, eventually pinching the cell in half.

The mechanism of cytokinesis is different for higher plant cells, which are surrounded by rigid cell walls. Rather than being pinched in half by a contractile ring, these cells divide by forming new cell walls and plasma membranes inside the cell (Figure 14.29). In early telophase, vesicles carrying cell wall precursors from the Golgi apparatus associate with spindle

(A)　　　Contractile ring　　　(B)

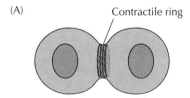

*Figure 14.28*
**Cytokinesis of animal cells**   (A) Cytokinesis results from contraction of a ring of actin and myosin filaments, which pinches the cell in two. (B) Scanning electron micrograph of a frog egg undergoing cytokinesis. (B, David M. Phillips/Visuals Unlimited).

*Figure 14.29*
**Cytokinesis in higher plants** Golgi vesicles carrying cell wall precursors associate with polar microtubules at the former site of the metaphase plate. Fusion of these vesicles yields a membrane-enclosed, disclike structure (the early cell plate) that expands outward and fuses with the parental plasma membrane.

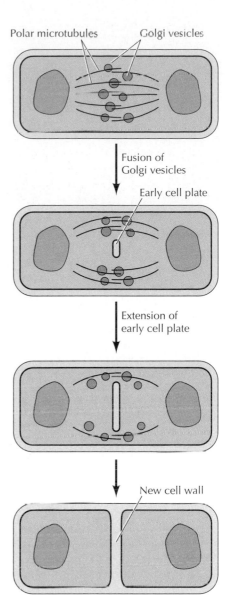

microtubules and accumulate at the former site of the metaphase plate. These vesicles then fuse to form a large, membrane-enclosed, disclike structure, and their polysaccharide contents assemble to form the matrix of a new cell wall (called a **cell plate**). The cell plate expands outward, perpendicular to the spindle, until it reaches the plasma membrane. The membrane surrounding the cell plate then fuses with the parental plasma membrane, dividing the cell in two.

## MEIOSIS AND FERTILIZATION

The somatic cell cycles discussed so far in this chapter result in diploid daughter cells with identical genetic complements. Meiosis, in contrast, is a specialized kind of cell cycle that reduces the chromosome number by half, resulting in the production of haploid daughter cells. Unicellular eukaryotes, such as yeasts, can undergo meiosis as well as reproducing by mitosis. Diploid *Saccharomyces cerevisiae*, for example, undergo meiosis and produce spores when faced with unfavorable environmental conditions. In multicellular plants and animals, however, meiosis is restricted to the germ cells, where it is key to sexual reproduction. Whereas somatic cells undergo mitosis to proliferate, the germ cells undergo meiosis to produce haploid gametes (the sperm and the egg). The development of a new progeny organism is then initiated by the fusion of these gametes at fertilization.

### The Process of Meiosis

In contrast to mitosis, **meiosis** results in the division of a diploid parental cell into haploid progeny, each containing only one member of the pair of homologous chromosomes that were present in the diploid parent (Figure 14.30). This reduction in chromosome number is accomplished by two sequential rounds of nuclear and cell division (called meiosis I and meiosis II), which follow a single round of DNA replication. Like mitosis, meiosis I initiates after S phase has been completed and the parental chromosomes have replicated to produce identical sister chromatids. The pattern of chromosome segregation in meiosis I, however, is dramatically different from that of mitosis. During meiosis I, homologous chromosomes first pair with one another and then segregate to different daughter cells. Sister chromatids remain together, so completion of meiosis I results in the formation of daughter cells containing a single member of each chromosome pair (consisting of two sister chromatids). Meiosis I is followed by meiosis II, which resembles mitosis in that the sister chromatids separate and segregate to different daughter cells. Completion of meiosis II thus results in the production of four haploid daughter cells, each of which contains only one copy of each chromosome.

The pairing of homologous chromosomes after DNA replication is not only a key event underlying meiotic chromosome segregation, but also allows recombination between chromosomes of paternal and maternal origin. This critical pairing of homologous chromosomes takes place during an extended prophase of meiosis I, which is divided into five stages (**leptotene**, **zygotene**, **pachytene**, **diplotene**, and **diakinesis**) on the basis of

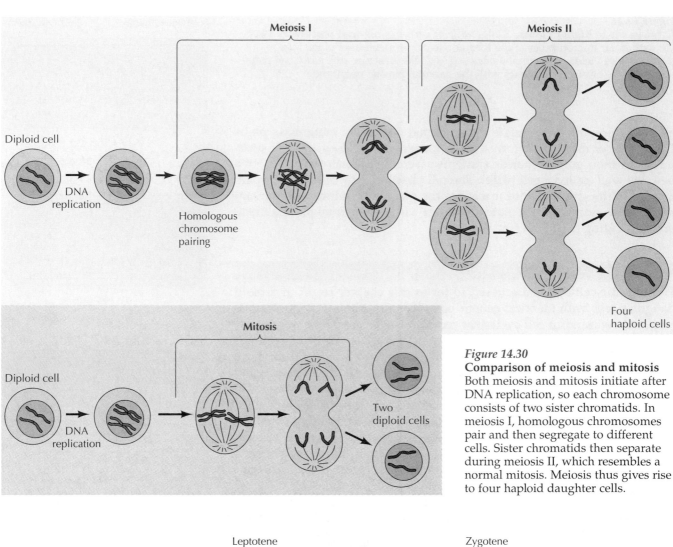

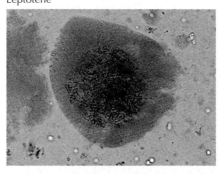

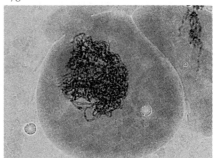

*Figure 14.30*
**Comparison of meiosis and mitosis**
Both meiosis and mitosis initiate after
DNA replication, so each chromosome
consists of two sister chromatids. In
meiosis I, homologous chromosomes
pair and then segregate to different
cells. Sister chromatids then separate
during meiosis II, which resembles a
normal mitosis. Meiosis thus gives rise
to four haploid daughter cells.

Leptotene

Zygotene

*Figure 14.31*
**Stages of the prophase of meiosis I**
Micrographs illustrating the mor-
phology of chromosomes of the lily.
(C. Hasenkampf/Biological Photo
Service.)

Pachytene

Diplotene

Diakinesis

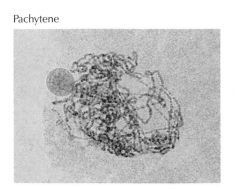

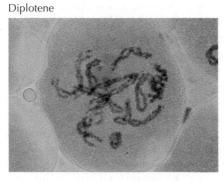

*Figure 14.32*
**The synaptonemal complex**   Chromatin loops are attached to the lateral elements, which are joined to each other by a zipperlike central element.

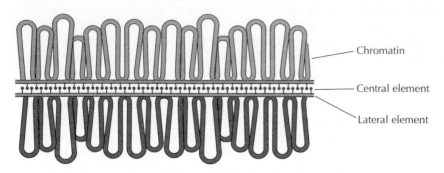

— Chromatin

— Central element

— Lateral element

*Figure 14.33*
**A bivalent chromosome at the diplotene stage**   The bivalent chromosome consists of paired homologous chromosomes. Sister chromatids of each chromosome are joined at the centromere. Chromatids of homologous chromosomes are joined at chiasmata, which are the sites at which genetic recombination has occurred. (John Cabisco/Visuals Unlimited.)

chromosome morphology (Figure 14.31). The initial association of homologous chromosomes is thought to be mediated by base pairing between complementary DNA strands during the leptotene stage, before the chromatin becomes highly condensed. The close association of homologous chromosomes (**synapsis**) begins during the zygotene stage. During this stage, a zipperlike protein structure, called the **synaptomenal complex**, forms along the length of the paired chromosomes (Figure 14.32). This complex keeps the homologous chromosomes closely associated and aligned with one another through the pachytene stage, which can persist for several days. Recombination between homologous chromosomes is completed during their association at the pachytene stage, leaving the chromosomes linked at the sites of crossing over (**chiasmata**; singular, chiasma). The synaptonemal complex disappears at the diplotene stage and the homologous chromosomes separate along their length. Importantly, however, they remain associated at the chiasmata, which is critical for their correct alignment at metaphase. At this stage, each chromosome pair (called a bivalent) consists of four chromatids with clearly evident chiasmata (Figure 14.33). Diakinesis, the final stage of prophase I, represents the transition to metaphase, during which the chromosomes become fully condensed.

At metaphase I, the bivalent chromosomes align on the spindle. In contrast to mitosis (see Figure 14.25), the kinetochores of sister chromatids are adjacent to each other and oriented in the same direction, while the kinetochores of homologous chromosomes are pointed toward opposite spindle poles (Figure 14.34). Consequently, microtubules from the same pole of the spindle attach to sister chromatids, while microtubules from opposite poles attach to homologous chromosomes. Anaphase I is initiated by disruption of the chiasmata at which homologous chromosomes are joined. The homologous chromosomes then separate, while sister chromatids remain associated at their centromeres. At completion of meiosis I, each daughter cell has therefore acquired one member of each homologous pair, consisting of two sister chromatids.

Meiosis II initiates immediately after cytokinesis, usually before the chromosomes have fully decondensed. In contrast to meiosis I, meiosis II resembles a normal mitosis. At metaphase II, the chromosomes align on the spindle with microtubules from opposite poles of the spindle attached to the kinetochores of sister chromatids. The link between the centromeres of sister chromatids is broken at anaphase II, and sister chromatids segregate to opposite poles. Cytokinesis then follows, giving rise to haploid daughter cells.

*Figure 14.34*
**Chromosome segregation in meiosis I**   At metaphase I, the kinetochores of sister chromatids are either fused or adjacent to one another. Microtubules from the same pole of the spindle therefore attach to the kinetochores of sister chromatids, while microtubules from opposite poles attach to the kinetochores of homologous chromosomes. Chiasmata are disrupted at anaphase I, and homologous chromosomes move to opposite poles of the spindle.

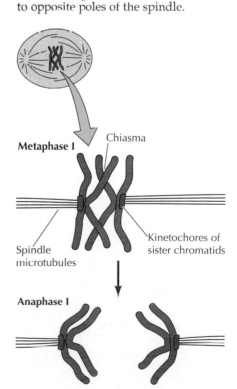

**Metaphase I**   Chiasma

Spindle microtubules

Kinetochores of sister chromatids

**Anaphase I**

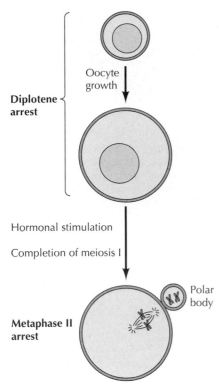

**Figure 14.35**
**Meiosis of vertebrate oocytes**    Meiosis is arrested at the diplotene stage, during which oocytes grow to a large size. Oocytes then resume meiosis in response to hormonal stimulation and complete the first meiotic division, with asymmetric cytokinesis giving rise to a small polar body. Most vertebrate oocytes are then arrested again at metaphase II.

## Regulation of Oocyte Meiosis

Vertebrate oocytes (developing eggs) have been particularly useful models for research on the cell cycle, in part because of their large size and ease of manipulation in the laboratory. A notable example, discussed earlier in this chapter, is provided by the discovery and subsequent purification of MPF from frog oocytes. Meiosis of these oocytes, like those of other species, is regulated at two unique points in the cell cycle, and studies of oocyte meiosis have illuminated novel mechanisms of cell cycle control.

The first regulatory point in oocyte meiosis is in the diplotene stage of the first meiotic division (Figure 14.35). Oocytes can remain arrested at this stage for long periods of time—up to 40 to 50 years in humans. During this diplotene arrest, the oocyte chromosomes decondense and are actively transcribed. This transcriptional activity is reflected in the tremendous growth of oocytes during this period. Human oocytes, for example, are about 100 µm in diameter (more than a hundred times the volume of a typical somatic cell). Frog oocytes are even larger, with diameters of approximately 1 mm. During this period of cell growth, the oocytes accumulate stockpiles of materials, including RNAs and proteins, that are needed to support early development of the embryo. As noted earlier in this chapter, early embryonic cell cycles then occur in the absence of cell growth, rapidly dividing the fertilized egg into smaller cells (see Figure 14.2).

Oocytes of different species vary as to when meiosis resumes and fertilization takes place. In some animals, oocytes remain arrested at the diplotene stage until they are fertilized, only then proceeding to complete meiosis. However, the oocytes of most vertebrates (including frogs, mice, and humans) resume meiosis in response to hormonal stimulation and proceed through meiosis I prior to fertilization. Cell division following meiosis I is asymmetric, resulting in the production of a small **polar body** and an oocyte that retains its large size. The oocyte then proceeds to enter meiosis II without having re-formed a nucleus or decondensed its chromosomes. Most vertebrate oocytes are then arrested again at metaphase II, where they remain until fertilization.

Like the M phase of somatic cells, the meiosis of oocytes is controlled by MPF. The regulation of MPF during oocyte meiosis, however, displays

**Figure 14.36**
**Activity of MPF during oocyte meiosis**    Hormonal stimulation of diplotene oocytes activates MPF, resulting in progression to metaphase I. MPF activity then falls at the transition from metaphase I to anaphase I. Following completion of meiosis I, MPF activity again rises and remains high during metaphase II arrest.

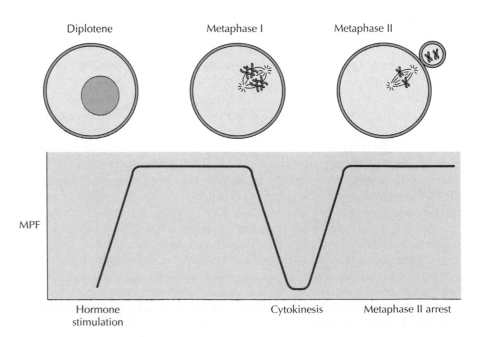

**Figure 14.37**
**Identification of cytostatic factor** Cytoplasm from a metaphase II egg is microinjected into one cell of a two-cell embryo. The injected embryo cell arrests at metaphase, while the uninjected cell continues to divide. A factor in metaphase II egg cytoplasm (cytostatic factor) therefore has induced metaphase arrest of the injected embryo cell.

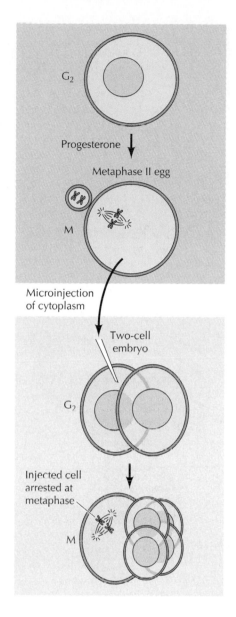

unique features that are responsible for metaphase II arrest (Figure 14.36). Hormonal stimulation of diplotene-arrested oocytes initially triggers the resumption of meiosis by activating MPF, as at the $G_2$ to M transition of somatic cells. As in mitosis, MPF then induces chromosome condensation, nuclear envelope breakdown, and formation of the spindle. Activation of the ubiquitin-mediated proteolysis system that degrades cyclin B then leads to the metaphase to anaphase transition of meiosis I, accompanied by a decrease in the activity of MPF. Following cytokinesis, however, MPF activity again rises and remains high while the egg is arrested at metaphase II. A regulatory mechanism unique to oocytes thus acts to maintain MPF activity during metaphase II arrest, preventing the metaphase to anaphase transition of meiosis II and the inactivation of MPF that would result from cyclin B proteolysis during a normal M phase.

The factor responsible for metaphase II arrest was first identified by Yoshio Masui and Clement Markert in 1971, in the same series of experiments that led to the discovery of MPF. In this case, however, cytoplasm from an egg arrested at metaphase II was injected into an early embryo cell that was undergoing mitotic cell cycles (Figure 14.37). This injection of egg cytoplasm caused the embryonic cell to arrest at metaphase, indicating that metaphase arrest was induced by a cytoplasmic factor present in the egg. Because this factor acted to arrest mitosis, it was called **cytostatic factor** (**CSF**).

More recent experiments have identified a protein-serine/threonine kinase known as **Mos** as an essential component of CSF. Mos is specifically synthesized in oocytes around the time of completion of meiosis I and is then required both for the increase in MPF activity during meiosis II and for the maintenance of MPF activity during metaphase II arrest. The action of Mos results from activation of MAP kinase, which plays a central role in the cell signaling pathways discussed in the previous chapter. In oocytes, however, MAP kinase plays a different role; its activation inhibits the ubiquitin proteolysis pathway responsible for cyclin B degradation, thereby arresting meiosis at metaphase II (Figure 14.38). Oocytes can remain arrested at this point in the meiotic cell cycle for several days, awaiting fertilization.

## Fertilization

At **fertilization**, the sperm binds to a receptor on the surface of the egg and fuses with the egg plasma membrane, initiating the development of a new diploid organism containing genetic information derived from both parents (Figure 14.39). Not only does fertilization lead to the mixing of paternal and maternal chromosomes, but it also induces a number of changes in the egg cytoplasm that are critical for further development. These alterations activate the egg, leading to the completion of oocyte meiosis and initiation

**Figure 14.38**
**Maintenance of metaphase II arrest by the Mos protein kinase** The Mos protein kinase maintains metaphase II arrest by inhibiting the ubiquitin proteolysis system responsible for cyclin B degradation. The action of Mos is mediated by MAP kinase.

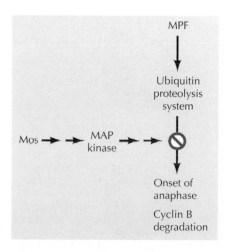

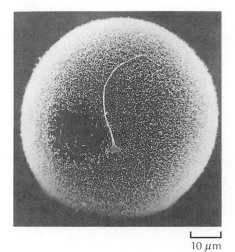

*Figure 14.39*
**Fertilization**   Scanning electron micrograph of a human sperm fertilizing an egg. (David M. Philips/Visuals Unlimited.)

10 μm

of the mitotic cell cycles of the early embryo.

A key signal resulting from the binding of a sperm to its receptor on the plasma membrane of the egg is an increase in the level of Ca²⁺ in the egg cytoplasm, probably as a consequence of stimulation of the hydrolysis of phosphatidylinositol 4,5-bisphosphate (PIP₂) (see Figure 13.28). One effect of this elevation in intracellular Ca²⁺ is the induction of surface alterations that prevent additional sperm from entering the egg. Because eggs are usually exposed to large numbers of sperm at one time, this is a critical event in ensuring the formation of a normal diploid embryo. These surface alterations are thought to result, at least in part, from the Ca²⁺-induced exocytosis of secretory vesicles that are present in large numbers beneath the egg plasma membrane. Release of the contents of these vesicles alters the extracellular coat of the egg so as to block the entry of additional sperm.

The increase in cytosolic Ca²⁺ following fertilization also signals the completion of meiosis (Figure 14.40). In eggs arrested at metaphase II, the metaphase to anaphase transition may result from activation of a Ca²⁺-dependent proteolysis system that degrades both cyclin B and Mos. The resultant inactivation of MPF leads to completion of the second meiotic division, with asymmetric cytokinesis (as in meiosis I) giving rise to a second small polar body.

Following completion of oocyte meiosis, the fertilized egg (now called a **zygote**) contains two haploid nuclei (called **pronuclei**), one derived from each parent. In mammals, the two pronuclei then enter S phase and replicate their DNA as they migrate toward each other. As they meet, the zygote enters M phase of its first mitotic division. The two nuclear envelopes break down and the condensed chromosomes of both paternal and maternal origin align on a common spindle. Completion of mitosis then gives rise to two embryonic cells, each containing a new diploid genome. These cells then commence the series of embryonic cell divisions that eventually lead to the development of a new organism.

(A)

Sperm

Metaphase II egg

Egg completes meiosis II

Zygote

Second polar body

Male and female pronuclei

DNA synthesis

Entry into M phase

Paternal and maternal chromosomes align on spindle

Two-cell embryo

*Figure 14.40*
**Fertilization and completion of meiosis**   (A) Fertilization induces the transition from metaphase II to anaphase II, leading to completion of oocyte meiosis and emission of a second polar body (which usually degenerates). The sperm nucleus decondenses, so the fertilized egg (zygote) contains two haploid nuclei (male and female pronuclei). In mammals, the pronuclei replicate DNA as they migrate toward each other. They then initiate mitosis, with male and female chromosomes aligning on a common spindle. Completion of mitosis and cytokinesis thus gives rise to a two-cell embryo, with each cell containing a diploid genome. (B) Micrographs of a mouse metaphase II egg, zygote, and two-cell embryo. (B, courtesy of Ann A. Kiessling, Harvard Medical School.)

(B) Metaphase II egg    Zygote    Two-cell embryo

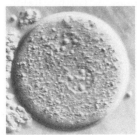
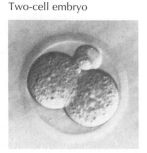

20 μm

## DEVELOPMENT, DIFFERENTIATION, AND PROGRAMMED CELL DEATH

Early development is characterized by the rapid proliferation of embryonic cells, which then differentiate to produce the many specialized types of cells that make up the tissues and organs of multicellular animals. As cells differentiate, their rate of proliferation usually decreases, and most cells in adult animals are arrested in the $G_0$ stage of the cell cycle. A few types of differentiated cells never divide again, but most cells are able to resume proliferation as required to replace cells that have been lost as a result of injury or cell death. In addition, some cells divide continuously throughout life to replace cells that have a high rate of turnover in adult animals.

Cell proliferation is thus carefully balanced with cell death to maintain a constant number of cells in adult tissues and organs. In many cases, cell death is a normal physiological event rather than the result of accidental trauma. Such physiological cell death (programmed cell death) not only serves to eliminate damaged cells and maintain constant cell numbers in adults, but also plays a key role in normal development by eliminating unwanted cells from developing tissues. A carefully regulated balance between cell proliferation and cell death is thus required for both the development and maintenance of animal tissues and organs.

### Cell Proliferation in Adults

The cells of adult animals can be grouped into three general categories with respect to cell proliferation. A few types of differentiated cells, including lens cells, nerve cells, and cardiac muscle cells in humans, are no longer capable of cell division. These cells are produced during embryonic development, differentiate, and are then retained throughout the life of the organism. If they are lost because of injury (e.g., the death of cardiac muscle cells during a heart attack), they can never be replaced.

In contrast, most cells in adult animals enter the $G_0$ stage of the cell cycle but resume proliferation as needed to replace cells that have been injured or have died. Cells of this type include skin fibroblasts, smooth muscle cells, the endothelial cells that line blood vessels, and the epithelial cells of most internal organs, such as the liver, pancreas, kidney, lung, prostate, and breast. One example of the controlled proliferation of these cells, discussed earlier in this chapter, is the rapid proliferation of skin fibroblasts to repair damage resulting from a cut or wound. Another striking example is provided by liver cells, which normally divide only rarely. However, if large numbers of liver cells are lost (e.g., by surgical removal of part of the liver), the remaining cells are stimulated to proliferate to replace the missing tissue. For example, surgical removal of two-thirds of the liver of a rat is followed by rapid cell proliferation, leading to regeneration of the entire liver within a few days.

Other types of differentiated cells, including blood cells, epithelial cells of the skin, and the epithelial cells lining the digestive tract, have short life spans and must be replaced by continual cell proliferation in adult animals. In these cases, the fully differentiated cells do not themselves proliferate. Instead, they are replaced via the proliferation of cells that are less differentiated, called **stem cells** (Figure 14.41). Stem cells divide to produce daughter cells that can either differentiate or remain as stem cells, thereby serving as a source for the production of differentiated cells throughout life.

A good example of the continual proliferation of stem cells is provided by blood cell differentiation. There are several distinct types of blood cells with specialized functions: Erythrocytes (red blood cells) transport $O_2$ and $CO_2$; granulocytes and macrophages are phagocytic cells; platelets (which

*Figure 14.41*
**Stem cell proliferation** Stem cells divide to form one daughter cell that remains a stem cell and a second that differentiates (e.g., to an intestinal epithelial cell).

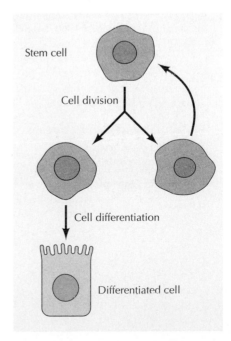

Stem cell

Cell division

Cell differentiation

Differentiated cell

# MOLECULAR MEDICINE

## In Vitro Fertilization

### The Disease

Reproduction is a basic drive of all species of animals, and humans are no exception. Indeed, producing children and raising a family is a fundamental aspect of the lives of most people. The inability to reproduce (infertility) is not only a devastating problem for many affected couples, but it is also common. For example, one frequently used clinical definition of subnormal fertility is the failure to achieve a pregnancy after a two-year period of trying. According to this criterion, it is estimated that approximately 20% of couples in the United States are subfertile.

Infertility can result from a variety of causes, which can affect either the female or the male partner. Many cases are due to anatomical defects in the Fallopian tubes, through which oocytes must travel from the ovary to the uterus. Some women are infertile (or subfertile) because of hormonal disorders that result in the failure to produce normal metaphase II eggs. Other cases of infertility or subfertility result from a failure of the male partner to produce or ejaculate adequate numbers of normal sperm. *In vitro* fertilization (IVF), which was originally developed as a treatment for tubal defects, is now widely used in the treatment of both male and female reproductive disorders.

### Molecular and Cellular Bases

The basic procedure of *in vitro* fertilization is to recover metaphase II

eggs from the ovary and culture them together with sperm. Fertilization is assessed 12 to 18 hours later by microscopic observation to detect the formation of two pronuclei. Fertilized eggs can then be returned to the Fallopian tube or uterus. Alternatively, they can be maintained in culture for a few additional days, during which the early embryonic cell divisions take place.

*In vitro* fertilization was initially developed to bypass defects in the Fallopian tube, which present an obvious anatomical block to a natural pregnancy. As already noted, however, IVF is also effective for other reproductive disorders. For example, hormonal defects resulting in the failure to produce metaphase II eggs can be treated by administration of appropriate hormones to the female partner. This treatment results in the production of multiple eggs, which can then be fertilized *in vitro*, followed by the return of successfully fertilized eggs to the uterus.

IVF is also effective in treating cases of infertility resulting from failure of the male partner to ejaculate sufficient functional sperm to achieve fertilization *in vivo*. Moreover, a new procedure in which a single sperm is directly injected into an egg (intracytoplasmic sperm injection, or ICSI) can be applied to cases in which fertilization does not result from the standard *in vitro* coculture of sperm and eggs. Interestingly, successful fertilization by the ICSI procedure

also requires stimulation of the surface of the egg in order to achieve egg activation.

The rate of pregnancy resulting from fertilization *in vitro* is similar to that observed *in vivo*. In particular, it is estimated that about 25% of ovulated human eggs naturally give rise to offspring in a normal fertile couple. Similar pregnancy rates are obtained following *in vitro* fertilization of eggs of young women with an anatomical block to pregnancy. Although lower pregnancy rates are obtained with eggs of older women or of women with hormonal defects, these reduced rates can be compensated for by the return of multiple fertilized eggs to the uterus. ICSI also yields a pregnancy rate of about 25%, suggesting that there are intrinsic limitations on the developmental capacity of human eggs, both *in vitro* and *in vivo*.

A human egg containing two pronuclei after fertilization *in vitro*. (Courtesy of Ann A. Kiessling, Harvard Medical School.)

are fragments of megakaryocytes) function in blood coagulation; and lymphocytes are responsible for the immune response. All these cells have limited life spans, ranging from less than a day to a few months, and are continually produced by the division of a common stem cell (the pluripotent stem cell) in the bone marrow (Figure 14.42). Descendants of the pluripotent stem cell then become committed to specific differentiation pathways.

## In Vitro Fertilization (continued)

### Prevention and Treatment

The first attempt at *in vitro* fertilization of human eggs was reported in 1942, but not until 1978 was the first IVF baby born. Since then, a variety of technical improvements have made IVF a standard treatment for infertility, and IVF has allowed tens of thousands of otherwise infertile couples to bear children. Although not successful for many couples, IVF has clearly had a major impact on a widespread medical problem.

In addition to its direct impact on the treatment of infertility, IVF offers new opportunities for the diagnosis and prevention of genetic disease. It is possible to remove one or two cells of an early embryo developing *in vitro* (usually at the eight-cell stage) without harming further embryonic development. Consequently, single embryo cells can be isolated and tested either for chromosomal abnormalities or for mutant alleles of any inherited disease gene that can be identified by PCR. It is then possible to select only the embryos carrying normal gene copies for transfer to the uterus. In combination with these methods of preimplantation embryo diagnosis, IVF thus allows couples carrying genes for inherited diseases to prevent the transmission of these diseases to their offspring.

### References

Wassarman, P. M. 1991. *Elements of Mammalian Fertilization II. Practical Applications.* Boca Raton, FL: CRC Press.

Winston, R. M. L. and A. H. Handyside. 1993. New challenges in human *in vitro* fertilization. *Science* 260: 932–936.

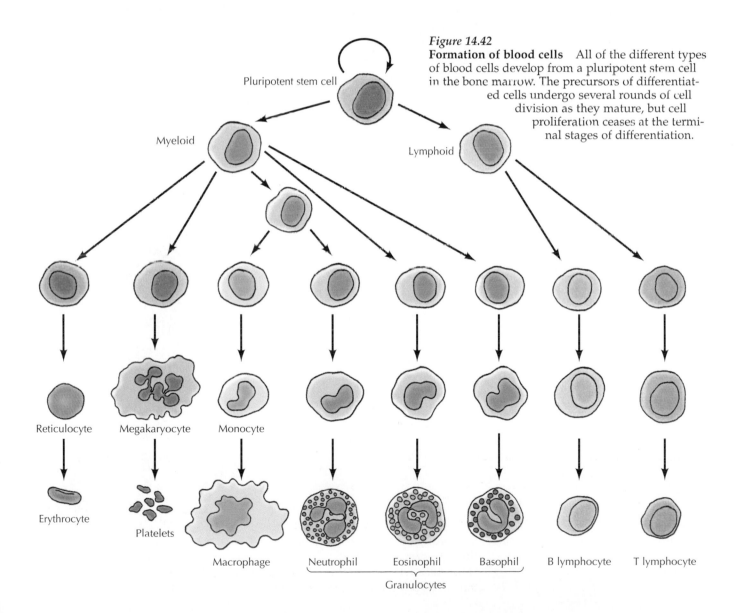

**Figure 14.42**
**Formation of blood cells**   All of the different types of blood cells develop from a pluripotent stem cell in the bone marrow. The precursors of differentiated cells undergo several rounds of cell division as they mature, but cell proliferation ceases at the terminal stages of differentiation.

Pluripotent stem cell

Myeloid

Lymphoid

Reticulocyte    Megakaryocyte    Monocyte

Erythrocyte

Platelets

Macrophage    Neutrophil    Eosinophil    Basophil    B lymphocyte    T lymphocyte

Granulocytes

These cells continue to proliferate and undergo several rounds of division as they differentiate. Once they become fully differentiated, however, they cease proliferation, so the maintenance of differentiated blood cell populations is dependent on continual proliferation of the pluripotent stem cell.

### Programmed Cell Death

**Programmed cell death** is a normal physiological form of cell death that plays a key role both in the maintenance of adult tissues and in embryonic development. In adults, programmed cell death is responsible for balancing cell proliferation and maintaining constant cell numbers in tissues undergoing cell turnover. For example, about $5 \times 10^{11}$ blood cells are eliminated by programmed cell death daily in humans, balancing their continual production in the bone marrow. In addition, programmed cell death provides a defense mechanism by which damaged and potentially dangerous cells can be eliminated for the good of the organism as a whole. Virus-infected cells frequently undergo programmed cell death, thereby preventing the production of new virus particles and limiting spread of the virus through the host organism. Other types of insults, such as DNA damage, also induce programmed cell death. In the case of DNA damage, programmed cell death may eliminate cells carrying potentially harmful mutations, including cells with mutations that might lead to the development of cancer.

During development, programmed cell death plays a key role by eliminating unwanted cells from a variety of tissues. For example, programmed cell death is responsible for the elimination of larval tissues during

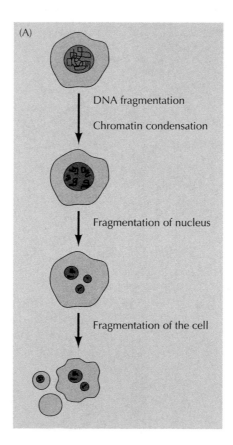

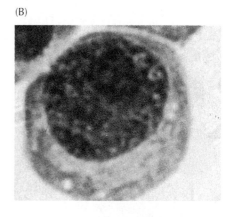

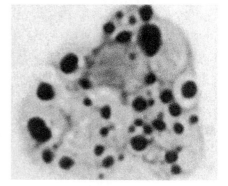

*Figure 14.43*
**Apoptosis** (A) Diagrammatic representation of the events of apoptosis. (B) Light micrographs of normal and apoptotic human leukemia cells, illustrating chromatin condensation and nuclear fragmentation during apoptosis. (B, courtesy of D. R. Green/La Jolla Institute for Allergy and Immunology.)

amphibian and insect metamorphosis, as well as for the elimination of tissue between the digits during the formation of fingers and toes. Another well-characterized example of programmed cell death is provided by development of the mammalian nervous system. Neurons are produced in excess, and up to 50% of developing neurons are eliminated by programmed cell death. Those that survive are selected for having made the correct connections with their target cells, which secrete growth factors that signal cell survival by blocking the neuronal cell death program. The survival of many other types of cells in animals is similarly dependent on growth factors or contacts with neighboring cells or the extracellular matrix, so programmed cell death is thought to play an important role in regulating the associations between cells in tissues.

In contrast to the accidental death of cells that results from an acute injury, programmed cell death is an active process characterized by a distinct morphological change known as **apoptosis** (Figure 14.43). During apoptosis, chromosomal DNA is usually fragmented as a result of cleavage between nucleosomes. The chromatin condenses and the nucleus then breaks up into small pieces. Finally, the cell itself shrinks and breaks up into membrane-enclosed fragments called apoptotic bodies. Such apoptotic cells and cell fragments are readily recognized and phagocytosed by both macrophages and neighboring cells, so cells that die by apoptosis are efficiently removed from tissues. In contrast, cells that die as a result of acute injury swell and lyse, releasing their contents into the extracellular space and causing inflammation.

Studies of programmed cell death during the development of *Caenorhabditis elegans* have identified three genes that play key roles in regulating and executing apoptosis. During normal nematode development, 131 somatic cells out of a total of 1090 are eliminated by programmed cell death. Two genes, *ced-3* and *ced-4*, are required for apoptosis to occur; if either of these genes is inactivated, the normal programmed cell deaths do not take place. A third gene, *ced-9*, functions as a negative regulator of apoptosis. If *ced-9* is inactivated by mutation, the cells that would normally survive fail to do so. Instead, they also undergo apoptosis, leading to death of the developing animal. Conversely, if *ced-9* is expressed at an abnormally high level, the normal programmed cell deaths fail to occur.

Genes related to both *ced-3* and *ced-9* have been identified in mammals and appear to encode proteins that represent conserved effectors and regulators of apoptosis induced by a variety of stimuli (Figure 14.44). Ced-3 is related to a family of mammalian cysteine proteases, known as the **ICE protease** family. Like Ced-3 in nematodes, members of the ICE protease family appear to play key roles in apoptosis in mammalian cells, indicating that induction of apoptosis involves proteolytic cleavage of one or more critical target proteins. The *ced-9* gene encodes a protein homologous to the **Bcl-2** family of proteins, which regulate apoptosis in mammalian cells. Like Ced-9, overexpression of Bcl-2 prevents apoptosis induced by a variety of agents. Genetic studies in *C. elegans*, as well as recent experiments in mammalian cells, suggest that Bcl-2 acts upstream of the ICE proteases and may control cell survival by regulating the activity of ICE family members. Although the molecular mechanisms remain to be elucidated, it thus appears that a conserved pathway regulates the fundamental process of programmed cell death in animals as diverse as nematodes and humans.

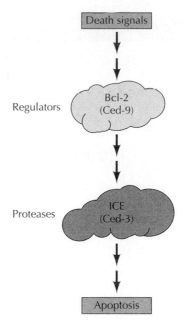

*Figure 14.44*
**Regulators and effectors of apoptosis** Many cell death signals induce apoptosis through a common conserved pathway. Apoptosis is regulated by members of the Bcl-2 family (Ced-9 in *C. elegans*), which act upstream of members of the ICE family of cysteine proteases (Ced-3 in *C. elegans*).

## KEY TERMS

mitosis, interphase, cytokinesis, M phase, $G_1$ phase, S phase, $G_2$ phase, $G_0$, flow cytometer, fluorescence-activated cell sorter

START, restriction point

p53

maturation promoting factor (MPF), Cdc2, cyclin

$G_1$ cyclin, Cln, Cdk, Cdk inhibitor (CKI)

Rb, tumor suppressor gene, E2F

TGF-$\beta$

prophase, metaphase, anaphase, telophase, centromere, kinetochore, centrosome, mitotic spindle, prometaphase

histone H1, lamin, polar wind, kinetochore microtubule, polar microtubule, astral microtubule

## *Summary*

### THE EUKARYOTIC CELL CYCLE

*Phases of the Cell Cycle:* Eukaryotic cell cycles are divided into four discrete phases: M, $G_1$, S, and $G_2$. M phase consists of mitosis, which is usually followed by cytokinesis. S phase is the period of DNA replication.

*Regulation of the Cell Cycle by Cell Growth and Extracellular Signals:* Extracellular signals and cell size regulate progression through specific control points in the cell cycle.

*Cell Cycle Checkpoints:* Checkpoints and feedback controls coordinate the events that take place during different phases of the cell cycle and arrest cell cycle progression if DNA is damaged.

*Coupling of S Phase to M Phase:* Once DNA replication has taken place, initiation of a new S phase is prevented until the cell has passed through mitosis.

### REGULATORS OF CELL CYCLE PROGRESSION

*MPF: A Dimer of Cdc2 and Cyclin:* MPF is the key molecule responsible for regulating the $G_2$ to M transition in all eukaryotes. MPF is a dimer of cyclin B and the Cdc2 protein kinase.

*Families of Cyclins and Cyclin-Dependent Kinases:* Distinct pairs of cyclins and Cdc2-related protein kinases regulate progression through different stages of the cell cycle. The activity of Cdk's is regulated by association with cyclins, activating and inhibitory phosphorylations, and the binding of Cdk inhibitors.

*Growth Factors and the D-Type Cyclins:* Growth factors stimulate animal cell proliferation by inducing synthesis of the D-type cyclins. Cdk4/cyclin D complexes then act to drive cells through the restriction point in $G_1$. A key substrate of Cdk4/cyclin D complexes is the tumor suppressor protein Rb, which regulates transcription of genes required for DNA synthesis and cell cycle progression.

*Inhibitors of Cell Cycle Progression:* DNA damage and a variety of extracellular signals inhibit cell cycle progression. These inhibitory factors frequently act by inducing synthesis of Cdk inhibitors.

### THE EVENTS OF M PHASE

*Stages of Mitosis:* Mitosis is conventionally divided into four stages: prophase, metaphase, anaphase, and telophase. The basic events of mitosis include chromosome condensation, formation of the mitotic spindle, nuclear envelope breakdown, and attachment of spindle microtubules to chromosomes at the kinetochore. Sister chromatids then separate and move to opposite poles of the spindle. Finally, nuclei re-form, the chromosomes decondense, and cytokinesis divides the cell in half.

*MPF and Progression to Metaphase:* M phase is initiated by activation of MPF, which phosphorylates other protein kinases, as well as some structural proteins, such as the nuclear lamins and histone H1. Activation of MPF is responsible for chromatin condensation, nuclear envelope breakdown, fragmentation of the endoplasmic reticulum and Golgi apparatus, and reorganization of microtubules to form the mitotic spindle. The attachment of spindle microtubules to the kinetochores of sister chromatids then leads to their alignment on the metaphase plate.

*Proteolysis and the Inactivation of MPF: Anaphase and Telophase:* Activation of a ubiquitin-mediated proteolysis system at the metaphase to anaphase transition leads to the degradation of cyclin B and inactivation of MPF. This proteolysis system also initiates anaphase by degrading a distinct target protein, which may maintain the linkage between sister chromatids. Inactivation of MPF leads to re-formation of nuclear envelopes, chromatin decondensation, and the return of microtubules to the interphase state.

*Cytokinesis:* Inactivation of MPF also triggers cytokinesis. In animal cells, cytokinesis results from contraction of a ring of actin and myosin filaments. In higher plant cells, cytokinesis results from the formation of a new cell wall and plasma membrane inside the cell.

**contractile ring, cell plate**

## MEIOSIS AND FERTILIZATION

*The Process of Meiosis:* Meiosis is a specialized cell cycle that gives rise to haploid daughter cells. A single round of DNA synthesis is followed by two sequential cell divisions. During meiosis I, homologous chromosomes first pair and then segregate to different daughter cells. Meiosis II then resembles a normal mitosis, in which sister chromatids separate.

**meiosis, leptotene, zygotene, pachytene, diplotene, diakinesis, synapsis, synaptonemal complex, chiasma**

*Regulation of Oocyte Meiosis:* Meiosis of vertebrate oocytes is regulated at two unique points in the cell cycle: the diplotene stage of meiosis I and metaphase of meiosis II. Metaphase II arrest results from inhibition of the ubiquitin-mediated proteolysis system that degrades cyclin B by a protein kinase expressed in oocytes.

**polar body, cytostatic factor (CSF), Mos**

*Fertilization:* Fertilization triggers the resumption of oocyte meiosis by activating a $Ca^{2+}$-dependent proteolysis system that initiates the transition from metaphase to anaphase and degrades cyclin B. The fertilized egg then contains two haploid nuclei, which form a new diploid genome and initiate embryonic cell divisions.

**fertilization, zygote, pronucleus**

## DEVELOPMENT, DIFFERENTIATION, AND PROGRAMMED CELL DEATH

*Cell Proliferation in Adults:* Most cells in adult animals are arrested in the $G_0$ stage of the cell cycle. A few types of differentiated cells are no longer capable of proliferation, but most are able to proliferate as required to replace cells that have been lost because of injury or cell death. Some differentiated cells have short life spans and are continually replaced via the proliferation of stem cells.

**stem cell**

*Programmed Cell Death:* Programmed cell death plays a key role both in the maintenance of adult tissues and in embryonic development. In contrast to the accidental death of cells resulting from an acute injury, programmed cell death takes place by the active process of apoptosis. Genes responsible for both the regulation and execution of apoptosis are conserved from nematodes to humans.

**programmed cell death, apoptosis, ICE protease, Bcl-2**

## QUESTIONS

1. Consider a mammalian cell that divides every 30 hours. Microscopic observation indicates that 3.3% of the cells are in mitosis at any given time. Analysis in the flow cytometer establishes that 53.3% of the cells have DNA contents of 2n, 16.7% have DNA contents of 4n, and 30% have DNA contents ranging between 2n and 4n. What are the lengths of the $G_1$, S, $G_2$, and M phases of the cycle of these cells?

2. The metaphase checkpoint delays the onset of anaphase until all chromosomes are properly aligned on the spindle. What would be the result if a fail-ure of this checkpoint permitted ana-phase to initiate while one chromosome was still attached to microtubules from only a single centrosome?

3. The Cdk inhibitor p16 binds specifically to Cdk4/cyclin D complexes. What would be the predicted effect of overexpression of p16 on cell cycle progression? Would overexpression of p16 affect a tumor cell lacking functional Rb protein?

4. *In vitro* mutagenesis of cloned lamin cDNAs has been used to generate mutants that cannot be phosphorylated by MPF. How would expression of these mutant lamins affect nuclear envelope breakdown at the end of prophase?

5. Mutants of cyclin B that are resistant to degradation by the cyclin B protease have been generated. How would expression of these mutant cyclin B's affect the events at the metaphase to anaphase transition?

6. Homologous recombination has been used to inactivate the *mos* gene in mice. What effect would you expect this to have on oocyte meiosis?

## REFERENCES AND FURTHER READING

### General Reference

Murray, A. and T. Hunt. 1993. *The Cell Cycle: An Introduction.* New York: W. H. Freeman.

### The Eukaryotic Cell Cycle

Baserga, R. 1985. *The Biology of Cell Reproduction.* Cambridge, MA: Harvard University Press.

Forsburg, S. L. and P. Nurse. 1991. Cell cycle regulation in the yeasts *Saccharomyces cerevisiae* and *Schizosaccharomyces pombe*. *Ann. Rev. Cell Biol.* 7: 227–256. [R]

Hartwell, L. H. and M. B. Kastan. 1994. Cell cycle control and cancer. *Science* 266: 1821–1828. [R]

Hartwell, L. H. and T. A. Weinert. 1989. Checkpoints: Controls that ensure the order of cell cycle events. *Science* 246: 629–634. [R]

Murray, A. W. 1992. Creative blocks: Cell-cycle checkpoints and feedback controls. *Nature* 359: 599–604. [R]

Norbury, C. and P. Nurse. 1992. Animal cell cycles and their control. *Ann. Rev. Biochem.* 61: 441–470. [R]

Nurse, P. 1994. Ordering S phase and M phase in the cell cycle. *Cell* 79: 547–550. [R]

Pardee, A. B. 1989. $G_1$ events and the regulation of cell proliferation. *Science* 246: 603–608. [R]

Rao, P. N. and R. T. Johnson. 1970. Mammalian cell fusion studies on the regulation of DNA synthesis and mitosis. *Nature* 225: 159–164. [P]

Su, T. T. S., P. J. Follette and P. H. O'Farrell. 1995. Qualifying for the license to replicate. *Cell* 81: 825–826. [R]

### Regulators of Cell Cycle Progression

Beach, D., B. Durkacz and P. Nurse. 1982. Functionally homologous cell cycle control genes in budding and fission yeast. *Nature* 300: 706–709. [P]

Evans, T., E. T. Rosenthal, J. Youngbloom, D. Distel and T. Hunt. 1983. Cyclin: A protein specified by maternal mRNA in sea urchin eggs that is destroyed at each cleavage division. *Cell* 33: 389–396. [P]

Hartwell, L. H., R. K. Mortimer, J. Culotti and M. Culotti. 1973. Genetic control of the cell division cycle in yeast: V. Genetic analysis of *cdc* mutants. *Genetics* 74: 267–287. [P]

Heichman, K. A. and J. M. Roberts. 1994. Rules to replicate by. *Cell* 79: 557–562. [R]

Hunter, T. and J. Pines. 1994. Cyclins and cancer II: Cyclin D and CDK inhibitors come of age. *Cell* 79: 573–582. [R]

King, R. W., P. K. Jackson and M. W. Kirschner. 1994. Mitosis in transition. *Cell* 79: 563–571. [R]

Kirschner, M. 1992. The cell cycle then and now. *Trends Biochem. Sci.* 17: 281–285. [R]

La Thangue, N. B. 1994. DRTF1/E2F: An expanding family of heterodimeric transcription factors implicated in cell-cycle control. *Trends Biochem. Sci.* 19: 108–114. [R]

Lohka, M. J., M. K. Hayes and J. L. Maller. 1988. Purification of maturation-promoting factor, an intracellular regulator of early mitotic events. *Proc. Natl. Acad. Sci. USA* 85: 3009–3013. [P]

Masui, Y. and C. L. Markert. 1971. Cytoplasmic control of nuclear behavior during meiotic maturation of frog oocytes. *J. Exp. Zool.* 177: 129–146. [P]

Morgan, D. O. 1995. Principles of CDK regulation. *Nature* 374: 131–134. [R]

Murray, A. 1995. Cyclin ubiquitination: The destructive end of mitosis. *Cell* 81: 149–152. [R]

Peter, M. and I. Herskowitz. 1994. Joining the complex: Cyclin-dependent kinase inhibitory proteins and the cell cycle. *Cell* 79: 181–184. [R]

Riley, D. J., E. Y.-H. Lee and W.-H. Lee. 1994. The retinoblastoma gene: More than a tumor suppressor. *Ann. Rev. Cell Biol.* 10: 1–29. [R]

Sherr, C. J. 1994. G1 phase progression: Cycling on cue. *Cell* 79: 551–555. [R]

Sherr, C. J. 1995. D-type cyclins. *Trends Biochem. Sci.* 20: 187–190. [R]

Smith, L. D. and R. E. Ecker. 1971. The interaction of steroids with *Rana pipiens* oocytes in the induction of maturation. *Dev. Biol.* 25: 232–247. [P]

Swenson, K. I., K. M. Farrell and J. V. Ruderman. 1986. The clam embryo protein cyclin A induces entry into M phase and the resumption of meiosis in *Xenopus* oocytes. *Cell* 47: 861–870. [P]

Weinberg, R. A. 1995. The retinoblastoma protein and cell cycle control. *Cell* 81: 323–330. [R]

### The Events of M Phase

Earnshaw, W. C. and A. F. Pluta. 1994. Mitosis. *BioEssays* 16: 639–643. [R]

Fuller, M. T. 1995. Riding the polar winds: Chromosomes motor down east. *Cell* 81: 5–8. [R]

Holloway, S. L., M. Glotzer, R. W. King and A. W. Murray. 1993. Anaphase is initiated by proteolysis rather than by the inactivation of maturation-promoting factor. *Cell* 73: 1393–1402. [P]

Holm, C. 1994. Coming undone: How to untangle a chromosome. *Cell* 77: 955–957. [R]

King, R. W., P. K. Jackson and M. W. Kirschner. 1994. Mitosis in transition. *Cell* 79: 563–571. [R]

Kirschner, M. 1992. The cell cycle then and now. *Trends Biochem. Sci.* 17: 281–285. [R]

Koshland, D. 1994. Mitosis: Back to the basics. *Cell* 77: 951–954. [R]

McIntosh, J. R. and G. E. Hering. 1991. Spindle fiber action and chromosome movement. *Ann. Rev. Cell Biol.* 7: 403–426. [R]

McIntosh, J. R. and M. P. Koonce. 1989. Mitosis. *Science* 246: 622–628. [R]

Murray, A. 1995. Cyclin ubiquitination: The destructive end of mitosis. *Cell* 81: 149–152. [R]

Peterson, C. L. 1994. The SMC family: Novel motor proteins for chromosome condensation? *Cell* 79: 389–392. [R]

Rieder, C. L. and E. D. Salmon. 1994. Motile kinetochores and polar ejection forces dictate chromosome position on the vertebrate mitotic spindle. *J. Cell Biol.* 124: 223–233. [R]

Staehelin, L. A. and P. K. Hepler. 1996. Cytokinesis in higher plants. *Cell* 84: 821–824. [R]

Warren, G. 1993. Membrane partitioning during cell division. *Ann. Rev. Biochem.* 323–348. [R]

## Meiosis and Fertilization

Austin, C. R. and R. V. Short, eds. 1982. *Reproduction in Mammals I. Germ Cells and Fertilization*, 2d ed. Cambridge, England: Cambridge University Press.

Carpenter, A. T. C. 1994. Chiasma function. *Cell* 77: 959–962. [R]

Gilbert, S. F. 1994. *Developmental Biology*, 4th ed. Sunderland, MA: Sinauer Associates.

Hawley, R. S. and T. Arbel. 1993. Yeast genetics and the fall of the classical view of meiosis. *Cell* 72: 301–303. [R]

Longo, F. J. 1987. *Fertilization*. London: Chapman and Hall.

McKim, K. S. and R. S. Hawley. 1995. Chromosomal control of meiotic cell division. *Science* 270: 1595–1601. [R]

Moens, P. B. 1994. Molecular perspectives of chromosome pairing at meiosis. *BioEssays* 16: 101–106. [R]

Orr-Weaver, T. L. 1995. Meiosis in *Drosophila*: Seeing is believing. *Proc. Natl. Acad. Sci. USA* 92: 10443–10449. [R]

Roeder, G. S. 1995. Sex and the single cell: Meiosis in yeast. *Proc. Natl. Acad. Sci. USA* 92: 10450–10456. [R]

Sagata, N. 1996. Meiotic metaphase arrest in animal oocytes: Its mechanisms and biological significance. *Trends Cell Biol.* 6: 22–28. [R]

Snell, W. J. and J. M. White. 1996. The molecules of mammalian fertilization. *Cell* 85: 629–637. [R]

Wassarman, P. M. 1987. The biology and chemistry of fertilization. *Science* 235: 553–560. [R]

## Development, Differentiation, and Programmed Cell Death

Baserga, R. 1985. *The Biology of Cell Reproduction*. Cambridge, MA: Harvard University Press.

Fawcett, D. W. 1994. *Bloom and Fawcett: A Textbook of Histology*, 12th ed. New York: Chapman and Hall.

Hengartner, M. O. and H. R. Horvitz. 1994. Programmed cell death in *Caenorhabditis elegans*. *Curr. Opin. Genet. Dev.* 4: 581–586. [R]

Korsmeyer, S. J. 1995. Regulators of cell death. *Trends Genet.* 11: 101–105. [R]

Kumar, S. 1995. ICE-like proteases in apoptosis. *Trends Biochem. Sci.* 20: 198–202. [R]

Martin, S. J. and D. R. Green. 1995. Protease activation during apoptosis: Death by a thousand cuts? *Cell* 82: 349–352. [R]

Martin, S. J., D. R. Green and T. G. Cotter. 1994. Dicing with death: Dissecting the components of the apoptosis machinery. *Trends Biochem. Sci.* 19: 26–30. [R]

Metcalf, D. 1992. Hemopoietic regulators. *Trends Biochem. Sci.* 17: 286–289. [R]

Raff, M. C. 1992. Social controls on cell survival and cell death. *Nature* 356: 397–400. [R]

Raff, M. C., B. A. Barres, J. F. Burne, H. S. Coles, Y. Ishizaki and M. D. Jacobson. 1993. Programmed cell death and the control of cell survival: Lessons from the nervous system. *Science* 262: 695–700. [R]

Steller, H. 1995. Mechanisms and genes of cellular suicide. *Science* 267: 1445–1449. [R]

Vaux, D. L. and A. Strasser. 1996. The molecular biology of apoptosis. *Proc. Natl. Acad. Sci. USA* 93: 2239–2244. [R]

# 15

# *Cancer*

ANCER IS A PARTICULARLY APPROPRIATE TOPIC for the concluding chapter of this book because it results from a breakdown of the regulatory mechanisms that govern normal cell behavior. As discussed in preceding chapters, the proliferation, differentiation, and survival of individual cells in multicellular organisms are carefully regulated to meet the needs of the organism as a whole. This regulation is lost in cancer cells, which grow and divide in an uncontrolled manner, ultimately spreading throughout the body and interfering with the function of normal tissues and organs.

Because cancer results from defects in fundamental cell regulatory mechanisms, it is a disease that ultimately has to be understood at the molecular and cellular levels. Indeed, understanding cancer has been an objective of molecular and cellular biologists for many years. Importantly, studies of cancer cells have also illuminated the mechanisms that regulate normal cell behavior. In fact, many of the proteins that play key roles in cell signaling, regulation of the cell cycle, and control of programmed cell death were first identified because abnormalities in their activities led to the uncontrolled proliferation of cancer cells. The study of cancer has thus contributed significantly to our understanding of normal cell regulation, as well as vice versa.

## THE DEVELOPMENT AND CAUSES OF CANCER

The fundamental abnormality resulting in the development of cancer is the continual unregulated proliferation of cancer cells. Rather than responding appropriately to the signals that control normal cell behavior, cancer cells grow and divide in an uncontrolled manner, invading normal tissues and organs and eventually spreading throughout the body. The generalized loss of growth control exhibited by cancer cells is the net result of accumulated abnormalities in multiple cell regulatory systems and is reflected in several aspects of cell behavior that distinguish cancer cells from their normal counterparts.

(A)

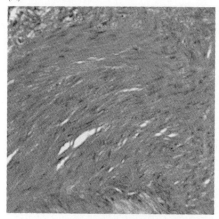

(B)

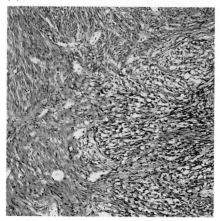

*Figure 15.1*
**A malignant tumor of the uterus**
Micrographs of normal uterus (A) and
a section of a uterine sarcoma (B).
Note that the cancer cells (darkly
stained) have invaded the surrounding
normal tissue. (Cecil Fox/Molecular
Histology Inc.)

## Types of Cancer

**Cancer** can result from abnormal proliferation of any of the different kinds of cells in the body, so there are more than a hundred distinct types of cancer, which can vary substantially in their behavior and response to treatment. The most important issue in cancer pathology is the distinction between benign and malignant tumors (Figure 15.1). A **tumor** is any abnormal proliferation of cells, which may be either benign or malignant. A **benign tumor**, such as a common skin wart, remains confined to its original location, neither invading surrounding normal tissue nor spreading to distant body sites. A **malignant tumor**, however, is capable of both invading surrounding normal tissue and spreading throughout the body via the circulatory or lymphatic systems (**metastasis**). Only malignant tumors are properly referred to as cancers, and it is their ability to invade and metastasize that makes cancer so dangerous. Whereas benign tumors can usually be removed surgically, the spread of malignant tumors to distant body sites frequently makes them resistant to such localized treatment.

Both benign and malignant tumors are classified according to the type of cell from which they arise. Most cancers fall into one of three main groups: carcinomas, sarcomas, and leukemias or lymphomas. **Carcinomas**, which include approximately 90% of human cancers, are malignancies of epithelial cells. **Sarcomas**, which are rare in humans, are solid tumors of connective tissues, such as muscle, bone, cartilage, and fibrous tissue. **Leukemias** and **lymphomas**, which account for approximately 8% of human malignancies, arise from the blood-forming cells and from cells of the immune system, respectively. Tumors are further classified according to tissue of origin (e.g., lung or breast carcinomas) and the type of cell involved. For example, fibrosarcomas arise from fibroblasts, and erythroid leukemias from precursors of erythrocytes (red blood cells).

Although there are many kinds of cancer, only a few occur frequently (Table 15.1). More than a million cases of cancer are diagnosed annually in the United States, and more than 500,000 Americans die of cancer each year. Cancers of 13 different body sites account for approximately 85% of this total cancer incidence. The four most common cancers, accounting for

*Table 15.1*   Most Frequent Cancers in the United States

| Cancer site | Cases per year | Deaths per year |
|---|---|---|
| Prostate | 317,000 (23%) | 41,000 (7.4%) |
| Breast | 186,000 (14%) | 45,000 (8.1%) |
| Lung | 177,000 (13%) | 159,000 (29%) |
| Colon/rectum | 134,000 (10%) | 55,000 (9.9%) |
| Lymphomas | 60,000 (4.4%) | 25,000 (4.5%) |
| Bladder | 53,000 (3.9%) | 12,000 (2.2%) |
| Uterus | 50,000 (3.7%) | 11,000 (2.0%) |
| Skin (melanoma) | 38,000 (2.8%) | 7,000 (1.3%) |
| Kidney | 31,000 (2.3%) | 12,000 (2.2%) |
| Oral cavity | 29,000 (2.1%) | 8,000 (1.4%) |
| Leukemias | 28,000 (2.1%) | 21,000 (3.8%) |
| Ovary | 27,000 (2.0%) | 15,000 (2.7%) |
| Pancreas | 26,000 (1.9%) | 28,000 (5.0%) |
| Subtotal | 1,156,000 (85%) | 439,000 (79%) |
| All sites | 1,359,000 (100%) | 555,000 (100%) |

Source: American Cancer Society, *Cancer Facts and Figures—1996.*

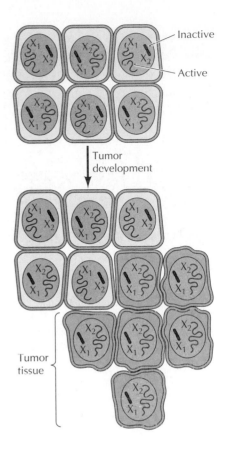

**Figure 15.2**
**Tumor clonality** Normal tissue is a mosaic of cells in which different X chromosomes ($X_1$ and $X_2$) have been inactivated. Tumors develop from a single initially altered cell, so each tumor cell displays the same pattern of X inactivation ($X_1$ inactive, $X_2$ active).

more than half of all cancer cases, are those of the prostate, breast, lung, and colon/rectum. Lung cancer, by far the most lethal, is responsible for nearly 30% of all cancer deaths.

## The Development of Cancer

One of the fundamental features of cancer is tumor clonality, the development of tumors from single cells that begin to proliferate abnormally. The single-cell origin of many tumors has been demonstrated by analysis of X chromosome inactivation (Figure 15.2). As discussed in Chapter 8, one member of the X chromosome pair is inactivated by being converted to heterochromatin in female cells. X inactivation occurs randomly during embryonic development, so one X chromosome is inactivated in some cells, while the other X chromosome is inactivated in other cells. Thus, if a female is heterozygous for an X chromosome gene, different alleles will be expressed in different cells. Normal tissues are composed of mixtures of cells with different inactive X chromosomes, so expression of both alleles is detected in normal tissues of heterozygous females. In contrast, tumor tissues generally express only one allele of a heterozygous X chromosome gene. The implication is that all of the cells constituting such a tumor were derived from a single cell of origin, in which the pattern of X inactivation was fixed before the tumor began to develop.

The clonal origin of tumors does not, however, imply that the original progenitor cell that gives rise to a tumor has initially acquired all of the characteristics of a cancer cell. On the contrary, the development of cancer is a multistep process in which cells gradually become malignant through a progressive series of alterations. One indication of the multistep development of cancer is that most cancers develop late in life. The incidence of colon cancer, for example, increases more than tenfold between the ages of 30 and 50, and another tenfold between 50 and 70 (Figure 15.3). Such a dramatic increase of cancer incidence with age suggests that most cancers develop as a consequence of multiple abnormalities, which accumulate over periods of many years.

At the cellular level, the development of cancer is viewed as a multistep process involving mutation and selection for cells with progressively increasing capacity for proliferation, survival, invasion, and metastasis (Figure 15.4). The first step in the process, **tumor initiation**, is thought to be the result of a genetic alteration leading to abnormal proliferation of a single cell. Cell proliferation then leads to the outgrowth of a population of clonally derived tumor cells. **Tumor progression** continues as additional mutations occur within cells of the tumor population. Some of these mutations confer a selective advantage to the cell, such as more rapid growth, and the descendants of a cell bearing such a mutation will consequently

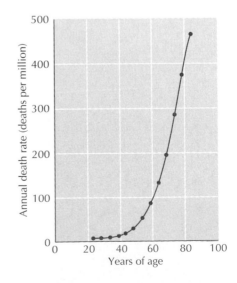

**Figure 15.3**
**Increased rate of colon cancer with age** Annual death rates from colon cancer in the United States. (Data from J. Cairns, 1978, *Cancer: Science and Society*. New York, W. H. Freeman.)

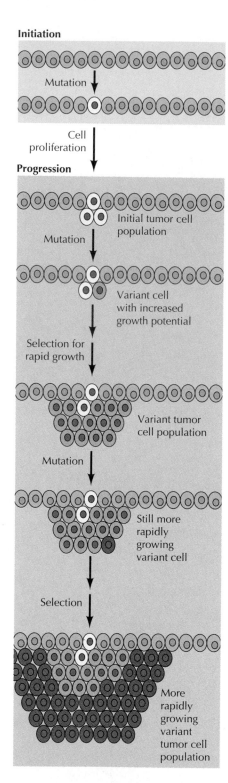

**Initiation**

Mutation

Cell proliferation

**Progression**

Initial tumor cell population

Mutation

Variant cell with increased growth potential

Selection for rapid growth

Variant tumor cell population

Mutation

Still more rapidly growing variant cell

Selection

More rapidly growing variant tumor cell population

*Figure 15.4*
**Stages of tumor development**   The development of cancer initiates when a single mutated cell begins to proliferate abnormally. Additional mutations followed by selection for more rapidly growing cells within the population then result in progression of the tumor to increasingly rapid growth and malignancy.

become dominant within the tumor population. The process is called clonal selection, since a new clone of tumor cells has evolved on the basis of its increased growth rate or other properties (such as survival, invasion, or metastasis) that confer a selective advantage. Clonal selection continues throughout tumor development, so tumors continuously become more rapid-growing and increasingly malignant.

Studies of colon carcinomas have provided a clear example of tumor progression during the development of a common human malignancy (Figure 15.5). The earliest stage in tumor development is increased proliferation of colon epithelial cells. One of the cells within this proliferative cell population is then thought to give rise to a small benign neoplasm (an **adenoma** or **polyp**). Further rounds of clonal selection lead to the growth of adenomas of increasing size and proliferative potential. Malignant carcinomas then arise from the benign adenomas, indicated by invasion of the tumor cells through the basal lamina into underlying connective tissue. The cancer cells then continue to proliferate and spread through the connective tissues of the colon wall. Eventually the cancer cells penetrate the wall of the colon and invade other abdominal organs, such as the bladder or small intestine. In addition, the cancer cells invade blood and lymphatic vessels, allowing them to metastasize throughout the body.

## Causes of Cancer

Substances that cause cancer, called **carcinogens**, have been identified both by studies in experimental animals and by epidemiological analysis of cancer frequencies in human populations (e.g., the high incidence of lung cancer among cigarette smokers). Since the development of malignancy is a complex multistep process, many factors may affect the likelihood that cancer will develop, and it is overly simplistic to speak of single causes of most cancers. Nonetheless, many agents, including radiation, chemicals, and viruses, have been found to induce cancer in both experimental animals and humans.

Radiation and many chemical carcinogens (Figure 15.6) act by damaging DNA and inducing mutations. These carcinogens are generally referred to as initiating agents, since the induction of mutations in key target genes is thought to be the initial event leading to cancer development. Some of the initiating agents that contribute to human cancers include solar ultraviolet radiation (the major cause of skin cancer), carcinogenic chemicals in tobacco smoke, and aflatoxin (a potent liver carcinogen produced by some molds that contaminate improperly stored supplies of peanuts and other grains). The carcinogens in tobacco smoke (including benzo(a)pyrene, dimethylnitrosamine, and nickel compounds) are the major identified causes of human cancer. Smoking is the undisputed cause of 80 to 90% of lung cancers, as well as being implicated in cancers of the oral cavity, pharynx, larynx, esophagus, and other sites. In total, it is estimated that smoking is responsible for nearly one-third of all cancer deaths—an impressive toll for a single carcinogenic agent.

Other carcinogens contribute to cancer development by stimulating cell proliferation, rather than by inducing mutations. Such compounds are

*Figure 15.5*
**Development of colon carcinomas** A single initially altered cell gives rise to a proliferative cell population, which progresses first to benign adenomas of increasing size and then to malignant carcinoma. The cancer cells invade the underlying connective tissue and penetrate blood and lymphatic vessels, thereby spreading throughout the body.

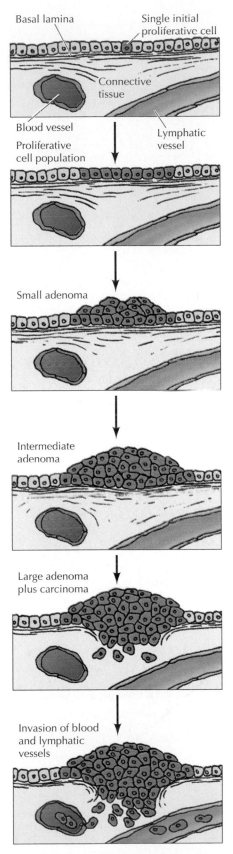

referred to as **tumor promoters**, since the increased cell division they induce is required for the outgrowth of a proliferative cell population during early stages of tumor development. The phorbol esters that stimulate cell proliferation by activating protein kinase C (see Figure 13.26) are classic examples. Their activity was defined by studies of chemical induction of skin tumors in mice (Figure 15.7). Tumorigenesis in this system can be initiated by a single treatment with a mutagenic carcinogen. Tumors do not develop, however, unless the mice are subsequently treated with a tumor promoter (usually a phorbol ester) to stimulate proliferation of the mutated cells.

Hormones, particularly estrogens, are important as tumor promoters in the development of some human cancers. The proliferation of cells of the uterine endometrium, for example, is stimulated by estrogen, and exposure to excess estrogen significantly increases the likelihood that a woman will develop endometrial cancer. The risk of endometrial cancer is therefore substantially increased by long-term postmenopausal estrogen replacement therapy with high doses of estrogen alone. Fortunately, this risk is minimized by administration of progesterone to counteract the stimulatory effect of estrogen on endometrial cell proliferation.

In addition to chemicals and radiation, some viruses induce cancer both in experimental animals and in humans. The common human cancers caused by viruses include liver cancer and cervical carcinoma, which together account for 10 to 20% of worldwide cancer incidence. These viruses are important not only as causes of human cancer; as discussed later in this chapter, studies of tumor viruses have played a key role in elucidating the molecular events responsible for the development of cancers induced by both viral and nonviral carcinogens.

*Figure 15.6*
**Structure of representative chemical carcinogens**

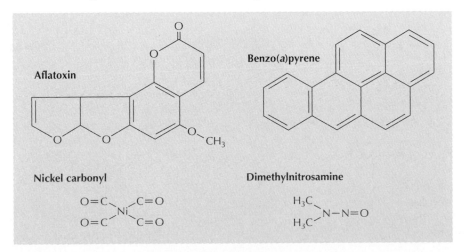

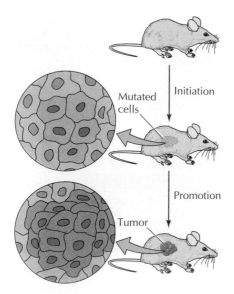

**Figure 15.7**
**Induction of tumors in mouse skin**
Tumors are initiated by mutations induced by a carcinogen. Development of a tumor then requires treatment with a tumor promoter to stimulate proliferation of the mutated cells.

## Properties of Cancer Cells

The uncontrolled growth of cancer cells results from accumulated abnormalities affecting many of the cell regulatory mechanisms that have been discussed in preceding chapters. This relationship is reflected in several aspects of cell behavior that distinguish cancer cells from their normal counterparts. Cancer cells typically display abnormalities in the mechanisms that regulate normal cell proliferation, differentiation, and survival. Taken together, these characteristic properties of cancer cells provide a description of malignancy at the cellular level.

The uncontrolled proliferation of cancer cells *in vivo* is mimicked by their behavior in cell culture. A primary distinction between cancer cells and normal cells in culture is that normal cells display **density-dependent inhibition** of cell proliferation (Figure 15.8). Normal cells proliferate until they reach a finite cell density, which is determined in part by the availability of growth factors added to the culture medium (usually in the form of serum). They then cease proliferating and become quiescent, arrested in the $G_0$ stage of the cell cycle (see Figure 14.6). The proliferation of most cancer cells, however, is not sensitive to density-dependent inhibition. Rather than responding to the signals that cause normal cells to cease proliferation and enter $G_0$, tumor cells generally continue growing to high cell densities in culture, mimicking their uncontrolled proliferation *in vivo*.

A related difference between normal cells and cancer cells is that many cancer cells have reduced requirements for extracellular growth factors. As discussed in Chapter 13, the proliferation of most cells is controlled, at least in part, by polypeptide growth factors. For some cell types, particularly fibroblasts, the availability of serum growth factors is the principal determinant of their proliferative capacity in culture. The growth factor requirements of these cells are closely related to the phenomenon of density-dependent inhibition, since the density at which normal fibroblasts become quiescent is proportional to the concentration of serum growth factors in the culture medium.

The growth factor requirements of many tumor cells are reduced compared to their normal counterparts, contributing to the unregulated proliferation of tumor cells both *in vitro* and *in vivo*. In some cases, cancer cells produce growth factors that stimulate their own proliferation (Figure 15.9). Such abnormal production of a growth factor by a responsive cell leads to continuous autostimulation of cell division (**autocrine growth stimula-**

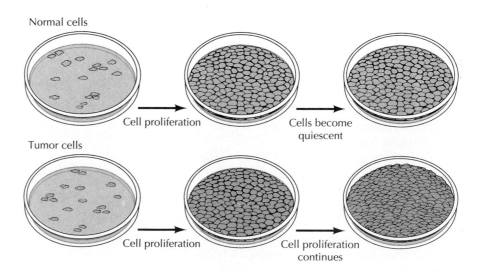

Normal cells

Cell proliferation → Cells become quiescent

Tumor cells

Cell proliferation → Cell proliferation continues

**Figure 15.8**
**Density-dependent inhibition** Normal cells proliferate in culture until they reach a finite cell density, at which point they become quiescent. Tumor cells, however, continue to proliferate independent of cell density.

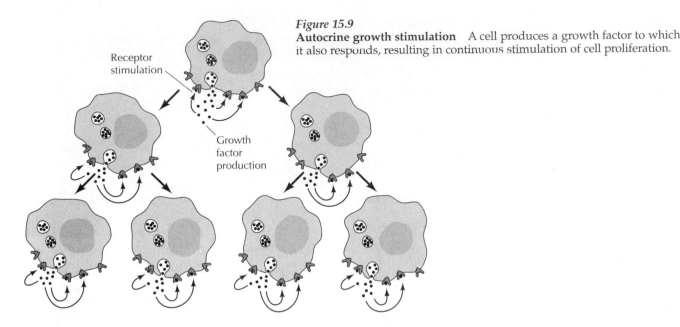

*Figure 15.9*
**Autocrine growth stimulation**   A cell produces a growth factor to which it also responds, resulting in continuous stimulation of cell proliferation.

Receptor
stimulation

Growth
factor
production

tion), and the cancer cells are therefore less dependent on growth factors from other, physiologically normal sources. In other cases, the reduced growth factor dependence of cancer cells results from abnormalities in intracellular signaling systems, such as unregulated activity of growth factor receptors or other proteins (e.g., Ras proteins or protein kinases) that were discussed in Chapter 13 as elements of signal transduction pathways leading to cell proliferation.

Cancer cells are also less stringently regulated than normal cells by cell-cell and cell-matrix interactions. Most cancer cells are less adhesive than normal cells, often as a result of reduced expression of cell surface adhesion molecules. Consequently, cancer cells are comparatively unrestrained by interactions with other cells and tissue components, contributing to the ability of malignant cells to invade and metastasize. The reduced adhesiveness of cancer cells also results in morphological and cytoskeletal alterations: Many tumor cells are rounder than normal, in part because they are less firmly attached to either the extracellular matrix or neighboring cells.

A striking difference in the cell-cell interactions displayed by normal cells and those of cancer cells is illustrated by the phenomenon of **contact inhibition** (Figure 15.10). Normal fibroblasts migrate across the surface of a culture dish until they make contact with a neighboring cell. Further cell migration is then inhibited, and normal cells adhere to each other, forming an orderly array of cells on the culture dish surface. Tumor cells, in contrast, continue moving after contact with their neighbors, migrating over adjacent cells, and growing in disordered, multilayered patterns. Not only the movement but also the proliferation of many normal cells is inhibited by cell-cell contact, and cancer cells are characteristically insensitive to such contact inhibition of growth.

Two additional properties of cancer cells affect their interactions with other tissue components, thereby playing important roles in invasion and metastasis. First, malignant cells generally secrete proteases that digest extracellular matrix components, allowing the cancer cells to invade adjacent normal tissues. Secretion of collagenase, for example, appears to be an important determinant of the ability of carcinomas to digest and penetrate through basal laminae to invade underlying connective tissue (see Figure

*Figure 15.10*
**Contact inhibition** Light micrographs (left) and scanning electron micrographs (right) of normal fibroblasts and tumor cells. The migration of normal fibroblasts is inhibited by cell contact, so they form an orderly side-by-side array on the surface of a culture dish. Tumor cells, however, are not inhibited by cell contact, so they migrate over one another and grow in a disordered, multilayered pattern. (Courtesy of Lan Bo Chen, Dana-Farber Cancer Institute.)

Normal cells

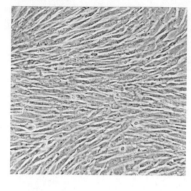
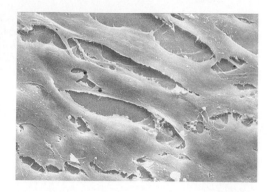

Tumor cells

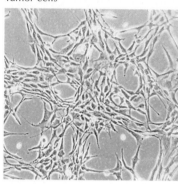
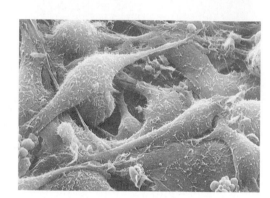

15.5). Second, cancer cells secrete growth factors that promote the formation of new blood vessels (**angiogenesis**). Angiogenesis is needed to support the growth of a tumor beyond the size of about a million cells, at which point new blood vessels are required to supply oxygen and nutrients to the proliferating tumor cells. Such blood vessels are formed in response to growth factors, secreted by the tumor cells, that stimulate proliferation of endothelial cells in the walls of capillaries in surrounding tissue, resulting in the outgrowth of new capillaries into the tumor. The formation of such new blood vessels is important not only in supporting tumor growth, but also in metastasis. The actively growing new capillaries formed in response to angiogenic stimulation are easily penetrated by the tumor cells, providing a ready opportunity for cancer cells to enter the circulatory system and begin the metastatic process.

Another general characteristic of most cancer cells is that they fail to differentiate normally. Such defective differentiation is closely coupled to abnormal proliferation, since, as discussed in Chapter 14, most fully differentiated cells either cease cell division or divide only rarely. Rather than carrying out their normal differentiation program, cancer cells are usually blocked at an early stage of differentiation, consistent with their continued active proliferation.

The leukemias provide a particularly good example of the relationship between defective differentiation and malignancy. All of the different types of blood cells are derived from a common stem cell in the bone marrow (see Figure 14.43). Descendants of these cells then become committed to specific differentiation pathways. Some cells, for example, differentiate to form erythrocytes whereas others differentiate to form lymphocytes, granulocytes, or macrophages. Cells of each of these types undergo several rounds of division as they differentiate, but once they become fully differ-

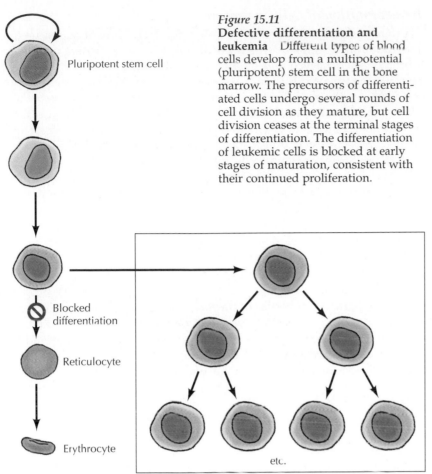

*Figure 15.11*
**Defective differentiation and leukemia** Different types of blood cells develop from a multipotential (pluripotent) stem cell in the bone marrow. The precursors of differentiated cells undergo several rounds of cell division as they mature, but cell division ceases at the terminal stages of differentiation. The differentiation of leukemic cells is blocked at early stages of maturation, consistent with their continued proliferation.

Pluripotent stem cell

Blocked differentiation

Reticulocyte

Erythrocyte

etc.

Leukemic cells fail to differentiate and continue to divide

entiated, cell division ceases. Leukemic cells, in contrast, fail to undergo terminal differentiation (Figure 15.11). Instead, they become arrested at early stages of maturation at which they retain their capacity for proliferation and continue to reproduce.

As discussed in Chapter 14, **programmed cell death**, or **apoptosis**, is an integral part of the differentiation program of many cell types, including blood cells. Coincident with their failure to differentiate normally, many cancer cells fail to undergo apoptosis, and therefore exhibit increased life spans compared to their normal counterparts. This failure of cancer cells to undergo programmed cell death contributes substantially to tumor development. For example, the survival of many normal cells is dependent on signals from growth factors or from the extracellular matrix that prevent apoptosis. In contrast, tumor cells are often able to survive in the absence of growth factors required by their normal counterparts. Such a failure of tumor cells to undergo apoptosis when deprived of normal environmental signals may be important not only in primary tumor development but also in the survival and growth of metastatic cells in abnormal tissue sites. Normal cells also undergo apoptosis following DNA damage, while many cancer cells fail to do so. In this case, the failure to undergo apoptosis contributes to the resistance of cancer cells to irradiation and many chemotherapeutic drugs, which act by damaging DNA. Abnormal cell survival, as well as cell proliferation, thus plays a major role in the unrelenting growth of cancer cells in an animal.

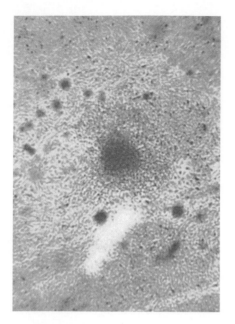

*Figure 15.12*
**The focus assay** A focus of chicken embryo fibroblasts induced by Rous sarcoma virus. (From H. M. Temin and H. Rubin, 1958. *Virology* 6: 669.)

## Transformation of Cells in Culture

The study of tumor induction by radiation, chemicals, or viruses requires experimental systems in which the effects of a carcinogenic agent can be reproducibly observed and quantitated. Although the activity of carcinogens can be assayed in intact animals, such experiments are difficult to quantitate and control. The development of *in vitro* assays to detect the conversion of normal cells to tumor cells in culture, a process called **cell transformation**, therefore represented a major advance in cancer research. Such assays are designed to detect transformed cells, which display the *in vitro* growth properties of tumor cells, following exposure of a culture of normal cells to a carcinogenic agent. Their application has allowed experimental analysis of cell transformation to reach a level of sophistication that could not have been attained by studies in whole animals alone.

The first and most widely used assay of cell transformation is the focus assay, which was developed by Howard Temin and Harry Rubin in 1958. The focus assay is based on the ability to recognize a group of transformed cells as a morphologically distinct "focus" against a background of normal cells on the surface of a culture dish (Figure 15.12). The focus assay takes advantage of three properties of transformed cells: altered morphology, loss of contact inhibition, and loss of density-dependent inhibition of growth. The result is the formation of a colony of morphologically altered transformed cells that overgrow the background of normal cells in the culture. Such foci of transformed cells can usually be detected and quantified within a week or two after exposure to a carcinogenic agent. In general, cells transformed *in vitro* are able to form tumors following inoculation into susceptible animals, supporting *in vitro* transformation as a valid indicator of the formation of cancer cells.

## TUMOR VIRUSES

Members of six distinct families of animal viruses, called **tumor viruses**, are capable of causing cancer in either experimental animals or humans (Table 15.2). Viruses belonging to five of these families have DNA genomes and are referred to as DNA tumor viruses. Members of the sixth family of tumor viruses, the retroviruses, have RNA genomes in virus particles but replicate via synthesis of a DNA provirus in infected cells. The viruses that cause human cancer include hepatitis B virus (liver cancer), papillomaviruses (cervical and other anogenital cancers), Epstein-Barr virus (Burkitt's lymphoma and nasopharyngeal carcinoma), and human T-cell lymphotropic virus

*Table 15.2* Tumor Viruses

| Virus family | Human tumors | Genome size (kb) |
|---|---|---|
| **DNA tumor viruses** | | |
| Hepatitis B viruses | Liver cancer | 3 |
| SV40 and polyomavirus | None | 5 |
| Papillomaviruses | Cervical carcinoma | 8 |
| Adenoviruses | None | 35 |
| Herpesviruses | Burkitt's lymphoma, nasopharyngeal carcinoma | 100–200 |
| **RNA tumor viruses** | | |
| Retroviruses | Adult T-cell leukemia | 9 |

(adult T-cell leukemia). In addition, HIV is indirectly responsible for the cancers that develop in AIDS patients as a result of immunodeficiency.

As already noted, tumor viruses not only are important as causes of human disease but have also played a critical role in cancer research by serving as models for cellular and molecular studies of cell transformation. The small size of their genomes has made tumor viruses readily amenable to molecular analysis, leading to the identification of viral genes responsible for cancer induction and paving the way to our current understanding of cancer at the molecular level.

### Hepatitis B Viruses

The **hepatitis B viruses**, which have the smallest genomes (approximately 3 kb) of all animal DNA viruses, specifically infect liver cells of several species, including ducks, woodchucks, squirrels, and humans. Infection with hepatitis B virus usually results in acute liver damage. In 5 to 10% of cases, however, the acute infection is not resolved and a chronic infection of the liver develops. Such chronic infection is associated with more than a hundredfold increased risk of liver cancer. Hepatitis B virus infection is particularly common in parts of Asia and Africa, where it is associated with up to a million cases of liver cancer annually (approximately 10% of worldwide cancer incidence).

Although hepatitis B viruses are causative agents of a major human cancer, the mechanism by which they induce cell transformation is not yet clear. It is possible that tumors result from expression of a viral gene (the X gene) that stimulates transcription of cellular genes that drive cell proliferation. Alternatively, cancers may develop simply as a result of continual cell proliferation resulting from chronic damage to the liver.

### SV40 and Polyomavirus

The best studied DNA tumor viruses, from the standpoint of molecular biology, are probably **simian virus 40 (SV40)** and **polyomavirus**. Although neither of these viruses is associated with human cancer, they have been critically important as models for understanding the molecular basis of cell transformation. The utility of these viruses in cancer research has stemmed from the availability of good cell culture assays for both virus replication and transformation, as well as from the small size of their genomes (approximately 5 kb).

SV40 and polyomavirus do not induce tumors or transform cells of their natural host species—monkeys and mice, respectively. In cells of their natural hosts (permissive cells), infection leads to virus replication, cell lysis, and release of progeny virus particles (Figure 15.13). Since a permissive cell is killed as a consequence of virus replication, it cannot become transformed. The transforming potential of these viruses is revealed, however, by infection of nonpermissive cells, in which virus replication is blocked. In this case, the viral genome sometimes integrates into cellular DNA, and expression of specific viral genes results in transformation of the infected cell.

The SV40 and polyomavirus genes that lead to cell transformation have been identified by detailed molecular analyses. The viral genomes and mRNAs have been completely sequenced, viral mutants that are unable to induce transformation have been isolated, and the transforming potentials of individual viral genes have been determined by gene transfer assays. Transformation by these viruses has thus been found to result from expression of the same viral genes that function in early stages of lytic infection. The genomes of SV40 and polyomavirus are divided into early and late regions. The early region is expressed immediately after infection and is

*Figure 15.13*
**SV40 replication and transformation**
Infection of a permissive cell results in virus replication, cell lysis, and release of progeny virus particles. In a nonpermissive cell, virus replication is blocked, allowing some cells to become permanently transformed.

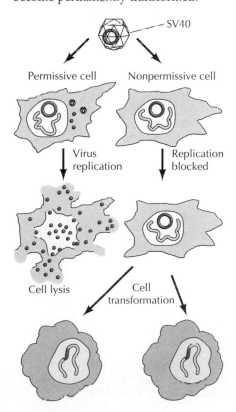

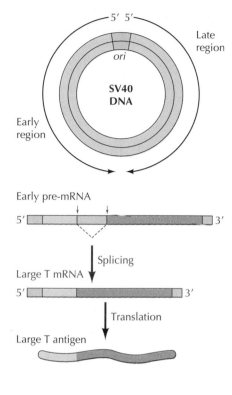

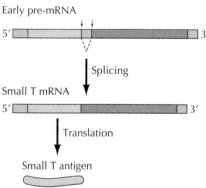

*Figure 15.14*
**The SV40 genome**   The genome is divided into early and late regions. Large and small T antigens are produced by alternative splicing of early-region pre-mRNA.

required for synthesis of viral DNA. The late region is not expressed until after viral DNA replication has begun, and includes genes encoding structural components of the virus particle. The early region of SV40 encodes two proteins, called small and large T antigens, of about 17 kd and 94 kd, respectively (Figure 15.14). Their mRNAs are generated by alternative splicing of a single early-region primary transcript. Polyomavirus likewise encodes small and large T antigens, as well as a third early-region protein of about 55 kd, designated middle T. Transfection of cells with cDNAs for individual early-region proteins has shown that SV40 large T is sufficient to induce transformation, whereas middle T is primarily responsible for transformation by polyomavirus.

During lytic infection, these early-region proteins fulfill multiple functions required for virus replication. SV40 T antigen, for example, binds to the SV40 origin and initiates viral DNA replication (see Chapter 5). In addition, the early-region proteins of SV40 and polyomavirus stimulate host cell gene expression and DNA synthesis. Since virus replication is dependent on host cell enzymes (e.g., DNA polymerase), such stimulation of the host cell is a critical event in the viral life cycle. Most cells in an animal are nonproliferating, and therefore must be stimulated to divide in order to induce the enzymes needed for viral DNA replication. This stimulation of cell proliferation by the early gene products can lead to transformation if the viral DNA becomes stably integrated and expressed in a nonpermissive cell.

As discussed later in this chapter, both SV40 and polyomavirus early-region proteins induce transformation by interacting with host proteins that regulate cell proliferation. For example, SV40 T antigen binds to and inactivates the host cell tumor suppressor proteins Rb and p53, which are key regulators of cell proliferation and cell cycle progression (see Figures 14.19 and 14.20).

## Papillomaviruses

The **papillomaviruses** are small DNA viruses (genomes of approximately 8 kb) that induce both benign and malignant tumors in humans and a variety of other animal species. Approximately 60 different types of human papillomaviruses, which infect epithelial cells of several tissues, have been identified. Some of these viruses cause only benign tumors (such as warts), whereas others are causative agents of malignant carcinomas, particularly cervical and other anogenital cancers. The mortality from cervical cancer is relatively low in the United States, in large part as a result of early detection and curative treatment made possible by the Pap smear. In other parts of the world, however, cervical cancer remains common; it is responsible for 5 to 10% of worldwide cancer incidence.

Cell transformation by human papillomaviruses results from expression of two early-region genes, *E6* and *E7* (Figure 15.15). The E6 and E7 proteins act analogously to SV40 T antigen by interfering with the function of the cellular Rb and p53 proteins. In particular, E7 binds to Rb, and E6 stimulates the degradation of p53 by ubiquitin-mediated proteolysis.

## Adenoviruses

The **adenoviruses** are a large family of DNA viruses with genomes of about 35 kb. In contrast to the papillomaviruses, the adenoviruses are not associated with naturally occurring cancers in either humans or other animals. However, they are widely studied and important models in experimental cancer biology.

Like SV40 and polyomaviruses, the adenoviruses are lytic in cells of their natural host species, but can induce transformation in nonpermissive

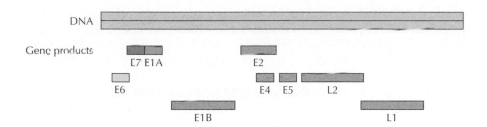

*Figure 15.15*
**The genome of a human papilloma-virus** Gene products are designated E (early) or L (late). Transformation results from the action of E6 and E7.

hosts. Transformation by the adenoviruses results from expression of two early genes, *E1A* and *E1B*, which are required for virus replication in permissive cells (Figure 15.16). These transforming proteins inactivate the Rb and p53 tumor suppressor proteins, with E1A binding to Rb and E1B binding to p53. It thus appears that SV40, papillomaviruses, and adenoviruses all induce transformation by a common pathway, in which altering regulation of the cell cycle by interfering with the activities of Rb and p53 plays a central role.

## Herpesviruses

The **herpesviruses** are among the most complex animal viruses, with genomes of 100 to 200 kb. Several herpesviruses induce tumors in animal species, including frogs, chickens, and monkeys. In addition, one member of the herpesvirus family, **Epstein-Barr virus**, is associated with the development of some human malignancies, including Burkitt's lymphoma in some regions of Africa, B-cell lymphomas in AIDS patients and other immunosuppressed individuals, and nasopharyngeal carcinoma in China.

In addition to its association with these human malignancies, Epstein-Barr virus is able to transform human B lymphocytes in culture. Partly because of the complexity of the genome, however, the molecular biology of Epstein-Barr virus replication and transformation remains to be fully understood. Several viral genes required to induce transformation of lymphocytes have been identified, but their functions have not been established.

## Retroviruses

Members of one family of RNA viruses, the **retroviruses**, cause cancer in a variety of animal species, including humans. One human retrovirus, human T-cell lymphotropic virus type I (HTLV-I), is the causative agent of adult T-cell leukemia, which is common in parts of Japan, the Caribbean, and Africa. A related virus (HTLV-II) may be associated with a rare form of leukemia called hairy T-cell leukemia. AIDS is caused by another retrovirus, HIV. In contrast to HTLV-I and HTLV-II, HIV does not cause cancer by directly converting a normal cell into a tumor cell. However, AIDS patients suffer a high incidence of some malignancies, particularly lymphomas and Kaposi's sarcoma. These cancers, which are also common among other immunosuppressed individuals, apparently develop as a secondary consequence of immunosuppression in AIDS patients.

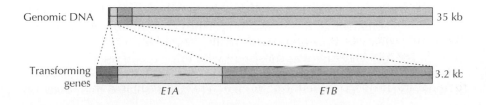

*Figure 15.16*
**The adenovirus genome** Two early-region genes, *E1A* and *E1B*, are responsible for induction of transformation.

*Figure 15.17*
**A typical retrovirus genome**   The DNA provirus, integrated into cellular DNA, is transcribed to yield genome-length RNA. This primary transcript serves as the genomic RNA for progeny virus particles, and as mRNA for the *gag* and *pol* genes. In addition, the full-length RNA is spliced to yield mRNA for *env*. The *gag* gene encodes the viral protease and structural proteins of the virus particle, *pol* encodes reverse transcriptase and integrase, and *env* encodes envelope glycoproteins.

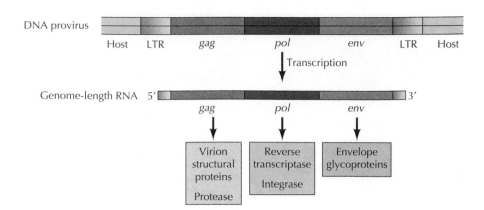

Different retroviruses differ substantially in their oncogenic potential. Most retroviruses contain only three genes (*gag*, *pol*, and *env*) that are required for virus replication but play no role in cell transformation (Figure 15.17). Retroviruses of this type induce tumors only rarely, if at all, as a consequence of mutations resulting from the integration of proviral DNA within or adjacent to cellular genes.

Other retroviruses, however, contain specific genes responsible for induction of cell transformation and are potent carcinogens. The prototype of these highly oncogenic retroviruses is **Rous sarcoma virus (RSV)**, first isolated from a chicken sarcoma by Peyton Rous in 1911. More than 50 years later, studies of RSV led to identification of the first viral oncogene, which has provided a model for understanding many aspects of tumor development at the molecular level.

## ONCOGENES

Cancer results from alterations in critical regulatory genes that control cell proliferation, differentiation, and survival. Studies of tumor viruses revealed that specific genes (called **oncogenes**) are capable of inducing cell transformation, thereby providing the first insights into the molecular basis of cancer. However, the majority (approximately 80%) of human cancers are not induced by viruses and apparently arise from other causes, such as radiation and chemical carcinogens. Therefore, in terms of our overall understanding of cancer, it has been critically important that studies of viral oncogenes also led to the identification of cellular oncogenes, which are involved in the development of non-virus-induced cancers. The key link between viral and cellular oncogenes was provided by studies of the highly oncogenic retroviruses.

*Figure 15.18*
**Cell transformation by RSV and ALV**   Both RSV and ALV infect and replicate in chicken embryo fibroblasts, but only RSV induces cell transformation.

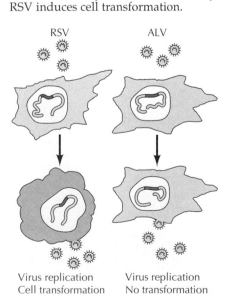

RSV          ALV

Virus replication     Virus replication
Cell transformation   No transformation

### Retroviral Oncogenes

Viral oncogenes were first defined in RSV, which transforms chicken embryo fibroblasts in culture and induces large sarcomas within 1 to 2 weeks after inoculation into chickens (Figure 15.18). In contrast, the closely related avian leukosis virus (ALV) replicates in the same cells as RSV without inducing transformation. This difference in transforming potential suggested the possibility that RSV contains specific genetic information responsible for transformation of infected cells. A direct comparison of the genomes of RSV and ALV was consistent with this hypothesis: The genomic RNA of RSV is about 10 kb, whereas that of ALV is smaller, about 8.5 kb.

In the early 1970s, Peter Vogt and Steven Martin isolated both deletion

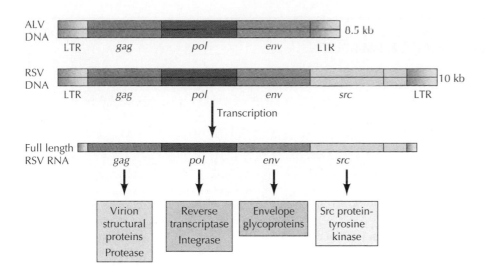

mutants and temperature-sensitive mutants of RSV that were unable to induce transformation. Importantly, these mutants still replicated normally in infected cells, indicating that RSV contains genetic information that is required for transformation but not for virus replication. Further analysis demonstrated that both the deletion and temperature-sensitive RSV mutants define a single gene responsible for the ability of RSV to induce tumors in birds and transform fibroblasts in culture. Because RSV causes sarcomas, its oncogene was called *src*. The *src* gene is an addition to the genome of RSV; it is not present in ALV (Figure 15.19). It encodes a 60-kd protein that was the first protein-tyrosine kinase to be identified (see the Key Experiment in Chapter 13).

More than 40 different highly oncogenic retroviruses have been isolated from a variety of animals, including chickens, turkeys, mice, rats, cats, and monkeys. All of these viruses, like RSV, contain at least one oncogene (in some cases two) that is not required for virus replication but is responsible for cell transformation. In some cases, different viruses contain the same oncogenes, but more than two dozen distinct oncogenes have been identified among this group of viruses (Table 15.3). Like *src*, many of these genes (such as *ras* and *raf*) encode proteins that are now recognized as key components of signaling pathways that stimulate cell proliferation (see Figure 13.32).

## Proto-Oncogenes

An unexpected feature of retroviral oncogenes is their lack of involvement in virus replication. Since most viruses are streamlined to replicate as efficiently as possible, the existence of viral oncogenes that are not an integral part of the virus life cycle

*Table 15.3* Retroviral Oncogenes

| Oncogene | Virus | Species |
|----------|-------|---------|
| *abl* | Abelson leukemia | Mouse |
| *akt* | AKT8 virus | Mouse |
| *cbl* | Cas NS-1 | Mouse |
| *crk* | CT10 sarcoma | Chicken |
| *erbA* | Avian erythroblastosis-ES4 | Chicken |
| *erbB* | Avian erythroblastosis-ES4 | Chicken |
| *ets* | Avian erythroblastosis-E26 | Chicken |
| *fes* | Gardner-Arnstein feline sarcoma | Cat |
| *fgr* | Gardner-Rasheed feline sarcoma | Cat |
| *fms* | McDonough feline sarcoma | Cat |
| *fos* | FBJ murine osteogenic sarcoma | Mouse |
| *fps* | Fujinami sarcoma | Chicken |
| *jun* | Avian sarcoma-17 | Chicken |
| *kit* | Hardy-Zuckerman feline sarcoma | Cat |
| *maf* | Avian sarcoma AS42 | Chicken |
| *mos* | Moloney sarcoma | Mouse |
| *mpl* | Myeloproliferative leukemia | Mouse |
| *myb* | Avian myeloblastosis | Chicken |
| *myc* | Avian myelocytomatosis | Chicken |
| *qin* | Avian sarcoma 31 | Chicken |
| *raf* | 3611 murine sarcoma | Mouse |
| *rasH* | Harvey sarcoma | Rat |
| *rasK* | Kirsten sarcoma | Rat |
| *rel* | Reticuloendotheliosis | Turkey |
| *ros* | UR2 sarcoma | Chicken |
| *sea* | Avian erythroblastosis-S13 | Chicken |
| *sis* | Simian sarcoma | Monkey |
| *ski* | Avian SK | Chicken |
| *src* | Rous sarcoma | Chicken |
| *yes* | Y73 sarcoma | Chicken |

## KEY EXPERIMENT

# The Discovery of Proto-Oncogenes

**DNA Related to the Transforming Gene(s) of Avian Sarcoma Viruses Is Present in Normal Avian DNA**

Dominique Stehelin, Harold E. Varmus, J. Michael Bishop and Peter K. Vogt

Department of Microbiology, University of California, San Francisco (DS, HEV, and JMB) and Department of Microbiology, University of California, Los Angeles (PKV)

*Nature,* Volume 260, 1976, pages 170–173

## The Context

Genetic analysis of RSV defined the first viral oncogene (*src*) as a gene that was specifically responsible for cell transformation but was not required for virus replication. The origin of highly oncogenic retroviruses from tumors of infected animals then suggested the hypothesis that retroviral oncogenes are derived from related genes of host cells. Consistent with this suggestion, normal cells of several species were found to contain retrovirus-related DNA sequences that could be detected by nucleic acid hybridization. However, it was unclear whether these sequences were related to the retroviral oncogenes or to the genes required for virus replication.

Harold Varmus, J. Michael Bishop, and their colleagues resolved this critical issue by exploiting the genetic characterization of the *src* oncogene. In particular, Peter Vogt had previously isolated transformation-defective mutants of RSV that had sustained

deletions of approximately 1.5 kb, corresponding to most or all of the *src* gene. Stehelin and collaborators used these mutants to prepare a cDNA probe that specifically represented *src* sequences. The use of this defined probe in nucleic acid hybridization experiments allowed them to definitively demonstrate that normal cells contain *src*-related DNA sequences.

## The Experiments

First reverse transcriptase was used to synthesize a radioactive cDNA probe composed of short single-stranded DNA fragments complementary to the entire genomic RNA of RSV. This probe was then hybridized to an excess of RNA isolated from a transformation-defective deletion mutant. Fragments of cDNA that were complementary to the viral replication genes hybridized to the transformation-defective RSV RNA, forming RNA-DNA duplexes. In contrast, cDNA fragments that were complementary to *src* were unable to hybridize and

remained single-stranded. This single-stranded DNA was then isolated to provide a specific probe for *src* oncogene sequences. As predicted from the size of deletions in the transformation-defective RSV mutants, the *src*-specific probe was homologous to about 1.5 kb of RSV RNA.

The radioactive *src* cDNA was then used as a hybridization probe to attempt to detect related DNA sequences in normal avian cells. Strikingly, the *src* cDNA hybridized extensively to normal chicken DNA, as well as to DNA of other avian species (see figure). These experi-

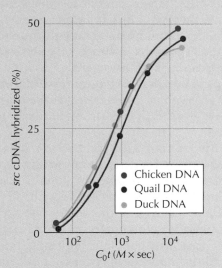

Hybridization of *src*-specific cDNA to normal chicken, quail, and duck DNA.

seems paradoxical. Scientists were thus led to question where the retroviral oncogenes had originated and how they had become incorporated into viral genomes—a line of investigation that ultimately led to the identification of cellular oncogenes in human cancers.

The first clue to the origin of oncogenes came from the way in which the highly oncogenic retroviruses were isolated. The isolation of Abelson leukemia virus is a typical example (Figure 15.20). More than 150 mice were inoculated with a nontransforming virus containing only the *gag, pol,*

## The Discovery of Proto–Oncogenes (continued)

ments thus demonstrated that normal cells contain DNA sequences that are closely related to the *src* oncogene, supporting the hypothesis that retroviral oncogenes originated from cellular genes that became incorporated into viral genomes.

### The Impact

Stehelin and colleagues concluded their 1976 paper by raising the possibility that the cellular *src* sequences are "involved in the normal regulation of cell growth and development or the transformation of cell behavior by physical, chemical or viral agents." This prediction has been strikingly substantiated, and the discovery of cellular *src* sequences has opened new doors to understanding both the regu-

lation of normal cell proliferation and the molecular basis of human cancer. Studies of oncogene and proto-oncogene proteins, including Src itself, have proven critical in unraveling the signaling pathways that control the proliferation and differentiation of normal cells. The discovery of the *src* proto-oncogene further suggested that non-virus-induced tumors could arise as a result of mutations in related cellular genes, leading directly to the discovery of oncogenes in human tumors. By unifying studies of tumor viruses, normal cells, and non-virus-induced tumors, the results of Varmus, Bishop, and their colleagues have had an impact on virtually all aspects of cell regulation and cancer research.

J. Michael Bishop

Harold Varmus

and *env* genes required for virus replication. One of these mice developed a lymphoma from which a new, highly oncogenic virus (Abelson leukemia virus), which now contained an oncogene (*abl*), was isolated. The scenario suggested the hypothesis that the retroviral oncogenes are derived from genes of the host cell, and that occasionally such a host cell gene becomes incorporated into a viral genome, yielding a new, highly oncogenic virus as the product of a virus-host recombination event.

The critical prediction of this hypothesis was that normal cells contain genes that are closely related to the retroviral oncogenes. This was definitively demonstrated in 1976 by Harold Varmus, J. Michael Bishop, and their colleagues, who showed that a cDNA probe for the *src* oncogene of RSV hybridized to closely related sequences in the DNA of normal chicken cells. Moreover, *src*-related sequences were also found in normal DNAs of a wide range of other vertebrates (including humans) and thus appeared to be highly conserved in evolution. Similar experiments with probes for the oncogenes of other highly oncogenic retroviruses have yielded comparable results, and it is now firmly established that the retroviral oncogenes were derived from closely related genes of normal cells.

The normal-cell genes from which the retroviral oncogenes originated are called **proto-oncogenes**. They are important cell regulatory genes, in many cases encoding proteins that function in the signal transduction path-

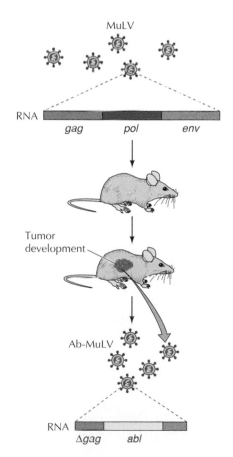

*Figure 15.20*
**Isolation of Abelson leukemia virus** The highly oncogenic virus Ab-MuLV was isolated from a rare tumor that developed in a mouse that had been inoculated with a nontransforming virus (Moloney murine leukemia virus, or MuLV). MuLV contains only the *gag*, *pol*, and *env* genes required for virus replication. In contrast, Ab-MuLV has acquired a new oncogene (*abl*), which is responsible for its transforming activity. The *abl* oncogene replaced some of the viral replicative genes and is fused with a partially deleted *gag* gene (designated Δ*gag*) in the Ab-MuLV genome.

**Raf proto-oncogene protein**

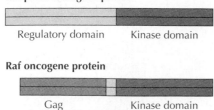

Regulatory domain    Kinase domain

**Raf oncogene protein**

Gag    Kinase domain

*Figure 15.21*
**The Raf oncogene protein** The Raf proto-oncogene protein consists of an amino-terminal regulatory domain and a carboxy-terminal protein kinase domain. In the viral Raf oncogene protein, the regulatory domain has been deleted and replaced by viral Gag sequences. As a result, the Raf kinase domain is constitutively active, causing cell transformation.

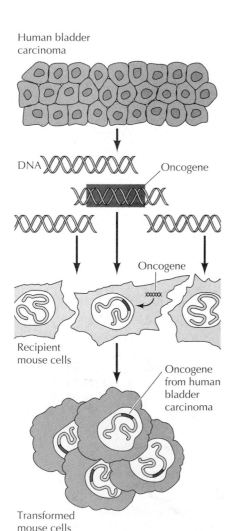

Human bladder carcinoma

DNA

Oncogene

Oncogene

Recipient mouse cells

Oncogene from human bladder carcinoma

Transformed mouse cells

ways controlling normal cell proliferation (e.g., *src*, *ras*, and *raf*). The oncogenes are abnormally expressed or mutated forms of the corresponding proto-oncogenes. As a consequence of such alterations, the oncogenes induce abnormal cell proliferation and tumor development.

An oncogene incorporated into a retroviral genome differs in several respects from the corresponding proto-oncogene. First, the viral oncogene is transcribed under the control of viral promoter and enhancer sequences, rather than being controlled by the normal transcriptional regulatory sequences of the proto-oncogene. Consequently, oncogenes are usually expressed at much higher levels than the proto-oncogenes and are sometimes transcribed in inappropriate cell types. In some cases, such abnormalities of gene expression are sufficient to convert a normally functioning proto-oncogene into an oncogene that drives cell transformation.

In addition to such alterations in gene expression, oncogenes frequently encode proteins that differ in structure and function from those encoded by their normal homologs. Many oncogenes, such as *raf*, are expressed as fusion proteins with viral sequences at the amino terminus (Figure 15.21). Recombination events leading to the generation of such fusion proteins often occur during the capture of proto-oncogenes by retroviruses, and sequences from both the amino and carboxy termini of proto-oncogenes are frequently deleted during the process. Such deletions may result in the loss of regulatory domains that control the activity of the proto-oncogene proteins, thereby generating oncogene proteins that function in an unregulated manner. For example, the viral *raf* oncogene encodes a fusion protein in which amino-terminal sequences of the normal Raf protein have been deleted. These amino-terminal sequences are critical to the normal regulation of Raf protein kinase activity, and their deletion results in unregulated constitutive activity of the oncogene-encoded Raf protein. This unregulated Raf activity drives cell proliferation, resulting in transformation.

Many other oncogenes differ from the corresponding proto-oncogenes by point mutations, resulting in single amino acid substitutions in the oncogene products. In some cases, such amino acid substitutions (like the deletions already discussed) lead to unregulated activity of the oncogene proteins. An important example of such point mutations is provided by the *ras* oncogenes, which are discussed in the next section in terms of their role in human cancers.

### Oncogenes in Human Cancer

Understanding the origin of retroviral oncogenes raised the question as to whether non-virus-induced tumors contain cellular oncogenes that were generated from proto-oncogenes by mutations or DNA rearrangements during tumor development. Direct evidence for the involvement of cellular oncogenes in human tumors was first obtained by gene transfer experiments in the laboratories of Robert Weinberg and of the author in 1981. DNA of a human bladder carcinoma was found to efficiently induce transformation of recipient mouse cells in culture, indicating that the human tumor contained a biologically active cellular oncogene (Figure 15.22). Both gene transfer assays and alternative experimental approaches have since

*Figure 15.22*
**Detection of a human tumor oncogene by gene transfer** DNA extracted from a human bladder carcinoma induced transformation of recipient mouse cells in culture. Transformation resulted from integration and expression of an oncogene derived from the human tumor.

*Table 15.4*   Representative Oncogenes of Human Tumors

| Oncogene | Type of cancer | Activation mechanism |
|---|---|---|
| *abl* | Chronic myelogenous leukemia, acute lymphocytic leukemia | Translocation |
| *bcl 2* | Follicular B-cell lymphoma | Translocation |
| *F2A/pbx1* | Acute lymphocytic leukemia | Translocation |
| *erbB-2* | Breast and ovarian carcinomas | Amplification |
| *gip* | Adrenal cortical and ovarian carcinomas | Point mutation |
| *gli* | Glioblastoma | Amplification |
| *gsp* | Pituitary and thyroid tumors | Point mutation |
| *hox-11* | Acute T-cell leukemia | Translocation |
| *lyl* | Acute T-cell leukemia | Translocation |
| *c-myc* | Burkitt's lymphoma | Translocation |
| *c-myc* | Breast and lung carcinomas | Amplification |
| *L-myc* | Lung carcinoma | Amplification |
| *N-myc* | Neuroblastoma, lung carcinoma | Amplification |
| *PML/RARα* | Acute promyelocytic leukemia | Translocation |
| *PRAD1* | Parathyroid adenoma | Translocation |
| *PRAD1* | Breast carcinoma | Amplification |
| *ras*H | Thyroid carcinoma | Point mutation |
| *ras*K | Colon, lung, pancreatic, and thyroid carcinomas | Point mutation |
| *ras*N | Acute myelogenous and lymphocytic leukemias, thyroid carcinoma | Point mutation |
| *ret* | Thyroid carcinoma | DNA rearrangement |

led to the detection of active cellular oncogenes in human tumors of many different types (Table 15.4).

Some of the oncogenes identified in human tumors are cellular homologs of oncogenes that were previously characterized in retroviruses, whereas others are new oncogenes first discovered in human cancers. The first human oncogene identified in gene transfer assays was subsequently identified as the human homolog of the *ras*H oncogene of Harvey sarcoma virus (see Table 15.3). Three closely related members of the **ras** gene family (*ras*H, *ras*K, and *ras*N) are the oncogenes most frequently encountered in human tumors. These genes are involved in approximately 15% of all human malignancies, including about 50% of colon and 25% of lung carcinomas.

The *ras* oncogenes are not present in normal cells; rather, they are generated in tumor cells as a consequence of mutations that occur during tumor development. The *ras* oncogenes differ from their proto-oncogenes by point mutations resulting in single amino acid substitutions at critical positions. The first such mutation discovered was the substitution of valine for glycine at position 12 (Figure 15.23). Other amino acid substitutions at position 12, as well as at positions 13 and 61, are also frequently encountered in

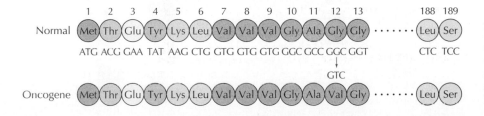

*Figure 15.23*
**Point mutations in *ras* oncogenes**   A single nucleotide change, which alters codon 12 from GGC (Gly) to GTC (Val) is responsible for the transforming activity of the *ras*H oncogene detected in bladder carcinoma DNA.

*ras* oncogenes in human tumors. In animal models, it has been shown that mutations that convert *ras* proto-oncogenes to oncogenes are caused by chemical carcinogens, providing a direct link between the mutagenic action of carcinogens and cell transformation.

As discussed in Chapter 13, the *ras* genes encode guanine-nucleotide binding proteins that function in transduction of mitogenic signals from a variety of growth factor receptors. The activity of the Ras proteins is controlled by GTP or GDP binding, such that they alternate between active (GTP-bound) and inactive (GDP-bound) states (see Figure 13.33). The mutations characteristic of *ras* oncogenes have the effect of maintaining the Ras proteins constitutively in the active GTP-bound conformation. In large part, this effect is a result of nullifying the response of oncogenic Ras proteins to GAP (GTPase-activating protein), which stimulates hydrolysis of bound GTP by normal Ras. Because of the resulting decrease in their intracellular GTPase activity, the oncogenic Ras proteins remain in the active GTP-bound state and drive unregulated cell proliferation.

Point mutations are only one of the ways in which proto-oncogenes are converted to oncogenes in human tumors. Many cancer cells display abnormalities in chromosome structure, including translocations, duplications, and deletions. The gene rearrangements resulting from chromosome translocations frequently lead to the generation of oncogenes. In some cases, analysis of these rearrangements has implicated already known oncogenes in tumor development. In other cases, novel oncogenes have been discovered by molecular cloning and analysis of rearranged DNA sequences.

The first characterized example of oncogene activation by chromosome translocation was the involvement of the **c-myc** oncogene in human Burkitt's lymphomas and mouse plasmacytomas, which are malignancies of antibody-producing B lymphocytes (Figure 15.24). Both of these tumors are

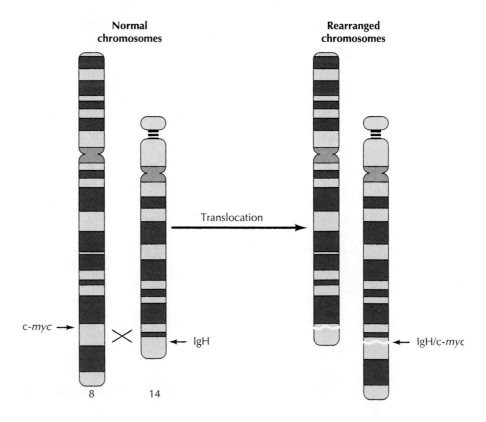

*Figure 15.24*
**Translocation of c-*myc***   The c-*myc* proto-oncogene is translocated from chromosome 8 to the immunoglobulin heavy-chain locus (IgH) on chromosome 14 in Burkitt's lymphomas, resulting in abnormal c-*myc* expression.

characterized by chromosome translocations involving the genes that encode immunoglobulins. For example, virtually all Burkitt's lymphomas have translocations of a fragment of chromosome 8 to one of the immunoglobulin gene loci, which reside on chromosomes 2 ($\kappa$ light chain), 14 (heavy chain), and 22 ($\lambda$ light chain). The fact that the immunoglobulin genes are actively expressed in these tumors suggested that the translocations activate a proto-oncogene from chromosome 8 by inserting it into the immunoglobulin loci. This possibility was investigated by analysis of tumor DNAs with probes for known oncogenes, leading to the finding that the c-*myc* proto-oncogene was the chromosome 8 translocation break point in Burkitt's lymphomas. These translocations inserted c-*myc* into an immunoglobulin locus, where it was expressed in an unregulated manner. Such uncontrolled expression of the c-*myc* gene, which encodes a transcription factor normally induced in response to growth factor stimulation, is sufficient to drive cell proliferation and contribute to tumor development.

Translocations of other proto-oncogenes frequently result in rearrangements of coding sequences, leading to the formation of abnormal gene products. The prototype is translocation of the ***abl*** proto-oncogene from chromosome 9 to chromosome 22 in chronic myelogenous leukemia (Figure 15.25). This translocation leads to fusion of *abl* with its translocation partner, a gene called *bcr*, on chromosome 22. The result is production of a Bcr/Abl fusion protein in which the normal amino terminus of the Abl proto-oncogene protein has been replaced by Bcr amino acid sequences. The fusion of Bcr sequences results in aberrant activity and altered subcellular localization of the Abl protein-tyrosine kinase, leading to cell transformation.

A distinct mechanism by which oncogenes are activated in human tumors is gene amplification, which results in elevated gene expression. Gene amplification (see Figure 5.54) is common in tumor cells, occurring more than a thousand times more frequently than in normal cells, and amplification of oncogenes may play a role in the progression of many tumors to more rapid growth and increasing malignancy. Indeed, novel oncogenes have been identified by molecular cloning and characterization of DNA sequences that are amplified in tumors.

A prominent example of oncogene amplification is the involvement of

*Figure 15.25*
**Translocation of *abl*** The *abl* oncogene is translocated from chromosome 9 to chromosome 22, forming the Philadelphia chromosome in chronic myelogenous leukemias. The *abl* proto-oncogene, which contains two alternative first exons (1A and 1B), is joined to the middle of the *bcr* gene on chromosome 22. Exon 1B is deleted as a result of the translocation. Transcription of the fused gene initiates at the *bcr* promoter and continues through *abl*. Splicing then generates a fused Bcr/Abl mRNA, in which *abl* exon 1A sequences are also deleted and *bcr* sequences are joined to *abl* exon 2. The Bcr/Abl mRNA is translated to yield a recombinant Bcr/Abl fusion protein.

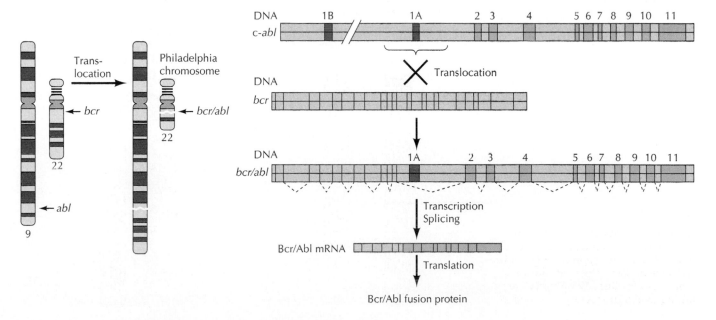

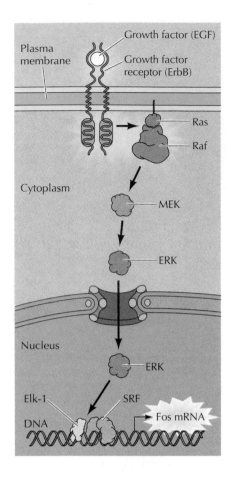

*Figure 15.27*
**Mechanism of *erb*B oncogene activation** The *erb*B proto-oncogene encodes the receptor for epidermal growth factor (EGF). Protein-tyrosine kinase activity of the proto-oncogene protein is controlled by EGF binding. In the ErbB oncogene protein, deletion of the amino-terminal ligand-binding domain results in constitutive kinase activity.

*Figure 15.26*
**Oncogenes and signal transduction** Oncogene proteins act as growth factors (e.g., EGF), growth factor receptors (e.g., ErbB), and intracellular signaling molecules (Ras and Raf). Ras and Raf activate the ERK MAP kinase pathway (see Figures 13.32 and 13.35), leading to the induction of additional genes (e.g., *fos*) that encode potentially oncogenic transcriptional regulatory proteins. Proteins with known oncogenic potential are highlighted with a yellow glow.

the **N-*myc*** gene, which is related to c-*myc*, in neuroblastoma (a childhood tumor of embryonal neuronal cells). Amplified copies of N-*myc* are frequently present in rapidly growing aggressive tumors, indicating that N-*myc* amplification is associated with the progression of neuroblastomas to increasing malignancy. Amplification of another oncogene, ***erb*B-2**, which encodes a receptor protein-tyrosine kinase, is similarly related to progression of breast and ovarian carcinomas.

### Functions of Oncogene Products

The viral and cellular oncogenes have defined a large group of genes (more than 70 in total) that can contribute to the abnormal behavior of malignant cells. As already noted, many of the proteins encoded by proto-oncogenes regulate normal cell proliferation; in these cases, the elevated expression or activity of the corresponding oncogene proteins drives the uncontrolled proliferation of cancer cells. In addition, some oncogene products contribute to other aspects of the behavior of cancer cells, such as defective differentiation and failure to undergo programmed cell death.

The majority of oncogene proteins function as elements of the signaling pathways that regulate cell proliferation in response to growth factor stimulation. These oncogene proteins include polypeptide growth factors, growth factor receptors, elements of intracellular signaling pathways, and transcription factors (Figure 15.26).

The action of growth factors as oncogene proteins results from their abnormal expression, leading to a situation in which a tumor cell produces a growth factor to which it also responds. The result is autocrine stimulation of the growth factor-producing cell (see Figure 15.9), which drives abnormal cell proliferation and contributes to the development of a wide variety of human tumors.

A large group of oncogenes encode growth factor receptors, most of which are protein-tyrosine kinases. These receptors are frequently converted to oncogene proteins by deletion of their amino-terminal domains, which would normally bind extracellular growth factors (Figure 15.27). Such deletions result in constitutive activity of the intracellular kinase

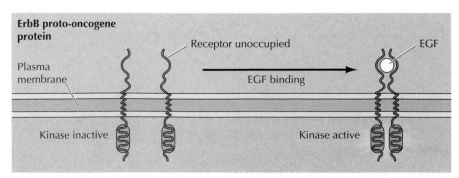

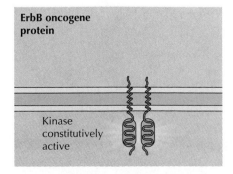

domain, leading to unregulated production of a proliferative signal from the oncogene protein. Alternatively, genes that encode some receptor protein-tyrosine kinases, such as *erb*B-2, are activated by gene amplification. Other oncogenes (including *src* and *abl*) encode nonreceptor protein-tyrosine kinases that are constitutively activated by deletions or mutations of regulatory sequences.

The Ras proteins play a key role in mitogenic signaling by coupling growth factor receptors to activation of the Raf protein-serine/threonine kinase, which initiates a protein kinase cascade leading to activation of the ERK MAP kinase (see Figure 13.32). As discussed earlier, the mutations that convert *ras* proto-oncogenes to oncogenes result in constitutive Ras activity, which leads to activation of the MAP kinase pathway. The *raf* gene can similarly be converted to an oncogene by deletions that result in loss of the amino-terminal regulatory domain of the Raf protein (see Figure 15.21). These deletions result in unregulated activity of the Raf protein kinase, which also leads to constitutive MAP kinase activation.

The MAP kinase pathway ultimately leads to the phosphorylation of transcription factors and alterations in gene expression. As might therefore be expected, many oncogenes encode transcriptional regulatory proteins that are normally induced in response to growth factor stimulation. For example, transcription of the *fos* proto-oncogene is induced as a result of phosphorylation of Elk-1 by the ERK MAP kinase (see Figure 15.26). **Fos** and the product of another proto-oncogene, **Jun**, are components of the AP-1 transcription factor, which activates transcription of a number of target genes in growth factor-stimulated cells (Figure 15.28). Constitutive activity of AP-1, resulting from unregulated expression of either the Fos or Jun oncogene proteins, is sufficient to drive abnormal cell proliferation, leading to cell transformation. The Myc proteins similarly function as transcription factors regulated by mitogenic stimuli, and abnormal expression of *myc* oncogenes contributes to the development of a variety of human tumors. Other transcription factors are frequently activated as oncogenes by chromosome translocations in human leukemias and lymphomas.

G protein-coupled receptors and G proteins also act as oncogenes in some human tumors (Figure 15.29). For example, mutations of the gene encoding the thyrotropin receptor convert it to an oncogene in thyroid tumors. Thyrotropin is a pituitary hormone that stimulates proliferation of thyroid cells through a G protein-coupled receptor that activates adenylyl cyclase. Mutations of the thyrotropin receptor in thyroid tumors result in constitutive activity of the receptor, which then drives cell proliferation via activation of the cAMP signaling pathway. Likewise, the genes encoding G

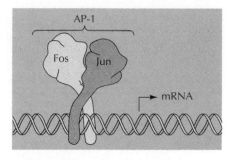

*Figure 15.28*
**The AP-1 transcription factor** Fos and Jun dimerize to form AP-1, which activates transcription of a variety of growth factor-inducible genes.

*Figure 15.29*
**Oncogenic activity of G protein-coupled receptors and G proteins** The thyrotropin receptor is coupled to adenylyl cyclase by $G_s$. The genes encoding both the receptor and the $G_s$ $\alpha$ subunit ($G_s\alpha$) can act as oncogenes by stimulating thyroid cell proliferation.

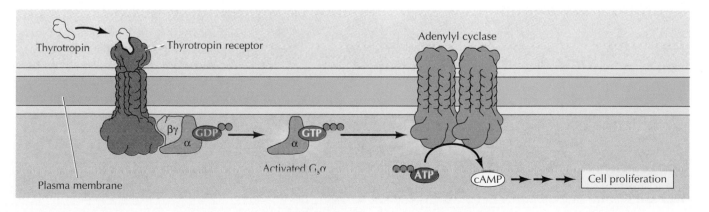

proteins can act as oncogenes in some cell types. The *gsp* oncogene, which encodes the $\alpha$ subunit of $G_s$, is generated by point mutations similar to those found in *ras*. These mutations result in constitutive activity of the $G_s$ $\alpha$ subunit, leading to unregulated stimulation of adenylyl cyclase. As might be expected, the *gsp* oncogene is involved in thyroid and pituitary tumors, where cAMP stimulates cell proliferation. Similar mutations convert the genes encoding other G protein $\alpha$ subunits to oncogenes in other cell types, including adrenal and ovarian tumors.

The intracellular signaling pathways activated by growth factor stimulation ultimately regulate components of the cell cycle machinery that promote progression through the restriction point in $G_1$. The D-type cyclins are induced in response to growth factor stimulation and play a key role in coupling growth factor signaling to cell cycle progression (see Figure 14.18). Perhaps not surprisingly, the gene encoding cyclin D1 is a proto-oncogene, which can be activated as an oncogene (called **PRAD1**) by chromosome translocation or gene amplification. These alterations lead to constitutive expression of cyclin D1, which then drives cell proliferation in the absence of normal growth factor stimulation.

Although most oncogenes stimulate cell proliferation, the oncogenic activity of some transcription factors instead results from inhibition of cell differentiation. As noted in Chapter 13, thyroid hormone and retinoic acid induce differentiation of a variety of cell types. These hormones diffuse through the plasma membrane and bind to intracellular receptors that act as transcriptional regulatory molecules. Mutated forms of both the thyroid hormone receptor (**ErbA**) and the retinoic acid receptor (**PML/RAR$\alpha$**) act as oncogene proteins in chicken erythroleukemia and human acute promyelocytic leukemia, respectively. In both cases, the mutated oncogene receptors appear to interfere with the action of their normal homologs, thereby blocking cell differentiation and maintaining the leukemic cells in an actively proliferating state (Figure 15.30). In the case of acute promyelocytic leukemia, high doses of retinoic acid can overcome the effect of the PML/RAR$\alpha$ oncogene protein and induce differentiation of the leukemic cells. This biological observation has a direct clinical correlate: Patients with acute promyelocytic leukemia can be treated effectively by administration of retinoic acid, which induces differentiation and blocks continued cell proliferation.

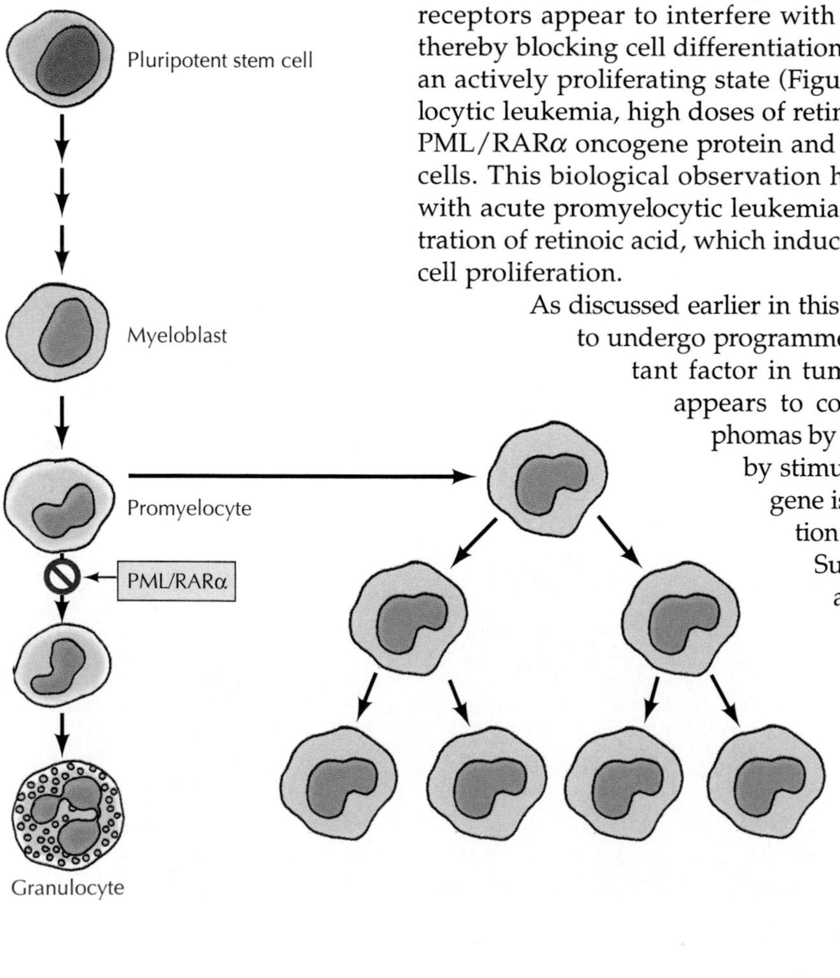

As discussed earlier in this chapter, the failure of some cancer cells to undergo programmed cell death, or apoptosis, is an important factor in tumor development. The *bcl*-2 oncogene appears to contribute to the development of lymphomas by protecting against apoptosis, rather than by stimulating cell proliferation. The *bcl*-2 oncogene is generated by a chromosome translocation that results in elevated gene expression. Such abnormal expression of Bcl-2 blocks apoptosis and maintains cell survival under conditions that normally induce cell death, such as growth factor deprivation. Although the mechanism of action of the Bcl-2 protein is unknown, the role of *bcl*-2 as an oncogene clearly illustrates the significance of programmed cell death in the development of malignancy.

*Figure 15.30*
**Action of the PML/RAR$\alpha$ oncogene protein** The PML/RAR$\alpha$ fusion protein blocks the differentiation of promyelocytes to granulocytes.

## TUMOR SUPPRESSOR GENES

The activation of cellular oncogenes represents only one of two distinct types of genetic alterations involved in tumor development; the other is inactivation of tumor suppressor genes. Oncogenes drive abnormal cell proliferation as a consequence of genetic alterations that either increase gene expression or lead to uncontrolled activity of the oncogene-encoded proteins. **Tumor suppressor genes** represent the opposite side of cell growth control, normally acting to inhibit cell proliferation and tumor development. In many tumors, these genes are lost or inactivated, thereby removing negative regulators of cell proliferation and contributing to the abnormal proliferation of tumor cells.

### Identification of Tumor Suppressor Genes

The first insight into the activity of tumor suppressor genes came from somatic cell hybridization experiments, initiated by Henry Harris and his colleagues in 1969. The fusion of normal cells with tumor cells yielded hybrid cells containing chromosomes from both parents (Figure 15.31). In most cases, such hybrid cells were not capable of forming tumors in animals. Therefore, it appeared that genes derived from the normal cell parent acted to inhibit (or suppress) tumor development. Definition of these genes at the molecular level came, however, from a different approach—the analysis of rare inherited forms of human cancer.

The first tumor suppressor gene was identified by studies of retinoblastoma, a rare childhood eye tumor. Provided that the disease is detected early, retinoblastoma can be successfully treated, and many patients survive to have families. Consequently, it was recognized that some cases of retinoblastoma are inherited. In these cases, approximately 50% of the children of an affected parent develop retinoblastoma, consistent with Mendelian transmission of a single dominant gene that confers susceptibility to tumor development (Figure 15.32).

Although susceptibility to retinoblastoma is transmitted as a dominant trait, inheritance of the susceptibility gene is not sufficient to transform a normal retinal cell into a tumor cell. All retinal cells in a patient inherit the susceptibility gene, but only a small fraction of these cells give rise to

*Figure 15.31*
**Suppression of tumorigenicity by cell fusion**  Fusion of tumor cells with normal cells yields hybrids that contain chromosomes from both parents. Such hybrids are usually nontumorigenic.

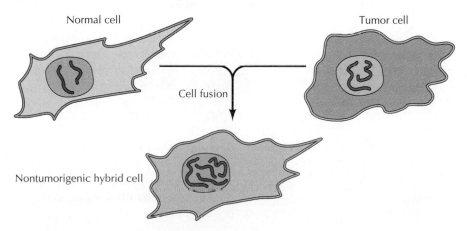

*Figure 15.32*
**Inheritance of retinoblastoma**  Susceptibility to retinoblastoma is transmitted to approximately 50% of offspring. Affected and normal individuals are indicated by purple and green symbols, respectively.

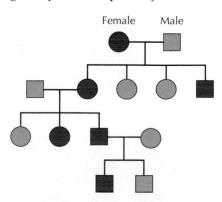

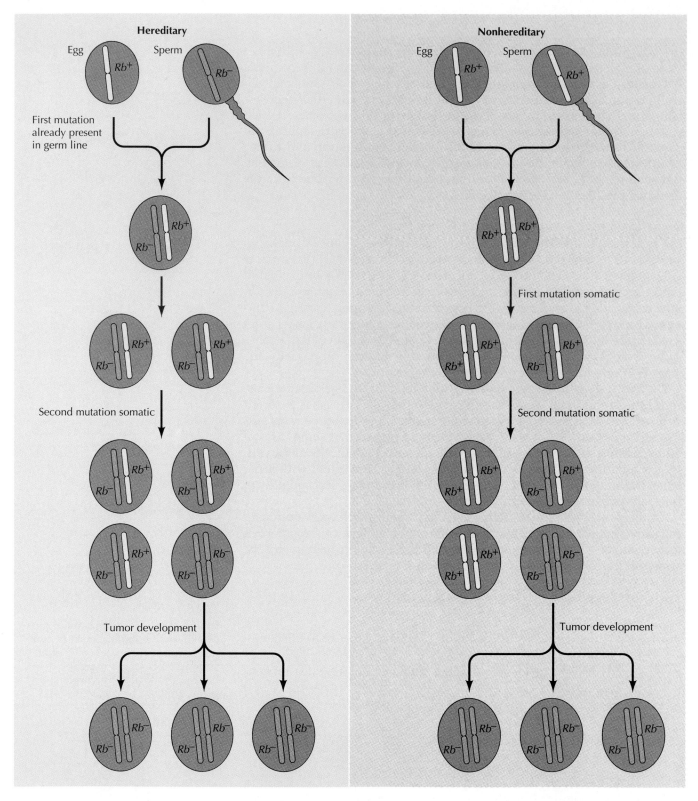

*Figure 15.33*
**Mutations of *Rb* during retinoblastoma development**   In hereditary retinoblastoma, a defective copy of the *Rb* gene (*Rb⁻*) is inherited from the affected parent. A second somatic mutation, which inactivates the single normal *Rb⁺* copy in a retinal cell, then leads to the development of retinoblastoma. In nonhereditary cases, two normal *Rb⁺* genes are inherited, and retinoblastoma develops only if two somatic mutations inactivate both copies of *Rb* in the same cell.

tumors. Thus, tumor development requires additional events beyond inheritance of tumor susceptibility. In 1971, Alfred Knudson proposed that the development of retinoblastoma requires two mutations, which are now known to correspond to the loss of both of the functional copies of the tumor susceptibility gene (the **Rb** tumor suppressor gene) that would be present on homologous chromosomes of a normal diploid cell (Figure 15.33). In inherited retinoblastoma, one defective copy of *Rb* is genetically transmitted. The loss of this single *Rb* copy is not by itself sufficient to trigger tumor development, but retinoblastoma almost always develops in these individuals as a result of a second somatic mutation leading to the loss of the remaining normal *Rb* allele. Noninherited retinoblastoma, in contrast, is rare, since its development requires two independent somatic mutations to inactivate both normal copies of *Rb* in the same cell.

The functional nature of the *Rb* gene as a negative regulator of tumorigenesis was initially indicated by observations of chromosome morphology. Visible deletions of chromosome 13q14 were found in some retinoblastomas, suggesting that loss (rather than activation) of the *Rb* gene led to tumor development (Figure 15.34). Gene-mapping studies further indicated that tumor development resulted from loss of normal *Rb* alleles in the tumor cells, consistent with the function of *Rb* as a tumor suppressor gene. Isolation of the *Rb* gene as a molecular clone in 1986 then firmly established that *Rb* is consistently lost or mutated in retinoblastomas. Gene transfer experiments also demonstrated that introduction of a normal *Rb* gene into retinoblastoma cells reverses their tumorigenicity, providing direct evidence for the activity of *Rb* as a tumor suppressor.

Although *Rb* was identified in a rare childhood cancer, it is also involved in some of the more common tumors of adults. In particular, studies of the cloned gene have established that *Rb* is lost or inactivated in many bladder, breast, and lung carcinomas. The significance of the *Rb* tumor suppressor gene thus extends beyond retinoblastoma, apparently contributing to development of a substantial fraction of more common human cancers. In addition, as noted earlier in this chapter, the Rb protein is a key target for the oncogene proteins of several DNA tumor viruses, including SV40, adenoviruses, and human papillomaviruses, which bind to Rb and inhibit its activity (Figure 15.35). Transformation by these viruses thus results, at least in part, from inactivation of Rb at the protein level rather than from mutational inactivation of the *Rb* gene.

Characterization of *Rb* as a tumor suppressor gene served as the prototype for the identification of additional tumor suppressor genes that contribute to the development of many different human cancers (Table 15.5). Some of these genes were identified as the causes of rare inherited cancers, playing a role similar to that of *Rb* in hereditary retinoblastoma. Other tumor suppressor genes have been identified as genes that are frequently deleted or mutated in common noninherited cancers of adults, such as colon carcinoma. In either case, it appears that most tumor suppressor genes are involved in the development of both inherited and noninherited forms of cancer. Indeed, mutations of some tumor suppressor genes appear to be the most common molecular alterations leading to human tumor development.

The second tumor suppressor gene to have been identified is *p53*, which is frequently inactivated in a wide variety of human cancers, including leukemias, lymphomas, sarcomas, brain tumors, and carcinomas of many tissues, including breast, colon, and lung. In total, mutations of *p53* may play a role in up to 50% of all cancers, making it the most common target of genetic alterations in human malignancies. It is also of interest that inherited mutations of *p53* are responsible for genetic transmission

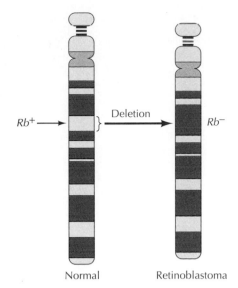

**Figure 15.34**
***Rb* deletions in retinoblastoma**
Many retinoblastomas have deletions of the chromosomal locus (13q14) that contains the *Rb* gene.

**Figure 15.35**
**Interaction of Rb with oncogene proteins of DNA tumor viruses**  The oncogene proteins of several DNA tumor viruses (e.g., SV40 T antigen) induce transformation by binding to and inactivating Rb protein.

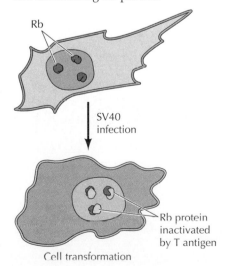

*Table 15.5* Tumor Suppressor Genes

| Gene | Type of cancer |
| --- | --- |
| APC | Colon/rectum carcinoma |
| BRCA1 | Breast and ovarian carcinomas |
| BRCA2 | Breast carcinoma |
| DCC | Colon/rectum carcinoma |
| DPC4 | Pancreatic carcinoma |
| INK4 | Melanoma, lung carcinoma, brain tumors, leukemias, lymphomas |
| NF1 | Neurofibrosarcoma |
| NF2 | Meningioma |
| p53 | Brain tumors; breast, colon/rectum, esophageal, liver, and lung carcinomas; sarcomas; leukemias and lymphomas |
| Rb | Retinoblastoma; sarcomas; bladder, breast, and lung carcinomas |
| VHL | Renal cell carcinoma |
| WT1 | Wilms' tumor |

of a rare hereditary cancer syndrome, in which affected individuals develop any of several different types of cancer. In addition, the p53 protein (like Rb) is a target for the oncogene proteins of SV40, adenoviruses, and human papillomaviruses.

Like *p53*, the **INK4** tumor suppressor gene plays a role in several common cancers, including lung cancer. Two other tumor suppressor genes (**APC** and **DCC**) are frequently deleted or mutated in colon cancers. In addition to being involved in noninherited cases of this common adult cancer, inherited mutations of the *APC* gene are responsible for a rare hereditary form of colon cancer, called familial adenomatous polyposis. Individuals with this condition develop hundreds of benign colon adenomas (polyps), some of which almost inevitably progress to malignancy.

Additional tumor suppressor genes have been implicated in the development of breast, ovarian, and pancreatic carcinomas, as well as in several rare inherited cancer syndromes, such as Wilms' tumor. The number of identified tumor suppressor genes is rapidly expanding, and the characterization of these genes is one of the most active areas of current cancer research.

### Functions of Tumor Suppressor Gene Products

In contrast to proto-oncogene and oncogene proteins, the proteins encoded by most tumor suppressor genes inhibit cell proliferation. Inactivation of tumor suppressor genes therefore leads to tumor development by eliminating negative regulatory proteins. In several cases, tumor suppressor proteins inhibit the same cell regulatory pathways that are stimulated by the products of oncogenes.

The protein encoded by the **NF1** tumor suppressor gene is an interesting example of antagonism between oncogene products and tumor suppressor gene products. The NF1 protein down-regulates Ras proto-oncogene proteins by acting as a GTPase-activating protein (see Figure 13.33). High levels of the active GTP-bound form of Ras are therefore present in tumor cells in which *NF1* has been inactivated, and inactivation of the *NF1* tumor suppressor gene appears to stimulate cell proliferation as a result of deregulation of the Ras signaling pathway.

Several tumor suppressor genes encode transcriptional regulatory proteins. A good example is provided by the product of **WT1**, which is fre-

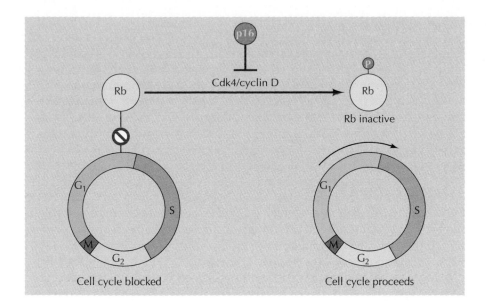

*Figure 15.36*
**Inhibition of cell cycle progression by Rb and p16** Rb inhibits progression past the restriction point in $G_1$. Cdk4/cyclin D complexes promote passage through the restriction point by phosphorylating and inactivating Rb. The activity of Cdk4/cyclin D is inhibited by p16.

*Figure 15.36*
**Inhibition of cell cycle progression by Rb and p16** Rb inhibits progression past the restriction point in $G_1$. Cdk4/cyclin D complexes promote passage through the restriction point by phosphorylating and inactivating Rb. The activity of Cdk4/cyclin D is inhibited by p16.

quently inactivated in Wilms' tumors (a childhood kidney tumor). The WT1 protein is a repressor that appears to suppress transcription of a number of growth factor-inducible genes. One of the targets of WT1 is thought to be the gene that encodes insulin-like growth factor II, which is overexpressed in Wilms' tumors and may contribute to tumor development by acting as an autocrine growth factor. Inactivation of WT1 may thus lead to abnormal growth factor expression, which in turn drives tumor cell proliferation.

The products of the *Rb* and *INK4* tumor suppressor genes regulate cell cycle progression at the same point as that affected by cyclin D1 (Figure 15.36). Rb inhibits passage through the restriction point in $G_1$ by repressing transcription of a number of genes involved in cell cycle progression and DNA synthesis (see Figure 14.19). In normal cells, passage through the restriction point is regulated by Cdk4/cyclin D complexes, which phosphorylate and inactivate Rb. Mutational inactivation of *Rb* in tumors thus removes a key negative regulator of cell cycle progression. The *INK4* tumor suppressor gene, which encodes the Cdk inhibitor p16, also regulates passage through the restriction point. As discussed in Chapter 14, p16 inhibits Cdk4/cyclin D activity. Inactivation of *INK4* therefore leads to elevated activity of Cdk4/cyclin D complexes, resulting in uncontrolled phosphorylation of Rb.

The *p53* gene product regulates both cell cycle progression and apoptosis (Figure 15.37). DNA damage leads to rapid induction of p53, which activates transcription of the Cdk inhibitor p21 (see Figure 14.20). The inhibitor p21 blocks cell cycle progression, both by acting as a general inhibitor of Cdk/cyclin complexes and by inhibiting DNA replication by binding to PCNA (proliferating cell nuclear antigen). The resulting cell cycle arrest presumably allows time for damaged DNA to be repaired before it is replicated. Loss of p53 prevents this damage-induced cell cycle arrest, leading to increased mutation frequencies and a general instability of the cell genome. Such genetic instability is a common property of cancer cells, and it may contribute to further alterations in oncogenes and tumor suppressor genes during tumor progression.

In addition to mediating cell cycle arrest, p53 is required for apoptosis induced by DNA damage. Unrepaired DNA damage normally induces apop-

*Figure 15.37*
**Action of p53** Wild-type p53 is required for both cell cycle arrest and apoptosis induced by DNA damage.

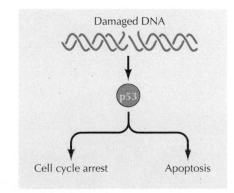

**Figure 15.38**
**Genetic alterations in colon carcinomas** Inactivation of *APC* is an early event in tumor development, giving rise to a proliferative cell population. Mutations of *ras*K then frequently occur and are found in early-stage adenomas. Subsequent mutations of *DCC* and *p53* are associated with later stages of tumor progression.

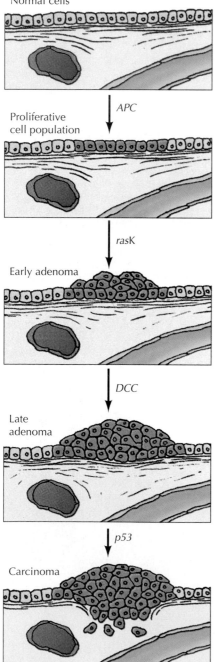

Normal cells

Proliferative cell population
*APC*

*ras*K

Early adenoma

*DCC*

Late adenoma

*p53*

Carcinoma

tosis of mammalian cells, a response that is presumably advantageous to the organism because it eliminates cells carrying potentially deleterious mutations (e.g., cells that might develop into cancer cells). Cells lacking p53 fail to undergo apoptosis in response to agents that damage DNA, including radiation and many of the drugs used in cancer chemotherapy. This failure to undergo apoptosis in response to DNA damage contributes to the resistance of many tumors to chemotherapy. In addition, loss of p53 appears to interfere with apoptosis induced by other stimuli, such as growth factor deprivation and oxygen deprivation. These effects of p53 inactivation on cell survival are thought to account for the high frequency of *p53* mutations in human tumors.

The proteins encoded by two tumor suppressor genes (*APC* and *DCC*) may regulate cell-cell interactions. *DCC* encodes a transmembrane protein related to the immunoglobulin superfamily of cell adhesion molecules, which mediate selective interactions between cells in tissues (see Table 12.3). It therefore appears possible that loss of DCC contributes to the development of malignancy as a result of alterations in such cell-cell interactions, releasing cancer cells from the regulatory constraints imposed by contact with their neighbors. Similarly, the APC protein may be involved in interactions between the cytoskeleton and plasma membrane at stable cell-cell junctions. In particular, APC binds β-catenin, which links actin filaments to transmembrane proteins (cadherins) at adherens junctions (see Figure 12.63). Although the function of APC is unknown, its interaction with β-catenin suggests that it regulates some aspect of cytoskeletal organization at these sites of cell-cell adhesion.

### Roles of Oncogenes and Tumor Suppressor Genes in Tumor Development

As discussed earlier, the development of cancer is a multistep process in which normal cells gradually progress to malignancy. The complete sequence of events required for the development of any human cancer is not yet known, but it is becoming increasingly clear that both the activation of oncogenes and the inactivation of tumor suppressor genes are critical steps in tumor initiation and progression. Accumulated damage to multiple genes eventually results in the increased proliferation, invasiveness, and metastatic potential that are characteristic of cancer cells.

The role of multiple genetic defects is best understood in the case of colon carcinomas, which have been studied extensively by Bert Vogelstein and his colleagues. These tumors frequently involve mutation of *ras*K oncogenes and inactivation or deletion of three distinct tumor suppressor genes—*APC*, *DCC*, and *p53*. Lesions representing multiple stages of colon cancer development are regularly obtained as surgical specimens, so it has been possible to correlate these genetic alterations with discrete stages of tumor progression (Figure 15.38).

These studies indicate that inactivation of *APC* is an early event in tumor development. Genetic transmission of mutant *APC* genes in patients with familial adenomatous polyposis results in abnormal colon cell proliferation, leading to the outgrowth of multiple adenomas in the colons of affected patients. Mutations of *APC* also occur frequently in patients with noninherited colon carcinomas and are generally detected at early stages of the disease process. Mutations of *ras*K genes then appear to occur, and *ras*K oncogenes are also frequently present in small and intermediate-size adenomas. In contrast, the *DCC* and *p53* tumor suppressor genes are inactivated usually at later stages of tumor progression. Mutations in these genes are only rarely found in early-stage adenomas, but they are frequently present in advanced adenomas and malignant carcinomas. The loss of these tumor suppressor genes in colon cancers thus appears to be involved in later stages of progression to malignancy, rather than in the initial stages of tumor formation.

Although these genetic alterations most often occur in the order described here, this is not an obligatory sequence of events. For example, mutations in *p53* are sometimes detected in early adenomas. More important than the order in which mutations occur is the fact that colon cancer ultimately results from accumulated damage to multiple genes. Accumulated damage to both oncogenes and tumor suppressor genes similarly appears to be responsible for the development of other types of cancer, including breast and lung carcinomas. The progressive loss of growth control that is characteristic of cancer cells is thus thought to be the end result of abnormalities in the products of multiple genes that normally regulate cell proliferation, differentiation, and survival.

## APPLICATIONS OF MOLECULAR BIOLOGY TO CANCER PREVENTION AND TREATMENT

A great deal has been learned about the molecular defects responsible for the development of many human cancers. However, cancer is more than a topic of scientific interest. It is a dread disease that claims the lives of nearly one out of every four Americans. It is therefore appropriate to ask whether our growing understanding of cancer will contribute to its prevention and treatment. The potential application of molecular biology to these practical issues represents a major challenge for future research.

### Prevention and Early Detection

The most effective way to deal with cancer would be to prevent development of the disease. A second-best, but still effective, alternative would be to reliably detect early premalignant stages of tumor development that can be easily treated. Many cancers can be cured by localized treatments, such as surgery or radiation, if they are detected before they spread through the body by metastasis. For example, early premalignant stages of colon cancer (adenomas) are usually completely curable by relatively minor surgical procedures (Figure 15.39). The cure rate for early carcinomas that remain localized to their site of origin is also high, about 90%. However, survival rates drop to about 50% for patients whose cancers have spread to adjacent tissues and lymph nodes, and to less than 10% for patients with metastatic colon cancer. Early detection of cancer can thus be a critical determinant of the outcome of the disease.

The major application of molecular biology to prevention and early detection may lie in the identification of individuals with inherited susceptibilities to cancer development. Such inherited cancer susceptibilities can result from mutations in tumor suppressor genes, in at least one oncogene (*ret*), and in DNA repair genes, such as the mismatch repair genes responsible for development of hereditary nonpolyposis colon cancer (see Molecular Medicine in Chapter 5). Mutations in these genes can be detected by genetic testing, allowing the identification of high-risk individuals before disease develops.

In addition to contributing to family planning decisions, careful monitoring of high-risk individuals may allow early detection and more effective treatment of some types of cancer. For example, colon adenomas can be detected by colonoscopy and removed prior to the development of malignancy. Patients with familial adenomatous polyposis (resulting from

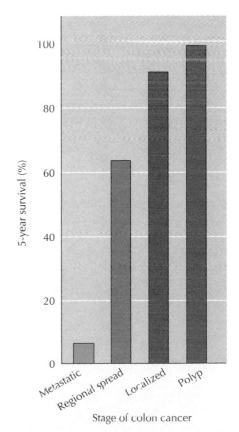

*Figure 15.39*
**Survival rates of patients with colon carcinoma** Five-year survival rates are shown for patients diagnosed with adenomas (polyps), with carcinoma still localized to its site of origin, with carcinoma that has spread regionally to adjacent tissues and lymph nodes, and with metastatic carcinoma.

inherited mutations of the *APC* tumor suppressor gene) typically develop hundreds of adenomas within the first 20 years of life, so the colons of these patients are usually removed before the inevitable progression of some of these polyps to malignancy. However, patients with hereditary nonpolyposis colon cancer develop a smaller number of polyps later in life and may therefore benefit from routine colonoscopy and possible pharmacologic measures to inhibit colon cancer development.

The direct inheritance of cancers resulting from defects in known genes is a rare event, constituting a small fraction of total cancer incidence. The most common inherited cancer susceptibility is hereditary nonpolyposis colon cancer, which accounts for about 15% of colon cancers and approximately 2% of all cancers in the United States. However, still-unidentified genes conferring cancer susceptibility may contribute to the development of a larger fraction of common adult malignancies, such as breast, colon, and lung cancers. The continuing isolation of cancer susceptibility genes is thus an important undertaking with clear practical implications. Individuals with such inherited susceptibility genes could be appropriately advised to avoid exposure to relevant carcinogens (e.g., tobacco smoke in the case of lung cancer) and carefully monitored to detect tumors at early stages that are more readily treated. The reliable identification of susceptible individuals, if it were followed by appropriate preventive and early detection measures, might ultimately make a significant impact on cancer mortality.

### Molecular Diagnosis

Molecular analysis of the oncogenes and tumor suppressor genes involved in particular types of tumors has the potential of providing information that is useful in the diagnosis of cancer and in monitoring the effects of treatment. Indeed, several applications of such molecular diagnosis are already being put into clinical practice. In some cases, mutations in oncogenes have provided useful molecular markers for monitoring the course of disease during treatment. The translocation of *abl* in chronic myelogenous leukemia is a good example. As previously discussed, this translocation results in the fusion of *abl* with the *bcr* gene, leading to expression of the Bcr/Abl oncogene protein (see Figure 15.25). The polymerase chain reaction (see Figure 3.27) provides a sensitive method of detecting the recombinant *bcr/abl* oncogene in leukemic cells and is therefore used to monitor the response of patients to treatment.

In other cases, the detection of mutations in specific oncogenes or tumor suppressor genes may provide information pertinent to choosing between different therapeutic options. For example, amplification of N-*myc* in neuroblastomas and *erb*B-2 in breast and ovarian carcinomas predicts rapid disease progression. Therefore, it might be appropriate to treat patients with such amplified oncogenes more aggressively. Because of the role of p53 in apoptosis induced by DNA damage, analysis of *p53* mutations may help predict the response of tumors to radiation and chemotherapy. As further information is accumulated on the roles of specific genetic abnormalities in determining tumor behavior, it is likely that molecular diagnosis will become increasingly important in dealing with cancer.

### Treatment

The most critical question, however, is whether the discovery of oncogenes and tumor suppressor genes will allow the development of new drugs that act selectively against cancer cells. Most of the drugs currently used in cancer treatment either damage DNA or inhibit DNA replication. Consequently, these drugs are toxic not only to cancer cells but also to normal

## MOLECULAR MEDICINE

# *Acute Promyelocytic Leukemia and Retinoic Acid*

### The Disease

Acute promyelocytic leukemia results from the proliferation of a specific type of immature blood cell. There are several different types of leukemia, which together account for approximately 28,000 cases and 21,000 deaths annually in the United States. These leukemias are divided about evenly between the chronic and acute leukemias. Chronic leukemias progress slowly, and many patients have no symptoms of disease for years. In contrast, acute leukemias progress rapidly and are fatal if untreated.

The acute leukemias are further divided into lymphocytic and non-lymphocytic types. The acute lymphocytic leukemias, which arise from immature lymphocytes, account for about 5000 cases per year. About 2000 of these occur in children, and acute lymphocytic leukemia is the most common type of childhood cancer. Fortunately, acute lymphocytic leukemia responds well to chemotherapy, and survival rates of affected children are greater than 75%.

The acute nonlymphocytic leukemias, approximately 9000 cases per year, occur primarily in adults. These leukemias are generally less responsive to chemotherapy than the acute lymphocytic leukemias, and only about 20% of patients are cured by current treatments. The acute non-lymphocytic leukemias arise from precursors of myeloid cells, including erythrocytes, monocytes, macrophages, megakaryotes, and granulocytes. Acute promyelocytic leukemia is one form of acute nonlymphocytic

leukemia, which arises from granulocyte precursors called promyelocytes.

### Molecular and Cellular Basis

Acute promyelocytic leukemia is characterized by a translocation between chromosomes 15 and 17. In 1990, three groups of researchers found that this translocation fuses the retinoic acid receptor gene *RARα* from chromosome 17 to the *PML* gene on chromosome 15, forming the *PML/RARα* oncogene. Although its mechanism of action remains to be fully understood, the PML/RARα protein appears to block cell differentiation, resulting in the continual proliferation of promyelocytic leukemic cells that fail to mature normally to granulocytes.

### Prevention and Treatment

Retinoic acid was first used to treat acute promyelocytic leukemia in 1986, before the *PML/RARα* oncogene had been identified. In these studies, high doses of retinoic acid were found to induce differentiation of the promyelocytic leukemia cells, leading to the elimination of all signs of leukemia in about 90% of patients. Although striking, this favorable response was temporary; patients relapsed within several months, and their leukemias were then resistant to further treatment with retinoic acid. However, current use of retinoic acid in combination with other chemotherapeutic drugs appears to reduce the incidence of relapse and significantly improve survival rates.

The therapeutic activity of retinoic

acid was observed prior to identification of the *PML/RARα* oncogene, so its effectiveness against leukemic cells expressing this oncogene protein was discovered by chance rather than by rational drug design. Moreover, the molecular mechanism by which high doses of retinoic acid overcome the block to differentiation resulting from expression of the PML/RARα oncogene protein is not understood. It is possible that high doses of retinoic acid stimulate the activity of other normal retinoid receptors in addition to binding directly to the PML/RARα fusion protein. Nonetheless, the use of retinoic acid for treatment of acute promyelocytic leukemia is noteworthy as the first example of a cancer therapy targeted against an oncogene product, and may serve as a model for the design of novel therapeutic agents designed to correct specific regulatory defects in cancer cells.

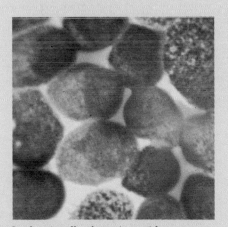

Leukemic cells of a patient with acute promyelocytic leukemia. (Courtesy of James Griffin, Dana-Farber Cancer institute.)

cells, particularly those normal cells that are undergoing rapid cell division (e.g., hematopoietic cells, epithelial cells of the gastrointestinal tract, and hair follicle cells). The action of anticancer drugs against these normal cell populations accounts for most of the toxicity associated with these drugs and limits their effective use in cancer treatment. Do oncogenes provide unique targets, against which drugs more specific for cancer cells could be designed?

Unfortunately, from the standpoint of cancer treatment, oncogenes are not unique to tumor cells. Since proto-oncogenes play important roles in normal cells, general inhibitors of oncogene expression or function are likely to act against normal cells as well as tumor cells. The exploitation of oncogenes as targets for anticancer drugs is therefore not a straightforward proposition, but there are reasons to hope that it will not ultimately be an impossible one.

Importantly, a therapeutic regimen targeted against a specific oncogene is already in use for the treatment of acute promyelocytic leukemia. This leukemia is characterized by a chromosome translocation in which the gene that encodes the retinoic acid receptor (*RARα*) is fused to another gene (*PML*) to form the *PML/RARα* oncogene. The PML/RARα protein is thought to function, at least in part, by blocking cell differentiation. These leukemic cells, however, differentiate in response to treatment with high doses of retinoic acid, which appear to overcome the inhibitory effect of the PML/RARα oncogene protein. Such treatment with retinoic acid results in remission of the leukemia in most patients, although this favorable response is temporary and patients eventually relapse. However, combined treatment with retinoic acid and standard chemotherapeutic agents significantly reduces the incidence of relapse. The use of retinoic acid is thus of substantial benefit in the treatment of acute promyelocytic leukemia, providing the first example of a clinically useful drug targeted against an oncogene protein.

The development of drugs targeted against Ras proteins is an area of active exploration, consistent with the prevalence of *ras* oncogenes in human tumors. Ras proteins are targeted to the plasma membrane by addition of a farnesyl isoprenoid to their carboxy terminus (see Figure 7.30). Because this is a relatively uncommon type of protein modification, inhibitors of farnesylation have been investigated as potential therapeutic drugs. Such compounds are surprisingly selective inhibitors of cells expressing oncogenic Ras proteins, and their potential clinical utility in the treatment of human cancer is currently being evaluated.

Inhibitors of a variety of other oncogene proteins, including protein-tyrosine kinases and transcription factors, are actively being investigated as new areas of drug development. The possibility also exists of developing drugs that interact with components of the cell cycle machinery, potentially compensating for defects in tumor suppressor genes such as *Rb* or *INK4*. Yet another area of drug development is the identification of compounds that might induce apoptosis of tumor cells, possibly by inhibiting Bcl-2 or compensating for the loss of p53. Although the eventual impact of molecular biology on the treatment of cancer remains to be seen, the exploitation of oncogenes and tumor suppressor genes as potential targets for therapeutic drugs clearly represents both an opportunity and a challenge for future research on the molecular biology of cancer.

# Summary

## THE DEVELOPMENT AND CAUSES OF CANCER

*Types of Cancer:* Cancer can result from the abnormal proliferation of any type of cell. The most important distinction for the patient is between benign tumors, which remain confined to their site of origin, and malignant tumors, which can invade normal tissues and spread throughout the body.

cancer, tumor, benign tumor, malignant tumor, metastasis, carcinoma, sarcoma, leukemia, lymphoma

*The Development of Cancer:* Tumors develop from single altered cells that begin to proliferate abnormally. Additional mutations lead to the selection of cells with progressively increasing capacities for proliferation, survival, invasion, and metastasis.

tumor initiation, tumor progression, adenoma, polyp

*Causes of Cancer:* Radiation and many chemical carcinogens act by damaging DNA and inducing mutations. Other chemical carcinogens contribute to the development of cancer by stimulating cell proliferation. Viruses also cause cancer in both humans and other species.

carcinogen, tumor promoter

*Properties of Cancer Cells:* The uncontrolled proliferation of cancer cells is reflected in reduced requirements for extracellular growth factors and lack of inhibition by cell-cell contact. Many cancer cells are also defective in differentiation, consistent with their continued proliferation *in vivo*. The characteristic failure of cancer cells to undergo apoptosis also contributes substantially to tumor development.

density-dependent inhibition, autocrine growth stimulation, contact inhibition, angiogenesis, programmed cell death, apoptosis

*Transformation of Cells in Culture:* The development of *in vitro* assays for cell transformation has allowed the conversion of normal cells into tumor cells to be studied in cell culture.

cell transformation

## TUMOR VIRUSES

*Hepatitis B Viruses:* The hepatitis B viruses cause liver cancer in several species, including humans.

tumor virus, hepatitis B viruses

*SV40 and Polyomavirus:* Although neither SV40 nor polyomavirus causes human cancer, they are important models for studying the molecular biology of cell transformation. SV40 T antigen induces transformation by interacting with the cellular Rb and p53 tumor suppressor proteins.

simian virus 40 (SV40), polyomavirus

*Papillomaviruses:* Papillomaviruses induce tumors in a variety of animals, including cervical carcinoma in humans. Like SV40 T antigen, the transforming proteins of papillomaviruses interact with Rb and p53.

papillomavirus

*Adenoviruses:* The adenoviruses do not cause naturally occurring cancers in either humans or other species but are important models in cancer research. Their transforming proteins also interact with Rb and p53.

adenovirus

*Herpesviruses:* The herpesviruses, which are among the most complex animal viruses, cause cancer in several species, including humans. The molecular biology of transformation by these viruses is not well understood.

herpesvirus, Epstein-Barr virus

*Retroviruses:* Retroviruses cause cancer in humans and a variety of other animals. Some retroviruses contain specific genes responsible for inducing cell transformation, and studies of these highly oncogenic retroviruses have led to the characterization of both viral and cellular oncogenes.

retrovirus, Rous sarcoma virus (RSV)

**ONCOGENES**

oncogene, *src, ras, raf*

*Retroviral Oncogenes:* The first oncogene to be identified was the *src* gene of RSV. Subsequent studies have identified more than two dozen distinct oncogenes in different retroviruses.

proto-oncogene

*Proto-Oncogenes:* Retroviral oncogenes originated from closely related genes of normal cells, called proto-oncogenes. The oncogenes are abnormally expressed or mutated forms of the corresponding proto-oncogenes.

*ras,* c-*myc,* abl, N-*myc,* erbB-2

*Oncogenes in Human Cancer:* A variety of oncogenes are activated by point mutations, DNA rearrangements, and gene amplification in human cancers. Some of these human tumor oncogenes, such as the *ras* genes, are cellular homologs of oncogenes that were first described in retroviruses.

Fos, Jun, *PRAD*1, ErbA, PML/RARα, *bcl*-2

*Functions of Oncogene Products:* Most oncogene proteins function as elements of signaling pathways that stimulate cell proliferation. The gene that encodes cyclin D1 is also a potential oncogene, which stimulates cell cycle progression. Other oncogene proteins interfere with cell differentiation, and Bcl-2 inhibits apoptosis.

**TUMOR SUPPRESSOR GENES**

tumor suppressor gene, *Rb, p53, INK4, APC, DCC*

*Identification of Tumor Suppressor Genes:* In contrast to oncogenes, tumor suppressor genes inhibit tumor development. The prototype tumor suppressor gene, *Rb*, was identified by studies of inheritance of retinoblastoma. Loss or mutational inactivation of *Rb* and other tumor suppressor genes, including *p53*, contributes to the development of a wide variety of human cancers.

*NF1, WT1*

*Functions of Tumor Suppressor Gene Products:* The proteins encoded by most tumor suppressor genes act as inhibitors of cell proliferation. The Rb, INK4, and p53 proteins are negative regulators of cell cycle progression. In addition, p53 is required for apoptosis induced by DNA damage and other stimuli, so its inactivation contributes to enhanced tumor cell survival. The proteins encoded by the *DCC* and *APC* tumor suppressor genes may regulate cell-cell interactions and organization of the cytoskeleton.

*Roles of Oncogenes and Tumor Suppressor Genes in Tumor Development:* Mutations in both oncogenes and tumor suppressor genes contribute to the progressive development of human cancers. Accumulated damage to multiple such genes is thought to result in the abnormalities of cell proliferation, differentiation, and survival that characterize the cancer cell.

**APPLICATIONS OF MOLECULAR BIOLOGY TO CANCER PREVENTION AND TREATMENT**

*Prevention and Early Detection:* Many cancers can be cured if they are detected at early stages of tumor development. Genetic testing to identify individuals with inherited cancer susceptibilities may allow early detection and more effective treatment of high-risk patients.

*Molecular Diagnosis:* Detection of mutations in oncogenes and tumor suppressor genes may be useful in diagnosis and in monitoring response to treatment.

*Treatment:* The development of drugs targeted against specific oncogenes or tumor suppressor genes may lead to the discovery of new therapeutic agents that act selectively against cancer cells.

## QUESTIONS

**1.** What is the role of clonal selection in the development of cancer?

**2.** You have constructed a mutant of SV40 T antigen that fails to induce cell transformation because it no longer binds Rb. Would this mutant T antigen induce transformation if introduced into cells together with a papillomavirus cDNA that encodes E7? How about E6?

**3.** The proliferation of fibroblasts is generally inhibited by cAMP. Would *gsp* act as an oncogene in these cells?

**4.** What effect would overexpression of the *INK4* tumor suppressor gene product be expected to have on tumor cells in which *Rb* has been inactivated by mutation?

**5.** Which would you expect to be more sensitive to treatment with irradiation— tumors with wild-type *p53* genes or tumors with mutant *p53* genes?

## REFERENCES AND FURTHER READING

### General References

Brugge, J., T. Curran, E. Harlow and F. McCormick, eds. 1991. *Origins of Human Cancer.* Plainview, NY: Cold Spring Harbor Laboratory Press.

Cooper, G. M. 1992. *Elements of Human Cancer.* Boston: Jones and Bartlett.

Cooper, G. M. 1995. *Oncogenes.* 2nd ed. Boston: Jones and Bartlett.

Varmus, H. and R. A. Weinberg. 1993. *Genes and the Biology of Cancer.* New York: Scientific American Library.

### The Development and Causes of Cancer

Doll, R. and R. Peto. 1981. *The Causes of Cancer: Quantitative Estimates of Avoidable Risks of Cancer in the United States Today.* New York: Oxford University Press.

Fialkow, P. J. 1979. Clonal origin of human tumors. *Ann. Rev. Med.* 30: 135–143. [R]

Fidler, I. J. 1990. Critical factors in the biology of human cancer metastasis. *Cancer Res.* 50: 6130–6138. [R]

Henderson, B. E., R. Ross and L. Bernstein. 1988. Estrogens as a cause of human cancer. *Cancer Res.* 48: 246–253. [R]

Liotta, L. A., P. S. Steeg and W. G. Stetler-Stevenson. 1991. Cancer metastasis and angiogenesis: An imbalance of positive and negative regulation. *Cell* 64: 327–336. [R]

Miller, J. A. 1970. Carcinogenesis by chemicals: An overview. *Cancer Res.* 30: 559–576. [R]

Nowell, P. C. 1986. Mechanisms of tumor progression. *Cancer Res.* 46: 2203–2207. [R]

Raff, M. C. 1992. Social controls on cell survival and cell death. *Nature* 356: 397–400. [R]

Sawyers, C. L., C. T. Denny and O. N. Witte. 1991. Leukemia and the disruption of normal hematopoiesis. *Cell* 64: 337–350. [R]

Sporn, M. B. and A. B. Roberts. 1985. Autocrine growth factors and cancer. *Science* 313: 745–747. [R]

Stetler-Stevenson, W. G., S. Aznavoorian and L. A. Liotta. 1993. Tumor cell interactions with the extracellular matrix during invasion and metastasis. *Ann. Rev. Cell Biol.* 9: 541–573. [R]

Temin, H. M. and H. Rubin. 1958. Characteristics of an assay for Rous sarcoma virus and Rous sarcoma cells in culture. *Virology* 6: 669–688. [P]

Thompson, C. B. 1995. Apoptosis in the pathogenesis and treatment of disease. *Science* 267: 1456–1462. [R]

### Tumor Viruses

Botchan, M., T. Grodzicker and P. A. Sharp, eds. 1986. *Cancer Cells,* Vol. 4, *DNA Tumor Viruses.* New York: Cold Spring Harbor Laboratory Press.

DeCaprio, J. A., J. W. Ludlow, J. Figge, J.-Y. Shew, C.-M. Huang, W.-H. Lee, E. Marsilio, E. Paucha and D. M. Livingston. 1988. SV40 large tumor antigen forms a specific complex with the product of the retinoblastoma susceptibility gene. *Cell* 54: 275–283. [P]

Dyson, N., P. M. Howley, K. Munger and E. Harlow. 1989. The human papillomavirus-16 E7 oncoprotein is able to bind to the retinoblastoma gene product. *Science* 243: 934–937. [P]

Fields, B. N., D. M. Knipe, P. M. Howley, R. M. Chanock, J. L. Melnick, T. P. Monath, B. Roizman and S. E. Straus, eds. 1996. *Fundamental Virology.* 3rd ed. New York: Lippincott-Raven.

Scheffner, M., B. A. Werness, J. M. Huibregtse, A. J. Levine and P. M. Howley. 1990. The E6 oncoprotein encoded by human papillomaviruses types 16 and 18 promotes the degradation of p53. *Cell* 63: 1129–1136. [P]

Weiss, R., N. Teich, H. Varmus and J. Coffin, eds. 1985. *Molecular Biology of Tumor Viruses: RNA Tumor Viruses,* 2nd ed. New York: Cold Spring Harbor Laboratory Press.

Zur Hausen, H. 1991. Viruses in human cancer. *Science* 254: 1167–1173. [R]

### Oncogenes

Aaronson, S. A. 1991. Growth factors and cancer. *Science* 254: 1146–1153. [R]

Alitalo, K. and M. Schwab. 1986. Oncogene amplification in tumor cells. *Adv. Cancer Res.* 47: 235–281. [R]

Bos, J. L. 1989. *Ras* oncogenes in human cancer: A review. *Cancer Res.* 49: 4682–4689. [R]

Cantley, L. C., K. R. Auger, C. Carpenter, B. Duckworth, A. Graziani, R. Kapeller and S. Soltoff. 1991. Oncogenes and signal transduction. *Cell* 64: 281–302. [R]

Cleary, M. L. 1991. Oncogenic conversion of transcription factors by chromosomal translocations. *Cell* 66: 619–622. [R]

Daum, G., I. Eisenmann-Tappe, H.-W. Fries, J. Troppmair and U. R. Rapp. 1994. The ins and outs of Raf kinases. *Trends Biochem. Sci.* 19: 474–480. [R]

Der, C. J., T. G. Krontiris and G. M. Cooper. 1982. Transforming genes of human bladder and lung carcinoma cell lines are homologous to the *ras* genes of Harvey and Kirsten sarcoma viruses. *Proc. Natl. Acad. Sci. USA* 79: 3637–3640. [P]

Hunter, T. and J. Pines. 1994. Cyclins and cancer II: Cyclin D and CDK inhibitors come of age. *Cell* 79: 573–582. [R]

Korsmeyer, S. J. 1995. Regulators of cell death. *Trends Genet.* 11: 101–105. [R]

Krontiris, T. G. and G. M. Cooper. 1981. Transforming activity of human tumor DNAs. *Proc. Natl. Acad. Sci. USA* 78: 1181–1184. [P]

Leder, P., J. Battey, G. Lenoir, C. Moulding, W. Murphy, H. Potter, T. Stewart and R. Taub. 1983. Translocations among antibody genes in human cancer. *Science* 222: 765–771. [R]

Lowy, D. R. and B. M. Willumsen. 1993. Function and regulation of Ras. *Ann. Rev. Biochem.* 62: 851–891. [R]

Marcu, K. B., S. A. Bossone and A. J. Petel. 1992. Myc function and regulation. *Ann. Rev. Biochem.* 61: 809–860. [R]

Martin, G. S. 1970. Rous sarcoma virus: A function required for the maintenance of the transformed state. *Nature* 227: 1021–1023. [P]

Ransome, L. J. and I. M. Verma. 1990. Nuclear proto-oncogenes *FOS* and *JUN*. *Ann. Rev. Cell Biol.* 6: 539–557. [R]

Reed, J. C. 1994. Bcl-2 and the regulation of programmed cell death. *J. Cell Biol.* 124: 1–6. [R]

Sherr, C. J. 1995. D-type cyclins. *Trends Biochem. Sci.* 20: 187–190. [R]

Shih, C., L. C. Padhy, M. Murray and R. A. Weinberg. 1981. Transforming genes of carcinomas and neuroblastomas introduced into mouse fibroblasts. *Nature* 300: 539–542. [P]

Solomon, E., J. Borrow and A. D. Goddard. 1991. Chromosome aberrations and cancer. *Science* 254: 1153–1160. [R]

Stehelin, D., H. E. Varmus, J. M. Bishop and P. K. Vogt. 1976. DNA related to the transforming gene(s) of avian sarcoma viruses is present in normal avian DNA. *Nature* 260: 170–173. [P]

Sugden, B. 1993. How some retroviruses got their oncogenes. *Trends Biochem. Sci.* 18: 233–235. [R]

Tabin, C. J., S. M. Bradley, C. I. Bargmann, R. A. Weinberg, A. G. Papageorge, E. M. Scolnick, R. Dhar, D. R. Lowy and E. H. Chang. 1982. Mechanism of activation of a human oncogene. *Nature* 300: 143–149. [P]

Thompson, C. B. 1995. Apoptosis in the pathogenesis and treatment of disease. *Science* 267: 1456–1462. [R]

Vogt, P. K. 1971. Spontaneous segregation of nontransforming viruses from cloned sarcoma viruses. *Virology* 46: 939–946. [P]

## Tumor Suppressor Genes

Fearon, E. R. and B. Vogelstein. 1990. A genetic model for colorectal tumorigenesis. *Cell* 61: 759–767. [R]

Fisher, D. E. 1994. Apoptosis in cancer therapy: Crossing the threshold. *Cell* 78: 539–542. [R]

Friend, S. H., R. Bernards, S. Rogelj, R. A. Weinberg, J. M. Rapaport, D. M. Albert and T. P. Dryja. 1986. A human DNA segment with properties of the gene that predisposes to retinoblastoma and osteosarcoma. *Nature* 323: 643–646. [P]

Harris, H., O. J. Miller, G. Klein, P. Worst and T. Tachibana. 1969. Suppression of malignancy by cell fusion. *Nature* 223: 363–368. [P]

Hartwell, L. H. and M. B. Kastan. 1994. Cell cycle control and cancer. *Science* 266: 1821–1828. [R]

Hollstein, M., D. Sidransky, B. Vogelstein and C. C. Harris. 1991. *p53* mutations in human cancers. *Science* 253: 49–53. [R]

Hunter, T. and J. Pines. 1994. Cyclins and cancer II: Cyclin D and CDK inhibitors come of age. *Cell* 79: 573–582. [R]

Knudson, A. G. 1976. Mutation and cancer: A statistical study of retinoblastoma. *Proc. Natl. Acad. Sci. USA* 68: 820–823. [P]

Knudson, A. G. 1993. Antioncogenes and human cancer. *Proc. Natl. Acad. Sci. USA* 90: 10914–10921. [R]

Lee, W.-H., R. Bookstein, F. Hong, L.-J. Young, J.-Y. Shew and E. Y.-H. P. Lee. 1987. Human retinoblastoma susceptibility gene: Cloning, identification, and sequence. *Science* 235: 1394–1399. [P]

Levine, A. J. 1993. The tumor suppressor genes. *Ann. Rev. Biochem.* 62: 623–651. [R]

Marshall, C. J. 1991. Tumor suppressor genes. *Cell* 64: 313–326. [R]

Riley, D. J., E. Y.-H. P. Lee and W.-H. Lee. 1994. The retinoblastoma gene: More than a tumor suppressor. *Ann. Rev. Cell Biol.* 10: 1–29. [R]

Sherr, C. J. 1994. $G_1$ phase progression: Cycling on cue. *Cell* 79: 551–555. [R]

Stanbridge, E. J. 1990. Human tumor suppressor genes. *Ann. Rev. Genet.* 24: 615–657. [R]

Vogelstein, B. and K. W. Kinzler. 1993. The multistep nature of cancer. *Trends Genet.* 9: 138–141. [R]

Weinberg, R. A. 1991. Tumor suppressor genes. *Science* 254: 1138–1146. [R]

Weinberg, R. A. 1995. The retinoblastoma protein and cell cycle control. *Cell* 81: 323–330. [R]

## Applications of Molecular Biology to Cancer Prevention and Treatment

Degos, L., H. Dombret, C. Chomienne, M.-T. Daniel, J.-M. Miclea, C. Chastang, S. Castaigne and P. Fenaux. 1995. All-*trans*-retinoic acid as a differentiating agent in the treatment of acute promyelocytic leukemia. *Blood* 85: 2643–2653. [P]

Gibbs, J. B., A. Oliff and N. E. Kohl. 1994. Farnesyltransferase inhibitors: Ras research yields a potential cancer therapeutic. *Cell* 77: 175–178. [R]

Grignani, F., M. Fagioli, M. Alcalay, L. Longo, P. P. Pandolfi, E. Donti, A. Biondi, F. Lo Coco, F. Grignani and P. G. Pelicci. 1994. Acute promyelocytic leukemia: From genetics to treatment. *Blood* 83: 10–25. [R]

Karp, J. E. and S. Broder. 1994. New directions in molecular medicine. *Cancer Res.* 54: 653–665. [R]

Levitzki, A. and A. Gazit. 1995. Tyrosine kinase inhibition: An approach to drug development. *Science* 267: 1782–1788. [R]

Winawer, S. J. with 15 others and the National Polyp Study Workshop. 1993. Prevention of colon cancer by colonoscopic polypectomy. *N. Engl. J. Med.* 329: 1977–1981. [P]

# Answers to Questions

## Chapter 1

1. $O_2$ became abundant in Earth's atmosphere as a result of photosynthesis.

2. Yeasts are surrounded by cell walls; animal cells are not.

3. The refractive index of air is 1.0; the refractive index of oil is approximately 1.4. Since resolution = $0.61\lambda/\eta \sin \alpha$ ($\eta$ is the refractive index), viewing a specimen through air rather than through oil changes the limit of resolution from about 0.2 $\mu$m to about 0.3 $\mu$m.

4. You would use velocity centrifugation. In an equilibrium gradient the two organelles would band at the same position.

5. The genome of Rous sarcoma virus (10,000 base pairs) is 100,000 times smaller than that of chicken cells (1.2 billion base pairs).

## Chapter 2

1. The formation of stable membranes depends on the amphipathic character of phospholipids.

2. Since phenylalanine residues are hydrophobic, they would probably be located in $\alpha$ helices or $\beta$ sheets within the interior of the protein. The loop regions connecting these elements of secondary structure would be expected to contain hydrophilic amino acids.

3. Aspartate is an acidic amino acid that interacts with basic amino acids in the substrates of trypsin.

Substitution of aspartate with lysine (a basic amino acid) would therefore interfere with substrate binding and catalysis.

4. Substituting the given values into the equation

$$\Delta G = \Delta G + RT \ln \frac{[B][C]}{[A]}$$

gives the following result: $\Delta G - 1.93$ kcal/mol. The reaction will therefore proceed from left to right, with A being converted to B plus C within the cell.

5. Since saturated fatty acids are more highly reduced, more free energy is obtained from their oxidation than from that of unsaturated fatty acids.

6. In order to drive the pathway in the direction of glucose synthesis, the energy-yielding reactions of glycolysis must be bypassed by alternative reactions that are coupled to the utilization of ATP and NADH.

7. The $\alpha$-helical structure allows the CO and NH groups of peptide bonds to form hydrogen bonds with each other, thereby neutralizing their polar character.

## Chapter 3

1. The genetic map is $z\underline{\quad 0.5 \quad}x\underline{\quad 0.2 \quad}y$

2. UGC also encodes cysteine, so this mutation would have no effect on enzyme function. UGA, however, is a stop codon, so this mutation would abolish enzyme activity.

3. The restriction map is

*Eco*RI

2.5 kb        0.5 kb    1 kb

*Hin*dIII

4. The haploid sperm contains a single starting copy of the DNA sequence. Each cycle of PCR amplifies the starting material two-fold, so 10 cycles yields $2^{10}$, or approximately a thousand, copies. Amplification for 30 cycles yields $2^{30}$, or more than a billion, copies.

5. Antibodies against the protein could be raised either by using the predicted amino acid sequence to generate synthetic peptides or by expressing the protein in bacteria. Fluorescence microscopy could then be used to determine subcellular localization.

6. Mutations of the amino acids of interest could be produced by *in vitro* mutagenesis. The effects of these mutations on catalytic activity could then be determined following expression of the mutated protein in bacteria or eukaryotic cells.

## Chapter 4

1. DNA reassociation is a bimolecular reaction, so the reaction rate is proportional to the square of the DNA concentration. A sequence repeated 1000 times in the genome would therefore reassociate with a rate $10^6$-fold greater than that of single-copy DNA.

2. Given one nucleosome every 200 base pairs, there are $3 \times 10^7$ nucleosomes in a diploid human cell. Each nucleosome contains two molecules of the four core histones and one molecule of histone H1, so there are nearly $3 \times 10^8$ histone molecules in the cell. Taking the average molecular weight of the histones as approximately 15,000 daltons, the total mass of histones is approximately 0.008 ng—nearly 1% of total cell protein.

3. Approximately 15%.

4. Because they can accommodate inserts containing thousands of kilobases, YACs can be used for mapping large genomes, which may contain thousands of megabases of DNA. Telomeres maintain the ends of YACs, allowing them to replicate as linear, chromosome-like molecules in yeasts.

5. Markers are currently spaced at approximately 1-Mb intervals, so you would need to clone approximately 1 Mb of DNA in order to cover the region that might contain your gene. A minimum of 25 cosmid clones with inserts about 40 kb long would be needed to span this region, assuming minimal overlap between different cosmids.

## Chapter 5

1. This property of DNA polymerases is necessary for proofreading because it enables the polymerase to recognize and excise mismatched bases that are not hydrogen-bonded to the template strand.

2. Primase, like other RNA polymerases, is not capable of proofreading, so errors occur at a comparatively high frequency in RNA primers. However, since the primers are later removed, the overall fidelity of replication is not compromised.

3. The high frequency of skin cancer results from DNA damage induced by solar UV irradiation, which is subject to repair by the nucleotide-excision repair system. The lack of elevated incidence of other cancers may suggest that similar types of damage are not frequent in internal organs and that most cancers of these organs result from other types of mutations (e.g., the incorporation of mismatched bases during DNA replication).

4. RecA mutants are sensitive to UV irradiation because they are deficient in recombinational repair of DNA damage.

5. The mouse would be immunodeficient, lacking both B and T lymphocytes, as a result of being unable to rearrange its immunoglobulin and T cell receptor genes.

6. The cellular reverse transcriptases that might be affected by these drugs include telomerase and the reverse transcriptases encoded by LINE sequences. Inhibition of the LINE reverse transcriptases would not be toxic to the cell, but inhibition of telomerase might interfere with the replication of chromosome ends.

## Chapter 6

1. The promoter containing the sequence TATGAT, which more closely resembles the consensus sequence, will be transcribed more efficiently.

2. The wild-type gene will be regulated normally, but the temperature-sensitive gene will be expressed constitutively. $\beta$-Galactosidase will therefore be produced at the permissive but not the nonpermissive temperature.

3. The promoter containing the TATA box can be transcribed *in vitro* in the presence of either TBP or TFIID. However, the Inr promoter requires TFIID, since the Inr sequence is recognized by TAFs rather than by TBP.

4. The sequence element would be a candidate binding site for a tissue-specific repressor.

5. Two different DNA-binding domains could be joined to the factor's activation domain by tissue-specific alternative splicing.

## Chapter 7

1. A Shine-Delgarno sequence is needed.

2. Polyadenylation is an important translational regulatory mechanism in early development. Its inhibition would block the translation of many oocyte mRNAs following fertilization.

3. An Hsp70 chaperone is expected to be required for the selective uptake of proteins into lysosomes from the cytosol, since this process involves the transfer of unfolded polypeptide chains across the lysosomal membrane. Such a chaperone would not be required for the fusion of membrane vesicles with lysosomes during autophagy.

4. Phospholipase treatment would release a GPI-anchored protein, but not a transmembrane protein, from the cell surface.

5. The mutation from serine to threonine would be expected to have no effect, since both serine and threonine residues are phosphorylated by the same protein kinases. However, the mutation from serine to alanine would prevent phosphorylation and block activation of the enzyme.

## Chapter 8

1. The transcription factor could no longer be phosphorylated at these sites, so it would be constitutively imported into the nucleus and would activate target gene expression.

2. Inactivating the nuclear export signal would result in retention of the protein in the nucleus.

3. An insulator prevents enhancers from acting on promoters in separate chromosomal domains. Insertion of an insulator between E and P1 would therefore inhibit transcription from both P1 and P2, whereas insertion of an insulator between P1 and P2 would only inhibit transcription from P2.

4. Because formation of a nucleolus requires transcription of the rRNA genes, it is blocked by treatment with actinomycin D. However, because the rRNA genes are transcribed by RNA polymerase I, $\alpha$-amanitin has no effect.

5. The B-type lamins are modified by farnesylation. The attached lipid causes them to remain associated with membrane vesicles following nuclear envelope breakdown. In contrast, the A- and C-type lamins are released in the cytosol.

## Chapter 9

1. The signal sequence would direct the growing polypeptide chain to the ER and the protein would then be secreted via the bulk flow pathway.

2. These proteins are unable to enter the ER and therefore remain in the cytosol.

3. Carbohydrate groups are added within the lumen of the ER and Golgi apparatus, which are topologically equivalent to the exterior of the cell.

4. The normally cytosolic protein lacks a signal sequence and does not enter the ER. Therefore, addition of a lysosome-targeting signal would have no effect. In contrast, such an addition would direct a normally secreted protein to lysosomes from the Golgi apparatus.

5. In the absence of mannose-6-phosphate formation in the Golgi apparatus, normally lysosomal proteins would be secreted.

## Chapter 10

1. The voltage component of the electrochemical gradient drives membrane insertion and translocation of the positively charged presequences of mitochondrial proteins.

2. Under these conditions, $\Delta G$ is approximately $-1.4$ kcal/mol. Approximately nine protons would therefore have to be transferred across the membrane for each molecule of ATP synthesized.

3. They are topologically equivalent compartments.

4. In contrast to mitochondria, there is no electric potential across the chloroplast membrane. Therefore, the charge of transit peptides does not contribute to protein translocation.

5. Two high-energy electrons are required to split each molecule of $H_2O$, so 24 high-energy electrons are required for the synthesis of each molecule of glucose. The passage of these electrons through the two photosystems generates 12 molecules of NADPH and between 16 and 24 molecules of ATP, depending on the stoichiometry of proton pumping at the cytochrome *bf* complex. Since 18 molecules of ATP are required for the Calvin cycle, the synthesis of glucose may require additional ATPs produced by cyclic electron flow.

6. Three out of four carbons converted to glycolate are returned to choroplasts and reenter the Calvin cycle.

## Chapter 11

1. Cytochalasin inhibits actin polymerization. It would therefore inhibit the movement of cells by pseudopodia but would not affect the microtubule-based movement of flagella.

2. The polarity of actin filaments defines the direction of myosin movement. If actin filaments were not polar, the unidirectional movement of myosin that results in the sliding of actin and myosin filaments could not take place.

3. No. Keratins are expressed only in epithelial cells.

4. Colchicine inhibits microtubule polymerization, so it would affect chromosome segregation. It would not affect cytokinesis, which is driven by an actin-myosin contractile ring.

5. By linking the microtubule doublets of cilia together, nexin converts the sliding of individual microtubules to a bending motion that leads to the beating of cilia. If it were eliminated, the microtubules would simply slide past one another.

## Chapter 12

1. The presence of porins in the plasma membrane would allow the free diffusion of ions and small molecules between the cytosol and extracellular fluids—a disaster for the cell.

2. Given $C_o/C_i = 10$, the $K^+$ equilibrium potential calculated from the Nernst equation is $-58$ mV. The actual resting membrane potential differs from the $K^+$ equilibrium potential because the plasma membrane is not completely impermeable to other ions.

3. The plant cell wall prevents cell swelling and allows the buildup of turgor pressure.

4. Because gap junctions allow ions to diffuse freely between adjacent cells, expression of normal CFTR in one epithelial cell would provide a functional $Cl^-$ channel to its neighbors.

5. The correct localization of transporters mediating active transport and facilitated diffusion is necessary for the polarized function of epithelial cells in transferring glucose from the intestinal lumen to the blood supply. Tight junctions prevent the diffusion of these transporters between domains of the plasma membrane, as well as sealing the spaces between cells of the epithelium.

## Chapter 13

1. Inhibition of cAMP phosphodiesterase would result in elevated levels of cAMP, which would stimulate cell proliferation.

2. The recombinant molecule would function as an epinephrine receptor coupled to $G_i$. Epinephrine would therefore inhibit adenylyl cyclase, lowering intracellular cAMP levels. Acetylcholine would have no effect, since it would not bind to the recombinant receptor.

3. PDGF monomers would not induce receptor dimerization. Since this is the first critical step in signaling from receptor protein-tyrosine kinases, they would be unable to stimulate the PDGF receptor.

4. Protein phosphatase 1 dephosphorylates serine residues that are phosphorylated by protein kinase A. Cyclic AMP-inducible genes are activated by CREB, which is phosphorylated by protein kinase A, so overexpression of protein phosphatase 1

would inhibit their induction. However, protein phosphatase 1 would not affect the activity of cAMP-gated ligand channels, since these channels are opened directly by cAMP binding rather than by protein phosphorylation.

5. Hydrolysis of $PIP_2$ by phospholipase C yields both diacylglycerol and $IP_3$, which signals the release of $Ca^{2+}$ from the endoplasmic reticulum. $PIP_2$ hydrolysis can therefore activate both PKC-$\alpha$ and PKC-$\varepsilon$. Hydrolysis of phosphatidylcholine yields diacylglycerol but not $IP_3$; consequently, phosphatidylcholine hydrolysis is sufficient to activate PKC-$\varepsilon$ but not PKC-$\alpha$.

6. Raf acts downstream of Ras in the MAP kinase pathway. Activated Raf can therefore bypass the effects of dominant negative Ras, but activated Ras cannot overcome the inhibitory effects of dominant negative Raf. MEK acts downstream of Raf, so activated MEK can overcome the effects of dominant negative Ras or Raf.

## Chapter 14

1. $G_1 = 16$ hours; $S = 9$ hours; $G_2 = 4$ hours; $M = 1$ hour.

2. One daughter cell would receive two copies of the misaligned chromosome; the other daughter cell would receive none.

3. In a normal cell, overexpression of p16 would inhibit cell cycle progression at the restriction point in $G_1$. Because Rb is the principal target of Cdk4/cyclin D complexes, a tumor cell lacking functional Rb would be unaffected by p16 overexpression.

4. The nuclear lamina would fail to break down.

5. Anaphase would initiate normally. However, MPF would remain active, so re-formation of nuclei, chromosome decondensation, and cytokinesis would not occur.

6. Oocytes of these mice would fail to arrest at metaphase II.

## Chapter 15

1. Clonal selection drives tumor progression by favoring the outgrowth of cells within the tumor population that are proliferating more rapidly.

2. E7 binds Rb, so it would be able to induce transformation in combination with the mutant T antigen. E6 would be unable to do so, since it interacts with p53 but not with Rb.

3. The *gsp* gene encodes the $\alpha$ subunit of $G_s$, which stimulates adenylyl cyclase. It would therefore not act as an oncogene in cells whose proliferation is inhibited by cAMP.

4.  *INK4* encodes the Cdk inhibitor p16, which inhibits Cdk4/cyclin D complexes. Since phosphorylation of Rb is the critical target of Cdk4/cyclin D, over-expression of p16 would have no effect on cells in which Rb has already been inactivated.

5.  Cells with wild-type p53 would be more sensitive because p53 is required for apoptosis induced by DNA damage.

# Glossary

**α helix**   A coiled secondary structure of a polypeptide chain formed by hydrogen bonding between amino acids separated by four residues.

**ABC transporters**   A large family of membrane transport proteins characterized by a highly conserved ATP binding domain.

**actin**   An abundant 43-kd protein that polymerizes to form cytoskeletal filaments.

**actin bundle**   Actin filaments that are crosslinked into closely packed arrays.

**actin network**   Actin filaments that are crosslinked into loose three-dimensional meshworks.

**activation energy**   The energy required to raise a molecule to its transition state to undergo a chemical reaction.

**active site**   The region of an enzyme that binds substrates and catalyzes an enzymatic reaction.

**active transport**   The transport of molecules in an energetically unfavorable direction across a membrane coupled to the hydrolysis of ATP or other source of energy.

**adaptin**   A protein that binds to membrane receptors and mediates the formation of clathrin-coated vesicles.

**adenine**   A purine that base-pairs with either thymine or uracil.

**adenoma**   A benign tumor arising from glandular epithelium.

**adenylyl cyclase**   An enzyme that catalyzes the formation of cyclic AMP from ATP.

**adherens junction**   A region of cell-cell adhesion at which the actin cytoskeleton is anchored to the plasma membrane.

**allele**   One copy of a gene.

**allosteric regulation**   The regulation of enzymes by small molecules that bind to a site distinct from the active site, changing the conformation and catalytic activity of the enzyme.

**alternative splicing**   The generation of different mRNAs by varying the pattern of pre-mRNA splicing.

**amino acid**   Monomeric building blocks of proteins, consisting of a carbon atom bound to a carboxyl group, an amino group, a hydrogen atom, and a distinctive side chain.

**aminoacyl tRNA synthetase**   An enzyme that joins a specific amino acid to a tRNA molecule carrying the correct anticodon sequence.

**amphipathic**   A molecule that has both hydrophobic and hydrophilic regions.

**anaphase**   The phase of mitosis during which sister chromatids separate and move to opposite poles of the spindle.

**anaphase A**   The movement of daughter chromosomes toward the spindle poles during mitosis.

**anaphase B**   The separation of the spindle poles during mitosis.

**angiogenesis**   The formation of new blood vessels.

**antibody**   A protein produced by B lymphocytes that binds to a foreign molecule.

**anticodon**   The nucleotide sequence of transfer RNA that forms complementary base pairs with a codon sequence on messenger RNA.

**antigen**   A molecule against which an antibody is directed.

**antiport**   The transport of two molecules in opposite directions across a membrane.

**AP endonuclease**   A DNA repair enzyme that cleaves next to apyrimidinic or apurinic sites in DNA.

**apical domain**   The exposed free surface of a polarized epithelial cell.

**apoptosis**   An active process of programmed cell death, characterized by cleavage of chromosomal DNA, chromatin condensation, and fragmentation of both the nucleus and the cell.

*Arabidopsis thaliana*   A small flowering plant used as a model for plant molecular biology and development.

**archaebacteria**   One of two major groups of prokaryotes; many species of archaebacteria live in extreme conditions similar to those prevalent on primitive Earth.

**ARF**   A GTP-binding protein required for vesicle budding from the *trans*-Golgi network.

**astral microtubules**   Microtubules of the mitotic spindle that extend to the cell periphery.

**ATP (adenosine 5′-triphosphate)**   An adenine-containing nucleoside triphosphate that serves as a store of free energy in the cell.

**ATP synthase**   A membrane spanning protein complex that couples the energetically favorable transport of protons across a membrane to the synthesis of ATP.

**autocrine signaling**   A type of cell signaling in which a cell produces a growth factor to which it also responds.

**autophagy**   The degradation of cytoplasmic proteins and organelles by their enclosure in vesicles from the endoplasmic reticulum that fuse with lysosomes.

**autoradiography**   The detection of radioisotopically labeled molecules by exposure to X-ray film.

**axonemal dynein**   The type of dynein found in cilia and flagella.

**axoneme**   The fundamental structure of cilia and flagella composed of a central pair of microtubules surrounded by nine microtubule doublets.

**β sheet**   A sheetlike secondary structure of a polypeptide chain, formed by hydrogen bonding between amino acids located in different regions of the polypeptide.

**bacteriophage**   A bacterial virus.

**basal body**   A structure similar to a centriole that initiates the growth of axonemal microtubules and anchors cilia and flagella to the surface of the cell.

**basal lamina**   A sheetlike extracellular matrix that supports epithelial cells and surrounds muscle cells, adipose cells, and peripheral nerves.

**basal transcription factors**   Transcription factors that are part of the general transcription machinery.

**base-excision repair**   A mechanism of DNA repair in which single damaged bases are removed from a DNA molecule.

**basement membrane**   See basal lamina.

**basolateral domain**   The surface region of a polarized epithelial cell that is in contact with adjacent cells or the extracellular matrix.

**benign tumor**   A tumor that remains confined to its site of origin.

**cadherins**   A group of cell adhesion molecules that form stable cell-cell junctions at adherens junctions and desmosomes.

*Caenorhabditis elegans*   A nematode used as a simple multicellular model for development.

**calmodulin**   A calcium-binding protein.

**Calvin cycle**   A series of reactions by which six molecules of $CO_2$ are converted into glucose.

**cAMP phosphodiesterase**   An enzyme that degrades cyclic AMP.

**cAMP-dependent protein kinase**   See protein kinase A.

**cancer**   A malignant tumor.

**carbohydrate**   A molecule with the formula $(CH_2O)_n$. Carbohydrates include both simple sugars and polysaccharides.

**carcinogen**   A cancer-inducing agent.

**carcinoma**   A cancer of epithelial cells.

**cardiolipin**   A phospholipid containing four hydrocarbon chains.

**carrier proteins**   Proteins that selectively bind and transport small molecules across a membrane.

**catalase**   An enzyme that decomposes hydrogen peroxide.

**caveolae**   Small invaginations of the plasma membrane that may be involved in endocytosis.

**Cdc2**   A protein-serine/threonine kinase that is a key regulator of mitosis in eukaryotic cells.

**Cdk inhibitor (CKI)**   A family of proteins that bind Cdks and inhibit their activity.

**Cdks**   Cyclin dependent protein kinases that control the cell cycle of eukaryotes. See also Cdc2

**cDNA library**   A collection of recombinant cDNA clones.

**cell adhesion molecules**   Transmembrane proteins that mediate cell-cell interactions.

**cell cortex**   The actin network underlying the plasma membrane.

**cell wall**   A rigid, porous structure forming an external layer that provides structural support to bacteria, fungi, and plant cells.

**cellulose**   The principal structural component of the plant cell wall, a linear polymer of glucose residues linked by $\beta(1\rightarrow4)$ glycosidic bonds.

**central dogma** The concept that genetic information flows from DNA to RNA to proteins.

**centriole** A cylindrical structure consisting of nine triplets of microtubules in the centrosomes of most animal cells.

**centromere** A specialized chromosomal region that connects sister chromatids and attaches them to the mitotic spindle.

**centrosome** The microtubule-organizing center in animal cells.

**cGMP phosphodiesterase** An enzyme that degrades cGMP.

**channel proteins** Proteins that form pores through a membrane.

**chaperone** A protein that facilitates the correct folding or assembly of other proteins.

**chaperonin** A family of heat-shock proteins within which protein folding takes place.

**chemiosmotic coupling** The generation of ATP from energy stored in a proton gradient across a membrane.

**chiasmata** Sites of recombination that link homologous chromosomes during meiosis.

**chitin** A polymer of *N*-acetylglucosamine residues that is the principal component of fungal cell walls.

**chlorophyll** The major photosynthetic pigment of plant cells.

**chloroplast** The organelle responsible for photosynthesis in the cells of plants and green algae.

**cholesterol** A lipid consisting of four hydrocarbon rings. Cholesterol is a major constituent of animal cell plasma membranes and the precursor of steroid hormones.

**chromatin** The fibrous complex of eukaryotic DNA and histone proteins. See histones, nucleosome, and chromatosome.

**chromatosome** A chromatin subunit consisting of 166 base pairs of DNA wrapped around a histone core and held in place by a linker histone.

**chromosomes** The carriers of genes, consisting of long DNA molecules and associated proteins.

**cilium** A microtubule-based projection of the plasma membrane that moves a cell through fluid or fluid over a cell.

***cis*-acting control element** A regulatory DNA sequence that serves as a protein binding site and controls the transcription of adjacent genes.

**citric acid cycle** A series of reactions in which acetyl CoA is oxidized to $CO_2$. The central pathway of oxidative metabolism.

**clathrin** A protein that coats the cytoplasmic surface of cell membranes and assembles into basketlike lattices that drive vesicle budding.

**clone** See recombinant molecule.

**codon** The basic unit of the genetic code; one of the 64 nucleotide triplets that code for an amino acid or stop sequence.

**coenzyme A (CoA)** A coenzyme that functions as a carrier of acyl groups in metabolic reactions.

**coenzyme Q** A small lipid-soluble molecule that carries electrons between protein complexes in the mitochondrial electron transport chain.

**coenzymes** Low-molecular-weight organic molecules that work together with enzymes to catalyze biological reactions.

**colcemid** A drug that inhibits the polymerization of microtubules.

**colchicine** A drug that inhibits the polymerization of microtubules.

**collagen** The major structural protein of the extracellular matrix.

**collenchyma** Plant cells characterized by thick cell walls; they provide structural support to the plant.

**complementary DNA (cDNA)** A DNA molecule that is complementary to an mRNA molecule, synthesized *in vitro* by reverse transcriptase.

**contact inhibition** The inhibition of movement or proliferation of normal cells that results from cell-cell contact.

**contractile ring** A structure of actin and myosin II that forms beneath the plasma membrane during mitosis and mediates cytokinesis.

**corticosteroids** Steroid hormones produced by the adrenal gland.

**cosmid** A vector that contains bacteriophage λ sequences, antibiotic resistance sequences, and an origin of replication. It can accomodate large DNA inserts of up to 45 kb.

**crista** A fold in the inner mitochondrial membrane extending into the matrix.

**cyanobacteria** The largest and most complex prokaryotes in which photosynthesis is believed to have evolved.

**cyclic AMP (cAMP)** Adenosine monophosphate in which the phosphate group is covalently bound to both the 3′ and 5′ carbon atoms, forming a cyclic structure; an important second messenger in the response of cells to a variety of hormones.

**cyclic electron flow** An electron transport pathway associated with photosystem I that produces ATP without the synthesis of NADPH.

**cyclic GMP (cGMP)** Guanosine monophosphate in which the phosphate group is covalently bound to both the 3′ and 5′ carbon atoms, forming a cyclic structure; an important second messenger in the response of cells to a variety of hormones and in vision.

**cyclins** A family of proteins that regulate the activity of Cdks and control progression through the cell cycle.

**cytochalasin**    A drug that blocks the elongation of actin filaments.

**cytochrome oxidase**    A protein complex in the electron transport chain that accepts electrons from cytochrome *c* and transfers them to $O_2$.

**cytokines**    Growth factors that regulate blood cells and lymphocytes.

**cytokinesis**    Division of a cell following mitosis or meiosis.

**cytosine**    A pyrimidine that base-pairs with guanine.

**cytoskeleton**    A network of protein filaments that extends throughout the cytoplasm of eukaryotic cells. It provides the structural framework of the cell and is responsible for cell movements.

**density gradient centrifugation**    A method of separating particles by centrifugation through a gradient of a dense substance, such as sucrose or cesium chloride.

**deoxyribonucleic acid (DNA)**    The genetic material of the cell.

**desmosome**    A region of contact between epithelial cells at which keratin filaments are anchored to the plasma membrane. See also hemidesmosome.

**diacylglycerol**    A secondary messenger formed from the hydrolysis of $PIP_2$ that activates protein kinase C.

**diakinesis**    The final stage of the prophase of meiosis I during which the chromosomes fully condense and the cell progresses to metaphase.

***Dictyostelium discoideum***    A unicellular eukaryote used for studies of cell movement and cell-cell signaling.

**diploid**    An organism or cell that carries two copies of each chromosome.

**diplotene**    The stage of mieosis I during which homologous chromosomes separate along their length but remain associated at chiasmata.

**DNA glycosylase**    A DNA repair enzyme that cleaves the bond linking a purine or pyrimidine to the deoxyribose of the backbone of a DNA molecule.

**DNA ligase**    An enzyme that seals breaks in DNA strands.

**DNA polymerase**    An enzyme catalyzing the synthesis of DNA.

**dolichol phosphate**    A lipid molecule in the endoplasmic reticulum upon which oligosaccharides are assembled for the glycosylation of proteins.

**domains**    Compact, globular regions of proteins that are the basic units of tertiary structure.

**dominant**    The allele that determines the phenotype of an organism when more than one allele is present.

**dominant inhibitory mutant**    A mutant that interferes with the function of the normal allele of the gene.

***Drosophila melanogaster***    A species of fruit fly commonly used for studies of animal genetics and development.

**dynein**    A motor protein that moves along microtubules towards the minus end.

***E. coli (Escherichia coli)***    A species of bacteria used as a model for biochemistry and molecular biology.

**E2F**    A family of transcription factors that regulate the expression of genes involved in cell cycle progression and DNA replication.

**ecdysone**    An insect steroid hormone that triggers metamorphosis.

**ectoderm**    The outer germ layer; gives rise to tissues that include the skin and nervous system.

**eicosanoid**    A class of lipids, including prostaglandins, prostacyclins, thromboxanes, and leukotrienes, that act in autocrine and paracrine signaling.

**electrochemical gradient**    A difference in chemical concentration and electric potential across a membrane.

**electron microscopy**    A type of microscopy that uses an electron beam to form an image. In transmission electron microscopy, a beam of electrons is passed through a specimen stained with heavy metals. In scanning electron microscopy, electrons scattered from the surface of a specimen are analyzed to generate a three-dimensional image.

**electron transport chain**    A series of carriers through which electrons are transported from a higher to a lower energy state.

**embryonal stem (ES) cells**    Cells cultured from early embryos.

**endocrine signaling**    A type of cell-cell signaling in which endocrine cells secrete hormones that are carried by the circulation to distant target cells.

**endocytosis**    The uptake of extracellular material in vesicles formed from the plasma membrane.

**endoderm**    The inner germ layer; gives rise to internal organs.

**endoplasmic reticulum (ER)**    An extensive network of membrane-enclosed tubules and sacs involved in protein sorting and processing as well as in lipid synthesis.

**endosome**    A vesicular compartment involved in the sorting and transport to lysosomes of material taken up by endocytosis.

**endosymbiosis**    A symbiotic relationship in which one cell resides within a larger cell.

**enhancer**    A transcriptional regulatory sequence that can be located at a site distant from the promoter.

**enzymes**    Proteins or RNAs that catalyze biological reactions.

**epidermal cells**    Cells forming a protective layer on the surfaces of plants and animals.

**epidermal growth factor (EGF)** A growth factor that stimulates cell proliferation.

**epithelial cells** Cells forming sheets (epithelial tissue) that cover the surface of the body and line internal organs.

**equilibrium centrifugation** The separation of particles on the basis of density by centrifugation to equilibrium in a gradient of a dense substance.

**erythrocytes** Red blood cells.

**estrogen** A steroid hormone produced by the ovaries.

**eubacteria** One of two major groups of prokaryotes, including most common species of bacteria.

**euchromatin** Decondensed, transcriptionally active interphase chromatin.

**eukaryotic cells** Cells that have a nuclear envelope, cytoplasmic organelles, and a cytoskeleton.

**exon** A segment of a gene that contains a coding sequence.

**exonuclease** An enzyme that hydrolyzes DNA molecules in either the 5′ to 3′ or 3′ to 5′ direction.

**expression vector** A vector used to direct expression of a cloned DNA fragment in a host cell.

**extracellular matrix** Secreted proteins and polysaccharides that fill spaces between cells and bind cells and tissues together.

**facilitated diffusion** The transport of molecules across a membrane by carrier or channel proteins.

**FAK (focal adhesion kinase)** A nonreceptor protein-tyrosine kinase that plays a key role in integrin signaling.

**fats** See triacylglycerols.

**fatty acids** Long hydrocarbon chains usually linked to a carboxyl group ($COO^-$).

**feedback inhibition** A type of allosteric regulation in which the product of a metabolic pathway inhibits the activity of an enzyme involved in its synthesis.

**fibroblast** A cell type found in connective tissue.

**fibronectin** The principal adhesion protein of the extracellular matrix.

**filiopodium** A thin projection of the plasma membrane supported by actin bundles.

**flagellum** A microtubule-based projection of the plasma membrane that is responsible for cell movement.

**flavin adenine dinucleotide ($FADH_2$)** A coenzyme that functions as an electron carrier in oxidation/reduction reactions.

**flow cytometer** An instrument that measures the fluoresence intensity of individual cells.

**fluid mosaic model** A model of membrane structure in which proteins are inserted in a fluid phospholipid bilayer.

**fluid-phase endocytosis** The nonselective uptake of extracellular fluids during endocytosis.

**fluorescence *in situ* hybridization (FISH)** A method used to localize genes on chromosomes using fluorescent probes.

**fluorescence microscopy** Type of microscopy in which molecules are detected based on the emission of flourescent light.

**focal adhesion** A site of attachment of cells to the extracellular matrix at which integrins are linked to bundles of actin filaments.

**freeze fracture** Method of electron microscopy in which specimens are frozen in liquid nitrogen and then fractured to split the lipid bilayer, revealing the interior faces of cell membranes.

**G-protein** A family of cell signaling proteins regulated by guanine nucleotide binding.

**G protein-coupled receptor** A receptor characterized by seven membrane-spanning $\alpha$ helices. Ligand binding causes a conformational change that activates a G protein.

**$G_0$** A quiescent state in which cells remain metabolically active but do not proliferate.

**$G_1$ phase** The phase of the cell cycle between the end of mitosis and the begining of DNA synthesis.

**$G_2$ phase** The phase of the cell cycle between the end of S phase and the begining of mitosis.

**gap junction** A plasma membrane channel forming a direct cytoplasmic connection between adjacent cells.

**gene** A segment of DNA that encodes a polypeptide chain or an RNA molecule.

**gene amplification** An increase in the number of copies of a gene resulting from the repeated replication of a region of DNA.

**gene transfer** The introduction of foreign DNA into a cell.

**general homologous recombination** The exchange of segments between DNA molecules that share extensive sequence homology.

**genetic code** The correspondence between nucleotide triplets and amino acids in proteins.

**genomic imprinting** The regulation of genes whose expression depends on whether they are maternally or paternally inherited, apparently controlled by DNA methylation.

**genomic library** A collection of recombinant DNA clones that collectively contain the genome of an organism.

**genotype** The genetic composition of an organism.

**Gibbs free energy (*G*)** The thermodynamic function that combines the effects of enthalpy and entropy to predict the energetically favorable direction of a chemical reaction.

**gluconeogenesis**   The synthesis of glucose.

**glycerol phospholipids**   Phospholipids consisting of two fatty acids bound to a glycerol molecule.

**glycocalyx**   A carbohydrate coat covering the cell surface.

**glycogen**   A polymer of glucose residues that is the principal storage form of carbohydrates in animals.

**glycolipid**   A lipid consisting of two hydrocarbon chains linked to a polar head group containing carbohydrates.

**glycolysis**   The anaerobic breakdown of glucose.

**glycoprotein**   A protein linked to oligosaccharides.

**glycosaminoglycan (GAG)**   A gel-forming polysaccharide of the extracellular matrix.

**glycosidic bond**   The bond formed between sugar residues in oligosaccharides or polysaccharides.

**glycosylation**   The addition of carbohydrates to proteins.

**glycosylphosphatidylinositol (GPI) anchor**   Glycolipids containing phosphatidylinositol that anchor proteins to the external face of the plasma membrane.

**glyoxylate cycle**   The conversion of fatty acids to carbohydrates in plants.

**Golgi apparatus**   A cytoplasmic organelle involved in the processing and sorting of proteins and lipids. In plant cells, it is also the site of the synthesis of cell wall polysaccharides.

**growth factors**   Polypeptides that control animal cell growth and differentiation.

**guanine**   A purine that base-pairs with cytosine.

**guanylyl cyclase**   An enzyme that catalyzes the formation of cyclic GMP from GTP.

**haploid**   An organism or cell that has one copy of each chromosome.

**heat-shock proteins**   A highly conserved group of chaperone proteins expressed in cells exposed to elevated temperatures or other forms of environmental stress.

**helicase**   An enzyme that catalyzes the unwinding of DNA.

**hemicellulose**   A polysaccharide that crosslinks cellulose microfibrils in plant cell walls.

**hemidesmosome**   A region of contact between cells and the extracellular matrix at which keratin filaments are attached to integrin.

**heterochromatin**   Condensed, transcriptionally inactive chromatin.

**high-energy bonds**   Chemical bonds that release a large amount of free energy when they are hydrolyzed.

**histones**   Proteins that package DNA in eukaryotic chromosomes.

**HMG-14 and HMG-17**   Nonhistone chromosomal proteins associated with decondensed transcriptionally active chromatin.

**Holliday junction**   The central intermediate in recombination, consisting of a crossed-strand structure formed by homologous base pairing between strands of two DNA moleucles.

**homeobox**   Conserved DNA sequences of 180 base pairs that encode homeodomains.

**homeodomain**   A type of DNA binding domain found in transcription factors that regulate gene expression during embryonic development.

**homologous recombination**   Recombination between segments of DNA with homologous nucleotide sequences.

**hormones**   Signaling molecules produced by endocrine glands that act on cells at distant body sites.

**hydrophilic**   Soluble in water.

**hydrophobic**   Not soluble in water.

**immunoblotting**   A method that uses antibodies to detect proteins separated by SDS-polyacrylamide gel electrophoresis.

**immunoglobulin**   See antibody.

**immunoprecipitation**   The use of antibodies to isolate proteins.

***in situ* hybridization**   The use of radioactive or flourescent probes to detect RNA or DNA sequences in cell extracts, chromosomes, or intact cells.

***in vitro* mutagenesis**   The introduction of mutations into cloned DNA *in vitro*.

***in vitro* translation**   Protein synthesis in a cell-free extract.

**inositol 1,4,5-trisphosphate ($IP_3$)**   A second messenger, formed from the hydrolysis of $PIP_2$, that signals the release of calcium ions from the endoplasmic reticulum.

**integral membrane proteins**   Proteins embedded within the lipid bilayer of cell membranes.

**integrin**   A transmembrane protein that mediates the adhesion of cells to the extracellular matrix.

**intermediate filament**   A cytoskeletal filament about 10 nm in diameter that provides mechanical strength to cells in tissues. See also keratins and neurofilaments.

**interphase**   The period of the cell cycle between mitoses that includes $G_1$, S, and $G_2$ phases.

**intracellular signal transduction**   A chain of reactions that transmits chemical signals from the cell surface to their intracellular targets.

**intron**   A noncoding sequence that interrupts exons in a gene.

**ion channel**   A protein that mediates the rapid passage of ions across a membrane by forming open pores through the phospholipid bilayer.

**ion pump**   A protein that couples ATP hydrolysis to the transport of ions across a membrane.

**Janus kinase (JAK)**   A family of nonreceptor protein-tyrosine kinases associated with cytokine receptors.

**keratin**   A type of intermediate filament protein of epithelial cells.

**kilobase (kb)**   One thousand nucleotides or nucleotide base pairs.

**kinesin**   A motor protein that moves along microtubules towards the plus end.

**kinetochore**   A specialized structure consisting of proteins attached to a centromere that mediates the attachment and movement of chromosomes along the mitotic spindle.

**kinetochore microtubules**   Microtubules of the mitotic spindle that attach to condensed chromosomes at their centromeres.

**Krebs cycle**   See citric acid cycle.

**lagging strand**   The strand of DNA synthesized opposite to the direction of movement of the replication fork by ligation of Okazaki fragments.

**lamellipodium**   A broad, actin-based extension of the plasma membrane involved in the movement of fibroblasts.

**laminin**   The principal adhesion protein of basal laminae.

**lamins**   Intermediate filament proteins that form the nuclear lamina.

**leading strand**   The strand of DNA synthesized continuously in the direction of movement of the replication fork.

**leptotene**   The initial stage of the extended prophase of meiosis I during which homologous chromosomes pair before condensation.

**leucine zipper**   A protein dimerization domain containing repeated leucine residues; found in many transcription factors.

**leukemia**   Cancer arising from the precursors of circulating blood cells.

**ligand**   A molecule that binds to a receptor.

**light reactions**   The reactions of photosynthesis in which solar energy drives the synthesis of ATP and NADPH.

**lignin**   A polymer of phenolic residues that strengthens secondary cell walls.

**lipids**   Hydrophobic molecules that function as energy storage molecules, signaling molecules, and the major components of cell membranes.

**long terminal repeat (LTR)**   DNA sequences found at the ends of retroviral and retrotransposon DNA that are direct repeats of several hundred nucleotides resulting from reverse transcriptase activity.

**lymphocyte**   A blood cell that functions in the immune response. B lymphocytes produce antibodies and T lymphocytes are responsible for cell mediated immunity.

**lymphoma**   A cancer of lymphoid cells.

**lysogeny**   A viral infection leading to integration of an inactive copy of viral DNA into the cell genome.

**lysosome**   A cytoplasmic organelle containing enzymes that break down biological polymers.

**M phase**   The mitotic phase of the cell cycle.

**macrophage**   A type of white blood cell specialized for phagocytosis.

**malignant tumor**   A tumor that invades normal tissue and spreads throughout the body.

**MAP kinases**   A family of mitogen-activated protein-serine/threonine kinases that are ubiquitous regulators of cell growth and differentiation.

**matrix**   The inner mitochondrial space.

**maturation promoting factor (MPF)**   A complex of Cdc2 and cyclin B that promotes entry into the M phase of either mitosis or meiosis.

**megabase (Mb)**   One million nucleotides or nucleotide base pairs.

**meiosis**   The division of diploid cells to haploid progeny, consisting of two sequential rounds of nuclear and cellular division.

**membrane-anchored growth factors**   Growth factors associated with the plasma membrane that function as signaling molecules during cell-cell contact.

**mesoderm**   The middle germ layer; gives rise to connective tissues and the hematopoietic system.

**messenger RNA (mRNA)**   An RNA molecule that serves as a template for protein synthesis.

**metaphase**   The phase of mitosis during which the chromosomes are aligned on a metaphase plate in the center of the cell.

**metastasis**   Spread of cancer cells through the blood or lymphatic system to other organ sites.

**5′ methylguanosine cap**   A structure consisting of GTP and methylated sugars that is added to the 5′ ends of eukaryotic mRNAs.

**microfilament**   A cytoskeleton filament composed of actin.

**microspike**   See filopodium.

**microtubule**   A cytoskeletal component formed by the polymerization of tubulin into rigid, hollow rods about 25 nm in diameter.

**microtubule-organizing center**   An anchoring point near the center of the cell from which most microtubules extend outward.

**microvillus**   An actin-based protrusion of the plasma membrane, abundant on the surfaces of cells involved in absorption.

**mismatch repair**   A repair system that removes mismatched bases from newly synthesized DNA strands.

**mitochondria**   Cytoplasmic organelles responsible for synthesis of most of the ATP in eukaryotic cells by oxidative phosphorylation.

**mitosis**   Nuclear division.

**mitotic spindle**   An array of microtubules extending from the spindle poles that is responsible for separating daughter chromosomes during mitosis. See also kinetochore microtubules, polar microtubules, and astral microtubules.

**molecular motor**   A protein that generates force and movement by converting chemical energy to mechanical energy.

**monocistronic**   Messenger RNAs that encode a single polypeptide chain.

**monoclonal antibody**   An antibody produced by a clonal line of B lymphocytes.

**monosaccharides**   Simple sugars with the basic formula of $(CH_2O)_n$.

**mutagen**   A chemical that induces a high frequency of mutations.

**mutation**   A genetic alteration.

**myosin**   A protein that interacts with actin as a molecular motor.

**N-myristoylation**   The addition of myristic acid (a 14-carbon fatty acid) to the N-terminal glycine residue of a polypeptide chain.

**Na$^+$-K$^+$ pump**   An ion pump that transports Na$^+$ out of the cell and K$^+$ into the cell.

**Nernst equation**   The relationship between ion concentration and membrane potential.

**neurofilament**   A type of intermediate filament that supports the axons of nerve cells.

**neuron**   A nerve cell specialized to receive and transmit signals throughout the body.

**neurotransmitter**   A small, hydrophilic molecule that carries a signal from a stimulated neuron to a target cell at a synapse.

**nexin**   A protein that links microtubule doublets to each other in the axoneme.

**nicotinamide-adenine dinucleotide (NAD$^+$)**   A coenzyme that functions as an electron carrier in oxidation/reduction reactions.

**nitric oxide (NO)**   A simple gas, synthesized from arginine, that is a major paracrine signaling molecule in the nervous, immune, and circulatory systems.

**nitrogen fixation**   The reduction of atmospheric nitrogen ($N_2$) to $NH_3$.

**nonreceptor protein-tyrosine kinase**   An intracellular protein-tyrosine kinase.

**nuclear envelope**   The barrier separating the nucleus from the cytoplasm, composed of an inner and outer membrane, a nuclear lamina, and nuclear pore complexes.

**nuclear lamina**   A meshwork of lamin filaments providing structural support to the nucleus.

**nuclear localization signal**   An amino acid sequence that targets proteins for transportation from the cytoplasm to the nucleus.

**nuclear membranes**   Membranes forming the nuclear envelope; the outer nuclear membrane is continuous with the endoplasmic reticulum and the inner nuclear membrane is adjacent to the nuclear lamina.

**nuclear pore complex**   A large structure forming a transport channel through the nuclear envelope.

**nucleic acid hybridization**   The formation of double stranded DNA and/or RNA molecules by complementary base pairing.

**nucleolus**   The nuclear site of rRNA transcription, processing, and ribosome assembly.

**nucleoside**   A purine or pyrimidine base linked to a sugar (ribose or deoxyribose).

**nucleosome**   The basic structural unit of chromatin consisting of DNA wrapped around a histone core.

**nucleosome remodeling factors**   Proteins that disrupt chromatin structure, allowing transcription factors to bind nucleosomal DNA.

**nucleotide**   A phosphorylated nucleoside.

**nucleotide excision repair**   A mechanism of DNA repair in which oligonucleotides containing damaged bases are removed from a DNA molecule.

**nucleus**   The most prominent organelle of eukaryotic cells; contains the genetic material.

**Okazaki fragments**   Short DNA fragments that are joined to form the lagging strand of DNA.

**oligonucleotide**   A short polymer of only a few nucleotides.

**oligosaccharide**   A short polymer of only a few sugars.

**oncogene**   A gene capable of inducing one or more characteristics of cancer cells.

**operator**   A regulatory sequence of DNA that controls transcription of an operon.

**operon**  A group of adjacent genes transcribed as a single mRNA.

**origin of replication**  A specific DNA sequence that serves as a binding site for proteins that initiate replication.

**origin replication complex (ORC)**  A protein complex that initiates DNA replication at yeast origins.

**oxidative phosphorylation**  The synthesis of ATP from ADP coupled to the energetically favorable transfer of electrons to molecular oxygen as the final acceptor in an electron transport chain.

**p53**  A transcription factor (encoded by the *p53* tumor suppressor gene) that arrests the cell cycle in $G_1$ in response to damaged DNA and is required for apoptosis induced by a variety of stimuli.

**pachytene**  The stage of meiosis I during which recombination takes place between homologous chromosomes.

**palmitoylation**  The addition of palmitic acid (a 16-carbon fatty acid) to cysteine residues of a polypeptide chain.

**paracrine signaling**  Local cell-cell signaling in which a molecule released by one cell acts on a neighboring target cell.

**parenchyma cell**  A type of plant cell responsible for most metabolic activities.

**passive diffusion**  The diffusion of small hydrophobic molecules through a phospholipid bilayer.

**passive transport**  The transport of molecules across a membrane in the energetically favorable direction.

**pectin**  A gel-forming polysaccharide in plant cell walls.

**peptide bond**  The bond joining amino acids in polypeptide chains.

**peptidoglycan**  The principal component of bacterial cell walls consisting of linear polysaccharide chains crosslinked by short peptides.

**peptidyl prolyl isomerase**  An enzyme that facilitates protein folding by catalyzing the *cis-trans* isomerization of prolyl peptide bonds.

**pericentriolar material**  The material in the centrosome that initiates microtubule assembly.

**peripheral membrane proteins**  Proteins indirectly associated with cell membranes by protein-protein interactions.

**peroxisome**  A cytoplasmic organelle specialized for carrying out oxidative reactions.

**phagocytosis**  The uptake of large particles, such as bacteria, by a cell.

**phalloidin**  A drug that binds to actin filaments and prevents their disassembly.

**phenotype**  The physical appearance of an organism.

**phorbol esters**  A class of tumor promoters that stimulate protein kinase C by acting as analogs of diacylglycerol.

**phosphatidylinositide 3-kinase (PI 3-kinase)**  An enzyme that phosphorylates $PIP_2$ yielding the second messenger phosphatidylinositol 3,4,5-trisphosphate ($PIP_3$).

**phosphatidylinositol 4,5-bisphosphate ($PIP_2$)**  A minor phospholipid component of the inner leaflet of the plasma membrane. Hormones and growth factors stimulate its hydrolysis by phospholipase C, yielding the second messengers diacylglycerol and inositol trisphosphate.

**phosphodiester bond**  A bond between the 5′-phosphate of one nucleotide and the 3′-hydroxyl of another.

**phospholipase C**  An enzyme that hydrolyzes $PIP_2$ to form the second messengers diacylglycerol and inositol trisphosphate.

**phospholipid bilayer**  The basic structure of biological membranes, in which the hydrophobic tails of phospholipids are buried in the interior of the membrane and their polar head groups are exposed to the aqueous solution on either side.

**phospholipid transfer protein**  A protein that transports phospholipid molecules between cell membranes.

**phospholipids**  The principal components of cell membranes, consisting of two hydrocarbon chains (usually fatty acids) joined to a polar head group containing phosphate.

**phosphorylation**  The addition of a phosphate group to a molecule.

**photoreactivation**  A mechanism of DNA repair in which solar energy is used to split pyrimidine dimers.

**photosynthesis**  The process by which cells harness energy from sunlight and synthesize glucose from $CO_2$ and water.

**pinocytosis**  The uptake of fluids or molecules into a cell by small vesicles.

**plant hormones**  A group of small molecules that coordinate the responses of plant tissues to environmental signals.

**plasma membrane**  A phospholipid bilayer with associated proteins that surrounds the cell.

**plasmalogens**  A family of phospholipids that have an ether bond and an ester bond.

**plasmid**  A small, circular DNA molecule capable of independent replication in a host cell.

**plasmodesma**  A cytoplasmic connection between adjacent plant cells formed by a continuous region of the plasma membrane.

**plastids**  A family of plant organelles including chloroplasts, chromoplasts, leucoplasts, amyloplasts, and elaioplasts.

**platelet-derived growth factor (PDGF)**  A growth factor released by platelets during blood clotting to stimulate the proliferation of fibroblasts.

**polar microtubules**   Microtubules of the mitotic spindle that overlap in the center of the cell and push the spindle poles apart.

**poly-A tail**   A tract of about 200 adenine nucleotides added to the 3′ ends of eukaryotic mRNAs.

**polyadenylation**   The process of adding a poly-A tail to a pre-mRNA.

**polycistronic**   Messenger RNAs that encode multiple polypeptide chains.

**polymerase chain reaction (PCR)**   A method for amplifying a region of DNA by repeated cycles of DNA synthesis *in vitro*.

**polynucleotide**   A polymer containing up to millions of nucleotides.

**polyp**   A benign tumor projecting from an epithelial surface.

**polypeptide**   A polymer of amino acids.

**polysaccharide**   A polymer containing hundreds or thousands of sugars.

**polysome**   A series of ribosomes translating a messenger RNA.

**pre-mRNA**   The primary transcripts that are processed to form messenger RNAs in eukaryotic cells.

**pre-rRNA**   The primary transcript, which is cleaved to form individual ribosomal RNAs (the 28S, 18S, and 5.8S rRNAs of eukaryotic cells).

**pre-tRNA**   The primary transcript, which is cleaved to form transfer RNAs.

**prenylation**   The addition of specific types of lipids (prenyl groups) to C terminal cysteine residues of a polypeptide chain.

**primary structure**   The sequence of amino acids in a polypeptide chain.

**primase**   An RNA polymerase used to intiate DNA synthesis.

**processed psuedogene**   A psuedogene that has arisen by reverse transcription of mRNA.

**progesterone**   A steroid hormone produced by the ovaries.

**programmed cell death**   A normal physiological form of cell death characterized by apoptosis.

**prokaryotic cells**   Cells lacking a nuclear envelope, cytoplasmic organelles, and a cytoskeleton (primarily bacteria).

**prometaphase**   A transition period between prophase and metaphase during which the microtubules of the mitotic spindle attach to the kinetochores and the chromosomes shuffle until they align in the center of the cell.

**promoter**   A DNA sequence to which RNA polymerase binds to initiate transcription.

**pronuclei**   Two haploid nuclei in a newly fertilized egg.

**proofreading**   The selective removal of mismatched bases by DNA polymerase.

**prophase**   The beginning phase of mitosis, marked by the appearance of condensed chromosomes and the development of the mitotic spindle.

**prosthetic groups**   Small molecules bound to proteins.

**proteasome**   A large protease complex that degrades proteins tagged by ubiquitin.

**protein disulfide isomerase**   An enzyme that catalyzes the formation and breakage of disulfide (S–S) linkages.

**protein kinase**   An enzyme that phosphorylates proteins by transferring a phosphate group from ATP.

**protein kinase A**   A protein kinase regulated by cyclic AMP.

**protein kinase C**   A family of protein-serine/threonine kinases that are activated by diacylglycerol and $Ca^{2+}$ and function in intracellular signal transduction.

**protein phosphatase**   An enzyme that reverses the action of protein kinases by removing phosphate groups from phosphorylated amino acid residues.

**protein-serine/threonine kinase**   A protein kinase that phosphorylates serine and threonine residues.

**protein-tyrosine kinase**   A protein kinase that phosphorylates tyrosine residues.

**protein-tyrosine phosphatase**   An enzyme that removes the phosphate groups from phosphotyrosine residues.

**proteins**   Polypeptides with a unique amino acid sequence.

**proteoglycan**   A protein linked to glycosaminoglycans.

**proteolysis**   Degradation of polypeptide chains.

**proto-oncogene**   A normal cell gene that can be converted into an oncogene.

**pseudogene**   A nonfunctional gene copy.

**psuedopodium**   An actin-based extension of the plasma membrane responsible for phagocytosis and amoeboid movement.

**pyrimidine dimer**   A common form of DNA damage caused by UV light in which adjacent pyrimidines are joined to form a dimer.

**quaternary structure**   The interactions between polypeptide chains in proteins consisting of more than one polypeptide.

**Raf**   A protein-serine/threonine kinase (encoded by the *raf* oncogene) that is activated by Ras and leads to activation of MAP kinase.

**Ras**   A family of small GTP binding proteins (encoded by the *ras* oncogenes) that couple growth factor receptors to intracellular targets, including the Raf protein-serine/threonine kinase and the MAP kinase pathway.

**Rb**  A transcriptional regulatory protein encoded by a tumor suppressor gene that was identified by the genetic analysis of retinoblastoma.

**receptor protein-tyrosine kinase**  Membrane-spanning protein-tyrosine kinases that are receptors for extracellular ligands.

**receptor-mediated endocytosis**  The selective uptake of macromolecules that bind to cell surface receptors that concentrate in clathrin-coated pits.

**recessive**  An allele that is masked by a dominant allele.

**recombinant DNA library**  A collection of genomic or cDNA clones.

**recombinant molecule**  A DNA insert joined to a vector.

**recombination**  The exchange of genetic material.

**replication fork**  The region of DNA synthesis where the parental strands separate and two new daughter strands elongate.

**repressor**  A regulatory molecule that blocks transcription.

**restriction endonuclease**  An enzyme that cleaves DNA at a specific sequence.

**restriction map**  The locations of restriction endonuclease cleavage sites on a DNA molecule.

**restriction point**  A regulatory point in animal cell cycles that occurs late in $G_1$. After this point a cell is committed to entering S and undergoing one cell division cycle.

**retroposon**  A transposable element that moves via reverse transcription of an RNA intermediate.

**retrovirus**  A virus that replicates by making a DNA copy of its RNA genome by reverse transcription.

**reverse transcriptase**  A DNA polymerase that uses an RNA template.

**rhodopsin**  A G protein-coupled photoreceptor in retinal rod cells that activates transducin in response to light absorption.

**ribonucleic acid (RNA)**  A polymer of ribonucleotides.

**ribosomal RNA (rRNA)**  The RNA component of ribosomes.

**ribosomes**  Particles composed of RNA and proteins that are the sites of protein synthesis.

**ribozyme**  An RNA enzyme.

**RNA editing**  RNA processing events other than splicing that alter the protein coding sequences of mRNAs.

**RNA polymerase**  An enzyme that catalyzes the synthesis of RNA.

**RNA splicing**  The joining of exons in a precursor RNA molecule.

**RNase H**  An enzyme that degrades the RNA strand of RNA-DNA hybrid molecules.

**RNase P**  A ribozyme that cleaves the 5′ end of pre-tRNAs.

**rough endoplasmic reticulum**  The region of the endoplasmic reticulum covered with ribosomes and involved in protein metabolism.

**Rous sarcoma virus (RSV)**  An acutely transforming retrovirus, in which the first oncogene was identified.

**ryanodine receptors**  Calcium channels in muscle and nerve cells that open in response to changes in membrane potential.

**S phase**  The phase of the cell cycle during which DNA replication occurs.

**sarcoma**  A cancer of cells of connective tissue.

**sarcomere**  The contractile unit of muscle cells composed of interacting myosin and actin filaments.

**sarcoplasmic reticulum**  A specialized network of membranes in muscle cells that stores a high concentration of $Ca^{2+}$.

**scanning electron microscopy**  See electron microscopy.

**sclerenchyma cells**  Plant cells characterized by thick cell walls that provide structural support to the plant.

**SDS-polyacrylamide gel electrophoresis (SDS-PAGE)**  A commonly used method to separate proteins by gel electrophoresis on the basis of size.

**second messenger**  A compound whose metabolism is modified as a result of a ligand-receptor interaction; it functions as a signal transducer by regulating other intracellular processes.

**secondary structure**  The regular arrangement of amino acids within localized regions of a polypeptide chain. See $\alpha$ helix and $\beta$ sheet.

**secretory vesicles**  Membrane-enclosed sacs that transport proteins from the Golgi apparatus to the cell surface.

**selectins**  Cell adhesion molecules that recognize oligosaccharides exposed on the cell surface.

**self-splicing**  The ability of some RNAs to catalyze the removal of their own introns.

**SH2 domain**  A protein domain of approximately 100 amino acids that binds phosphotyrosine-containing peptides.

**Shine-Delgarno sequence**  The sequence prior to the initiation site that correctly aligns bacterial mRNAs on ribosomes.

**signal patch**  A recognition determinant formed by the three-dimensional folding of a polypeptide chain.

**signal peptidase**  An enzyme that removes the signal sequence of a polypeptide chain by proteolysis.

**signal recognition particle (SRP)**  A particle composed of proteins and 7SL RNA that binds to signal sequences and targets polypeptide chains to the endoplasmic reticulum.

**signal sequence** A hydrophobic sequence at the amino terminus of a polypeptide chain that targets it for secretion in bacteria or incorporation into the endoplasmic reticulum in eukaryotic cells.

**site-specific recombination** Recombination mediated by proteins that recognize specific DNA sequences.

**smooth endoplasmic reticulum** The major site of lipid synthesis in eukaryotic cells.

**Southern blotting** A method in which radioactive probes are used to detect specific DNA fragments that have been separated by gel electrophoresis.

**spectrin** A major actin-binding protein of the cell cortex.

**sphingomyelin** A phospholipid consisting of two hydrocarbon chains bound to a polar head group containing serine.

**spliceosomes** Large complexes of snRNAs and proteins that catalyze the splicing of pre-mRNAs.

**Src** A nonreceptor protein-tyrosine kinase encoded by the oncogene (*src* ) of Rous sarcoma virus.

**starch** A polymer of glucose residues that is the principal storage form of carbohydrates in plants.

**START** A regulatory point in the yeast cell cycle that occurs late in $G_1$. After this point a cell is committed to entering S and undergoing one cell division cycle.

**STAT proteins** Trancription factors that have an SH2 domain and are activated by tyrosine phosphorylation, which promotes their translocation from the cytoplasm to the nucleus.

**stem cell** A cell that divides to produce daughter cells that can either differentiate or remain as stem cells.

**stereocilium** A specialized microvillus of auditory hair cells.

**steroid hormones** A group of hydrophobic hormones that are derivatives of cholesterol.

**steroid receptor superfamily** A family of transcription factors that regulate gene expression in response to steroids and related hormones.

**substrate** A molecule acted upon by an enzyme.

**symport** The transport of two molecules in the same direction across a membrane.

**synapse** The junction between a neuron and another cell, across which information is carried by neurotransmitters.

**synapsis** The association of homologous chromosomes during meiosis.

**synaptic vesicle** A secretory vesicle that releases neurotransmitters at a synapse.

**synaptomenal complex** A zipperlike protein structure that forms along the length of paired homologous chromosomes during meiosis.

**T cell receptor** A T lymphocyte surface protein that recognizes antigens expressed on the surface of other cells.

**TATA box** A regulatory DNA sequence found in the promoters of many eukaryotic genes transcribed by RNA polymerase II.

**TATA-binding protein (TBP)** A basal transcription factor that binds directly to the TATA box.

**taxol** A drug that binds to and stabilizes microtubules.

**telomerase** A reverse transcriptase that synthesizes telomeric repeat sequences at the ends of chromosomes from its own RNA template.

**telomeres** Repeats of simple-sequence DNA that maintain the ends of linear chromosomes.

**telophase** The final phase of mitosis, during which the nuclei re-form and chromosomes decondense.

**temperature-sensitive mutant** A cell expressing a protein that is functional at one temperature but not at another, whereas the normal protein is functional at both temperatures.

**tertiary structure** The three-dimensional folding of a polypeptide chain that gives the protein its functional form.

**testosterone** A steroid hormone produced by the testis.

**thylakoid membrane** The innermost membrane of chloroplasts that is the site of electron transport and ATP synthesis.

**thymine** A pyrimidine found in DNA that base-pairs with adenine.

**thyroid hormone** A hormone synthesized from tyrosine in the thyroid gland.

**tight junction** A continuous network of protein strands around the circumference of epithelial cells, sealing the space between cells and forming a barrier between the apical and basolateral domains.

**topoisomerase** An enzyme that catalyzes the reversible breakage and rejoining of DNA strands.

*trans*-**splicing** The joining of exons from different RNA molecules.

**transcription factor** A protein that regulates the activity of RNA polymerase.

**transcription** The synthesis of an RNA molecule from a DNA template.

**transcriptional attenuation** A regulatory mechanism that controls elongation of an mRNA molecule past a certain site on the DNA template.

**transcytosis** The sorting and transport of proteins to different domains of the plasma membrane following endocytosis.

**transducin** A G protein that stimulates cGMP phosphodiesterase when it is activated by rhodopsin.

**transfection** The introduction of a foreign gene into eukaryotic cells.

**transfer RNA (tRNA)**   RNA molecules that function as adaptors between amino acids and mRNA during protein synthesis.

**transforming growth factor β (TGF-β)**   A polypeptide growth factor that generally inhibits animal cell proliferation.

**transgenic mouse**   A mouse that carries foreign genes incorporated into the germ line.

**translation**   The synthesis of a polypeptide chain from an mRNA template.

**transmembrane proteins**   Integral membrane proteins that span the lipid bilayer and have portions exposed on both sides of the membrane.

**transmission electron microscopy**   See electron microscopy.

**transposition**   The movement of DNA sequences throughout the genome.

**transposon**   A DNA sequence that can move to different positions in the genome.

**triacylglycerol**   Three fatty acids linked to a glycerol molecule.

**tropomyosin**   A fibrous protein that binds actin filaments and regulates contraction by blocking the interaction of actin and myosin.

**tubulin**   A cytoskeletal protein that polymerizes to form microtubules.

**tumor**   Any abnormal proliferation of cells.

**tumor promoter**   A compound that leads to tumor development by stimulating cell proliferation.

**tumor suppressor gene**   A gene whose inactivation leads to tumor development.

**tumor virus**   A virus capable of causing cancer in animals or humans.

**turgor pressure**   The internal hydrostatic pressure within plant cells.

**ubiquinone**   See coenzyme Q.

**ubiquitin**   A highly conserved protein that acts as a marker to target other cellular proteins for rapid degradation.

**uniport**   The transport of a single molecule across a membrane.

**uracil**   A pyrimidine found in RNA that base-pairs with adenine.

**vacuole**   A large membrane-enclosed sac in the cytoplasm of eukaryotic cells. In plant cells, vacuoles function to store nutrients and waste products, to degrade macromolecules, and to maintain turgor pressure.

**vector**   A DNA molecule used to direct the replication of a cloned DNA fragment in a host cell.

**vinblastine**   A drug that inhibits microtubule polymerization.

**vincristine**   A drug that inhibits microtubule polymerization.

**Western blotting**   See immunoblotting.

**X-ray crystallography**   A method in which the diffraction pattern of X rays is used to determine the arrangement of individual atoms within a molecule.

*Xenopus laevis*   An African clawed frog used as a model system for developmental biology.

**yeast artificial chromosome (YAC)**   A vector that can replicate as a chromosome in yeast cells and can accomodate very large DNA inserts (hundreds of kb).

**yeasts**   The simplest unicellular eukaryotes. Yeasts are important models for studies of eukaryotic cells.

**zinc finger domain**   A type of DNA binding domain consisting of loops containing cysteine and histidine residues that bind zinc ions.

**zygote**   A fertilized egg.

**zygotene**   The stage of meiosis I during which homologous chromosomes become closely associated.

# Index

## About the Book

*Editors:* Patrick J. Fitzgerald, Andrew D. Sinauer

*Project Manager:* Carol J. Wigg

*Copy Editor:* Stephanie Hiebert

*Production Manager:* Christopher Small

*Book Layout and Production:* Jefferson Johnson

*Art Editing and Illustration Program:* Patrick Lane of J/B Woolsey Associates

*Photo Research:* Jane Potter

*Book and Cover Design:* Susan Brown Schmidler

*Color Separations:* Lanman Lithotech and Vision Graphics, Inc.

*Cover Manufacture:* Henry N. Sawyer Company, Inc.

*Book Manufacture:* The Courier Companies, Inc.